1st International Conference on Engineering and Applied Technology (ICEAT 2017)

IOP Conference Series: Materials Science and Engineering
Volume 403

Mataram, Indonesia
29 – 30 November 2017

ISBN: 978-1-5108-7297-4
ISSN: 1757-8981

Printed from e-media with permission by:

Curran Associates, Inc.
57 Morehouse Lane
Red Hook, NY 12571

Some format issues inherent in the e-media version may also appear in this print version.

Printed by Curran Associates, Inc. (2018)

For permission requests, please contact the Institute of Physics
at the address below.

Institute of Physics
Dirac House, Temple Back
Bristol BS1 6BE UK

Phone: 44 1 17 929 7481
Fax: 44 1 17 920 0979

techtracking@iop.org

Additional copies of this publication are available from:

Curran Associates, Inc.
57 Morehouse Lane
Red Hook, NY 12571 USA
Phone: 845-758-0400
Fax: 845-758-2633
Email: curran@proceedings.com
Web: www.proceedings.com

Table of contents

Volume 403

2017 1st International Conference on Engineering and Applied Technology

(ICEAT)29–30 November 2017, Mataram, Indonesia

Accepted papers received: 1 August 2018

Preface

Papers

IOP Conf. Series: Materials Science and Engineering **403** (2018) 011001 doi:10.1088/1757-899X/403/1/011001

Introduction

Assalamualaikum wr. wb.

On behalf of the organizing committee, I would like to welcome all of you to the 1st International Conference on Engineering and Applied Technology (ICEAT) and the 5th National Seminar on Applied Technology (SNTT), to be held from 29th – 30th November 2017.

The sustainability issues emerge and influence the human life, on the other side technology plays a significant role to accelerate the achievement of sustainable development goals. These two factors have motivated the researchers to engage themselves with continuous discussion on sustainability and technology. This is timely placed with the theme of our conference 'Bridging the gap from emerging technology to sustainable and smart technology'. It is hoped that the conference will lead to cross fertilization of ideas and act as a catalyst for collaborations at both the national and international levels.

I am pleased to inform you that there will be 150 papers presented across 7 parallel sessions. Paper presenters are from Japan, Malaysia, and Indonesia. We have very diverse presenter from Indonesia represents western and eastern part of Indonesia (from Aceh to Papua). I would like to convey our deepest gratitude to the keynote speakers: Professor Masaharu Fujita from Kyoto University Japan, Professor Michele Rosano from Curtin University Western Australia, and Ibu Inten Sripeni from PT Indonesia Power Jakarta. I also take this opportunity to thank our sponsors namely Garuda Indonesia, Tourism Office of West Nusa Tenggara, and PT. Amman Mineral Nusa Tenggara.

Lastly, I hope the conference will bring a new exciting experience, provide a good platform to make friends and learn from each other. I hope your stay in Lombok will be enjoyable and a memorable one.

Wabillahi Taufik Walhidayah Wassalamualaikum Warahmatullahi Wabarakatuh.

With best wishes,

Organizing Committee Chairman
Joni Safaat Adiansyah, Ph.D

IOP Conf. Series: Materials Science and Engineering **403** (2018) 011001 doi:10.1088/1757-899X/403/1/011001

Welcome Message from the Rector of Universitas Muhammadiyah Mataram

Assalamu alaikum wr. Wb
Peace be upon you

Honorable the three keynote speakers for the First International Conference on Engineering and Applied Technology:

1. A/Professor Michele Rosano: from Sustainable Engineering Group (SEG) Curtin University Australia,
2. Professor Masaharu Fujita from Disaster Prevention Research Institute Kyoto University of Japan, and
3. Mrs. Sripeni Inten Cahyani from PT Indonesia Power
4. All Deans of Association of Muhammadiyah University from all over Indonesia
5. Honorable all ladies and gentlemen.

All thanks only delivered to the one and only, the Almighty Allah SWT who always give us his blessing, mercy, healthy and opportunity to allow us to come and attend this special occasion.
All ladies and gentlemen, welcome to West Nusa Tenggara especially at Muhammadiyah University of Mataram, it is a pride to be trusted by the members of Forum Group Discussion of Technology's Forum (FGDT), as a committee and executor of annual meeting of Dean of Engineering Faculty at the 8th Association of Muhammadiyah University (PTM) all over Indonesia held together with the National Seminar on Applied V Technology, and the International Conference on Engineering and Applied Technology I.

It must be acknowledged that the issues in development and technology become an important part of human civilization. Technology takes a significant role in the dynamics and activities of modern society. This is because the technology offers easiness in facilities, communications, transportation and so forth. Therefore, humans as designers as well as users are required to continue to keep on innovative in developing the utilization and application of technology, in order to improve the quality of human life. And I think these scientific meetings are an absolutely must be done, because this is such a good opportunity for researchers and scientists all over Indonesia with different backgrounds of expertise to meet each other, and to share knowledge in maximizing the efficient and effective technology, for mankind by carrying the local wisdom and human nature as a social creature.

The efficiency and effectiveness of using technology has a fundamental effect on the acceleration of development. Admittedly or not, discourse on sustainable development has become one of the world's top topics today and next in the future. The successful of implementation related to the paradigm of sustainable development will provide the changes of value in all aspects of life, including; social, economic, cultural, and environment. Although, in fact we cannot deny that there are still many points of views that cast doubt on the crucial role of technology in promoting sustainable development. So, I think this extraordinary meeting is very fitting with the current theme, "Bridging the gap from emerging technology to sustainable and smart technology".

I hope that the vision to be achieved in this meeting can be accomplished in the form of developing sustainable and humane technological engineering research, then the results of these technological innovations can be published globally and inspire the new emergence and more advanced technologically. And the most important thing to be noted is through this meeting, I hope that good communication is established and strong networking among researchers, institutions, and related stakeholders would be built up regularly, which will facilitate the development of technology, applications, and new innovations.

Rector of Universitas Muhammadiyah Mataram
Drs. H. Mustamin H. Idris, MS

IOP Conf. Series: Materials Science and Engineering **403** (2018) 011002 doi:10.1088/1757-899X/403/1/011002

ICEAT 2017 Committee

Steering Committee

- Moehamad Aman (Universitas Muhammadiyah Magelang, Indonesia)
- Eko Budi Leksono (Universitas Muhammadiyah Gresik, Indonesia)
- M. Taufiq Tamam (Universitas Muhammadiyah Purwokerto, Indonesia)
- Sri Sunarjono (Universitas Muhammadiyah Surakarta, Indonesia)
- Hamzah Al Imran (Universitas Muhammadiyah Makasar, Indonesia)
- Jazaul Ikhsan (Universitas Muhammadiyah Yogyakarta, Indonesia)
- Kartika Firdausy (Universitas Ahmad Dahlan, Indonesia)
- Yun Arifatul Fatimah (Universitas Muhammadiyah Magelang, Indonesia)
- Sudarman (Universitas Muhammadiyah Malang, Indonesia)

Organizing Committee

Honorary Board

- Majlis DIKTI LITBANG PP Muhammadiyah
- Mustamin H. Idris (Rector of Universitas Muhammadiyah Mataram, Indonesia)

Executive Board

- Isfanari (Universitas Muhammadiyah Mataram, Indonesia)
- Anwar Ma'ruf (Universitas Muhammadiyah Purwokerto, Indonesia)

General Chair

- Tole Sutikno (Universitas Ahmad Dahlan, Indonesia)

General Co-Chair

- Herry Purnama (Universitas Muhammadiyah Surakarta, Indonesia)

Chairman

- Joni Safaat Ardiansyah (Universitas Muhammadiyah Mataram, Indonesia)

Secretary

- Ima Rahmawati (Universitas Muhammadiyah Mataram, Indonesia)

Treasury

- Diah Rahmawati (Universitas Muhammadiyah Mataram, Indonesia)

Local Committee

- Titik Wahyuningsih (Universitas Muhammadiyah Mataram, Indonesia)
- Salahudin (Universitas Muhammadiyah Mataram, Indonesia)
- Humaira (Universitas Muhammadiyah Mataram, Indonesia)

- Sri Hardiningsih (Universitas Muhammadiyah Mataram, Indonesia)
- M. Ghazali (Universitas Muhammadiyah Mataram, Indonesia)
- Sarman (Universitas Muhammadiyah Mataram, Indonesia)
- Muslihin (Universitas Muhammadiyah Mataram, Indonesia)
- Nuraini (Universitas Muhammadiyah Mataram, Indonesia)
- Sri Maryati (Universitas Muhammadiyah Mataram, Indonesia)
- Alpiana (Universitas Muhammadiyah Mataram, Indonesia)
- Bedy Fara Aga Matrani (Universitas Muhammadiyah Mataram, Indonesia)
- Rifqi Anra Wijaya (Universitas Muhammadiyah Mataram, Indonesia)
- Fariz Primadi Hirsan (Universitas Muhammadiyah Mataram, Indonesia)
- Agus Kurniawan (Universitas Muhammadiyah Mataram, Indonesia)
- Rudy Razak (Universitas Muhammadiyah Mataram, Indonesia)
- Nurul Qiyaam (Universitas Muhammadiyah Mataram, Indonesia)
- Maria Ulfa (Universitas Muhammadiyah Mataram, Indonesia)
- M. Ramdhani (Universitas Muhammadiyah Mataram, Indonesia)
- Rasyid Ridha (Universitas Muhammadiyah Mataram, Indonesia)
- Maya Saridewi Pascanawati (Universitas Muhammadiyah Mataram, Indonesia)
- Hidayati (Universitas Muhammadiyah Mataram, Indonesia)

Technical Program Committee

General

- Dwi Winarti
- M. Islamy Rusyda (Universitas Muhammadiyah Mataram, Indonesia)
- Tole Sutikno (Universitas Ahmad Dahlan, Indonesia)
- Aris Widyo Nugroho (Universitas Muhammadiyah Yogyakarta, Indonesia)
- Andri Pranolo (Universitas Ahmad Dahlan, Indonesia)
- Adhi Prahara (Universitas Ahmad Dahlan, Indonesia)
- Ahmad Azhari (Universitas Ahmad Dahlan, Indonesia)
- Herry Purnama (Universitas Muhammadiyah Surakarta, Indonesia)
- Tri Widayatno (Universitas Muhammadiyah Surakarta, Indonesia)

Technical

- Slamet Riyadi (Universitas Muhammadiyah Yogyakarta, Indonesia)
- Husni Thamrin (Universitas Muhammadiyah Surakarta, Indonesia)
- Gunawan Ariyanto (Universitas Muhammadiyah Surakarta, Indonesia)
- Andri Pranolo (Universitas Ahmad Dahlan, Indonesia)
- Zulfatman (Universitas Muhammadiyah Malang, Indonesia)
- Yun Arifatul Fatimah (Universitas Muhammadiyah Magelang, Indonesia)
- Hari Prasetyo (Universitas Muhammadiyah Surakarta, Indonesia)
- Eko Setiawan (Universitas Muhammadiyah Surakarta, Indonesia)
- Ilyas Mas'udin (Universitas Muhammadiyah Malang, Indonesia)
- Siti Mahsanah (Universitas Ahmad Dahlan, Indonesia)

IOP Conf. Series: Materials Science and Engineering **403** (2018) 011002 doi:10.1088/1757-899X/403/1/011002

- Aris Widyo Nugroho (Universitas Muhammadiyah Yogyakarta, Indonesia)
- Sudarisman (Universitas Muhammadiyah Yogyakarta, Indonesia)
- Muji Setiyo (Universitas Muhammadiyah Magelang, Indonesia)
- Marwan Effendy (Universitas Muhammadiyah Surakarta, Indonesia)
- Tri Widayatno (Universitas Muhammadiyah Surakarta, Indonesia)
- Denny Vitasari (Universitas Muhammadiyah Surakarta, Indonesia)
- Haryanto (Universitas Muhammadiyah Purwokerto, Indonesia)
- Nurul H Fitriyah (Universitas Muhammadiyah Jakarta, Indonesia)
- Jazaul Ikhsan (Universitas Muhammadiyah Yogyakarta, Indonesia)
- Agus Setyo Munthohar (Universitas Muhammadiyah Yogyakarta, Indonesia)
- Sri Sunarjono (Universitas Muhammadiyah Surakarta, Indonesia)
- Nurul Hidayati (Universitas Muhammadiyah Surakarta, Indonesia)
- Samin (Universitas Muhammadiyah Malang, Indonesia)
- Wisnu Setiawan (Universitas Muhammadiyah Surakarta, Indonesia)
- Gunawan (Universitas Muhammadiyah Surabaya, Indonesia)

ICEAT 2017 Reviewer

- Hamid Abdillah (Universitas Muhammadiyah Surakarta, Indonesia)
- Leon A. Abdillah (Bina Darma University, Indonesia)
- Agus Dwi Anggono (Universitas Muhammadiyah Surakarta, Indonesia)
- Erna Astuti (Universitas Ahmad Dahlan, Indonesia)
- Gunawan Rudi Cahyono (Politeknik Negeri Banjarmasin, Indonesia)
- Rafał Dreżewski (AGH University of Science and Technology, Poland)
- Siti Fatimah (Universitas Muhammadiyah Surakarta, Indonesia)
- Yun Arifatul Fatimah (Universitas Muhammadiyah Magelang, Indonesia)
- Rois Fatoni (Universitas Muhammadiyah Surakarta, Indonesia)
- Lukas G. Swan (Dalhousie University, Canada)
- Nur Hidayati (Universitas Muhammadiyah Surakarta, Indonesia)
- Pedro Hokama (University of Campinas, Brasil)
- Shi-Jinn Horng (National Taiwan University of Science and Technology, Taiwan)
- Jazaul Ikhsan (Universitas Muhammadiyah Yogyakarta, Indonesia)
- Dewi Pramudi Ismi (Universitas Ahmad Dahlan, Indonesia)
- Wolfgang Keller (University of Colorado, Colorado, US)
- Nuning Kurniasih (Universitas Padjadjaran, Indonesia)
- Ilyas Masudin (Universitas Muhammadiyah Malang, Indonesia)
- Darmawan Napitupulu (Indonesian Institute of Sciences, Indonesia)
- E.P. Nowicki (University of Calgary, Canada)
- Aris Widyo Nugroho (Universitas Muhammadiyah Yogyakarta, Indonesia)
- Adhi Prahara (Universitas Ahmad Dahlan, Indonesia)
- Utomo Pujianto (State University of Malang, Indonesia)
- Herry Purnama (Universitas Muhammadiyah Surakarta, Indonesia)
- Denny Vitasari (Aberystwyth University, Australia)
- Aji Prasetya Wibawa (Universitas Negeri Malang, Indonesia)
- Tri Widayatno (Universitas Muhammadiyah Surakarta, Indonesia)
- Iwan Tri Riyadi Yanto (Universitas Ahmad Dahlan, Indonesia)

IOP Conf. Series: Materials Science and Engineering **403** (2018) 011004 doi:10.1088/1757-899X/403/1/011004

Organizers and Sponsors

Organized by

FGDT PTM (Engineering Forum of Muhammadiyah Higher Education), Indonesia

Hosted by

Universitas Muhammadiyah Mataram, Indonesia

Technical Co-Sponsored by

IOPscience, IOP

Supported by

Majelis Pendidikan Tinggi Penelitian dan Pengembangan, PP Muhammadiyah, Indonesia
Universitas Ahmad Dahlan, Indonesia
Universitas Muhammadiyah Magelang, Indonesia
Universitas Muhammadiyah Malang, Indonesia
Universitas Muhammadiyah Purwokerto, Indonesia
Universitas Muhammadiyah Surakarta, Indonesia
Universitas Muhammadiyah Yogyakarta, Indonesia

Peer review statement

All papers published in this volume of *IOP Conference Series: Materials Science and Engineering* have been peer reviewed through processes administered by the proceedings Editors. Reviews were conducted by expert referees to the professional and scientific standards expected of a proceedings journal published by IOP Publishing.

Planning and design approach in Islamic green city towards sustainable city: the case of Martapura

H M Caesarina[1] and N Aina[2]

[1] Urban and Regional Planning Department, University of Muhammadiyah Banjarmasin, Jl. S. Parman Komp. RS. Islam Banjarmasin
[2] Architecture Department, University of Muhammadiyah Banjarmasin, Jl. S. Parman Komp. RS. Islam Banjarmasin

hanny.planarch@gmail.com, pwkumbjm.hanny@umbjm.ac.id, nooraina1986@umbjm.ac.id

Abstract. Global climate is changing and the green dimension of sustainable development has becoming an imperious agenda for all countries to pursue. Indonesia as part of Asia (which is the most populous Muslim majority country in the world) faces serious environmental problem, and uses the green city concept in most of its Islamic cities. Martapura, as one of the capital city in Indonesia is famous as an Islamic city with great vision and mission for green city. This study analyses the context of urban planning and design approach in Islamic Green city to help improving the quality of environment towards sustainable city. This is a descriptive-analysis study, and data collection conducted by documents and field survey. It concludes lessons learned from the practice of planning and design in Martapura and presents recommendations in planning and design approach for urban planner and designer to make a city more sustainable from an Islamic green city context.

1. Introduction

Global climate is changing and the green dimension of sustainable development has becoming an imperious agenda for all countries to pursue. According to Asian Development Bank, cities in Asia consume nearly 75 percent of total resource use and are responsible for more than 75 percent of greenhouse gas (GHG) emissions. Environmental degradation and social challenges such as polluted air, contaminated river and many other problems have been linked to urbanization. Recognizing the problems are concentrated on urban regions as the key driver of economic growth, ADB's Urban Operational Plan 2012-2020 emphasizes balanced urban development in an effort to promote liveable cities across Asia [1].

The ADB uses the green city approach and focusing on integrated and balanced growth. The green city approach is used to realize a city that is liveable for all its residents by promoting a paradigm shift where cities make a concerted effort to pursue integrated urban development and environmental planning in response to urbanization and other challenges [2]. With respect to environment, the focus is on managing land, water, and air in an efficient manner to maintain the long term sustainability.

Indonesia as part of Asia (which is the most populous Muslim majority country in the world) faces serious environmental problem, and uses the green city concept to overcome the ecological challenges in most of its Islamic cities [2]. There are so many cities in Indonesia that can be categorized as Islamic cities and faces so many urban problems. The need of proper planning and design of green city is required to accommodate urban problems [3]. With respect to environment, the focus of green city is on managing land, water, and air in an efficient manner to maintain the long term sustainability [1] by following the Green City Action Plan as a step towards the vision of sustainable city. It includes some elements which are water management, energy efficiency and renewable energy, green transportation, zero waste, cultural heritage and tourism, and urban forestry and agriculture.

As observed by [1] and [4], integrative and comprehensive approaches that base on three Es (Environment, Economy, Equity) will lead to a greener city. The three Es and green cities concept called the GrEEEn cities operational framework forms the basis of a comprehensive Green City Action Plan which is a program linked to goals for environmental quality improvement [5]. According to [6] and [7], other than physical planning and design, the eco-communities approach also holds an important role in planning and design approach to improve the quality of city and environment. In context of Islamic city, there are some important elements that support the concept of green city such as 'Eco Islamic Belief' which is the expression 'Greening Islam' that may lead to an Islamic perception of sustainability based on some [8] principles and goals [9]. Eco-Islamic beliefs community often advocated by social movements for sustainability may increase chances of long term success in environmental policy making [7, 10, 11].

This paper explores and presents the planning and design approach that have been used in Islamic green city, Martapura. It is expected that this paper could provide the basic idea of how an Islamic green city might lead to sustainable city in Indonesia.

2. Study area

The planning and design approach of Islamic green city has been discussed as an effort to help improving the quality of environment towards sustainable city in the urban context of the study area: Martapura, the capital city of Banjar Regency. Martapura is located 40 km from Banjarmasin (the capital of South Kalimantan). As stated in [12], Martapura is identified as an Islamic city due to its history as the capital of Banjar Empire (1859-1863). *Ulama* (Muslim leader) and Muslim Scholars had the main point in politic. Muhammad Arshad al-Banjari is one of the famous *Ulama* in Banjar Empire. This historical background explains the identity of Martapura which has famously known as the "Veranda of Mecca". The influence of Islam on the form and design of the city is literally seen in Martapura. In RAKH [13], the Mission of Banjar Regency is to increase the role of stakeholders in order to establish Martapura as a green city, based on Islamic Green City development guidelines by improving the quality of natural resources and environment.

3. Methodology

This paper focuses on the planning and design approach that has been implemented in urban planning and urban design in Martapura by combining the concept of green city and Islamic city principles. The following discussion has been carried out by stimulating various literatures using some contexts of Islamic city and green city to define an Islamic green city. Elements of Islamic city and green city are well-defined as a unity and were used as the combined variable to indicate planning and design in Martapura which have the vision of Islamic green city.

4. Planning and design approach in Martapura as an Islamic Green City

Islamic city has been discussed in only several literatures and few papers. Islamic city means a city formed base on the Quran and prophetic principles and teachings. Those principles are implemented through planning and design process for the city. Regulations and policies are based on Islamic principles, such as the differentiation of spaces by gender (Male and Female should be placed in separated zone) [14]. Some important design principles of Islamic city are: Natural laws; Religious and cultural beliefs; Design principles stemmed from *Sharia law* (Islamic law based on the teachings of Al-Qur'an and the prophet); and Social principles as mentioned in [15]. The first principle which is natural laws expressed in the adaptation of the build form and plan of the city to the weather conditions and topography. The second principle, Religious and natural beliefs gives the mosque the central position in spatial and institutional hierarchies. The third principle reflected how the Islamic city linked with *Sharia law* by determining physical and social relations between public and private zone. The fourth principle is dealing with how the social urban society grouped into some organizations based on ethnic origin and cultural perspectives. The design principles of Islamic city have some specific elements, therefore, the personnel in planning and urban design should understand both traditional trends and current techniques suitable for shaping the city as mentioned in [16].

According to [15], [17] and [18], Islamic city has some typical physical features: the main mosque (Islamic city should have a congregational Friday mosque and a market/chief bazaar nearby); *suqs* (market); citadel/office park (Islamic city often lacked of Municipal government and institution. This lack of municipal government and institutions was recompense by some *sheikhs* (Muslim Leader in Middle East); residential quarters; street network; wall; exterior space. One of these Islamic city features, exterior space or open space is the heart of green city concepts. Open space includes green urban spaces, such as urban green park, gardens and green belt have the main point to help reducing CO_2 and make a healthier living environment.

The form of Martapura as an Islamic City started in the 1990s when most of the public spaces in Martapura were built based on Islamic guidelines. According to [12], There were a lot of *Pengajian* (Islamic study) which developed rapidly into the center of religion, politics, economy and social of Muslim in South Kalimantan. The influence of Islam on the form and design of the city is literally seen in Martapura. As the capital of Banjar Regency, its mission is to increase the role of stakeholders in order to establish Martapura as a green city, based on Islamic Green City development guidelines by improving the quality of natural resources and environment as seen in [13]. Martapura has implemented the integrated and comprehensive approaches in form of green city action plan which base on three main elements of sustainability, which are environment, society and economy -also known as three Es (Environment, Economy, Equity) [1].

Green city according to P2KH [8] is a city that able to adapt to climate changes, environmental friendly by using natural sources efficiently, reducing waste, green transportation system, balancing build and natural environment based on sustainable planning and design. Green city concept is a good step in achieving a long term vision of sustainable city, a rich quality of life and strong connections between humans and nature. Green city has 8 (eight) main elements, which are: green planning and design; green water; green open space ; green waste; green transportation, green energy; green community; and green building as shown in [8]. As can be seen from analysis in figure 1, Martapura also incorporated Islamic guidelines into those eight elements of green city and forms a new approach for Islamic green cities. In Eco Islamic belief, humans and environment are connected each other [14]. Sustainable development elements are clearly stated in *Al-Qur'an* (central religious text of Islam), such as natural resources, water, material, construction, green land, sky healthy environment, etc.

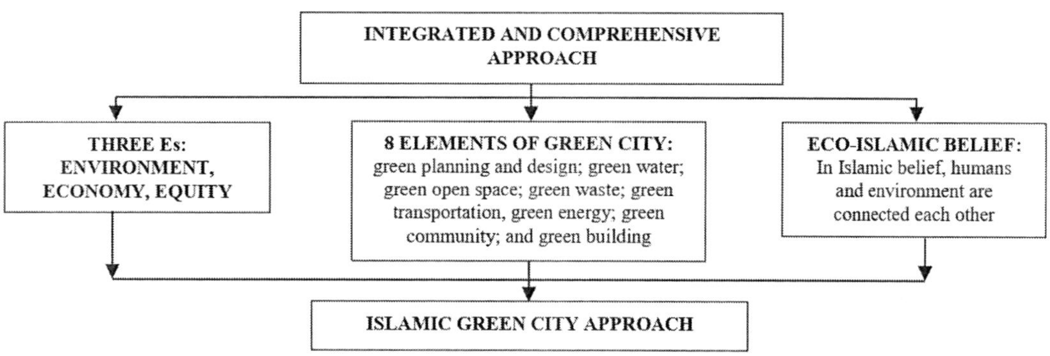

Figure 1. Analysis of planning and design approach in Martapura as an Islamic Green City.

IOP Conf. Series: Materials Science and Engineering **403** (2018) 012001 doi:10.1088/1757-899X/403/1/012001

Martapura has tried to implement some approach in planning and design based on *Al-Qur'an* and prophetic principles and teachings through some specific regulation, policy and advisory. The analysis of application of Islamic green city approach in Martapura can be seen from table 1. Islamic atmosphere can be seen in any corner of Martapura especially around *Kubah Sekumpul*. Many Muslims from all over Indonesia come to *Kubah Sekumpul* for *ziarah* (a pilgrimage to holy grave), especially during special events such as *Haul Guru Sekumpul* (Death anniversary of *Guru Sekumpul*), *Maulid Nabi* (Celebration of Prophet Muhammad's birthday), etc. Those celebrations take place at *Kubah Sekumpul* Mosque and the aftermath are some environmental problems such as serious traffic problem as mentioned in [19] and [20], lack of parking space, the damage of public facilities like grass and shrubs in the park, litter ends up everywhere and producing a huge amount of garbage.

Martapura implement so many Islamic values as part of green cities concept. Table 1 shows forbids any forms of living creature's statues in the city as part of green open spaces. Besides being the "Veranda of Mecca", Martapura also known as *Santri* City where a lot of *Santri* (students or followers of Islamic school) wears white clothes around the city to study Islam and often using bicycle as their main transportation which supports the concept of green transportation. Nonetheless, it needs to be detailed more as a specific guideline for an Islamic green city. Green city concept in Martapura is implemented towards Islamic atmosphere but mostly in forms of advisory, not guidelines. The awareness of environment and ecology conditions need to be improved with definite role from the Government, institutions, academic, and *Ulama* or other religious leader.

Table 1. Analysis: Application of Islamic green city approach in Martapura.

Elements	Application in Planning (Martapura)	Application in Design (Martapura)
Regulations and Policy	Government's goal of the improvement of *Al-Qur'an* literacy rate; Government's goal of the improvement of Arab –Melayu literacy rate; Special policies during *Ramadhan* –Restaurant can't open before 3 p.m.; Middle schools have special subjects regarding Islam and Environment.	Any form of living creatures statues are forbidden in the city (advisories); Cinema and Clubs are forbidden to be built in the city (advisories); *Dzikr* (Praise to Allah) signage's; No specific policy and guidelines for green design.
The Main Mosque	Mosque only functioned for praying and *Pengajian* regarding the hereafter. There's no *Pengajian* in the mosque which is related to the socialization of green city (Ulama has not been involved in planning process).	• Main Mosque "*Al Karamah*" is located in the heart of the City • '*Kubah Sekumpul'* has become the point of interest in the heart of Martapura, and could be the focus in arranging the Central Business District.
***Suqs* (Market)**	Special policy is applied during *Ramadhan* time, restaurants are not allowed to open their business before 3 pm.	*Suqs* or the market is located around the main mosque (*Pertokoan Cahaya Bumi Selamat* near *Al Karamah* mosque and Street Market near *Kubah Guru Sekumpul*).
Citadel	There is an advisory for government employee to dress in Muslim clothes,	*Kantor Bupati* (Governmental office) is located near the market.
Residential Quarters	Residential quarters should consider cleanliness, etc.	Residential is located around the city and are not grouped by their ethnicity, so people can reach other facilities by foot.
Street Network	Accessibility and street networks are based on function in each zone.	Narrow streets and connecting each zoning.
Wall	There are no regulations regarding neighborhood's wall.	There are limited number of gates, mostly people live peacefully.
Exterior	Improvement in quality and quantity of green open spaces, urban parks, garden, etc. by arranging proper sites and location.	Limited private garden, limited green open spaces.
Green Planning and design	Urban Design Documents of Banjar Regency Areas based on green city guidelines; Public policy and regulation for green spaces, green buildings, and green office park; No specific and detailed guidelines for green planning and design..	Green Urban Design, Encouraging Middle Schools for *Adiwiyata* (Go green); program to build green house and plant trees/providing green area.
Green Open	The development of urban park, green belt, green	Urban park with Islamic themes and names

Elements	Application in Planning (Martapura)	Application in Design (Martapura)
Space	open space, etc. around the city.	(advisory); Sculpture is not allowed in living creatures forms (advisory).
Green Waste	Waste management programme, Waste water management programme, sanitation and cleanliness program.	Waste bins are separated into recycled and non-recycled items; Islamic sign "*Kebersihan itu sebagian dari iman*/Cleanliness is part of our faith" can be seen everywhere.
Green Transportation	Cycling path in some of the main roads; No regulation in promoting cycle for people; Methane collection systems, Solar Cell installations.	Cycling path is poorly applied, because the main roads are already narrow.
Green Energy		Schools are the main institutions who are active in promoting methane collection systems.
Green Water	Installations of Clean water network, Management of clean water supply.	-
Green Building	Green house modelling; Green building policy.	Inspectorate building was designed in green concepts with a lot of open spaces, passive ventilation systems (1st green building pilot project in Martapura).
Green Community	Improvement of *Kampung Hijau* (Green Neighbourhood) program in some areas.	Not all of Martapura's citizen is aware of green city concept, hence a lot of socialization and discussion need to be conducted.

5. Conclusion

Islamic green city has so many significant elements that should be integrated into planning and design process. Therefore, planning and design for Islamic green city need a particular approach. Urban planner and urban designer in Islamic green city should have the ability to understand Islamic principles and Green city concept for shaping the build environment and the city. Significant role of *Ulama* (Muslim leader) has the opportunity to be the main point to socialize the concept of green city to citizens.

Martapura still lacks of many important elements as an Islamic green city. However, Martapura has started to implement the integrated and comprehensive approaches in form of green city action plan which base on three main elements of sustainability, which are environment, society and economy - also known as three Es (Environment, Economy, Equity); eight elements of green city; and Eco-Islamic Belief. This three combination made a new approach to be combined and integrated in Islamic green cities.

References

[1] Asian Development Bank 2012 *Green Cities – A Sustainable Urban Future in Southeast Asia*
[2] Steinberg B V F and Lindfield M 2012 *Green cities* (Mandaluyong City, Philippines: Asian Development Bank)
[3] Pankaja M S, Student P D and Nagendra H N 2015 Green city concept – as new paradigm in urban planning *Int. J. Eng. Sci.* pp 55–60
[4] El Ghorab H K and Shalaby H A 2016 Eco and Green cities as new approaches for planning and developing cities in Egypt *Alexandria Eng. J.* vol 55 pp 495–503
[5] Sandhu S C 2014 *Enabling GrEEEn Cities: An operational framework for integrated urban development in southeast asia* ADB Southeast Asia Work. Pap. Ser.
[6] Villumsen S 2012 Habermas in the garden : broadening the scope of social movements for sustainability *Master Thesis* Ser. Environ. Stud. Sustain. Sci.
[7] Johnston L F 2013 Religion and sustainability social movements and the politics of the environment *J. Ecol. Anthropol.* p 273
[8] Kementrian Pekerjaan Umum dan Perumahan Rakyat 2017 Panduan penyelenggaraan Program Pengembangan Kota Hijau, VI (Jakarta, Indonesia: Kemeterian Pekerjaaan Umum dan

Perumahan Rakyat, Direktorat Jenderal Cipta Karya, Direktorat Bina Penataan Bangunan)
[9] Abedi-sarvestani A and Shahvali M 2008 Environmental ethics : toward an islamic perspective *Am. J. Agric. Environ. Sci.* vol 3 pp 609–17
[10] Vincenti D 2016 Sustainability transitions in arab-islamic countries: egypt as a case study *Agric. Agric. Sci. Procedia* vol 8 pp 135–40
[11] Kunreuther H, Gupta S, Bosetti V, Cooke R, Dutt V, Ha-Duong M, Held H, Llanes-Regueiro J, Patt A, Shittu E and Weber E 2014 Chapter 2 - Integrated risk and uncertainty assessment of climate change response policies (Cambridge: Cambridge University Press) pp 151–206
[12] Van Klinken G and Berenschot W 2014 *In Search of Middle Indonesia* (Leiden, Boston: BRILL)
[13] BAPPEDA Kabupaten Banjar 2015 Rencana Aksi Kota Hijau (Banjar, Indonesia: Kabupaten Banjar)
[14] Jamalinezhad M, Talakesh S M and Soltani S H K 2012 Islamic principles and culture applied to improve life quality in Islamic cities *Procedia - Soc. Behav. Sci.* vol 35 pp 330–4
[15] Saoud R 2002 Introduction to the Islamic city found. Sci. Technol. Civilis. Pp 1–9
[16] Abu-Lughod J 1980 Contemporary relevance of Islamic urban principles (extract) *Ekistics* vol 47 pp 6–10
[17] Abu-Lughod J L 1987 The Islamic City--Historic myth, islamic essence, and contemporary relevance *Int. J. Middle East Stud.* vol 19 pp 155–76
[18] Eickelman D F 1974 Is there an islamic city? the making of a quarter in a moroccan town *Int. J. Middle East Stud.* vol 5 pp 274–94
[19] Banjarmasin Post 2017 Jemaah Haul Guru Sekumpul Diimbau Jalan Kaki (Banjarmasin, Indonesia: Banjarmasin Post) accessed on March 26th, 2017
[20] Banjarmasin Post 2017 Gundukan Sampah Masih Berserakan di Martapura (Banjarmasin, Indonesia: Banjarmasin Post) accessed on April 3rd, 2017

Study comparative of stability performance between PVC fishing boat and wooden traditional fishing boat

D Chrismianto, P Manik and G Rindo

Department of Naval Architecture, Faculty of Engineering, Diponegoro University, Semarang, Indonesia

deddychrismianto@yahoo.co.id

ABSTRACT: The use of PVC materials as an alternative material for fishing boat is being developed in Pekalongan, Central Java, Indonesia. Therefore, further analysis is needed to assist this development. In this paper, the analysis of ship stability of PVC fishing boat is investigated and be compared with wooden traditional fishing boat. The results of this analysis indicate that the PVC fishing boat has better stability which the PVC boat has GM values greater than traditional fishing boat. The GM values of traditional fishing boat is 3.988 m on normal load conditions, while the PVC fishing boat has a GM value of 5.828 m.

1. Introduction

In ancient era, the water transport is often used in Indonesian island is a raft. The used material is many bamboos which is arranged in such a way as to assemble a raft. This transportation is the beginning of the creation of modern ships. The raft is applied as a crossing tool between an island to other one or used by fisherman to catch a fish. In Taiwan, the rafts are mostly found, but using different materials. As a substitute for bamboo, PVC pipes are used to make rafts. With a ship length about 30-40 feets, the PVC boat is used to catching fish. The driving machine used is a diesel engine.

Actually, the fishing boat in Indonesia mostly use wood as raw materials. The wood used is not just any wood, but the kind of wood that has certain conditions, such as resistant to underwater environment, and has good properties of material strength. The wood used is usually old age and has a long size. Recently, the wood for boat material is decreasing and difficult to be obtained. In the long term, the logging for boat material can damage the environment.

In Indonesia on 2016, there is already fishing boat that use an alternative material instead of wood, one of which is Poly Vinyl Chloride (PVC). The fishing boat uses PVC material for hull part and expected to minimize the cost of manufacture and shorten the working time. The cost used to make a ship made of PVC ranges from IDR 800 million, while the fishing boat in general reach IDR 1.2 billion. The production time of the PVC fishing boat is also relatively faster about five months compared with the wooden boats production usually takes up to seven months. However, from the ship performance side, the stability of PVC fishing boat need to be assessed and compared with IMO regulation. In this research, the analysis of ship stability of PVC fishing boat is investigated and compared with wooden traditional fishing boat.

IOP Conf. Series: Materials Science and Engineering **403** (2018) 012002 doi:10.1088/1757-899X/403/1/012002

2. Literature review

PVC (Polyvinyl Chloride) was discovered accidentally by Henri Viktor Regnault in 1835 and Eugen Baumann in 1872. At the beginning of the 20th century, Russian chemist Ivan Ostromislensky and Fritz Klatte of the German chemical company Griesheim- Electron tried to use PVC as commercial products. But the difficulty of material pruning hinders their efforts. In 1926, Waldo Semon and the B.F Goodrich company developed the method of making PVC become truly plastic by adding some additives. As a result, PVC is more flexible and easier to process which then reaches global usage.

Polyvinyl chloride (PVC) is a pipe made of plastic and some other vinyl combinations. It is a third generation of thermoplastic polymer in terms of world usage, after polyethylene and polypropylene. Around the world more than 50% of PVC produced is used as a construction material. PVC is relatively cheap, durable, and easy to assemble. PVC can be made more elastic and flexible by adding plasticizer.

The mandatory stability requirements of ships shall refer to the standards established by the local Indonesian Bureau of Classification or the Marine Authority such as International Maritime Organization (IMO). Thus the process of stability analysis carried out must be in accordance with IMO (International Maritime Organization) standards. Code A.749 (18) Ch 3 - design criteria applicable to all ships Section A.749 (18), Chapter 3.1.2.3: The area under the curve GZ at oscillation angle 0°- 30° ≥ 3,101 m.deg, area 0°- 40° (deg) ≥ 5,157 m.deg and area 30°- 40° ≥ 1,719 m.deg. Chapter 3.1.2.2: Maximum GZ value occurring at an angle of 30°-180° ≥ 0.2 m. Chapter 3.1.2.2: angle at the maximum GZ value ≥ 25 °. Chapter 3.1.2.4: Early GM at 0° ≥ 0.15 m [1]. For ship stability analysis, the principle dimension of wooden traditional fishing boat and PVC fishing boat can be seen in table 1.

Table 1. The principal dimension.

The wooden traditional fishing boat		
1	Length (LOA)	15 m
2	Breadth	5.2 m
3	Height	1.8 m
4	Draught	1 m
5	Tonnage	17 GT
The PVC fishing boat		
1	Length (LOA)	18.3 m
2	Breadth	4.5 m
3	Height	0.75 m
4	Draught	0.75 m
5	Tonnage	17 GT

3. Results and discussion

First, the wooden traditional fishing boat and PVC fishing boat is developed in 3D model for the stability calculation (figure 1) [2, 3]. The stability analysis calculated under various loading conditions. The loading conditions that are investigated such as: departure condition, condition at sea when catching fish, and arrive condition [4-6]. The stability performance of wooden traditional fishing boat is obtained as in table 2 and figure 2. While the stability performance of PVC fishing boat can be seen in table 3 and figure 3.

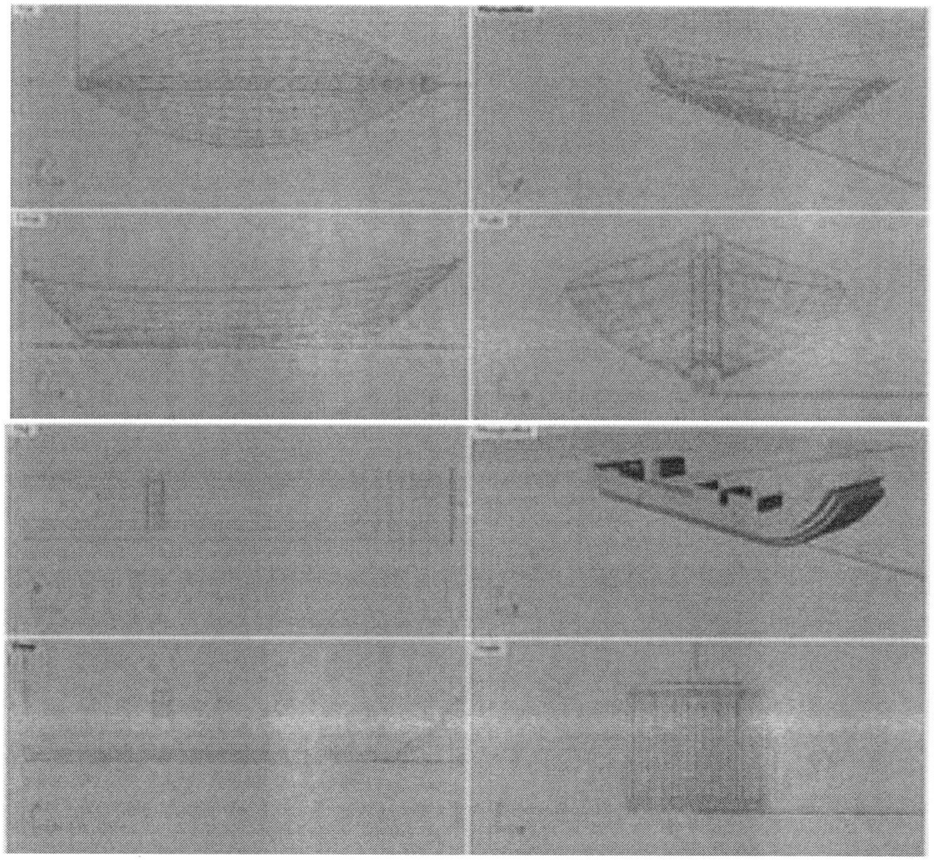

Figure 1. 3D Model of wooden traditional fishing boat and PVC fishing boat.

Table 2. Stability analysis for the wooden traditional fishing boat.

No	Criteria	IMO	Units	Load Condition		
				I	II	III
1	Area 0 to 30	3.151	m.deg	26.817	23.600	21.655
2	Area 0 to 40	5.157	m.deg	44.040	38.424	35.055
3	Area 30 to 40	1.719	m.deg	17.222	14.824	13,400
4	GZ on 30 and more	0.2	m	1.952	1.650	1. 460
5	Max GZ angle	25	deg	50.9	49.1	49.1
6	Initial GMt value	0.15	m	3.988	3.510	3.221
7	Initial GMt value for ships ≥ 24 m	0.35	m	3.988	3.510	3.221
8	Initial GMt value for ships ≥ 70 m	0.15	m	3.988	3.510	3.221
	Status			Pass	Pass	Pass

IOP Conf. Series: Materials Science and Engineering **403** (2018) 012002 doi:10.1088/1757-899X/403/1/012002

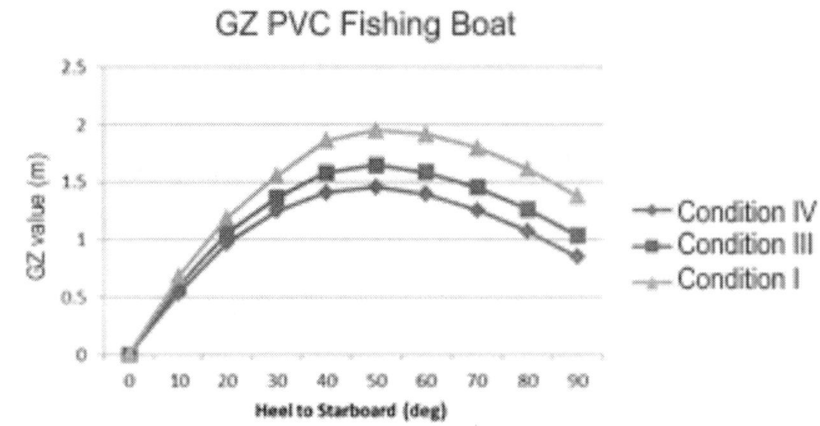

Figure 2. Graph of GZ Values for the wooden traditional fishing boat

Table 3. Stability analysis for the PVC fishing boat.

No	Criteria	IMO	Units	Load Condition		
				I	II	III
1	Area 0 to 30	3.151	m.deg	26.765	16.803	10.733
2	Area 0 to 40	5.157	m.deg	38.567	22.839	14.524
3	Area 30 to 40	1.719	m.deg	11.802	6.036	3.791
4	GZ on 30 and more	0.2	m	1.187	0.652	0.401
5	Max GZ angle	25	deg	31.8	19.1	17.3
6	Initial GMt value	0.15	m	5.828	3.705	3.062
7	Initial GMt value for ships ≥ 24 m	0.35	m	5.828	3.705	3.062
8	Initial GMt value for ships ≥ 70 m	0.15	m	5.828	3.705	3.062
	Status			Pass	Pass	Pass

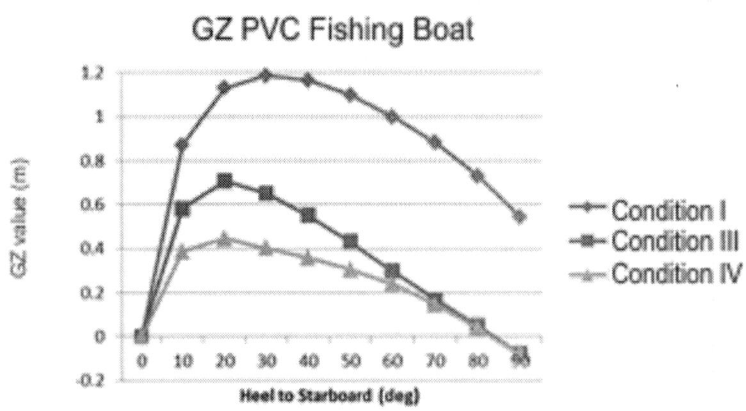

Figure 3. Graph of GZ Values for the PVC fishing boat.

The calculation of ship stability performed in this study analyzes the 3 conditions of each model that have been made, in accordance with IMO A.749 (18) Chapter 3.5. As for the 3 conditions are:

a. Condition I: Represents a ship condition before departing from port to fishing ground. Under these conditions, the vessel in a consumable charge state is empty.
b. Condition II: The ship is already on the fishing ground. In this condition, the ship has been given additional cargo ship (mackerel) 50%.
c. Condition III: The ship is still on the fishing ground. In this condition, the boatload (mackerel) is 100%.

Table 2 and table 3 show that the value of the area under the GZ curves at points 1, 2, and 3 for wooden traditional fishing boats and PVC fishing boats are still above the IMO standard. At the heel angle that assumed as down flooding point of the ship, that is between 0 - 30 degrees, 0 - 40 degrees and 30 - 40 degrees. Both of boats are still in the stable conditions because they have a large righting moment. While IMO standard on points 4 and 5 mentions that minimum distance and minimum angle at maximum GZ value should not less than 0.2 m and 25 degrees. From the calculations on all conditions stated that the GZ value for wooden traditional fishing boat and PVC fishing boat are still above the IMO standard.

IMO rule on point 6 states that the distance of metacenter gravity (MG) for minimum fishing boat is 0.35 m and the results show that MG values for wooden traditional fishing boats and PVC fishing boat under all conditions are above the standard set of requirements set by IMO. This condition can be declared stable because it has a positive MG value.

4. Conclusion

The GZ value of PVC fishing boat at condition I of 1.577 m; condition II of 1.266 m; the condition of III is 1.181 m which has fulfilled the criteria specified by IMO. The MG value of PVC fishing boat on condition I amounted to 3.624 m; condition II of 2.654 m; condition III of 2.700 m which already fulfill the criteria specified by IMO. While, The GZ value of wooden traditional fishing boat at condition I of 1.820 m; condition II of 1.573 m; condition III of 1.398 m where the value has also met the criteria specified by IMO. The MG value of wooden traditional fishing boat condition I is 3.617 m; condition II of 3.244 m; condition of III equal to 2.981 m where the value already fulfill the criteria specified by IMO.

The model of PVC fishing boat has better the stability performance than wooden traditional fishing boats, because it has larger MG distance than the wooden traditional fishing boat model. The large MG value makes the ability to reverse the ship in its original position become better.

Acknowledgement

The researcher would like to highly appreciate the Directorate of Research and Community Service, Directorate General of Research Strengthen and Development, Ministry of Research, Technology and Higher Education according to Research Contract Number 007/SP2H/LT/DRPM/V/2017.

References

[1] Barras B and Derrett C D 2011 *Ship Stability for Masters and Mates* Elsevier
[2] Chrismianto D and Dong J K 2014 Parametric bulbous bow design using the cubic Bezier curve and curve-plane intersection method for the minimization of ship resistance in CFD *J Mar Sci Technol* vol 19 pp 479-92
[3] Chrismianto D, Zakki A F, Arswendo B and Dong J K 2015 Development of cubic bezier curve and curve-plane intersection method for parametric submarine hullform design in order to optimize the hull resistance by using CFD *Journal of Marine Science and Application* vol 14 issue 4 pp 399-405
[4] Robert F B 1998 *Motion In Waves and Controllability, Principles of Naval Architecture Volume III* (USA: The Society of Naval Architecture and Marine Engineers)

[5] Hind J A 1982 *Stability and Trim Of Fishing Vesels And Other Small Ships. Second Edition.* (Famham, Surrey. England: Fishing News Book Ltd.)
[6] Iqbal M dan Rindo G 2016 Pengaruh anti Slamming Bulbous Bow Terhadap Gerakan Slamming Pada Kapal Perintis 200 DWT *KAPAL* vol 13 no 1

IOP Conf. Series: Materials Science and Engineering **403** (2018) 012003 doi:10.1088/1757-899X/403/1/012003

Green architecture concept in Chinese region at Clan Jetties area, Penang Malaysia

R D Nur'aini[1], Windarta[2], F Firdaus[1] and I Makinudin[1]

[1]Architecture Department, [2]Mechanical Engineering Department
Universitas Muhammadiyah Jakarta, Jl. Cempaka Putih Tengah 27 Jakarta Pusat, 10510

ratnadewina@ftumj.ac.id, ratnadewina@gmail.com

Abstract. Penang is part of Malaysia's territory where there is an area recognized by UNESCO as the world heritage, George Town. As a center of the conservation area, it is a port that became the center of commerce during the colonial period. It makes the city inhabited by multi ethnics from different parts of the world, such as Malay ethnic, Chinese ethnic, and Indian ethnic. This multi ethnic settlement becomes the main attraction to be made as the research material. This study aims to find the concept of green architecture of clan jetties area, describe the use of natural materials in its buildings and residences and explain the utility system on the region. This research uses qualitative method with analysis approach using general-specific method. Based on the research we conducted with the method of observation and direct observation of the research object, it can be concluded that clan jetties is a special settlement for Chinese ethnic who are migrants. The residential location is along the coastline adjacent to the Weld Quay harbor. The settlement is divided into 7 areas based on family or clan; Ong Jetty, Lim Jetty, Chew Jetty, Tan Jetty, Lee Jetty, New Jetty and Yeoh Jetty. Each clan has a similar pattern of settlement that extends to the sea with one main road and the side of the road is occupied by the houses of the citizens. The utility system of the area is broadly divided according to the utility of clean water for sewage and fire, lighting, and fire-fighting utilities. Each clan is already equipped with the utility. The main material used for building houses is wood. The placement of vegetation for greening using a flower pot is either placed at the bottom or hung on top.

1. Introduction

George Town is one of the historic areas in Penang, Malaysia. This conservation area is protected by UNESCO since 2008. More than 500 years ago this city was the center of trade and port so that now the region has many cultures, a blend of eastern and western cultures. This can be seen from the image of the buildings that dominate the region is a blend of European, Malay and Chinese ethnic styles.

Weld Quay and Clan Jetties are a coastal area of George Town which is located on the beach so this area becomes a port and residential area. Communities that live in the ports commonly have a livelihood as fishermen and workers. This area is protected as a historic conservation area with various ethnic groups including Malay, Chinese, Indian and Turkish. Each of these tribes occupies different areas so as to create a settlement area according to the tribe.

In ancient times, the main mode of transport for shipments of goods was sea vessels that made the port area became the attraction of the community as a place of livelihood. Therefore, there are many trading companies and warehouses here. In George Town area, there is a residential area of clan Jetties which is mainly inhabited by Chinese ethnics who originally worked as fishermen or port workers. Inside it, there are several clans which are distinguished by family ties, such as Ong Jetty, Lim Jetty, Chew Jetty, Tan Jetty, Lee Jetty, New Jetty and Yeoh Jetty.

According to [1], Green Architecture is a minimal architecture consuming resources: natural resources, including energy, water and materials, and minimal negative impact on the environment. Meanwhile, according to [2] formulates the definition of Sustainable Development (sustainable development) as development that can meet the needs of today's people without sacrificing the potential of future generations.

Green Architecture is a concept of planning a building that is friendly to the environment. This is an attempt to solve the problem of earth global warming. The implementation of Green Architecture is not only in Indonesia but other countries also begin to organize environmentally friendly buildings [3].

The World Green Building Council defines the rating system in Green Building as Greenship as a rating system divided into six aspects consisting of:

a) Appropriate Site Development (ASD)
b) Energy Efficiency & Refrigerant (EER)
c) Water Conservation (WAC)
d) Material Resources & Cycle (MRC)
e) Indoor Air Health & Comfort (IHC)
f) Building Environment Management (BEM)

2. Methodology

This research used descriptive qualitative method. The research object was located in the clan Jetties area of George Town Penang Malaysia with an area of 259.42 hectares (figure 1). Data collection techniques were conducted in several ways: literature review, field observation, and interviews with local residents.

Data collected from the survey in Penang, Malaysia were then selected and classified into some categories. Those that were not suitable with the result were eliminated. The data analysis was done from of sketches and documentation of observation result.

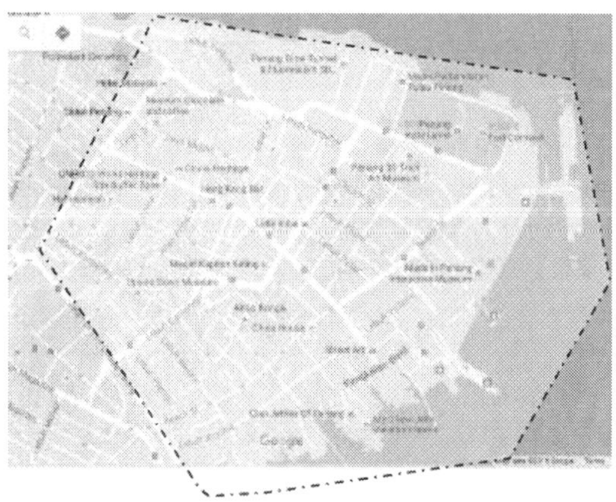

Figure 1. Research area.

3. Result and discussion

Clan Jetty consists of 7 (seven) clans residing in the port area [4]. The clan division is based on lineage or family. However, this research only focused on Lim Jetty and Chew Jetty [5].

3.1. Lim Jetty

Lim Jetty was previously a much larger settlement with 42 buildings before World War II. Unfortunately, it was burned and destroyed during the Japanese bombing in December 1941. The reconstruction began slowly after the war and new temples appeared to meet the spiritual needs of the people. The houses with zinc roofs are divided into two separate main areas with one row of wooden shacks lined along one path facing the sea and the other cottage facing each other on the other side (figure 2).

(a)

(b)

Figure 2. Lim Jetty area (a) main gate; (b) settlement.

3.2. Chew Jetty

Chew Jetty is the largest and the most visited water village among the other clans. The closed-tied Chew community played an important role in the settlement of Chew clan immigrants from the same hometown in the old days and has largely been spared the war damage for years. The area offers a public shuttle to reach the ships parked in the center of the sea lane and it is a favorite spot for shutterbugs and setting of movies with its friendly villagers, along with small shops, hair salons, beverage outlets and souvenir stores lined the narrow alleys (figure 3).

(a)

(b)

Figure 3. Chew Jetty area (a) front area; (b) central area (souvenir area).

The annual prayer ceremony of Thi Kong She held to commemorate the Emperor's birthday on the 9th day of the Lunar New Year. It is especially celebrated on a grand scale here and remains as one of the highlights on the Penang calendar for the festival. They will prepare food like roast pork, pastries and fruits and put them on a 50-meter-long altar with main road sides. The side of the table is tied with sugarcane, a symbol of gratitude to God and favorable pieces of red paper hung from sugarcane stems

IOP Conf. Series: Materials Science and Engineering **403** (2018) 012003 doi:10.1088/1757-899X/403/1/012003

and pasted on food offerings. The altar is surrounded by huge tusces of colorful giant incense and midnight fireworks that turn the road into a magical landscape.

3.3. *Green architecture*

Green Architecture in clan Jetty area covers various aspects such as natural lighting and the use of natural materials in residential buildings and supporting infrastructure of the area.

3.3.1. Natural lighting. Lim Jetty is an area close to the Quay weld harbor, where the area is mostly a dock for ships - docked. Ventilation in this settlement is using air vents located above the door of the building where the open ventilation is wide enough to circulate the air (figure 4(a)). Natural lighting in this settlement is using windows of a house like a house in general. The types of windows are various - naco glass windows and ordinary glass window.

Chew Jetty is a souvenir central area in clan Jetty. The area is mostly stalls merged with the home residents. Like other clan Jetty settlements, houses in Chew Jetty area are also made of wood, so for home air circulation they use ventilation and windows as the main air circulation (figure 4(b)). The main lighting in the settlement is obtained from the windows and doors of the house where the windows are blown with glass and woodwork so that the lighting and air conditioner can be maximized.

(a)

(b)

Figure 4. Natural lighting (a) Lim Jetty; (b) Chew Jetty.

3.3.2. Building materials. Clan Jetty in Penang area is a settlement established on the sea, where its wall and main roads of the settlement are made of natural material, wood (figure 5).

(a)

(b)

Figure 5. Wood and concrete foundations (a) Chew Jetty; (b) Lee Jetty.

3.4. Vegetation
Reforestation in clan Jetty area utilizes the media in the form of pots and hanging plants. This is influenced by the location of a relatively large area located over the sea, so harnessing pot for planting area is very profitable and become solution of greening problem. The plant plots are in various positions such as on the roadside and in the home garden area so that the treatment of the plants is done by each house (figure 6).

(a)

(b)

Figure 6. vegetation plots in the area of citizen, rear (a) and along the road (b).

4. Conclusion
Based on the research, it can be concluded that clan Jetty is an area that belongs to a heritage protected by UNESCO. This area is included in the scope of George Town area of Penang Island Malaysia. It is a residential area inhabited by Chinese migrants ethnic where the area is located near to the harbor of the weld base. At first, the migrants from China came to Penang Island to trade when George Town was a well-known port city at the time. Now, the migrants occupy the coastal areas and work on average at the base of the ship in various professions.

Clan Jetty is a residential area of Chinese ethnic based on clan so that each region is a collection of clan's descendants. The clan in the area consists of 7 districts and 6 clans and one area which is a mixed area of the entire clan. Clans which are as follows: clan Ong Jetty, clan Lim Jetty, clan Chew Jetty, clan Tan Jetty, clan Lim Jetty, clan New Jetty and clan Yeoh Jetty. From the seven clans, the clan and new Jetty are clans of various clans.

The layout of each clan has the same characteristics that are both located above sea level and jutting toward the sea except Ong Jetty and Yeoh Jetty areas which are on the land. The pattern of settlement is linear in which the road as its center and the settlement is on the left and right of the road, except Chew Jetty residential area where its settlement pattern is forked and divided into a residential area.

Lighting and carriage in every house of the area uses natural stimulation and is assisted with artificial lighting during the night. Air circulation in the residents is using windows and ventilation, and assisted with the use of air conditioner.

The utilization of vegetation for greening in this area is using plants in pots and placed in the area of the road and yard of the house.

References
[1] Karyono T H 2010 *Green architecture; Pengantar pemahaman arsitektur hijau di Indonesia* (Jakarta: Rajawali Pers)

[2] World Commission on Environment and Development 1987 *Report of the World Commission on Environment and Development: Our Common Future* (Oxford: Oxford University Press)

[3] Iyengar K 2015 *Sustainable Architectural Design: An Overview* (Routledge)
[4] -- *Clan* Jetties *Floating Village in Penang, wonderful Malaysia* [Online] Available at http://www.wonderfulmalaysia.com/ (accessed on 1 January 2017)
[5] -- 2009 Chew *Jetties, A Stroll Down a Living Heritage Community* [Online] Available at http://www.visitpenang.gov.my/ (accessed on 3 January 2017)

IOP Conf. Series: Materials Science and Engineering **403** (2018) 012004 doi:10.1088/1757-899X/403/1/012004

Shifting of air pollutants distribution during car free day event

H S Huboyo, I W Wardhana, E Sutrisno and S H Budisulistiorini

Environmental Engineering Department, Faculty of Engineering,
Diponegoro University, Jl. Prof. H. Soedarto SH, Tembalang, Semarang, Indonesia

huboyo@gmail.com, huboyo@undip.ac.id

Abstract. According to Decree of Semarang Mayor No. 22/2011, car free day activities is addressed to give clean air for facilitating citizens activities. This car free day event is held every Sunday in the morning in the city center of Semarang i.e. located at Simpang Lima square. This research is aimed at identifying the shifting of pollutant during car fee day event by comparing ambient air pollutant concentration represented by carbon monoxide during car free day event and non-car free day event. About 14 streets had been measured its ambient CO concentration during Saturday (non-car free day event) and Sunday (car free day event). We also modeled (using Caline4) the CO dispersion at the certain area on those streets to know the spatial distribution of concentration during those two events. The ambient CO concentration, in general, during car free day event were somewhat increase for certain roads. The emission load of vehicles emission during CFD event was 1.37 times of non-CFD event. Nevertheless, based on spatial distribution of ambient CO concentration at the area of roads of interest, its concentrations were below the ambient CO concentration standard (PP.41/99).

1. Introduction

As the city becoming bigger due to urbanization, transportation matter is important part of city growth. Uncontrolled growth of vehicles owned privately in one side and limited growth of road inner the city in other side make unbalanced growth and further it makes congestion and traffic jam. Based on this issue, it is indispensable for many cities to eagerly try to reduce the private vehicles use in city centers. Nowadays, cities from around the world seem to become more willing to try car free initiatives [1] particularly in developed countries. These initiatives supported by city government should be participated by the citizen to run it effectively. The measures on reduction of motorized traffic is deemed to give benefit to public health, both in the short and long-term [2].

Car sharing, carpooling and car free days are several measures for minimizing private cars. In Netherland, car sharers drive around 15%–20% fewer car kilometers than before they started car sharing and emit between 240 and 390 fewer kilograms of CO_2 per person, per year [3].

Based on the research in Italy, carpooling with certain strategy could reduce the emission by 22 – 28% [4]. Changing habitual commuting by bicycle in the area of Stockholm, theoretically, could reduce RR of NOx by 8% associated by 10 μgm^{-3} decrease [5]. Other study by Scheepers reviewed the effectiveness measure by shifting car use to cycling and its impact to mortality [6]. Car free day (CFD) is the foremost popular measure in developing world due to less costly and attract economical aspect. However, this CFD face big challenge as awareness and participation of the public are still in question.

IOP Conf. Series: Materials Science and Engineering **403** (2018) 012004 doi:10.1088/1757-899X/403/1/012004

Clearly the awareness and event quality, in CFD event, were not the main driving factor for the people to participate at the CFD as regular event [7]. In Hanoi, Vietnam, car free city and city-of-short-distance concepts had difficulties in the management level of authority, awareness and public participation [8]. This CFD is claimed to clean the ambient air in the CFD area although it is temporary. The context of reducing pollutant by implementing CFD event is true only in the CFD area. Whilst in the surrounding area, it is predicted that level of air pollutant is higher than before CFD implementation. This study is aimed at knowing the impact of air pollutant distribution due to car free day event particularly at surrounding area. Emission load calculation and shifting distribution of air pollutant are the main topic of this study. The discrepancies of implementing CFD in the road segment could be handled by implementing CFD at wider area (comprising multiple roads, not operating CFD event at a road segment only).

2. Methodology

This section explains the methodology used in this research. The location of study, data collection method, and data analysis are elaborated in the following subsection.

2.1. Location of study and data collection

This study was conducted during car free day event (Sunday) and non-car free day event (Saturday) in 2016 and 2017. The location of the study and the sampling points are depicted in figure 1. We collected the data on number of vehicles passing through the measured road (hourly basis), CO concentration (Extech) and other micrometeorological aspects such as wind speed, wind direction and ambient temperature. Measurements were conducted during car free day event i.e. 06.00 – 09.00 AM. We also measured several ambient CO backgrounds during early morning (before 06.00 AM). We assume, during this time, the background CO was not affected by traffic emission. Totally 14 roads around the car free day area (Simpang Lima Square) were measured.

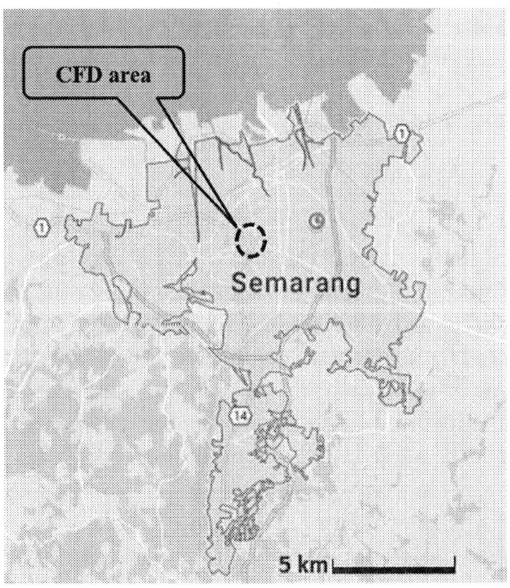

Sampling points	*Lat Long Coordinates*	Sampling points	*Lat Long Coordinates*
Kusumawardhani	6°59'46"S 110°25'16"E	Pandanaran_2	6°59'29"S 110°25'12"E
Undip Pleburan	6°59'40"S 110°25'26"E	Menteri Supeno	6°59'36"S 110°25'09"E
Erlangga	6°59'31"S 110°25'24"E	Gergaji	6°59'43"S 110°25'07"E
Ahmad Yani	6°59'31"S 110°25'36"E	Imam Bonjol	6°58'59"S 110°24'35"E
Ahmad Dahlan	6°59'22"S 110°25'30"E	Imam Bonjol – Pierre Tendean	6°58'43"S 110°24'44"E
Gajah Mada	6°59'18"S 110°25'23"E	Pierre Tendean	6°58'41"S 110°24'51"E
Pandanaran_1	6°59'20"S 110°25'12"E	MH Thamrin	6°58'42"S 110°25'01"E

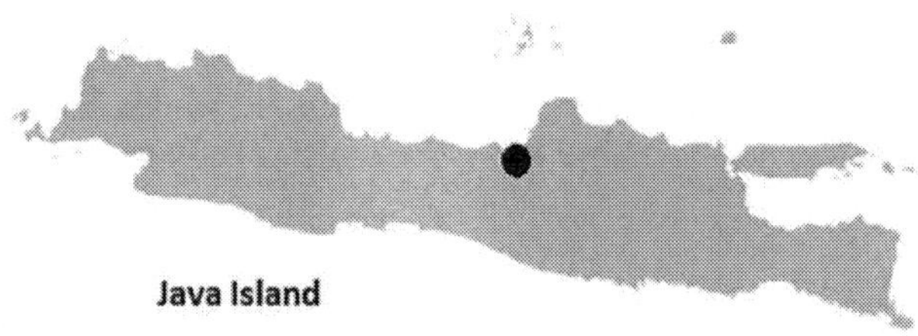

Figure 1. Location of the study.

2.2. Data analysis

After collecting the ambient CO concentration and number of vehicles (in category), we analyze the distribution of CO concentration in the concerned area using Caline4 in Calroad View. We compare it during car free day event and non-car free day event. Based on the modelled map, we are able to know the difference of CO distribution during CFD and non-CFD event. For knowing the impact of increasing number of vehicles in concerned roads we estimated the CO emission load (g/h). To do so, we used guideline for emission inventory for road transportation from Ministry of Environment.

3. Results and discussion

This section discusses the research results. Authors analyze the traffic activities and emission load during CFD and non-CFD event and estimate the distribution of ambient CO. The discussion of the result is elaborated in the following subsection.

3.1. Traffic activities and emission load

During CFD event, the number of vehicles in the road surrounding the CFD arena showed higher than during non-CFD event (see table 1). This is true to all mode vehicles i.e. gasoline car, diesel car and motorcycle. However, several few roads showed lower number of vehicles. The highest increase of vehicle number took place at small roads (i.e. alternative roads) such as Kusumawardhani road, Gergaji road. Ahmad Yani was the main road which had high increase of vehicles.

Quantitatively, as featured in figure 2, the increased of CO concentration at the roads were identified in Menteri Supeno street (208.7%), Gergaji street (145.5%), Kusumawardhani street (228.6%), Imam Bardjo street (85.7%), Erlangga street (63.6%), Pandanaran street section 2 (143.75%), Ahmad Dahlan street (79.4%), Pierre Tendean street (13,15%), Imam Bonjol (12,62%) and Pemuda street (18,94%). At contrast, at Gajah Mada street, it decreased by -13.3%, Pandanaran street (-34.3%). While at A. Yani street there was no increased or decreased. Due to this rise, commonly the ambient CO concentration at many roads surrounding CFD arena during CFD event were much higher than those at non-CFD event.

Table 1. Number of vehicles during CFD and non-CFD events.

Sampling points	Non-CFD Event (units)			CFD Event (Units)			CFD/Non-CFD		
	Gasoline		Motorcycle	Gasoline		Motorcycle	Gasoline		Motorcycle
	Gasoline	Diesel		Gasoline	Diesel		Gasoline	Diesel	
Kusumawardhani	235	42	1640	808	107	4801	3.4	2.5	2.9
Undip Pleburan	508	40	2032	720	44	3155	1.4	1.1	1.6
Erlangga	479	78	2040	145	29	2280	0.3	0.4	1.1
Ahmad Yani	442	62	1875	1981	170	7150	4.5	2.7	3.8
Ahmad Dahlan	1409	262	4266	1214	132	5125	0.9	0.5	1.2
Gajah Mada	2436	255	5919	1289	95	6204	0.5	0.4	1.0
Pandanaran_1	411	55	1684	449	53	2609	1.1	1.0	1.5
Pandanaran_2	935	127	3371	1037	163	5447	1.1	1.3	1.6
Menteri Supeno	806	86	4890	820	45	6972	1.0	0.5	1.4
Gergaji	65	8	655	195	28	798	3.0	3.5	1.2
Imam Bonjol	1657	253	6385	1947	508	7038	1.2	2.0	1.1
Imam Bonjol –Pierre Tendean	1571	361	6287	2676	403	7060	1.7	1.1	1.1
Pierre Tendean	1809	192	6449	2386	256	7087	1.3	1.3	1.1
MH Thamrin	1216	187	4516	1452	283	5172	1.2	1.5	1.1

There is a fluctuation of measured ambient CO concentrations (both during CFD event and non-CFD event) indicating that the number of vehicles vary with hour during CFD event or non-CFD event. There is no exact pattern of ambient CO concentration with time which may indicate dissipation of CO concentration in ambient air prevent accumulation of ambient CO concentration.

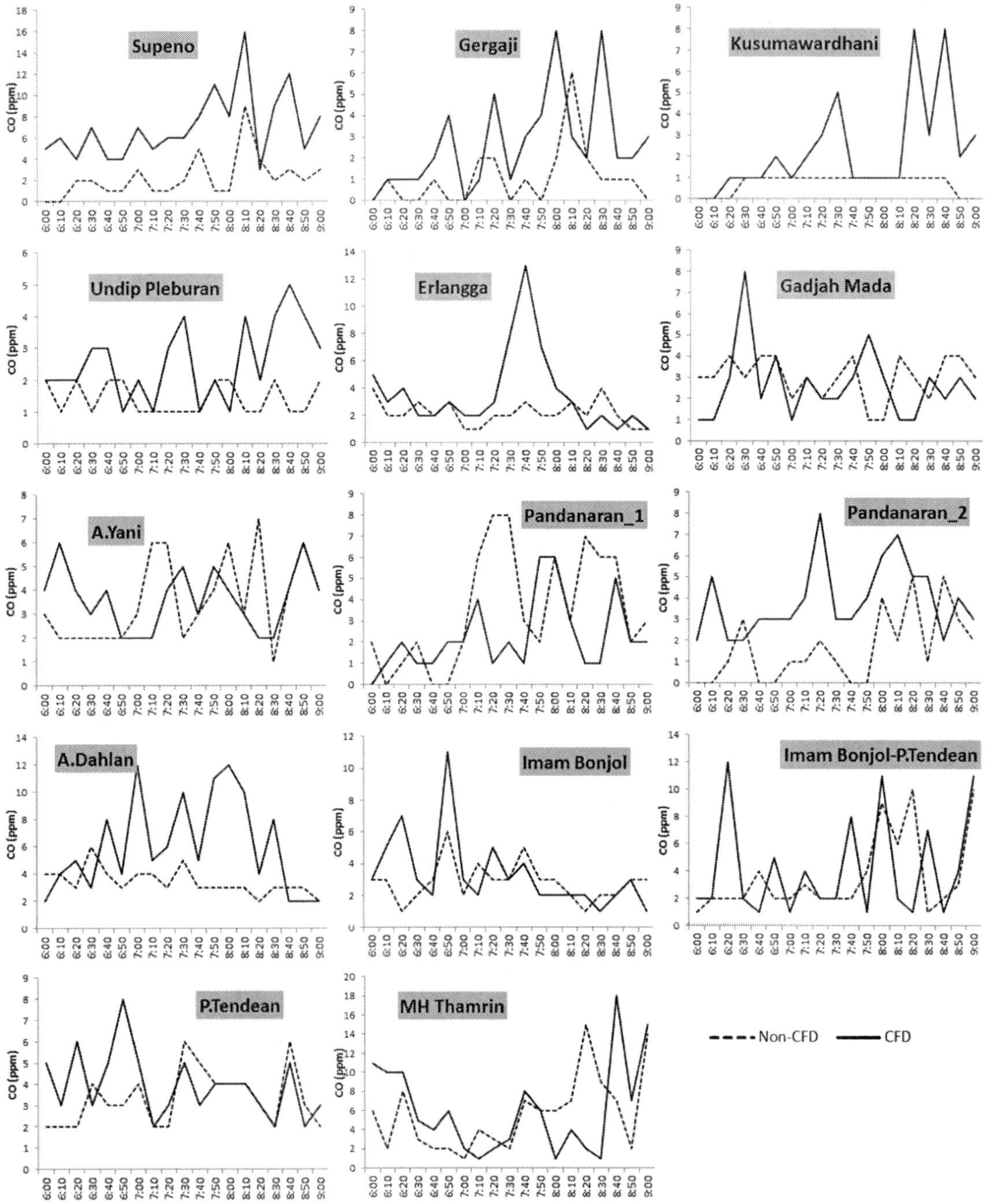

Figure 2. Varying concentration of ambient CO during CFD and non-CFD event.

As derived from figure 3, the emission load of vehicles during CFD event in total roads about 16.5 ton CO_2/year (if we assume the CFD take place every two weeks). This value is 1.37 times of non-CFD event. Thus this CFD objective to reduce the emission from its activity will be irrelevant. It is important to apply this CFD as an area CFD (not just for CFD implication for certain roads). No more attracted activities should be delivered in the CFD arena to prevent people to buy things at CFD arena which may be interested for CFD users.

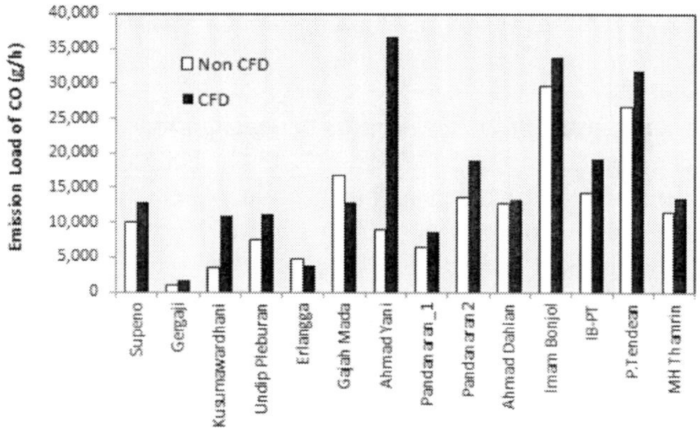

Figure 3. Ratio of emission load during CFD and non-CFD.

3.2 Estimated distribution of ambient CO

In modeling air pollutant distribution, we used Caline4 in Callroad View in concerned roads with several assumption parameters as follows: wind direction deviation 20°, atmospheric stability class 7, mixing height 450 m and mixing zone with 10 m. While the CO emission factor for gasoline car vehicle, diesel vehicle car and motorcycle are 64.37 g/mil, 4.5 g/mil, 22.53 g/mil respectively. We accommodate field measurement of temperature, wind (speed and direction) and traffic volume to be inputted in the model. Receptor sites then could be plotted to know the CO distribution, in this study we select 7 receptor sites (around Pemuda road) and 20 receptor sites (around Simpang Lima square). After defining link activities for each related road, then we could get the results of Caline4 model as depicted in figure 4 where (A) and (C) for non-CFD event and (B) and (D) for CFD event.

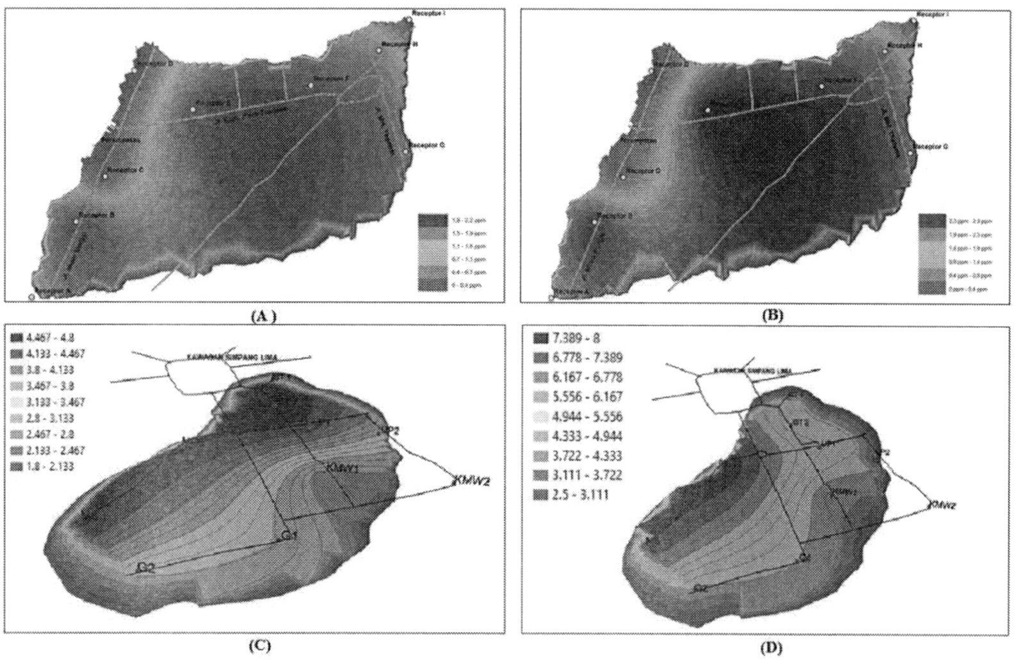

Figure 4. Increasing ambient CO concentration surrounding CFD area.

Based on the model results, the majority of the roads surrounding CFD area had higher ambient CO concentration than those during non CFD event. The reason for this incidence CFD event attracted more vehicles to come the area, avoiding vehicles driving through Simpang Lima square to surrounding available roads. Nevertheless, this shifting of CO distribution from CFD area to surrounding CFD area were still less than CO ambient standard (Government Regulation 41/1999).

4. Conclusion

About 14 streets had been measured its ambient CO concentration during Saturday (non-car free day event) and Sunday (car free day event). We also modeled (using Caline4) the CO dispersion at the certain area on those streets to know the spatial distribution of concentration during those two events. The ambient CO concentration, in general, during car free day event were somewhat increase for certain roads. Based on the model results, the majority of the roads surrounding CFD area had higher ambient CO concentration than those during non CFD event. The reason for this incidence CFD event attracted more vehicles to come the area, avoiding vehicles driving through Simpang Lima square to surrounding available roads. Nevertheless, based on spatial distribution of ambient CO concentration at the area of roads of interest, its concentrations were below the ambient CO concentration standard (PP.41/99).

Acknowledgement

This study was funded by DIPA Faculty of Engineering Diponegoro University 2017. Field measurements by Raditya and Diken are greatly appreciated.

References

[1] Nieuwenhuijsen M J and Khreis H 2016 Car free cities: Pathway to healthy urban living *Environ. Int.* vol 94 pp 251–62

[2] de Nazelle A, Nieuwenhuijsen M J, Antó J M, Brauer M, Briggs D, Braun-Fahrlander C, Cavill N, Cooper A R, Desqueyroux H, Fruin S, Hoek G, Panis L I, Janssen N, Jerrett M, Joffe M, Andersen Z J, van Kempen E, Kingham S, Kubesch N, Leyden K M, Marshall J D, Matamala J, Mellios G, Mendez M, Nassif H, Ogilvie D, Peiró R, Pérez K, Rabl A, Ragettli M, Rodríguez D, Rojas D, Ruiz P, Sallis J F, Terwoert J, Toussaint J-F, Tuomisto J, Zuurbier M and Lebret E 2011 Improving health through policies that promote active travel: A review of evidence to support integrated health impact assessment *Environ. Int.* vol 37 pp 766–77

[3] Nijland H and van Meerkerk J 2017 Mobility and environmental impacts of car sharing in the Netherlands *Environ. Innov. Soc. Transitions* vol 23 pp 84–91

[4] Bruck B P, Incerti V, Iori M and Vignoli M 2017 Minimizing CO2 emissions in a practical daily carpooling problem *Comput. Oper. Res.* vol 81 pp 40–50

[5] Johansson C, Lövenheim B, Schantz P, Wahlgren L, Almström P, Markstedt A, Strömgren M, Forsberg B and Sommar J N 2017 Impacts on air pollution and health by changing commuting from car to bicycle *Sci. Total Environ.* vol 584–585 pp 55–63

[6] Schepers P, Fishman E, Beelen R, Heinen E, Wijnen W and Parkin J 2015 The mortality impact of bicycle paths and lanes related to physical activity, air pollution exposure and road safety *J. Transp. Heal.* vol 2 pp 460–73

[7] Hussein A S 2016 How event awareness, event quality and event image creates visitor revisit intention? a lesson from car free day event *Procedia Econ. Financ.* vol 35 pp 396–400

[8] Minh N Q 2016 Application of "Car-Free City" and "City of Short Walks" to living quarters in Hanoi towards sustainable mobility and logistics *Procedia Eng.* vol 142 pp 284–91

Estimating methane emission from solid waste landfill using various different methods

R Mahful and S Managi
Kyushu University, Japan

rmahful@gmail.com

Abstract. Solid waste management is one of biggest problems for big cities in Indonesia. Currently, open dumping method is applied for managing the waste generated and produces methane emission as one of potential global warming gasses. This paper examined the methane that generated by Benowo landfill in Surabaya and Antang landfill in Makassar. Four methods were applied, namely IPCC default method, zero order, landGEM, and Order decay method. The result showed that methane emission quantity generated by these two landfills was significantly influenced by two local climate parameters: temperature and rainfall intensity. In addition, the analysis method used was also contributing to the volume of methane emission emitted.

1. Introduction

As a developing country, Indonesia has a main problem in waste management particularly municipal organic waste. This typical waste is predicted to increase in yearly basis align with the population growth. The Indonesian Statistical Bureau (2015) revealed that Surabaya city with 2,870,200 inhabitants generated 539,342.25 tons of waste while Makassar city with 1,369,606 populations generated 1,632,045 ton of waste. These figures indicate that the ratio between waste and residents in Surabaya and Makassar cities were 0.19 ton/person and 1.19 ton/person respectively and it increases every year. The waste management issues can be found globally particularly in Southeast Asia [1]. In addition, [2] and [3] noted that the waste management study is the most crucial issue that should be considered.

Methane (CH_4), Hydrogen Sulphide (H_2S), and Carbon dioxide (CO_2) are three main gasses that generated from organic waste landfill. These gasses contribute to the global warming and should be managed properly. There are three tools that can be used to calculate those gasses emissions from landfill operations, namely IPCC, CLEEN Model, and Land GEM [4, 5]. Zero and order-decay method are used as the basic reference.

The aim of this study is to calculate the Greenhouse Gasses (GHG) emission that generated by two landfills in Indonesia, namely Benowo landfill in Surabaya and Antang landfill in Makassar. Four methods that used are as follows: zero order, IPCC method and First Order, LandGEM Sofware.

2. Materials and methodology

2.1. The landfill site

Two landfills in Indonesia were investigated in this paper. One is located in Surabaya and another one is located in Makassar. Both of the landfills are managed the huge amount of municipal waste. Therefore, these two landfills are suitable to be used as a case study to determine and evaluate the feasibility of landfill energy projects.

2.1.1. Benowo Landfill (LF 1). LF 1 is ± 37.4 hectares in size and its final height will be between 5 – 12 m. The Benowo landfill has been operating since November 2001 and manages all wastes generated by Surabaya city. The LF1 closure plan is predicted on 2030. The total municipal waste that managed by LF1 is approximately 539 thousand tons per year as shown in table 1.

Table 1. Quantity of solid waste disposed into Benowo landfill.

Year	Quantity (ton/year)
2003	204,000.00
2004	335,618.50
2005	467,237.00
2006	598,855.50
2007	540,200.00
2008	459,425.50
2009	448,741.95
2010	463,779.05
2011	478,816.14
2012	493,853.24
2013	508,890.33
2014	531,403.50
2015	539,342.25

Source: central bureau of statistic, Indonesia

2.1.2. Antang Landfill (LF 2). LF 2 is ± 14.3 hectares in size and its final height will be between 5 – 12 m. Most of the areas have been capped and almost reach its final design capacity. Antang landfill was operated in 1993 and covers all wastes generated by Makassar city. The LF2 closure stage is planned on 2032 with total wastes disposed on 2015 were 1.6 million tons as shown in table 2.

Table 2. Quantity of solid waste disposed into Antang Landfill.

Year	Quantity (ton)
2003	1,027,495
2004	1,077,874
2005	1,128,254
2006	1,178,633
2007	1,229,012
2008	1,279,391
2009	1,329,770
2010	1,380,149
2011	1,432,085
2012	1,480,907
2013	1,531,287
2014	1,581,666
2015	1,632,045

Source: central bureau of statistic, Indonesia

2.2. Landfill gas generation model

Two methods were used in this study for estimating the emission rates of total landfill gas generated by municipal solid waste landfill. These methods are Landfill Gas Emission Model (LandGEM) software and Default Model that developed by The United State Environmental Protection Agency (US EPA) and IPCC respectively. The model determines the mass of methane generated by using the methane generation capacity and the mass of dumped waste. Table 3 shows the comparison of CHG emission estimation methods.

Table 3. Comparison of CHG emission estimation methods.

Features	Methodology	
Reaction Kinetics	Default method zero order	LandGEM first order
Model equation	$QCH_4 = \left\{ \frac{MSW_t \; x \; MSW_f \; x \; MCF \; x \; DOC \; x \; DOC_f \; x \; F \; x}{\left(\frac{16}{12} - R\right) \cdot (1 - OX)} \right\}$	$QCH_4 = \sum_{i=1}^{n} \sum_{j=0.1}^{n} K \, L_o \left(\frac{M_i}{10}\right) e^{-KT_{ij}}$
	Where QCH_4: annual CH_4 generation (Gg/yr.); MSWT: total MSW generated (Gg/yr); MSW_f: Fraction of MSW disposed at SWDS (generally 80%); MCF: CH_4 correction factor (fraction); DOC: Degradable organic carbon[$Fraction\ (CGgMSW)$]; DOC_f: Fraction DOC dissimilated; F: Fraction by volume of CH_4 in Landfill gas; R: recovered CH_4 (Gg/yr.); OX: oxidation factor (fraction).	Where QCH_4: annual CH_4 generation (mg/yr); i=1 year time increment; n: (year of the calculation) – (initial year of waste acceptance); j: 0.1 year time increment; k: CH_4 generation rate(year-1); Lo: potential CH_4 generation capacity (m3/Mg); Mi: mass of waste accepted in the ith year (Mg); tij: age of the Jth section of waste mass Mi accepted in the i-th year.

The study initiated by EPA defines the electricity generation potential from landfill between 75-85% of the total methane produced. With the calorific value of methane is 4.5 kWh/year then the formula to estimate the potential electricity as shown in equation (1).

$$Electricity\ per\ year = \frac{CH_4}{year} \cdot 75\% \cdot 4.5kWh \tag{1}$$

The CH_4 generation rate constant (K) is the rate of waste decay and It can be calculated by using the equation (2).

$$K = 3.2 \cdot 10^{-5}(R) + 0.01 \tag{2}$$

Where R is an average annual precipitation in mm. Value of K for Benowo and Antang landfills are presented in table 4 and some assumptions made for Benowo and Antang landfills are shown in table 5.

Table 4. K values for Benowo and Antang landfills.

Location	Average precipitation (mm/year)	K ($year^{-1}$)
Benowo	141.1 mm	0.01
Antang	22,186.0 mm	0.08

Table 5. Assumption made for Benowo and Antang landfills.

Model parameter	Benowo Landfill		Antang Landfill	
	IPCC Default Method	LandGEM Method	IPCC Default Method	LandGEM Method
Methane Correction factor (MCF)	0.4	-	0.4	-
Fraction of CH4 in landfill gas (F)	0.5	0.5	0.5	0.5

Model parameter	Benowo Landfill		Antang Landfill	
	IPCC Default Method	LandGEM Method	IPCC Default Method	LandGEM Method
CH4 recovery (R)	0	-	0	-
Oxidation Factor (OX)	0	-	0	-
The CH4 generation rate constant (K) (year-1)	-	0.01	-	0.08
CH4 Generation potential (L0) (m3/mg)	-	170 (arid area)	-	170

3. Result and discussion

Based on the characteristics, Benowo landfill is located in arid area with the low annual precipitation and high temperature as presented in table 6. These two parameters resulted in the L_0 difference of Antang and Benowo landfills. Therefore, the data analysis used different reference LandGEM where Antang landfill use CAA Conventional and Benowo landfill uses CAA Arid Area.

Table 6. Operational parameters of Benowo and Antang Landfills, Indonesia.

Features	Detail	
	Benowo	Antang
Location: Latitude	7°13'9, 70'' S 112°37'52.18 T	5°10'36.02'' S 119°29'25,35'' T
Longitude		
Year of start	2001	1993
Year of closure	2032	2030
Area (Ha)	37.4	14.3
Annual precipitation	141.1 mm	2186 mm
Temperature (c)	35	27

In this study, the $CH_{4\ emission}$ calculation for Antang and Benowo Landfills showed the differences result between IPCC Default method and LandGem. The result generated by IPCC was 20 times higher than LandGEM. The similar result was also shown in [1], when the author calculated $CH_{4\ emission}$ from open dump site in Kapur, India. The author found that the difference due to the elimination of temperature and rainfall fall parameters from the analysis. However, this study was considering use the higher result of CH_4 Produced to manage the risks that might occurred properly.

$CH_{4\ emission}$ generated by Benowo Landfill using LandGEM showed that it increased in yearly bases until the landfill closure period. In addition, the analysis indicated that the estimation was two times higher compared to IPCC calculation. The result also showed that methane emission generated by landfill in arid area was higher compared to landfill in other areas. The calculation of methane emission using various methods is presented in table 7 and figure 1 and figure 2.

Table 7. CH_4 emission from Benowo and Antang Landfills using various models.

Year	Waste quantity (mg/year)		Benowo	Antang	Benowo	Antang
	Benowo	Antang	IPCC(Default Method)	IPCC (Default Method)	Land GEM	Land GEM
2004	305,108	979,886	9,021.12	318,996.40	458.6	5,180
2005	424,761	1,025,685	12,558.91	304,215.70	1,135.0	10,360
2006	544,414	1,071,484	16,914.97	318,434.40	2,068.0	15,540
2007	491,091	1,117,284	15,573.15	332,653.20	3,251.0	20,730

Year	Waste quantity (mg/year)		Benowo	Antang	Benowo	Antang
	Benowo	Antang	IPCC(Default Method)	IPCC (Default Method)	Land GEM	Land GEM
2008	417,660	1,163,083	13,244.54	346,872.30	4,290.0	25,910
2009	407,947	1,208,882	12,936.55	361,091.10	5,144.0	31,100
2010	421,617	1,254,681	12,150.29	29,815.42	5,959.0	36,280
2011	435,287	1,301,895	12,544.24	31,365.01	6789.0	41,470
2012	448,957	1,346,279	12,938.19	32,777.35	7,633.0	46,670
2013	462,628	1,392,079	13,332.13	74,907.72	8,491.0	51,860
2014	483,094	1,437,878	13,921.94	77,456.05	9,363.0	57,050
2015	490,311	1,483,677	14,129.93	80,004.33	10,260.0	62,240
2016	489,784	1,454,126	14,114.73	82,552.62	11,160.0	67,430
2017	493,163	1,470,557	14,212.12	242,725.10	12,040.0	72,210
2018	496,566	1,487,175	14,310.18	409,113.10	12,910.0	76,840
2019	499,992	1,503,980	14,408.93	579,230.70	13,770.0	81,340
2020	503,442	1,520,975	14,508.35	753,140.60	14,620.0	85,710
2021	506,916	1,538,162	14,608.45	930,906.90	15,470.0	89,960
2022	510,414	1,555,544	14,709.25	1,112,595.00	16,300.0	94,110
2023	513,936	1,573,121	14,810.75	1,298,270.00	17,130.0	98,140
2024	517,482	1,590,897	14,912.94	1,487,999.00	17,940.0	102,100
2025	521,052	1,608,875	15,015.84	1,681,850.00	18,750.0	105,900
2026	524,648	1,627,055	15,119.45	1,879,893.00	19,550.0	109,700
2027	528,268	1,645,440	15,223.77	2,082,196.00	20,340.0	113,400
2028	531,913	1,664,034	15,328.82	2,288,830.00	21,130.0	116,900
2029	535,583	1,682,837	15,434.59	2,499,870.00	21,900.0	120,500
2030	539,279	1,701,854	15,541.08	2,715,386.00	22,670.0	123,900
2031	543,000	-	15,648.32	-	23,440.0	-
2032	546,746	-	15,756.29	-	24,190.0	-

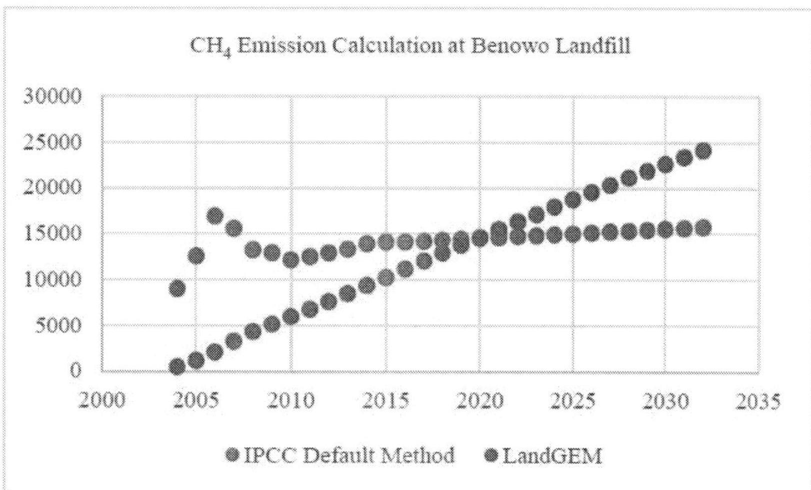

Figure 1. CH_4 emission calculation at Benowo landfill.

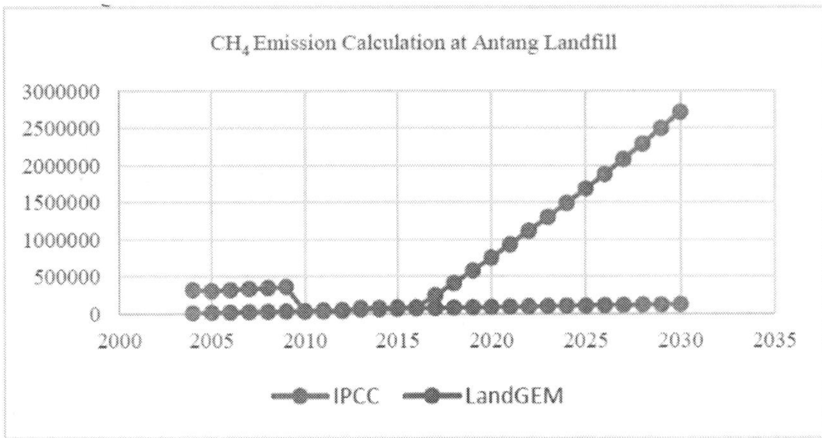

Figure 2. CH4 emission calculation at Antang landfill.

Waste management is required to minimize the CH_4 emission environmental impact. Application of proper waste management system including sanitary landfill method would assist the CH_4 emission reduction from landfill area.

4. Conclusions

The study shows that temperature and rainfall intensity are two main parameters for estimating CH_4 emission from landfill. These two parameters resulted in high methane emission emitted by Antang and Benowo landfill. In addition, the calculation method used is also contributing to the analysis results. IPCC and LandGEM method generate different results in estimating CH_4 emission from Benowo and Antang landfill. The difference occurred due to temperature and rainfall intensity variation.

Acknowledgement

The author is thankful to *Lembaga Pengelolaan Dana Pendidikan* (LPDP) for providing financial during my master of urban and environmental engineering in Kyushu University.

References

[1] Kaushal A and Sharma M P 2016 Methane Emission from Panki Open Dump Site of Kanpur, India *Procedia Environmental Sciences* vol 35 pp 337-47

[2] Meidiana C and Gamse T 2010 Development of waste management practices in Indonesia *European journal of scientific research* vol 40 no 2 pp 199-210

[3] Kawai M, Purwanti I F, Nagao N, Slamet A, Hermana J and Toda T 2012 Seasonal variation in chemical properties and degradability by anaerobic digestion of landfill leachate at Benowo in Surabaya, Indonesia *Journal of environmental management* vol 110 pp 267-75

[4] Kumar S, Gaikwad S A, Shekdar A V, Kshirsagar P S and Singh R N 2004 Estimation method for national methane emission from solid waste landfills *Atmospheric Environment* vol 38 no 21 pp 3481-7

[5] Kalantarifard A and Yang G S 2012 Estimation of methane production by LANDGEM simulation model from Tanjung Langsat municipal solid waste landfill, Malaysia *Int. J. Sci.*

IOP Conf. Series: Materials Science and Engineering **403** (2018) 012006 doi:10.1088/1757-899X/403/1/012006

Development of porous ceramic membrane from natural zeolite – clay for microfiltration

A Ma'ruf[1] and M A S Al Fathoni[2]

[1] Chemical Engineering Department, [2] Civil Engineering Department
Universitas Muhammadiyah Purwokerto
Jl. Raya Dukuh Waluh Kembaran, Purwokerto, Jawa Tengah, Indonesia 53182

anwarump@yahoo.com

Abstract. Indonesia has a large amount of natural zeolite and clay. These materials are potential for ceramic membrane production. The aims of this study were to describe the characteristic of ceramic membrane development from natural zeolite and clay. With an addition of TiO_2, ceramic membrane was produced by dry process by pressing at 5 tons for 30 minutes. From the results of the investigation, it can be concluded that the natural zeolite and clay from Banyumas may be utilized for obtaining ceramic membranes for microfiltration at low sintering temperature of 900°C. Optimum addition of TiO_2 was 5%wt. At this condition, the densities of ceramic membranes were 1.44 gr/cm^3, 1.8 gr/cm^3, and 1.5 gr/cm^3. The porosities of ceramic membranes were 53.29%, 31.099%, and 35.42%. The water absorption values of ceramic membranes were 36.83%, 17.06%, and 23.61% for natural zeolite, clay, and natural zeolite – clay respectively.

1. Introduction

Membrane separation process widely used for many purposes. There are two types of membrane that are organic (polymeric membrane) and non-organic membrane (metal and ceramic membranes). Each membrane has advantages and weaknesses. The ceramic membranes have many advantages such as high thermal and chemical stability, pressure resistance, long lifetime, good resistance to fouling, and easy to clean [1]. The weakness of the ceramic membrane is fragile and high-cost production due to the materials used. Materials commonly used for ceramic membranes are Al_2O_3, TiO_2, ZrO_2, SiO_2 etc. or a combination of these materials [2].

Several researchers used the natural materials for ceramic membrane development. The aim of utilization of natural material is to decrease the production cost while good characteristics are also achieved. For instance, kaolin-clay ceramic membrane for microfiltration is used in [1], apatite material for microfiltration and ultrafiltration is used in [3], clay for microfiltration of oily waste water is used in [4] and fly ash for microfiltration is used in [5].

The characteristics and properties of the ceramic membranes are mainly determined by their composition, the pore-former content and the sintering temperature. Based on pore of membrane, membranes can be classified into 2 main groups, porous and non-porous membranes. Separation mechanism in porous membranes is by molecular sieving. Mechanism of transport in non-porous membranes involve permeating molecules dissolve into a membrane, diffuse, and desorb from the membrane [6].

The definition of porous membranes is more in agreement with the definitions adopted by the International Union of Pure and Applied Chemistry (IUPAC). The pore size classification given refers to pore diameter: (i) Macropores > 50 nm. (ii) Mesopores 2 nm < pore size < 50 nm. (iii) Micropores < 2 nm [7]. Transport occurs through porous membranes rather than the dense matrix and ideal gas separation membranes possess a high flux and a high selectivity.

Indonesia has a large of natural zeolite and clay. These materials are potential for ceramic membrane production. The aims of this study were to describe the characteristic of ceramic membrane developed from natural zeolite and clay, with an addition of TiO_2.

2. Materials and methods

This section explains the materials and methods used in this research. The materials, methods, and characterizations are elaborated in the following subsections.

2.1. Materials

Natural zeolite and starch were purchased from local market at Purwokerto, Indonesia. Clay was obtained from Kebumen district, Indonesia. TiO_2 was purchased from Merck.

2.2. Methods

Natural zeolite and clay were crushed to the size of < 45 μm. The materials were mixed with TiO_2 at various composition. The composition of mixture is given in table 1. The mixture was then placed at O-ring casting with the diameter of 3.2 cm and the thickness of 1 cm and pressed at 5 tons for 30 minutes. The mixture was then calcinated at the temperature of 900°C for 6 hours.

Table 1. Composition of sample.

Sample	Natural zeolite (%w)	Clay (%w)	TiO_2 (%w)
1	100	0	0 – 5 – 10
2	0	100	0 – 5 – 10
3	50	50	0 – 5 – 10

2.3. Characterizations

The composition of raw materials was analyzed using X-Ray Diffraction (XRD). Morphology of ceramic membrane was analyzed using Scanning Electron Microscopy (SEM). Bulk density, porosity, and water absorption of membrane was analyzed using gravimetric methods.

3. Results and Discussion

The discussion of the result is explained in the following subsections.

3.1. Materials characterization

XRD was used to determine the characteristic of natural zeolite and clay. The Crystallinity of natural zeolite and clay were 84.6354% and 54.41% respectively. Figure 1 shows the XRD pattern of natural zeolite and clay. SiO_2 content in natural zeolite is 75.62% while in clay is 53.07%.

3.2. Ceramic membrane characterization

The characteristic of ceramic membrane is explained in the following subsections.

3.2.1. Bulk density. The characteristics of ceramic membrane are described into three parameters such as bulk density, porosity and water absorption. Figure 2 shows the effluence of addition of TiO_2 on bulk density of ceramic membrane. Ceramic membrane developed from clay has higher bulk density, while from natural zeolite has lower bulk density. Addition of TiO_2 will increase the bulk density of ceramic membrane. The optimum addition of TiO2 is 5%w. At this condition, the densities of ceramic membranes are 1.44 gr/cm^3, 1.8 gr/cm^3, and 1.5 gr/cm^3 for natural zeolite, clay, and natural zeolite – clay respectively. The density of ceramic membrane developed from natural zeolite was 1.86 g/cm^3 at temperature calcination 900°C according to the report in [8].

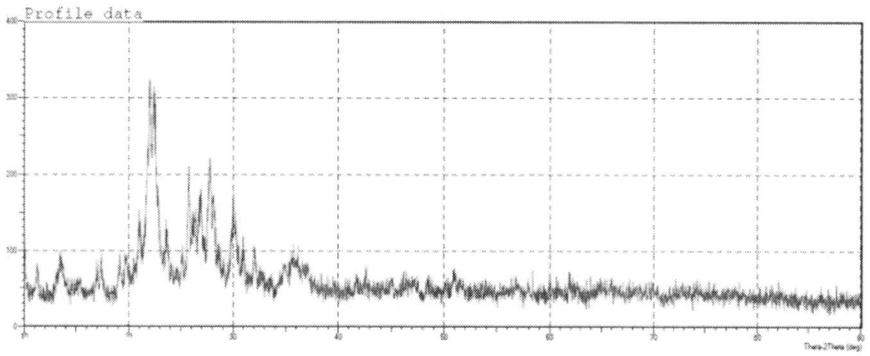

(a)

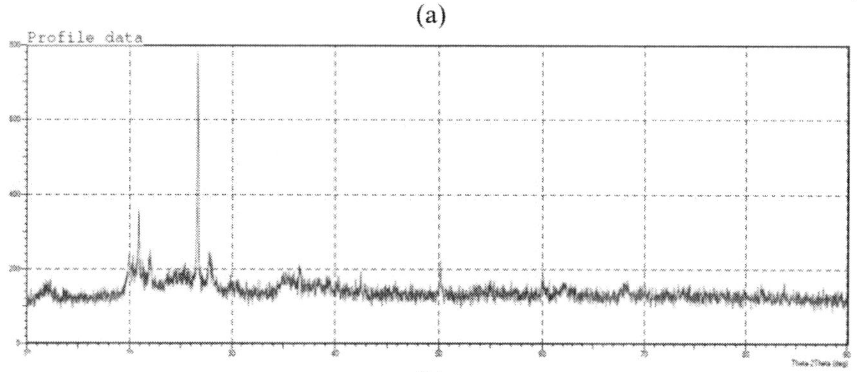

(b)

Figure 1. XRD pattern of materials: (a) natural zeolite, (b) clay.

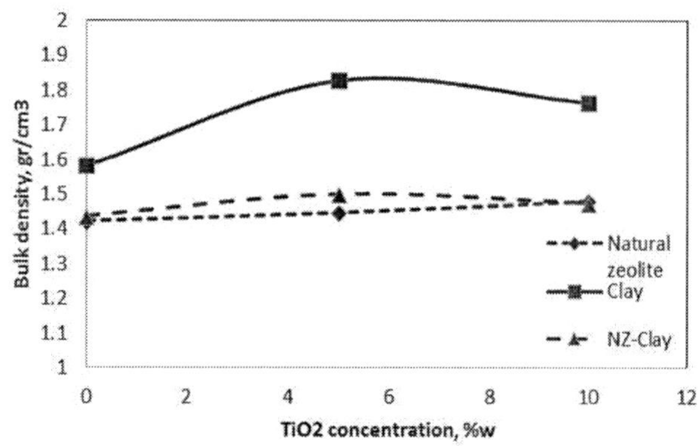

Figure 2. The effluence of TiO_2 concentration on bulk density.

3.2.2. Porosity. Figure 3 describes the porosity of the ceramic membrane. Porosity of Ceramic membrane developed from clay is higher than the other. Addition of TiO_2 will increase the porosity of ceramic membrane. The porosities of ceramic membranes are 53.29%, 31.099%, and 35.42% for natural zeolite, clay, and natural zeolite – clay respectively at addition 5% of TiO_2. The porosities of ceramic membranes are 53.29%, 31.099% and 35.42%.

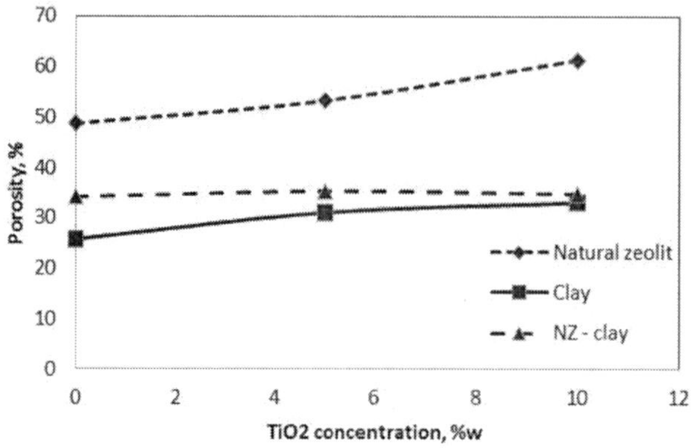

Figure 3. The effluence of TiO_2 concentration on porosity.

3.2.3. Water absorption. Figure 4 describes the porosity of the ceramic membrane. Porosity of Ceramic membrane developed from clay is higher than the other. Addition of TiO_2 will increase the porosity of ceramic membrane. The water absorption values of ceramic membranes are 36.83%, 17.06%, and 23.61% for natural zeolite, clay, and natural zeolite – clay respectively at addition 5% of TiO_2.

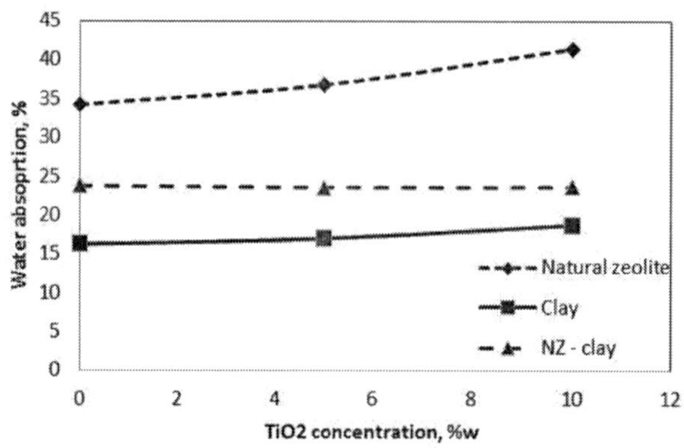

Figure 4. The effluence of TiO_2 concentration on water absorption.

3.3. Morphology of membrane

Figure 5 shows the morphology of ceramic membrane developed. Ceramic membrane from natural zeolite shows more pores. The pore size of ceramic membrane from clay has good unity of pore size.

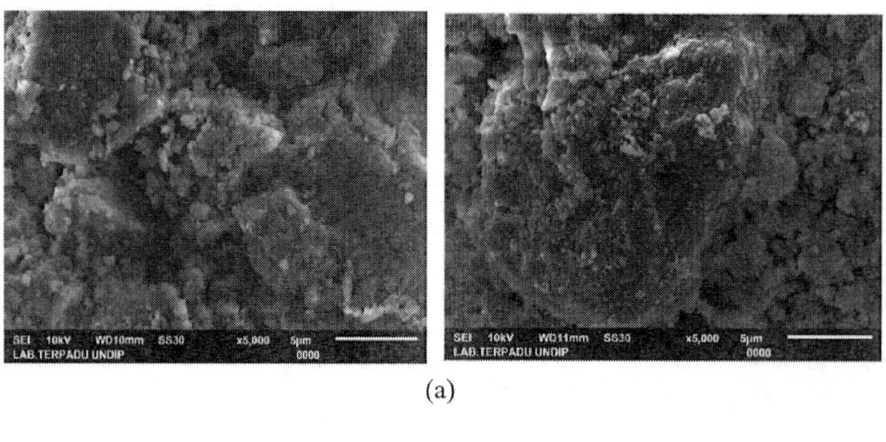

(a)

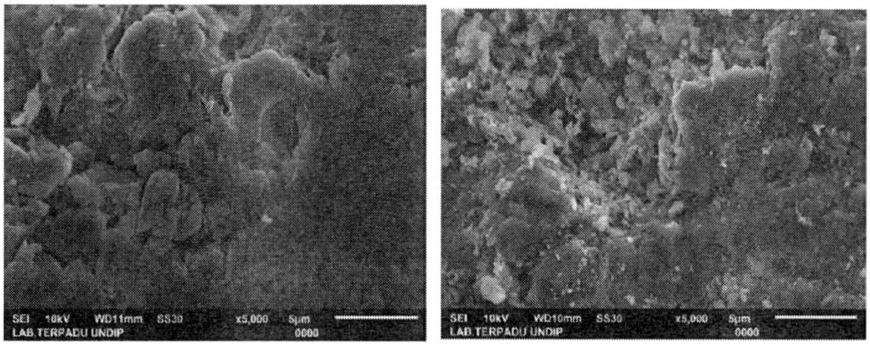

(b)

Figure 5. The morphology of ceramic membrane: (a) natural zeolite, (b) clay.

4. Conclusion

From the results of the investigation, it can be concluded that the natural zeolite and clay from Banyumas may be utilized for obtaining ceramic membranes for microfiltration at low sintering temperature of 900°C. Optimum addition of TiO_2 is 5%wt. At this condition, the densities of ceramic membranes were 1.44 gr/cm^3, 1.8 gr/cm^3, and 1.5 gr/cm^3, the porosities of ceramic membranes were 53.29%, 31.099%, and 35.42%, the water absorption values of ceramic membranes were 36.83%, 17.06%, and 23.61% for natural zeolite, clay, and natural zeolite – clay respectively.

Acknowledgments

The authors are grateful to Directorate of Higher Education, Ministry of National Education of Indonesia in providing a grant to conduct this research.

References

[1] Rekik S B, Bouaziz J, Deratani A and Beklouti S 2017 Study of ceramic membrane from naturally occurring-kaolin clays for microfiltration applications *Period. Polytech. Chem. Eng.* vol 61 p 206

[2] Akbarnezhad S, Mousavi S M and Sarhaddi R 2010 Sol-gel synthesis of alumina-titania ceramic membrane: preparation and characterization *Indian J. Sci. Technol.* vol 3 pp 1048–51

[3] Masmoudi S, Larbot A, Feki H El and Amar R Ben 2007 Elaboration and characterisation of apatite based mineral supports for microfiltration and ultrafiltration membranes *Ceram. Int.* vol 33 pp 337–44

[4] Parma S and Chowdhury P 2014 Preparation and characterization of microfiltration ceramic membrane for oily waste water treatment *Int. J. Res. Eng. Technol.* vol 3 no 3 pp 725–30

[5] Singh G and Bulasara V K 2015 Preparation of low-cost microfiltration membranes from fly ash *Desalin. Water Treat.* vol 53 no 5 pp 1204-12

[6] Khemakhem S, Larbot A and Ben Amar R 2009 New ceramic microfiltration membranes from Tunisian natural materials: Application for the cuttlefish effluents treatment *Ceram. Int.* vol 35 pp 55–61

[7] Zdravkov B D, Čermák J J, Šefara M and Jank\ru J 2007 Pore classification in the characterization of porous materials: A perspective *Cent. Eur. J. Chem.* vol 5 pp 385–95

[8] Hristov P, Yoleva A, Djambazov S, Chukovska I and Dimitrov D 2012 Preparation and characterization of porous ceramic membranes for micro-filtration from natural zeolite *J. Univ. Chem. Technol. Metall.* vol 47 pp 476–80

The characterization of Merapi volcanic ash as adsorbent for dyes removal from batik wastewater

S Salamah[1] and E T Wahyuni[2]

[1]Department of Chemical Engineering, University of Ahmad Dahlan, Prof. Soepomo Street, Yogyakarta 55164
[2]Department of Chemistry, University of Gadjah Mada, Sekip Utara Street, Yogyakarta 55281

[1]sitisalamah@che.uad.ac.id, [2]endtriw@yahoo.com

Abstract. In this paper, characterization, acid and base treatments, and adsorption assessment of the ash have been carried out to verify the ability of Merapi volcanic ash as an adsorbent. The characterizations were performed to determine the chemical composition by using ICP and surface area and pore volume by using surface area analyzer. The adsorbent ability was tested for removal of dyes, represented by naphtolis standard, and Batik waste water. The adsorption was carried by batch technique, in which adsorbent dose and adsorption time were varied to obtain the optimum condition. From the characterization, it is obtained that Merapi volcanic ash contains Si and Al as mayor elements, Fe, Ca, Mg, Na, and K with low percentage, and As, Mo, Cd, Cr, Cu, Mn and Pb with trace level. The adsorption study indicates that the Merapi volcanic ash can adsorbs the batik waste water and the adsorption increase by the increase in the weight of 49.6 % and stirrer time 50 minute can adsorb dyes waste of 3.9 gr/ml. Volcano ash is less suitable if it is used as adsorbent compared to natural zeolite.

1. Introduction

Merapi volcano eruption has exploded ash in very high amount that led to air pollution and made high pile of the ash promoting the environmental problem. It has been reported that Merapi volcanic ash contains mayor elements including SiO_2 and Al_2O_3 [1] that are around 60 % and 20 % respectively, accompanied with minor as well as trace level elements. In addition, radioactive elements are also found in the ash [2]. Based on the chemical composition, the effect of the ash contaminating soil on the plant growth has been studied and concluded that small amount of ash could induce the plant growth, while large pale ash on the soil inhibited the growth [3, 4].

The high content of SiO_2 and Al_2O_3 in the volcanic ash, as found in clays [5] and zeolites [6], enables the ash to act as adsorbent. In addition, the ash also has large surface area that is ~ 10 m^2 per gram weight volcanic ash [7] that may give high adsorption capacity. The material high composition of silica and other metal oxides has the possibility to remove dye from wastewater [8]. However, the use of Merapi volcanic ash as adsorbent has not been explored yet. Furthermore, the eruption of Merapi Volcano was in rainy season, so that the volcanic ash distributed in open place was rinsed by the rain water. Such rinsing may remove some metals from the ash. This may cause the change composition and change the adsorption capacity of the ash.

Adsorption is a process by which a solid adsorbent attracts a component in solution its surface through a physical or chemical interaction [9] Based on the fact reported above, the adsorption ability of the volcanic ash, before and after rinsing, is interesting to be studied. Batik waste water was selected as a waste model that was treated with volcanic ash adsorbent, because many Batik industries are existing in Yogyakarta.

The large number of Batik industries create very high volume of the waste water, contents high level of dyes. The contamination of dyes in the environment including rivers leads to the environmental damage and health problem [8, 10]. The removal of dyes waste water from Batik industry has to be carried out before being disposed to the environment. Several methods have been used for removal dyes in Batik waste water [11]. Several adsorbents such as zeolite [11], modified clay [11], and activated carbon [12] can be used to manage the waste of Batik production in Yogyakarta [13]. Effectiveness of waste management has been studied for removal dyes from Batik waste water. However, removing the dyes from the waste by using volcanic ash has not been examined. In this paper, a study on dyes removal from Batik waste water by using volcanic ash as adsorbent, whether fresh / before rinsed and after rinsed is reported.

2. Experimental

This section explains the procedure of proposed experiment.

2.1. Material

Volcanic ash obtained from the eruption of Merapi Volcano on Oct 26, 2010 at radius ± 28 km from the peak of southern side, Sinduadi village, Mlati, Sleman (before and after, Naphthol dyes used in batik coloring, Naphthol α powder, Ethanol, Aquadest, HCl pro analyst, NaOH crystal, Filter paper Whatman 42 and pH paper were used in this research without any purification.

2.2. Instrumentation

Surface area analyzer NOVA 100, ICP (Ion coupled plasma), UV-Vis Spectrophotometer, Glass tools, Sieve (100 mesh size) model RX-86, Heater and magnetic stirrer, Dryer oven (Precision model 17), Digital scale Mettler AE 163, Grinder, Buncher funnel and vacuum pump, Vacuum oven (Yamamoto Scientific, LTD). Thermometer and a set of reflux tools and stopwatch were used in this research.

2.3. Procedure

2.3.1. Preparation of naphthol dyes solution. Naphthol dye powder, as a standard dye, as much as 0.4 gram; 0.3 gram; 0.2 gram and 0.1 gram was dissolved in 10 ml of cold water, while naphthol salt as a dye model was dissolved in 1 L of boiling water.

2.3.2. Determination of the chemical composition in Merapi volcanic ash. Merapi volcanic was dried at 100°C then sieved with 100 mesh sieve. The 100 mesh volcanic ash powder then was analyzed by ICP Analyzer.

2.3.3. Preparation of Merapi volcanic ash for adsorbent. The 100 mesh volcanic ash powder as much as 150 grams was mixed with 1 M HCl 125 ml and refluxed with three-neck flask for 4 hours at 60 °C by stirring gently using a magnetic stirrer. The ash was separated from the mixture by filtering and then rinsed with water to remove the HCl excess. Then, the rinsed ash was neutralized using by stirring for 4 h using 200 mL of 1 M NaOH. The Merapi volcanic ash was rinsed again with aquadest until the pH is 7 and dried in the oven for 2 hours at 100°C. These steps of preparation for volcanic ash were repeated in natural zeolite preparation to prepare the natural zeolite as adsorbent.

2.3.4. Adsorption of batik dyes using Merapi volcanic ash. Make a standard curve for various concentrations of dyes and analyzed the absorbance by spectrophotometer. Then making naphthol dyes solution and then measuring the concentration using spectrophotometer. Then pour the Merapi volcanic ash with variation of a certain weight (10 grams, 20 grams, 30 grams, 40 grams, and 50 grams) in 200 mL batik waste solution, and then stirred using a mercury stirrer for 10 minutes. After stirring for 10 minutes, the mixture was precipitated for 24 hours, and then centrifuged for 10-15 minutes. The absorbance of the sample was then measured by using a spectrophotometer and the concentration was calculated with the aid of a standard curve that was created.

2.3.5. The influence of stirring time. Taking the volcanic ash with the most optimal datum of data that has been obtained through the previous experiments. Stirring the ashes for 4 minutes. After that, repeat steps 1-2 for stirring time 10 minutes, 20 minutes, 30 minutes, 40 minutes, and 50 minutes. After stirring, the mixture was precipitated for 24 hours, and then centrifuged for 10-15 minutes. Then, measure the absorbance of the sample with a spectrophotometer. Measure the concentration of the sample with the aid of a standard curve has been made.

3. Result and discussion

This section presents and discusses the experiment results.

3.1. Chemical composition of the Merapi volcanic ash

The chemical composition of the volcanic ash determined by ICP (Inductivity Coupled Plasma spectrometers), both before rinsed/fresh and that of after rinsed is presented in table 1 for mayor elements and that of for heavy metals is in table 2.

Table 1. The content of mayor elements in Merapi volcanic ash.

No	Element	volcanic ash (mg/kg)→ %w	volcanic ash + water (mg/kg)→ %w
1	Al	62890	39590
2	Ca	39500	16010
3	Fe	54040	31660
4	Mg	9773	4944
5	Na	22440	13640
6	P	765.4	113.1
7	S	907.7	815.6

Table 2. The content of heavy hazard metals in the volcanic ash.

No	Element	volcanic ash (mg/kg)	volcanic ash+ water (mg/kg)
1	As	8.948	8.302
2	Cd	1.295	0.839
3	Cr	3.88	2.262
4	Cu	7.407	2.492
5	Mn	1310	796
6	Mo	1.436	1.436
7	Ni	2.767	1.623
8	Pb	10.14	11.70
9	Zn	160.6	72.34

Table 1 and [5] showed that Merapi volcanic ash contains SiO_2 and Al_2O_3 in high level accompanied by other mayor elements. Based on this data, it is clear that the volcanic ash has the potential to be used as an adsorbent by providing silanol and aluminol as the active adsorption sites. According to [8], charcoal has a structure like aluminum silicate which has the possibility to be used as an adsorbent. In addition, as seen in table 2, some hazardous metals in very low quantities were found in the volcanic ash.

The presence of the hazardous metals made the ash harmful to the body as it can cause lung problems when the ash-polluted air is inhaled. Moreover, when the ash contaminates water including open well and rivers, biota may be dead and people may get health problems. Furthermore, data in both tables indicate that the rain water has made the content of all metals, except for Pb and Si, decreased. The decrease is caused by releasing the metals in the rain water. Such releasing makes the content of Si increased, that may provide more active adsorption sites.

Merapi volcanic ash contains hazardous metals (As, Cd, Pb, and sulfur) in small quantities, also contain other metals which is in large enough quantities e. g. aluminum (Al), calcium (Ca), iron (Fe), and sodium also phosphorus (P). These results also indicate the metal content in the volcanic ash exposed to rain is mostly decline, it was likely due to the metals dissolve in water, and pouring rain water can dissolve some of the metals contained in the ash. The metal content of volcanic ash can be utilized such as a mixing compound cement manufacturing. According to [5] in addition to Al and Ca also contains silica which is quite high, thus allowing the use of ash as a basis for the manufacture of cement-based material.

3.2. The surface characters of Merapi volcanic ash

Measurement of the specific surface area, total pore volume and pore radius average conducted to determine how large specific surface area and pore distribution of ash that can be used as an adsorbent. This measurement was done because the specific surface area of the sample will indirectly affect the adsorption ability of the sample. According to [14] with increasing adsorption on the surface of the powders, the possibility of formation of a product which is expressed as the concentration of reactants that can be adsorbed will increase. Data specific surface area, total pore volume and pore radius average volcanic ash are presented in table 3.

Table 3. The surface characters of Merapi volcanic ash.

Sample	Parameter	Test result	Unit
MerapiVolcano ash	Specific surface area	4.100	m^2/g
	Total pore volume	5.234×10^{-3}	cc/g
	Pore radius average	25.531	Angstrom
Merapi Volcano ash + water	Specific surface area	4.297	m^2/g
	Total pore volume	5.241×10^{-3}	cc/g
	Pore radius average	24.39	Angstrom
Natural zeolite	Specific surface area	29.366	m^2/g
	Total pore volume	5.241×10^{-3}	cc/g
	Pore radius average	18.183	Angstrom

Table 3 showed that the Merapi volcanic ash specific surface area and specific pore area are quite large and thus have the potency as an adsorbent. As a comparison, Y-zeolite which is zeolite synthesis produced in Japan [15] has a specific surface area of 475.94 m^2/g and specific pore area of 29.366 m^2/g and the fly ash which is a product of coal combustion has a specific surface area between 170 to 1000 m^2/kg or 0.170 to 1 m^2/g [7], both are the adsorbents widely used nowadays. The specific surface area of Merapi volcanic ash increased after affected by rain water. It is probably because some of the metals are dissolved in rain water and thus it increases its specific surface area. This is also supported by the reduction in the metal content of the volcanic ash exposed to rain.

3.3. volcanic ash utilization as dyes adsorbent

The weight of volcanic ash used as an adsorbent is 10 grams, 20 grams, 30 grams, 40 grams and 50 grams, minutes. Once adsorbed by stirring, the solution was precipitated for 24 hours and centrifuged for 15 minutes and then analyzed for absorbance.

IOP Conf. Series: Materials Science and Engineering **403** (2018) 012007 doi:10.1088/1757-899X/403/1/012007

The adsorption experimental was carried out by batch technique, in which the influences of adsorbent mass and stirring time are studied to obtain the optimum adsorption conditions. The influence of the volcanic ash mass on the dyes adsorption is presented as table 4 and 5.

Table 4. The influence of the volcanic ash on the dyes adsorption.

No	volcanic ash weight (gr)	Adsorption Concentration	Adsorption (%)
1	10	1.639	20.594
2	20	1.797	22.572
3	30	2.434	30.576
4	40	3.301	41.458
5	50	3.952	49.641

Table 4 showed that the more Merapi volcanic ash used as adsorbent represented by the increase in volcanic ash weight, the more wastewater that is adsorbed. This is because the more materials used as adsorbent, the more the pore area increased, and thus the more wastewater color being adsorbed.

Table 5. The influence of the natural zeolite on the dyes adsorption.

No	Natural Zeolite weight (gr)	Adsorption Concentration	Adsorption (%)
1	10	4.654	50.454
2	20	7.876	61.241
3	30	5.678	71.314
4	40	6.186	77.699
5	50	6.286	78.958

Table 5 showed that the more zeolite being used as adsorbent represented by the increase in zeolite weight, the more wastewater being adsorbed compared with wastewater adsorbed by Merapi volcanic ash. This is because the zeolite has a specific pore area of 29.366 m^2/g larger than volcanic ash which only has a specific pore area of 4.1 m^2/g hence the wastewater adsorbed is more than those adsorbed by volcanic ash. If it is compared with volcanic ash adsorbent, zeolite adsorbed 30% even up to 60% more than volcanic ash. This is because of specific pore area of zeolite which is larger than volcanic ash.

3.4. The influence of the stirring time to adsorption

The influence of the stirring time was examined by running the adsorption experiment with time variation by volcanic ash and natural zeolite. The result of the adsorption in various stirring times is displayed as figure 1.

Figure 1 shows that the longer the time or stirring, the more wastewater being adsorbed. The optimal amount of wastewater absorbed by volcanic ash is 3.98 gr/ml with the percentage of 50 %, while zeolite absorbed more amount of wastewater which is 6.3 gr/ml with the percentage of 79.4 %. This is because by prolonging the stirring time until the optimal time, the longer the contact between the adsorbent and the wastewater and the closer to perfect the adsorption process.

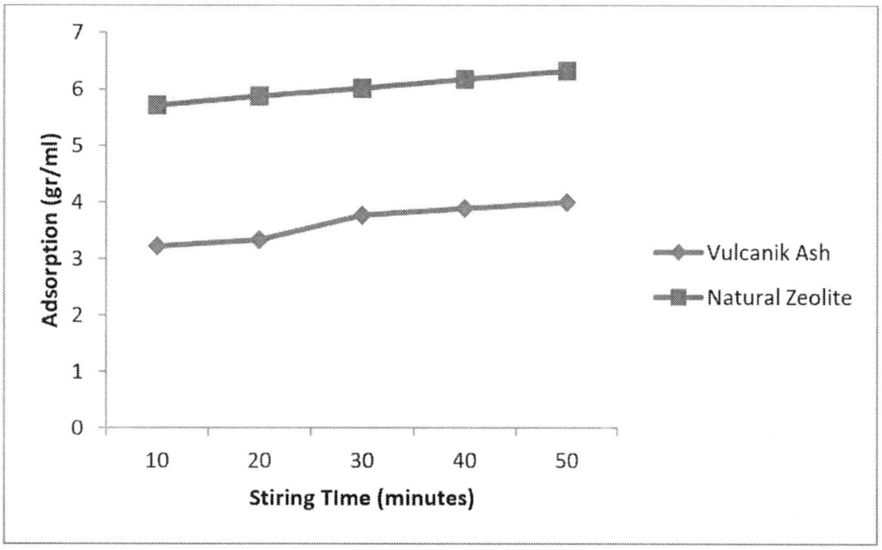

Figure 1. The influence of the stirring time on the dyes adsorption.

The comparison of adsorbed color concentration between stirring time of volcanic ash and zeolite showed that zeolite has higher adsorption ability which reaches 79.4%. This is because zeolite has larger specific surface area than volcanic ash. Natural zeolite has a specific surface area of 29.366 m^2/g whereas Merapi volcanic ash has a specific surface area of 4.100 m^2/g. The research showed that Merapi volcanic ash is less suitable if used as adsorbent if it is compared to other adsorbent which has larger specific surface area.

4. Conclusion

Merapi volcanic ash contains heavy metals such as Arsenic (As), Cadmium (Cd), Chromium (Cr), Copper (Cu), Manganese (Mn), Molybdenum (Mo), and Lead (Pb) with varying concentrations and a maximum of 10 mg/kg. It also contains other metal elements such as Aluminum (Al), Calcium (Ca), Iron (Fe), and Sodium (Na). Merapi volcanic ash has a surface area of 4.100 m^2/g, pore volume of 214.47 x $10^{-3}$and the mean pore of 214.47 x 10^{-3}. Based on this result, it can be concluded that Merapi volcanic ash can be used as an adsorbent and when the volcanic ash is used as an absorbent of batik waste water, it works optimally at a weight of 50 grams can adsorbent dyes of 49.6% and stirring time 50 minutes can adsorbent dyes of 3.98 gr/ml. Merapi Volcano ash is less suitable if used as adsorbent if it is compare natural zeolite.

Acknowledgement

This research was supported by Research and Development Institute of Ahmad Dahlan University grant No: M-75 / LPP-UAD / I/2014.

References

[1] Delmelle P 2005 Surface Area, Porosity and Water Adsorption Properties of Fine Volcanic Ash Particles. *Bull. Volcanology* vol 67 pp 160-9
[2] Clift P D and Fitton J G 1998 Trace and rare earth element chemistry of volcanic ashes from sites 918 and 919: Implication for Icelandic Volcanism *Proc. of the Ocean Drilling*
[3] Brotopuspito K 2012 Physics of the Volcano *Inaugural Speech Professor of Geophysics* (Yogyakarta: University of Gadjah Mada)
[4] Wilson T M, Stewart C, and Cole J 2007 Impact on the 2006 Eruption of Merapi Volcano, Indonesia on Agriculture and Infrastructure (Indonesia: GNS Science Report)
[5] Wahyuni E T 2012 Determination of volcanic Ash Chemical Composition from Merapi

Volcano Eruption *J. Human and Environmental* vol 19 pp 150-9
[6] Salamah S 2016 Utilization of Kelut's Volcanic Ash as the Aggregate Mixture of Concrete Brick *J. Key Engineering Material* vol 718 pp 196-200
[7] Pratama Y, Putranto H T 2007 Fly Ash Coal Conversion to Zeolite for Removal of Chromium and Nickel from Wastewaters (www.majarikanayakan.com)
[8] Kusmiyati, Listiyanto P A and Vitasari D 2017 Coal Bottom Ash and Activated Carbon for Removal of Vertigo Blue Dye in Batik Textile Waste Water: Adsorbent Characteristic, Isotherm and Kinetic Studies *J. WALAILAK Sci & Tech* vol 14 **no** 5 pp 427-39
[9] Bediako J K, Wei W, Y-S Yun 2016 Conversion of Waste Textile Cellulose Fibers into Heavy Metal Adsorbents *Journal of Industrial and Engineering Chemistry* vol 43 pp 61-8
[10] Hernaningsih T 2016 Industrial Wastewater Treatment Technology Review by Electrocoagulation Process *J. Environmental Engineering* vol 9 **pp** 31-46
[11] Riyanto 2015 Discovery New Technique for Batik Wastewater Treatment *J. Research Outcome in Sleman Regency* vol 2 pp 65-77
[12] Setyaningsih H 2007 Batik Wastewater Treatment with Chemical Process and Activated Carbon Adsorption *Post Graduate Thesis* (Jakarta: University of Indonesia)
[13] Al Rasyid M R and Asri R W P 2017 Waste Prevention Effectiveness of Batik Production in Yogyakarta, Indonesia, *Proceeding of The 2nd International Conference on Sustainable Innovation* pp 437-81
[14] Gasser R P H 1985 *An Introduction to Chemisorption and Catalysis by Metal* (Oxford: Clarendon Press)
[15] Salamah S, Trisunaryanti W and Triyono 2011 Influence of total metal contents in catalyst $N_{(3)}$-Pd/zeolite Y on fuel hydrocracking asphalt fraction from Buton asphalt with semi bath reactor, *Proceeding of The 7th Indonesia Zeolite Conference* ITS Surabaya

Consumer's valuation of the traditional fish handling practices using importance-performance analysis

T Juwitaningtyas[1], W Supartono[2] and I W F Aziz[2]

[1] Department of Food Technology, Industrial Technology Faculty, Universitas Ahmad Dahlan. Jl. Dr. Sopemomo, Umbulharjo, Yogyakarta. 55164
[2]Department of Agroindustrial Technology, Faculty of Agricultural Technology, Universitas Gadjah Mada. Jl. Flora No. 1, Bulaksumur, Sleman, Daerah Istimewa Yogyakarta 55281

titisari.juwitaningtyas@tp.uad.ac.id

Abstract. Fishery products, especially fish, have special characteristics on the diversity of species, inconsistent supplies, and short shelf-life. Most of the businesses in the fishery sector in Indonesia are still running using traditional methods. Lack of infrastructure and knowledge of Good Handling Practices (GHP) in fish, resulted in the decreasing of quality and often losses. This research tries to explore perception and consumer appraisal of fresh fish products to fishery business practices in Indonesia, especially in the coastal area of Yogyakarta. By using Importance-Performance Analysis (IPA) tools, we can measure the relationship between consumer perceptions with the priority of improvement that should be done. The method of this research is by using a questionnaire, where the questionnaire that has been valid and reliable (level of confidence 95%) then distributed to 101 respondents (fresh fish consumers). Based on the data processing, we got result that there are 3 attributes that enter in Quadrant I that is existence of garbage near market area, the hygiene of display table, and existence of dirty water puddle. The attributes included in Quadrant I indicate that these attributes are given the highest priority of improvement according to the consumer.

1. Introduction

Fishery is one of agriculture sub-sectors that play an important role in economic growth of Indonesian society. Based on data from BPS that is processed in [1], mentioned that fishery sub-sector holds the second largest contribution after the food crop sub-sector. This is triggered by increased consumption of fresh fish and processed fish products. According to BPS observation data from 2013-2014, there was an increase in fish supply coupled with an increase in fish consumption by 8.44% and 8.32%. Province that experienced the highest increase of fish consumption is Special Region of Yogyakarta with increasing value 30,96%. One of the factors that causes the increase in fish consumption is the support from the government both in the form of campaigns and events about fond of eating fish to the community.

The potency in supply and demand of fishery products is unfortunately not balanced with the quality of handling. According to [2], the handling of inappropriate fishery products is the main cause of the declining quality of fresh fish products that are often under the established requirements. Fresh fish products that are often found in the market are mostly handled with traditional and simple techniques. Environmental conditions around the place of fish seller were still less awake hygiene. Juwitaningtyas [2] also mentioned that the practice of handling fish that takes place in traditional markets is below the standard where found in fresh fish found microbial contamination that exceeds the threshold, breakdown of cold chain system, and minimum sanitation. Therefore, to ensure the ongoing food security system to consumers, this study aims to measure consumer ratings of traditional fish handling practices so that the priority of improvement on any problems that arise can be done.

2. Materials and methods

This study uses an analysis of *Importance-Performance Analysis* (IPA) to interpret consumer assessment and give priority to the improvement of quality attributes based on consumer ratings. IPA is a tool in marketing science that is used to measure the level of customer satisfaction on the quality attributes of an item of goods or services. This tool brings consumer judgment on the performance of items of goods / services to the level of importance attributes contained in the item. This method is generally an easy tool to apply in a variety of interests in order to improve the marketing performance as well as decision making improvements to the required quality attributes [3-8].

Therefore, this study involved respondents who are fish consumers in Fish Market of Depok Beach, Yogyakarta. The following steps are taken in this research:

1) Creating an initial questionnaire distributed to 35 respondents.
2) Testing the validity and reliability of the initial questionnaire. Validity and reliability test is performed on the initial questionnaire to determine the questionnaire's ability as a valid and reliable measurement tool. A valid item is an item that has a value of $r_{calc} > r_{table}$, where r_{calc} is derived from equation (1).

$$r = \frac{\left(n\sum_{i=1}^{n} X_i Y_{ii}\right) - \left(\sum_{i=1}^{n} X_i\right)\left(\sum_{i=1}^{n} Y_i\right)}{\left(\sqrt{n\sum_{i=1}^{n} X_i^2 - \left(\sum_{i=1}^{n} X_i\right)^2}\right)\left(\sqrt{n\sum_{i=1}^{n} Y_i^2 - \left(\sum_{i=1}^{n} Y_i\right)^2}\right)} \quad (1)$$

3) The reliable items are items that have Cronbach's alpha > r_{table} values. This reliability test can be done by using Spearman-Brown formula as shown in equation (2).

$$\boldsymbol{ri} = \frac{2\,(\boldsymbol{rb})}{1+\boldsymbol{rb}} \quad (2)$$

4) Calculates the adequacy of the sample size. The sample required for this study was calculated using a 95% confidence level and error rate of 0.1 using equation (3).

$$n = p(1-p)\left[\frac{Z_{\alpha/2}}{E}\right]^2 \quad (3)$$

5) Creating final questionnaires distributed to 101 respondents. The final questionnaire distributed is a selection of validity and reliability test results so that 13 questions are obtained.
6) Analyze satisfaction test and interest test. This analysis is done by calculating the average value of each attribute and then calculating the average of the overall value of satisfaction level and importance level. The formula used to calculate such averages is:

$$\overline{\overline{Xi}} = \frac{\sum_{t=1}^{k} \overline{Xi}}{n} \text{ and } \overline{\overline{Yi}} = \frac{\sum_{t=1}^{k} \overline{Yi}}{n} \quad (4)$$

7) Creating IPA Quadrant. The IPA quadrant is a quadrant constructed from Cartesian diagrams divided into 4 areas as shown in figure 1.

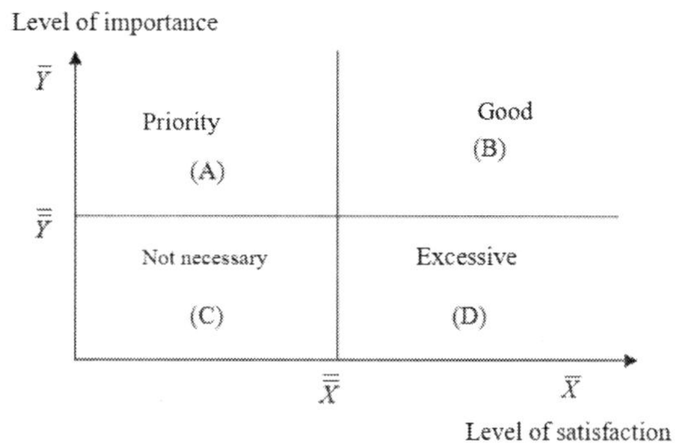

Figure 1. The theory of IPA diagram.

The average value of degree of importance and degree of satisfaction calculated based on the results in step 5 becomes the coordinate point which is then plotted on the quadrant.

3. Result and discussion

Based on the value of the validity and reliability test results of all questions that include the attribute of satisfaction / interest and performance attributes on the questionnaire, obtained the result that there are 13 items that meet the valid and reliable criteria with the value of $r_{calc} > r_{table}$ and the value of Cronbach's alpha > r_{table}. The final questionnaire was distributed to 101 respondents to assess the level of importance / satisfaction and satisfaction attributes, with the results shown in table 1.

Table 1. The questionnaire result of level of important / satisfaction assessment.

Code	Attributes	Average satisfaction	Average interest
Q1	Fish eyes are clear and fresh color	1,800	2,136
Q2	The fish' gill are fresh red	1,827	2,164
Q3	There is no bad odor	1,982	2,500
Q4	Fish meat is not slimy	1,936	2,291
Q5	Comparison of the use of ice is 1 Kg ice for 1 Kg fish (to keep the freshness of the fish)	1,464	2,136
Q6	The display tables are clean	1,555	2,327
Q7	The market floor does not look seedy	1,436	2,218
Q8	There is no stagnant water	1,618	2,264
Q9	No garbage piling near the sales area	1,618	2,382
Q10	There is a sufficient supply of clean water	1,755	2,355
Q11	Merchants use clean clothes	1,600	2,064
Q12	Merchants use gloves while serving purchase	1,200	2,045
Q13	Fish storage containers are clean	1,691	2,418

Based on the value of each attribute in the table above, then calculated the average value and obtained the result that is $\bar{\bar{Xi}} = 1.652$ (degree of importance) and $\bar{\bar{Yi}} = 2{,}254$ (degree of satisfaction). The point ($\bar{\bar{Xi}}$, $\bar{\bar{Yi}}$) becomes the point of intersection on the Cartesian diagram which then becomes the

IOP Conf. Series: Materials Science and Engineering **403** (2018) 012008 doi:10.1088/1757-899X/403/1/012008

dividing point of the plotting area (Quadrant I, Quadrant II, Quadrant III, and Quadrant IV). The pair of degree of importance and degree of satisfaction on each attribute becomes the coordinate point (X1, Y1; X2, Y2; ..... X13, Y13). All these points are then plotted in Cartesian diagrams that are spread across the areas of Quadrant I up to Quadrant IV (figure 2).

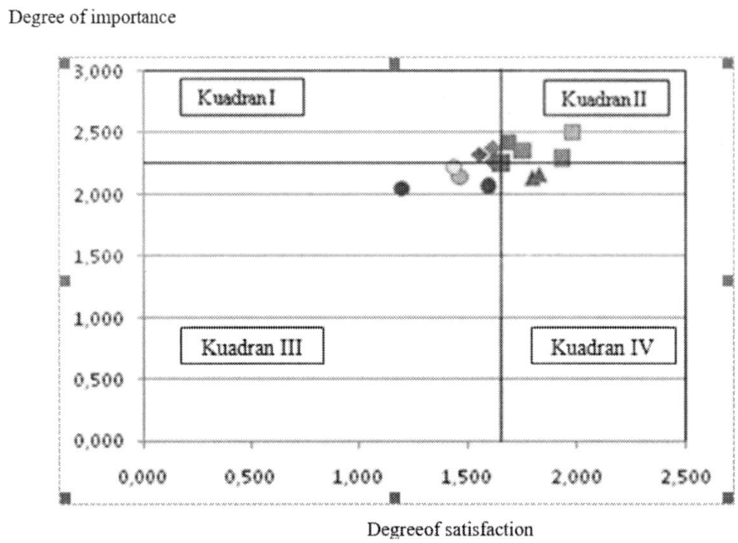

Figure 2. Data spread in IPA diagram.

Based on the results of plotting on Quadrant Importance-Performance Analysis (IPA) above obtained the result that:

1) Attributes that are in Quadrant I are the presence of garbage near the market area, display table hygiene, and the presence of stagnant water. Attributes spread in Quadrant I fall into the priority category for immediate correction.
2) Attributes that are in Quadrant II are the sensory assessment of fish are the odor and appearance of fish skin, clean water supply, and cleanliness of fish storage container. Attributes that occupy in Quadrant II means attributes that have shown good performance so that has met the expectations and customer satisfaction.
3) Attributes that are in Quadrant III are attributes of the use of ice to maintain freshness of fish, fish market floor conditions, and personal hygiene of traders that is related to clothing and the use of gloves. Attributes that are in Quadrant III mean that the attributes are actually considered not too important by consumers but has shown good performance.
4) Attributes that are in Quadrant IV are attributes related to the product directly in the form of assessment of the eye and fish gills. Attributes that fall into this category have been considered good by consumers and the improvement will be felt to be excessive.

4. Conclusion

Based on the research on consumer's assessment of traditional fish handling practices in Depok Fish Market which analyzed using *Importance-Performance Analysis* (IPA) method, it is concluded that consumers feel the environmental hygiene factor in the market area is very necessary to be improved. The quality attribute concerning the quality of the direct product (fresh fish) is considered good and meets the expectations and satisfaction of the consumers.

References

[1] Fathurrohman Y E 2016 Model Kerjasama Kelompok Pembudidaya Ikan (POKDAKAN) Pada Pemasaran Agribisnis Ikan Gurami di Kabupaten Banyumas *Agriekonomika* Vol 5 No 2

[2] Juwitaningtyas T and Supartono W 2011 Evaluasi dan Rekomendasi Penerapan Good Handling Practices Pada Ikan di Pasar Ikan Tradisional Pantai Depok, Bantul, DIY *Proceeding of Seminar Nasional APTA 2011* pp 64-72

[3] Abalo J, Varela J and Manzano V 2007 Importance Values for Importance-Performance Analysis: A Formula For Spreading Out Values Derived From Preference Rankings *Journal of Business Research* 60 2 pp 115-121

[4] Kitcharoen K 2004 The Importance-Performance Analysis of Service Quality in Administrative Departments of Private Universities in Thailand *ABAC Journal* 24 3 pp 20-46

[5] Matzler K, Saurwein E and Heischmidt K 2003 Importance-Performance Analysis Revisited: The Role of The Factor Structure of Customer Satisfaction *The Service Industries Journal* 23 2 pp 112-129.

[6] Silva F and Fernandez O 2010 Using Importance-Performance Analysis in Evaluating of Higher Education: A Case Study *ICEMT 2010 International Conference on Education and Management Technology. IEEE* pp 121-123

[7] Silva F J H and Fernandez P O 2011 Importance-Performance Analysis As A Tool In Evaluating Higher Education Service Quality: The Empirical Results Of ESTiG *Proceeding in IBIMA Proceeding (Scopus Indexed)*

[8] Slack N 1994 The Importance-Performance Matrix as a Determinant of Improvement Priority *International Journal of Operations & Production Management* 14 5 pp 59-75

IOP Conf. Series: Materials Science and Engineering **403** (2018) 012009 doi:10.1088/1757-899X/403/1/012009

Recovery of oryzanol from rice bran oil by silica-based batch adsorption: Radial diffusivity inside particle

A D Susanti[1,2], W B Sediawan[2], S K Wirawan[2] and Budhijanto[2]

[1]Department of Chemical Engineering, Universitas Sebelas Maret, Jl. Ir. Sutami No. 36 A, Surakarta, Indonesia 57126
[2]Department of Chemical Engineering, Universitas Gadjah Mada, Jl. Grafika No. 2, Yogyakarta, Indonesia 55281

aridiana@staff.uns.ac.id

Abstract. In this study, quantitative description on oryzanol batch adsorption from rice bran oil via mathematical model has been developed. The accuracies of the model proposed were tested based on experimental data from literature. The main idea is that the effective radial diffusivity of oryzanol in the adsorbent particle be dependent on the distance to the center of the adsorbent particle. The oryzanol mass transfer rate from the bulk of the liquid to the surface of silica gel particle follows the film theory. Meanwhile the distribution coefficient correlation was used to describe the adsorption equilibrium. The values of the involving parameters were obtained by visual inspection. The model proposed works quantitatively well but in order to get general values of parameters for various particle sizes, further improvement is still required.

1. Introduction

Oryzanol, a valuable compound, is naturally contained in rice bran oil and is an important value-added co-product of rice bran processing. Oryzanol is a mixture of at least five trans ferulic acid esters with phytosterols [1-7]. Literature studies show that oryzanol is proficient to prevent cardiovascular disorder because of the ability of lowering cholesterol levels [8-13].

It had been reported that adsorption-desorption systems are believed as one of promising methods to recover valuable components such as oryzanol from rice bran oil, provided that appropriate mobile phases are available [14, 15]. Quantitative studies of the adsorption equilibrium and kinetics of oryzanol isolation need to be further studied for the purposes of engineering process design and operation. The studies conduct mathematical model simulation to obtain quantitative description of the process as well as to approximate the values of the involving parameters. The experimental data from literature were applied to verify the accuracy of the proposed mathematical model. It is imagined that the adsorption consists of several processes in series which are (1) the mass transfer of oryzanol from the bulk of liquid to the surface of silica gel particle, (2) the intra-particle diffusion through the liquid inside pore from the surface of silica gel particle to the inside part, and (3) the adsorption from the liquid inside pore to the adjacent pore surface. The third process is considered to be fast so the both first two processes affect the rate of oryzanol adsorption. Usually, the adsorption at the pore surface is assumed to be instantaneous, so local adsorption equilibrium always attained.

Experimental studies on batch oryzanol adsorption from rice bran oil have been conducted by [16]. They also proposed mathematical model by assuming oryzanol effective radial diffusivity in the silica gel and distribution coefficient for the local equilibrium of adsorption be constant (called as Model 1). Model 1 works quantitatively well, but discrepancies were encountered. The studies show that larger particles better adsorb oryzanol than the smaller one.

The effective radial diffusivities of oryzanol inside silica gel with average diameter of 0.0815 mm and 0.35 mm were 8 x 10^{-5} cm^2/s and 4.2 x 10^{-4} cm^2/s, respectively. It means that the smaller particle has smaller value than the larger one. These phenomena are not conceivable since smaller particle are usually better adsorb solute from solution because the internal diffusion for small particle is faster than the one of larger particle. This study intends to quantitatively explain these discrepancies by assuming that the effective radial diffusivity depends on the distance to the center of the particle. The larger the distance to the center of particle, the higher effective radial diffusivity would be. In this paper the effective radial diffusivity is assumed to be proportional to the distance to the center of particle (Model 2).

2. Mathematical model development

This section explains the proposed mathematical model. The model is elaborated into several subsections as follow.

2.1. *Material balance in silica gel particle*

Oryzanol material balance in the volume element in the adsorbent particle resulted in partial differential equation as shown in equation (1). In equation (1), the accumulation of oryzanol in the volume element includes the accumulation of oryzanol in the liquid inside the pore and the accumulation of oryzanol adsorbed at the surface of the pore.

$$\frac{\partial}{\partial r}\left[D_e r^2\left(\frac{\partial C_{As}}{\partial r}\right)\right]=\varepsilon r^2\frac{\partial C_{As}}{\partial t}+a_s r^2\frac{\partial X_{As}}{\partial t} \tag{1}$$

By assuming that local oryzanol adsorption equilibrium at the surface of pore is always attained, and the equilibrium follows distribution coefficient model, equation (1) becomes equation (2).

$$\frac{\partial}{\partial r}\left[D_e r^2\left(\frac{\partial C_{As}}{\partial r}\right)\right]=\left(\varepsilon+a_s K_f\right)r^2\frac{\partial C_{As}}{\partial t} \tag{2}$$

Equation (2) can be manipulated into equation (3). In equation (3), D_e is radial effective diffusivity of oryzanol through the pore of the particle (cm/s) and responsible to the oryzanol concentration gradient in silica gel particle.

$$\frac{\partial C_{As}}{\partial t}=\frac{1}{\left(\varepsilon+a_s K_f\right)}\frac{1}{r^2}\frac{\partial}{\partial r}\left[D_e r^2\left(\frac{\partial C_{As}}{\partial r}\right)\right] \tag{3}$$

In the previous work, the developed model assumed constant D_e and constant K_f [16]. The model well describes the dynamics of oryzanol concentration inside silica gel at various time, but cannot logically explain the phenomenon that the larger particles better adsorb oryzanol than the smaller ones.

The model proposed in this study is intended to logically describe the discrepancies found in the previous study. The basic idea is that the effective radial diffusivity is proportional to the distance to the center of particle (r). The argument of this idea is based on the fact that silica gel particles was made by granulation. The initial solid formed (the inner part of silica gel particle) is more compact than the later one (the outer part of silica gel particle) so the porosity in the inner part is smaller than the one of the outer part. As a result, the effective radial diffusivity of the outer part is larger than the one of the inner part. Mathematically the effective radial diffusivity is modeled by equation (4).

$$D_e=\mathfrak{D}r \tag{4}$$

with $\mathfrak{D}$ is a constant. Combination of equation (3) and (4) gives equation (5).

$$\frac{\partial C_{As}}{\partial t} = \frac{\mathfrak{D}}{(\varepsilon + a_s K_f)} \left[r \frac{\partial^2 C_{As}}{\partial r^2} + 2 \frac{\partial C_{As}}{\partial r} \right] \tag{5}$$

2.2. Material balance in the bulk of the liquid

Oryzanol concentration inside the adsorbent particle was not uniform and was assumed to be depending on the distance to the center of the silica gel particle. The mass of oryzanol in the particle (m_{As}) was calculated by integrand and mathematically formulated as shown in equation (6).

$$m_{As} = N_b \int_0^R 4\pi r^2 (\varepsilon C_{As} + a_s X_{As}) dr \tag{6}$$

Since the local oryzanol adsorption equilibrium at the surface of pore is always attained, and the equilibrium follows distribution coefficient model, equation (6) can be written as equation (7).

$$m_{As} = 4\pi N_b (\varepsilon + a_s K_f) \int_0^R C_{As} r^2 dr \tag{7}$$

The concentration of oryzanol in the liquid can be computed using equation (8).

$$C_{Al} = C_{Al0} - \frac{4\pi N_b}{V} (\varepsilon + a_s K_f) \int_0^R C_{As} r^2 dr \tag{8}$$

where C_{Al} is oryzanol concentration in the bulk of the liquid (mg/ml), C_{Al0} is initial oryzanol concentration in the bulk of the liquid (mg/ml), N_b is number of adsorbent particles, V is volume of liquid (ml), ε is porosity of the adsorbent particles, a_s is pore specific surface area of the adsorbent particles (cm^2/cm^3), K_f is distribution coefficient, C_{As} is oryzanol concentration inside the pore of particle (mg/ml), and r is distance to the center of the particle (cm).

In order to solve the mathematical models proposed, initial and boundary conditions are required. Initially, the oryzanol concentration in the bulk of the liquid is C_{Al0} while the oryzanol concentration in the silica gel particle is zero. Mathematically it can be written as: $C_{Al}(0) = C_{Al0}$ and $C_{As}(r, 0) = 0$. The first boundary condition is that at every time the value of C_{As} at the center of silica gel particle is minimum or mathematically: $\frac{\partial}{\partial r} C_{As}(0, t) = 0$. The second boundary condition is that at any time, the flux of oryzanol at the surface of silica gel particle is continuous or mathematically: $-D_e \frac{\partial}{\partial r} C_{As}(R, t) = \mathrm{k}_c (C_{As}(R, t) - C_{Al}(t))$.

2.3. Numerical calculations

The equations were numerically solved using Finite Difference Approximation (FDA). The implicit method was applied [17]. Meanwhile, the integral for calculating the amount of oryzanol in the particle was conducted by Simpson's rule.

3. Results and discussions

The previous experiment was performed for two different sizes of silica gel particles (diameter of 0.0815 mm and 0.35 mm) at the same initial oryzanol concentration in the liquid which was 0.011288 mg/mL and at the same of ratio of mass of adsorbent to the volume of solution (140 mg adsorbent per 7 mL liquid).

The results studies from [16] are presented in figure 1 and table 1. In figure 1, fractions of remaining oryzanol in the liquid (C_t/C_0) for both of particle sizes were plotted versus time.

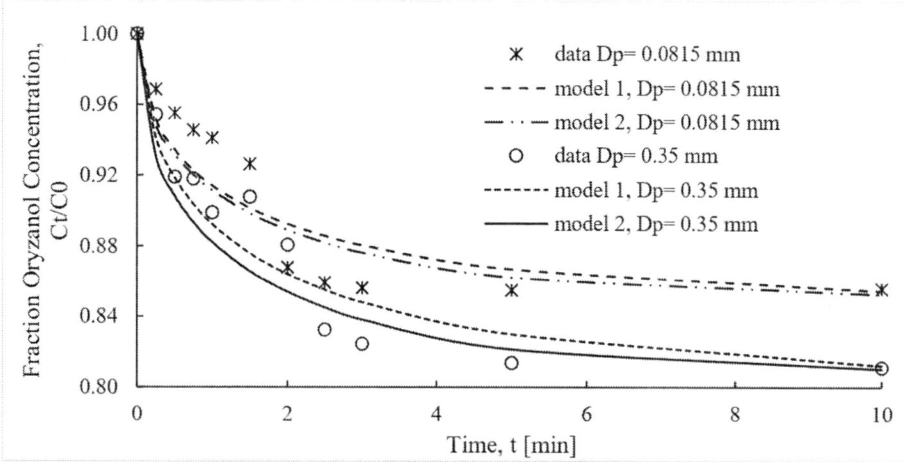

Figure 1. Fraction of remaining oryzanol mass in the bulk of the liquid in batch adsorption.

Model 2 was then applied to the data from [16]. In these calculations the values of the involving parameters are obtained by visual inspection. The computational results using Model 2 are shown also in figure 1. The values of the parameters obtained for Model 2 are also listed in table 1.

Table 1. The adsorption parameters for Model 1 and Model 2.

Parameters	Values			
Model	Model 1	Model 2	Model 1	Model 2
Average particle diameter, mm	0.0815	0.0815	0.35	0.35
Effective radial diffusivity, D_e, cm^2/s	8×10^{-5}	-	4.2×10^{-4}	-
Constant of effective radial diffusivity, $\mathfrak{D}$, cm^2/s	-	6×10^{-5}	-	4×10^{-4}
Distribution coefficient, K_f, g silica/ml solution	1.2×10^{-6}	1.2×10^{-6}	1.7×10^{-6}	2.1×10^{-6}
Mass transfer coefficient, k_c,1/s	0.825	0.825	0.825	0.825
Sum of Squares of Errors, SSE,%	2.8×10^{-5}	2.5×10^{-7}	2.5×10^{-5}	2.6×10^{-7}

Even though the values of Sum of Squares of Errors for Model 2 are slightly smaller but the two values are almost comparable. Furthermore, figure 1 shows that the accuracies of Model 1 and Model 2 are visually comparable. Hence, it can be concluded that the computational results as of Model 1 and Model 2 are comparable.

The gain of Model 2 is that the model can logically explain the phenomena that larger particle can adsorb oryzanol from solution faster than the one of smaller particle. However, the values of constant of effective radial diffusivity ($\mathfrak{D}$) as well as the distribution coefficient (K_f) for smaller particle differ to the ones of larger particle. Further improvement of the model is still needed and we are still working on that study.

The comparison of time consumed for the calculation using Model 1 and Model 2 at the same computer showed that the time of calculation for Model 2 was approximately six time to the one of Model 1. It implies that Model 2 is significantly more complicated than Model 1.

4. Conclusion

This study concluded that Model 2 which assumes that the effective radial diffusivity is proportional to the distance to the center of the particle works well for adsorption of oryzanol from solution system using silica gel adsorbent. However further improvement is still required to obtain the general values of parameters that can be applied for particles at different sizes.

Acknowledgement
The authors greatly appreciate the Directorate of Research and Public Services, Ministry of Research, Technology, and Higher Education (DRPM – Ristek Dikti), for providing funding by the *Penelitian Disertasi Doktor (*PDD) 2017 Grant.

References

[1] Chen M H and Bergman C J 2005 A Rapid Procedure for Analyzing Rice Bran Tocopherol, Tocotrienol and γ-oryzanol Contents *Journal of Food Composition and Analysis* vol 18 pp 139-51
[2] Danielski L, Zetzl C, Hense H and Brunner G 2005 A Process Line for the Production of Raffinated Rice Oil from Rice Bran *Journal of Supercritical Fluids* vol 34 pp 133–41
[3] Da Silva M A, Sanches C and Amante E R 2006 Prevention of Hydrolytic Rancidity in Rice Bran *Journal of Food Engineering* vol 75 pp 487–91
[4] Devi R R and Arumughan C 2007 Phytochemical Characterization of Defatted Rice Bran and Optimization of a Process for Their Extraction and Enrichment *Bioresource Technology* vol 98 pp 3037-43
[5] Stoggl W, Huck C, Wongyai S, Scherz H and Bonn G 2005 Simultaneous Determination of carotenoids, tocopherols, and γ-oryzanol in crude RBO by liquid chromatography coupled to diode array and mass spectrometric detection employing silica C30 stationary phases *Journal of Separation Science* vol 28 pp 1712–18
[6] Xu Z and Godber J S 2000 Comparison of Supercritical Fluid and Solvent Extraction Methods in Extracting γ-Oryzanol from Rice Bran *JAOCS* **77** pp 547-51
[7] Zigoneanu I G, Williams L, Xu Z and Sabliov C M 2008, Determination of Antioxidant Components in Rice Bran Oil Extracted by Microwave-assisted Method *Bioresource Technology* vol 99 pp 4910-18
[8] Crevel R W R, Kerkhoff M A T and Konong M M G 2000 Allergenicity of Refined Vegetable Oils *Food Chem. Toxicology* vol 38 pp 385-7
[9] Imsanguan P, Roaysubtawee A, Orirak R, Pongamphai S, Douglas S and Douglas P L 2008 Extraction of α–tocopherol and γ–Oryzanol from Rice Bran *LWT* vol 41 pp 1417-24
[10] Lloyd B J, Siebenmorgen T J and Beers K W 2000 Effects of Commercial Processing on Antioxidants in Rice Bran *Cereal Chem.*vol 77 pp 551–5
[11] Patel M and Naik S N 2004 Gamma-oryzanol from Rice Bran Oil-A Review *Journal of Scientific & Industrial Research* vol 63 pp 569-78
[12] Seetharamaiah G S and Chandrasekhara N 1989 Studies on Hypocholesterolemic Activity of Rice Bran Oil *Atherosclerosis* vol 78 pp 219–23
[13] Xu Z, Hua N and Godber J S 2000 Antioxidant Activity of Tocopherols, Tocotrienols, and γ–Oryzanol Components from Rice Bran Against Cholesterol Oxidation Accelerated by 2,2-Azobis (2methylpropionamidine) Dihydrochloride *J. Agric. Food Chem.* vol 49 pp 2077- 81
[14] Diack M and Saska M 1994 Separation of Vitamin E and γ-Oryzanols from Rice Bran by Normal-Phase Chromatography *JAOCS* vol 71 pp 1211-7
[15] Lai P, Li K Y, Lu S and Chen H H 2009 Analytical Method: Phytochemicals and Antioxidant Properties of Solvent Extracts from Japonica Rice Bran *Food Chemistry* vol 117 pp 538-44
[16] Susanti A D, Sediawan W B, Wirawan S K and Budhijanto 2017 Mathematical Modelling of Oryzanol Separation from Rice Bran Oil by Silica-Based Batch Adsorption: Equilibrium and Mass Transfer Rate *Material Sciences Forum* vol 901 pp 190-6
[17] Burden R L, Faires J D and Reynolds A C 1981 Numerical Analysis, Weber & Schmidt

Grafted flocculant based on the modified taro starch varied by length of polyacrylamide chain

M Kaavessina, S Distantina, Fadilah and Margono

Chemical Engineering Dept., Universitas Sebelas Maret, Surakarta 57126, Indonesia

mkaavessina@staff.uns.ac.id

Abstract. The grafted flocculant was prepared from modified taro starch as a backbone. This flocculant was utilizing the combination properties of the synthetic and natural polymer. Taro starch was enzymatic modified using α amylase before grafting. Non terminated polyacrylamide (nt-PAM) with different chain length was grafted gently into the starch in a nitrogen (N_2) atmosphere. Further, this grafted flocculant was called starch-g-Pam. Mark Houwink equation was used to calculate the average molecular weight (Mv) of product that correlates directly with the intrinsic viscosity (η). Fourier Transform Infra-Red (FTIR) analysis was also carried out to depict its molecular structures. The flocculation performance test was evaluated to diminish the turbidity of the simulated wastewater (about 200 NTU). This artificial wastewater was created by dissolving kaolin in water (10 g of kaolin/1000 ml). The results explained that the grafted flocculant has good performance as well as some commercial organic flocculants. It can be highlighted that the longer chain of polyacrylamide, the better performance flocculation will be exhibited. Although fewer polyacrylamide is grafted into the backbone of the modified taro starch.

1. Introduction

The availability of clean water is now a hot issue due to many people lack access it in a sufficient amount of potable water, sanitation water, etc. To tackle this problem, a large amount of waste water should be processed in a treatment unit. This treatment is a terraced process to reduce or remove pollutants or contaminants from wastewater, such as minerals, organic matters, solids, nutrients, micro-organisms. The purpose of this treatment is to recover/improve the quality of water from wastewater sources to be utilized in particular need.

Flocculation is an efficient, eco-friendly and economical process in particular of wastewater treatment for disposal/reduction of suspended solid particles such as solid minerals, toxic heavy metals, etc. This process agglomerates colloidal particles, that difficult to be sedimented, into a bigger one. This agglomeration causes the easiness of further separation, such as filtration or sedimentation [1]. There are two types of flocculants that have been developed and available in the market, i.e. polymeric and non-polymeric flocculants. The polymeric flocculant exhibits the forming of large flocs and strongly bound to restrain the forces of shear degradation [2]. Thus, some researchers consistently devoted to developing this flocculant either from a synthetic polymer, natural polymer or combination both of them.

Many reports explained that the synthetic ones are showing better performance to tie up the colloidal particles than that of natural ones [2, 3]. However, the drawbacks of synthetic polymers are inadequate biodegradability and shear stability. In contrast, those weaknesses are not found in the natural polymers.

The natural ones have shear stable, highly biodegradable and non-toxic [4]. Thus, it is interesting to combine the excellences of properties both synthetic and natural polymer in a new flocculant. Grafting copolymerization is widely conducted to modify structure and properties of the polymer by inserting a certain functional group [5]. This method can be done through redox grafting [6], γ-ray irradiation [7], microwave irradiation [8], and electron beam irradiation [9].

Our previous work has successfully copolymerized the polyacrylamide and varied starches (i.e., taro starch, cassava starch and their modification) through redox grafting [10]. We grafted polyacrylamide in the form of free radicals into the starches. Free radicals were formed when the polymerization of acrylamide initiated by potassium persulfate was not terminated. The non-terminated polyacrylamide would substitute the hydroxyl groups in the starches. The modified starches both taro and cassava were easier to be replaced with non-terminated polyacrylamide than that of neat starches. Modified starch can be undertaken through chemical processes (esterification, oxidation, etc.), enzymatic process, or physical processes (pre-gelatinization and hydrothermal treatment). This process is to revamp the hydroxyl groups of starch. Generally, the flocculation performance of this flocculant is comparable to that of Tawas (alum). However, it still needs to enhance its effectivity.

There are some reports about the factors influencing the effectiveness of flocculant in the flocculation of colloidal particles, such as dosage of flocculant, type of suspended particles, etc. However, the characteristic effects of flocculant are found in few articles [11, 12]. In the present article, the polymerization time of non-terminated polyacrylamide before grafted into the starches was studied. This time represents the length of polyacrylamide chains. We used modified taro starch as a backbone. Taro starch was modified enzymatically by adding α-amylase. After grafted flocculants were obtained, we analyzed their molecular structure, average molecular weight and its flocculation performance in simulated wastewater.

2. Experimental details

Taro was available from the traditional market in Surakarta and made into taro starch. The taro starch was modified by enzyme α-amylase (from Sigma Aldrich). Some chemicals, i.e., Acrylamide, $K_2S_2O_8$, TMEDA, acetone, methanol and hydroquinone were also kindly ordered by Sigma Aldrich. Further, these chemicals were directly used without any specific treatment.

The formation of Starch-g-Pam was initiated with redox reaction to synthesize non terminated polyacrylamide (nt-PAM). This free radical polymer would be grafted into the modified taro starch by substituting the hydroxyl (-OH) groups. In this investigation, the polymerization time of acrylamide was varied at 0.5, 1 and 1.5 hours. Whereas, the other conditions of both synthesizing and grafting of nt-PAM were accorded to our previous work [10].

We used an infrared spectrophotometry (Perkin-Elmer 630 IR) to determine the molecular structure of products. The wavenumber range of IR spectra was recorded at 4000-400 cm^{-1}. The preparation of IR samples was using a KBr pellet method in a solid state. The Mark Houwink equation as shown in equation (1) was involved in determining the average molecular weight (Mv) of the grafted flocculants where, the *K* and a constant are 6.31 x 10^{-3} g/ml and 0.8 for PAm in water, respectively [13].

$$\eta = KMv^a \tag{1}$$

The performance of grafted flocculant was tested to agglomerate kaolin particles in colloidal size. The concentration of kaolin in this simulated wastewater is 10 g in 1 L of solution. The parent solution of flocculant was prepared by dissolving 10 mg grafted flocculant in 20 ml of water. The dosage of flocculant in this test is 4 ml of parent solution decanted into 500 ml of the artificial wastewater. The mixture was then shaken at 120 rpm (for 2 minutes) and 50 rpm (for 10 minutes), sequentially. Finally, the turbidity of mixture before and after adding the grafted flocculant was calculated by EUTECH TN100.

3. Results and discussion

The taro starch was modified enzymatically using α-Amylase prior to use as a backbone of grafted flocculant. The modified starch was visually clearer in water solution and had lower viscosity compared to the native one. However, the IR spectrum showed this modification was not changing the molecular structure taro starch [10]. The schematic representation of starch-g-polyacrylamide varied in polymerization time of acrylamide 0.5 h (A) and 1.5 h (C) is shown in figure 1.

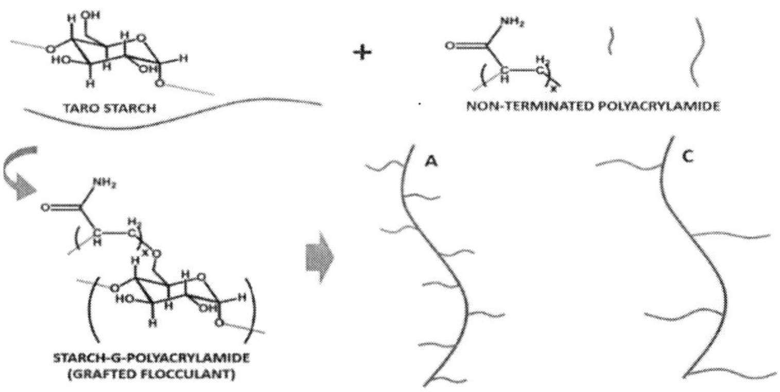

Figure 1. Schematic representation of starch-g-polyacrylamide varied in polymerization time of acrylamide: 0.5 h (A) and 1.5 h (C).

In this article, we study the effect of polymerization time of acrylamide toward flocculation performance of grafted flocculants. The same amount of acrylamide monomer was polymerized without any termination processes in the different time, i.e., 0.5 (A), 1 (B) and 1.5 hours (C). Further, we call the obtained flocculants as sample A, sample B and sample C representing those times, respectively. We proposed that it would produce the different length of radical chains. The amount of shorter chains was more than that of the longer chains. For 1.5 hours, it had enough time to propagation step and resulted in the longer chains of non-terminated polyacrylamide (nt-polyacrylamide). However, these chains were fewer formed due to the limited availability of acrylamide monomer. The IR spectrum of starch-g-polyacrylamide varied in polymerization time of acrylamide for sample A and sample C are shown in figure 2.

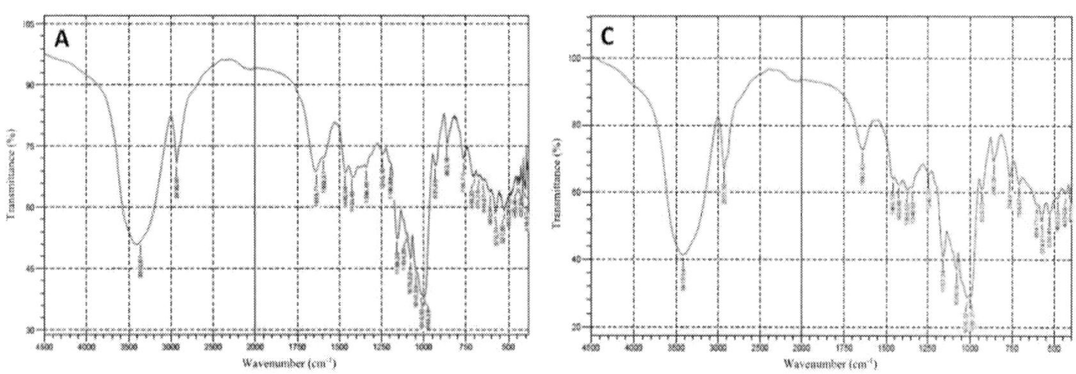

Figure 2. IR spectrum of starch-g-polyacrylamide varied in polymerization time of acrylamide: sample A and sample C.

The obtained nt-polyacrylamide with different length of chains was then grafted into the modified taro starch as the backbone. The schematic representation of the obtained flocculants was depicted clearly in figure 1. It can be seen that the backbone of sample A was more grafted by the shorter chains of nt-polyacrylamide. It means that the hydroxyl groups in this sample were more substituted. Compared to the sample C, the hydroxyl groups were fewer substituted by the longer chains of nt-polyacrylamide. This statement will be confirmed by Fourier-transform infrared spectroscopy (FTIR) analysis.

The similar IR spectra of grafted flocculants varied their length of polyacrylamide chains were shown in figure 2. The absorption band data of samples was recapitulated in table 1. It can be underlined at wavenumber around 3419.94 cm^{-1} and 3392.89 cm^{-1} that indicating the hydroxyl (-OH) groups. The different intensity is clearly appearance which the stronger intensity is seen for sample C. It was confirming our previous statement that the hydroxyl groups in modified taro starch of sample C were still more exist and unsubstituted with nt-polyacrylamide.

Table 1. Comparison infrared spectra of Sample A and Sample C.

Wavenumbers		Chemical Bonding	Mode	Notification
Sample A	Sample C			
3419.94	3418.92	–OH	Stretching	stronger intensity for sample C
1643.42	1643.40	–C-O	Stretching	
1340.58	1340.58	–CH–	Deformation	asymmetric, symmetric
1368.55	1368.55			
1420.63	1420.60			
1460.18	1460.15			
1193.99	1190.49	C-N	Bending	Similar intensity for sample A and C
1240.32	1241.02			
1242.21	1268.21	–C-OH	Bending	
890.00	921.10	–C–C–	Stretching	

Figure 2 also shows two similar peaks around 1193.99 cm^{-1} and 1240.32 cm^{-1} indicating the existence of C-N bond. This bond refers to the appearance of chemical bonds between starch and polyacrylamide in the obtained products as illustrated in figure 1. This result was confirming that starch-g-polyacrylamide has been successfully formed through "grafting to" method.

Table 2. Tabulation of intrinsic viscosity $|\eta|$ and average molecular weight (Mv).

Grafted Flocculant	Backbone, gram	Acrylamide, M (4 ml)	Polymerization time, h	$\|\eta\|$, ml/g	Mv.10^3
Sample A	2.5	0.1	0.5	7.4 ± 0.2	6.862
Sample B	2.5	0.1	1.0	7.6 ± 0.1	7.095
Sample C	2.5	0.1	1.5	7.7 ± 0.2	7.212

Table 2 summarizes the measurement of intrinsic viscosity and the calculation of molecular weight of starch-g-PAm. As seen in this table, the molecular weight of grafted flocculants has almost the same. In agreement, their intrinsic viscosity was also on the same trend. It may be explained that the utilization of the same starch gives the same opportunity of free radical to substitute the hydroxyl groups of the starch. Whereas, the free radical was added in the same amount of monomer although resulting in the different of nt-polyacrylamide chains.

Table 3 recapitulates flocculation test in the simulated wastewater containing a kaolin suspension. It shows that sample C is more efficient to reduce the water turbidity, although the average molecular weight is almost same. It might be concluded that this phenomenon was related to the length of polyacrylamide grafted in the backbone. The longer chains may be easier to get hold of colloidal kaolin particles becoming agglomerates. However, it needs further analysis to verify this explanation.

Table 3 also shows the performance of the grafted flocculants may be compared to the synthetic commercial flocculant (Alum/$Al_2(SO_4)$). The percentage residual turbidity of sample C was almost the same value with the synthetic one, i.e., 12.46 NTU and 6.23 NTU, respectively. The acidity of the simulated wastewater after flocculation tended to decrease. However, the larger drop in acidity occurred when alum was used. This phenomenon showed the advantages of the grafted flocculant.

Table 3. Flocculation test of grafted flocculant for 1 hour.

Criteria	Sample			
	A	B	C	Alum
Colloidal particle	Kaolin	Kaolin	Kaolin	Kaolin
Temperature, °C	30	30	30	30
pH before/after flocculation	7.1/6.6	7.1/6.5	7.1/6.6	7.1/6
Volume of the simulated wastewater, L	0.5	0.5	0.5	0.5
Dosage of flocculant, mg/L	2.67	2.67	2.67	2.67
Initial Turbidity, NTU	200	200	200	200
Residual Turbidity, NTU	19.83	16.51	12.46	6.23
Removal, %	90.08	91.74	93.77	96.88

4. Conclusion

The different length of non-terminated polyacrylamide can be obtained by varying the polymerization time. The obtained nt-polyacrylamide was successfully grafted into the modified taro starch. The average molecular weight related to the intrinsic viscosity was observed in the similar value. However, the longer of polyacrylamide chains grafted in the starch the more effective of its performance to decrease turbidity of kaolin suspension in water. It means that the agglomeration of colloidal particles was easily achieved when the length of polyacrylamide chains was getting longer. Although, the number of chains is fewer. Generally, the grafted flocculant consisting of modified taro starch and polyacrylamide is potential to be applied in a part of wastewater treatments.

Acknowledgement

This research was financially supported by Universitas Sebelas Maret through research grant of "*Hibah Peneliti Utama sebagai Rujukan Hibah* MRG-UNS PNBP 2017".

References

[1] Singh R P, Tripathy T, Karmakar G P, Rath S K, Karmakar N C, Pandey S R, Kannan K, Jain S K and Lan N T 2000 *Curr. Sci.* **78** 6
[2] Nayak B R, Biswal D R, Karmakar N C and Singh R P 2002 Bull. Mater. Sci. 25 537
[3] Adhikary P, Tiwari K N and Singh R P 2007 J. Appl. Polym. Sci. 103 773
[4] Biswal D R and Singh R P 2004 Carbohydr. Polym. 57 379
[5] Kaavessina M, Ali I, Elleithy R and Al-Zahrani S 2012 J. Polym. Res. 19 1
[6] Kang H-M, Cai Y-L and Liu P-S 2006 Carbohydr. Res. 341 2851
[7] Iskandar S 2011 Atom Indonesia 37 5
[8] Alfaifi A Y A, El-Newehy M H, Abdel-Halim E S and Al-Deyab S S 2014 *Carbohydr. Polym* **106** 440
[9] Zhang S, Wang W, Wang H, Qi W, Yue L and Ye Q 2014 Carbohydr. Polym 101 798
[10] Kaavessina M, Distantina S and Fadilah 2017 MATEC Web Conf. 101 01003
[11] Liu Z, Wei H, Li A and Yang H 2017 Water Res. 118 160

[12] Salehizadeh H, Yan N and Farnood R 2017 Biotechnol. Adv. INPRESS
[13] Qudsieh I Y, l-Razi A F, Kabbashi N A, Mirghani M E S, Fandi K G, Alam M Z, Muyibi S A and Nasef M M 2008 *J. Appl. Polym. Sci.* **109** 3140

Optimization of biodiesel production by various plant sources and micro-reactor simulation

R Fatoni[1], M M Rahman[2] and A Elkamel[2]

[1] Chemical Engineering Department, Universitas Muhammadiyah Surakarta, Jl. Ahmad Yani Tromol Pos No. 1, Pabelan, Surakarta, Jawa Tengah 57102, Indonesia
[2] Chemical Engineering Department, University of Waterloo, 200 University Avenue West, Waterloo N2GL, Ontario, Canada

rois.fatoni@ums.ac.id

Abstract. This paper discusses the alternative ways to optimize biodiesel production. First, different oils from different plant sources were compared to see the degree of conversion in trans-esterification reaction and the purity of bio-diesel produced. Second, micro-reactor is proposed as an alternative to conventional, batch reactor in producing biodiesel. The results show that soybean oil gives the highest conversion among other oils. Process simulation using COMSOL shows that using micro-reactor can cut the reaction time to 25 minutes; much shorter than the time needed by batch reactor which is 60 minutes. The simulation is performed at the temperature, reactant ratio, and acid catalyst concentration of 60°C, 6:1, and 1%, respectively.

1. Introduction

Biodiesel is considered as the best alternative to petroleum based fuel. It is produced from biological sources which are renewable. The main sources are animal fats and vegetable oils. Therefore, biodiesel is biodegradable and non-toxic. When it burns, biodiesel produces less smoke. That is because biodiesel contains 11% of oxygen or more. That helps in combustion process, leading to decreasing the emissions of unburned hydrocarbon up to 90%. The emission of carcinogenic substance such as polycyclic aromatic hydrocarbon (PAH) is decreased significantly when biodiesel is used. The emission of CO_2 is 50% less compared to petro-diesel [1].

Despite those benefits, however, biodiesel has some disadvantages compared with the fossil fuel. The cost of production of biodiesel is still higher than petro-diesel cost of production. In sub-tropical countries, where temperature drops significantly during winter, biodiesel needs more additives. This is because the higher cloud point possessed by biodiesel. Another problem is storage issues. Biodiesel is relatively unstable compared to petro-diesel. Biodiesel has higher rancidity because of oxidation with atmospheric, bacterial air. Furthermore, certain types of elastomers and natural rubber can be degraded rapidly due to higher percentage of biodiesel blends. Hence, it raises higher concerns in materials in fueling system. These disadvantages of biodiesel drive the researchers to find the optimum ways to produce biodiesel in order to increase its utilization [1–3].

Activities in biodiesel research dated back in late 1970's and early 1980's. Biodiesel can be produced in different ways. Some of the methods are common than others, for example direct use and blending, trans esterification and other recently developed methods with supercritical methanol.

Catalytic trans esterification is by far is the most common method in producing biodiesel. Due to its nature of reaction, which is mass-transfer-controlled reaction, the continuous stirred reactor is used widely [4,5].

Biodiesel can be produced from different source of plant. The availability of the plant and the degree of oil yield which can be produced by the plant farming per unit areas vary from place to place. Typical data of the information can be seen in table 1. It can be seen that some source plant has higher yield than the other. Oil palm has the highest yield with 5950 litres/ha; followed by the likes of coconut, avocado, and macadamia nut which produce around more than 2000 litres/ha. The list is followed by jatropha, jojoba, castor vean, rapeseed, peanut, and cocoa. Soybean has the lowest yield among the plants with 446 litres/ha. Surprisingly, soybean is the largest biodiesel feedstock in US biodiesel production. It accounts for about 55% of total production, followed by corn oil, canola oil, and palm oil. The choice of biodiesel feedstock is not only by considering the oil yield per farming area, but also by considering the cost of farming and the degree of trans esterification of each oil feedstock [1].

Table 1. Typical oil yield of plant farming for biodiesel feedstock

Crop	Litres oil/ha
Oil palm	5950
Coconut	2689
Avocado	2638
Macadamia nut	2246
Jatropha	1892
Jojoba	1818
Castor vean	1413
Rapeseed	1190
Peanut	1059
Cocoa	1026
Mustard seed	572
Soybean	446
Corn (maize)	172

2. The scope of work

The objective of this work is to obtain information which will be beneficial in optimizing biodiesel from plant oil. The information comes from two areas: the comparison of biodiesel yield in trans-esterification of different plant oil and a computer-aided simulation comparing conventional and micro-reactor performance of biodiesel production.

Biodiesel production can be carried by reacting fat or oil with alcohol to produce esters and glycerol. This reaction is well-known as trans-esterification. Sometimes, it also called as alcoholysis. Simple alcohol can be used for trans-esterification. Ethanol and methanol are used most frequently. Methanol is primarily used because of physical and chemical advantages. Methanol is a polar molecules and the chain is shortest among other alcohols. Those features make methanol is very suitable for this. Furthermore, methanol is inexpensive and can react with triglycerides quickly [6].

Catalytic trans-esterification method is used to produce biodiesel for this research project. The catalyst used is sodium hydroxide (NaOH); a catalyst which is easily dissolved in methanol. Theoretically, a 3:1 molar ratio of alcohol to triglycerides is required to complete a trans-esterification reaction [4,5].

2.1. Biodiesel yield comparison among different oil source

There are five feedstock oil in this work that was being compared their biodiesel conversion in trans-esterification reaction: vegetable oil, corn oil, canola oil, soybean oil, and sunflower oil. Each oil was treated with the same operating condition in trans-esterification reaction performed in lab-scale batch reactor. The degree of conversion obtained will be useful information in choosing feedstock in biodiesel production.

2.2. Simulation of conventional and micro-reactor in biodiesel production.
Many researchers have suggested the use of micro-reactor in trans-esterification reaction. Unlike conventional batch reactor, micro-reactor is usually a continuous flow reactor. It is a device where chemical reactions occur in a confinement such as micro-channels with lateral dimension below 1 millimeter. The structure of micro-reactor usually contains an extensive number of parallel channels with common inlet flow regions. These micro-channels offer great advantages over conventional batch reactors, such as increase in reaction yield and speed, more efficient and reliable with much enhanced degree of process control [7,8].

Micro-channel usually produces small amount of material. Therefore, in order to increase production capacity in fluid throughout in micro-reactors, it is facilitated by numbering-up approach, rather than by scaling-up approach. This way of increasing production quantity gives the desired feature of a basic unit remains the same when the total capacity is increasing.

There are some advantages of micro-reactors which also known as miniaturized systems compared to large-scale process. Micro-reactor needs less space. This is because the batch nature of operation in conventional, large-scale process can be replaced by a continuous flow operation. The cost is kept low by parallel micro-fabrication and automated assembly. It needs less energy, it is safer operation, it is easier to scale up of production capacity, and it offers more flexibility in response to the fluctuated demands of market [9].

To compare the performance of micro-reactor, a computer-aided simulation was performed in this work. The reaction time needed to achieve 100% conversion was compared between micro-reactor and batch reactor.

3. Experimental method
This section explains the experimental method used in this research. The method is elaborated in the following subsections.

3.1. Experimental procedure of trans-esterification
All experiments were performed at room temperature and atmospheric pressure. A typical experiment with a 50 mL Erlenmeyer flask and add 14 ml of pure methanol. Slowly 0.50 g of NaOH is added (Caution: the mass cannot go over 0.50 g). Once the NaOH is dissolved completely, a very strong base is produced sodium methoxide (Step 1). Simultaneously 60 mL of vegetable oil is heated to 50°C. Sodium methoxide is then added slowly to the warm vegetable oil (Step 2). The solution turns cloudy. The solution is stirred for 20 minutes under medium agitation. The solution is then allowed to settle for another 20 minutes in a separatory funnel. Biodiesel is formed at the top and the glycerol layer settles down at the bottom. The glycerol is then removed first and then biodiesel is drained from the separatory funnel.

3.2. Computer-aided simulation of micro-reactor
The micro-reactor system consists of the following components: syringe pump, micro-reactor, and cold trap for the sample collection. The process can be seen in figure 1 and the COMSOL diagram for trans-esterification micro reactor can be seen in figure 2. The step of dissolving methanol with NaOH was replaced with stock solution of sodium methoxide for micro-reactor reaction to reduce process time and it also decreases the exposure to the dangerous base sodium methoxide. Since stock solution of methanol with dissolved NaOH can be found off the shelf, this reduces the reaction steps significantly. This completely eliminates the Step1 from the experimental procedure.

IOP Conf. Series: Materials Science and Engineering **403** (2018) 012011 doi:10.1088/1757-899X/403/1/012011

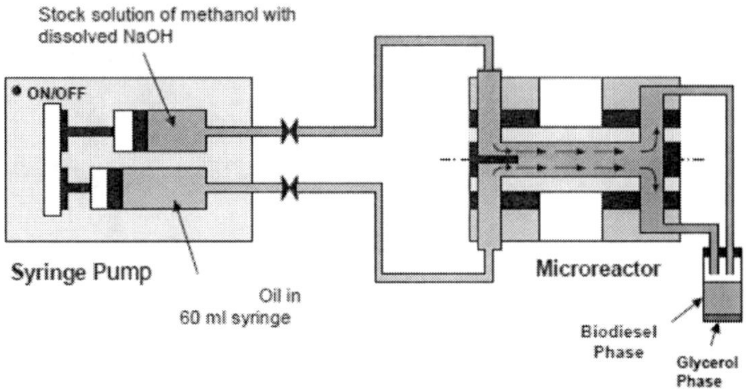

Figure 1. Schematic of trans-esterification micro-reactor.

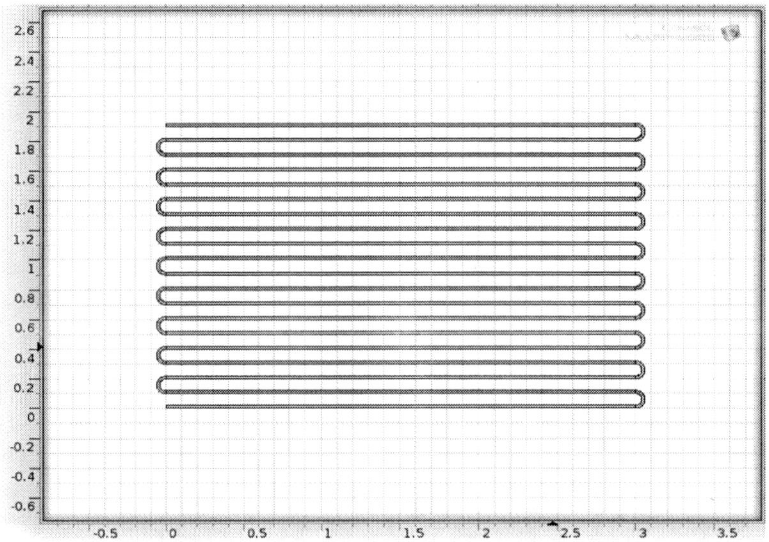

Figure 2. COMSOL diagram for trans-esterification micro reactor.

The oil conversions were calculated from trans-esterification of vegetable oil to biodiesel by using a micro-tube reactor. Step 2 of the reaction process was done in the micro-reactor which directly affects the reaction yield. Reaction engineering module was used for the simulation and it was assumed to be stationary plug flow with no slip condition. To compare the performance of micro-reactor, simulation of batch esterification is also carried out in this simulation.

4. Results and discussion

This section discuss the result of the experiments conducted in Section 3. The discussion is elaborated in the following subsections.

4.1. Effect of different feedstock for biodiesel production

Five different oils have been used to in this project to determine which has the greater percentage yield. Figure 3 shows that soybean has the highest conversion. This shows soybean oil can be used for biodiesel can be used most effectively.

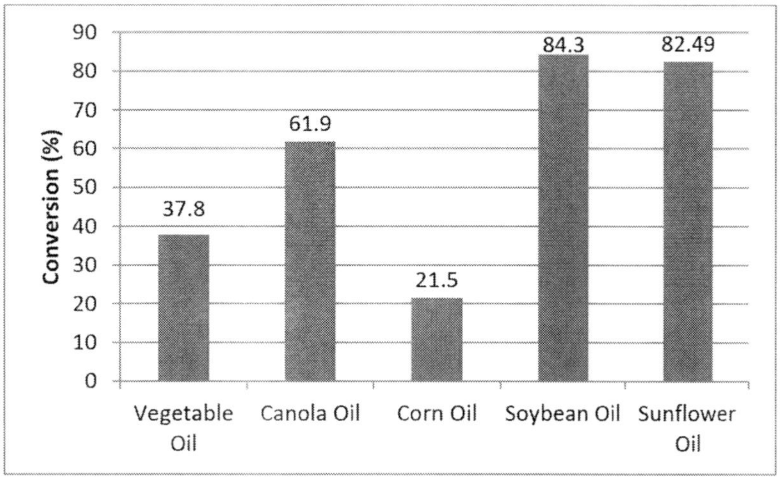

Figure 3. Biodiesel conversion from various oil feedstock.

4.2. Performance comparison between micro reactor and batch reactor
The reaction formula was set to be A+3B=>3C+D. The reaction type is irreversible reaction. Although in practice it is a equilibrium reaction but for simplicity it was assumed to be irriversible. Reactor was designed to have a volume of 1.885x10 m^3 with a diameter of 200 microns and 20 tubes each of length of 3 cm. The volumetric flow rate is set to 20 microL/min. The ambient condition was set to room temperature 293 Kelvin and 1 atmospheric pressure. Reaction engineering module was used to simulate the trans esterification with the data provided in table 2.

Table 2. Data for trans-esterification simulation by using micro reactor.

Species	Density [kg/m^3]	Molecular weight[kg/mol]	Initial Conc [mol/m^3]
Vegetable Oil (A)	920	0.900	1.050
Sodium Methoxide (B)	945	0.054	0.893
Biodiesel (C)	880	0.292	-
Glycerol (D)	1260	0.093	-

The result of simulation showing the reaction time needed to achieve 100% conversion can be seen in figure 4. The graph clearly shows for batch reaction the oil conversion was 100% after a reaction time of 1hr. However, under the same condition, while the reaction is performed in a micro-reactor the 100% conversion is reached at 0.62hr.

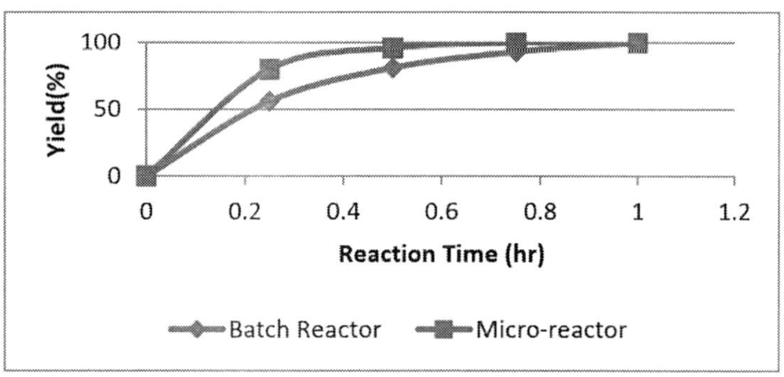

Figure 4. Batch reactor vs micro reactor reaction time of trans esterification.

5. Conclusion

Investigation of effects of oil feedstock, operating conditions and the use of micro-reactor has been carried out in this work to obtain information which can be used to optimize biodiesel production Soybean oil gives the highest yield of biodiesel compared to other oil feedstock. Reaction simulation by using engineering module in COMSOL shows that micro-reactor is better in reaction time needed to achieve 100% conversion (0.62 hour) compared to the conventional batch reactor (1 hour).

References

[1] Sales A 2011 *Production of biodiesel from sunflower oil and ethanol by base catalyzed transesterification* (Royal Institute of Technology (KTH) Stockholm, Swedia)
[2] Avhad M R and Marchetti J M 2015 A review on recent advancement in catalytic materials for biodiesel production *Renew. Sustain. Energy Rev.* vol 50 pp 696–718
[3] Lee S J 2010 Process simulation, economic analysis and synthesis of biodiesel from waste vegetable oil using supercritical methanol
[4] Liu X, Piao X, Wang Y and Zhu S 2010 Model Study on Transesterification of Soybean Oil to Biodiesel with Methanol Using Solid Base Catalyst *J. Phys. Chem. A* vol 114 pp 3750–5
[5] El-Gendy N S, Abu Amr S S and Aziz H A 2014 The Optimization of Biodiesel Production from Waste Frying Sunflower Oil Using a Heterogeneous Catalyst Energy Sources, Part A *Recover. Util. Environ. Eff.* vol 36 pp 1615–25
[6] Martínez S L, Romero R, Natividad R and González J 2014 Optimization of biodiesel production from sunflower oil by transesterification using Na2O/NaX and methanol *Catal. Today* vol 220–222 pp 12–20
[7] Richard R, Thiebaud-Roux S and Prat L 2013 Modelling the kinetics of transesterification reaction of sunflower oil with ethanol in microreactors *Chem. Eng. Sci.* vol 87 pp 258–69
[8] Bhoi R, Sen N, Singh K K, Mahajani S M, Shenoy K T, Rao H and Ghosh S K 2014 Transesterification of sunflower oil in microreactors *Int. J. Chem. React. Eng.* vol 12 pp 47–62
[9] Šalić A and Zelić B 2011 Microreactors - Portable Factories for Biodiesel Fuel Production Main pp 98–110

IOP Conf. Series: Materials Science and Engineering **403** (2018) 012012 doi:10.1088/1757-899X/403/1/012012

Evaluation of moisture content in drying of grated coconut meat using grain moisture meter

E Sulistiawati, I Santosa and A Aktawan

Chemical Engineering Departement, Faculty of Industrial Technology, Universitas Ahmad Dahlan, Jl. Prof. Soepomo, Warungboto, Yogyakarta, Indonesia.

Abstract. Drying the coconut meat is one of the process in the coconut oil industry. The objective of this research was to evaluate the moisture content that measured by grain moisture meter (GMM), and it was compared to gravimetric results. Grated coconut meat was put into the dryer that carried out at 80°, 85°, 90°, 95°, and 100°C. It needed 150 to 110 minutes until zero moisture content. Measuring the moisture content of grain particles (i.e. grated coconut meat) by Grain Moisture Meter (GMM) is one of a simple way to understand how much the real time moisture content, but it must be calibrated with gravimetric method.

1. Introduction

Most methods of measuring moisture content in solid particles using gravimetric methods. This is done by some researchers. The weakness of this way is not known the current water content. The measurement also takes a long time. Therefore, it is need to find a way to measure the water content in real time.

Aregbelosa *et al.* studied the thin layer drying characteristics of *dika* kernels and nuts. It was investigated at four drying temperatures of 50°, 60°, 70° and 80°C and the data was fitted to drying models. The moisture content of the kernels and nuts was determined by gravimetric method. The samples were oven dried at 130°C until the mass didn't changes between two weighing succession. The sample's weight was taken quickly to obviate interference with the drying process. The final moisture content was specified by oven drying method [1].

Experiments of drying copra were conducted by Deepa *et al.* [2]. In the determination of water content, samples (10 g) were chopped from randomly selected 5 cups and kept in a convective electrical oven, maintained at 105° ± 1°C for 5 h. Initial and final mass at time t of samples were recorded using electronic balance and repeated every 1 h interval till end of drying.

Drying of grated coconut was investigated by Ginting *et al.* [3]. The research was aimed at testing the temperature drying (100°C, 110°C, 120°C and 130°C) on the quality of the desiccated coconut dried in desiccated coconut dryer. Water content was known by weighing the dried material in the oven at 105°C for 4 hours. The difference between initial and final mass is the water content. Pestano and Jose studied improvement coconut processing systems in the Philippines [4]. Samples of air-dried coconut meat was weighed using a precision balance with an accuracy of 0.01 g. Drying take place at 60°, 70° and 80°C.

Rosa *et al* investigate the kinetics of drying orange seed, waste products from juice processing, that was studied in the temperatures of 40°, 50°, 60° and 70 °C and drying velocities of 0.6, 1.0 and 1.4 m/s. The experiments were obtained using a convective air forced dryer. Drying was stopped when the weight of the test sample reached a constant value. Moisture contents at reach time interval was calculated in dry basis (d.b.) from both weight loss data and dry solid weight of the sample [5].

Grain moisture meter is used to measure the moisture content in solid granules and the water content is read directly on the screen. Is the result of the water content read from the grain moisture meter (GMM) equal to the water content of the gravimetric way? In this research, drying of fresh grated coconut is done. The purpose of this research was to evaluate the moisture content that indicated by grain moisture meter compared to the gravimetric results.

2. Materials and methods

Water equilibrium in solid phase (x) and in air (y) could be approximated by Henry Equation (1).

$$y = \mathrm{H.x} \tag{1}$$

Experimentally, the coconut was de-husked, de-shelled, peeled, grated, and dried in the batch dryer. Drying was carried out in the batch dryer equipped with a motor, an agitator and a heater from a fire derived from LPG gas. The dryer was also equipped with oil bath to regulate its temperature. The area of the dryer was 29.5 cm x 40 cm. Grated coconut 3 kilograms was put into the dryer that operates from 80°C to 100°C, and every 10 minutes was measured its moisture content (y) using the Grain Moisture Meter - Hold Peak - HP-70326. The results were compared with the solid water content (x) which measured gravimetrically.

3. Results and discussion

This section explains and discusses the experiment results.

3.1. Evaluation of moisture content from GMM and gravimetrically method

The samples of 30 ± 0.01 g of grated coconut meat were put into porcelain dish. The grated coconut was dried in the oven, and every ten minutes was measured the moisture using GMM and also measured by electrical balance. The moisture content of grated coconut from the grain moisture meter (y, g/g) was compared with water content from gravimetrically method (x, g water/g solid dry basis). The results can be seen at figure 1 and figure 2.

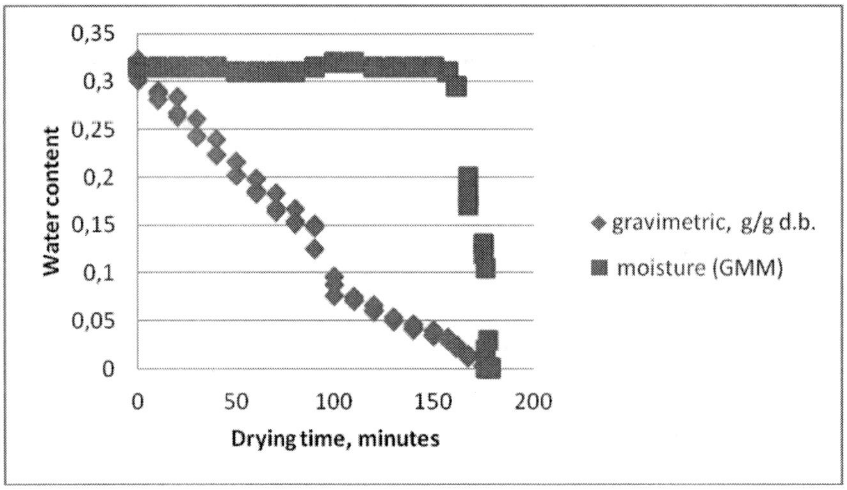

Figure 1. Measuring water content using Grain Moisture Meter (GMM) and Gravimetric Method.

There were two periods of drying, constant rate period (for x = 0.024 - 0.32), and falling rate period (x = 0 - 0.024). For constant drying period, the moisture content at the solid surface was 0.31 to 0.32, and it needed more than 150 minutes for oven drying. For the falling rate period, the relation between y and x was expressed on figure 3, and the Henry constant was 12.957.

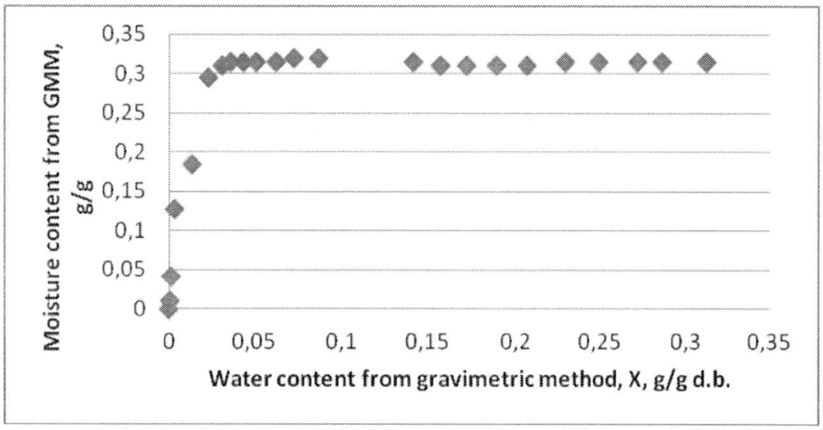

Figure 2. The relation between moisture content from grain moisture meter (y) and water content in solid gravimetrically (x).

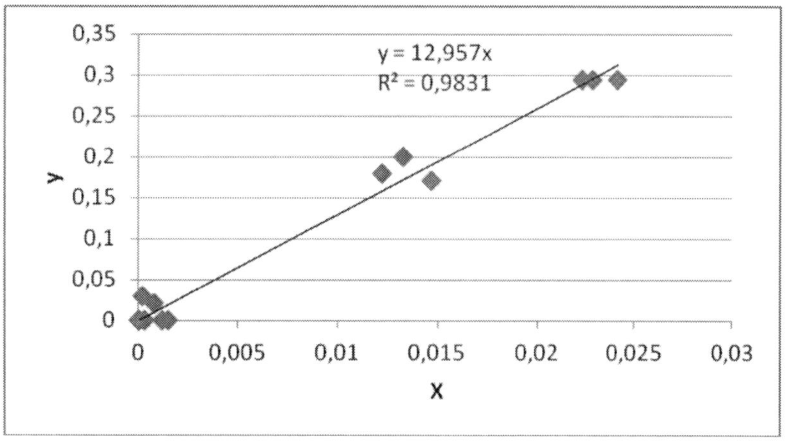

Figure 3. Relation between y and x in the falling rate period.

The measurement using grain moisture meter (GMM) is simple method, but need to compare with the gravimetric method. In the beginning of drying process, the water covers the solid surface, so that the moisture content that measured is constant, until critically x. After that, the moisture content (y) is falling until zero. Grain moisture meter is a simple to understand how much the moisture content, but it must be checked with the gravimetric method to get a certain value of water content in the solid. Grain moisture meter only measure the moisture at the air near the surface of the solid. In the constant rate period, the measurement or moisture is not valid, but in the falling rate period, under 0.024 (g/g dry basis) it's almost valid. It is needed to make a relationship between drying time (t) and water content in dry basis by gravimetric method (x), and is shown as figure 4 and equation (2). It is needed to evaluate the true moisture content over 0.024 g/g d.b. So, for $x > 0.024$ it is applied the equation (2) and for $x \leq 0.02$ equation (3) is applied.

$$x = 0.2951 - 0.0017t \tag{2}$$

Where, x = water content, g/g solid dry basis and t = drying time, minutes.

$$x = y/1295.7 \tag{3}$$

Where, y = moisture content by grain moisture meter, %.

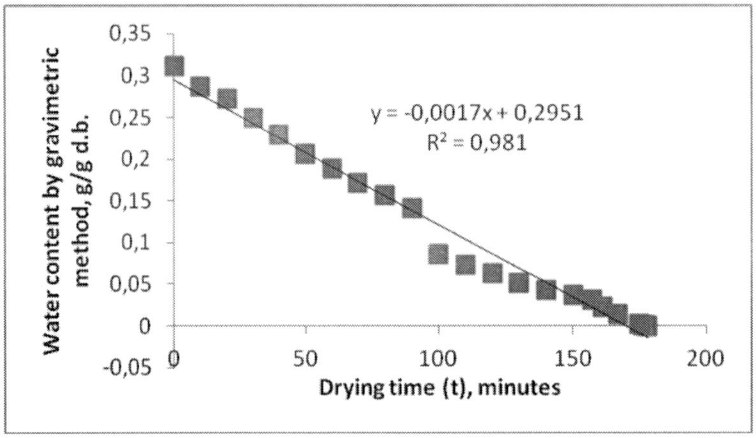

Figure 4. Relationship between water content by gravimetric method against drying time (t).

3.2. *Application on a batch modified screw mixed dryer*

Grated coconut as much as 3 kg was put into the modified screw mixed dryer, and batch operation. The dryer was operated at 80°C and atmospheric pressure. Every ten minutes the water content of grated coconut was measured using GMM. The drying operation was continued until zero moisture content. The same operation was carried out at 85°, 90°, 95°, and 100°C. The results of measurement of water content using grain moisture meter and applied calculation of equation (2) and equation (3) is shown in table 1.

Table 1. The average moisture content of grated coconut meat at different temperature operation observed by grain moisture meter.

	Average Moisture Content, g water/g solid dry basis									
t, min	T=80°C		T=85°C		T=90°C		T=95°C		T=100°C	
	GMM	x	GMM	x	GMM	x	GMM	x	GMM	x
0	0.320	0.295	0.320	0.295	0.320	0.295	0.320	0.295	0.320	0.295
10	0.320	0.278	0.320	0.278	0.320	0.278	0.320	0.278	0.315	0.278
20	0.320	0.261	0.320	0.261	0.315	0.261	0.315	0.261	0.315	0.261
30	0.320	0.244	0.315	0.244	0.315	0.244	0.315	0.244	0.315	0.244
40	0.320	0.227	0.315	0.227	0.315	0.227	0.315	0.227	0.315	0.227
50	0.320	0.210	0.315	0.210	0.315	0.210	0.315	0.210	0.315	0.210
60	0.315	0.193	0.315	0.193	0.315	0.193	0.315	0.193	0.315	0.193
70	0.315	0.176	0.315	0.176	0.315	0.176	0.315	0.176	0.215	0.017
80	0.315	0.159	0.315	0.159	0193	0.015	0.213	0.016	0.120	0.009
90	0.315	0.142	0.220	0.017	0.120	0.009	0.120	0.009	0.067	0.005
100	0.315	0.125	0.130	0.010	0.047	0.004	0.078	0.006	0.027	0.002
110	0.172	0.013	0.087	0.007	0.022	0.002	0.037	0.003	0.000	0.000
120	0.120	0.009	0.037	0.003	0.000	0.000	0.000	0.000		
130	0.050	0.004	0.000	0.000						
140	0.020	0.002								
150	0.000	0.000								

It can be seen from the table 1 that the higher operating temperature will reduce the drying time. The drying temperature of 100°C took only 110 minutes to reach zero moisture content, whereas the drying time 120 minutes at 90°C and 95°C. At 85°C and 80°C it needed 130 minutes and 150 minutes, respectively. The higher operating temperature will reduce the humidity of air surrounding the dryer, and eventually will increase the driving force. This will increase the rate of mass transfer of water from solid surface to the air. Finally, it reduces the drying time.

The results of this research is better than the Wijose Process, because the Wijose Process needs more time, 188.3 minutes for the same temperature [4]. This results also better than another researcher which used cabinet dryer [2], that needs 24 hours.

Grain moisture meter is not suitable for measuring water content in drying operation of grated coconut meat, so it must be corrected by equation (2) and (3) to get the current water content gravimetrically. It is shown as x (gram of water/gram of solid dry basis) in table 1 and figure 5. The calibration of GMM results can be seen at figure 5.

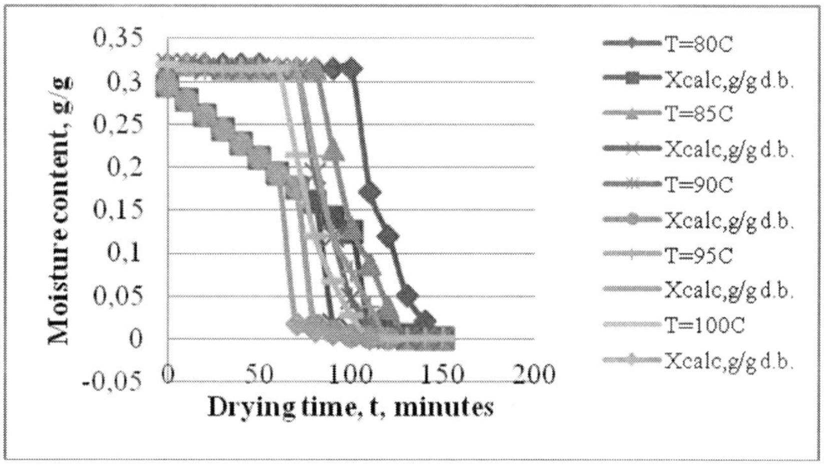

Figure 5. The calibrated results of GMM that applied in batch modified screw mixed dryer.

4. Conclusion

Measuring the moisture content of grain particles (i.e. grated coconut meat) by Grain Moisture Meter (GMM) is one of a simple way to understand how much the real time moisture content, but it must be calibrated with gravimetric method.

Acknowledgements

The authors wish to thank Indonesian Government, Kemenristek-Dikti, for the research grant. This research would not have been possible without the support and cooperation of LPP UAD and Laboratory staff of Chemical Engineering Department, M. Tamrin, and also Dewi Wulandari, Mita Prastika, Budiharjo, Handika Bayu and Donianto as technical staffs.

References

[1] Aregbelosa O A, Ogunsina B S, Sofolahan A E and Chime N N 2015 Mathematical modeling of thin layer drying characteristics of dika (*Irvingia gabonensis*) nuts and kernels *Nigerian Food Journal* vol 33 pp 83–9

[2] Deepa J, Rajkumar P and Arumuganathan T 2015 Quality analysis of copra dried at different drying air temperatures *IJASR* vol 5 no 4 pp 1-6

[3] Ginting W L, Harahap W A and Rohanah A 2015 The Effect of Temperature Variation on Quality of Desiccated Coconut Dried in Desiccated Coconut Dryer *J.Rekayasa Pangan dan Pert.* vol 3 no 3 pp 407-11

[4] Pestano L D B and Jose W I 2016 Engineering an Improved Coconut Processing System in the Philippines at the Farm-Level *Journal of Advanced Agricultural Technologies* vol 3 no 1 pp 58 - 62

[5] Rosa D P, Cantu-Lozano D, Luna-Solano G, Polachini TC and Telis-Romero J 2015 Mathematical modeling of orange seed drying kinetics *Cienc. Agrotec.* vol 39 no 3 pp 291 - 300

Formulation process making of *Aloe vera* mask with variable percentage of *Aloe vera* gel extract

T Y Hendrawati[1], R A Nugrahani[1], S Utomo[1] and A I Ramadhan[2]

[1]Chemical Engineering Department, [2]Mechanical Engineering Department
Faculty of Engineering, Universitas Muhammadiyah Jakarta, Indonesia

yuni.hendrawati@ftumj.ac.id

Abstract. The purpose of this research is to get the effect of adding *Aloe vera* extract to mask formulation and get the best mask formula. The *Aloe vera* mask formulation is done with the amount of *Aloe vera* extract added. The mask composition consists of Polyvinyl Pirolidon K 30 (7.55%), Polyvinyl alcohol (1.51%), Methyl Paraben (0.10%), Propyl Paraben (0.10%), BHT (0.12%) and Water (90.61%). The amount of *Aloe vera* extract added as much as 5%, 10%, 15% and 20%. The properties of the mask tested include density, pH, vitamin C content, drying time and favorite test. The conclusions of this research were the effect of the addition of *Aloe vera* extract to the density, pH, vitamin C and drying time. The best content was 0.15%, 1.066 g / ml density, vitamin C 2.2 gram, pH 6.47, Drying time 25 minutes and viscosity 24,000 CP. In a favorite test with odor criteria, a soft feeling on the face, a clean feel on the face and color done on 20 panelists found the most preferred is a mask with 0.15% *Aloe vera* extract content.

1. Introduction

The *Aloe vera* plant is included in the *liliaceae* family. Distribution areas cover the whole world. *Aloe vera* itself has more than 350 types of plants [1]. The most widely developed species in Asia, including Indonesia, are *Aloe Chinensis Baker*, originally from China. This species in Indonesia has been grown commercially in West Kalimantan and is better known as *Aloe vera Pontianak* [2]. *Aloe vera* is used in the cosmetic, food, and pharmaceutical industries. In the cosmetic and toilet industry, it is used as a base material for mask, skin moisturizers, soaps, shampoos, sun lotions, make up creams, perfumes, shaving creams, bath aids, and many other products [3]. Determination of phenolic constituents of *Aloe vera* such as Aloin A and B, Aloenin (B), Aloesin and Chrysophanol [4].

Aloe vera fluid contain the main elements of aloin, emodin, gum and other elements such as essential oils. Aloin is an active ingredient that is antiseptic and antibiotic [5]. Carbohydrates in *Aloe vera* gel consist of three components: 16.2% of the liquid contains mannan, 0.7% microparticles containing polysaccharides (rich in galactose), and 83.1% of cell walls containing galacticonic acid [6]. Polysaccharides isolated from *Aloe vera* with alcohol, composed of mannose: glucose: galactose: galactose A: fructose: arabinose: silosa with a ratio of 120: 9: 6: 3: 2: 2: 1. The active components of polysaccharides, and barbaloin of the *aloe vera* gel have maximum stability at 70 °C and a decrease in stability at a temperature higher or lower than the normal temperature. Barbaloin is an unstable component when dissolved in methanol, and undergoes transformation into a series of unidentified components [7].

APS 1 is a polysaccharide isolated from *Aloe veravarchinensis*. APS 1 is composed of manosa and glucose (ratio 18: 5) with a molecular weight of about $2.1x10^5$. Testing of APS 1 shows that APS 1 can be an antioxidant active ingredient for the prevention and treatment of health problems associated with free radicals, such as coronary heart disease [8]. *Aloe vera* can support the immune system. Amino acids serve as a builder of cell and tissue cells, so that *Aloe vera* is good in helping the regeneration of new cells. Vitamin C, vitamin E and zinc (Zn) serves as a natural antioxidant [9].

Aloe vera has unique uses. *Aloe vera* polysaccharide component is responsible for penetration and lubrication ability. Enzymes are useful in the processing of nutrients in food. Protein is useful for repairing body tissues. As a cosmetic, *Aloe vera* is excellent in maintaining moisture, tightening and smoothing the skin. Pulp and yellow fluid fraction from aloe leaves efficacious as antifungal. Both materials may inhibit the growth of pathogenic fungi *F. oxysporum*, *R. solani*, *C.* [10].

Masks are cosmetic preparations for facial skin care. Masks are applied to the facial skin in relatively thick layers and removed some time later, usually 15-30 minutes [11]. This type of cosmetics works to maintain healthy skin including cleaning, moisture protection, protection from UV hazards, antioxidants, whitening, prevent skin aging, prevent wrinkles, prevent lax and acne on the skin. In the formulation of a gel-peel off face mask type, the composition of the ingredients used is gelling agent, viscosity enhancing agent, and humectant will affect the physical and chemical properties of the peel off gel face mask.

The commonly used film coating polymers are polyvinyl alcohol (PVA) and polyvinyl pyrrolidone (PVP). The film layer is formed by the hydration process of the solvent component and the polymer chain which then coalesces into a film coating when it dries. In the gel peel offset formulation the viscosity-enhancing agent that can be used is HPMC, carbomer, gom, guar and CMC Na. Humectant serves to maintain stability by absorbing moisture from the environment and reducing water evaporation from the preparation. In addition to maintaining the stability of dosage, humectan indirectly also can maintain skin moisture so that the skin is not dry. The most commonly used types of humectants are glycerine, propylene glycol, and sorbitol [12].

Based on the content of these chemical compounds, the diversification of *Aloe vera* products is very broad, among others, required for cosmetic ingredients. Even *Aloe vera* commodity is also called as a plant of a million benefits. These factors cause *Aloe vera* products have opportunities and prospects of global marketing and very promising in the future. *Aloe vera* contains various substances in the leaves such as vitamins, minerals, enzymes and amino acids. The various substances that cause *Aloe vera* can be utilized in various fields such as cosmetics, pharmaceutical and food industries [13]. Amino acids and their properties and concentrations in the gel conditions vary depending on the conditions of storing aloe leaves in dark places or exposed to light. Given the quinone and anthraquinone contained in it, the presence of light will cause reddish changes in the gel eventually form a brown to brown color. Therefore, since light or heat is the catalyst for the reaction, the best way to avoid this color change is to store in dark, cold and dry places [14].

Efficacy of *Aloe vera* is quite diverse, such as antibiotics, antiseptic, antibacterial, antiviral, antifungal, anti-infection, anti-inflammatory and anti-swelling. *Aloe vera* can inhibit the growth of skin disease-causing organisms. In the invitro test, it is known that *Aloe vera* can inhibit the growth of Dermatophiluscongolensis. In the presence of such capabilities, *Aloe vera* may be used as a cosmetic for the treatment and skin care. *Aloe vera* gel is able to hold skin moisture so as not easy to dry, this is because the lignin level can penetrate and seep into the skin, and hold too much fluid loss. The gel also contains 30% crystal aloin ingredients, consisting of barbaloin and isobarbaloin, resin, aloe emodin and amarphousaloin that stimulate hair growth [15]. *Aloe vera* works by combining several mechanisms. The gel, which consists mainly of polysaccharides, plays a role in blocking moisture. *Aloe vera* gel contains a variety of antibiotics and anti-fungi that have the potential to slow or block the disease-causing microorganisms [14].

Mask is one type of facial skin care used to improve the level of hygiene, healthy and beauty of the skin and improve and stimulate the return of dead cell activity. Masks include depth cleansing cosmetics because they can lift dead horn cells. The use of masks is usually done after the massage, applied to the entire face except the eyebrows, eyes, lips so it will appear to wear a facial mask. In general, this mask aims to refresh, tighten the skin and as an anti-oxidant. Currently a natural face mask is often used as an alternative choice because it has no risk or harmful side effects [12]. The purpose of this research is to get the effect of adding *Aloe vera* extract to mask formulation and get the best mask formula. Masks are cosmetic preparations for facial skin care. Masks applied to the skin in the form of layers and removed in some time later usually 15-30 minutes.

2. Experimental method

The research was conducted at Applied Chemistry Laboratory, Chemical Engineering Department, Faculty of Engineering, Universitas Muhammadiyah Jakarta. The *Aloe vera* mask formulation is done with the amount of *Aloe vera* extract added. The mask composition consists of Polyvinyl Pirolidon K 30 (7.55%), Polyvinyl alcohol (1.51%), Methyl Paraben (0.10%), Propyl Paraben (0.10%), BHT (0.12%) and Water (90.61%). Mask Base Making All materials are weighed, Polyvinyl alcohol 72000 coupled with distilled water six times and then heated in a cup, stirred until the color is clear and homogeneous. Polyvinyl Pirolidon K30 is stirred in the mortar with the addition of a small amount of distilled water. The two masses are mixed and added propylenglycol, stirred until homogeneous. Methylparaben and propylparaben are added and stirred until homogeneous. The mask preparation mixes *Aloe vera* gel extracts in the base of the stirring mask until homogeny. The total of *Aloe vera* extract added as much as 5%, 10%, 15% and 20%. Mask properties tested include density, pH, vitamin C content, drying time and favorite test. On a favorite test with odor criteria, a gentle feeling on the face, a clean feel on the face and colors done on 20 panelists.

3. Results and discussion

This section presents and discuss the experimental results.

3.1. Effect of comparison of Aloe vera extract on vitamin C content

In the amount of *Aloe vera* extract added to the mask formula, vitamin C in the mask increased to 15 grams in 100 grams of mask formula (15% w / w) and decreased at 20% w / w. This can be seen in figure 1.

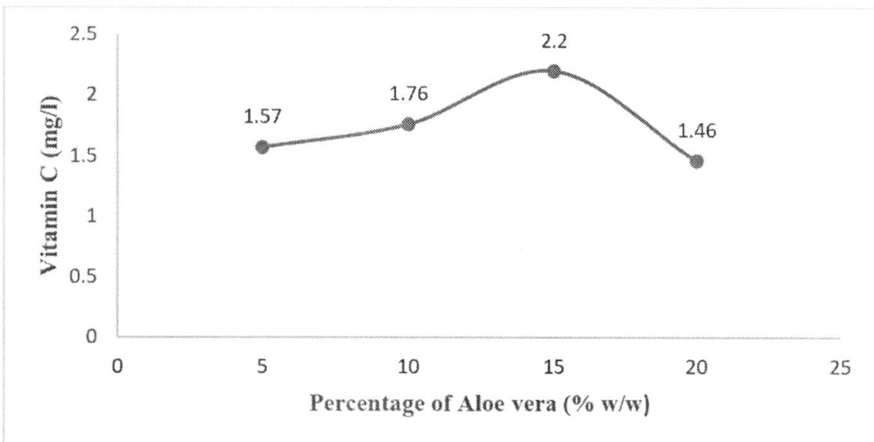

Figure 1. Effect of *Aloe vera* extract on vitamin C mask content.

Vitamin C in the 5% mask formula showed vitamin C 1.57 mg / liter, for 10% vitamin C 1.76 mg / litre, at 15% vitamin C 2,2 mg / liter and at 20% vitamin C decreased at 1, 46 mg / liter. High vitamin C is needed in the mask to give the effect of brightening and cleaning the face. From this analysis obtained the highest vitamin C in the mask with the content of *Aloe vera* extract 15% (w / w). In the content of 15% *Aloe vera* extract, vitamin C levels show 1.066 g/ml and is the highest level.

3.2. Effect of comparison of Aloe vera extract on the pH of the mask
Analysis of pH mask to determine the conformity of skin physiology to the mask. In the amount of *Aloe vera* extract added to the mask formula, the pH of the mask increased to 15 grams in 100 grams of mask formula (15% w / w) and decreased at 20% w / W. This can be seen in figure 2. From pH masks to all formulas are still within the standard for physiological pH for the skin of 4.5-7.5.

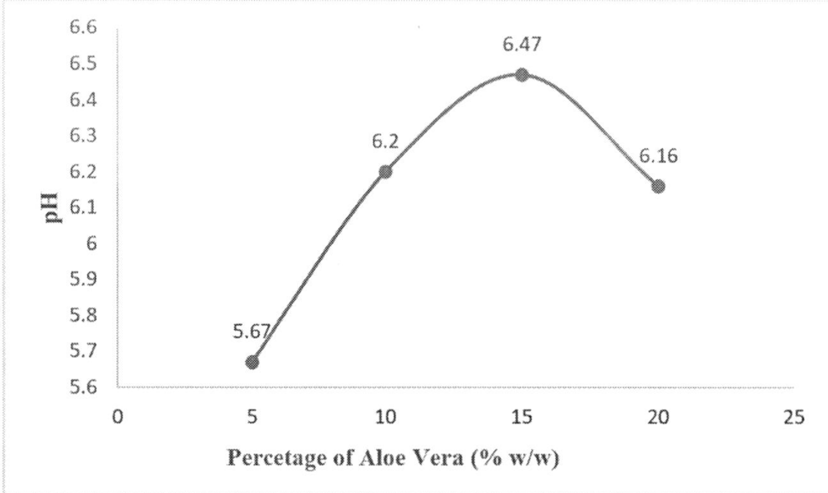

Figure 2. Effect of *Aloe vera* extract on pH mask.

3.3. Influence of Aloe vera extract on density of mask
In the amount of *Aloe vera* extract added to the mask formula, vitamin C in the mask increased to 15 grams in 100 grams of mask formula (15% w / w) and decreased at 20% w / w. This can be seen in figure 3.

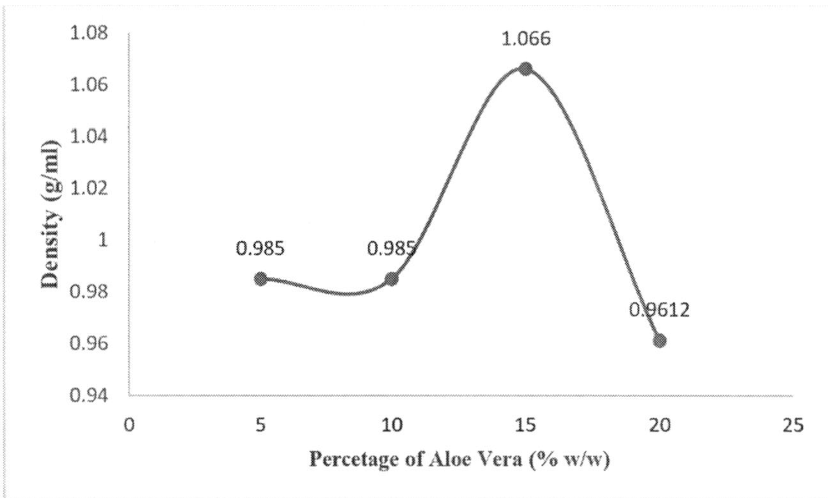

Figure 3. Effect of *Aloe vera* extract on mask density.

3.4. *Effect of comparison of Aloe vera extract on mask drying time*

In the amount of *Aloe vera* extract added to the mask formula, the drying time in the mask increases up to 15 grams in 100 grams of mask formula (15% w / w) and remains at 20% w / w. This can be seen in figure 4. Mask drying time is about 30 minutes longest and for all formula have been qualified. The standard for drying time for masks is 30 minutes [12].

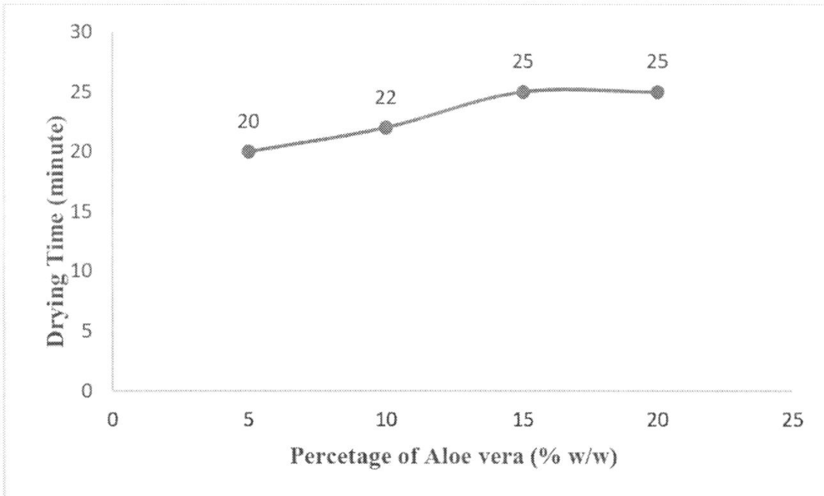

Figure 4. Effect of *Aloe vera* extract on dry mask time.

3.5. *Results of favorite test*

In the analysis of favorite test conducted on 20 panelists for odor criteria, a soft feeling on the face, a sense of cleanness in the face and color then the results can be presented as follows. Assessment given value 1 for dislike, 2 less like, 3 likes and 4 very likes. The average result of the assessment on each criterion is multiplied by the weights so as to obtain the average weighting rating. For the highest appraisal results it is the preferred panelist mask formula. The results of this study show *Aloe vera* mask with 15% (w / w) *Aloe vera* extract is the best formula favored by panelists. The results are presented in the following table 1.

Table 1. Results of favorite test on *Aloe vera* mask.

Panelists	Average rating of Odor criteria	Average grading criteria on Soft Face	Average assessment of net taste criteria on the face	Average color criteria assessment	Average weight assessment
Weight Criteria	25%	25%	30%	20%	-
Mask of 5% *Aloe vera*	3.50	3.75	2.75	2.50	3.1375
Mask of 10 % *Aloe vera*	3.25	4.00	3.25	3.50	3.4875
Mask of 15% *Aloe vera*	4.00	4.00	3.75	3.75	3.875
Mask of 20% *Aloe vera*	3.25	2.50	3.50	3.50	3.1875

On testing the viscosity of *Aloe vera* mask with Brookfield Viscometer. The spindle used is the number 7 spindle or the smallest spindle because the spindle is best suited for a sample that has a mass of cream. The size of the viscosity in the sample depends on the speed of the spin and the multiplier factor on the spindle used. From the viscosity testing that has been done, obtained viscosity of 24,000 CP for all formulas.

4. Conclusion

The conclusion of this research is the effect of *Aloe vera* extract on density, pH, vitamin C and drying time. The best content is 15% of *Aloe vera* Extract, 1.066 g / ml density, vitamin C 2.2 gram, pH 6.47, time Drying 25 minutes and viscosity 24,000 CP. On the favorite test with the odor criteria, the soft feeling on the face, the clean feeling on the face and the color done on 20 panelists found the most preferred is the mask with the content of *Aloe vera* extract 15%.

Acknowledgment

We would like to thanks for the LPPM Universitas Muhammadiyah Jakarta and Ministry of Research, Technology and Higher Education Republic of Indonesia for financial support of this work through research grant of PUPT on year 2017.

References

[1] Boudreau M D and Beland F A 2006 An evaluation of the biological and toxicological properties of Aloe barbadensis (miller), Aloe vera *J. Environ Sci Health* C. vol 24 pp 103–54
[2] Bozzi A, Perrin C, Austin S and Vera F A 2007 Quality and authenticity of commercial aloe vera gel powders *Food Chem.* vol 103 pp 22–30
[3] Eshun K and He Q 2004 Aloe vera: a valuable ingredient for the food, pharmaceutical and cosmetic industries--a review *Crit Rev Food Sci Nutr.* vol 44 p 91
[4] Grindlay D and Reynolds T 1986 The Aloe vera phenomenon: a review of the properties and modern uses of the leaf parenchyma gel *J. Ethnopharmacol.* vol 16 p 117
[5] Hamman J H 2008 Composition and Applications of Aloe vera Leaf Gel *Molecules* **13** p 1599
[6] Hendrawati T Y, Rukmayadi D and Wahjono E 2012 Rekayasa program perancangan industri tepung aloe vera dengan pengering semprot *Prosiding Seminar Nasional Teknik Kimia* Universitas Indonesia
[7] Okamura N, Asai M, Hine N and Yagi A 1996 High-performance liquid chromatographic determination of phenolic compounds in Aloe species *Journal of Chromatography A.* vol 746 pp 225-231
[8] Rahyuni T, Hadijah S dan Kesumadewi Y S 2002 Teknologi pengolahan tepung lidah buaya (aloe vera) sebagai alternatif komoditi eksport Universitas Tanjungpura Kalimantan
[9] Ross S A, Elsohly M A, Wilkins S P 1997 Quantitative analysis of Aloe vera mucilaginous polysaccharide in commercial Aloe vera products *J AOAC Int.* vol 80 p 455
[10] Saccu D, Bogoni P and Procida G 2001 Aloe exudate: Characterization by reversed phase HPLC and head-space GC-MS *J Agric Food Chem.* vol 49 p 4526
[11] Shai A, Maibach H I and Baran R 2009 *Handbook of Cosmetics skin care, second edition* (London: Informa Heathcare)
[12] Masluhiya AF S , Widodo, and S Widyarti S 2016 Formulasi masker alami berbahan dasar bengkoang dan jintan hitam untuk mengurangi kerutan pada kulit wajah *Jurnal Care* vol 4 no 2 pp 1-11
[13] Turner C, Williamson D A, Stroud P A, Talley D J 2004 Evaluation and comparison of commercially available Aloe vera L. products using size exclusion chromatography with refractive index and multiangle laser light scattering detection *Int Immunopharmacol.* vol 4 no 14 pp 1727–37
[14] Hendrawati T Y 2015 Aloe Vera Powder Properties Produced from Aloe Chinensis Baker, Pontianak, Indonesia *Journal of Engineering Science and Technology* January pp 47-59

[15] Wang Y, Strong K 1995 A two-year study monitoring several physical and chemical properties of field-grown Aloe barbadensis Miller leaves *Subtropical Plant Sci.* pp 4734–8

IOP Conf. Series: Materials Science and Engineering **403** (2018) 012014 doi:10.1088/1757-899X/403/1/012014

Hydrogel film of polyethylene oxide-polypropylene glycol diacrylate for wound dressing application

Haryanto and A N Aini

Chemical Engineering Department, Faculty of Engineering, Universitas Muhammadiyah Purwokerto, Indonesia

harymsl@gmail.com

Abstract. Hydrogel polymer receives more attention in medical and other applications due to its biocompatibility properties and swelling ratio. Hydrogel has an ability to absorb and retain water, a property which is very useful for medical applications. We developed polyethylene oxide (PEO)-polypropylene glycol diacrylate (PPGDA) hydrogel film through electron beam irradiation for wound dressing application. Various compositions of PPGDA/PEO (0%, 5%, 10%, 15% and 20%) were irradiated using electron beam to obtain the optimum composition for a better hydrogel characteristic. The PPGDA content in the hydrogel film influences the level of swelling capacity, gel fraction and mechanical strength. The results show that 20% PPGDA content of hydrogel film has the highest mechanical strength (6,2 N/mm^2). The composition of 20% PPGDA hydrogel film also has the highest gel fraction (79%). However, the hydrogel prepared using that composition has the lowest swelling ratio (300%). Based on the results, 20% PPGDA hydrogel film is very potential for wound dressing application with high mechanical strength and ideal water vapor transmission rate.

1. Introduction

Wound dressing accelerates wound healing and protect the wound surface. It absorbs the exudate to speed up its healing. Nowadays, hydrogel material especially hydrogel film has been used as a wound dressing materials. Hydrogel is a perfect dressing to accelerate autolytic debridement of necrosis and slough due to the ability to absorb and retain water inside [1-4]. An ideal wound dressing must initiate an appropriate environment to speed up healing process, easy to be applied, and require fewer dressing changes, so it take shorter nursing time [2]. Hydrogel film has low mechanical strength caused the limited use of this type of wound dressing in wound healing applications. Therefore, the hydrogel film from polyethylene oxide (PEO) and polypropylene glycol diacrylate (PPGDA) produced via electron beam irradiation is expected to have a better tensile strength and vapor transmission rate.

Various polymers can be polymerized by crosslinking to form hydrogel film [5]. Polyethylene oxide demonstrates the excellence as a good biocompatible material [6]. However, PEO hydrogel without any addition materials is easily be broken and has low tensile strength [7-9]. As a non-polar polymer, PEO shows a strong interaction with protein [10]. It has been applied for various medical applications [11, 12] and it has recently been used as a wound dressing [13-15].

PPGDA is a polymer which is soluble in water and non-hazardous to the body. It has been used in various biomedical applications, including oral insulin delivery and bone regeneration applications [16, 17]. Nevertheless, very few researches have noted the application of this polymer as a wound dressing matter.

There are several functions of an electron beam which is irradiated on the polymers, namely crosslinking, grafting and degradation [18]. Moreover, in the electron beam system, the change of the intensity of energy or dose can be run without any difficulties. Furthermore, irradiated polymers are safe and free from unwanted initiators or accelerators in the cross-linked polymers [15]. The mechanical strength of cross-linked hydrogel and the water vapor transmission rate of PPGDA/PEO cross-linked hydrogels are expected to be higher than pure PEO hydrogel films.

2. Materials and methods

2.1. Materials

Polyethylene oxide and Polypropylene glycol diacrylate were bought from Sigma-Aldrich, Singapura.

2.2. Production of hydrogel films

Polyethylene oxide (PEO) (Mn 6×10^5) and Polyropylene glycol diacrylate (PPGDA) (M_n 750) were used for production of the hydrogels. PPGDA was purified by passing through Alumina column to remove monomethyl ether hydroquinon inhibitor. The mixture of PPGDA and PEO (5 % w/w) was stirred at ambient temperature for one day. Various percentages of PPGDA (0%, 5%, 10%, 15% and 20%) were prepared in the same way. The certain thickness of hydrogel films was made by pouring the PPGDA/PEO aqueous solution into a petri dish. The solution was dried for one day at 50°C in the oven. Finally, the dry films were irradiated using electron beam (EB) at 300 kGy after vacuum sealing with polyethylene bag.

2.3. Gel fraction

The gel fraction of hydrogel film was measured from the weight ratio of the insoluble dry film after soaking in the water and the initial weight of dry film. The dry PPGDA/PEO film was soaked into distilled water at 50°C for one day after irradiation. The gel fraction can be calculated using equation (1).

$$\text{Gel fraction} = (Wc/Wo) \times 100\ [\%] \quad (1)$$

where Wc and Wo are the weight of insoluble dry gel and the initial weight of a dry film [19].

2.4. Swelling ratio

The weight of swollen hydrogel was measured at certain time intervals until the hydrogel reached the equilibrium. The dry PPGDA/PEO film was weight immediately after irradiation as an initial weight. The swelling ratio (SR) was measured using equation (2).

$$SR = (Wt/Wo) \times 100\ [\%] \quad (2)$$

where Wt and Wo are the weight of swollen hydrogel at time t and that of the dry PPGDA/PEO film, respectively.

2.5. Mechanical strength

The mechanical properties of PPGDA/PEO hydrogel film were measured using universal tensile instrument (UTM Taiwan) with an extension rate of 10 mm/min.

2.6. Water vapor transmission rate

The transmission rate of water vapor of cross-linked hydrogel film was measured using equation (3).

$$WVTR = ((W_2\text{-}W_1)/S)\ [g\ m^{-2}h^{-1}] \quad (3)$$

where W_1 and W_2 are the weight of the whole cup at the first and the second hours, respectively, and S is the surface area of the sample. A round shape of dry film was put on the top of a cup and placed in an incubator of 90% RH at 40°C. Then the whole cup was weight at the first and second hours.

3. Results and discussion

3.1. The effect of PPGDA percentage on gel fraction and swelling properties

Figure 1(a) indicates that the gel fraction of cross-linked hydrogel increased linearly with increasing the weight percentage of the PPGDA in a mixture after 5 minutes. It shows that PPGDA enhance the degree of crosslinking due to the lower viscosity of PPGDA. Figure 1(b) indicates that all composition of hydrogel films absorb water very quickly and get an equilibrium state within 5 minutes. The swelling ratio decreased with increasing content of PPGDA from 800% at 0% of PPGDA to 300% at 20% of PPGDA. This result reveals that the higher content of PPGDA can decrease the swelling ratio due to the crosslinking of PPGDA together with radicals in PEO backbones via radical polymerization.

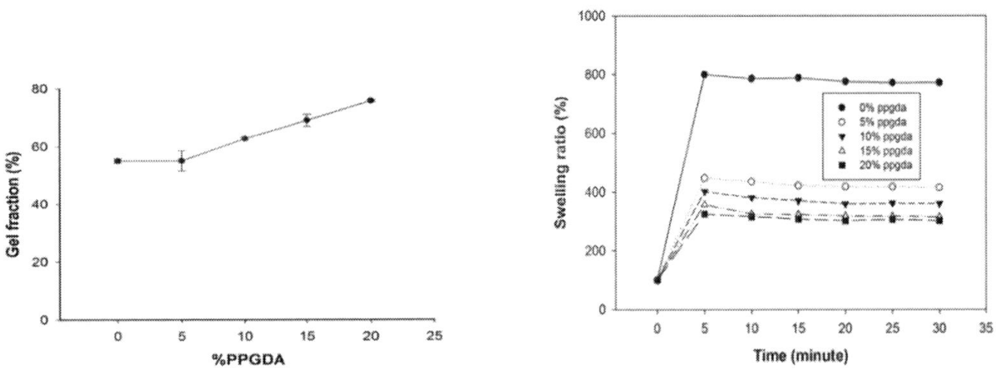

(a) (b)

Figure 1. The effect of PPGDA percentage on gel fraction (a) the swelling ratio over time (b).

3.2. Mechanical strength

Figure 2 shows the tensile strength of PPGDA/PEO cross-linked hydrogel film. The tensile strength increased linearly from 5% to 20% content of PPGDA and reached the maximum value (0.65 MPa) at 20% of PPGDA. The result indicates that mechanical strength of PPGDA/PEO cross-linked hydrogel film are high enough to be applied as a wound dressing.

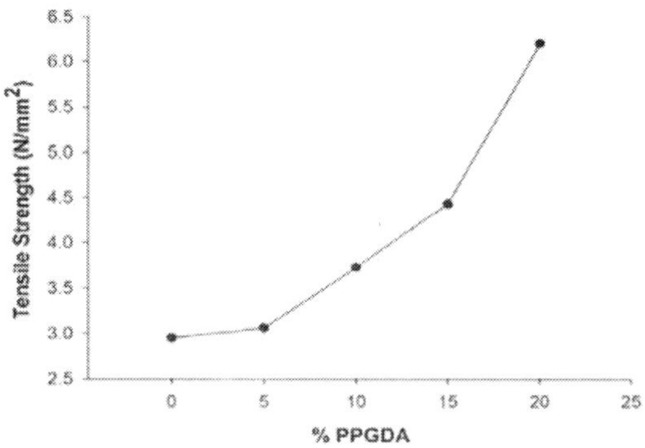

Figure 2. The effect of percentage of PPGDA on tensile strength.

3.3. Water vapor transmission rate (WVTR)

Figure 3 demonstrates that the WVTR is remain stable for all compositions of PPGDA. However, the average of WVTR of various compositions of PPGDA/PEO (25 g m^{-2} h^{-1}) is higher than of pure PEO hydrogel film (15.89 g m^{-2} h^{-1}). The different type of wound need different type of dressing. High exudate wound need the wound dressing which has high water vapor transmission rate to evaporate more exudate from wound. The ideal dressing should have WVTR higher than normal skin (8.5 g m^{-2} h^{-1}).

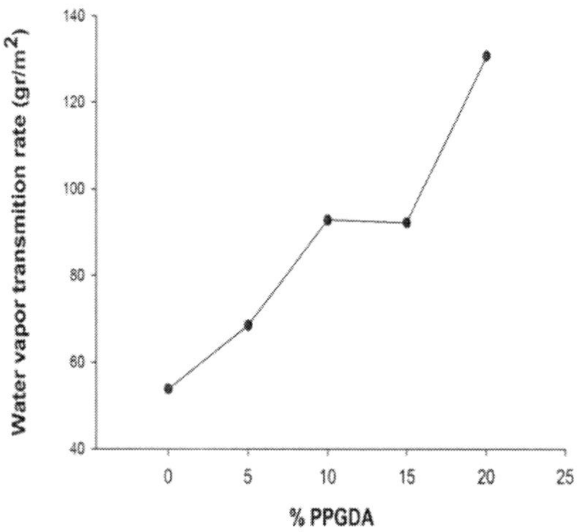

Figure 3. Water vapor transmission rate of PPGDA/PEO hydrogel film.

4. Conclusions

The PPGDA/PEO hydrogel film could be prepared by e-beam irradiation at 300 kGy. The increasing content of PPGDA in PPGDA/PEO hydrogel film was able to increase the mechanical strength. These results proved that the content of PPGDA influence the properties of cross-linked hydrogel film. The findings also proved that confirmed that PPGDA/PEO hydrogel film has a great potential for future use in wound dressing application.

References

[1] Cartmell J V, Sturtevant W R, Valadez M and Wolf M L 1991 Hydrogel wound dressing product *US Patent 5059424*
[2] Rajendran S and Anand S C 2002 Applications of medical textiles, in Development in Medical Textiles vol 32 The Textile Institute, Manchester pp 10-17
[3] Peppas N A and Scott J E *J Controll Release* vol 18 pp 95-100
[4] Abdelrahman T and Newton H 2011 *Surgery* vol 29 pp 491-495
[5] Ikada Y, Mita T, Horii F and Sakurada I 1977 *Radiat Phys Chem* vol 9 pp 633-45
[6] Manfred N and Stefan S 1997 *J. Surg. Res.* vol 67 pp 113-8
[7] Paul B M, James M M, Helen E G, Brian E G, James H N and Bradely S H 2004 *Fertil. Steril* vol l8 pp 1-5
[8] Haryanto, Singh D, Han S S, Son J H and Kim S C 2015 *Mater. Sci. Eng. C* vol 46 pp 195-201
[9] Wu M, Bao B, Yoshii F and Makuuchi K 2011 *J. Radioanal, Nucl. Chem.* vol 250 pp 391-5
[10] Jonathan E E 2007 *World J. Surg.* vol 31 p 2132
[11] Hellebrekers B W J, Trimbos-Kemper G C M, Van Blitterswijk C A, Bakkum E A and Trimbos J B M Z 2007 *Hum. Reprod.* vol 15 pp 1358-63
[12] Kathleen E R, Herbert E S, Norma R, Thornton M, Kobak W and Dizerega G S 2000 *Fertil. Steril.* vol 73 pp 831-38
[13] Jang S S, Goddard W A and Kalani M Y S 2007 *J Phys Chem B.* vol 111 pp 1729-37
[14] Haryanto, Kim S C, Kim J H, Kim J O, Ku S K, Cho H, Han D H and Huh P 2014 *Macromol Res* vol 22 pp 131-8
[15] Yoshi F, Zhanshan Y, Isobe K, Shinozaki K and Makuuchi K 1999 *Radiation Physic and Chemistry* vol 55 pp 133-8
[16] Arakawa T and Timasheff S N 1985 *Biochemistry* vol 24 pp 6756-62
[17] Ranger W R, Halpin D, Sawhney A S, Lyman M and Locicero J 1997 *Am. Surg.* vol 63 pp 788-95
[18] Dunn R, Lyman M D, Edelman P G and Campbell P K 2001 *Fertil Steril* vol 75 pp 411-16
[19] Amit K, Sitanshu S L and Harpal S 2006 *International Journal of Pharmaceutics* vol 323 pp 117-24

The effect of flow velocity on local scaling around hexagonal pillars (laboratory model test)

H Al Imran and Nenny

Department of Civil Engineering, University of Muhammadiyah Makassar, Makassar, Sulawesi Selatan, Indonesia

hamzah_alimran@yahoo.co.id

Abstract. The river has a dynamic nature that subject to change in the dimensions of space and time of. At the time of the balanced condition, the flow would be troubled by a pillars of the bridge and will form a more balanced condition caused scours the base. Scour around bridge pillars caused by the vortex system. This research will study the local scour depth at piers with two types of models (pillar 2 and pillar 3). Scour depth around pillars were observed for 15 minutes, 30 minutes and 50 minutes is done on a set of recirculating sediment on land lines with a length of 20 meters, width of 0.5 meters and height of 0.20 meters with a permanent uniform flow conditions. The model used is the type pillar with a hexagonal cross-section dimensions of 10 cm long, 5 cm width and 3.5 cm sloping sides. Scour depth measured each consisting of running time variation. The flow velocity was measured at each pillar around which causes the variation of maximum scour occurs.

1. Introduction

The bridge is one of the important infrastructure, maintaining the function and ability of the bridge in serving the traffic flow to be the key of the smoothness of the economy, therefore continuous inspection of the bridge condition must be an integral part in the bridge management system. Inspection of the condition of the bridge is intended to identify as soon as possible damage so that effective and efficient handling can be done in accordance with the conditions of damage that occurred. Damage to bridges can be caused by load factors, environmental and natural disasters.

The collapse of a bridge is largely due to the failure of the stability of the bridge pillar in transferring the bridge loads to the ground. The failure of the pillar is caused by scouring on the river bed or around the bridge pillar beyond the security level, thus endangering the construction of the bridge [1].

Scouring occurs usually lasts for long periods of time, as this process occurs gradually bit by bit. The process will look more real in the event of a major flood, it is based on when the floods occur, water fluctuations are no longer predictable. The basic scouring becomes even greater if the effective width of the river is reduced, this may result in the flow of water becoming focused towards a point.

Many cases of the collapse of bridge buildings are not only caused by construction factors, but the scouring problem around the pillars can be another cause, this is indicated by the continuous scouring process resulting in a decline in the base of the pillar [2].

Local scouring occurs due to the turbulence of water caused by the disruption of the flow, both large and its direction, resulting in the drift of basic materials or river cliffs [3, 4]. Turbulence is caused by changes in velocity to place and time. Local suction on the base material can occur directly by the flow velocity so that the material's endurance is exceeded. Theoretically the shear stress that occurs is greater than the critical shear stress of the bed materials.

In order to overcome this phenomenon, it is necessary to study the laboratory about the total scouring that occurs around the pillar to know the scouring process that occurs, and to know the flow parameters that affect the local scour around the bridge pillar so that further can be searched for the control and prevention of scouring on the pillar for damage and collapse of construction can be avoided.

The purpose of this study is to analyze the effect of the flow velocity on scouring occurring around the pillar, as well as the form of scour changes occurring on the river bed caused by flow velocity and flow. This result is expected to be a reference in planning the pillar of the bridge so that the failure of the stability of the lower pillar of the bridge can be avoided.

2. Methodology

The research was conducted at the River Laboratory Civil Engineering of Hasanuddin University and Soil Mechanics Laboratory Muhammadiyah University of Makassar. The pillar model used is a Hexagonal-shaped pillar model made of stone ash and cement that is pressed on a plate-made steel plate with a size of 2 cm x 5 cm x 40 cm with a 3.54 cm slope. The pillar is placed in the center of the channel at a distance between the 2.6 m pillar as shown in figure 1(a) for pillar type 1 (2 pillars) and figure 1(b) 1.6 m for pillar type 2 (3 pillars).

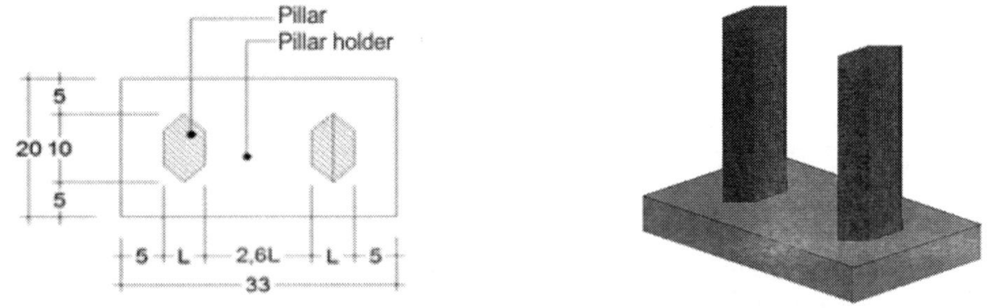

(a) Perspective type-1

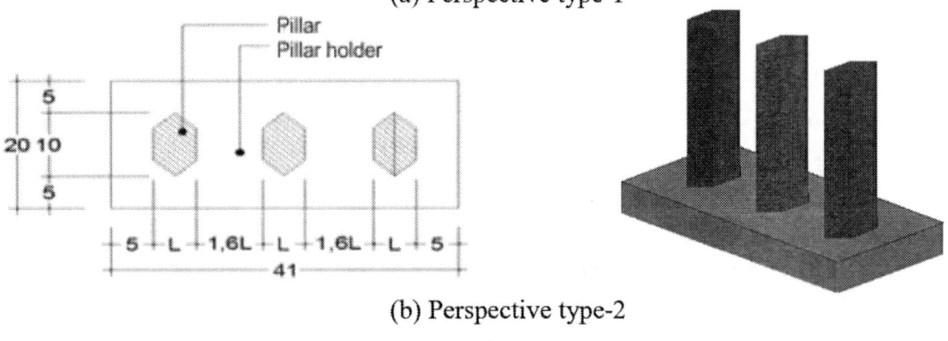

(b) Perspective type-2

Figure 1. The hexagonal pillar shape.

Determination of independent variables and bound variables, namely: In this running the free variable used is the time variation for drainage: $t_1 = 15$ minutes, $t_2 = 30$ minutes, $t_3 = 50$ minutes, channel width (*B*), channel length, channel slope (*S*), shape and size of pillar. The dependent variable is discharge (*Q*), streaming rate (*v*), and flow depth (*h*). In this running will be placed 2 (two) pillars with the distance between the pillars 2.6 cm in the middle of the channel with three variations of discharge (*Q*), namely: *Q1*, *Q2* and *Q3* with each of three variations of time (*t*), namely: $t_1 = 15$ minutes, $t_2 = 30$ minutes, $t_3 =$ 50 minutes. In this run will be placed 3 (three) pillars with the distance between pillars 1.6L in the middle of the channel with three variations of discharge (*Q*), namely: *Q1, Q2* and *Q3*, with each of three variations of time (*t*), namely: $t_1 = 15$ Minutes, $t_2 = 30$ minutes, $t_3 = 50$ minutes.

3. Results and discussion

This section explains and discuss the result into several subsections as follow.

3.1. Froude numbers

Froude numbers is a comparison between the force of inertia and the force of gravity. Thus, the Froude number is a function of all the event flow patterns in the channel. It is true that the Froude number is very important in determining the flow conditions in critical, subcritical, or supercritical streams. From the calculations it is seen that the Froude number is less than one, meaning that the current state of flow produces a subcritical flow condition can be seen figure 2.

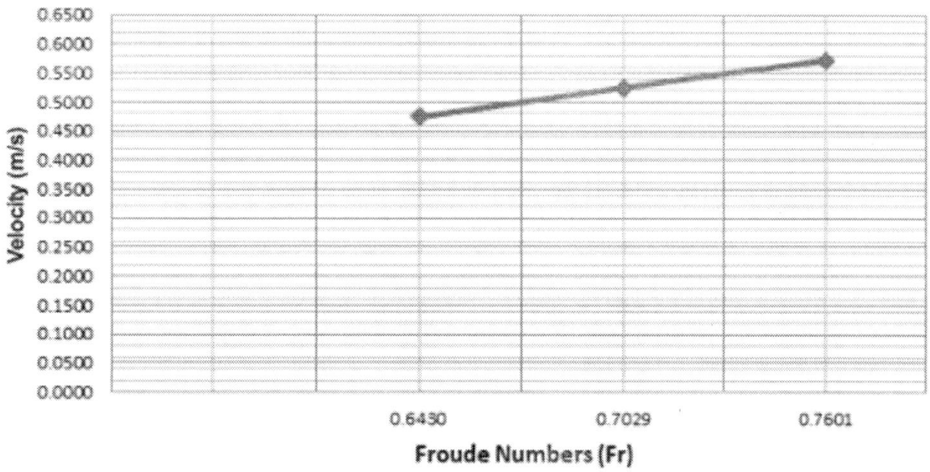

Figure 2. The relationship between flow velocity and Froude numbers.

3.2. Observation of depth of scouring on debit

Measurement of scour depth around the pillar is done on three sections: front, side and back of pillar with Total observation point 375 observation points, 15 grid / yarn pias with each thread marked 25 point with distance between point two centimeter, to get Accurate contour data. Observation points around the pillar and the distance of pillar placement can be seen in figure 3 and figure 4.

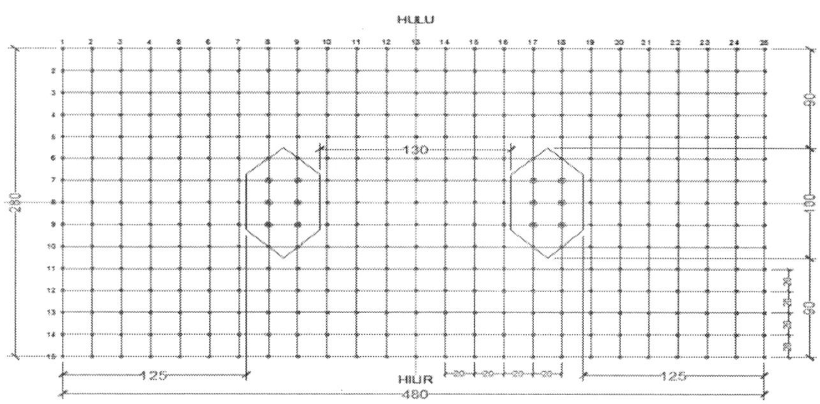

Figure 3. Pattern of point measurement for type 1 (2 pillars).

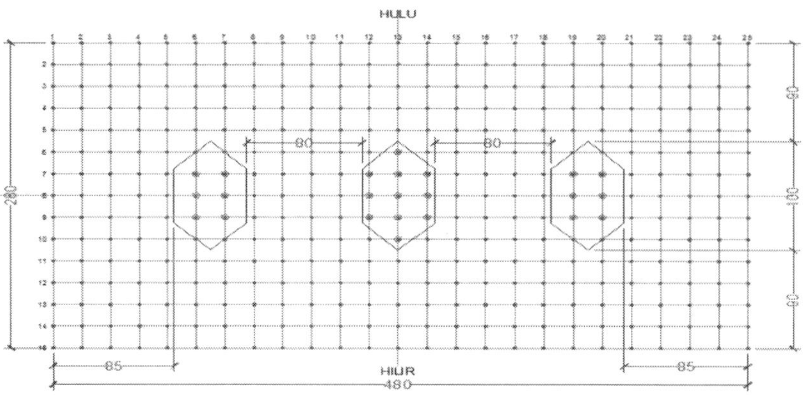

Figure 4. Pattern of point measurement for type 2 (3 pillars).

Flow pattern that occurs in the middle of the channel there is a barrier in the form of pillars will result in local scouring (local scouring) and the decrease of base elevation (degradation) around the pillar. At the point of observation pillars as a model in obtaining scour conditions on each observation. Namely in the following picture, can be seen figure 5, figure 6 and figure 7.

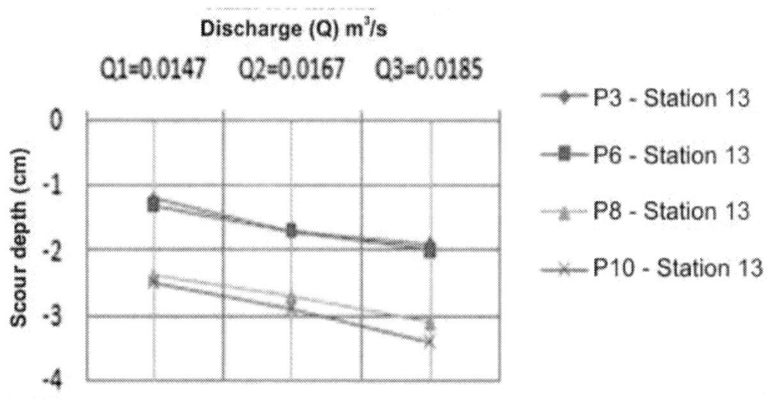

Figure 5. Relationship discharge (Q) with scour depth at time (t) = 15 minutes.

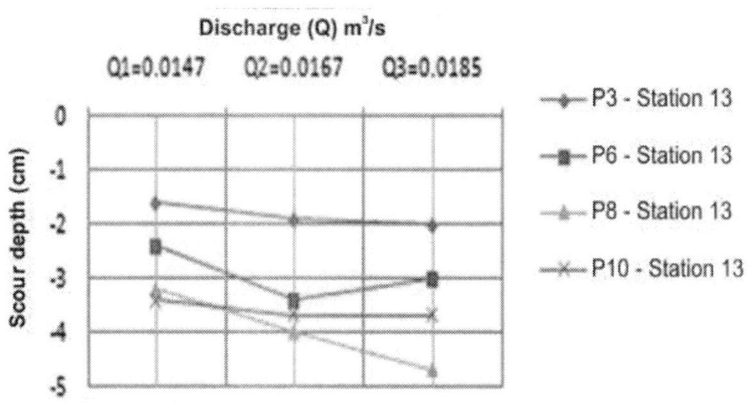

Figure 6. Relationship discharge (Q) with scour depth at time (t) = 30 minutes.

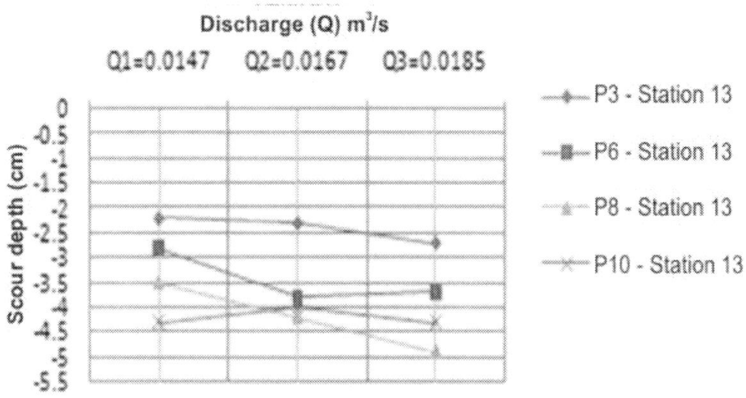

Figure 7. Relationship discharge (Q) with scour depth at time (t) = 50 minutes.

Three images above, variations of flowing discharge at the same time indicate the local scour rate around the pillar. In the observation of figure 7 with a flow time of 50 min, the maximum scaling depth was greater than 4.9 cm, while the observation in figure 5. with a flow time of 15 min, experienced a smaller scouring depth of 2.7 cm. To change the scour depth at each different discharge, we can see on one of the picture above, in figure 7. visible depth scour which happened at variation of debit with the same running time that is, at P3 Pars of maximum scour depth which happened for 2, 7 cm, at P6 pore the maximum scour depth that happened at 3.7 cm and at P8 pala the maximum scour depth which happened 4.9 cm. It can be assumed that of the three pias, the scour depth increases with increasing or increasing variation of discharge.

3.3. *Observation of scour depth (ds) to flow velocity (v)*

The effect of flow velocity variation on the local scour depth around the pillar can be seen in the following figure 8, figure 9 and figure 10.

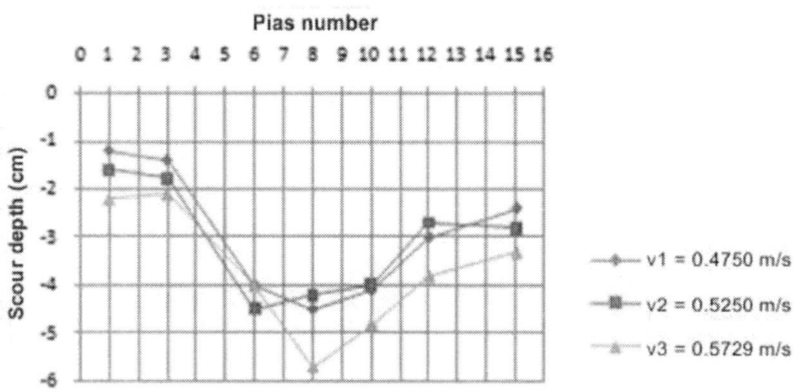

Figure 8. Effect of flow velocity on local scour at station 10 (t = 15 min).

The effect of velocity variation due to different discharge with the scour depth is shown in figure 9. At time (t) 30 minutes with velocity (v) = 0.475 m/s obtained by a scour depth value of 4.6 cm at station 10 grid to 8, at velocity (v) = 0.525 m/s obtained scaled depth value of 6.2 cm at station 10 grid 8 and at velocity (v) = 0.573 m/s obtained scaled depth value of 6.9 cm at station 10 grid 8. The observation of the scour depth process shows the magnitude of the addition of scour depth at the beginning of the drainage, and then the addition of scour depth decreases as it approaches the balance into the scour.

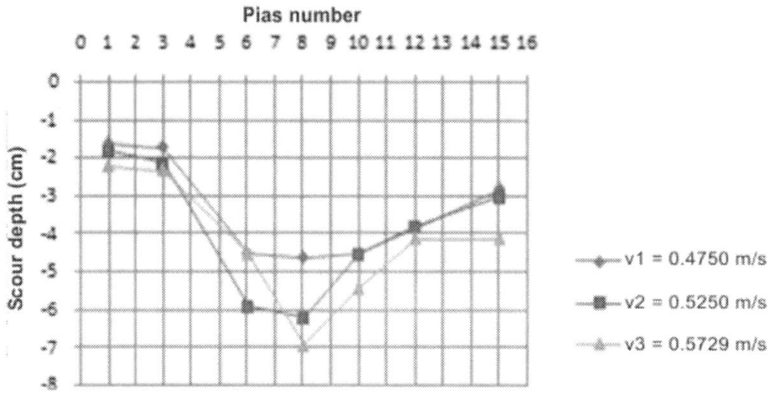

Figure 9. Effect of flow velocity on scouring at station 10 ($t = 30$ min).

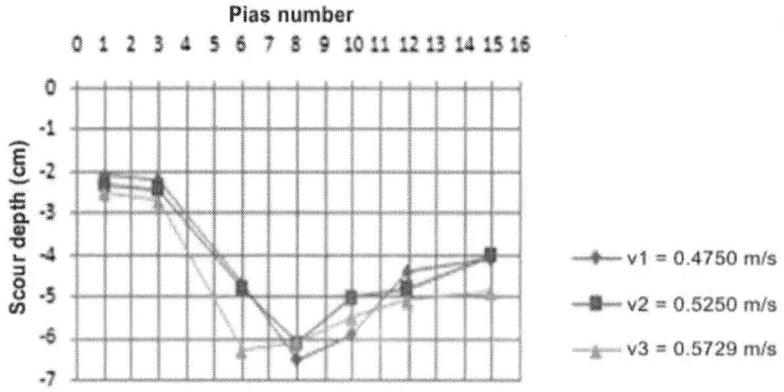

Figure 10. Effect of flow velocity on local scour at station 10 ($t = 50$ min).

4. Conclusions

Flow velocity will result in local scouring and decrease of base elevation (degradation) around the pillar which is a direct result of interactions between pillars, flow velocity, and bed materials. The existence of the pillar resulted in the stacking of the pressure (stagnation pressure) on the pillar, consequently the two-dimensional flow into three dimensions that have two streams namely downward flow along the upstream side of the pillar and horseshoe vortex around the pillar. The scour depth increased rapidly in the early minutes of drainage and the depth of the scour depth decreased to near balance.

References

[1] Khwairakpam P and Mazumdar A 2009 Local scour around hydraulic structures *International Journal of Recent Trends in Engineering* vol 1 no 6
[2] Kandasamy K J and Melville W B 1998 Maximum local scour depth at bridge piers and abutments *Journal of Hydraulic Research* vol 36 no 2 pp 183–98
[3] Barbhuiya, Karim A and Subhasish 2003 Vortex flow field in a scour hole around abutments *International Journal of Sediment Research* vol 18 no 4 pp 310–25
[4] Nenny, Pallu S, Thaha M A and Maricar F 2016 The model of pillar scouring protective using concave-sided curtain *International Journal on Development Research* vol 6 no 3 pp 7065–70

IOP Conf. Series: Materials Science and Engineering **403** (2018) 012016 doi:10.1088/1757-899X/403/1/012016

Shear behavior of slender geometry RC coupling beams in seismic response

Y Nurchasanah, M Ujianto and F M Annas

Civil Engineering, Universitas Muhammadiyah Surakarta, Indonesia

yenny.nurchasanah@ums.ac.id

Abstract. Coupling beams have geometric sizes with special heights known as deep beams, and the design usually takes up more space. Focus of this research is to analyze slender coupling beams with a diagonal reinforcement angle design of less than 20 degrees to the longitudinal reinforcement. New alternative design for slender coupling beams was experimentally investigated where the parameter was the span-to-depth ratios (2.75; 3.0; 3.25). Coupling beams should not be considered as load receiver, but instead as connector between the couple shear walls for working together. The result of the test indicated that coupling beams with a slenderness ratio of 3.25 has capacity to behave as coupling beams which function is to distribute the energy.

1. Introduction

In high-rise buildings, it is common to construct couple shear wall to function as structure wall. To meet the needs of openings on the structural wall, a structural element above the opening position that is able to behave resisting to lateral load is required. Coupling beams should also be used on short beams that receive horizontal loads, for instance corridor beams, since generally the design in such area is not considered as coupling beams. However, coupling beams are only designed as conventional flexural beams with longitudinal and shear reinforcement on the concrete so that fail frequently occurs, especially in diagonal areas. Currently, coupling beams with high geometry ratios are in demand along with the need for shortening the height of a floor on the building so that the dimensional need of the coupling beams must be able to adjust. Figure 1 shows the coupling beams element.

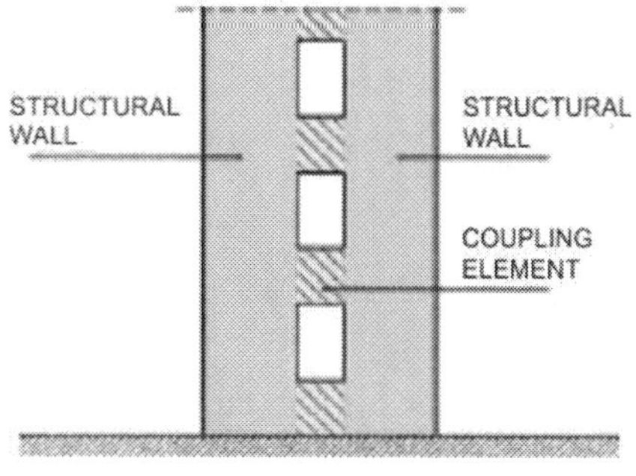

Figure 1. Coupling beams element [1].

Damages occur on conventional coupling beams can be identified by its incapability to withstand shear failure and its less optimal seismic performance. Observation on coupling beam with diagonal reinforcement group with confinement indicated the failure of coupling beams in the tension diagonal mode and yet it is able to maintain its seismic behavior [2]. The problem of the existence of transverse reinforcement on the diagonal reinforcement group the installation of the reinforcement.

Santhakumar examined the seven-story structure model with the couple shear wall. Laboratory results show that the failure of the coupling beam is the shear failure and buckling which occurs in the diagonal reinforcement [3]. Barney and Tassios tested eight reinforced concrete coupling beams with ratio 2.5 and 5.0 under reverse cyclic loading [4, 5]. Coupling beams with a ratio of 5.0 show an increased strength, although relatively low, but yet provide good seismic performance. Some disadvantages that may arise are the existence of diagonal reinforcement in a group leading to the broader beam width and thicker walls. In addition, the transverse reinforcement for diagonal reinforcement is quite difficult in the installation.

Galano carried out testing on short coupling beams with a ratio 1.5, specifically diagonal reinforcement without confining and with rhombic configuration [6]. The results showed that beams with diagonal reinforcement or rhombic configuration can provide better performance. Rhombic configuration is effective than diagonal reinforcement based on the value of ductility and strength.

1.1. Reinforce concrete coupling beams

Coupling beam is defined as a beam between two shear walls or a structural wall that serves to transfer horizontal force from the structural walls during lateral load occurrence, for example during an earthquake. The diagonal reinforcement concrete coupling beams (DRCCB) was initially developed after the 1964 Alaska earthquake to solve considerable damage to conventional coupling beams that have weaknesses in their shear capacity and seismic performance.

Provision that refers to ACI 318M-11 [1], coupling beams with $l_n/h < 2$ and with V_u exceeding $0.33.\lambda.\sqrt{f'_c}.A_{cw}$, shall be reinforced with two groups of intersecting diagonal bars symmetrically across the center of the span [1]. Coupling beams with two groups of intersected diagonal bar must satisfy to $V_u = 2A_{vd} \cdot f_y \cdot \sin\alpha \ll 0.83\sqrt{f_c \cdot A_{cw}}$, where α is the angle between the diagonal reinforcement bar and the longitudinal axis of the coupling beams. Figure 2 shows the confinement of individual diagonals.

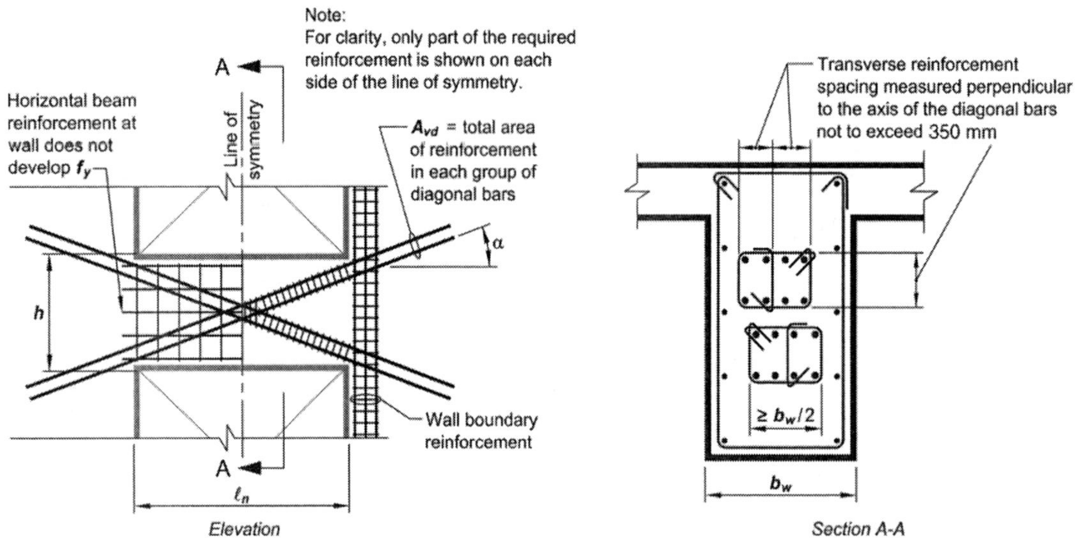

Figure 2. Confinement of individual diagonals [1].

1.2. Analysis of coupling beam

Analysis of coupling beam to the flexure and shear stresses to the behavior of structures formed by shear failure can be idealized by the force system as in figure 3, which is the half part of the triangle-shaped beam [7]. The force system is calculated using equation (1) – (5).

Vertical force equations, $P_u = 2.V + f_{tc}.b.a + P_v$ (1)

Horizontal force equations, $P_{st} = f_{tc}.b.h' + P_h + C$ (2)

Moment force equations (M=0), $P_{st}h' = V.a + f_{tc}.\frac{b.(h'^2 + a^2)}{2} + P_h.\frac{h'}{2} + P_v.\frac{a}{2}$ (3)

from equations (1), (2), (3), the ultimate load of the beam can be formulated as follows:

$$P_u = (f_{tc}.b.h' + 2.C + P_h).\frac{h'}{a} \quad (4)$$

One of the important deformation criteria in a coupling beam is when the concrete crush at the end of the compression area at $\frac{(h-h')}{2}$ when the greatest force occurs.

Compression force, $C = 0.67.f_{cu}.b.\frac{(h-h')}{2}$ (5)

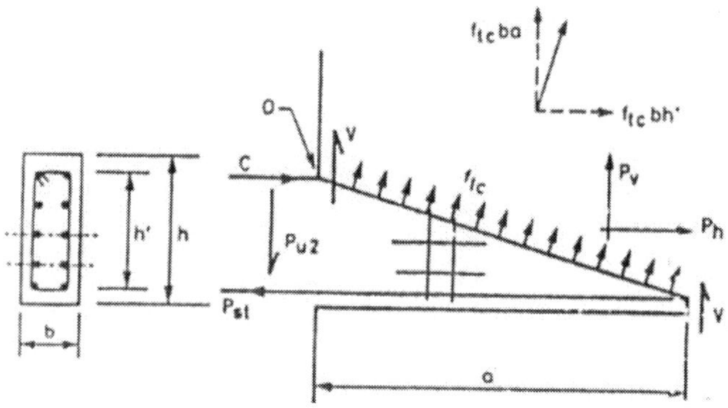

Figure 3. Idealization of triangular equations on coupling beams [7].

2. Experimental details

This section explains the detail of the experiment conducted on laboratory at Universitas Muhammadiyah Surakarta.

2.1. Specimens design

The design of specimens is shown in figure 4 where the span-to-depth ratio: $l_n/h = 2.75$ ($\alpha = 19.98°$); $l_n/h = 3.00$ ($\alpha = 18.43°$); $l_n/h = 3.25$ ($\alpha = 17.10°$).

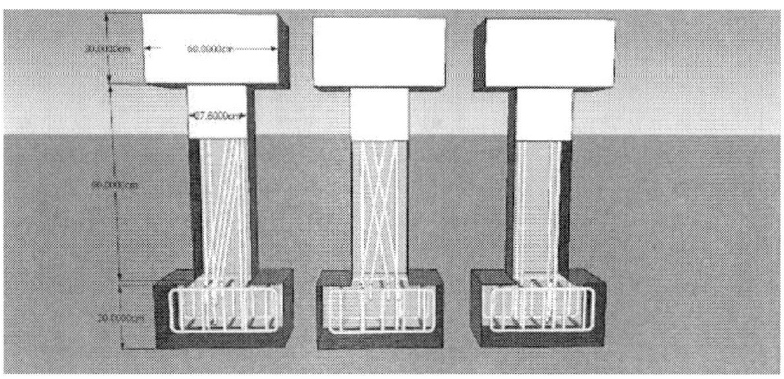

Figure 4. Detail of specimen reinforcement.

2.2. Specimens construction

The specimens were constructed at structures laboratory at Universitas Muhammadiyah Surakarta. Prior to the establishment, quality checks were performed on all materials (cement, sand, gravel, water and steel). These materials must meet the quality requirements of the regulations. The concrete mixture design method was according to Indonesian standard regulation with 0.5 water cement ratio. The quality of concrete was tested through compression test of concrete cylinder 15 x 30 cm. A day before the test of concrete compressive strength, the cylindrical test object was removed from the immersion bath for drying. Concrete cylinders were lifted and placed centrally on the Universal Testing Machine (UTM), specifically on testing machine holder. Once ready, it started loading with loading speed set 15 MPa/min. During the test, the size of the load and the shortness of the specimen were recorded.

The coupling beams test object was divided into four variables, which consisted of a non-diagonal reinforcement and a coupling beam with diagonal reinforcement with an angular slope below 20 degrees against its longitudinal reinforcement. The specimens were carried out after the final computation of the mixed design and the preparation of test instruments and materials were all endured in good condition. In order, concrete curing was carried out in order to keep the concrete surfaces always fresh in humid conditions. Concrete weight testing was intended to determine the density of a concrete.

2.3. Coupling beams testing

Preparation testing was done a few days before the main testing, which was intended to ensure all the equipment were ready and could be directly utilized during the testing. Placement of the test object on the Loading frame, it was positioned according to the setting up so that it had enough space to deform. Preparation of tables of plans for loading and recording of observations was also performed. LVDT and Yoke extensometer to measure the deformation must in zero position as well as to record the initial value on the strain indicator that showed the strain gauge's initial strain.

The test is performed in accordance with the designed load interval. The result of deformation and strain of each measuring instrument according to the load interval was read visually. During the testing process, observation of the crack pattern was started at first crack up to the maximum load and marked them with markers on the test specimen and noted the value of load at each of the crack stages. The test specimen was marked on both sides of the beam and the grid lines as a guide to determine the location of the crack was also done. The record of the loads and deformations that worked out and the observation at crucial moments, such as first crack and spalling (peeling of concrete cover), were also conducted. Before, during and after test, the entire results on each specimen that has been coded according to its characteristics were documented with photographs and camcorders. Subsequently, the devices were removed, to be installed on the next specimen.

3. Results and discussion

As composite materials, the structure of reinforced concrete coupling beams has several elements designed to withstand shear forces, such as V_c which is the contribution of the concrete material; V_s which is the contribution of transverse reinforcement; and V_d which is the contribution of diagonal reinforcement. So, $V_n = V_c + V_s + V_d$. Table 1 shows the coupling beams capacity.

At ratio 3.0, the load capacity that can be retained by coupling beams without diagonal reinforcement was 100 kN. After the installation of diagonal reinforcement into coupling beams structure, the load capacity was 113.33 kN, or there was an increase by 13.33 %. At ratio 2.75, coupling beams with diagonal reinforcement contributed to the load capacity of 146.67 kN while at ratio 3.25, the diagonal reinforcement group of steel contributed to 100 kN load capacity. Coupling beams with a slenderness ratio of 3.25 was still able to maintain its behavior according to the function of the coupling beams as shown in table 1.

Table 1. Coupling beams capacity.

Type of Test	Test objects	l_n (cm)	H (cm)	Ratio (l_n/h)	Load Capacity (kN)
Lateral Load	Coupling Beams without diagonal reinforcement	90	30.00	3.00	100.00
	Coupling Beams with diagonal reinforcement	90	32.75	2.75	146.67
			30.00	3.00	113.33
			27.60	3.25	100.00

3.1. Shear behavior

Shear stiffness, G' analysis was performed based on laboratory data, referring to ASTM E564, $G' = \frac{P}{d} \times \frac{a}{b}$. This stiffness analysis was carried out at load resistance where the wall had reached the maximum peak load. Ultimate shear strength, S_u analysis was also done based on the results of laboratory testing, referring to ASTM E564, $S_u = \frac{P_u}{b}$ as shown in table 2. The most dominant of shear behavior occurred in coupling beams with aspect ratio span-to-depth 2.75. Along with the shorter height of the coupling beams, the shear stiffness and shear strength were also getting weaker, but still in predicted behavior capacity of distributing energy.

Table 2. Shear stiffness and ultimate shear strength.

Type of objects	α (°)	Length (mm) a	Wide (mm) b	Deformation (mm) d	G' (kN/m)	S_u (kN/m)
without diagonal reinforcement	18.43	90	30.00	60.82	4.93	333.33
with diagonal reinforcement	19.98		32.75	58.27	6.86	444.44
	18.43		30.00	58.17	5.79	377.78
	17.10		27.60	58.74	5.55	362.32

The calculation with method analysis in [7] as in table 3 obtained two values of ultimate load P_u. Flexural failure was obtained from the flexural failure model formula, whereby this flexural strength was only held by its longitudinal reinforcement capacity alone. Shear failure is a value that represents the entire cross section. Thus, by comparing the two values of load above, it can be seen that shear failure has a lower value than flexural failure, or in other words, shear failure value is more decisive to predict the type of collapse that may occur.

The final failure model of a coupling beam is a shear failure, characterized by a diagonal directional crack followed by a split on the crack path. The coupling beam does not only indicate a shear phenomenon, but long before the shear cracking pattern occurs, there is also a flexural crack pattern characterized by a directional crack almost perpendicular to the longitudinal plane of the beam. However, as long as the flexural crack pattern is still within the load capacity of the service, it will not lead to the coupling beam flexure failure.

Table 3. Failure analysis.

Experiment Analysis			N. K. Subedi Analysis Method				
f_c' N/mm^2	Description of Failure	P_u kN	f_{cu} N/mm^2	f_{tc} N/mm^2	$P_{u,1}$ (Flexure Failure) kN	$P_{u,2}$ (Shear Failure) kN	Failure Prediction
21.073	Shear Failure, very wide of split, damage on concrete and bars at the tip of compress area.	113.33	26.60	1.267	115.093	71.769	Shear Failure at 71.769 kN. $P_{u,2}$ has the potential to fail earlier than $P_{u,1}$.

4. Conclusions

The following conclusions are drawn based on the experimental results. The coupling beam with diagonal reinforcement angle below 20 degrees to the longitudinal reinforcement is still capable of behaving according to the function of the coupling beam. The l_n/h ratios of 2.75, 3.0 and 3.25 as the geometry parameter of the coupling beam in this research are able to produce data that provide an alternative that the coupling beam with the slender geometry is still able to distribute the lateral force. It meets the purpose of shortening the height of the building floor, especially in buildings for business and apartment designations, so as to reduce the cost of construction.

References

[1] ACI-318M 2011 *Building Code Requirements for Structural Concrete (ACI 318M-11)* (Farming Hills: American Concrete Institute)

[2] Paulay J R B T 1974 Diagonally reinforced coupling beams of shear walls *ACI Special Publication* vol 42 pp 579–98

[3] Santhakumar The ductility of coupled shearwall (New Zealand: University of Cantebury) p 1074

[4] Barney G B, Shiu K N, Rabbit B G, Fiorato A E, Russel H G and Corley W G 1980 Behavioral of coupling beams under load reversals *RD068.01B Portl. Cem. Assoc.* Skokie, IL

[5] Tassios A B T P and Moretti M 1996 On the coupling behavior and ductility if reinforced concrete coupling beams of shear walls *ACI Struct. J.* vol 93 pp 711–20

[6] Galano A V L 2000 Seismic behavior of short coupling beams with different reinforcement layouts *ACI Structutal J.* vol 97 pp 876–85

[7] Subedi N K 1990 Ultimate Strength Analysis of Reinforced Concrete Coupling Beams *Struct. Eng.* vol 68 no 3 pp 45-50

IOP Conf. Series: Materials Science and Engineering **403** (2018) 012017 doi:10.1088/1757-899X/403/1/012017

Street vendor sustainable policy by learning characteristics of distribution model location in Losari beach line, Makassar

R Mahful[1], Ihsan[2] and I Sastrawati[3]

[1]Kyushu University, Japan
[2,3]Hasanuddin University, Indonesia

rmahful@gmail.com

Abstract. Lack of formal sector as essential problem in Indonesia, nowadays. The problem creates people to seek informal sector as one of solution in reducing possibility of social unrest. Build some business on the street that society know as Street Vendor is popular mode for lower income society due to not derived much fund. This paper attempts to reach problem from street vendor itself as contribution for government to create policy that can be acceptable in society and environment. By using agent-based modelling as the tool and Losari beach line as the case study, this paper attempts to show a real model based on location choices of street vendor.

1. Introduction

Jaya as secretary of Makassar in Tribun Timur Newspaper said Economic growth in tandem with population growth in the city of Makassar. he said during the past five years occurred a growing population of Makassar with an average increase of 1.56% each year of the amount Makassar City dwellers were estimated at 1.6 million. This is because the city became the center of the development of the formal sector, the city is deemed promising by village communities and led to the displacement of people from villages to cities. The condition is known as the theory of the origin of the driving factors and pull factors of the destination in the urbanization.

The rate of growth of street vendors in the city of Makassar so rapidly, until November 2013 estimated the number of street vendors around 14,000 inhabitants according to the head of the Department of Trade Industry and Commerce of Makassar. Lack of formal sector employment makes people look for other ways to survive, the informal sector is one of the solvers of this problem, so we can say the informal sector can reduce the possibility of social unrest as a result of the scarcity of employment opportunities.

Research about street vendors has been conducted by many researchers [1-7]. In Indonesia, street vendor is one of controversial things, since it has become a favorite destination for resident and tourist as a cheap play to buy something but there are still some problems with the illegality of the place. In Makassar City, development rate of street vendors in Makassar City becomes one of the factors causing behavioral problems of street vendors.

There are some researches about street vendor discussing about either the policy implementation or the design of street vendor itself. However, this research presents a different method, modeling the distribution of the street vendor have been implemented by using agent-based modelling as the tool to represent existing condition that actually happened into a form that spreads illustration based computing with certain rules drafted by character actors and their environment. Model shown as closely as possible

IOP Conf. Series: Materials Science and Engineering **403** (2018) 012017 doi:10.1088/1757-899X/403/1/012017

to the actual conditions, to determine the pattern and behavior of street vendors when determining the location of a trade in accordance with the characteristic / desires.

Topic of multi agent system has been discussed in [8-11]. Multi-agent system is a system consisting of several agents interact with each other from their environment. In such a system, not all agents are the same, each agent can have a uniqueness and abilities, goals also, which represent the real world counterpart. A multi-agent system is an assembly of different agents, with different roles, as well as the capabilities and objectives for different categories of each agent. Hence, it is hoped that, through this study, the policy of street vendor will be implemented and it will be acceptable in society and environment in Makassar City. In addition, this research is expected to be a benchmark in the development of other areas in Indonesia

2. Method

2.1. Method of collecting data

Data / information required in this study were obtained with a few ways below:

a. Observation, namely direct observation field conditions.
b. The study of literature, which explore a variety of important information such as literature and theories related work culture, organizational, human resource management, and the results of previous studies.
c. Interviews and questionnaires, namely the collection of facts and data by conducting interviews and questionnaires intensive and in-depth, structured and systematic.

The population in this study are vendors who are along the corridor Losari Makassar. Time survey is conducted on a Friday-Sunday with consideration on the day is a busy day at the normal time and breaks, while the time of the survey at 16:00 to 18:30 because at that time is the start time to trade hawkers along the corridors of Losari Beach, Makassar.

2.2. Data analysis method

There are three methods of analysis used in this research is descriptive, frequency distribution, and multi-agent simulation.

a. Descriptive analysis is used to identify the characteristics of the location of trade and distribution pattern of street vendors along the tourist corridor Losari, Makassar.
b. Frequency distribution analysis is used in the identification of factors influence vendors in selecting a location for trade along the tourist corridor Losari, Makassar City.
c. Multi-agent simulation used in Building a distribution model of the location of street vendors along the corridor Losari, Makassar City.

3. Analysis and interpretation

3.1. Existing

Losari beach is a tourism place which is located in Makassar, Sulawesi Selatan. This place is famous as tourism place since 1945. Losari beach as icon of Makassar becomes a magnet for people to run informal business to get finance, one of it is street vendor. Based on survey, there are 14 kinds of street vendor and 132 street vendors which are found in Losari beach corridor. Based on Survey that held by researcher, as well. The street vendor can be categorized by linear throught at Jl. Penghibur street, grouping in Losari corridor, or spreading in corridor Losari, grouping and spreading in Losari line as well.

Drink and *bakso* (traditional food) are those who seller their product grouping in Losari corridor. Others, such as *keripik ubi* (traditional snack), *gorengan* (traditional snack), burger, *gulali* (traditional snack), peanuts, sausage, accessories, watch, clothes, mixed are selling their product by spreading in

Losari line. On the other hand, *pisang epe* (traditional food) and toys are linear through at Jl. Penghibur Street and spreading and grouping in Losari corridor, perceptively.

3.2. Factors

Researchers find that there are eight factors from street vendor as the reason to choose their location. The factors are retribution, access, security, grouping, crowded, rivalry, no choice location, and kinship as shown in table 1. Frequency distribution analysis is the analysis that runs for finding the reason of street vendor to choose the location.

Table 1 shows crowded factor (76.5%) as very influential factor that street vendor said to choose the location. Kinship, access, security, grouping factors (more than 39%) are influential factor. And others such as no choice location, rivalry, and retribution are not as much as influences factors to street vendor in choosing their position.

Table 1. Eight factors to choose location.

		Frequency	Percent	Valid Percent	Cumulative Percent
Crowded					
valid	0	31	23.5	23.5	85.6
	1	101	76.5	76.5	100.0
	total	132	100.0	100.0	
Retribution					
valid	0	120	90.9	90.9	90.9
	1	12	9.1	9.1	100.0
	total	132	100.0	100.0	
No choice location					
valid	0	110	83.3	83.3	83.3
	1	22	16.7	16.7	100.0
	total	132	100.0	100.0	
Security					
valid	0	76	57.6	57.6	57.6
	1	56	42.4	42.4	100.0
	total	132	100.0	100.0	
Rivalry					
valid	0	113	85.6	85.6	85.6
	1	19	14.4	14.4	100.0
	total	132	100.0	100.0	
Access					
valid	0	71	53.8	53.8	90.9
	1	61	46.2	46.2	100.0
	total	132	100.0	100.0	
Kinship					
valid	0	80	60.6	60.6	60.6
	1	52	39.4	39.4	100.0
	total	132	100.0	100.0	
Grouping					
valid	0	78	59.1	59.1	59.1
	1	54	40.9	40.9	100.0
	total	132	100.0	100.0	

IOP Conf. Series: Materials Science and Engineering **403** (2018) 012017 doi:10.1088/1757-899X/403/1/012017

3.3. Simulations

Figure 1 shows the simulation of 1-10 multi-agents and table 2 shows the simulation result.

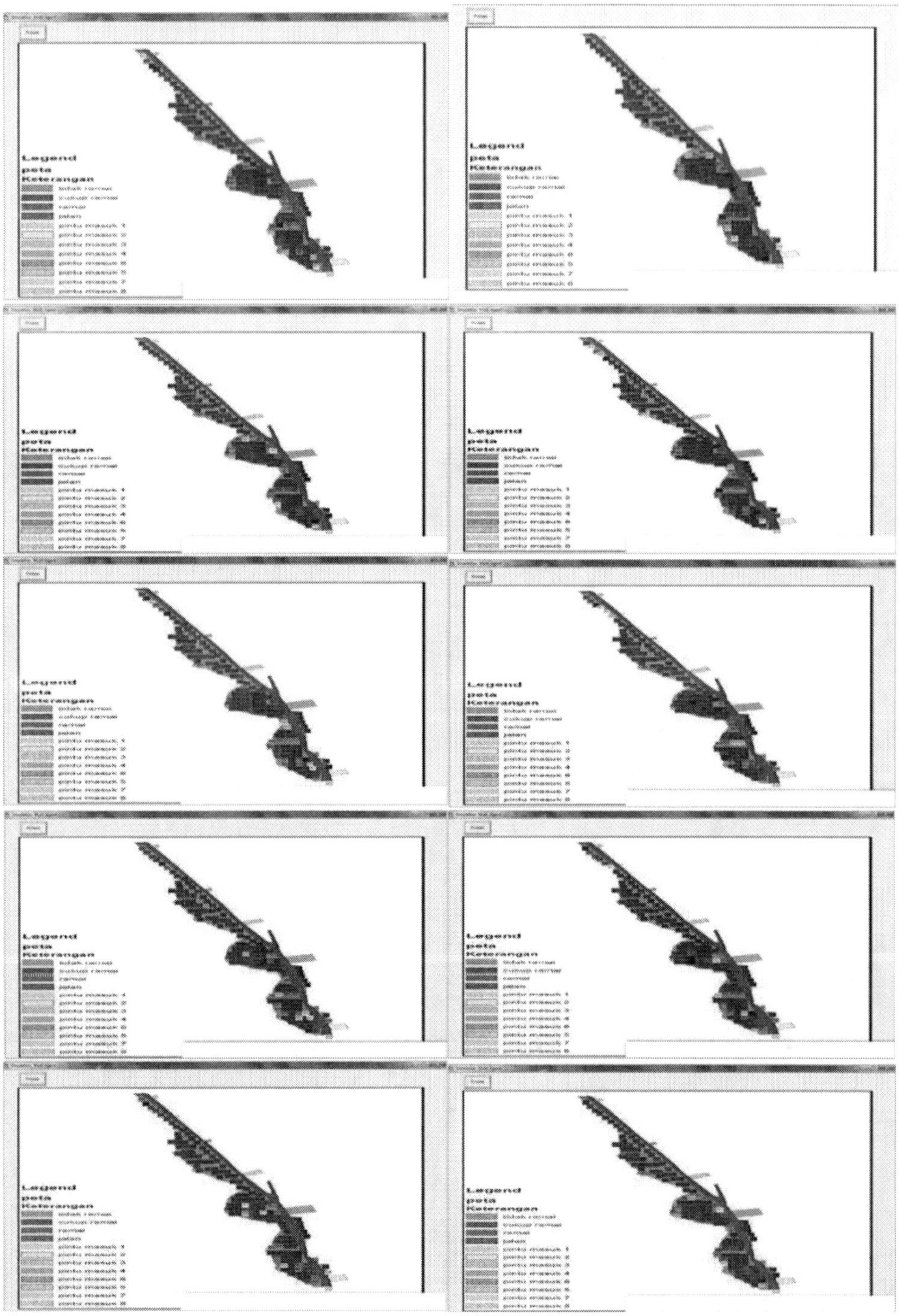

Figure 1. Multi-agent 1-10 simulation.

Table 2. Simulation result.

No	Type of seller	Simulation results	Result (compare to existing)
1	*Pisang epe* (traditional food)	*Pisang epe* sellers choose the location regarding on grouping with the similar kind, through at Penghibur street	Result from simulation and existing are similar due the pattern can interest and easy to be found by visitor.
2	Drink	Separation pattern of drink sellers choose the location, based on grouping with the similar kind and in Losari corridor	Similarity from the result of simulation and the existing because drinking sellers think that this pattern can make people easy to know their location.
3	Toys	The unique patterns come from toys sellers. Because some of them are grouping and others are separation in Losari corridor and through Penghibur beach	This is due to the different nature of each street vendors. There is a rivalry but also a sign of grouping. But all of them want to approach the customer so in the simulation sometimes they are in Penghibur street.
4	*Keripik Ubi* (traditional snack)	The patterns of *keribik ubi* seller are separation, mostly and group a few	The reason of this pattern, actually, is because the characteristic of seller. Special for this kind of seller they have characteristic as rival, mostly. Therefore, the patterns are separation one another
5	*Bakso* (traditional food)	To be short from the simulation, *Bakso* seller prefer to be group, although there are 3 simulation show spread.	That is caused by characteristic *bakso* seller. They do not stay in one place and choose place close consumer or held in separation one another
6	*Gorengan*	from the instance, *Gorengan* sellers disperse to inside or outside Losari corridor.	The result from simulation are similar with existing. It is caused by characteristics of *gorengan* sellers which do not settle and selling the product based on the consumer position.
7	Burger	Results of simulation, the burger sellers spread at Losari corridor.	The result from simulation are similar with existing. It is caused by characteristics of *gorengan* sellers which do not settle and selling the product based on the consumer position
8	*Gulali* (traditional snack)	Results of simulation, the *gulali* sellers spread at Losari corridor	The simulation is same as existing. It is caused by characteristics of *gulali* sellers which do not settle and selling the product based on the consumer position
9	Peanuts	The unique simulation result from peanuts sellers. From 10 simulations which be based on this simulation. It finds that 6 simulations shows the sellers are grouping and 4 others spread at Losari corridor	The results from simulation is most same as existing because the characteristic of this kind is selling product based on the consumer.
10	*Sosis*	Results of simulation, the *sosis* sellers spread at Losari corridor	That is caused by characteristic *sosis* seller. They do not stay in one place and choose place close consumer or held in crowded place

4. Discussion

From the result of simulation. We can discuss that some reason that street vendor said as reason in choosing their location are little bit miss. One factor that very influence in choosing place for some street vendor is rivalry. Rivalry is a factor as usually find in traders and it also finds in characteristics of street vendor in Losari beach corridor.

Researchers also hope that this finding can help Makassar government to make a plan for street vendor to increase their finance and also attract tourist. So, it can be symbiotic mutualism from government and society. Due to street vendor as good solution to decrease jobless, as well. For others governments, this finding also is able to be solution for make plan that have character or location like street vendor or Losari beach as tourism place.

References

[1] Bhowmik S K 2005 Street vendors in Asia: a review *Economic and political weekly* pp 2256-64
[2] Cross J 2000 Street vendors, and postmodernity: conflict and compromise in the global economy *International journal of sociology and social policy* vol 20 pp 29-51
[3] Bhowmik S 2012 Street vendors in the global urban economy (Taylor & Francis)
[4] Timothy D J and Wall G 1997 Selling to tourists: Indonesian street vendors *Annals of Tourism research* vol 24 no 2 pp 322-40
[5] Vollaard A M, Ali S, Van Asten H A, Ismid I S, Widjaja S, Visser L G, Surjadi Ch and Van Dissel J T 2004 Risk factors for transmission of foodborne illness in restaurants and street vendors in Jakarta, Indonesia *Epidemiology & Infection* vol 132 no 5 pp 863-72
[6] Yatmo Y A 2008 Street vendors as 'out of place' urban elements *Journal of Urban Design* vol 13 no 3 pp 387-402
[7] Illy H F 1986 Regulation and evasion: Street-vendors in Manila *Policy Sciences* vol 19 no 1 61-81
[8] Castiglione F 2006 Agent based modeling *Scholarpedia* vol 1 no 10 p 1562
[9] Getchell A 2008 Agent-based modeling *Physics* vol 22 no 6 pp 757-67
[10] Drogoul A, Vanbergue D and Meurisse T 2002 Multi-agent based simulation: Where are the agents? *International Workshop on Multi-Agent Systems and Agent-Based Simulation* (Springer, Berlin, Heidelberg)
[11] Jennings N R 2000 On agent-based software engineering *Artificial intelligence* vol 117 no 2 pp 277-296

IOP Conf. Series: Materials Science and Engineering **403** (2018) 012018 doi:10.1088/1757-899X/403/1/012018

Economic and environment feasibility of landfill gas project in Indonesia

R Mahful and S Managi

Kyushu University, Japan

rmahful@gmail.com

Abstract. Waste is a serious environmental problem in Indonesia. The waste based energy production using Waste Landfill Gas project is beneficial in the context of economic and environment aspect. The average electricity production cost from landfill gas project is 808.95 Rupiah for per KWh. This is comparatively cheaper than the production cost of gas turbine, diesel, combined cycle, geothermal sources. The emission of CO2 from the coal, oil, gas and natural gas is significantly higher than the waste landfill gas unit. This finding can benefit the policy making and manage the waste problem in Indonesia.

1. Introduction

Meidiana [1] and Kawai [2] stated that waste management is the most crucial issue that held in Indonesia. Haskarlinus [3] tries to make the best monitory on Landfill in Jakarta, Indonesia. Based on this issue, Indonesia needs method to deal with. LFG or landfill Gas as the one technique which can use to solve this problem. Many countries in the world, special in Southeast Asia apply this technique to deal with their problem of trash such as India, Malaysia, and Thailand. This technique also can overcome another issue such an electricity. To sum up, Landfill gas can be one of the best way to deal with waste problem in Indonesia.

On the other hand, to build Landfill gas system need more attention in many field such as the technology and the LFG technique. Methane, Hydrogen Sulfide, and Carbon dioxin are the emission which will occur in this project. It also the main exist emission as hot issue about environment in the world. Kumar [4] and Kalantarifard [5], used IPCC, CLEEN model and Land GEM to measure CH_4, CO_2, and H_2S in landfill site. Zero and order-decay method as the basic rules that they used to measure it. Furthermore, Dowling [6] measures the economic from landfill gas project. Therefore, this research attempts to know about the feasibility of LFG project, to show the advantages and disadvantages of the project and compare it with other electricity sources in Indonesia.

2. Materials and methodology

2.1. Landfill sites

Two landfills in Indonesia are investigated in this paper. One of them is in the large city of West Indonesia (Surabaya) and another one is in the metropolitan city of East Indonesia (Makassar). It is used to determine and evaluate the feasibility of landfill gas to energy projects.

2.1.1. Benowo Landfill (LF 1). LF 1 is ± 37.4 hectares in size and its final height will be between 5 – 12 m. The Benowo landfill begins on November 2001 and it accepts waste from whole Surabaya which is the part of East Java. The planning of Benowo landfill closure year is 2030. Table 1 shows yearly disposed quantities of Benowo landfill.

Table 1. Yearly disposed quantities of Benowo landfill.

Year	Quantity (ton/year)
2003	204,000.00
2004	335,618.50
2005	467,237.00
2006	598,855.50
2007	540,200.00
2008	459,425.50
2009	448,741.95
2010	463,779.05
2011	478,816.14
2012	493,853.24
2013	508,890.33
2014	531,403.50
2015	539,342.25

Source: central bureau of statistic, Indonesia

2.1.2. Antang Landfill (LF 2). LF 2 is ± 14.3 hectares in size and its final height will be between 5 – 12 m. Most of the areas/phases have been capped or are almost at full capacity. However, the city is currently in negotiations regarding expansion of this landfill. Antang landfill begins in 1993 and it accepts waste from whole Makassar which is the part of South Sulawesi. The planning of Antang Landfill closure year is 2032. Table 2 shows yearly disposed quantities of Antang landfill.

Table 2. Yearly disposed quantities of Antang landfill.

Year	Quantity (ton)
2003	1,027,495
2004	1,077,874
2005	1,128,254
2006	1,178,633
2007	1,229,012
2008	1,279,391
2009	1,329,770
2010	1,380,149
2011	1,432,085
2012	1,480,907
2013	1,531,287
2014	1,581,666
2015	1,632,045

Source: central bureau of statistic, Indonesia

2.2. Landfill gas generation model

Landfill Gas Emission Model (Land GEM) software is used by estimating emission rates for total landfill gas of municipal solid waste landfill. The United State Environmental Protection. Agency (US EPA) is the organization that develop this software. The model determines the mass of methane generated using the methane generation capacity and the mass of waste deposited.

Current research also calculates the electricity generation potential from the landfill. EPA defines between 75-85% of the produced methane and the calorific value of methane is 4.5 kWh/year and the form to find the potential electricity. Based on Land GEM rule, CO_2 emission is calculated from the production of methane and the methane content percentage.

2.3 Landfill Gas Energy Cost Model (LFGcost-Web)

LFGcost-Web is an LFG energy project cost estimating tool developed for EPA's LMOP. LFGcost-Web estimates LFG generation rates using a first-order decay equation. Variation in the gate and types of incoming waste, site operating conditions, and moisture and temperature conditions may provide substantial variations in the actual rates of generation.

The default inputs and cost estimated by LFGcost-Web are based on typical project designs and for typical landfill situations. The model attempts to include all equipment, site work, permits, operating activities, and maintenance that would normally be required for constructing and operating a typical project, however, Individual landfill may require unique design modifications which would add to the cost estimated by LFGcost-Web. First-Order Decay Equation for Average Annual Waste Acceptance Rate as shown in equation (1).

$$Q_t = (1/(CH_4/100)) \cdot L_o \cdot R \cdot [e^{-kc} - e^{-kt}] \quad (1)$$

Where Q_t is the landfill gas generation rate at time t (ft^3/year), CH_4 is the methane content of landfill gas (%), L_o is the potential methane generation capacity of waste (ft^3/ton), R is the average annual waste acceptance rate during active life (tons), k is the methane generation rate constant (1/year), c is the time since landfill closure (years), and t is the time since the initial waste placement (years). First-Order Decay Equation for Waste Disposal History is shown in equation (2).

$$Q_t = \Sigma_i[(1/(CH_4/100)) \cdot k \cdot L_o \cdot M_i \cdot e^{-kt_i}] \quad (2)$$

Where Q_t is the landfill gas generation rate at time t (ft^3/year), CH_4 is the methane content of landfill gas (%), k is the methane generation rate constant (1/year), L_o is the potential methane generation capacity of waste (ft^3/ton), M_i is the waste acceptance rate in the i-th section (tons) and t_i is the age of the i-th section (years).

After total of gas generation is calculated, it will calculate the gross power generation potential (NPGP) in year. Where is the collection system efficiency, typically 85%, is the energy content of landfill gas, typically 500 BTU/cf, and is the heat rate of equipment. Afterwards, the annual electricity is generated as shown in equation (3).

$$AEG = NPGP\ (24)\ (365)\ 85\% \quad (3)$$

Where net power generation potential (AEG) is estimated by subtracting the parasitic loads, 24 is hours per day, 365 is days per year, and 85% re represent the assumed average percentage of the time in a year that equipment is producing electricity at its rate capacity (net of maintenance, downtime, etc.).

3. Result and discussion

3.1. Generation Cost Landfill gas generation from Antang and Benowo landfills

Using the financial assumptions (table 3) and collection and flaring system assumptions (table 4), the waste-in-place from LFGCost-WebV3.0 as well, this research finds that an electricity price around $0.06 by operating cost, net income, and simple payback in table 5. Figure 1 describes landfill gas generation, collection and utilization curve.

Table 3. Financial assumptions.

Project life (years)	15
Project start year	2016
Down payment	20%
Loan rate	6%
Loan period	10 years

Depreciation	Straight line
Corporate tax	35%
Renewable energy tax credit	0 $/kWh
Discount rate	12%

Table 4. Collection and flaring system assumptions.

Cost Component	Cost (2013$'s)	Cost Unit
Drilling and pipe crew mobilization Installed cost of vertical	$20,000	per system
gas extraction wells	$4,675	per well
Installed cost of wellheads and pipe gathering system	$17,000	per well
Installed cost of knockout, blower, and flare system	$(x)^{0.61}$ * $4,600	$, x = ft^3/min
Engineering, permitting, and surveying	$700	per well
Annual O&M for collection (excluding energy)	$2,600	per well
Annual O&M for flare (excluding electricity)	$5,100	per flare
Electricity price (depends on type of project)	$0.090	per kWh with a
Project Component	**Quantity**	
Average depth of landfill waste (ft)	65	
Number of wells (1 well per acre)	10	
Number of flares (1 flare per system)	1	
Collected landfill gas design flow rate (ft3/min)	2,374	
Electricity usage by blowers (kWh/ft3)	0.002	
Installed Capital Costs:	2016	
Mobilization:	$21,224	
Extraction Wells:	$49,611	
Wellheads and Pipe Gathering System:	$180,405	
Knockout, Blower, and Flare System:	$559,314	
Engineering, Permitting, and Surveying:	$7,428	
Total Capital Costs Including Cost Contingency	$817,984	

Table 5. Operating cost, net income, and simple payback.

Year	Operating Cost	Maintenance Cost	Net Income
2017	$163,784	$1,807,823	$61,351
2018	$168,195	$1,853,018	$67,099
2019	$170,956	$1,899,344	$74,276
2020	$173,771	$1,946,827	$81,792
2021	$176,642	$1,995,498	$89,669
2022	$179,570	$2,045,386	$97,926
2023	$182,556	$2,096,520	$106,588
2024	$185,602	$2,148,933	$115,677
2025	$188,709	$2,202,657	$125,219
2026	$191,878	$2,257,723	$135,242
2027	$195,111	$2,314,166	$136,670
2028	$198,410	$2,372,020	$138,092
2029	$201,774	$2,431,321	$139,506
2030	$205,207	$2,492,104	$140,912
2031	$208,710	$2,554,406	$142,308
Simple payback			9 years

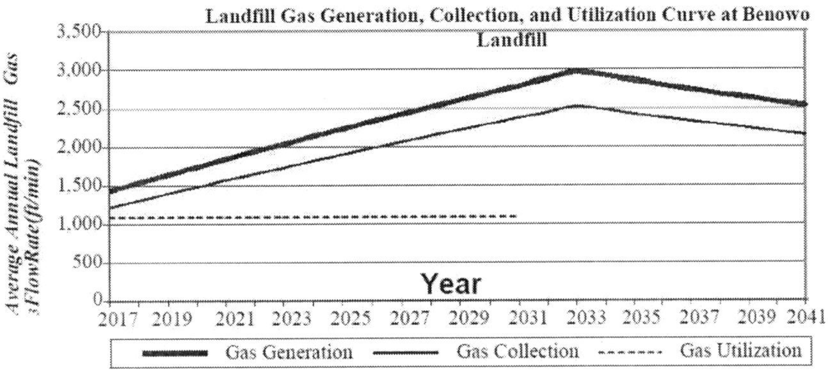

Figure 1. Landfill gas generation, collection and utilization curve.

3.2. *Comparison electricity sources in Indonesia*

This research attempts to compare Landfill gas CO2 emission in each power place that is used in Indonesia. Calculation of LFG CO2 Emission is got by result using LandGEM and the other sources are known by a review on the pattern of electricity generation and emission in Indonesia from 1998 to 2009 [7]. Table 6 shows CO2 comparison between LFG and other fuels.

Table 6. CO_2 Comparison between other fuels.

Fuels	CO2 Emission (kg/kwh)
LFG	0.00000000027
Gas	0.85
Fuel Oil	0.85
Coal	1.18

By the result of comparing to another source, the research find that LFG has CO2 emission lower (0.00000000027 kg/kWh) that others sources such as Gas (0.85 kg/kWh), Fuel Oil (0.85 kg/kWh), and Coal (1.18 kg/kWh). It means that LFG is one electricity generation who has possibility as electricity friendly generation for environment. It is based on CO2 as the biggest issue in Global warming.

LFGcost-Web as a software development from USA environmental protection agency is used to calculate the average generation cost if landfill gas generation is built and compare to other generation sources such as hydro generation, steam generation, diesel generation, turbine generation, geothermal generation, and combined cycle generation. Figure 2 shows comparison of different electricity production systems.

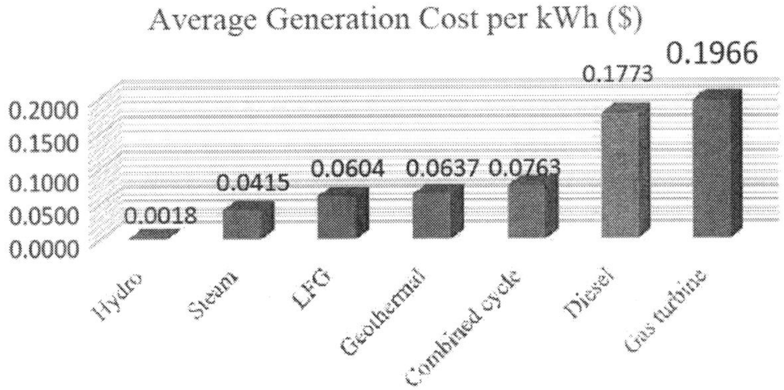

Figure 2. Comparing the different electricity production system.

Based on the figure 2, average generation of Landfill gas as the third lower generation which is \$0.0604 per kWh. It is bigger than Hydro (\$0.0018) and steam (\$0.0415) generation yet it is lower than geothermal (\$0.0637), Combined Cycle (\$0.0763), Diesel (\$0.1773), and Gas turbine (\$0.1966) generation.

4. Conclusion

From the result finding, Landfill gas project can be alternative way to deal waste problem. Change from waste to electricity could help government in electricity needs. The results show that landfill gas project more friendly in economic and environmental aspects rather than other recent electricity sources in Indonesia. To sum up, government can consider landfill gas project in Indonesia as solution for waste problem and also electricity.

Acknowledgement

The author thanks to *Lembaga Pengelolaan Dana Pendidikan* (LPDP) for providing financial during my master of urban and environmental engineering in Kyushu University.

References

[1] Meidiana C and Gamse T 2010 Development of waste management practices in Indonesia *European journal of scientific research* vol 40 issue 2 pp 199-210
[2] Kawai M, Purwanti I F, Nagao N, Slamet A, Hermana J and Toda T 2012 Seasonal variation in chemical properties and degradability by anaerobic digestion of landfill leachate at Benowo in Surabaya, Indonesia *Journal of environmental management* vol 110 pp 267-75
[3] Pasang H, Moore G A and Sitorus G 2007 Neighbourhood-based waste management: a solution for solid waste problems in Jakarta, Indonesia *Waste management* vol 27 no 12 pp 1924-38
[4] Kumar S, Gaikwad S A, Shekdar A V, Kshirsagar P S and Singh R N 2004 Estimation method for national methane emission from solid waste landfills *Atmospheric Environment* vol 38 no 21 pp 3481-7
[5] Kalantarifard A and Yang G S 2012 Estimation of methane production by LANDGEM simulation model from Tanjung Langsat municipal solid waste landfill, Malaysia *Int. J. Sci.*
[6] Dowling M, Kibaara S, Chowdhury S and Chowdhury S 2012 Economic feasibility analysis of electricity generation from landfill gas in South Africa *Power System Technology (POWERCON), 2012 IEEE International Conference on.*
[7] Hasan M H, Muzammil W K, Mahlia T M I, Jannifar A and Hasanuddin I 2012 A review on the pattern of electricity generation and emission in Indonesia from 1987 to 2009 *Renewable and Sustainable Energy Reviews* vol 16 no 5 pp 3206-19

Bearing capacity of circular skirted footing on clay soil

A B Listyawan, Renaningsih and N Kusumaningtyas

Civil Engineering Department, Universitas Muhammadiyah Surakarta, Jl. A. Yani Tromos Pos I Pabelan Surakarta 57102, Indonesia

anto.budi@ums.ac.id

Abstract. Footing foundation has been worldwide used in resisting light to medium load of structures. Footing's design has to consider the bearing capacity and the settlement of soil below. This research focuses on the investigation of the effect of skirt on circular footing to improve the footing performance in resisting vertical load on clay soil. There are nine samples of circular footing tested in vertical load system in laboratory in the diameter of 75 mm, 100 mm and 150 mm with length of skirt 100 mm and 150 mm on clay by keeping the similar water content and compaction method. The results show that the skirt effectively reduces the foundation settlement on clay which is observed on similar load 1 kN. The observations on L/D ratio on similar diameter also show that the higher L/D ratio the smaller settlement. On the observations of settlement 3 mm with similar diameter show that the longer the skirt the higher load retained by footing.

1. Introduction

Foundation is the part of structure which serves exclusively to transmit load from structure on to the sub-soil. One of the causes of the building collapse is the instability of the foundation, either the foundation design or the instability of soil. As known, clay is classified as one of the problematic soils due to its low shear strength and excessive settlement. Therefore, it is necessary to improve and to investigate either on the soil or the foundation design. In order to solve these problems, a vertical plates or skirts made out of steel are attached to the footing or called skirt foundation. Wakil [1] studied the skirt circular footing to improve the bearing capacity on sandy soil. The research was performing eighteen laboratory experiments on circular footing of different diameter and skirt length.

The investigation on the effect of skirt length and relative density of sand on ultimate load from the laboratory tests was found that the skirts are more beneficially in case of footing on loose sand that in case of medium and dense sand. Ashraf and Wasim [2] reported that the improvement of bearing capacity is remarkable; using both partially replaced sand pile with and without confinement by skirts. Golmoghani and Rowshanzamir [3] evaluated bearing capacity of skirted footing on sand. The test results were found that using structural skirts may improve the footing bearing capacity up to 3.68 times greater than the average value of calculated ultimate bearing capacity of foundations with the same depth of skirt. Satria [4] reported that the additional of skirt on the circular footing is very effective to improve the ultimate bearing capacity on the sand soil. With the same diameter of circular footing, as the length of skirt increases, the ultimate bearing capacity increases. There are also other researches related to bearing capacity of skirted foundation in sand [5-7].

2. Method of test

To study the behavior of the effect of skirted footing on clay, six laboratory tests of skirted footing and three other without skirts were conducted on small scale model of circular footing machined from steel plates having diameter (D) equal to 75 mm, 100 mm, 150 mm and of thickness 20 mm. The thin circumferential skirts are welded firmly and accurately around the periphery of footing, of thickness in 10 mm. Their lengths were 100 mm and 150 mm, measured after welding to footing. The skirt length (L) to the footing diameter ratio (L/D) are 1.33 (100/75), 1.00 (100/100), 0.67 (100/100), 2.00 (150/75), 1.50 (150/100), 1.00 (150/150) and three unskirted footing L/D are 0.00. The model footings have smooth face and notch at the center of top face for mounting piston to transfer the load when testing. There are two open hole drilled at the top face of footing to observe the top surface of the soil. The illustration of skirt circular footing is shown in figure 1.

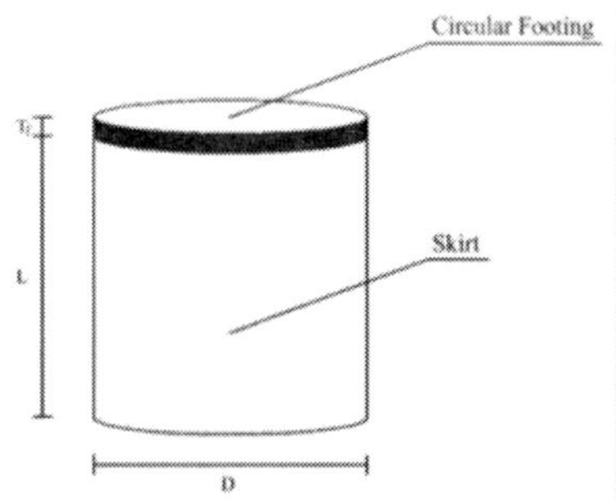

Figure 1. Skirt circular footing.

Where,

D = diameter of skirt circular footing
L = length of skirt
T_f = thickness of footing

The soil used in this research is clay coming from District Sukoharjo, Central of Java, Indonesia. The soil was keeping the same of water content and compaction method in each laboratory test. The bin was made of cylinder steel, having height 500 mm and diameter 600 mm. The vertical loadings are measured by frame load testing machine with hydraulic loading machine, it has maximum load of 250 kN. Two dial gauges installed to measure the vertical displacement. Both were placed on the top face of footing. The compaction was used the hammer with 5.5 lbs (2.5 kg) weight and 12 inch (305 mm) free fall height. The set-up of loading test is shown in figure 2.

First stage is preparing and setting up the equipment and clay that comes from same location. The soil is dried on the temperature (27-30°C) and passed no. 4 sieve. Second stage is investigating the air-dried water content. Third stage is conducting the laboratory test. The value of water content for this research must be kept in same condition for all nine models laboratory test. Adding the water into the soil and mix to produce homogenous water content in all area. Then, compact the 10 layers of soil and giving 100 blows for each layer. Set the soil bin on the hydraulic loading machine and place the skirt footing with L/D = 1.33 right on center of the bin, push it until all the skirt is already inside the soil. Observe the top surface of the clay inside the skirt via two open holes on the footing. Set the two dial gauges on the left and the right position to observe the vertical displacement. Then running the hydraulic loading machine, observe and record all of the magnitude that occurred on each dial. The settlement value is shown on the two dial gauges and the load value is shown on the digital device of the hydraulic loading machine. The last is stopping the loading when the load decrease as the indication that the foundation failure. The subsequent experiment, conduct the same test as sequence before but it just replaces the footing with value L/D = 0.0 (unskirted circular footing), 1.33, 1.00, 0.67, 2.00, 1.50, 1.00.

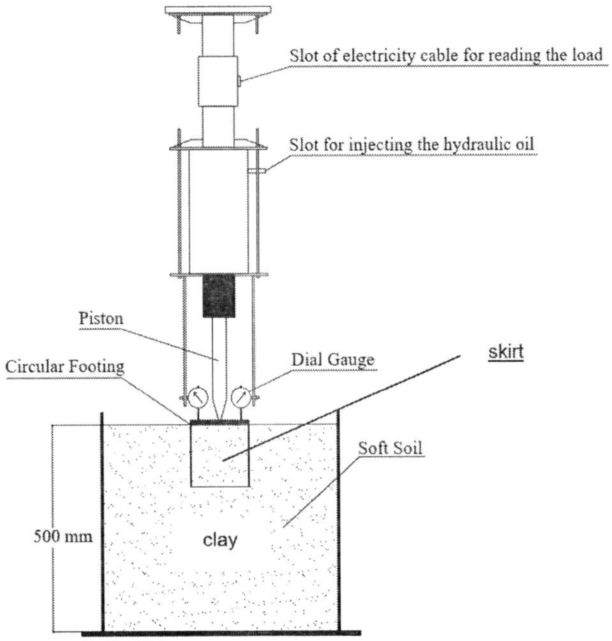

Figure 2. Set up of loading test.

3. Result and discussion

The research is conducted in similar water content and compaction method. The water content was investigated around 15%. This value kept for all testing model as reference. The result of the investigation shows load settlement curves for all model footing. The general result is presented in figure 3 - figure 5. The horizontal and vertical axes show the load and settlement value respectively.

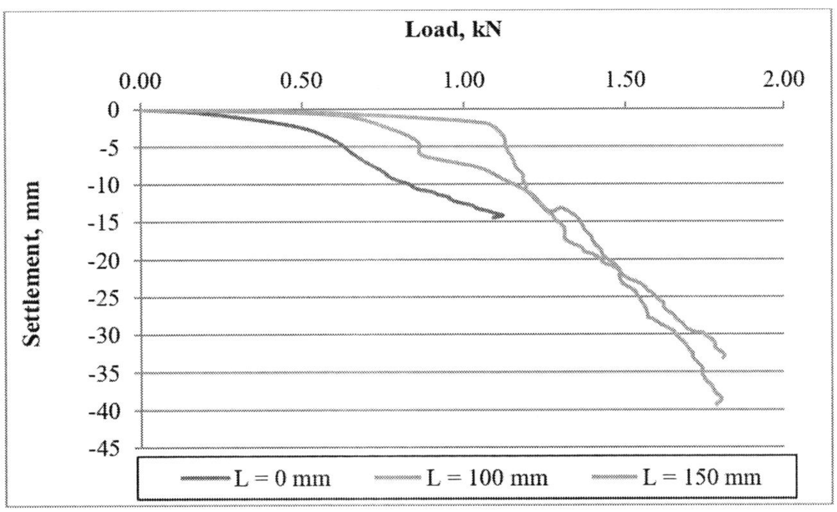

Figure 3. Load-settlement for footing diameter 75 mm.

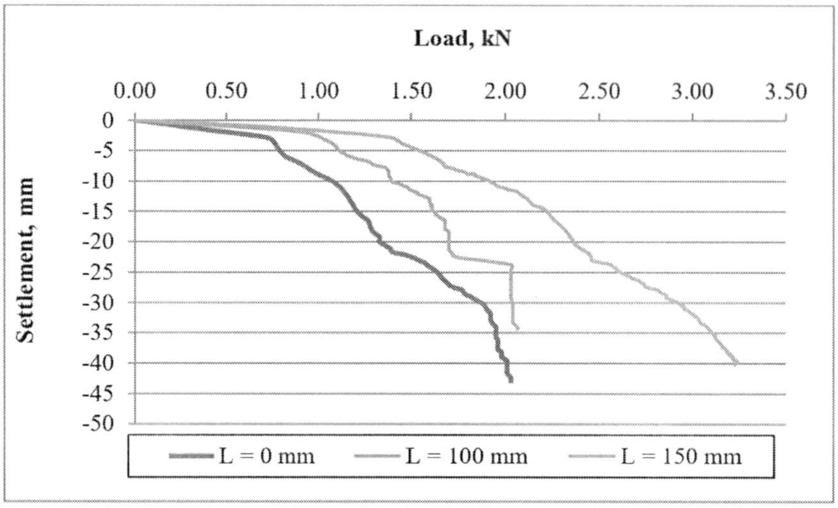

Figure 4. Load-settlement for footing diameter 100 mm.

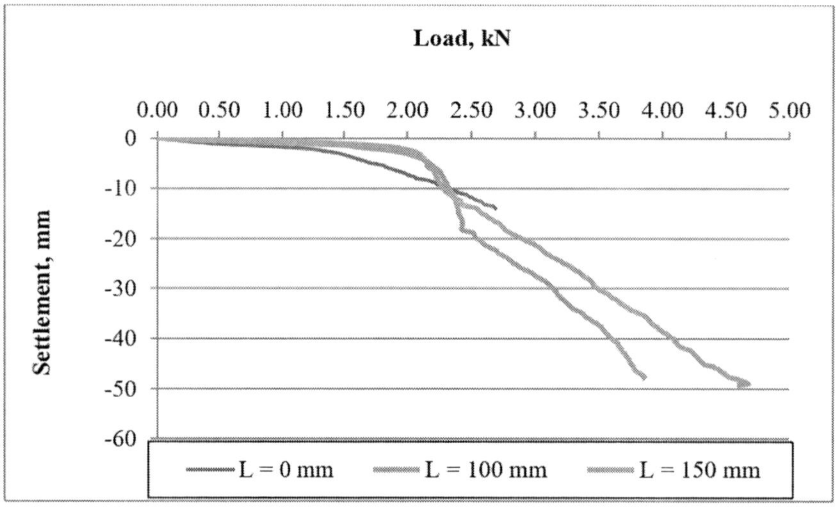

Figure 5. Load-settlement for footing diameter 150 mm.

It can be seen from figure 3 – figure 5 that the load-settlement relationship for all curves are fairly linear for small-load range, and then the relationship is nonlinear for large-load ranges and does not exhibit any peak value.

3.1 Settlement in similar load as reference

In order to investigate the magnitude of settlement due to the influence of the additional skirt to circular footing, it has to observed on the similar value of load (P, kN). It is taken 1 kN. The settlement on similar load (1 kN) is shown on table 1. The magnitude of settlement which is caused by the different skirt length indicated that the longer skirt the smaller settlement. It is shown on diameter (D) 150 mm with the length (L) 0.00 mm= 1.72 mm of settlement, (L) 100 mm= 0.90, (L) 150 mm= 0.52 mm. Table 1 also shows the different condition on L/D ratio, which is observed from the similar diameter and different skirt length. The higher L/D ratio the smaller settlement.

Figure 6 shows the smallest settlement of skirt circular footing is the footing that attached by the longest skirt 150 mm to the widest diameter 150 mm. The skirt gives resistant due to the sleeve friction between clay and shaft (i.e. skirt) of footing. This resistant contributes on minimizing the settlement duo to vertical loading. As the skirt gets longer, the area of shaft also larger and as consequent, the resistant raises to produce a smaller settlement.

Table 1. Settlement magnitude on load 1 kN.

Footing diameter D (mm)	Skirt length, L (mm)	L/D	Settlement (S) (mm)
75	0	0.00	12.63
75	100	1.33	7.35
75	150	2.00	1.59
100	0	0.00	8.90
100	100	1.00	2.59
100	150	1.50	1.70
150	0	0.00	1.72
150	100	0.67	0.90
150	150	1.00	0.52

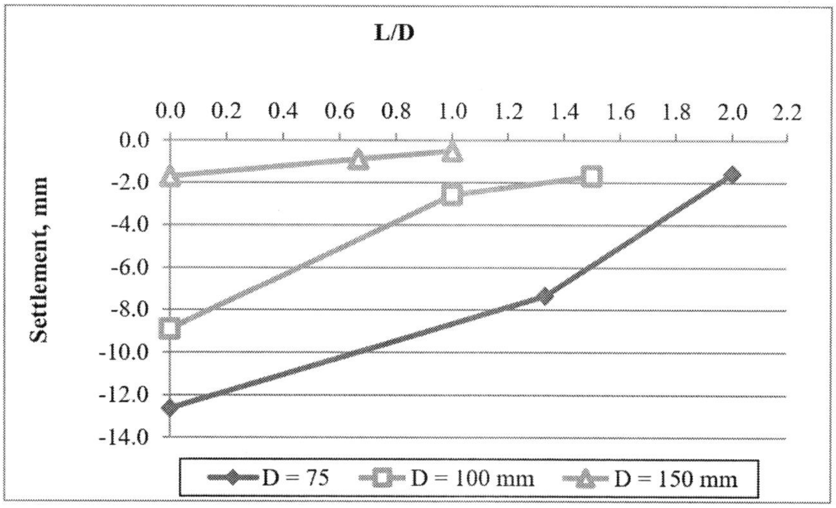

Figure 6. L/D Ratio-settlement relationship, different D.

3.2 Load in similar settlement as reference

Based on the test and investigation of six models of skirt footing and three models of unskirted footing to study the load magnitude due to the influence of the variant skirt length to the different diameter, it could be observed in the similar settlement in 3 mm and the results can be seen on table 2. Figure 7 shows the longer skirt the higher load in similar diameter. It also happens on L/D ratio which is observed in similar diameter with different skirt length. It obviously can be analyzed that the longer skirt the larger area of resistant due to sleeve friction between cohesive soil and the shaft and end up with the magnitude of load gets higher. As shown in figure 7, the higher L/D ratio the higher load in similar settlement 3 mm with different diameter.

Table 2. Load magnitude in similar settlement on 3 mm.

Footing diameter D (mm)	Skirt length, L (mm)	L/D	Area	Load (P) (kN)
75	0	0.00	4419.64	0.54
75	100	1.33	4419.64	0.80
75	150	2.00	4419.64	1.13
100	0	0.00	7857.14	0.74
100	100	1.00	7857.14	1.05
100	150	1.50	7857.14	1.42
150	0	0.00	17678.57	1.47
150	100	0.67	17678.57	1.99
150	150	1.00	17678.57	2.10

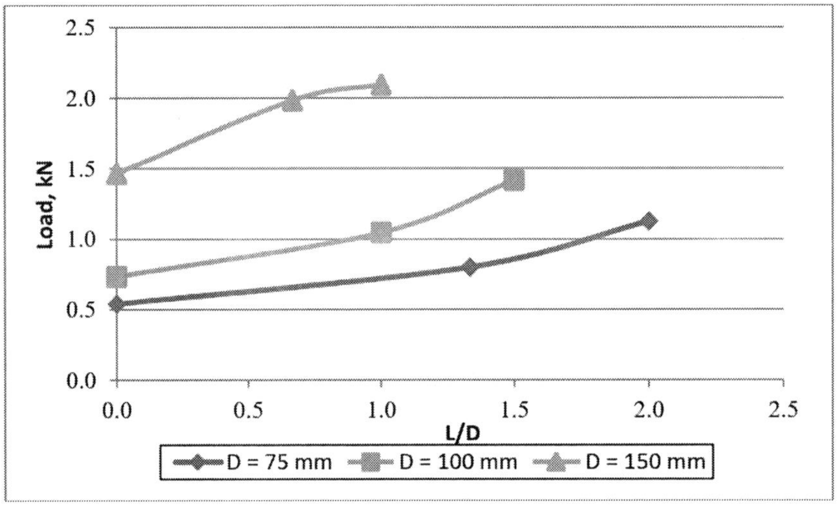

Figure 7. L/D ratio-load relationship in similar settlement 3 mm, different D.

4. Conclusion

From the accomplished laboratory tests, we derived some conclusions. The additional skirt on circular footing is effective to reduce the settlement on clay as can be seen when it is observed on the similar load on 1 kN load on the same diameter of circular footing, the longer skirt the smaller settlement. The magnitude of settlement for the influence of skirt length to the footing diameter, L/D ratio observed on the similar diameter of footing different length of skirt. It shown that the higher L/D ratio the smallest settlement. The magnitude of loading of the skirted and unskirted circular footing observed on similar settlement in 3 mm, it is explained that at the diameter of footing, the longer skirt the higher load.

References

[1] Wakil A Z E 2013 Bearing capacity of skirt circular footing on sand *Alexandria Engineering Journal* vol 52 no 3 pp 359-364

[2] Nazir A K and Azzam W R 2010 Improving the bearing capacity of footing on soft clay with sand pile with/without skirts *Alexandria Engineering Journal* vol 49 no 4 pp 371-7

[3] Golmoghani-Ebrahimi S and Rowshanzamir M A 2013 Experimental evaluation of bearing capacity of skirted footing *Civil Engineering and Architecture* vol 1 no 4 pp 103-108
[4] Satria I F and Listyawan A B 2016 *Effect of skirt on the bearing capacity of circular footing on sand* Doctoral dissertation, Universitas Muhammadiyah Surakarta, Indonesia
[5] Chandrawanshi S, Kumar R, Kaur D S and Jain D P Effect of skirt on pressure settlement behaviour of model circular footing in medium dense sand *International Journal of Advanced Engineering Technology* vol 5 no 2 pp 1-5
[6] Tripathy S 2013 *Load carrying capacity of skirted foundation on sand* Doctoral Dissertation National Institute of Technology, India
[7] Villalobos F 2007 Bearing capacity of skirted foundation in sand *VI Congreso Chileno de Geotecnia Valparaiso*

Mode technology assessment of mass transport master plan in Greater Jakarta (Jabodetabek)

Alvinsyah and E Hadian

Transport Research Group, Civil Engineering Department, University of Indonesia, Depok, Indonesia

alvinsyah2004@gmail.com, hadian@iutri.org

Abstract. The aim of this paper is to evaluate the appropriateness of the proposed mass transit mode technology in the Greater Jakarta (Jabodetabek) Public Transport Master Plan. The analysis is focused on the estimated demand on the proposed bus rapid transit (BRT) network. A transport demand model based on the four step modeling approaches is prepared to estimate the demand on each proposed BRT route. Prior conducting a simulation, the model developed from previous works is calibrated and validated with various primary data namely, trip length frequency distribution and bus passenger data obtained from the field. In parallel, some assumptions and operational scenarios were established by taking into account the total number of ridership, maximum passenger flow, operational headway and proposed mode capacity. The appropriateness of the proposed mode was assessed and analyzed. The results show that the ridership and operational headway exceed the standard commonly used for the operational aspects of BRT on some proposed BRT routes. In addition, the evaluation shows that the proposed routes should be shifted to a rail base mass transit technology in order to maintain best level of services.

1. Introduction

Referring to the draft of the *Badan Pengelola Transportasi Jabodetabek* (BPTJ) Transport Master Plan [1] seems that the most ambitious and potential program that should be implemented in a very short period is Bus Rapid Transit (BRT) network plan. Assuming that the definition of BRT [2, 3] is held, and taking into account the existing physical condition for its proposed network, it seems that BPTJ needs a special treatment and very extra effort to accomplished this Plan. Yet, regardless the difficulties found to implement the plan, some of the previous works [4-6] indicate that the potential public transport demand on some proposed BRT corridors is quite significant high. Although, choosing a proper type of mass transit technology can be a difficult and complex process [7], and the real intense controversies seem to be generated in making choice between rail and bus system [7, 8], this effort is clearly a worthwhile goal, since the choice will affect travel times, personal transport expenditures, and commuter comfort and safety.

Since implementing and operating an urban mass transport system will involve a very huge capital, a decision on adopting a specific mass transport technology must be done with a very careful, thorough and prudent process, hence the investment either from public or private fund for the infrastructure, system operation and maintenance will not be wasted for the whole life of the system. Therefore, prior continuing to a more detail work for implementation, it is necessary to conduct an assessment to the Plan, especially the appropriateness of the proposed mode technology.

Vuchic [9] indicates that the first step of mode evaluation should consider right of way (ROW) category, mode technology and type of service/operation on the basis of system requirements, where they are mutually independent. Yet, the basic one is ROW category, which influences the choice of technology and operations. The next phase is to evaluate each candidate mode in one of three basic ways that are monetary units, other quantitative units and qualitative (descriptive) terms.

Since passenger demand or ridership plays a dominant role in urban mass transit system, it is often used as a basis reference to determine the capacity of mass transit [7-9] either for the future plan or the implementation. Therefore, this research aims to assess the suitability of BRT system proposed in the aforementioned Plan based on quantitative units, and focused on the projected demand and the operational headway for each proposed corridors. The subsequent sections discuss research methods, model assumptions and operational scenarios, simulation and result analysis and finally the reached conclusion.

2. Research methods

This research is initiated with collecting and reviewing previous works [1, 6, 10-13] to prepare a transport model for the analysis which is based on the four-step method. The model adopted from previous works [1, 4, 5, 10, 12, 14] is then updated by a calibration and validation process where the updating procedure is explained in a more detail in [15]. Once the updated model, commonly called as base year model, is set, then it is utilized to estimate future motorized person trips.

Taking the actual mode share composition [13], this total motorized person trip is split into trips with private vehicle (i.e. car and motorcycle) and trips with public transport. Having obtained the total public transport trip, then, this trip is assigned to the existing and the planned public transport network through a transit assignment procedure. In this step, a macro planning software called EMME is utilized and a standard transit assignment based on the concept of optimal strategies which minimize transfer, waiting and in-vehicle time is adopted [16]. Further detail explanation on functions of this optimal strategy can be found in [17].

In transit assignment model, the monetary cost, which is fare applied to each particular line services are converted to 'time' and is weight-combined with total travel time which consists of in-vehicle, waiting at stops, access and egress time components in the specific form called 'generalized cost' function. The detail mathematical function can be found in [15].

Once the potential demand is obtained from the simulation, the estimated total boarding passenger for each BRT Corridor is then analyzed. Based on the magnitude of total boarding passenger, corridors having demand more than 8,000 pax/hr [7, 18, 19, 20], are identified as the potential mass transit corridors need to be served by a higher capacity technology (i.e. rail base system) and further analysis can proceed. To assure that the above potential corridors need a higher capacity mode, an operational parameter like service headway could be used as criteria [9, 21]. Therefore, to verify and assess the appropriateness of the proposed mode technology (i.e. BRT system), an initial headway need to be calculated [11, 19].

In order to know whether a transit mode can be operated safely and feasibly and also within the acceptance level of passenger waiting time, a standard headway need to be determined as a benchmark. For operation purpose, this standard headway is usually determined by the government as regulator (i.e. called policy headway). While, for planning or design purpose, a minimum and maximum headway can be defined based on technical aspects, safety operation, and passenger preference. An acceptable maximum headway for a mass rapid transit system should be less than 5-6 minutes [2, 9, 22], while the minimum headway varies from 12 seconds to 2 minutes [9, 19, 20, 23, 24] depends on ROW category, operational characteristics and the adopted policy or LOS standard for a particular transit mode. Yet, minimum headway commonly used in transit operation is 2 - 3 minutes [21]. In a more generic form, minimum headway for various urban mass rapid transit is described in detail in [9, 20] and [19].

It is important to note that the minimum (line) headway is usually adopted either from way headway or station headway [9]. Yet, in public transport, line or corridor capacity is mostly dictated

IOP Conf. Series: Materials Science and Engineering **403** (2018) 012020 doi:10.1088/1757-899X/403/1/012020

by station capacity [3], so consequently in vast majority cases the line minimum headway is determined by station headway [9]. Since the range of minimum headway is significantly wide, the designated value need to be verified by the station capacity with the formula as suggested in ITDP [3]. By comparing total passenger boarding per hour and initial headway, with that of commonly used in mass transit system operation [3, 21, 23], the appropriate mode technology could be determined as a preliminary assessment. Further, it is then verified with the standard of station or platform saturation level.

3. Model assumptions and operational scenarios

Referring to previous work [1] the trip growth for each district in Jabodetabek area is ranged from 0.77% to 1.99% for each local city in Jabodetabek with average of 1.38% in 2020. While, the public transport share adopted in this research is 24% [13], and the operational parameters and fare setting assumptions for each public transport is adopted from [1]. The simulation scenario which considers network plan, fare system and setting, and operational plan, is set for the year of 2020, a more detail the simulation scenario can be seen in [15].

4. Simulation and analysis

Simulation results show the estimated ridership for BPTJ-BRT corridors with high potential demand represented in total boarding passenger and the maximum station boarding-alighting passenger (in peak morning) are shown in table 1. Besides the boarding passenger, the maximum flow or commonly called as peak hour per direction (PHPD) volume occurs at each service route is also estimated. This PHPD figure is important for designing the operational plan of public transport such as number of fleet, service frequency, load factor, and the operation and maintenance cost.

Table 1. Potential demand on BPTJ-BRT.

BPTJ-BRT Service Route	Total Boarding (pass./peak-hr)	Maximum flow (pass./peak-hr)	Max. Total Boarding at Station (pass./peak-hr)	Max. Total Alighting at Station (pass./peak-hr)
Bekasi–Senen	32,114	12,344	6,463	5,372
Depok–BlokM	13,692	6,402	3,504	2,253
Bogor–Senen	14,512	3,624	2,945	2,409
Poris Plawad –BlokM	16,941	7,560	3,503	5,184
Cikarang–BlokM	47,811	19,276	7,373	6,325
Tn. Abang – Bekasi	25,700	9,093	3,406	2,092

From 20 (twenty) proposed BRT service routes simulated, there are 6 (six) routes have demand number that can be served either by rail system or high capacity BRT system. Also from table 1, two of the proposed BRT route have a very high demand (i.e. approximately around 30,000 passengers per hour) which close to the level that commonly served by heavy rail transit. Taking the projected demand as the initial criteria to select appropriate technology or system is quite common in practice [7, 8, 9]. Yet, using it as a dominant parameter for making decision can be misleading [7, 9]. This is due to some mass transit systems may blur the boundaries with the definition of other transit system when they, for instance, utilize same ROW category as a standard, hence it consequently affects the capacity boundaries as well [7, 9]. Therefore, specifically, for this research purpose, a service headway is used to verify whether the chosen technology or system based on the projected demand could be safely and feasibly operated under given ROW or other transit operational characteristics. By inputting all required parameters into the formula suggested in [11, 19], the initial headway for various transit mode technology for each proposed BRT routes are obtained as shown in table 2.

Table 2. Initial headway for various mode technology.

BPTJ-BRT Service Route	Max. Flow (x1,000 pax/hr)	MRT (C=1000 pax)	LRT (C=600 pax)	BRT Bi-Articulated (C=250 pax)	BRT Articulated (C=140 pax)	BRT Single Bus (C= 90 pax)
Bekasi–Senen	18,507	5.21	3.10	1.30	0.73	0.47
Depok–BlokM	9,102	8.84	5.30	2.21	1.24	0.80
Bogor–Senen	8,692	11.68	7.01	2.92	1.64	1.05
Poris Plawad –BlokM	11,443	6.42	3.85	1.61	0.90	0.58
Cikarang–BlokM	21,813	4.40	2.64	1.10	0.62	0.40
Tn. Abang – Bekasi	12,303	6.49	3.89	1.62	0.91	0.58

Looking at the calculated headway in table 2 and if 3 minutes is referred as the minimum standard headway [21], only two proposed routes in table 3 are still feasible served by BRT system with bi-articulated bus, while the others need to be served by a higher capacity system. In the other hand, if the minimum headway as described in [9, 21] and [19] is referred as the standard value, all proposed BPTJ-BRT routes are still could be served by BRT system [23, 24]. But minimum headway figures indicated in [9, 21] and [19], should be treated with caution, since it is in a generic form, especially for bus system operation. As explained previously for BRT system, the minimum headway depends on ROW category, operational characteristics such as safety standard operation, way and station capacity and also traffic signal cycle time. Therefore, it is necessary to conduct further assessment to make sure that a high demand corridor/route is still be able to be served by BRT system safely and feasibly. Referring to [3], in this case, the dwelling time for single bus, articulated bus, and bi-articulated are assumed 14 seconds, 13 seconds and 12 seconds respectively. While the average boarding and alighting time per passenger are assumed 0.3 seconds for articulated and bi-articulated bus and 3 seconds for single bus. The above values are adopted with assumptions that station and bus platform are level, and the transaction is done off board. Taking the headway values as shown in table 2, and utilizing formula for platform saturation level is in [3], the station saturation level is presented in table 3.

Table 3. Bus platform station saturation level.

BPTJ-BRT Service Route	Bi-Articulated (C=250 pax)	Articulated (C=140 pax)	Single Bus (C= 90 pax)
Bekasi–Senen	1.16	1.28	10.29
Depok–BlokM	0.58	0.65	5.05
Bogor–Senen	0.53	0.57	4.65
Poris Plawad –BlokM	0.87	0.96	7.58
Cikarang-BlokM	1.35	1.49	11.92
Tn. Abang–Bekasi	0.60	0.69	4.93

Referring to the calculated saturation level in table 3, and taking the maximum saturation level allowed is not more than 0.4 [3], it is shown that all proposed BPTJ-BRT routes in table 2 need a higher capacity mode such rail technology with ROW-A category. Based on the calculated headway in table 2, the alternative mode includes the operational headway that could be considered to serve the proposed service routes are described in table 4. From table 4, there are still two routes that can be served by bi-articulated bus with 2 minutes' headway. Yet, assuming that BRT is chosen to serve these two corridors, a precaution on platform design, Traffic Signal Priority (TSP) setting at intersection, and probability of providing passing lane need to be kept in mind.

It is important to notice, knowing that station saturation level presented in table 4 exceed that of the allowed level, that these proposed routes with high demand are still possible to be served by BRT system [3, 7, 23, 24].

Table 4. Alternative mode technology for BPTJ-BRT.

BPTJ-BRT Service Route	Mode Technology	Operational Headway[a] (minutes)
Bekasi–Senen	MRT or LRT	3.0 – 5.0
Depok–BlokM	Bi-Articulated Bus or LRT	2.0 – 5.0
Bogor–Senen	Bi-Articulated Bus or LRT	2.5 – 7.0[b]
Poris Plawad –BlokM	LRT or MRT	3.5 – 6.0
Cikarang-BlokM	LRT or MRT	2.5 – 4.0
Tn. Abang–Bekasi	LRT or MRT	3.5 – 6.0

[a] The values are rounded down
[b] Complies with prevailing standard value

But, special treatment or care need to be carried out, like passing lane provision at station, applying multiplatform or long station, dual lane provision, multirouting, good traffic engineering measure like TSP, along the corridor [9, 19, 22, 23, 24]. But if, these kind special treatment are almost impossible or very difficult to implement due to physical, political, legal or budget constraint, the minimum headway 2 or 3 minutes is applied, provided that the cycle time at each intersection along the corridor is not significantly higher than the above values.

5. Conclusion

An assessment to the proposed BPTJ-BRT network route by conducting a simulation through a transit model indicates that six routes have a relatively high demand reflected by total boarding passenger and a PHPD volume. These high demand corridors, at some extend might potentially be served by a higher capacity mode technology such as rail system. Yet, since the line capacity boundary between different transit modes is somewhat overlapped, further analysis on the minimum acceptable headway and station saturation level are necessary to be carried out. This further analysis yields that four proposed BRT corridors are need to be served by LRT with ROW-A or by MRT system. While the other two corridors still possible to be served by bi-articulated bus, with a very special care and attention to several aspects like proper platform design, Traffic Signal Priority (TSP) setting, and passing lane provision.

References

[1] IUTRI 2016 *Capacity and Demand Analysis of Jabodetabek Mass Transit Network* JICA Final Report
[2] TRB 2003 *Transit capacity and Quality of Service Manual (TCQSM)* TCRP Report 100 Transportation Research Board Washington D.C.
[3] ITDP 2007 *Bus Rapid Transit Planning Guide* (New York: Institute for Transport Development & Policy)
[4] BSTP 2009 *Studi Penyusunan Master Plan Pola Transportasi Makro (PTM) Jabodetabek tahap-1* Direktorat Jendral Perhubungan Darat Jakarta Final Report
[5] BSTP 2010 *Studi Pedoman Perencanaan Trayek Pengumpan (feeder) untuk Angkutan Massal berbasis Jalan* Direktorat Jendral Perhubungan Darat Jakarta Final Report
[6] JICA 2012 *A Study on Jabodetabek Public Transport Implementation Strategy (JapTraPIS)* Final Report
[7] Wright L 2004 *Bus Rapid Transit* Deutsche Gesellschaftfür Technische Zusammenarbeit (GTZ) GmbH
[8] Deen T B and Pratt R H 1992 *Evaluating Rapid Transit, in Public Transportation* ed Gray G G and Hoel L A (Englewood Cliffs, New Jersey, USA: Prentice Hall Inc.) pp 293–331
[9] Vuchic V R 2005 *Urban Transit; Operations, Planning, and Economics* (Hoboken, New Jersey, USA: Wiley &Sons Inc.)

[10] Alvinsyah and Hadian E 2016 Analisis Dampak Aktifitas Kawasan Reklamasi Pantura Terhadap Kinerja Jaringan Jalan di DKI Jakarta *MTI Journal* vol 1 pp 43-58
[11] BSTP 2014 *Pedoman Perencanaan Teknis Perhitungan Kebutuhan Sarana Angkutan Massal Perkotaan* Direktorat Jendral Perhubungan Darat Jakarta Final Report
[12] JICA 2004 *Study on an Integrated Transport Master Plan for Jabodetabek Area* JICA Final Report
[13] JICA 2011 *Study on Jabodetabek Urban Transportation Implementation Plan (Review on Sitramp-2004)* JICA 2nd Draft Final Report
[14] CTS 2004 *Pola Transportasi Makro Jakarta* Dishub DKI Jakarta- CTS UI Final Report
[15] Hadian E and Alvinsyah 2017 Impact on Ridership of New Railbase Transit Due to the Operation of Extensive Bus Semi Transit Network (Case Study: Greater Jakarta/Jabodetabek Public Transport Network) *Proceeding of the 15th International Conference on Quality in Research*
[16] INRO 2015 *EMME User Manual* EMME Suite Montreal
[17] Spiess H and Florian M 1989 Optimal Strategies: A New Assignment Model for Transit Networks *Transportation Research Part B: Methodological 23B* vol 2 pp 83-102
[18] Transcraft 2005 *Practical Guide for Improving Urban Bus Transport Systems, Volume 2; A Guide to High Capacity Modern Urban Bus Systems: Best Practice and Recommendations* Consia
[19] TRB 2007 *Bus Rapid Transit Practitioner's* Guide TCRP Report 118, Transportation Research Board Washington DC
[20] Vuchic V R 1992 *Urban Passenger Transportation Modes in Public Transportation* ed Gray G G and Hoel L A (Englewood Cliffs, New Jersey, USA: Prentice Hall Inc.) pp 79 – 113
[21] Ceder 2016, *Public Transportation Planning and Operation; Modeling, Practice and Behaviour 2nd ed.* (Boca Raton, Florida, USA: CRC Press)
[22] ITDP 2012 *the BRT Standard Ver. 1.0* Deutsche Gesellschaftfür Internationale Zusammenarbeit (GTZ) GmbH
[23] Cervero R 2013 Bus Rapid Transit (BRT): An Efficient and Competitive Mode of Public Transport *Working Paper 2013-01 Institute of Urban and Regional Development* (Berkeley USA: University of California)
[24] Fjellstrom K 2010 High capacity BRT planning, implementation & operation: Case study of the Guangzhou BRT *UNCRD EST Conference* Thailand vol 24

The optimum bus route selection for sustainable operation (case study: Tangerang bus lane)

Alvinsyah[1] **and U Halim**[2]

[1] Transport Research Group, Civil Engineering Department, University of Indonesia, Depok, Indonesia
[2] Institute for Transport Policy and Development-Indonesia, Jakarta 10340, Indonesia

alvinsyah2004@gmail.com

Abstract. The aim of this paper is to select the optimum route option on a bus lane corridor based on operational and economical characteristics. In this research the Tangerang Bus Lane is chosen as a case study. The operational characteristics analyzed are total boarding passenger, total number of bus, bus productivity, bus cost structure and estimated fare. Based on the previous works the operational cost and potential revenue are calculated for both route options, and then an incremental benefit cost ratio analysis is conducted to obtain the most feasible route to operate. The result shows that the alternative route still consistently in providing a better performance compared with the current route.

1. Introduction

The Government of Tangerang City operates a bus lane corridor which connect Poris Plawad Bus Terminal with Kali Deres Bus Terminal in DKI Jakarta. However, based on data recorded by the Tangerang City Transport Agency, the daily ridership is relatively very low compared with other busway systems. In the other hand, there is an alternative location which is more attractive to the potential user for the origin of this service [1], and it has been indicated that this alternative route shows a better operational performance.

Since the operational and maintenance cost are one of the most critical issues in a mass transit system, a right decision on selecting the best route will yield to a very high cost saving, maximum revenue, and low subsidy needed for the for the whole life of the system. Therefore, a further analysis on the operational cost and the potential revenue is an obligatory step to obtain the best route option.

In the following section, the basic theory, research framework, and current situation are described. The last section discusses the optimum route based on financial analysis of operational cost and potential revenue and the final section will be the conclusion of the research.

2. Methods

This section explains the methods used in this research into two subsections namely literature review and research framework.

2.1. Literature review

In the past, planning for public transport was focused on social equity where everybody has to be able to use public transport. This strategy resulted into long public transport routes with many stops, serving as much as people as possible. Nowadays, public transport routes have to be efficient. The focus of today's public transport planning is serving mainly areas with high potential of use. Therefore, the estimation of operational cost and potential revenue becomes important aspects in determining an operation route. These aspects are commonly derived from some operational characteristics such as ridership, route length, service frequency, vehicle occupancy, operating speed, and line capacity [2]. While demand estimates are critical to designing the system, planning, operation, and predicting the financial viability of the system. Referring to the previous work [1], the total bus required for service can be calculated using equation (1).

$$n_{op} = \frac{(2L/_{V}+60)+R_t}{h} = \frac{2L\times 60}{V\times h} + \frac{R_t}{h} \tag{1}$$

where,
n_{op} = Total number of operating bus
R_t = Lay over (or recovery) time (minutes)
L = One-way route length (km)
V = Average design speed (km/hr)
h = Bus headway (minutes)

Number of spares is usually 2 -3 buses or 10 – 20% of number of bus in operation, whichever is greater [3]. Bus productivity is an important parameter for vehicle operational cost calculation. It can be derived from the total number of round trip achieved by each bus during the designated operation hours. Therefore, the total productivity can be expressed in equation (2).

$$N_{rit} = \frac{T_{op} - T_{spbu}}{T_{rt} + R_t} \tag{2}$$

where,
N_{rit} = Total number of round trip
T_{op} = Total time of operation (minute)
T_{spbu} = Time for fuelling (minute)

In order to estimate service productivity for the whole system in operation, the distance-travelled by each bus needs to be estimated. There are three categories of distance-travelled, namely effective distance-travelled, empty distance-travelled and total distance-travelled which expressed in equation (3) – (5).

$$Eff_{dis} = N_{rit} \times 2\text{L} \tag{3}$$

$$Emp_{dis} = 3\% \times Eff_{dis} \tag{4}$$

$$Tot_{dis} = Eff_{dis} + Emp_{dis} \tag{5}$$

where,
Eff_{dist} = Effective distance travelled (km)
Emp_{dist} = Empty distance travelled (km)
Tot_{dis} = Total distance travelled (km)

Referring to the guidelines for bus route design [4], the un-productive (empty passenger) distance travelled is suggested 3% of the total effective daily distance travelled. While for the operational cost component, the guidelines suggested that the components need to be calculated based on actual condition. On the other hand, the potential revenue is determined by the daily ridership represented in equation (6) [1].

$$R_d = to \times \left[\left\{\left(\bar{F}_a^{p} + F_b^{p}\right) \times \bar{T}_p\right\} + \left\{\left(\bar{F}_a^{op} + \bar{F}_b^{op}\right) \times T_{op}\right\}\right] \quad (6)$$

where,

Rd	=	Daily ridership (pass./day)
$\bar{F}$	=	Average maximum flow (pass./hr)
a,b	=	Flow direction
to	=	turnover rate
p	=	Peak
op	=	Off peak
T	=	Operational service time

Referring to TRB [3], turnover rate ranges from 1.2 to 2.0 passengers per bus depending on the route structure and areas served. The financial feasibility of a bus route should be analyzed to achieve a feasible and sustainable operation. Therefore, Benefit-Cost (B/C) ratio method is used to evaluate the feasibility of each route option and then an incremental B/C ratio must be adopted as expressed in equation (7) [5].

$$\frac{\Delta B}{\Delta C_{(y\text{-}x)}} = \frac{Rev_y - Rev_x}{VOC_y - VOC_x} \quad (7)$$

where,

ΔB	=	Nett benefit (Rupiah)
ΔC	=	Nett cost (Rupiah)
Rev	=	Revenue (Rupiah)
VOC	=	Vehicle operational cost (Rupiah)
x,y	=	Route option

Since Bus operation is commonly consisted to an operator for a certain period, a financial analysis which considers future forecasting need to be adopted. One approach commonly used is by converting all forecasted value into present situation, called present value method and then continued with BC ratio analysis. The present value calculation can be expressed in equation (8) [5].

$$V\,z(n) = Z_1 + Z_2\left(\tfrac{P}{F}, i\%, 1\right) + Z_3\left(\tfrac{P}{F}, i\%, 2\right) + Z_4\left(\tfrac{P}{F}, i\%, 3\right) + \cdots + Z_n\left(\tfrac{P}{F}, i\%, n-1\right) \quad (8)$$

where,

PV	=	Present Value (Rupiah)
z	=	Revenue or Cost (Rupiah)
n	=	Concession period (year)
i	=	Interest rate (percentage)
P, F	=	Present & Future value

2.2. Research framework

The flow of this research consists of two consecutive stages namely demand and service productivity analysis and financial analysis. In the first stage, the potential demand, bus operational characteristics and bus service productivity is analyzed. Then, in the second stage, a benefit-cost ratio analysis based on the estimated bus operational cost and potential revenue is conducted. Adopting results and using similar method from previous work [1], a demand estimation is conducted first on each alternative route. The analysis is based on two main parameters namely total boarding passenger and some estimated operational parameters such as frequency, headway, cycle time and number of bus in operation. The total corridor demand is derived from bus passenger flow obtained from load factor and frequency survey. While for the demand of the proposed service is derived by factoring the corridor demand with the percentage of potential passenger willing to shift to the proposed service obtained from user opinion/preference survey.

Since the current bus lane system applies different operational headway, the potential demand should be adjusted with the prevailing headway. Consequently, a new split percentage needs to be derived from the result of preference survey conducted by previous work [1]. Therefore, based on this new split percentage and assuming that the corridor demand does not change significantly, the new potential demand is determined by applying similar procedure as explained in previous paragraph. Meanwhile, the supply analysis requires some operational parameters such as, bus frequency, roundtrip travel time, and number of bus and the operation are estimated based on the required minimum operational headway. Based on the estimated operational parameters, productivity measures such as total number of trips, total distance travelled are determined. In parallel, the operational cost structure is also calculated. Then, by utilizing the estimated demand, and the assumed tariff, the potential revenue is determined. While in the supply side, the total operational cost is also calculated based on the service productivity measures. Finally, by applying incremental benefit cost ratio analysis on the potential revenue and operational cost, the best alternative of route is determined.

3. Result and Discussion

Tangerang bus lane corridor has two route options and basically, both routes share the corridor more than half of its total length and there is only slight difference in term of route length. The one-way route length of route A, connecting Poris Plawad Terminal and Kali Deres Terminal, is about 10.8 km, and the alternative route (or Route B), connecting Cikokol Terminal and Kali Deres Terminal is, about 11.6 km. The observed average speed along the corridor are 21.2 km/hr (peak morning) and 26.1 km/hr 14 (peak evening). Referring to previous work [1], the assumed parameters of the proposed service is shown in table 1.

Table 1. Assumed parameter.

Parameter	Peak Hour	Off-Peak Hour
Average Design Speed	17 km/hr	20 km/hr
Bus Capacity	85 (seats & standing)	85 (seats & standing)
Service Time Window	6 hrs	11 hrs
Fueling Time	15 minutes	15 minutes
Lay over time	10 minutes	10 minutes
Seat Turnover rate	1.3	1.3

Adopting equation (6) and (8), the required minimum headway [6], the route length as explained previously, and the design speed shown in table 1, the adjusted bus frequency and roundtrip travel time are calculated as shown in table 2. While, by applying the equation (1) as in section 2, assumed parameters as in table 1, and the adjusted headway as in table 2, the number of bus in operation can be determined as in table 2.

Table 2. Estimated operational parameters.

Parameters	Route A		Route B	
	Peak	Off-Peak	Peak	Off-Peak
Adjusted Headway (minutes)	7.5	10.0	7.5	10.0
Adjusted Bus Frequency	8	5	8	5
Rndtrip travel time (minutes)	86	75	92	79
# of Bus in operation (bus)	11	7	12	8

Taking the assumed parameter as shown in table 1, the estimated operational parameters as shown in table 2, and equation (2) to equation (5) as described in section 2, the service productivity can be calculated as summarized in table 3. Referring to the actual cost for each component obtained from various resources the operational cost in term of bus-km for each alternative route is shown on table 4. By multiplying the service productivity in table 3 and bus operational cost per km in table 4, the total operational cost for each alternative route can be calculated as shown in table 5.

Table 3. Bus service productivity.

Service Productivity	**Route A**	**Route B**
Total service (Cycle) length (km)	21.5	23.1
Number of trip (peak-hour)	5	4
Number of trip (off-peak)	9	9
Distance-travelled (peak-km/day)	107.7	92.5
Distance-travelled (off peak-km/day)	193.9	208.1
Total Effective distance-travelled (km/day)	301.6	300.6
Empty distance travelled (km/day)	9.0	9.0
Daily Total distance-travelled (km)	310.6	309.6
Monthly distance-travelled (km)	9,319.3	9,288.2
Annual distance-travelled (km)	113,385.2	113,007.3

Table 4. Operational cost per bus-km (in rupiah).

Cost Component	**Route A**	**Route B**
Bus Investment	705.6	707.9
Air Conditioning (AC)	107.1	107.5
Crew	1,046	1,049.5
Vehicle License & Operational Permit	24.9	25.0
AC maintenance	21.4	21.5
Management	904.5	907.5
Fuel Consumption	1,500	1,500
Tires	116.7	116.7
Minor Maintenance	58.9	58.9
Major Maintenance	97.5	97.5
General Checking	11.7	11.7
Fuel consumption for AC	450	450
Non-productive fuel Consumption	17.6	17.0
Mechanic	82.6	82.8
Total Cost	5,144.8	5,153.1

Table 5. Total operational cost.

Cost	**Route A**	**Route B**
Daily Operational Cost (Rp/bus)	1,598,198.5	1,595,452.1
Daily Operational Cost (Rp/route)	15,068.729.0	16,322,702.7
Annual Operational Cost (Rp/route)	5,500,086,094.6	5,957,786,501.8

Since the system is operated as a bus lane, the potential user obtained from the opinion survey, is about 29% of the existing public transport user along the corridor. By modifying the estimated demand from the previous work [1] and applying equation (6) in section 2, the total estimated boarding for route A is 12,279 pax/day and for route B is 12,734 pax/day. Therefore, potential revenue for both route with assumption that the flat fare is IDR 3,500/trip can be estimated. The estimated potential revenue for both route is indicated in table 6.

Table 6. Total potential revenue.

Cost	Route A	Route B
Daily Revenue (Rp)	42,976,500	44,569,000
Annual Revenue (Rp)	15,686,422,500	16,267,685,000

As shown in table 5 the operational cost for route B is higher, but in contrary its potential revenue as shown in table 6 is higher than of route A. This situation resulted to a further analysis for selecting the best route option by utilizing an incremental benefit-cost ratio analysis for the base year operation.

In this research, the benefit is simply taken from the potential revenue and the cost is from the estimated operational cost. Therefore, by applying equation (7) as in section 2, the B/C ratio for Route A and Route B are 2.9 and 2.7 respectively. While the incremental B/C ratio for these two Route is 1.3

Taking a look solely from each B/C ratio value, seems that route A has a better performance, but having compared the B/C ratio value for both alternatives where the route A is taken as the reference, the route B finally shows better result. This is due to the difference of their potential revenue, while their operational cost is only slightly differed. Therefore, based on this analysis the result is confirmed with the results of previous work [1] which compares the performance indicator for each alternative route.

A further financial analysis by taking into account the assumed concession period (i.e. 7 yrs.), a consistency of previous analysis and sustainability of the operation of Route B could be checked. Therefore, for this purpose, assumptions are made for the following; annual passenger growth is assumed 3% [7]. Operational Cost is assumed increased 5% annually based on the ideal inflation rate.

Since there is no data for the elasticity of price to the demand, the fare is kept maintained as the basic year. In this analysis all forecasted value will be converted to the present value (PV), so by applying equation (8) as in section 2, the result is shown in table 7. Using similar approach as in the base year analysis, the B/C ratio for Route A and Route B are 2.13 and 2.07 respectively. While the incremental B/C ratio for these two Route is 1.2.

Table 7. Present Value Analysis

PV Benefit (Rp)		*PV Cost (Rp)*	
Route A	**Route B**	**Route A**	**Route B**
76,367,779,299	79,197,597,654	35,823,527,766	38,252,741,396

Based on the calculated B/C ratio, the results of financial analysis with 7 (seven) years concession period and some assumptions, the analysis indicates a consistence results with that of base year and furthermore a sustainable operation seems can be achieved.

4. Conclusion

The analysis consistently shows, as compared to previous work result, that the route B potentially provides a better performance. It is strongly recommended that, in complementary with the current operation, the Government of Tangerang City operates a bus lane from Cikokol to Kali Deres (i.e. Route B).

References

[1] Alvinsyah and Halim U 2012 The Impact of Routing Option on Bus Lane Corridor (Case Study: Tangerang Bus Lane) *Journal of FSTPT (University Transport Study Forum)* vol 12 pp 63-72

[2] Wright L and Hook W 2007 Bus Rapid Transit Planning Guide *Institute for Transport Development & Policy* New York

[3] Transportation Research Board 2007 *Bus Rapid Transit Practitioner's Guide* TCRP Report 118 Washington D.C

[4] DGLT 2002 *Guidelines for Urban Bus Transport with fixed route* Ministry of Transportation Republic of Indonesia

[5] De Garmo E P and Sullivan W G 1996 *Engineering Economy: Instructor's Manual (10th ed)* (New Jersey: Prentice-Hall)

[6] Permenhub 2012 *Minister Decree Number 10 on Minimum Level of Service for Road Base Mass Transit* Ministry of Transportation Republic of Indonesia

[7] Dishub Tangerang 2009 *A Study on Management & Operational Plan for Public Transport Service in Tangerang City* Dinas Perhubungan Tangerang Final Report

IOP Conf. Series: Materials Science and Engineering **403** (2018) 012022 doi:10.1088/1757-899X/403/1/012022

Seismic ground motion induced lateral earth pressures on basement walls

R Berangket and W A Prakoso

Civil Engineering Department, Universitas Indonesia, Depok

berangketr@gmail.com

Abstract. Research on the dynamic lateral earth pressures acting on basement walls induced by seismic ground motions has appeared in the literature in the last few years to examine the apparent inconsistency between the theoretical pressures and the pressures suggested by basement wall good performance of during large earthquakes. This inconsistency suggested that the current methods would yield excessive lateral earth pressures. In this paper, a two dimensional plane strain finite element model of a 2-story concrete basement wall was used to examine the dynamic lateral earth pressures. The seismic ground motion was the N-S component of the 1992 Erzincan earthquake motion record. The assumed soil profile was a sands deposit overlying a reflecting bedrock layer. The observed time history outputs were the wall acceleration in horizontal direction, the lateral earth pressures, and the associated lateral thrust. The outputs were examined further to evaluate the amplification of wall acceleration, the different time history pattern of wall acceleration and lateral thrust, the influence of limiting soil tensile strength, the time difference for peak wall acceleration and peak lateral thrust, and a discussion on phase difference. The results suggested that the complex interaction could not be represented by typical simple peak input acceleration – soil pressure relationships.

1. Introduction

Basement walls are to be designed against the increase in the lateral earth pressures induced by seismic ground motions, as required by the Indonesia seismic building code [1]. Although the code does not require any specific estimation methods, many design estimation methods can be found in the literature (e.g., [2]). It is noted that the typical estimation methods are based on relatively simple peak input acceleration – soil pressure relationships. At the international level in recent years (e.g., [2-6]), basement walls appeared to perform well during relatively larger-than-expected earthquakes. This observation has raised a major concern, not about the actual good performance, but about the apparent over-designed nature of the current estimation methods. This over-designed nature might consequently lead to unnecessary construction costs.

Sitar and his colleagues (e.g., [2, 4]) have shown through a series of laboratory tests and numerical studies that the actual lateral earth pressures induced by seismic ground motions were not as high as those estimated by the current estimation methods. Prakoso and his colleagues have built upon this suggestion by conducting a series of numerical analyses. Winner and Prakoso [5] found that, for basement structures without embedded walls, the peak value and frequency of the input acceleration would affect the earth pressures and wall bending moments. The limiting soil tensile strength provided by the soil cohesion would also affect the lateral pressures, but it would not have a significant effect on the wall bending moments. Prakoso *et al.* [6] found that the increase in horizontal wall acceleration and wall pressures was not linearly proportional to the increase in the peak input acceleration, and the embedment wall depths had a minor effect on the wall acceleration and wall pressures. It is noted that [5, 6] employed harmonic ground motions.

The paper objective is to examine numerically further the increase in dynamic lateral earth pressures acting on basement walls due to an actual earthquake ground motion record. This paper is an extension of [6]; the basement considered is a 2-level, 8 m deep basement without any pile foundations typically constructed in Jakarta area. The finite element model used and the case evaluated are described in the beginning, while the horizontal wall acceleration and lateral earth pressures are discussed in the latter part of the paper.

2. Research method

This research assumed that a concrete basement with dimensions of 8 m deep and 18 m wide was constructed in a 25 m thick sands deposit ($\phi = 35°$) overlying a reflecting engineering bedrock. The thickness of the basement concrete wall was 0.6 m, while the embedment depth was 6 m (total wall length 14 m). Two-dimensional, finite element models were used to examine the increased lateral earth pressures as shown as figure 1. The model used fifteen-node triangular finite elements for the soils and concrete walls; the total number of elements was 1,018 elements. To better capture the increased earth pressures and wall bending moments, finer elements were used for the soil directly behind the walls and the basement walls. Beam elements were used to model the basement structural elements (e.g., columns and beams). The Mohr-Coulomb constitutive soil model was used for all the soil elements, while the elastic model was used for basement wall elements and structural elements; details of the soil, wall, and structural element properties were given in [6]. Absorbent boundaries were specified on the both sides of the model to avoid spurious reflection during dynamic analyses, and the Rayleigh damping parameters adopted from [7] was used. The software used was Plaxis 2-D version 8 [8].

The input seismic motion used for this paper was the N-S component of the 1992 Erzincan earthquake motion record [9]. The considered record had a peak ground acceleration of 0.387 g and was 20 second long. The acceleration was applied using the prescribed displacement option at the base of the mesh representing the reflecting engineering bedrock. The number of dynamic calculation steps was 1,000 steps (Δt = 0.02 s). The dynamic calculation was performed after the static construction phases (excavation of basement, installation of wall and structure). The observed outputs include the lateral earth pressure behind wall and the wall acceleration.

Static and dynamic model verification results were reported in [6]. The static lateral earth pressures behind the basement wall at the end of construction stage were examined. In general, the pressures were about the same as the Rankine active pressures in the upper wall part, then increased to about the same as the at-rest pressure, and subsequently decreased to the Rankine active pressure at the wall tip level. The observed dynamic wall acceleration was compared to the peak input acceleration directly; for lower acceleration values, there was amplification of the ground motions, but for higher values, there was de-amplification of the ground motions, similar to the general trend shown in [10]. It is noted that a related dynamic model verification for a soil retaining system has been given in [11].

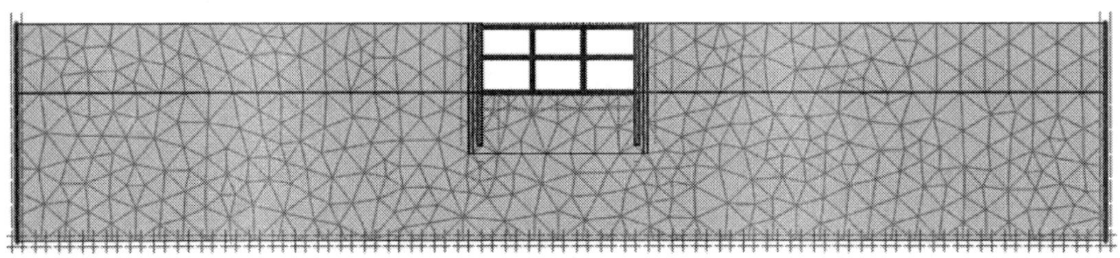

Figure 1. Finite element mesh.

3. Results

3.1. Horizontal wall acceleration

The time history of the basement wall acceleration in horizontal direction is shown in figure 2. The time history was monitored at three locations: 0.67 m (about top floor slab level), 4.0 m (middle floor slab level), and 8.0 m (bottom floor slab level). The peak wall acceleration values at these locations (> 0.6 g) were greater than the peak input acceleration (0.387 g). The amplification factors could subsequently be evaluated by calculating the ratio of peak wall acceleration to peak input acceleration; the factors for the three locations were 1.899, 1.658, and 1.570, respectively.

It can be observed clearly that there was a phase difference between the peaks of horizontal wall acceleration as shown in figure 2 (right) for the time window of 2.5 to 5.0 second. For the main peak (t = 3.1 – 3.5 s), the peaks at different depths occurred in slightly different patterns. However, for the subsequent local peaks, the peak at depth of 8.0 m (deepest monitoring point) occurred first, followed subsequently by that at depth of 4.0 m and 0.67 m (shallowest monitoring point). These different observed patterns suggest that, although it was of relatively high stiffness (t = 0.6 m, supported by 3 levels of floor slabs), the basement wall showed rather flexible behavior.

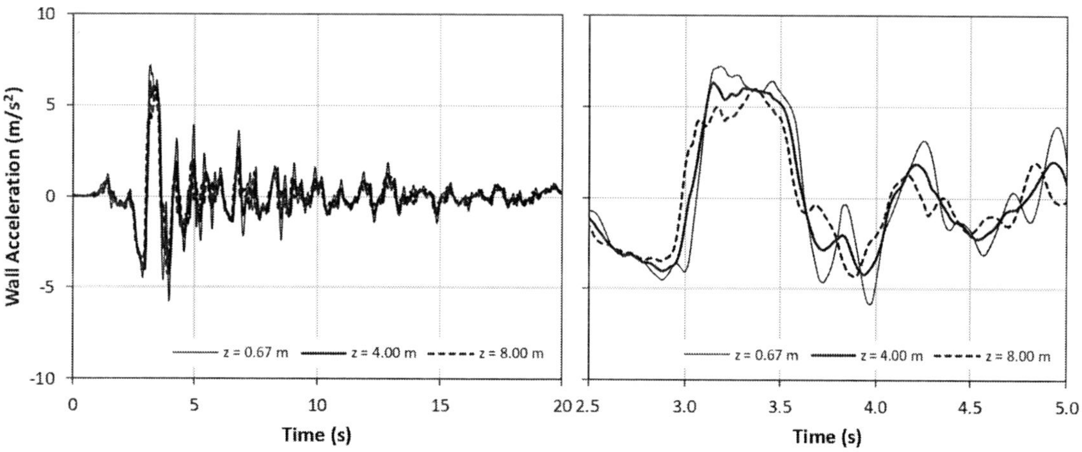

Figure 2. Horizontal wall acceleration at different depths.

3.2. Lateral earth pressures and thrust

The time history of the resulting dynamic lateral earth pressures (p_{AE}) acting on the basement wall is shown in figure 3. The time history of change in lateral earth pressures (= dynamic lateral earth pressure – static lateral earth pressure, Δp_{AE}) is also calculated and shown in figure 4. Note that a positive pressure is a compressive pressure. The earth pressures were monitored at five depths: 0.85 m (about ground surface level), 1.8 m (about midway between top and middle floor slabs), 4.0 m (middle slab level), 6.2 m (about midway between middle and bottom floor slabs), and 7.9 m (slightly above bottom floor slab). In all observation points, p_{AE} remained in compression due to limited soil tensile strength, consistent with observations reported in [5]. The p_{AE} and Δp_{AE} initially increased with an increase in the input acceleration. However, after the time window of the main peak (t = 3.1 – 3.5 s), p_{AE} and Δp_{AE} only decreased slightly, although the input acceleration was relatively small. Similar behavior was observed in centrifuge tests as reported in [12].

It can be observed clearly that there was a phase difference between the peaks of Δp_{AE} as shown in figure 4 (right) for the time window of 2.5 to 5.0 second. The peaks of Δp_{AE} for depths = 0.8 m and 1.8 m occurred earlier at dynamic time t of about 3.1 s, though at slightly different times. The Δp_{AE} subsequently decreased and then oscillated slightly (average of Δp_{AE}(t=8-20s) = 38.6 kPa and 19.5 kPa, respectively). The peaks of Δp_{AE} for depths = 4.0 m, 6.2 m, and 7.9 m occurred latter at t = 3.8 – 4.0 s, though again at slightly different times. The Δp_{AE} subsequently gradually decreased and then oscillated slightly (average of Δp_{AE}(t=8-20s) = 86.2 kPa, 98.3 kPa, and 110.9 kPa, respectively). Furthermore, at times, the local peaks for the shallow observed points occurred when Δp_{AE} for the deeper points reached their local minima, and vice versa. This different times for peaks at different depths suggest that the use of pressure envelope might lead to an excessively conservative design.

The Δp_{AE} can be integrated to obtain the dynamic lateral thrusts (P_{AE}), and figure 5 shows the time history of P_{AE}. The general trend of P_{AE} was similar that of Δp_{AE}; P_{AE} reached its peak (803 kN) at t about 3.8 – 4.0 s, then gradually decreased, and then oscillated slightly (average of P_{AE}(t=8-20s) = 581 kN.

It is of interest to relate the wall acceleration time history to Δp_{AE} and P_{AE}. It was observed that the time history of the wall acceleration had a different pattern compared to that of Δp_{AE} and P_{AE}. Furthermore, the main peak of wall acceleration occurred at t = 3.1 – 3.5 s. The main peaks of Δp_{AE} for shallow depths occurred at t about 3.1 s, while those for deeper depths occurred at t = 3.8 – 4.0 s. The main peaks of P_{AE} occurred at similar time as Δp_{AE} for deeper depths. All these observations suggest a complex dynamic soil-basement interaction which could not be represented by a simple peak input acceleration – soil pressure relationship as found in typical design approaches.

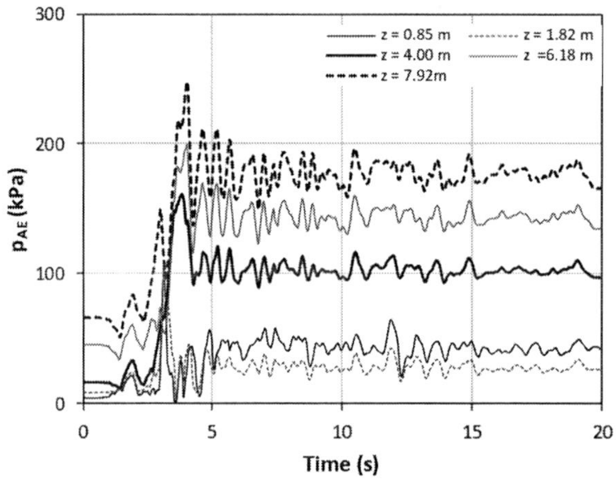

Figure 3. Lateral earth pressure acting on basement wall.

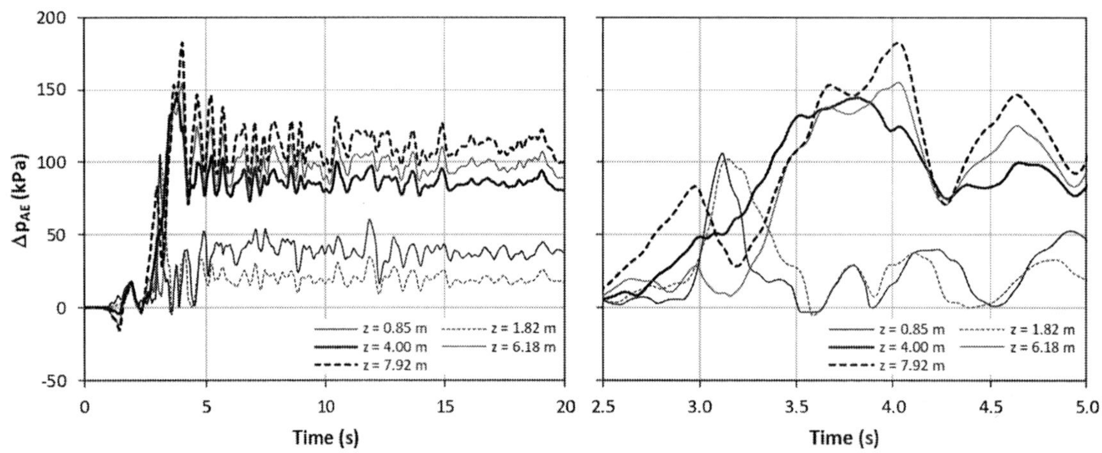

Figure 4. Change in lateral earth pressure on basement wall.

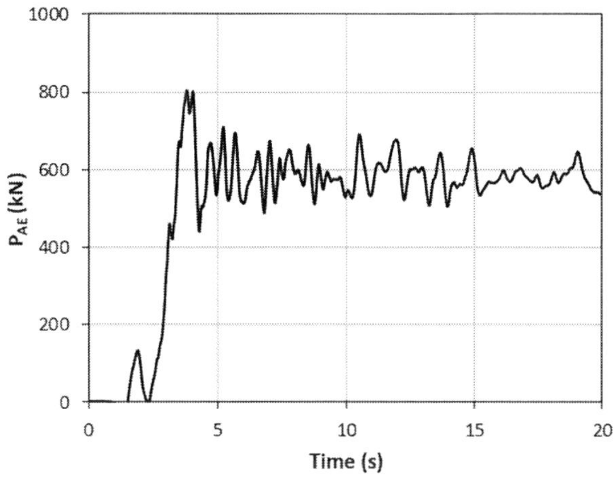

Figure 5. Lateral thrust on basement wall.

4. Conclusions

A two dimensional plane strain dynamic finite element model was used to evaluate the wall acceleration and the lateral earth pressures induced by the N-S component of the 1992 Erzincan earthquake motion record. The basement was a 2-story concrete basement. The basement was embedded in a sands deposit overlying an engineering bedrock layer. The highlighted observations are as follows:

- The amplification of horizontal acceleration was observed at different wall depths, and the amplification factors tend to decrease with greater wall depths.
- The time history of wall acceleration had a different general pattern compared to that of Δp_{AE} and P_{AE}.
- A phase difference was observed in horizontal wall acceleration and Δp_{AE}. The basement wall behavior was somewhat flexible, although it was of relatively high stiffness.
- The Δp_{AE} is affected by the limiting soil tensile strength.
- The time of the peak wall acceleration was different from that of the peak P_{AE}.

The above observations suggested that a complex dynamic soil-basement interaction occurred and it apparently could not be represented by a simple peak input acceleration – soil pressure relationship as found in typical design approaches.

Acknowledgment

The authors gratefully acknowledge the partial support provided by Universitas Indonesia through the 2017 PITTA research grant.

References

[1] Badan Standarisasi Nasional 2012 Tata cara perencanaan ketahanan gempa untuk struktur bangunan gedung dan non gedung (Indonesia seismic resistance building code) *SNI 1726* Indonesia

[2] Lew M, Sitar N, Al-Atik L, Pourzanjani M and Hudson M B 2010 Seismic earth pressures on deep building basements *Proc., 79th Annual Convention, Structural Engineers Association of California,* Indian Wells, CA

[3] Taiebat M, Amirzehni E and Finn W D L 2014 Seismic design of basement walls: evaluation of current practice in British Columbia *Canadian Geotech. J.* vol 51 no 9 pp 1004-20

[4] Mikola R G, Candia G and Sitar N 2016 Seismic earth pressures on retaining structures and basement walls in cohesionless soils *J. Geotech. Geoenviro. Engrg* vol 142 no 10

[5] Winner A B and Prakoso W A 2016 Behavior of basement wall subjected to synthetic harmonic

ground motions *The 3rd International Conference on Earthquake Engineering and Disaster Mitigation* Yogyakarta Indonesia

[6] Prakoso W A, Berangket R and Yulianda F 2017 *Proc. Intl. Conf. Earth Sciences and Engrg*

[7] Guler E, Cicek E, Demirkan M M and Hamderi M 2012 Numerical analysis of reinforced soil walls with granular and *cohesive* backfills under cyclic loads *Bull. Earthquake Engrg* vol 10 issue 3 pp 793-811

[8] Brinkgreve R B J 2002 *Plaxis: finite element code for soil and rock analyses: 2D – Version 8* Balkema

[9] PEER 2013 *Pacific Earthquake Engineering Research Center strong motion database*

[10] Athanasopoulos-Zekkos A, Vlachakis V S and Athanasopoulos G A 2013 Phasing issues in the seismic response of yielding, gravity-type earth retaining walls–Overview and results from a FEM study *Soil Dyn. Earthquake Engrg* vol 55 pp 59-70

[11] Prakoso W A and Kurniadi H 2015 Seismic Amplification of Double-Sided Geosynthetic-Reinforced Soil Retaining Walls *Proc. 6th Intl. Conf. Earthquake Geotech. Engrg* Christchurch, New Zaeland

[12] Al-Atik L and Sitar N 2010 Seismic earth pressures on cantilever retaining structures *J. Geotech. Geoenviro. Engrg* vol 136 no 10 pp 1324-1333

Resilient modulus of porous asphalt using oil palm fiber

T Syammaun and H A Rani

Faculty of Engineering, University Muhammadiyah Aceh

tamalkhani27@gmail.com

Abstract. In recent years, the use of waste fibers has been developing in many ways to increase the performance properties of asphalt mixture. Efforts to create a green technology which is more environmentally friendly that can produce economic value is also a consideration in the utilization of waste materials. The research is aimed to evaluate the resilient modulus of porous asphalt using oil palm fiber (OPF). Five different percentages of fiber (1%, 2%, 3%, 4% and 5%) were used by weight of bitumen. Several lab tests, such as characteristics, asphalts characteristics, and resilient modulus were carried out to compare and quantify performance of OPF-modified porous asphalt. In general, the results of experiments show that oil palm fiber addition to porous asphalt gives significant effect for its properties.

1. Introduction

Porous asphalt (PA) was developed in the early 50s in Europe and the United States. Usually, it is used for enhancing traffic and comforting drivers and occupants in the street of the region [1]. The utilization of PA in the top layers resulted on 20% air void content and interconnected micro voids are created due to higher proportions of coarse aggregates and lower sand content, which is in the wet weather cause the water depletes through a series of micro conduits [2].

PA also contains elimination of glare, splash and spray, improves visibility, increases skid resistance of the road surface [3, 4]. However, PA may experience problems which can affect its performance and service life due to the microstructure of PA. The open structure uncovered a large binder surface area to the oxidative impact of air and the damaging effect of water resulting from increased interfacial moisture content, leading to rapid aging of the binder, moisture damage of the bitumen aggregate bond and structural distress of the compounds [2].

Currently, many researchers have been used various types of fiber in asphalt mixtures, such as polyester fiber, cellulous fiber, glass fiber and mineral fiber to improve performances of pavements [5, 6]. Asphalt leakage during the material transportation and paving can be prevented through fiber stabilizing especially for the open-graded-friction-course (OGFC) and stone–mastic–asphalt (SMA) mixtures [7, 8]. The rheological properties of asphalt binder and the required optimum asphalt content for the mixture design are changed by the absorption of fiber in asphalt components, which is the important role in the formation of the interface bonding between fiber and asphalt [9].

Oil palm fiber (OPF) is squeezed out from empty fruit bunch fiber (EFB) [10]. The ability of the OPF in preventing the drain down of binders while improving the stability of the mix should lead to improving the stability of the mix which should lead to improved field performance [11]. Malaysia is one of major country in oil palm production which started about more than 70 years ago. It is reported by Palm Oil Board that there is 7.0 million tons of oil palm trunks, 26.2 million tons of oil palm fronds and 23% of Empty Fruit Bunch (EFB) per ton of Fresh Fruit Bunch (FFB) processed in oil palm mill. Based on these facts there will be ample amount of OPF available for use in suitable applications. However, there is not much study about use of OPF in asphalt road pavement. Therefore, it would be important to conduct an experimental laboratory study to investigate the use of this fiber in PA mixtures according to Malaysia condition.

The aim of this paper is to evaluate the effects of oil palm fiber (OPF) on some properties of porous asphalt (PA). Porous asphalt Grading-B using five different percentages of fiber were evaluated using laboratory tests. The performance of PA with respect to OPF additives were analyzed and evaluated using standard statistical techniques such as regression analysis and ANOVA.

2. Experimental program

2.1 Materials

Aggregates were employed from local crushed granite aggregate. The hydrated lime was added as filler. The aggregates were sieved into the selected size range according to the Malaysian Public Works Department for porous asphalt (PA-Grading B) as shown in table 1. The results of physical properties of aggregates tested are given in table 2. The conventional bitumen penetration-grade 80/100 was used. The physical properties of asphalt are presented in table 3. Oil palm fiber (OPF) produced by the palm oil industry in Malaysia is illustrated in figure 1 and the interior view of OPF at magnification 1.00KX is shown in figure 2. Table 4 illustrates the physical properties of OPF.

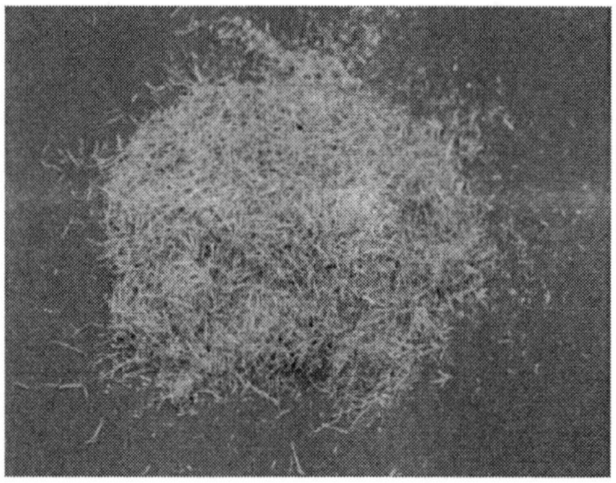

Figure 1. The oil palm fiber has to be washed to remove other materials before being sieved through a 425 μm sieve.

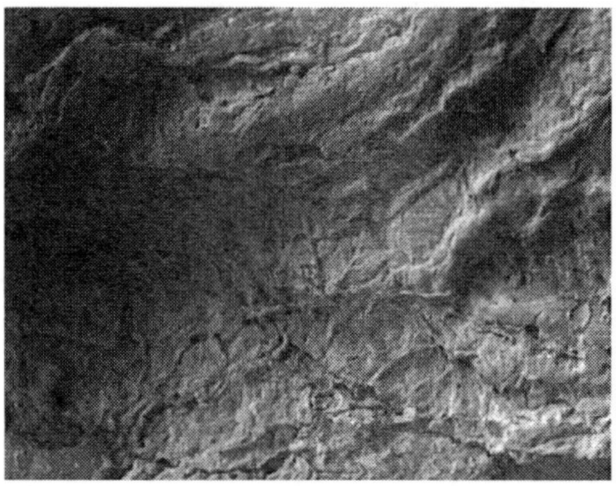

Figure 2. Interior View of OPF at Magnification 1.00KX.

Table 1. Gradation of porous asphalt (Grading-B) based on the Malaysian standard.

Sieve size (mm)	Specification passing range (%)	Used passing (%)
20.000	100	100.0
14.000	85-100	92.5
10.000	55-75	65.0
5.000	10-25	17.5
2.360	5-10	7.5
0.075	2-4	3.0

Table 2. Physical properties of crushed granite aggregate.

Properties	Test method	Test Value (%)
Aggregate crushing value	BS812:Part3	20.0
Elongation index	BS182:Part3	16.0
Flakiness index	BS182:Part3	4.0
Impact value	BS812:Part3	20.0
Los Angeles abrasion value	ASTM C-131	21.0
Soundness	BS812:Part3	4.1

Table 3. Physical properties of asphalt binder.

Properties	Test method	Unit	Test Value (%)
Ductility at 25°C	ASTM D-113	cm	>100
Penetration at 25°C	ASTM D 5-97	1/10 mm	84
Softening point	ASTM D-92	°C	46

Table 4. Physical properties of OPF.

Properties	Unit	Range
Density	g/cm³	0.7-1.55
Diameter	μm	150-500
Tensile strength	MPa	50-400

2.2 Specimen preparation

The oil palm fiber was washed to remove other materials and sieved through a 500 μm sieve. The binder contents (BC) used in the mix process were 4%, 4.5%, 5%, 5.5%, and 6% by weight of aggregate. Different percentages of fibers such as 1%, 2%, 3%, 4% and 5% by weight of bitumen were used to compare with control specimens (no fibers). Aggregates were mixed with asphalt binder at 160±5°C before adding fiber. The specimens were prepared using the standard Marshall hammer and tested to evaluate resilient modulus.

2.3 Test Method

Indirect tensile modulus test parameters are tabulated in table 4. According to the American Society Testing and Materials (ASTM) D4123-82, the Indirect Modulus test was carried out using the Universal Material Testing Apparatus (UMMATA) in order to apply compressive loads with acts parallel and along the vertical diametrical plane by two curves loading strips at 25°C. Following this, the information was sent to a computer program for data attainment. Horizontal tensile stress modulus of asphalt concrete mixtures can be obtained by equations (1).

$$\sigma_x\,(\max) = \frac{2\,x\,P}{\pi\,x\,d\,x\,t} \quad (1)$$

Where σ_x (max) is the maximum horizontal tensile stress in middle of specimen; P, applied vertical peak load, d, average diameter of specimen and t; average thickness of specimen.

3. Results and discussion

Resilient modulus is an important parameter to analysis and determines the performance of pavement response to traffic loading [12]. The indirect tensile test with repeated loading at 25°C was used to determine the stiffness modulus of porous asphalt.

3.1 Effect of different OPF content on resilient modulus

The results for different contents of OPF on the resilient modulus are shown in figure 3 and the result of resilient modulus value with the different binder contents is shown in figure 4. For all particular binder contents, there appeared to be a maximum resilient modulus value as the OPF content increased from 0 to 5%. As expected, for any value of OPF content the resilient modulus value decreased as the binder content increased. This implied that the higher amount of binder at any particular OPF content would result in a more pronounced viscous component of the binder when compared to the elastic component, hence reducing the resilient modulus value.

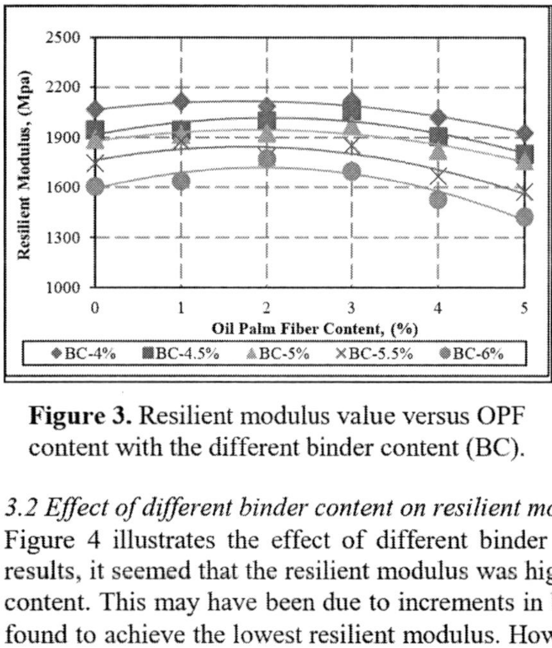

Figure 3. Resilient modulus value versus OPF content with the different binder content (BC).

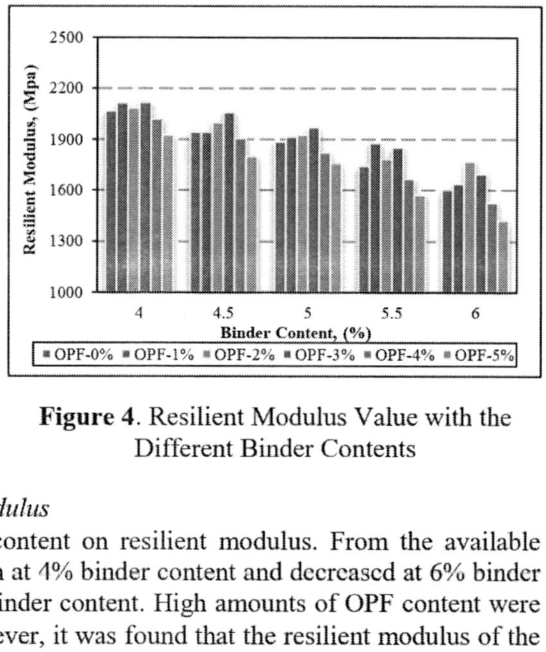

Figure 4. Resilient Modulus Value with the Different Binder Contents

3.2 Effect of different binder content on resilient modulus

Figure 4 illustrates the effect of different binder content on resilient modulus. From the available results, it seemed that the resilient modulus was high at 4% binder content and decreased at 6% binder content. This may have been due to increments in binder content. High amounts of OPF content were found to achieve the lowest resilient modulus. However, it was found that the resilient modulus of the mix with certain OPF contents had an inconsistent value. This indicated that the damage accumulation of the specimen is not a linear process.

3.3 Resilient Modulus at Optimum Binder Content

In this study, the optimum binder content was determined and selected as 5% for the oil palm fiber contents of 0%, 1%, 2%, 3%, 4% and 5%. The results of the resilient modulus at optimum binder contents are summarized in figure 4. As shown in figure 4, the results of the resilient modulus at the optimum binder content indicated that the mixes are generally higher than the specimen without fiber (0% OPF content) except for specimens made with 4% and 5% OPF content, with values of 1886Mpa, 1917Mpa, 1926Mpa, 1971Mpa, 1822Mpa, and 1761Mpa for fiber content 0%, 1%, 2%, 3%, 4% and 5% respectively.

4. Statistical analysis

Analysis of variance (ANOVA) was used as the statistical tool to evaluate the effect of OPF content on performance characteristics of porous asphalt. Two-factor ANOVA without replication and single factor ANOVA were adopted to evaluate the results of laboratory tests with confidence level 95% ($\alpha = 0.05$). Table 5 presents the ANOVA result on the effect of fiber content on certain parameters of the porous asphalt. For the resilient modulus, the analysis gave an F-value of 122.93 (F=122.93 > F-critical= 2.866) and a P-value smaller than 0.05. As such, the addition of OPF significantly affects the properties and characteristics of the porous asphalt mixes.

Table 5. Results of ANOVA: Two-Factor without replication.

Source of Variation	SS	df	MS	F	P-value	F crit
Resilient modulus						
Binder content	713047	4	178262	122.93	8.82412E-14	2.86608
OPF content	207898	5	41579.6	28.6733	1.79741E-08	2.71089
Error	29002.3	20	1450.11			
Total	949947	29				

5. Conclusion

Based on this preliminary laboratory study, a number of observations and conclusions may be drawn. The resilient modulus value decreased as the binder content increased for any value of OPF content. This implied that a higher amount of binder at any particular OPF content would result in a more pronounced viscous component of the binder when compared to the elastic component, hence reducing the resilient modulus value. The results showed that inconsistent value of some relationships between OPF contents and binder contents indicated that the damage accumulation of specimens was not a linear process. From the results of regression analysis and analysis of variance (ANOVA), it was evident that the addition of oil palm fiber-modified significantly affected the properties and characteristics of the porous asphalt mixes.

References

[1] Takahashi S, Poulikakus L D and Partl M N 2003 Evaluation of improved porous asphalt by various test methods *Proceedings of the 6th International RILEM symposium, Performance Testing and Evaluation of Bituminous Material* Zurich pp 230–236

[2] Poulikakos L D and Partl M N 2009 Evaluation of moisture susceptibility of porous asphalt concrete using water submersion fatigue tests *Construct Build Mater* vol 23 no 12 pp 3475-84

[3] Cooley Jr L A, Brown E R and Watson D E 2000 Evaluation of OGFC mixtures containing cellulose fibers *Transportation Research Record 1723* Transportation Research Board, national Research Council

[4] Moore L M, Hicks R G and Rogge D F 2001 Design, construction and maintenance guidelines for porous asphalt pavements *Transportat. Res. Rec.: J. Transportat. Res. Board.* vol 1778 pp 91-9.

[5] Chen H, Xu Q, Chen S and Zhang Z 2009 Evaluation and design of fiber-reinforced asphalt mixtures *Mater Design* vol 30 no 7 pp 2595-2603

[6] Moghaddam T B 2013 Dynamic properties of stone mastic asphalt mixtures containing waste plastic bottles *Construct Build Mater* vol 34 pp 236-242

[7] Hasan H F, Oraimi S A and Taha A 2005 Evaluation of open-graded friction course mixtures containing cellulose fibers and styrene butadiene rubber polymer *J Mater Civil Eng* vol 17 pp 416–22

[8] Sertass J P and Samanos J 1996 Fiber-modified asphalt concrete characteristics, applications and behavior *J Assoc Asphalt Paving Technol* vol 64 pp 193–230

[9] Chen H and Xu Q 2010 Experimental study of fibers in stabilizing and reinforcing asphalt binder *Fuel* vol 89 no 7 pp 1616-22

[10] Shinoj S, Visvanathan R, Panigrahi S and Kochubabu M 2011 Oil palm fiber (OPF) and its composites: A review *Indus Crops Prod* vol 33 no 1 pp 7-22

[11] Muniandy R, Ishak S B and Radin U R S 1999 Study on the use of oil palm fiber in rubberized stone mastic *REAAA Journal* vol 7 no 1

[12] Sangita, Khan T A, Sabina and Sharma D K 2011 Effect of waste polymer modifier on the properties of bituminous concrete mixes *Construct Build Mater* vol 25 no 10 pp 3841-8

IOP Conf. Series: Materials Science and Engineering **403** (2018) 012024 doi:10.1088/1757-899X/403/1/012024

Evaluation of thin square bending plate using IGA Galerkin

I Katili, K Octavianus, I J Maknun and J I Rastandi

Civil Engineering Department, Universitas Indonesia, Depok 16424, Indonesia

irwan.katili@gmail.com

Abstract. This paper presents an evaluation of thin bending plate problems using IGA Galerkin. It focuses on square plates under uniform load with different boundary conditions so that B-Spline is sufficient to generate basis functions. To demonstrate the performance of IGA Galerkin formulation, two numerical tests are presented.

1. Introduction

Researchers in worldwide have developed many kinds of numerical methods for structural engineering application. One of these numerical methods, Finite Element Analysis (FEA) has become the most popular and powerful numerical method that is widely used in many engineering and scientific disciplines. FEA now has been implemented in many commercial programs such as Computer Aided Engineering (CAE) for structural analysis. In FEA polynomial function is adopted as shape function to build discretization of structure. The results of our research in Finite Element Method have been published in [1-13].

In 2005, Hughes [14] introduced a new method called Isogeometric Analysis (IGA). Isogeometric Analysis (IGA) is a new development of FEA but it is not based on polynomial function as FEA. It uses Non-Uniform Rational B-Splines (called NURBS) as basis function. NURBS is a technology which has been used in Computer Aided Design (CAD) system to approximate the geometric shape with high accuracy. With the idea of using the same basis function NURBS in CAD and IGA, hopefully this can simplify mesh refinement. IGA also has flexibility in increasing of element number and polynomial degree independently, which FEA doesn't have.

Many researches about IGA and its applications have been developed, especially for beam, plate and shell. For plate bending problem, Kiendl [15] developed the solution of Reissner-Mindlin plate problem using Collocation Method in IGA, and Reali [16] developed the solution of Kirchhoff plate problem, also using Collocation Method in IGA. In 2017, the development of Isogeometric Galerkin based on UI approach have been proposed by Katili [17].

This paper will discuss evaluation of thin square bending plate problem under uniform load with different boundary conditions using IGA. The presentation will focus on square plate, and the problems will be solved by using Galerkin Method. Numerical test will be conducted to see the performance of IGA formulation for thin plate with different boundary conditions. The reference solution from [18] and [19], will be used to validate the output of this numerical test.

2. IGA discretization for rectangular thin plate

As described in previous section, the idea of Isogeometric Analysis (IGA) is to implement NURBS function as the replacement of shape function used in FEA. Equation (1) is NURBS equation in two dimension. For rectangular plate problem, $\omega_{i,j}=1$, so $\sum_{\hat{i}=1}^{n}\sum_{\hat{j}=1}^{m} N_{\hat{i},p}(\xi)M_{\hat{j},q}(\eta)\omega_{\hat{i},\hat{j}}=1$ [20]. Regarding this simplification, for rectangular plate problem, NURBS is as simple as B-Spline, which is the tensor product of two basis functions in one dimension, $N_{i,p}(\xi)$ and $M_{j,q}(\eta)$.

$$R_{i,j}^{p,q}(\xi,\eta)=\frac{N_{i,p}(\xi)M_{j,q}(\eta)\omega_{i,j}}{\sum_{\hat{i}=1}^{n}\sum_{\hat{j}=1}^{m} N_{\hat{i},p}(\xi)M_{\hat{j},q}(\eta)\omega_{\hat{i},\hat{j}}}=N_{i,p}(\xi)M_{j,q}(\eta) \tag{1}$$

B-Spline in one dimension is defined in equation (2), where ξ_i is knot-i.

$$N_{i,0}(\xi)=\begin{cases}1, \text{ if } \xi_i \le \xi < \xi_{i+1}\\ 0, \text{ otherwise}\end{cases} \text{ with } N_{i,p}(\xi)=\frac{\xi-\xi_i}{\xi_{i+p}-\xi_i}N_{i,p-1}(\xi)+\frac{\xi_{i+p+1}-\xi}{\xi_{i+p+1}-\xi_{i+1}}N_{i+1,p-1}(\xi) \tag{2}$$

As can be seen, knot is a component to build B-Spline. Knot depends on polynomial degree and number of element. Knot value is in parametric space $[0,1]$. Polynomial degree will give multiplicity to knot value 0 and 1, which amounts $p+1$. Number of element depends on the knot span number between the interval 0 and 1. The set of knot is called knot vector (Ξ) expressed in equation (3).

$$\Xi_\xi=\left[\xi_1,\xi_2,\cdots,\xi_{n+p+1}\right] \tag{3}$$

where $n+p+1$ is number of knot in knot vector, and n is number of basis function, which is determined after number of knot is known. Further explanation about knot vector and basis function can be seen in references. Analogous with shape function and degree of freedom $\{u_n\}$ in FEA are basis function and control variable in IGA. In IGA for two dimensional problems, displacement function can be expressed in equation (4), where $\langle R(\xi,\eta)\rangle$ is basis function and $\{\hat{w}_n\}$ is control variable.

$$w(\xi,\eta)=\langle R(\xi,\eta)\rangle\{\hat{w}_n\} \tag{4}$$

Control variable (CV) is not a real displacement as degree of freedom in FEA. Control variable number is equal to basis function number in parametric space. Example is given for $p=q=3$ and NELT $=2\times2$, which gives $r=m\times n=25$ basis functions, each element will have the same 25 control variables, $\hat{w}_1,\hat{w}_2,\hat{w}_3,\ldots\ldots\hat{w}_{25}$, as can be seen in figure 1. The transverse shear deformation is neglected in Kirchhoff plate theory, where $\gamma_x=\gamma_y=0$. Then, the vertical displacement will have a straight relation with rotation with $\beta_x=-w,_x$ and $\beta_y=-w,_y$. Based on equation (4), curvature $\{\chi\}$ can be discretized as equation (5).

$$\begin{Bmatrix}\chi_x(\xi,\eta)\\ \chi_y(\xi,\eta)\\ \chi_{xy}(\xi,\eta)\end{Bmatrix}=-\begin{Bmatrix}w(\xi,\eta),_{xx}\\ w(\xi,\eta),_{yy}\\ 2w(\xi,\eta),_{xy}\end{Bmatrix}=\begin{Bmatrix}-\sum_{i=1}^{r}R_i,_{xx}(\xi,\eta)\hat{w}_i\\ -\sum_{i=1}^{r}R_i,_{yy}(\xi,\eta)\hat{w}_i\\ -2\sum_{i=1}^{r}R_i,_{xy}(\xi,\eta)\hat{w}_i\end{Bmatrix} \tag{5}$$

And then, curvature can be expressed in bending strain matrix $\left[B_b\right]$ in equation (6).

$$\{\chi(\xi,\eta)\}=[B_b]\{\hat{w}_n\} \text{ with } [B_b]_{3\times r}=-\begin{bmatrix}R_1,_{xx} & R_2,_{xx} & \cdots & R_r,_{xx}\\ R_1,_{yy} & R_2,_{yy} & \cdots & R_r,_{yy}\\ 2R_1,_{xy} & 2R_2,_{xy} & \cdots & 2R_r,_{xy}\end{bmatrix}_{3\times r} \tag{6}$$

Galerkin Method adopted to solve the problem in this paper is developed by the principle of virtual work. Equation (7) represents the virtual work, which is based only on bending strain energy and distributed load as external load.

$$\int_A \langle\chi\rangle[H_b]\{\chi^*\}dA = \int_A w^* f_z dA \tag{7}$$

Applying the discretization in weak form equation (Galerkin) will yield stiffness matrix equation as shown in equation (8).

$$[k]\{\hat{w}_n\} = \{f_n\} \quad \text{with} \quad [k] = \int_A [B_b]^T [H_b][B_b]\, \mathrm{d}A \tag{8}$$

Isotropic elasticity matrix $[H_b]$ are calculated using equation (9).

$$[H_b] = D_b \begin{bmatrix} 1 & \upsilon & 0 \\ \upsilon & 1 & 0 \\ 0 & 0 & \dfrac{1-\upsilon}{2} \end{bmatrix} \; ; \; D_b = \frac{Eh^3}{12(1-\upsilon^2)} \tag{9}$$

D_b is the generalized bending constitutive matrix, E is the Young's Modulus, h is the thickness of plate and υ is the Poisson's ratio. Equivalent external load for distributed load can be expressed in equation (10).

$$\{f_n\} = \begin{Bmatrix} \vdots \\ f_{n_i} \\ \vdots \end{Bmatrix}_{r\times 1} \; ; \quad f_{n_i} = \int_A f_z R_i \, \mathrm{d}A \tag{10}$$

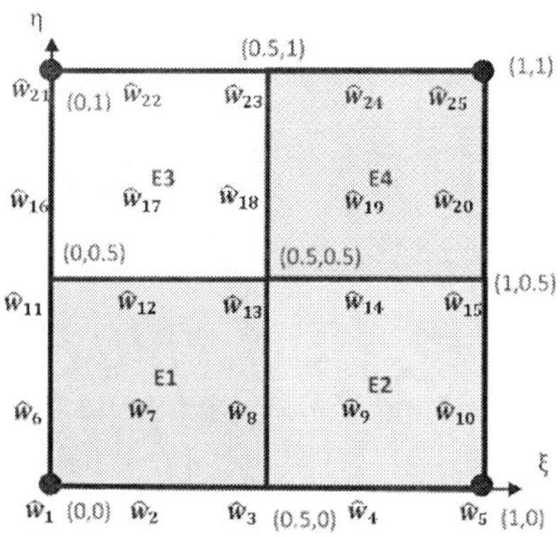

Figure 1. Control Variable in Parametric Space for $p = q = 3$ and NELT = 2×2.

In Isogeometric analysis using Galerkin Method, as in FEA, boundary conditions will reduce stiffness matrix and equivalent external load matrix. The difference between IGA and FEA is, in IGA, applying boundary conditions will yield the relations between control variables. These relations of control variables will be used to develop reduction matrix. The boundary conditions in Galerkin Method are vertical displacement and rotation as shown in equation (11).

$$w(\xi,\eta) = \langle R(\xi,\eta)\rangle\{\hat{w}_n\} \; ; \; \beta_x(\xi,\eta) = -\langle R_{,x}(\xi,\eta)\rangle\{\hat{w}_n\} \; ; \; \beta_y(\xi,\eta) = -\langle R_{,y}(\xi,\eta)\rangle\{\hat{w}_n\} \tag{11}$$

Analytical integration will face complexity issue and will take much longer time in computation. Numerical integration is employed to solve the stiffness matrix and equivalent load matrix for plate bending problem. Numerical integration is also used to solve the equivalent load matrix. Gauss integration will be used as numerical integration method. In IGA, the determination of Gauss point (Gauss Quadrature) is different with FEA. Nguyen [21] gives an alternative to use this method which is similar with FEA. According to Nguyen, the determination of number of Gauss point can be calculated using equation (12).

$$NPG_{\xi} = p+1 \text{ and } NPG_{\eta} = q+1 \tag{12}$$

Gauss points are in parent space (analogous with parametric space in FEA), while the basis functions are defined in parametric space, so the transformation from parent space to parametric space can be calculated using equation (13).

$$\begin{Bmatrix} \xi \\ \eta \end{Bmatrix} = \begin{Bmatrix} \frac{1}{2}\left[(\xi_{i+1}-\xi_i)\tilde{\xi}+(\xi_{i+1}+\xi_i)\right] \\ \frac{1}{2}\left[(\eta_{i+1}-\eta_i)\tilde{\eta}+(\eta_{i+1}+\eta_i)\right] \end{Bmatrix} \tag{13}$$

Numerical integration must be done in parent space, so using equation (13), we have transformation of Jacobian matrix from parent space to parametric space as shown in equation (14).

$$\left[\tilde{J}\right] = \begin{bmatrix} \frac{\partial \xi}{\partial \tilde{\xi}} & \frac{\partial \eta}{\partial \tilde{\xi}} \\ \frac{\partial \xi}{\partial \tilde{\eta}} & \frac{\partial \eta}{\partial \tilde{\eta}} \end{bmatrix} \text{ and } \left|\tilde{J}\right| = \frac{1}{4}(\xi_{i+1}-\xi_i)(\eta_{i+1}-\eta_i) \tag{14}$$

The stiffness matrix is expressed in equation (15).

$$[k] = [k_b] = \sum_{I}^{NPG} \omega_I^G |J||\tilde{J}|\left[B_b(\xi_I,\eta_I)\right]^T [H_b]\left[B_b(\xi_I,\eta_I)\right] \tag{15}$$

Based on the same principle, equivalent load matrix can be expressed in equation (16).

$$f_{n_i} = \sum_{I}^{NPG} \omega_I^G |J||\tilde{J}| f_z R_i(\xi_I,\eta_I) \tag{16}$$

3. Numerical Test

The conducted numerical test is a convergence test in Isogeometric Analysis for square thin plate under uniform load using Galerkin method. Parameters used in this convergence test are vertical displacement and bending moment. Refinements applied in IGA for convergence test are:

- Increasing in number of element (number of element is increased from 1×1, 2×2, 4×4 up to 8×8 in constant polynomial degree $p = 5$).
- Increasing in degree of polynomial (degree of polynomial is increased with p = 4, 6, 8, 10 in constant number of element NELT = 2×2).

Properties of geometry are presented in figure 2, with length L = 1000, thickness h = 10, material properties (E = 10.92 and υ = 0.3) and uniform distributed load f_z = -1. Square thin plate bending problems with two boundary conditions are presented: (a) full simply supported (SSSS) and (b) full clamped (CCCC).

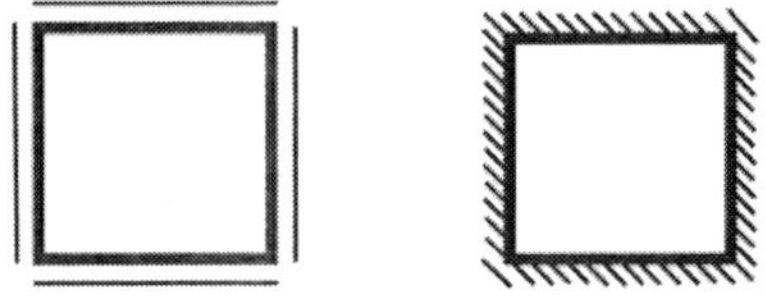

(a) SSSS (b) CCCC

Figure 2. SSSS (a) and CCCC (b) square plates with uniform load f_z.

3.1 Full Simply Supported (SSSS)

In full simply supported problem boundary conditions $w = \beta_s = 0$ are applied on all plate edges. The convergence of vertical displacement at the center of plate is shown in table 1 and figure 3. Table 2 and figure 3 present the convergence of bending moment at the center of plate. We found a good results converge to the reference solution.

Table 1. Convergence of displacement for SSSS problem.

p Escalation (using NELT = 2×2)				NELT Addition (using *p* = 5)			
p	CV	*w*	Error (%)	NELT	CV	*W*	Error (%)
4	6	-4140588	1.93	1×1	6	-4140586	1.93
6	8	-4066534	0.11	2×2	7	-4060986	0.02
8	10	-4066713	0.12	4×4	9	-4064282	0.06
10	12	-4066959	0.12	8×8	13	-4069792	0.19
Ref. -4062000 (Timoshenko & Wonowsky-Krieger [19])							

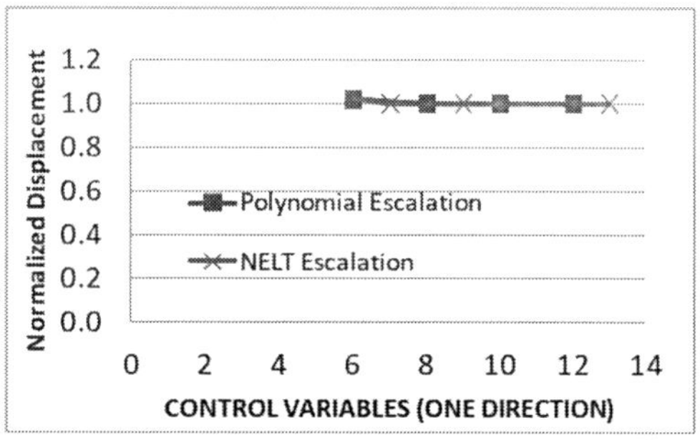

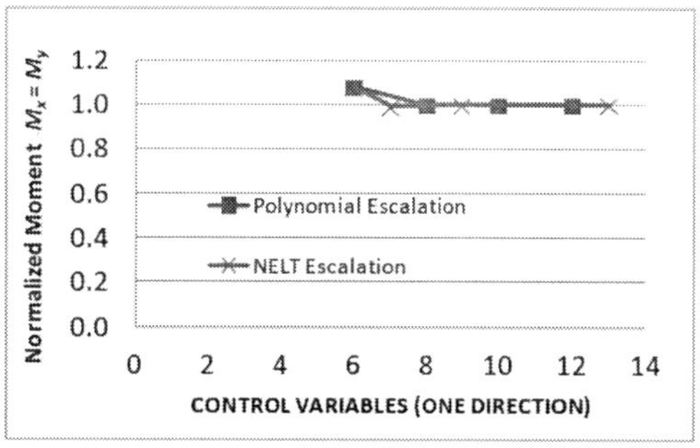

Figure 3. Convergence of Displacement and Bending moments $M_x = M_y$ for SSSS problem.

Table 2. Convergence of Bending Moment for SSSS problem.

p Escalation (using NELT = 2×2)				NELT Addition (using *p* = 5)			
p	CV	$M_x = M_y$	Error (%)	NELT	CV	$M_x = M_y$	Error (%)
4	6	-51722	7.98	1×1	6	-51722	7.98
6	8	-47934	0.07	2×2	7	-47463	0.91
8	10	-47902	0.00	4×4	9	-47824	0.16
10	12	-47946	0.10	8×8	13	-47935	0.07
Ref. -47900 (Timoshenko & Wonowsky-Krieger [19])							

3.2 Full Clamped (CCCC)

In full clamped problem boundary conditions $w = \beta_x = \beta_y = 0$ applies on all plate edges. The convergence of vertical displacement at the center of plate is shown in table 3 and figure 4. Table 4 and figure 4 present the convergence of bending moment at the center of plate. We found a good results converge to the reference solution.

Table 3. Convergence of displacement for CCCC problem.

p Escalation (using NELT = 2×2)				NELT Addition (using *p* = 5)			
p	CV	*w*	Error (%)	NELT	CV	*w*	Error (%)
4	6	-1329210	5.08	1×1	6	-1329209	5.08
6	8	-1266006	0.08	2×2	7	-1259755	0.41
8	10	-1271989	0.55	4×4	9	-1264963	0.00
10	12	-1268337	0.26	8×8	13	-1265304	0.02
Ref. -1265000 (Timoshenko & Wonowsky-Krieger [19])							

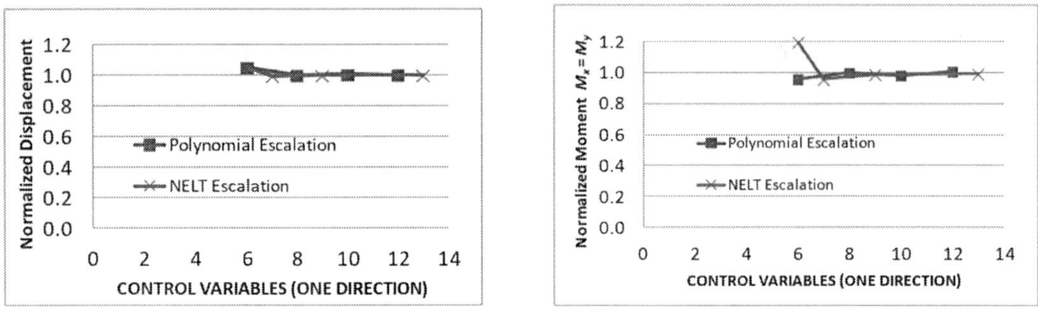

Figure 4. Convergence of Displacement and Bending moments $M_x = M_y$ for CCCC problem.

Table 4. Convergence of Displacement for CCCC problem.

p Escalation (using NELT = 2×2)				NELT Addition (using *p* = 5)			
p	CV	$M_x = M_y$	Error (%)	NELT	CV	$M_x = M_y$	Error (%)
4	6	-22110	4.28	1×1	6	-27648	19.69
6	8	-23064	0.16	2×2	7	-22088	4.38
8	10	-22656	1.92	4×4	9	-22772	1.42
10	12	-23254	0.67	8×8	13	-22889	0.92
Ref. -23100 (Timoshenko & Wonowsky-Krieger [19])							

4. Conclusion

Numerical test has been conducted in Isogeometric Analysis for square thin plate bending problem under uniform load with different boundary conditions using Galerkin method. The convergence tests show good correlation between the results and the reference solution for both *p* escalation and NELT addition, which means that Isogeometric Analysis using Galerkin method has good prospect to be an alternative computational method in solving the structural analysis problems.

Acknowledgement

The financial support from Universitas Indonesia through the *Hibah* PITTA program (No.786/UN2.R3.1/HKP.05.00/2017) is gratefully acknowledged.

References

[1] Katili I 1993 A new discrete Kirchhoff-Mindlin element based on Mindlin-Reissner plate theory and assumed shear strain fields - part I: An extended DKT element for thick-plate bending analysis *International Journal for Numerical Methods in Engineering* vol 36 pp 1859-83

[2] Batoz J-L and Katili I 1992 On a simple triangular Reissner/Mindlin plate element based on incompatible modes and discrete constraints *International Journal for Numerical Methods in Engineering* vol 35 pp 1603-32

[3] Katili I 1993 A new discrete Kirchhoff-Mindlin element based on Mindlin-Reissner plate theory and assumed shear strain fields - part II: An extended DKQ element for thick plate bending analysis *International Journal for Numerical Methods in Engineering* vol 36 pp 1885-1908

[4] Hamdouni A, Katili I, Maknun I J, Millet O 2013 Analytical and experimental analysis of an asymptotic thin-walled beam model *European Journal of Environmental and Civil Engineering* vol 17 no 1 pp 1-18

[5] Katili I, Batoz J-L, Maknun I J, Hamdouni A and Millet O 2014 The development of DKMQ plate bending element for thick to thin shell analysis based on Naghdi/Reissner/Mindlin shell theory

Finite Elements in Analysis and Design vol 100 pp 12-27
[6] Katili I, Maknun I J, Millet O and Hamdouni A 2015 Application of DKMQ element for composite plate bending structures *Composite Structures* vol 132 pp 166-174
[7] Maknun I J, Katili I and Purnomo H 2015 Development of DKMT element for error estimation in composite plate structures *International Journal of Technology* vol 6 no 5 pp 780 -9
[8] Maknun I J, Katili I, Millet O and Hamdouni A 2016 Application of DKMQ element for twist of thin-walled beams: comparison with Vlasov theory *International Journal for Computational Methods in Engineering Science and Mechanics* vol 17 pp 391-400
[9] Mahjudin M, Lardeur P, Druesne F and Katili I 2016 Stochastic finite element analysis of plates with the certain generalized stresses method *Structural Safety* vol 61 pp 12-21
[10] Irpanni H, Katili I and Maknun I J 2017 Development DKMQ Shell Element with Five Degrees of Freedom per Nodal *International Journal of Mechanical Engineering and Robotics Research* vol 6 pp 248-52
[11] Wong F T, Erwin, Richard A and Katili I 2017 Development of the DKMQ element for buckling analysis of shear-deformable plate bending *Procedia Engineering* vol 171 pp 805 – 12
[12] Katili I 2017 Unified and integrated approach in a new Timoshenko beam element. *European Journal of Computational Mechanics* vol 26 pp 282–308
[13] Katili I, Maknun I J, Tjahjono E and Alisjahbana I 2015 Error estimation for the DKMQ24 shell element using various recovery methods *International Journal of Technology* vol 8 no 6 pp 1060 - 9
[14] Hughes T J R, Cotrell J A and Bazilevs Y 2005 Isogeometric Analysis: CAD, Finite Elements, NURBS, Exact Geometry, and Mesh Refinement *Computer Method Applied Mechanics in Engineering* vol 194 pp 4135-95
[15] Kiendl J and Auricchio 2015 Isogeometric Collocation Methods for the Reissner-Mindlin Plate Problem *Computer Method Applied Mechanics in Engineering* vol 284 pp 489-507
[16] Reali A and Gomez H 2015 An Isogeometric Collocation Approach for Bernoulli-Euler Beams and Kirchhoff Plates *Computer Method Applied Mechanics in Engineering* vol 24 pp 623-36
[17] Katili I and Aristio R 2017 Isogeometric Galerkin in rectangular plate bending problem based on UI approach *European Journal of Mechanics - A/Solids* vol 67 pp 92-107
[18] Batista M 2010 Uniformly Loaded Rectangular Thin Plates with Symmetrical Boundary Conditions *arXiv preprint* arXiv: 1001.3016
[19] Timoshenko S and Woinowsky-Krieger S 1959 *Theory of plates and shells. Second edition*, (McGraw Hill)
[20] Calabro F and Manni C 2013 The Choice of Quadrature in NURBS-Based Isogeometric Analysis 3rd *South-East European Conference on Computational Mechanics*
[21] Nguyen V P, Bordas S P A and Rabezuk T 2012 *An Introduction to Isogeometric Analysis with MATLAB Implementation: FEM and XFEM formulations* Cardiff University

The change of value of soaked-CBR in embankment of subgrade due to water logging and repeated vehicles load

I Nurtjahjaningtyas[1], I B Mochtar[2*] and N E Mochtar[2]

[1]Department of Civil Engineering, Faculty of Engineering, University of Jember, Jember 68121, Indonesia.
[2]Department of Civil Engineering, Faculty of Civil Engineering, Environment and Earth, ITS Surabaya, Indonesia

indran.teknik@unej.ac.id, indramochtar@yahoo.com

Abstract. The value of soaked-CBR in subgrade soil is very important in the highway pavement thickness design [1]. If the soaked-CBR value of the subgrade does not meet the requirements, it is necessary to improve the soil by blasting with a better material. This is done to prevent damage to the road. Damage to the pavement surface may also cause damage to the subgrade. Alternatively, the damage to the pavement surface begins with damage on the subgrade. If the subgrade decreases in its power - caused by repetitive vehicle loads, otherwise known as repetition vehicle load, then it can collapse the foundation layer, in which - then continues on the surface damage. Based on the phenomenon, this research is conducted to know the change of soaked-CBR value on subgrade soil due to water logging and repetition of vehicle load passing. The result shows that the soaked-CBR value will increase along with the number of vehicles passing up to 25x passings, then down to 50x passings which is the lowest value of all load variations. After 50x passings, CBR value will increase slightly up to 250x passings which then decrease until the end of the research.

1. Introduction

The subgrade of highway is the base ground supporting the pavement construction design the road, where the CBR value in the subgrade has a large influence [1]. The CBR value used in pavement thickness design is the value of soaked-CBR. When the soaked-CBR value of the subgrade does not meet the requirements, it is necessary to treat the subgrade, using form of a better material or stabilization [2]. Treatment of subgrade with the better soil moisture is intended to increase the value of soaked-CBR, or to increase the capacity of the soil.

In highway pavement construction, repair subgrade is positioned under the foundation layer. When there is damage to the surface of the pavement continuously, it damages the foundation layer, then the subgrade is also damaged. Alternatively, the damage to the pavement surface is started by the damage on the subgrade.

While the subgrade decreases its strength caused by repetitive vehicle loads, otherwise known as vehicle load passing, it can collapse the foundation layer which then continues on the surface damage. Based on [3, 4] the damage that occurs in the subgrade is largely due to the water logging that cannot drain out, so that it continuously receives the repetition of the vehicle load, it will be easier to damage.

The condition of early damage to road pavement as mentioned above should not occur, if the traffic of passing vehicles and other factors is in accordance with the planning, because in the calculation of pavement construction planning, the CBR value for subgrade l using soaked CBR value. The CBR value represents the value of ground quality when compared to standard materials in the form of crushed stone having a 100% CBR value (AASHTO) [5], and the CBR value of soaking is the CBR value of soil in water-saturated condition whose collapse mechanism is due to the compressive load increases continuously in the function of time until the penetration decrease reaches 0.5". The load or compressive force which causes a decrease in penetration to 0.1" or 0.2" is compared to the standard load.

However, the example case in Pantura, East Java, Indonesia, there is still early damage especially if there is water logging continuously in the base and subbase layers. This condition was apparently answered by [6, 7] that unstable subgrade or soft subgrade conditions allow for settlement or deformation when subjected to repeated passing. Some images of road damage due to the presence of standing water are shown in figure 1.

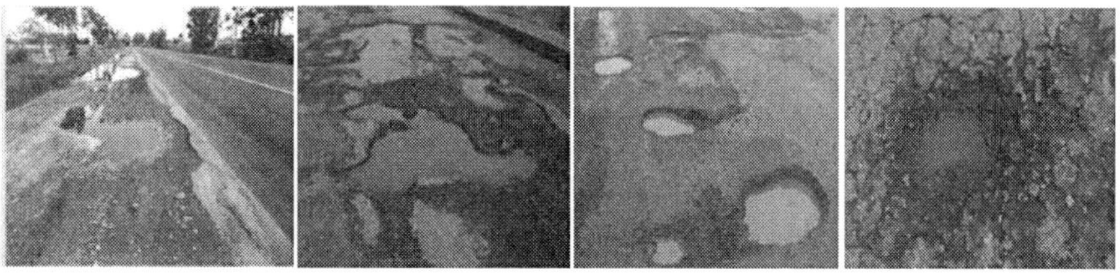

Figure 1. Example of road damage due to waterlogging.

The damage caused by water that occurs almost in all lowland areas in Indonesia, it is important to know the influence of water puddle on the degradation of CBR value immersion. In addition, the effects of excessive load and heavy vehicle reps also need to be studied, given that the largest vehicles in Indonesia have excessive payloads. On the basis of this thinking, this research is conducted in order to know the influence of waterlogging, overload and heavy vehicle reps on the degradation of CBR.

From all the above descriptions, it is found a problem that needs to be searched for and then sought the solution, that is whether the combination of water pool and repetitive load within a certain period of time in the foundation layer and subgrade cause the change of CBR value.

2. Material and method

2.1. Material

In this study embankment is used material taken from the district of Malang, East Java, Indonesia. The test results of physical properties as shown in table 1.

Table 1. Result of examination of physical properties of soil embankment.

Parameter	Result value	Unit
Liquid Limit	56.88	%
Plastic Limit	30. 98	%
Plasticity Index	25.89	%
Optimum Moisture Content	25.36	%
Max dry density	1.527	gr/cc

Material with parameters as shown in table 1 is included in the MH classification. Preliminary CBR values before some treatments have ranged from 6% to 10%, so the material is qualified to be used as an embankment material.

2.2. Method

The method used in this study is a modified soaked-CBR testing method. The modifications made in this study were to use cyclic loading prior to CBR testing. The purpose of this cyclic loading is to know the number of vehicle passing that weigh on the soil. Other modifications to the CBR test are also performed during the testing process by providing a water logging as well as – it is the initial purpose of the research, is to determine the effect of water logging on the value of soaked- CBR

With the presence of water logging during the testing CBR in the laboratory, it will have the same value as well as in the field when there is a damage by the stagnant water. In order to prepare the specimens until immersion test uses the applicable rules, namely by giving energy compacting as much as 56x blows per layer. The number of layers in each specimen is 5 layers. Then soaked for 4 days. The tool used for testing is also the same tool as per the common applicable rules.

3. Result and discussion

This research is conducted to know the change of soaked-CBR value on subgrade-soil caused by water logging and vehicle repetition passing. In the process of conducting the research, there are variations vehicle and their passing. Variation of vehicle load is intended to know the value of soaked-CBR when the soil burdened with several variations, namely 50, 100 and 150 Psi. While the variation of the number of passing as the result is shown in table 2. The variation in the amount of load is based on the pressure of the wheels of the light vehicle, which is assumed to be 50 Psi, the wheel pressure of the normal vehicle is 100 Psi, and the wheel pressure of the heavy vehicle is 150 Psi.

Part of the process of conducting the research is shown in figure 2. In figure 2 (a) it shows the condition of the specimen after loading and seen there are parts deformed. This proves the visual condition in the field that if the soil with the parameters as used in the study, subjected to repeated loading will experience a deformation. In figure 2 (b), shows the surface smoothing process of the specimen before CBR testing. And figure 2 (c) shows the CBR testing process.

From table 2 and figure 3, it can be seen that the three CBR test curves have a trend not very different at all vehicle wheel pressure, but have a change in the CBR value on each test of the number of vehicle paths. From the test results, for the three variations of wheel pressure, the change in CBR value increased on 10x the number of trajectories, then decreased greatly on 50x the number of trajectories. The largest decrease amount, up to 45% for wheel pressure 150 Psi. As for the pressure of 50 and 100 Psi wheels respectively by 20% and 30%. The CBR value at 50x the number of passes is the lowest CBR value in all wheel pressure variations. Furthermore, on the variation in the number of passes after 50x to 250x, the CBR value changes slightly and will change downward in the number of passes from 250x to 300x. Given the changing value of CBR, it is possible that this problem is one cause of premature damage to road pavement.

Table 2. Value of CBR (%) vs. Number of passing (n) and pressure variations (P) from soil.

n	P = 50 Psi;	P = 100 Psi;	P = 150 Psi;
0	5.20	9.32	8.00
10	7.62	9.91	16.33
25	5.50	8.65	14.44
50	4.50	4.75	7.49
70	7.49	8.66	10.00
100	10.07	9.32	11.99
150	11.15	10.16	12.82
250	14.00	11.70	15.00
300	12.74	9.91	14.44

(a) (b) (c)

Figure 2. Test specimens in the testing process.

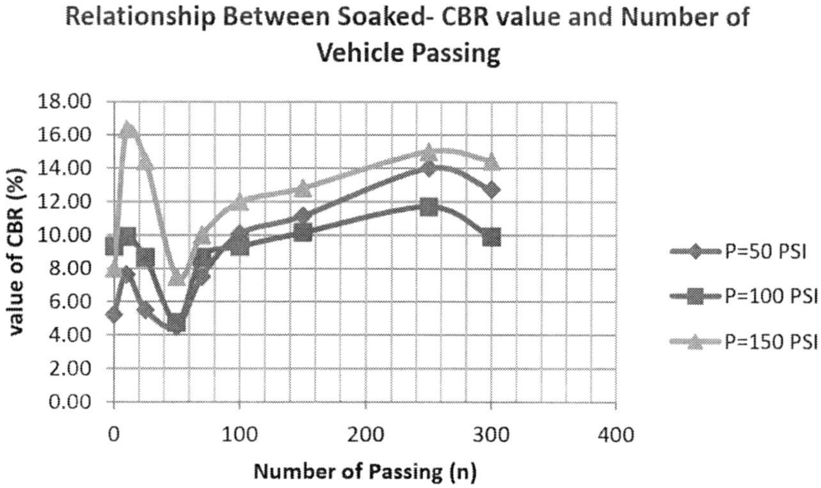

Figure 3. The soaked-CBR value relationship curve and the number of vehicle passing on the soil.

The change in immersion CBR value in this study has a difference with the change of CBR immersion value in the research that [8] and [9] have done. The difference in research being conducted with previous research lies in the modified methodology of CBR testing and its loading system. So the very important value obtained from this study is to know the change of CBR value when loaded with some variation of wheel pressure which can be interpreted that this research can be applied on the road with various classes according to the allowable wheel pressure limit. Another important result is that CBR values on the ground can be changed after being repeated by any vehicle. So it is very important to know by the planners and implementers of road pavement construction, especially the flexible road pavement, to be able to pay attention to the implementation starting from hoarding for subgrade treatment to foundation coating so as to maintain the CBR plan value until the end of the road plan. It is advisable to avoid the occurrence of early damage to road pavement

4. Conclusion

Based on the research, it can be concluded that the water logging and repetition of vehicle load can cause changes in the value of soaked-CBR. This has been proven by the results of research which states that for the three variations of wheel pressure, the change in CBR value increased on 10x the number of paths, then decreased greatly on 50x the number of trajectories. The largest decrease amount, up to 45% for wheel pressure 150 Psi. As for the pressure of 50 and 100 Psi wheels respectively by 20% and 30%. The CBR value at 50x the number of passes is the lowest CBR value in all wheel pressure variations. Furthermore, on the variation in the number of passes after 50x to 250x, the CBR value changes slightly and will change downward in the number of passes from 250x to 300x. Given the changing value of CBR, it is possible that the problem of water puddles and vehicle load reps as one cause of premature damage to road pavement.

References

[1] AASHTO 2008 *Standard Specifications for Transportation Materials and Methods of Sampling and Testing Part 2: Tests* (Washington D.C: AASHTO)
[2] Tjie-Liong G 2006 Konsep Desain Geosynthetics Untuk Konstruksi Jalan (Geosynthetics Design Concept for Road Construction) Road Construction in Indonesia with Special Reference to the Role of Geosynthetics Indonesian Chapter of IGS
[3] Mochtar I B 1999 *Penelitian dan Pemrosesan Data Teknik Jalan Bina Marga Jawa Timur* Research Report
[4] Sugiyanto M A 2007 Penurunan kinerja lapis pondasi bawah karena kandungan fraksi halus berlebih hasil kontaminasi material tanah dasar menyebabkan terjadinya kerusakan perkerasan jalan
[5] Austin R A and Gilchrist A J T 1996 Enhanced performance of asphalt pavements using geocomposites *Geotextiles and Geomembranes* vol 14 pp 175-186
[6] Yasuhara K, Yamanouchi T 1983 Approximate Prediction of soil deformation under drained-repeated loading *Soils and Foundations* vol 23 no 2 pp 13-25
[7] Hu A F, Zhang X D and Jia Y S 2014 Research on the permanent settlement of saturated soft subsoil *Advances in Soil Dynamics and Foundation Engineering* pp 168-177
[8] Etim R K, Eberemu A O and Osinubi K J 2017 Stabilization of black cotton soil with lime and iron ore tailings admixture *Transportation Geotechnics* vol 10 pp 85-95
[9] Kumar D, Nigam S, Nangia A and Tiwari S 2015 California Bearing Ratio Variations in Soil Reinforced with Natural Fibres (A Case Study Bhopal Bypass Road) *International Journal on Emerging Technologies* vol 6 no 2 pp 95-104

IOP Conf. Series: Materials Science and Engineering **403** (2018) 012026 doi:10.1088/1757-899X/403/1/012026

Optimization of superplasticizer MasterGlenium SKY 8614 with added materials fly ash, steel slag, and silica fume for high strength concrete

M Ujianto, E L Ardiansyah, G O Ilokana and M F Falah

Civil Engineering Department, Engineering Faculty, Universitas Muhammadiyah Surakarta Jl. *Ahmad Yani, TromolPos 1, Pabelan, Kartasura, Jawa Tengah*

ujianto@ums.ac.id

Abstract. Indonesia as a developing country continues build various infrastructures such as high storey building, bridge, highway, or fly over. Most of the construction project in Indonesia use concrete as the main material. Various problems arise as the ongoing developments such as reduced the natural resources to market demand require the concrete can be used one day after the project completion. The research is trying to solve this problem. The use of fly ash as substitute material to reduce the amount of cement use due to its chemical properties of silica and alumina, the use of steel slag instead of 50% coarse aggregate, and use of silica fume so greatly affect the compressive strength of concrete. The result of concrete compressive strength in 3 samples from first mix design with code AR were 29.16 MPa, 29.72 MPa, and 30.12 MPa. Fly ash, steel slag, and silica fume can increase the compressive strength which is much higher than design compressive strength of 15 MPa and with this mix design only use 51% of total cement requirement with normal mix design of ACI method.

1. Introduction

Development in the construction in Indonesia currently is heavy conducted. More than 60% construction project either from the simplest to the complicated project use concrete as the main construction material. Increasing the lower middle class by 56.5% from 240 million populations can be sure the development will be continuing to grow [1]. The purpose of self-development is to make people prosperous. Without realizing there is problem arising from uncontrolled development is reduction the natural resources that is the impact of massive development that occurred at this time. To solve this problem, so need for innovation. In this study the authors have conducted research on high quality concrete SCC concrete that can be hardened and can be used within 1 day. Concrete consisting of a mixture of cement, fine aggregate, coarse aggregate, and water is a material that cannot be separated in the world of construction.

Conventional concrete is considered less effective because in the process required a vibrator to compress fresh concrete. SCC is also otherwise called as High-Fluidity Concrete, Self-Levelling Concrete and Self- Consolidating Concrete. This concrete has a very good ability to flow inside the formwork in which there is a reinforcing steel relying on the effect of gravity force with a maintained homogeneity [2]. As a solution of the problem, is currently being developed concrete SCC (Self Compacting Concrete) that is self-flowing and self-compacted concrete, to support SCC's work is required also Admixture material that is Superplasticizer, in this opportunity using MasterGlenium SKY 8614 from BASF, added is required to meet the flow ability requirements.

Then, SCC development needs to be balanced with innovation to save natural resources. The cement production process mostly uses fossil fuels that are likely to cause a greenhouse effect. The residual ash from the cement production process is also directly mixed with the free air which makes it pollution for the environment. It should be noted that 1 ton of cement produces 1 ton of carbon dioxide as well, and cement production also accounts for 7% of overall CO2 generated from various sources That factor causes many innovations to reduce the amount of cement use in making SCC concrete. Some researchers have done research on the issue, such as the addition of materials as cementitious to reduce the amount of cement and increase the compressive strength of concrete.

Based on the above background, then the problem that became the focus of this research is what is the most optimal material for Superplasticizer MasterGlenium SKY 8614? How do other additives influence the compressive strength of the concrete? While the purpose of this research is to know the optimal added material for Superplasticizer MasterGlenium SKY 8614 and to know the effect of fly ash, steel slag and silica fume combined with Superplasticizer MasterGlenium SKY 8614 to the compressive strength of concrete. The benefits of this research are expected to become the potential developer material of steel slag, silica fume, and fly ash as added materials of SCC concrete construction and can be a solution of construction problems in buildings that will soon be used within 1 day.

2. Research methods

In the first stage of preparation based on study results, literature studies, preparations include materials and equipment to be used in the manufacture of test specimens. The sources of data from primary and secondary data. Primary data in this study was obtained through interviews and consultations with experts who are competent in their field. Secondary data in this study secondary data obtained from the observation and test samples of test specimens, and reference library related to this research. Some of the things observed and tested in this study are:

- Workability of Concrete Mixed Workability, this test is done by Slump Test.
- Compressive Test, this test is done on all samples after all samples reaches age of 24 hours.

In the second stage, the material characteristics and physical test were tested on each material used. This test used the ASTM C 40 assay standard for the ASTM C 127 as a mild aggregate assay test for specific gravity and absorbance tests on coarse aggregate ASTM C 128 for specific gravity and absorption tests on fine aggregate ASTM C 136 for fine aggregate filtration analysis and coarse aggregate.

In the third stage, batching is done with mix design composition 1 with AR code (Superplasticizer MasterGlenium SKY 8614, silica fume, slag steel, sand, gravel, cement, and water), mix design 2 with BS code (Superplasticizer MasterGlenium SKY 8614, fly ash, Slag of steel, sand, gravel, cement, and water). Mix design 3 with CT code (Superplasticizer MasterGlenium SKY 8614, fly ash, silica fume, sand, gravel, cement, and water), mix design 4 with DU code (Superplasticizer MasterGlenium SKY 8614, fly ash, silica fume, slag steel, sand, Gravel, cement, and water) and mix design 5 with the EV code (Superplasticizer MasterGlenium SKY 8614, steel slag, sand, gravel, cement, and water). Each mix design will be made into 3 specimens. Prior to printing into a 15 x 30 cm cylindrical test specimen, each of the fresh concrete results of each mix design is performed slump flow test which refers to the EFNARC testing standard (fresh concrete requirement is to have 55-85 cm slump flow). After the hardened concrete can be done capping (ASTM C 617 Standard) with sulfur material on each hardened concrete so that the surface of the concrete becomes flat [3 – 7].

In the fourth stage, the test of compressive strength of cylindrical concrete on each test specimen is 15 specimens. In the fifth stage data analysis and discussion of all test results that have been done on each specimen. In the sixth stage the conclusion of all test results, analysis and report writing.

2.1. Materials

SteelSlaag is the waste material from steel casting in the form of waste and has not maximized its use. Silica fume is a delicate pozzolan material, where the composition is mostly produced from the high blast furnace or the rest of the silicon or silicon iron alloy production. The use of silica fume in a concrete mix is intended to produce concrete with increasing strength from day to day. Superplasticizer is an added material used in the manufacture of SCC concrete intended to achieve its flow ability. The PCC (Portland Composite Cement) on this occasion uses a brand of Three Wheels of type I. The cement is often used for buildings in general, in addition to having lower hydration heat during the cooling process compared to other types of portland, the work will be easier and produce surface concrete is denser and smooth (SNI 157064-2004). Aggregates Larrad's (1990) study results show that the maximum grain size provides concrete evidence to make high quality concrete should not be more than 15 mm [8] List of materials are used in this research shown in table 1.

Table 1. Materials data.

No.	Materials	Density	Explanation
1	Cement	3.15 gr/cm^3	Tiga Roda tipe PCC
2	Sand	2.60 gr/cm^3	Merapi
3	Coarse	2.65 gr/cm^3	Kulon Progo
4	*Steel slag*	3.60 gr/cm^3	Ceper Klaten
5	*Silica fume*	2.20 gr/cm^3	SikaFume, PT. Sika
6	*Fly ash*	2.15 gr/cm^3	PT. Varia Usaha Beton
7	*Superplasticizer*	Brosur	PT. BASF

2.2. Mix design

AR code is used as designation for mix design type 1, BS code is used as designation for mix design type 2, CT code is used as designation for mix design type 3, DU code is used as designation for mix design type 4, and EV code is used as designation for mix design type 5. The above materials use fly ash 15% of cementitious, silica fume 10% of cementitious, superplasticizer 1.5% of cementitious, 50% steel slag of total coarse aggregate, fine aggregate ratio: coarse aggregate = 49:51. The concrete strength of the concrete plan is 15 MPa within 1 day. Mix design using American Concrete Institute (ACI) method, with strong press 15 MPa plan in 1 day and f'c 75 MPa in 28 days. As a result of normal concrete mix design, total cement requirement of 800 kg/m^3, gravel 672 kg/m^3, sand 650 kg/m^3, water 0,206 m^3. Then modified to mix design modification with cementitious deduction of 27.25% of cement content, to obtain total cement requirement of 550 kg/m^3, silica fume 5.5% from cementitious of 32 kg/m^3. The mix design is shown in table 2.

Table 2. Mix design.

No.	Material	Mix Design Code (Needs 3 cylinder)					Unit
		AR	BS	CT	DU	EV	
1	Cement	8.7	7.9	6.9	6.9	9.2	kg
2	*Silica fume*	0.5	-	0.9	0.9	-	kg
3	*Fly ash*	-	1.4	1.4	1.4	-	kg
4	Water	2.9	2.9	2.9	2.9	2.9	liter
5	*Superplasticizer*	0.2	0.1	0.2	0.2	0.1	liter
6	Sand	12	13	12	12	13	kg
7	Gravel	6.2	6.7	13	6.5	6.8	kg
8	*Steel Slag*	6.2	6.7	-	6.5	6.8	kg

IOP Conf. Series: Materials Science and Engineering **403** (2018) 012026 doi:10.1088/1757-899X/403/1/012026

3. Result of concrete cylindrical pressure test

The cylindrical concrete was tested for compressive strength at the age of 1 day (24 hours) with a compressive strength test machine. The following is the result of the cylinder concrete compressive test can be seen in the table 3.

Table 3. Result of concrete pressure test.

Mix Design Code	1	2	3	Average(MPa)
AR	29.14	29.71	30.10	29.65
BS	19.24	22.07	20.27	20.53
CT	4.00	4.44	4.44	4.27
DU	4.30	4.10	3.40	3.93
EV	26.60	23.80	30.60	27.00

4. Discussion

The result of compressive strength test is shown in figure 1 (a) AR concrete mix design (b) BS concrete mix design (c) CT concrete mix design (d) DU concrete mix design and (e) EV concrete mix design. Figure 2 shows the comparison result of all mix design and figure 3 shows the graph of concrete optimum compressive strength.

The results of compressive strength test of AR concrete mix design, AR1 cylinder has a compressive strength of 29.14 MPa, AR2 cylinder has a compressive strength of 29.71 MPa, and the AR3 cylinder has a compressive strength of 30.10 MPa. The average of AR concrete compressive strength is 29.65 MPa at 1 day (24 hours).

The results of compressive strength test from concrete BS design mix, BS1 cylinder have compressive strength of 19.24 MPa, BS2 cylinder have compressive strength 20.27 MPa, and BS3 cylinder have compressive strength equal to 22.07 MPa. The average of concrete strength of BS is 20.53 MPa at 1 day (24 hours).

The results of compressive strength test from concrete mix design CT, CT1 cylinder has a compressive strength of 4.00 MPa, cylinder CT2 has a compressive strength of 4.44 MPa, and CT3 cylinder has a compressive strength of 4.44 MPa. The average of compressive strength of CT concrete is 4.27 MPa at 1 day (24 hours).

The results of compressive strength test from DU concrete mix design, DU1 cylinder have compressive strength equal to 3.4 MPa, cylinder DU2 have compressive strength 4.1 MPa, and cylinder DU3 have compressive strength equal to 4.3 MPa. The average of concrete compressive strength of DU is 3.93 MPa at 1 day (24 hours).

The results of the compressive strength test of EV concrete mix design, EV1 cylinder has a compressive strength of 23.8 MPa, EV2 cylinder has a compressive strength of 26.6 MPa, and EV3 cylinder has a compressive strength of 30.6 MPa. The average of compressive strength of EV concrete is 27,00 MPa at 1 day (24 hours).

AR concrete has the highest average compressive strength compared to other mixed concrete that is 29.65 MPa in the age of 1 day (24 hours). Recommended additives for the SKY 8614 Superplasticizer MasterGlenium are silica fume and steel slag.

From the results of the compressive strength shown in the table 3 that the addition of Superplasticizer and fly ash is not effective in increasing the compressive strength of the concrete mixture. The use of steel slaags that have high hardness characteristics can support increased compressive strength.

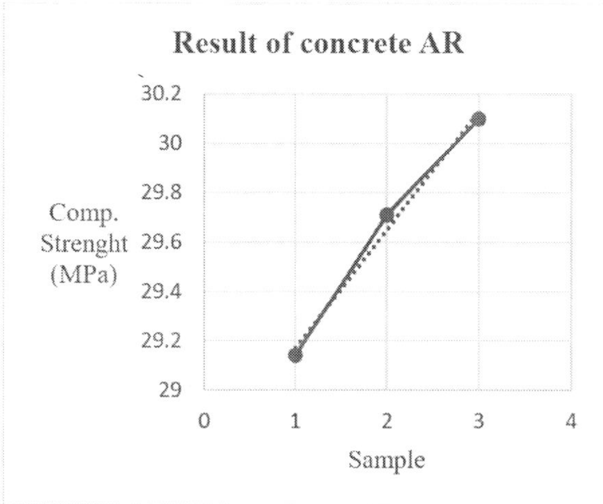

(a) Graph of AR concrete test result.

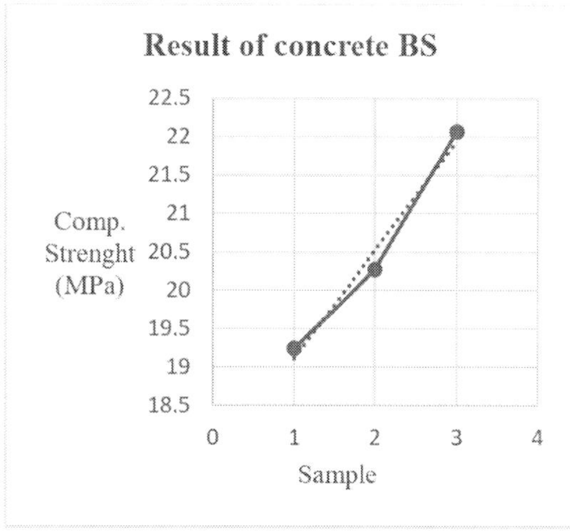

(b) Graph of BS concrete test result.

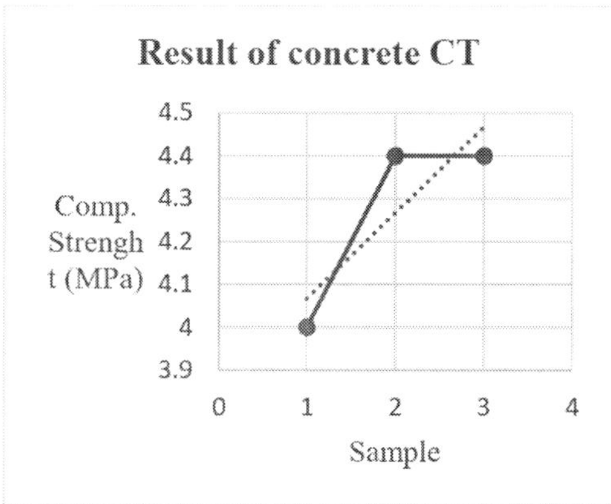

(c) Graph of CT concrete test result.

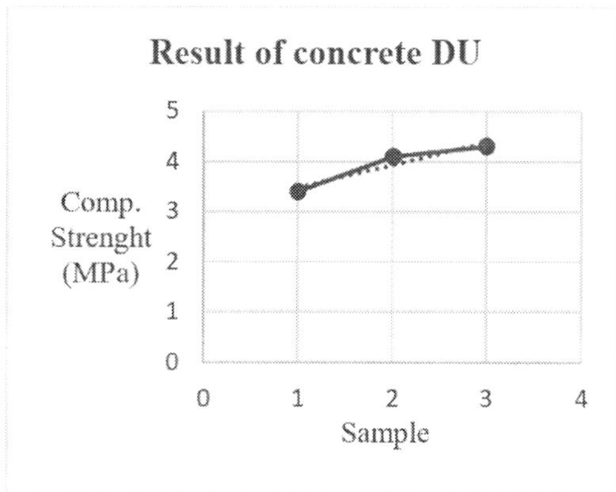

(d) Graph of DU concrete test result.

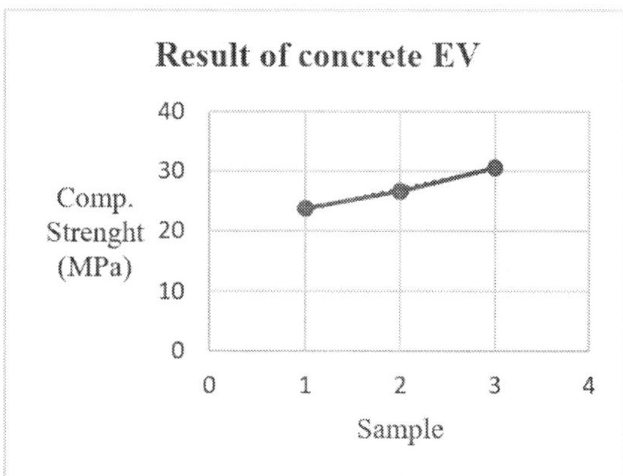

(e) Graph of EV concrete test result.

Figure 1. The result of compressive strength test.

IOP Conf. Series: Materials Science and Engineering **403** (2018) 012026 doi:10.1088/1757-899X/403/1/012026

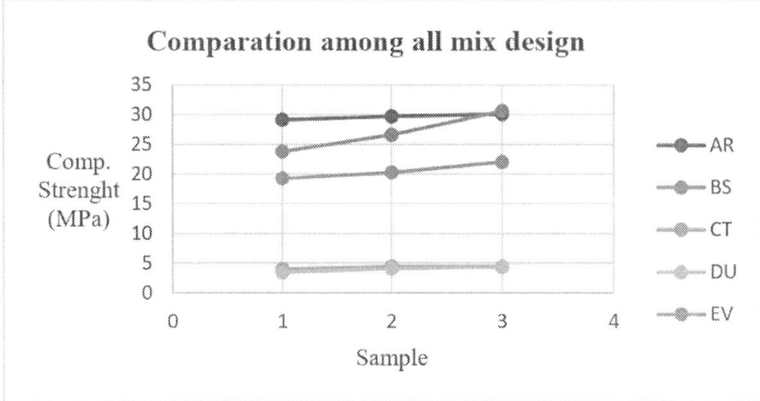

Figure 2. Result of all mix design.

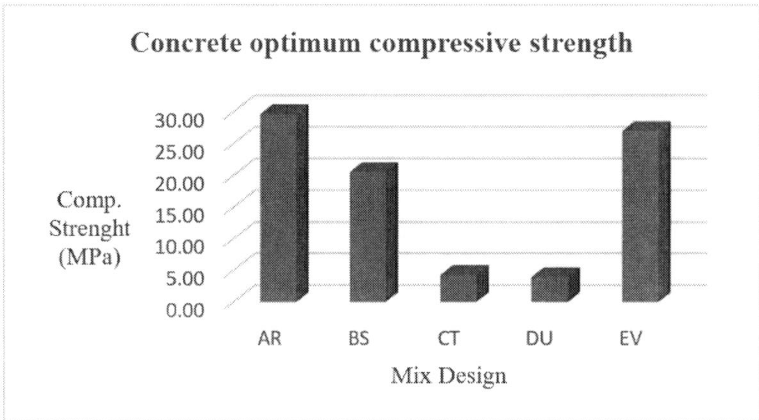

Figure 3. Graph of optimum compressive strength.

5. Conclusion

From the results and discussion can be concluded that the optimal added materials for Superplasticizer MasterGlenium SKY 8614 is silica fume and steel slag. And the addition of added materials will be optimum using silica fume with 5.5% content of cementitious, steel slag of 50% of the total coarse aggregate.

References

[1] Hediyanto 2013 Peranan Beton Dalam Pembangunan Infrastruktur Indonesia [Online] available at https://www.pu.go.id/berita/view/10400/peranan-beton-dalam-pembangunan-infrastruktur-indonesia

[2] Okamura H, Ozawa K and Ouchi M 2000 Self-compacting concrete *Structural concrete journal of the fib*. Vol 1 no 1 pp 3-17

[3] Annual Book of ASTM Standards 2004 *ASTM C40 Standard Test Method for Organic in Fine Aggregate for Concrete* (West, Conshohocken, PA: ASTM International)

[4] Annual Book of ASTM Standards 2015 *ASTM C127 Standard Test Method for Relativity Density (Specific Gravity) and Absorption of Coarse Aggregate* (West, Conshohocken, PA: ASTM International)

[5] Annual Book of ASTM Standards 2015 *ASTM C128 Standard Test Method for Relativity Density (Specific Gravity) and Absorption of Fine Aggregate* (West, Conshohocken, PA: ASTM

International)
[6] Annual Book of ASTM Standards 2014 *ASTM C136 Standard Test Method for Sieving Analysis of Fine and Coarse Aggregate* (West, Conshohocken, PA: ASTM International)
[7] EFNARC 2002 *EFNARC Specification and Guidelines for Self-Compacting Concrete* (Farnhamm, Surrey GU9 7EN, UK: EFNARC)
[8] Tjokrodimuljo K 1996 *Teknologi Beton* (Yogjakarta, Indonesia: Nafiri)

Mixture design consideration for foamed asphalt using RAP materials

S Sunarjono and N Hidayati

Centre for Transportation Studies, Civil Eng, Faculty of Engineering, Universitas Muhammadiyah Surakarta, Jl. A. Yani Pabelan Kartasura Tromol Pos I Surakarta 57101 Indonesia Ph. 0062271 717417 ext 2212

sri.Sunarjono@ums.ac.id

Abstract. The use of reclaimed asphalt pavement (RAP) in cold foamed asphalt mixture can decrease the need of fresh aggregate and bitumen, reduce fuel consumption and CO2 production. A cold foamed asphalt using RAP material is a unique asphalt mixture. Performance of these materials will be affected by mix design process. This paper discusses about mix design consideration for these mixture especially to set the aggregate and foamed bitumen properties properly. The method used to propose mix design consideration in this paper is by analyzing results of the long research work about foamed asphalt and RAP materials in laboratory. It is found that the optimum value of foamed bitumen properties is affected by bitumen grade, foam characteristic, water content at foaming, and temperature of bitumen. The best aggregate properties should consider the plasticity index, aggregate gradation, and the percentage of RAP proportion. The percentage of RAP proportion influence specific gravity, mixture density, volumetric characteristic, and mixture performance.

1. Introduction

The Report of TRL611 [1] published a classification for cold recycled materials. There are three families of cold recycled materials. Materials mixed with hydraulic are called as fully hydraulically bound. Materials mixed with visco-elastic binder are named as fully visco-elastically bound. Recycled materials mixed without binder are mentioned as unbound materials. Cold recycled mixture with different binder types and curing behavior can be categorized by areas within these three families. Four material types i.e. quick hydraulic (e.g. using cement), slow hydraulic (e.g. using lime), quick viscoelastic (e.g. using bitumen and cement), and slow viscoelastic (e.g. using bitumen and lime) fall into these three material families. Based on this TRL classification, asphalt using foamed bitumen and reclaimed asphalt pavement (RAP) materials will be closer to the viscoelastic behavior either slow or quick viscoelastic families.

The use of RAP in cold foamed asphalt mixture can decrease the need of fresh aggregate and bitumen, reduce fuel consumption and CO2 production, and overcome the addition of road surface level when it is conducted by overlay method. It is noted that RAP materials can replaced the fresh aggregate for road pavement construction with satisfied performance [2]. However, the properties of foamed asphalt mixture using RAP is unique. This mixture is superior in environment issues but inferior in properties issues. This mixture should be designed as properly as possible, included in the mix design process. This paper reports the results of research exploring the main consideration used for mix design of foamed asphalt using RAP. The objective of research is to open understanding factors affecting the mixture performance using RAP.

The laboratory mix design for bituminous materials is conducted to achieve the optimum proportions of material components that fulfill the mixture specification. The amount and properties of all mixture components should be combined economically to yield optimum mixture properties. According to the Asphalt Institute (1988) the important things to consider in the asphalt mix design are:

- The bitumen content to maintain durability and fatigue cracking resistance of mixture,
- The stability and stiffness modulus of mixture to withstand deformation caused by traffic loads.
- The void percentage to maintain the mixture from flushing, bleeding, or loss of stability during secondary compaction under traffic loading.
- Mix workability during mixing, laying and compaction process.

2. Research method

The method used to propose mix design consideration in this paper is by analyzing results of the long research work about foamed asphalt and RAP materials. There are two main researches with topics of the effect of foamed bitumen properties to cold mix asphalt performance, and study of mixture density and stability of cold mix asphalt using RAP materials.

This research focus on mix design consideration in term of two aspects namely determines the optimum foamed bitumen properties and aggregate properties considerations. First aspect was conducted by produce foamed bitumen using laboratory equipment i.e. Wirtgen WLB 10, and investigated the foamed bitumen properties e.g. the maximum expansion ration and half-life. Second aspect by investigated the properties of RAP materials using related testing, e.g. specific gravity, sieving analysis, compaction, Marshall test, and Pavement Facility Testing. The resulted data tests were then analyzed to explore the main consideration of mix design process for asphalt using foamed bitumen and RAP.

The considerations used in this mix design refer to these following procedures:

- CSIR procedure [3]. This procedure is widely implemented in many cold recycling projects in South Africa.
- Lee and Kim procedure [4]. This procedure developed specifically for cold-in-place rehabilitation (CIPR) using foamed bitumen.
- Wirtgen procedure [5]. This procedure developed the method to select the optimum properties of foamed bitumen in order to produce the best foamed asphalt mix performance.

3. Discussion

3.1. Consideration to select the optimum properties of foamed bitumen

Three important stages in order to select the optimum properties of foamed bitumen can be described as follows. Firstly, exploring the characteristic of foamed bitumen, i.e. maximum expansion ratio (ERm) and half-life (HL) values at some temperatures and water contents variations for several selected bitumen penetration grades; Secondly, determining the best bitumen penetration and temperature at foaming process using the ERm vs HL curve; Thirdly, determining the best foaming water content (FWC) using ER-HL vs FWC curve and selected foamed bitumen specification. It can use either CSIR [3], Lee and Kim [4], or Wirtgen procedures [5].

It has been widely understood that foamed bitumen with higher maximum expansion ratio and longer half-life will have better quality, and asphalt mix properties using it will have better performance. Unfortunately, maximum expansion ratio and half-life have inverse values, and this makes it difficult to determine the optimum properties of foamed bitumen. Based on the experiences, foamed bitumen which has good properties and can function effectively usually has ERm values between 10 and 15, which may be resulted by using foaming water content between 1 and 3 percent. table 1 shows various guidance of minimum application of ERm and HL limit based on empirical experiences.

Table 1. The Guidance of ERm and HL values.

Method	Minimum ERm	Minimum HL
CSIR [3]	10	12 seconds
TRL Report 386 [6]	10	10 seconds
Wirtgen [5]	8	6 seconds
Chiu and Huang [7]	8	8 seconds

Wirtgen [5] has proposed an easy method to determine the optimum properties of foamed bitumen. This method is not very rigorous but very suitable for field applications that require speed in making decisions. Wirtgen method use values data of ERm and HL in various FWC, and the optimum FWC is selected based on the minimum application of ERm and HL limit. Figure 1 shows the results of foaming characterization under laboratory foaming plant investigation using bitumen grade pen 50/70. The ERm and HL have been investigated at a range of temperature and foaming water content values. The foamed bitumen properties produced using bitumen pen 50/70 is investigated at temperature of 140°C, 160°C and 180°C and at FWC of 1%, 2%, 3%, 4%, and 5%. As shown in the figure 1, Foamed bitumen produced at a temperature of 160°C are determined as the best properties because the ERm values at temperature of 160°C for those 5 variation of FWC are higher than other temperatures, although the values of HL are not the highest. After determining the best temperature then the next step is to determine the optimum value of foaming water content at the best temperature as shown in figure 2.

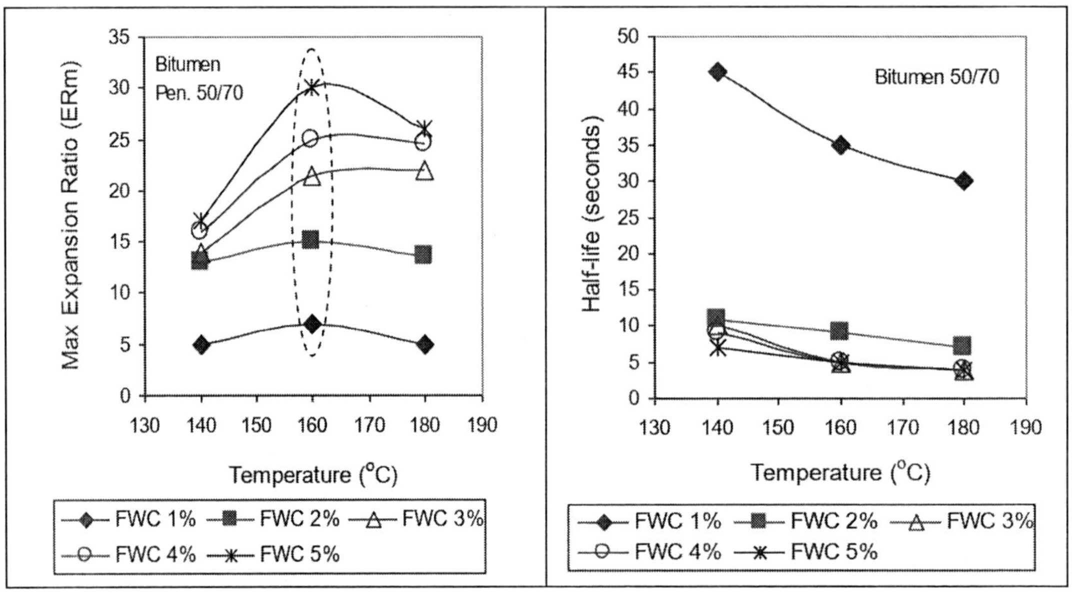

Figure 1. The ERm and HL values at various temperature and FWC using bitumen Pen. 50/70.

Jenkins [8] therefore introduced the Foam Index (FI) concept as a function of ERm and HL in order to find the optimum FWC. Values of FI will increase when values of ERm and HL increase. Optimum FWC could therefore be determined using this FI parameter. Foamed bitumen with high foam index value is indicated to have good properties. The optimum properties of foamed bitumen could therefore be determined from the full range of foaming water content variation at one temperature. However, as shown in figure 2, the relationship between FI and FWC has no optimum value, FI values rises continually with the increase of FWC values. In this case therefore the optimum FWC was selected using Wirtgen method. The optimum foaming water content is decided to be around 1.9% as shown in figure 2.

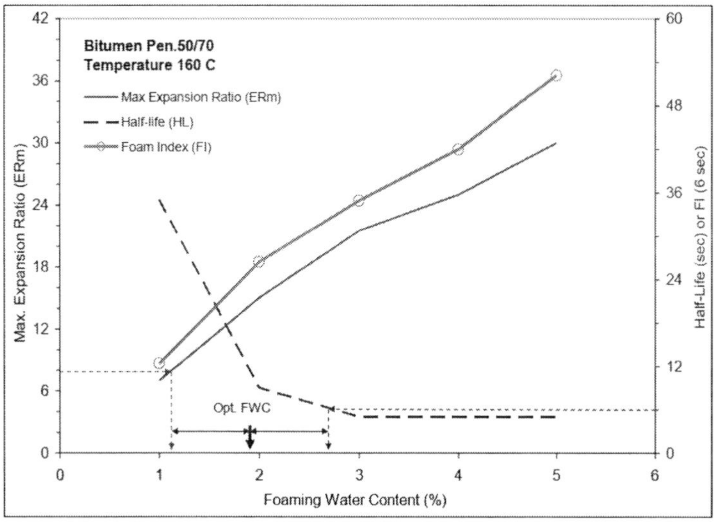

Figure 2. Selecting the optimum value of FWC.

The considerations in determining the bitumen grade for foamed asphalt is actually no different for hot mix asphalt. Both types of mixture are very concerned to the aspect of traffic load type (heavy or light traffic) and regional climate (hot or cold region). It is widely understood that for roads with high traffic load or roads in a hot climate requires bitumen with low penetration. It should be noted that in foamed asphalt mixture, the bitumen grade type will also significantly influence the binder distribution in the mixture during the mixing process, which affect to the stiffness value of foamed asphalt mixture. Thus, in selecting the usage of bitumen grade type of foamed asphalt must consider together the influence of bitumen penetration on the bitumen distribution and on the mixture performance in the field.

It should also be highlighted that bitumen grade types affect the values of viscosity and stiffness modulus. Bitumen with low penetration values will generally have high viscosity and stiffness modulus. Thus, a foamed asphalt generated using hard bitumen (low penetration) will have two sides different consequences. First, it will have negatively consequence because hard bitumen (high viscosity) will reduces workability during mixing. Second, it will have positive consequence because hard bitumen (high stiffness) will enhance layer stability in pavement structure. A foamed asphalt generated using soft bitumen (high penetration) will have opposite situation. Therefore, in the determination of bitumen grade type for foamed asphalt mixture must also consider two important things, i.e. the capability of mixer and the climate in where the road is located. A foamed asphalt produced using hard bitumen is recommended to use a high speed mixer so that the bitumen distribution is good and evenly distributed; but if there is no high speed mixer, it is recommended to use a soft bitumen grade. On the other hand, if a foamed asphalt mixture is to be implemented in a cold region, it is recommended to use higher bitumen penetration because the distribution of bitumen will be more significant to develop stiffness modulus value at low temperature; however, in the non-cold region, a harder bitumen grade type will more suitable because the stiffness modulus of mixture will more significant to develop the stability of foamed asphalt mixture at high temperature.

It should also be noted that consideration to determine the usage of bitumen grade type in the mixing process can be seen using different angles. In order to generate foamed bitumen with high ERm, it is recommended to implement high temperature foaming (160°C - 180°C) for hard bitumen (Pen 50/70) and to implement low temperature foaming (140°C - 160°C) for soft bitumen (Pen 160/220). In this case the influence of bitumen temperature foaming on the bitumen or foamed bitumen viscosity should also be considered properly.

3.2. Aggregate properties consideration

There are at least four steps to prepare aggregate in the mix design process. First, ensure that plasticity index of aggregate is low. Second, ensure that aggregate gradation meet the specification for foamed asphalt e.g. as developed by [9], and filler content should not be less than five percent. Third, the maximum dry density (MDD) and optimum water content (OWC) is determined using Proctor procedure both standard and modified type. Fourth, aggregate sample for about 3-5 batches is prepared carefully, check initial water content of RAP aggregate, and the usage of cement or lime (if needed) should replace the equivalent percentage of mineral filler.

It can be seen that The RAP gradation exhibits coarser according to the recommendation of ideal grading specification for foamed asphalt mixture [9]. The proportion of filler is approached to be 0.3% (using dry method sieving) or 2.1% (using washing method sieving). It is noted that minimum need of filler for foamed asphalt mixture is about 4.5%. As shown in the figure 3 the gradation of RAP was investigated using dry sieving and wet sieving method. Wirtgen [9] suggested using wet sieving system for RAP materials due to the attached fine aggregates problem. It was found that the wet sieving is finer than the dry sieving. The difference between these two sieving methods looks not significant. However, the recommendation of Wirtgen is true in terms of getting the real particle size distribution, not cluster size distribution.

The percentage of RAP proportion should be properly considered. First, the specific gravity of RAP is lower than fresh aggregate. Second, the density of RAP is lower than fresh aggregate. Third, the volumetric characteristic of asphalt mixture using RAP is worse than using fresh aggregate. Fourth, the performance of asphalt mixture using RAP is worse than using fresh aggregate.

Table 2 shows the specific gravity (SG) test of RAP, RAP aggregate and fresh aggregate (FA) materials. RAP aggregate is aggregate resulted from extraction test. In term of RAP materials which contain aggregate and aged bitumen, the density of RAP will be affected by their both component SG and their composition. It can be understood that the presence of aged bitumen will reduce RAP density. SG test is usually followed by absorption test i.e. the aggregate ability to absorb water into their pores. Absorption value of aggregate represents ability to absorb bitumen and adhesive strength of aggregate.

Table 3 shows density test results of several materials using standard and modified Proctor. The materials used for this test are RAP, modified gradation RAP, RAP aggregate, and fresh aggregate. It is supposed that the order of density values between those four materials from low to high is RAP, modified gradation RAP, RAP aggregate, and then fresh aggregate. RAP material has lowest value of density. Modification of the RAP gradation can elevate the density, but it is still lower than RAP aggregate or fresh aggregate.

Table 4 shows the comparison between volumetric performance of foamed asphalt specimens using RAP and fresh aggregate. It can be seen that the specimens using fresh aggregate have a lower void volume. It means that fresh aggregate makes the specimen more compactable than RAP aggregate. The excellence of fresh aggregate can be observed at both VMA and VIM values and at three compaction levels i.e. 15, 30, and 45 passes.

Figure 4 shows rutting performance of foamed asphalt using RAP under Pavement Facility Test. In this case, two different mixture types are compared, namely foamed asphalt using 50% of RAP (50% using fresh aggregate) and using 75% of RAP (25% using fresh aggregate). The surface rutting of the 7 point locations in average along the wheel path of the trial pavement sections has function to the load applications. Rutting was developed to varying degrees in these two types of section. It is found that in terms of the amount of rutting, the mixtures using 50% of RAP is better than using 75%.

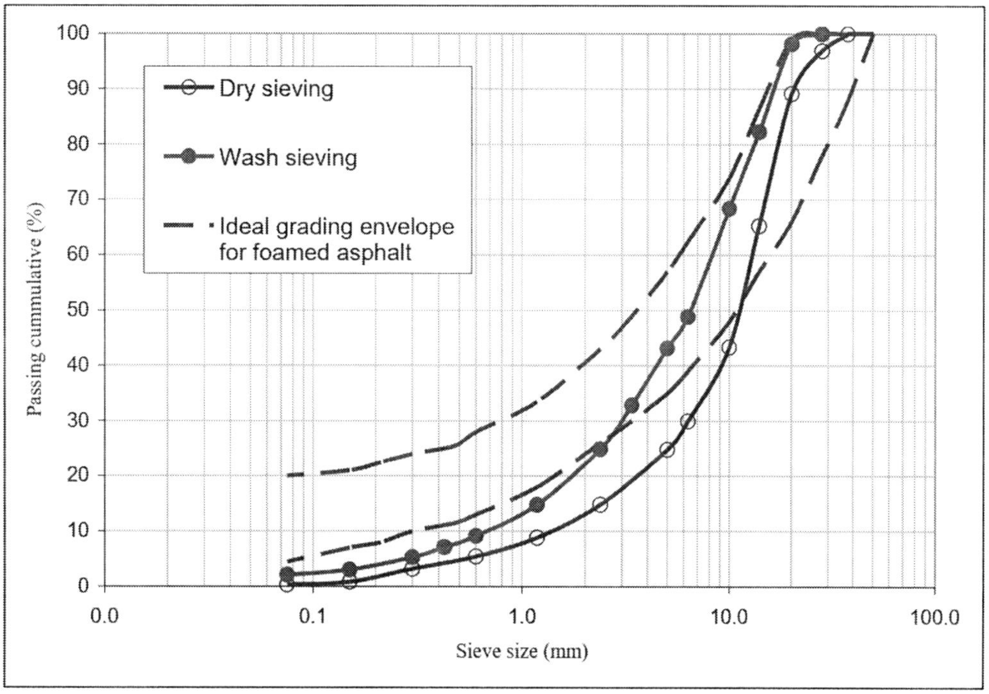

Figure 3. Gradation of RAP sieved using dry and wash method.

Table 2. Specific gravity test results of RAP, RAP aggregate and fresh aggregate.

Properties	Particle size								
	10-20 mm			5-10 mm			< 5mm		
	RAP	*RAP Agg*	*FA*	*RAP*	*RAP Agg*	*FA*	*RAP*	*RAP Agg*	FA
Bulk SG	2.060	2.243	2.937	2.109	2.192	2.762	1.937	1.968	2.565
SSD SG	2.090	2.277	3.014	2.133	2.227	2.841	1.976	2.016	2.688
Apparent SG	2.122	2.322	3.183	2.170	2.272	3.000	2.016	2.068	2.926
Absorption (%)	1.416	1.506	2.632	1.117	1.611	2.878	2.041	2.459	4.822

Note: RAP agg is aggregate from RAP after extracted, whereas FA is fresh aggregate

Table 3. Density of RAP using Standard and Modified Proctor.

No.	Method	Density (gr/cm^3)			
		RAP	**Mod grad RAP**	**RAP Agg**	**Fresh Agg**
1	Standard Proctor	1.640	1.701	NA	2.110
2	Modified Proctor	1.677	1.770	2.010	2.261

IOP Conf. Series: Materials Science and Engineering **403** (2018) 012027 doi:10.1088/1757-899X/403/1/012027

Table 4. Values of VIM, VMA, and VFA for Foamed Asphalt Specimen using RAP and Fresh Aggregate

Pass	RAP			Fresh Agg		
	VIM (%)	VMA (%)	VFA (%)	VIM (%)	VMA (%)	VFA (%)
15	14.24	19.13	25.56	12.8	17.83	28.21
30	13.53	18.45	26.67	12.04	17.01	29.22
45	12.58	17.56	28.36	10.74	15.92	32.54

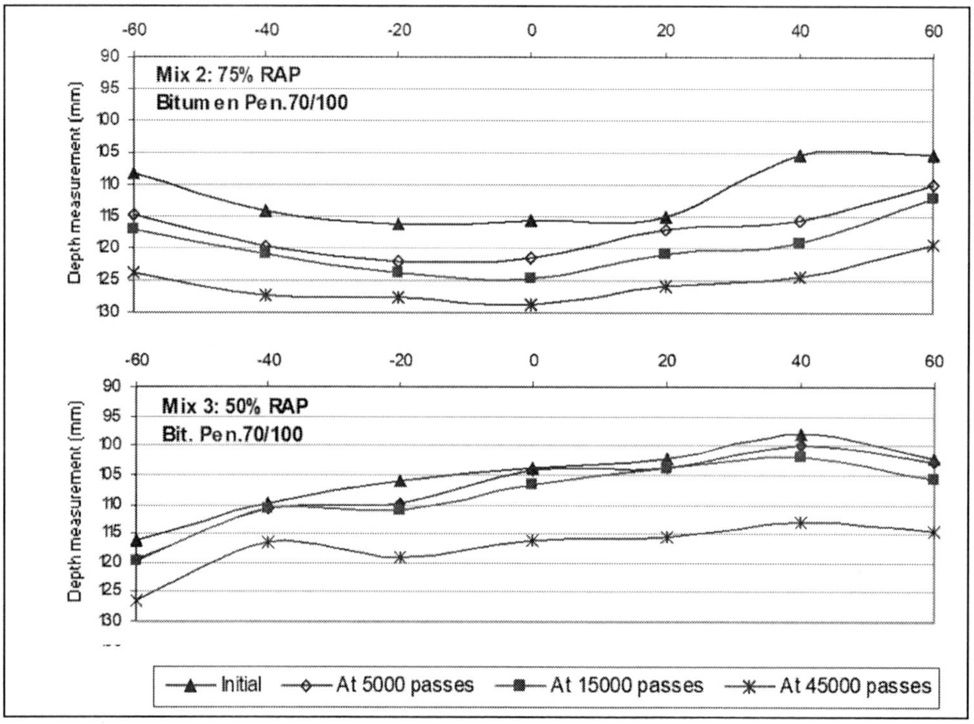

Figure 4. Rutting performance of Foamed Asphalt Using RAP in Longitudinal profiles.

4. Conclusions

The optimum foamed bitumen properties are affected by bitumen grade, maximum expansion ratio and half-life, foaming water content, and bitumen temperature when foaming. The best aggregate properties should consider the plasticity index, aggregate gradation, and the percentage of RAP proportion. The percentage of RAP proportion influence specific gravity, mixture density, volumetric characteristic, and mixture performance.

Acknowledgments

This project is financially supported by the *Direktorat Riset dan Pengabdian Masyarakat Direktorat Jenderal Penguatan Riset dan Pengembangan Kementerian Riset, Teknologi, dan Pendidikan Tinggi* (Indonesian Government) with contract number: 211.4/A.3-III/LPPM/V/2017, and the facilities support of Universitas Muhammadiyah Surakarta.

References

[1] Merill D, Nunn M and Carswell I 2004 A guide to the use and specification of cold recycled materials for the maintenance of road pavements *TRL Report TRL 611* (Wokingham, Berkshire United Kingdom: TRL)

[2] Tambake S O, Kumar D N and Manjunath K R 2014 laboratory investigation on hot mix asphalt using reclaimed asphalt pavement (rap) for bituminous concrete mix *International Journal of Research in Engineering and Technology* vol 03 issue 6

[3] Muthen K M 1999 Foamed Asphalt Mixes, Mix Design Procedure *Contract Report CR-98/077* CSIR Transportek, South Africa

[4] Lee H D and Kim Y J 2003 Developing of a mix design process for cold-in-place rehabilitation using foamed asphalt *Final Report for TR-474 Phase I* Iowa Department of Transportation, Public Policy Center, Civil and Environmental Engineering, The University of Iowa

[5] Wirtgen 2005 Foamed bitumen mix design procedure using the Wirtgen WLB 10 # 147 236.0001

[6] Milton L J and Earland M G 1999 Design guide and specification for structural maintenance of highway pavements by cold in-situ recycling *TRL Report TRL 386* Transport Research Laboratory, Crowthorne

[7] Chiu C T and Huang M Y 2002 A study on properties of foamed asphalt treated mixes Chung Hua University, Department of Civil Engineering, Taiwan

[8] Jenkins K J 2000 Characterisation of Foamed Bitumen. Chapter 3, Mix Design Considerations for Cold and Half-warm Bituminous Mixes with Emphasis on Foamed Bitumen *PhD Dissertation* University of Stellenbosch, South Africa

[9] Akeroyd F M L and Hicks B J 1988 Foamed Bitumen Road Recycling *Highways* vol 56 no 1933 pp 42, 43, 45

The utilization of fly ash in mix proportion of self-consolidation concrete

M Solikin, K O M Oraibi and Y Nurchasanah

Civil Engineering Department Universitas Muhammadiyah Surakarta, Jl. Ahmad Yani Tromol Pos 1 Pabelan Surakarta 57162

msolikin@ums.ac.id

Abstract. The application of Self-Consolidation concrete is needed for casting the concrete element with close-fitting reinforcement. This study was conducted to examine the effect of fly ash utilization on Self-Consolidation concrete properties and also to discover the optimum amount of fly ash as partially cement substitution in increasing mechanical properties of Self-Consolidation concrete. To coop with the problem three different mix design of Self-Consolidation concrete specimens were prepared to accomplish the laboratory experimental program. The mix proportion variation of fly ash replacement is 0%, 15% and 35% from cement weight. The specimens were prepared to test the properties of Self-Consolidation concrete, i.e.: slump flow test, flexural strength test, compressive strength test, and modulus of elasticity. Except slump flow test that was conducted for fresh concrete, the other tests were conducted after 28 days of curing time. After finishing testing programs, it is found that the utilization 15% of fly ash in production of Self-Consolidation concrete improve flexural strength of concrete whereas only a slightly decrease in compressive strength was found.

1. Introduction

One of development in concrete technology is called Self-Consolidation Concrete (SCC) which was developed by [1]. This genre of concrete can flow under its own weight to fill up the formwork and no external vibration is needed to compact it. Even though the fresh concrete is highly workable that flow easily trough restricted section, there is no segregation and bleeding effect to the concrete. In addition, homogeneity of fresh concrete should be maintained to ensure there is no change in quality during transportation.

There are two properties need to be balance in Self-Consolidation concrete i.e. deformability and stability that makes the concrete can be used for casting narrow concrete element or congested reinforcement condition without any external vibration. At the end of the process, a dense and homogeneous concrete will be produce and it has similar mechanical properties and durability as normal vibrated concrete.

Similar to normal concrete, the binder of Self-Consolidation concrete can be combined by using pozzolanic material. One of widely used pozzolanic material that can be used as partially cement replacement is fly ash [2]. Fly ash is non-combustible mineral material, a residual product of coal burned process in power plant which rise with flue gas and collected in dust collection system [3].

The fly ash that has spherical shape is beneficial to increase concrete workability as it would reduce the friction between the particle when it is used as partially cement replacement in Self-Consolidation concrete. Moreover, previous researcher also reported the positive effect of fly ash to reduce bleeding, reduce w/c ratio, and reduce heat hydration. At early stage the strength development of fly ash concrete is lower rather than normal concrete, however after longer period its strength could be higher rather than normal concrete. In addition to the advantage in improving mechanical properties, the reduce of cement consumption will reduce the production of green-house gas into the atmosphere [4-7].

A massive research about the application of fly ash in concrete as cementitious material, as fine aggregate replacement or as additional material has been conducted started in 1930's [8]. However, it needs a continuous study about the utilization of fly ash in Self-Consolidation concrete especially when the fly ash is produce from a local power plant. In addition, the compatibility of fly ash with other concrete ingredients such as superplasticizer and local aggregate need to be tested.

Therefore, this paper will report the mechanical properties of Self-Consolidation concrete by utilization of fly ash as partially cement replacement. The variation of fly ash replacement is design between 15 – 35% from cement content. The mechanical properties test consists of slump flow test, compressive strength test, and flexural strength test.

2. Research methodology

2.1. Material

The material to produce Self-Consolidation concrete in this research consist of cement, fly ash, fine aggregate and coarse aggregate which came from reliable sources were tested to ensure the quality of the material. According to ASTM standard for testing material, fine aggregate properties i.e. specific gravity, absorption and sand equivalent and also coarse aggregate properties i.e. abrasion test, absorption and specific gravity meet the standard as concrete material. In addition, the fly ash which is employed in this experimental program belongs to class F in which the over-all content of $SiO_2 + Fe_2O_3 + Al_2O_3$ more than 70%.

2.2. Mix design

Following the material test, a concrete mix design was prepared with total binder of 442 kg to possibly produce a 40 MPa design strength. Basically, the mix proportion of Self-Consolidation concrete is similar to normal concrete. However, some consideration must be taken i.e. the usage of superplasticizer is compulsory to enable the fresh concrete flow by its weight. Moreover, the content of fine aggregate is around 50% - 60% from total of aggregate, to reduce friction effect from the coarse aggregate.

Table 1. Mix design of concrete.

Mix type	Coarse aggregate (kg/m^3)	Fine aggregate (kg/m^3)	Water (kg/m^3)	Fly ash (kg/m^3)	Cement (kg/m^3)	Sp (% by weight of cement)	w/c
A (0)	866	884	208	0.0	442.0	2% (8.84 kg)	0.47
B (15)	866	884	208	66.3	375.7	2% (7.51 kg)	0.47
C (35)	866	884	208	154.7	287.3	2% (5.75 kg)	0.47

The mix design for this research in table 1 shows three concrete mix designs with different fly ash content i.e. 0%, 15% and 35%. Table 1 shows the reduction of cement content along with the increase of fly ash content. In addition, the mix proportion used 50.5% of fine aggregate to the total of aggregate used. Although 0.47 w/c was used, 2% of superplasticizer by the weight of cement was used to increase the workability of fresh concrete.

After preparing the mix proportion, a series of specimen was casted in regard to the appropriate experimental program. The specimens were removed from the mould after one day of casting and continued by immersing in water for curing process for 28 days. Table 2 shows the matrix of specimen for the experiment which shows the dimension and number of specimens.

Table 2. Specimens for experimental program.

No	Testing program	Specimen	Number of specimen		
			A	B	C
1	Compressive strength	Cube 15 x 15 x 15 cm	3	3	3
2	Flexural strength	beam 15 x 15 x 50 cm	3	3	3

3. Result and discussion

3.1. Slump flow test

Slump flow test was conducted to measure the workability and flowability of fresh concrete to check whether the concrete meet the criteria of Self-Consolidation concrete. The test was carried on by filling fresh concrete into a slump cone and lift up vertically to let the fresh concrete spread on a flat surface. The next step of test is measuring two perpendicular dimeter of the fresh concrete to find out the average diameter of slump flow. Based on ASTM C 1611 the slump flow of self-consolidating concrete in the laboratory or the field, the requirement diameter of slump flow test is 500 – 700 mm. The slump flow test illustration is shown in figure 1.

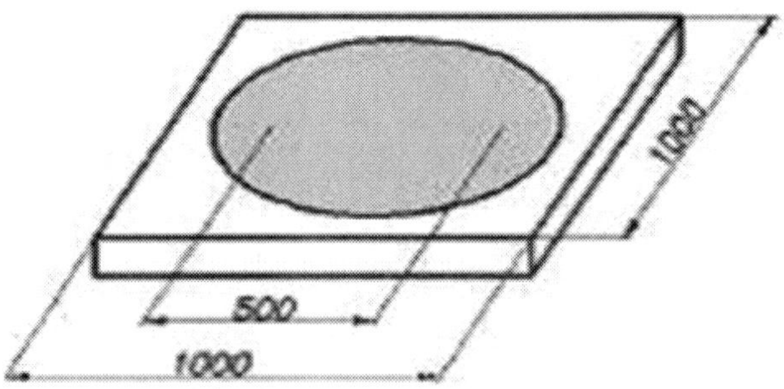

Figure 1. Slump flow test illustration [9].

The observed result as shown in table 3, the diameter of slump flow was increased following the increase of fly ash as partial cement substitution. Similar trend on increasing the slump flow as the amount of fly ash substitution increased also reported by [10-12]. The phenomena closed related to the spherical shape of fly ash.

Table 3. Diameter of slump flow (mm).

Fly Ash substitution	Slump Flow	Average
0 %	571.0	598.0
	625.0	
15 %	613.0	610.3
	658.0	
35 %	687.0	655.0
	640.0	

3.2. Concrete compressive strength test

Concrete compressive strength test was conducted on a concrete cylinder which has 15 cm of diameter and 30 cm of height to find out the strength of the concrete cylinder. The test was carried out after curing the specimen for 28 days and the strength can be found using deploying the formula in which the maximum load was divided by area of cylinder.

Compressive strength test result after curing process for 28 days shows a slightly reduction in compressive strength as the increase of fly ash amount, i.e.: 30.62 MPa for control concrete to 29.37 MPa for 15% of fly ash amount. However significant decrease of strength to 23.06 MPa is found when the amount of fly ash was increased to 35%, as revealed in table 4.

Table 4. Concrete compressive strength (28 days of curing).

Fly Ash	Compressive Strength
0%	30.62 MPa
15%	29.37 MPa
35%	23.06 MPa

Therefore, the concrete compressive strength result is significantly affected by percentage used of fly ash, in which highest compressive strength is belongs to normal Self-Consolidation concrete without Fly Ash in comparison to SCC modified with 15% and 35% of fly ash as partially cement substitution. Similar report was published by [13] who reported that the increase of fly ash used as substitution of cement lead to the lower of concrete compressive strength. The lower the compressive strength is possibly caused by the more the fly ash used the less the C_3A from cement content, a substance which responsible to the early strength of concrete [14]. However, the decrease of strength not too far as besides react with cement hydration product, some fly ash act as filler in concrete which can fill cavities in the concrete to make the concrete product more solid.

Previous researcher such as [15] reported, that fly ash concrete can produce high strength concrete, however it has similar result that the more of fly ash used the lower the compressive strength. In the research the decrease of compressive strength test is from 53.17 MPa at 10% of fly ash used to 39.25 MPa at 50% after 28 days of curing period.

The optimum amount of fly ash used in concrete production was reported by previous researcher, in which fly ash can be used until 30 % of fly ash in increasing self-consolidation concrete compressive strength. However, after the optimum amount of Fly Ash, the compressive strength of SCC is decrease following the increase in Fly Ash content. The difference result in optimum amount of Fly Ash content from this research is possibly affected by the difference content of binder and different w/c ratio. Previous research employs 550 kg/m^3 of total binder with w/c of 0.42 whereas in this research used less binder of 440 kg/m^3 and higher w/c of 0.47.

3.3. Flexural strength test

Flexural test investigates maximum bending load on a concrete beam with the dimension of 15 x 15 cm and clear span of 45 cm. The test is important to analyze the flexural behavior of concrete and to ensure the design mix proportion meet the ductility criteria of concrete. The flexural strength test was conducted based on ASTM C 78, a methods and procedures for unreinforced concrete bending test. The results of flexural strength are presented in table 5.

Table 5. Flexural strength test result (28 days of curing).

Fly ash	Flexural strength (Kg/cm^2)	Flexural Strength (MPa)
0 %	53.9	5.4
15 %	57.1	5.7
35 %	52.8	5.3

The flexural strength test result shows the effect of fly ash content to the flexural strength, which has slightly different trend from concrete compressive strength. The employment of 15% of fly ash is slightly increase the flexural strength, however addition fly ash content of 35% decrease the flexural strength. The fact of slightly flexural strength improvement at 15% of fly ash utilization is possible affected by the use of optimum content of fly ash in improving concrete compressive strength. However, the increase of fly ash content to 35% decrease the flexural strength. Early research conducted in 2011 by [16] reported the same trend of flexural strength result and obtained similarly result, that optimum content of fly ash to improve flexural strength is 15%.

4. Conclusion

After analyzing the result of experimental program on Self-Consolidation concrete with fly ash as partially cement replacement, it can be concluded that the utilization of fly ash is beneficial to improve the fresh concrete properties of Self-Consolidation concrete in which the more the use of fly ash the better the slump flow result. In addition, it is recommended to use 15% fly ash as cement replacement as optimum amount related to compressive strength and flexural strength of concrete.

Acknowledgments

The author expresses gratitude to the staff at laboratory of Civil Engineering Universitas Muhammadiyah Surakarta, who kindly provide the experimental tools and assistance during experimental program.

References

[1] Naik T R and Kumar R 2001 Self-consolidation concrete (SCC) or self leveling concrete (SLC) The University of Wisconsin - Milwaukee

[2] Kosmatka S H, Kerkhoff B and Panares W C 2003 *Design and Control of Concrete Mixtures 14th edition* (Illinois, USA: Portland Cement Association)

[3] Nawy E G 1996 *Fundamentals of High Strength High Performance Concrete* (London, UK: Longman Group Limited)

[4] Bilodeau A and Malhotra V M 2000 High-Volume Fly Ash System: Concrete Solution for Sustainable Development *ACI Materials Journal* vol January-February 2000 pp 41-50

[5] Malhotra V M 1999 Making concrete greener with fly ash *Concrete International* Detroit: American Concrete Institute vol 21 pp 61-6

[6] Oner A, Akyuz S and Yildiz R 2005 An experimental study on strength development of concrete containing fly ash and optimum usage of fly ash in concrete *Cement and Concrete Research* vol 35 pp 1165– 71

[7] Siddique R 2008 *Waste materials and by-products in concrete* (Berlin: London Springer)

[8] ACI 226-3R-87 1987 Use of Fly Ash in Concrete: reported by ACI Committee 226 *ACI* materials *journal* vol Sep-Oct 1987 pp 381-409

[9] Chockalingam M 2014 Experimental investigation on self-consolidation concrete using marble powder and silica fume *International journal & magazine of engineering, technology, management and research* vol 1 pp 15-21

[10] Solikin M, Setunge S and Patnaikuni I 2013 Experimental Design Analysis of Ultra Fine Fly Ash, Lime Water, and Basalt Fibre in Mix Proportion of High Volume Fly Ash Concrete *Pertanika Journal Science & Technology* vol. 21 pp 589 - 600

[11] Prihantoro T F and Solikin M 2015 Perkembangan Kuat Tekan Beton Mutu Tinggi dengan Memanfaatkan Teknologi High Volume Fly Ash Concrete *Simposium Nasional Rekayasa Aplikasi Perancangan Industri (RAPI XVI)* Surakarta pp S-301

[12] Kamran M and Mishra M 2014 Behavior of self-consolidation concrete using ppc and opc with different proportions of fly ash *IJRET: International Journal of Research in Engineering and Technology* vol 03 pp 342-5

[13] Ganeshwaran P A, Suji and Deepashri S 2012 Evaluation of mechanical properties of self-consolidation concrete with manufactured sand and fly ash *International journal of civil engineering and technology (IJCIET)* vol 3 pp 60-9

[14] Solikin M 2012 High Performance Concrete with High Volume Ultra Fine Fly Ash Reinforced

with Basalt Fibre *Doctor of Philosophy* School of Civil, Chemical and Environment Engineering Melbourne: RMIT University p 203.

[15] Ankush E G, Rajeev C and M K K 2014 Influence of Fine Aggregate Particle Size and Fly-ash on the Compressive Strength of Mortar for SCC *International Journal of Scientific and Research Publications* vol 4 pp 1-6

[16] Gaywala N R and Raijiwala D B 2011 Self-consolidation concrete: a concrete of next decade *Journal of Engineering Research and Studies* vol II pp 213-18

IOP Conf. Series: Materials Science and Engineering **403** (2018) 012029 doi:10.1088/1757-899X/403/1/012029

Analysis of pipe diameter variation in axial pumps for reducing head loss

E Widodo and R Y Pradhana

Universitas Muhammadiyah Sidoarjo, Indonesia

ediwidodo@umsida.ac.id

Abstract. Axial type water pumps are widely used in daily activities. The size of the output pipe has a significant influence on the efficiency and performance of the pump performance. By changing the size of the pump will cause changes in flow velocity and fluid pressure on the pipe. The PVC pipe used in this test is 1, 1/2 ', 1', 3/4 ', 1/2' 'inches. By varying the pipe press carried out measurements of flow speed generated, as well as the flow of the water in the pipe. The length of the pipe causes the loss of the head, because the friction of fluid with the pipe surface is wider. While the pipe size change that will cause fluid flow change. From the test results obtained fluid flow that occurs is turbulent flow with Reynold numbers 17.050 up to 59.884. The lowest head loss value occurs on the 1.5 "pipe size of 2.504, with a water debit of 41x10-5 m3 / s. The water flow through the pipeline causes the Reynolds' to be lower, if the water flow is low. The tendency of the liquid stream to be laminar, and the flow rate becomes lower.

1. Introduction

The axial pump is a positive displacement pump with rotary components such as lobes, gears, threads, vanes, and rollers. The way it works is to pump the liquids to the suction side, and push it into the gap or the press room between the pumping components, then pressed so that the smaller the gap further liquid is removed through the side of the exhaust. The fall in pump performance suddenly and instability in the operation would be a problem, an indication of the cause of the decline in the performance of the pump is one caused by cavitation [1]

The problem raised in this study was to find out whether there was any effect on the size of the rotary PVC pump pipe on the speed of the water thrust. There is also this study discussing the size of PVC pipe with a variety of diameters (1½", 1", ¾", ½") discusses the rotational force of the pump on the PVC pipe and discusses the speed of water pressure in each variation of PVC pipe diameter, but this research except the discussion about the rotary pump components of the machine.

This study was conducted with the aim of knowing on the size of PVC pipe required for a rotary pump speed of water thrust, knowing the thrust on the tested rotary pump and for knowing the value of the water compressive velocity on a predetermined PVC pipe. The benefit of this research to know about the size of PVC pipe with the use of rotary pump, to know the speed of water coming out with various sizes of PVC pipe and know the thrust of the rotary pump.

2. Methodology

PVC pipe used with diameter variation (1½", 1", ¾", ½"). The research steps on the pump which is assembled starting from the planning of PVC pipe installation, installation of PVC pipe, data collection, data processing, discussion and conclusion. Data analysis performed using SPSS for Windows 10 software.

2.1. Fluid flow in the pipe

In the drainage process inside the pipe there is a loss of the pump pressure which is often called head losses. The flow type in the pipe is a turbulent type, since the value of the Reynolds number [2] and [3]. There are several fluid streams in the pipeline that are steady flow and flow on a straight pipe. On straight pipeline flow is classified into laminar and turbulent flow. To determine whether a flow is laminar or turbulent a Reynold number as in table 1 can be used using equation (1).

$$\mathrm{Re} = \frac{V.D}{v} \tag{1}$$

Where:
Re = Reynold number.
V = average of velocity of flow in pipe (m/s).
D = inner diameter of pipe (m).
V = kinetic viscosity (m^2/s).

Table 1. Reynold number.

Reynold number	Characteristics
< 2000	Laminar flow
> 4000	Turbulent Flow

2.2. Regression

Regression in the modern sense [4, 5] is the study of the dependence of a variable, which is variables depend on one or more other variables or it called explanatory variables in order to make estimates and / or prediction about the average population or the mean value.Variable values are dependent in relation to know the values of their explanatory variables. According to [6] and [5], the regression is used to determine the nature and strength of the relationship between two unknown variables and the view on past observations of the variables and other variables. Since Levin & Rubin in defining the regression also use the term "correlation analysis", it is preferable in this section that the authors need to explain the difference between regression and correlation. According to [4] and [5] correlation analysis aims to measure the strength or degree of linear association between two variables.

2.3. Head losses

The compressive height is the height of the rising fluid column to obtain an amount of energy equal to that of one unit of fluid weight under the same conditions. A compressive loss is one of the unavoidable losses in a fluid flow in the form of reduced pressure on a stream, and the results on reduced flow velocity. One of the most common and irreversible disadvantages of piped water flow is compressive losses due to friction and cross-sectional change or pipe bends that interfere with normal flow. This causes the water flow to become weaker and smaller [3, 7]. Head Losses is obtained by using equation (2).

$$hl = f\frac{LV^2}{D2g} \tag{2}$$

Where
D = pipe diameter (m).
V = velocity average of flow in pipes (m/s).

L = pipe length (m).
G = acceleration of gravity (9.81 m/s^2).

2.4. *Coefficient of losses*

In a stream through a pipe system or installation there is a flow resistance. These barriers are caused by form-installation factors. These barriers can cause a decrease in energy from a fluid that is often called a head loss or pressure drop caused by the influence of friction losses and changes in flow patterns. In the laminar flow conditions, the frictional resistance is only affected by the viscosity of the fluid. However, in the turbulent flow the resistance is affected by the viscosity of the fluid and the pipe surface roughness. The effect of pipe surface roughness was first researched widely by Nikuradse. The results of his experiments show that surface roughness greatly affects the flow of high Reynolds, the coefficient of friction depends on the Reynolds number. Von Karman lowered the formula for turbulent flow by including surface roughness. The result of the decrease is shown in equation (3).

$$\frac{1}{\sqrt{f}} = 1.14 + 2\log\frac{D}{e} \tag{3}$$

Blasius equation also describes the coefficient of friction for turbulent flow as shown in equation (4).

$$f = 0.3164\, Re^{-\frac{1}{4}} \tag{4}$$

According to [8], Moody developed the experimental results of Nikurades into a mathematical model and succeeded in plotting a graph of friction coefficient relation with the Reynold number of turbulent flow with surface hardness variation. In this study as for materials and equipment used are as follows:

a) Centrifugal pump.
b) PVC pipe with variation diameter 1½ ", 1", ¾ "and ½".
c) Knee.
d) Chainsaw saws.
e) Meter.
f) Lock pipe.
g) Glue pipe.
h) Stopwatch.
i) Glass measuring cup.
j) Stop the faucet.
k) Tee.
l) Sok.
m) Water union.
n) Water reservoir.

After preparing the tools and starting materials with the installation of the pump installation, testing the flow velocity of the water fluid, the measurement of the flow of water flow until the calculation of thrust that can release water quickly. The operational procedure are:

a) Provide tools and materials needed in research.
b) Install the rotary pump used.
c) Install PVC pipe to know the thrust of the water.
d) Turn on the prepared pump.
e) Begin research with a comparison of PVC pipe with various sizes by opening the lid of the faucet.
f) Data retrieval and data processing required.

Figure 1 shows the research installation design.

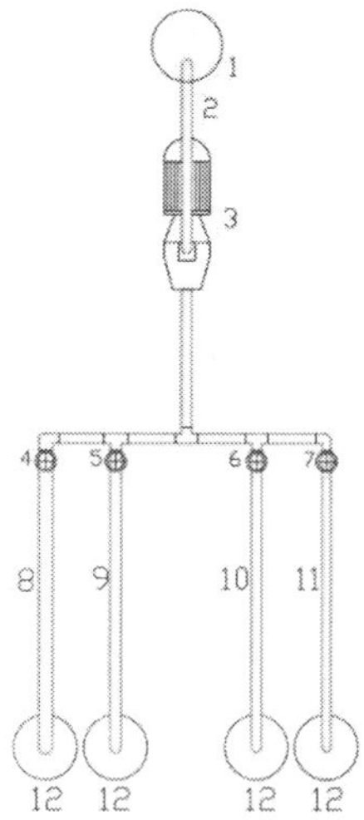

Where:
1. Water container
2. Suction pipe ¾"
3. Pump
4. Valve 1
5. Valve 2
6. Valve 3
7. Valve 4
8. Pipe 1½"
9. Pipe 1"
10. Pipe ¾"
11. Pipe ½"
12. Water reservoir

Figure 1. Research installation design.

3. Result and discussion

Average data time required to move 5 liters of water from tub A to tub B PVC press pipe on rotary pump with diameter ½" is 10.361 s, diameter ¾" is 10.695 s, diameter 1" is 10.944 s, and the diameter of 1½" is 19.931 s. Then from the obtained average it is continued with regression analysis using SPSS 20. The correlation is shown in table 2, regression is shown in table 3 and hypothesis is shown in table 4.

Table 2. Model Summary[b]

Model	R	R Square	Adjusted R Square	Std. Error of the Estimate
1	0.990[a]	0.980	0.970	0.116739

a. Predictors: (Constant), X1
b. Dependent Variable: Y1

Table 3. Anova[a]

Model		Sum of Squares	df	Mean Square	F	Sig.
	Regression	1.343	1	1.343	98.532	0.010[b]
1	Residual	0.027	2	0.014		
	Total	1.370	3			

a. Dependent Variable: Y1
b. Predictors: (Constant), X1

Table 4. Coefficients[a]

Model		Unstandardized Coefficients		Standardized Coefficients	t	Sig.
		B	Std. Error	Beta		
1	(Constant)	9.514	0.159		59.801	0.000
	X1	1.567	0.158	0.990	9.926	0.010

a. Dependent Variable: Y1

In the coefficients in table 4, the column B Constant (a) is 9.514, while the value (b) 1.567, so the regression can be written as Y = a + bX or Y = 9.514 + 1.567X.

3.1. Debit calculation

Obtained Q (discharge) with known volume of 5 liters which was first converted to 0.005 m^3 and to obtain the discharge calculated by using an equation (5).

$$\text{If } V = Q.t \text{ then } Q = \frac{V}{t} \quad (5)$$

Where:
V = Volume (m^3)
Q = Debit (m^3/s)
T = Time (s)

Then the average of 5 liters discharge is obtained, followed by manual calculation to know the head losses.

3.2. Calculation of cross-sectional area

Calculate the flow velocity (v) by using the formula Q = v. A, since the cross-sectional area is not known then, look for the cross-sectional area using the formula $A = \pi r^2$ or $A = \frac{1}{4}\pi D^2$ [6], where D is the diameter of pipe 1 inch = 0.0254 m.

3.3. Calculation of flow velocity

The flow velocity (v) is calculated using equation (6).

$$V = \frac{Q}{A} \quad (6)$$

Where:
Q = Debit (m^3/s)
V = flow velocity (ms)
A = Area of cross section (m^2)

3.4. Calculation of Reynolds number

There are several fluid streams pipe that is steady flow and flow on straight pipe. In straight stream pipes are classified into laminar flow, transition stream, and turbulent flow. To determine whether the flow is laminar or turbulent can be used Reynolds number

$$Re = \frac{V.D}{v} \quad (7)$$

Where:
Re = Reynold number
V = average velocity of flow in pipe (m/s)
D = inner diameter of pipe (m)
V = kinetic viscosity (for temperature $30^0 = 0.804 \times 10^{-6}\, m^2/_s$)

3.5. Calculation of coefficient losses

When calculating the losses head using turbulent flow because all of Re > 4000 which includes turbulent flow. The calculation of coefficient losses is shown in equation (8).

$$f = 0.3164\ Re^{-\frac{1}{4}} \quad (8)$$

Head Losses are obtained using equation (9) and the result of head loss calculation is shown in table 5 and figure 2.

$$hl = f\frac{LV^2}{D2g} \quad (9)$$

Where:

D = pipe diameter (m)
V = average velocity of flow in pipes (m/s)
L = pipe length (m)
G = acceleration of gravity (9.81 m/s^2)

Table 5. Calculation of head losses.

Diameter of Pipe	Head Losses	Debit (m^3/s)	Velocity of flow (m/s)	Coeff Losses (f)	Reynold number
½"	1,141.957	$48x10^{-5}$	3.791	$20x10^{-3}$	59,884.139
¾"	123.476	$46x10^{-5}$	1.615	$22x10^{-3}$	38,259.311
1"	6.488	$45x10^{-5}$	0.889	$24x10^{-3}$	28,070.690
1 ½"	2.504	$41x10^{-5}$	0.359	$28x10^{-3}$	17,050.345

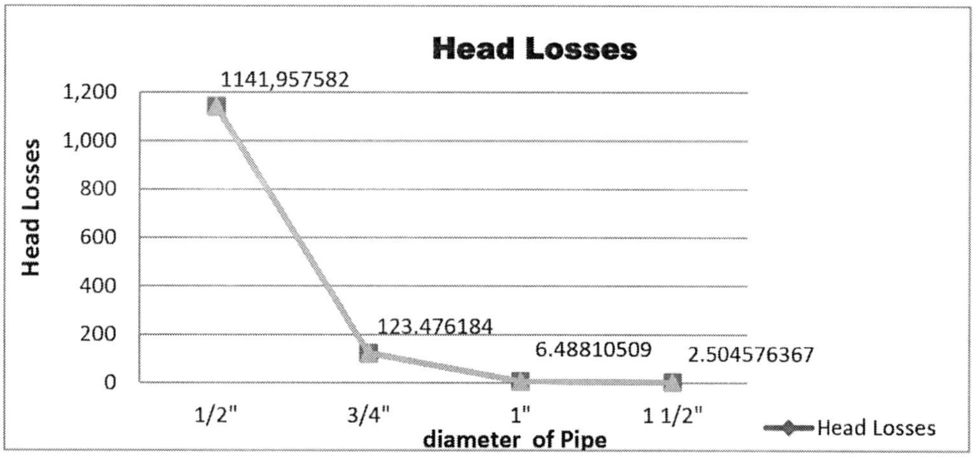

Figure 2. Graph of head losses.

Figure 2 shows the graph of data collection diameter PVC pipe with a variety of pipe size ½", ¾", 1", 1 ½". The graph in figure 2 explains the highest value of Head Losses has a value of 1,141.957582 with ½ " diameter pipe and the lowest value has a value of 2.504576367 with a diameter of 1 ½ " pipe. In the graph above shows the higher the value of Head Losses, the higher the thrust force in the pipe and the thrust force it is in the ½ "diameter pipe.

Figure 3 shows that the highest point of water discharge on a ½" PVC pipe with a value of 48 × 10-5 m^3/s and the lowest point of PVC pipe water discharge on a 1 ½" pipe size with a value of 41 × 10-5 m^3/s. Graph in figure 3 shows that the PVC pipe size ½" is higher or faster in moving water from tub A to tub B while the PVC pipe is 1 ½" slower, this is because PVC pipe diameter is very influential in the soon or later the water is flowing.

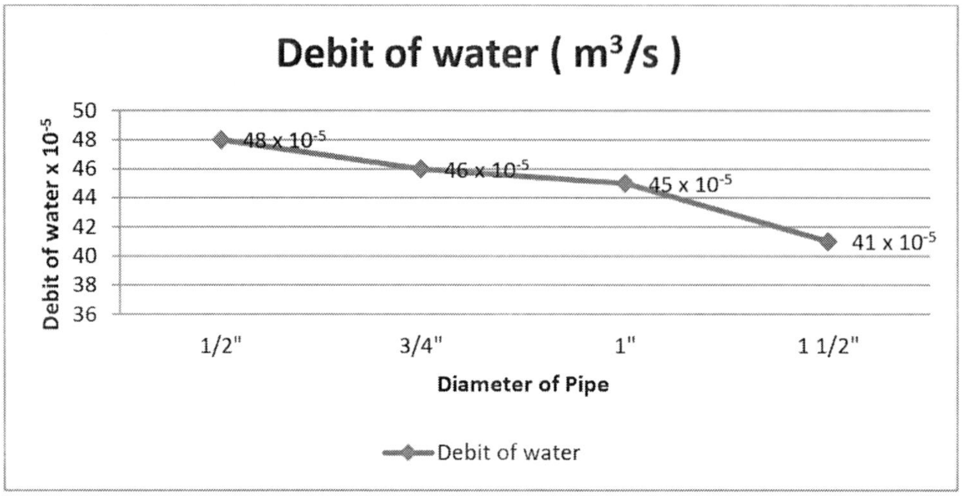

Figure 3. Graph of debit of water.

In figure 4, the comparison of the graph of the water discharge data to the Reynolds number explains that the smaller the diameter of the PVC pipe then the water debit value will be large and comparable with the Reynold number value the smaller the size of the PVC pipe then the Reynold number will be greater, The larger the diameter of PVC pipe then the smaller the value of the water debit and the value of the Reynold number.

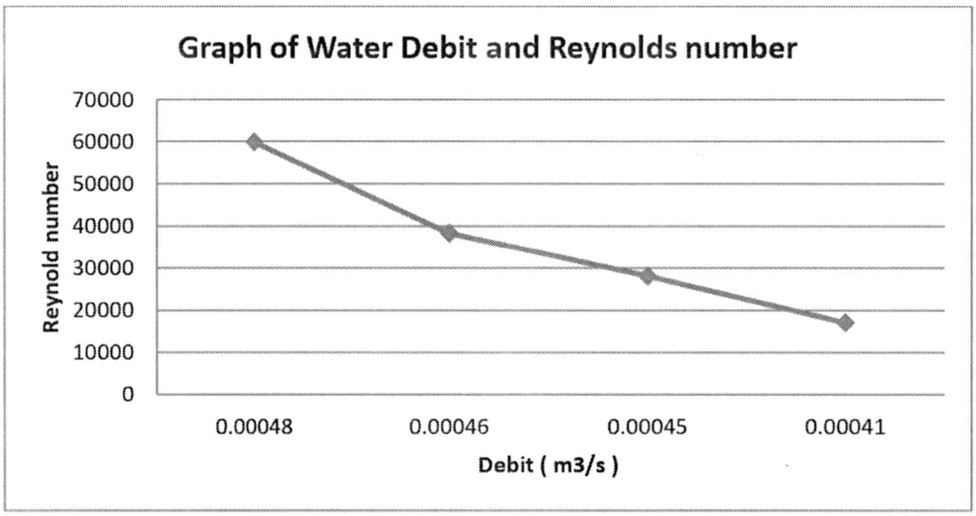

Figure 4. Graph of water debit and Reynolds number.

4. Conclusion

Based on the data obtained, the pipeline further accelerates the water push force on the rotary pump with the diameter of the pipe ½". In this research can be obtained data showing that PVC pipe diameter pipe ½" can accelerate the water push force of the rotary pump with the last time. This water-push acceleration study using PVC pipe with different variations in the rotary pump can be a good reference for the use of pvc pipe on rotary pumps.The smaller the size of the PVC pipe taken, the faster the resulting thrust force. The thrust energy generated on the PVC pipe diameter of 1/2" pipe in this study can be used to save the cost and time required to drive water using a rotary pump as a vacuum.

References

[1] Alfayez L, D Mba D and Dyson G 2005 The application of acoustic emission for detecting incipient cavitation and the best efficiency point of a 60kW centrifugal pump: Case study *Ndt & E International* vol 38 no 5 pp 354-8

[2] Ozbayoglu E M and Sorgun M 2010 Frictional pressure loss estimation of non-newtonian fluids in realistic annulus with pipe rotation *Journal of Canadian Petroleum Technology* vol 49 no 12 pp 57-64

[3] Chiaramonti D, Prussi M, Casini D, Tredici M R, Rodolfi L, Bassi N, Zittelli G C and Bondioli P 2013 Review of energy balance in raceway ponds for microalgae cultivation: re-thinking a traditional system is possible *Applied Energy* vol 102 pp 101-11

[4] Juana L, guez-Sinobas L R and Losada A 2004 Closure to "Determining Minor Head Losses in Drip Irrigation Laterals. I: Methodology" by Luis Juana, Leonor Rodr ı′ guez-Sinobas, and Alberto Losada *Journal of irrigation and Drainage Engineering* vol 130 no 3 pp 255-6

[5] Provenzano G, Pumo D and Dio P D 2005 Simplified procedure to evaluate head losses in drip irrigation laterals *Journal of irrigation and drainage engineering* vol 131 no 6 pp 525-32

[6] Geem Z W, Kim J H and Loganathan G V 2002 Harmony search optimization: application to pipe network design *International Journal of Modelling and Simulation* vol 22 no 2 pp 125-33

[7] Romeo E, Royo C and Monzón A 2002 Improved explicit equations for estimation of the friction factor in rough and smooth pipes *Chemical engineering journal* vol 86 no 3 pp 369-74

[8] Fang X, Xu Y and Zhou Z 2011 New correlations of single-phase friction factor for turbulent pipe flow and evaluation of existing single-phase friction factor correlations *Nuclear Engineering and Design* vol 241 no 3 pp 897-902

An analysis of intelligent LED emergency lamp with voltage and resistance activated sensor

E N Budisusila and J P Hapsari

Electrical Engineering Department, Sultan Agung Islamic University (UNISSULA)
Jl. Raya Kaligawe km.4 Semarang, Indonesia.

ekanbs@unissula.ac.id, jenny@unissula.ac.id

Abstract. LED as illumination lamps has become popular because of its efficiency to save energy, especially electricity. This lamp can be operated as normal lamp connected to AC power line but if the failure of power line electricity occurs, the lamp is still on for hours depends on battery capacity. This lamp circuits have voltage and resistance sensor in the end of circuit connected to threaded (screw) electrode of the lamp. This paper analyzes the principle and performance of the circuit, especially to voltage and resistance threshold activation, and the other factor that can cause success or failure operation. The analysis concludes that the voltage from power line installation can activate the lamp to bright up and to charge its battery at once. When the supply of power source is failure, it triggers the circuit to keep the lamp on with battery operated, by term and condition applied that the connection to line power installation established. It is caused of resistance sensor that detected no resistance value in loop circuit connection. This lamp will be able to operate correctly in single line load circuit, and only one lamp as line load, no other resistive or inductive load.

1. Introduction

LED as illumination lamps has become popular because of its efficiency to save energy, especially electricity. The LED consumes less voltage and current to produce more bright light, which means using minimal watt can produce optimal bright light. Finally, the LED lamps production becomes massive in the market. One of them is used for emergency lamps with DC or battery power supply and can be recharged with AC power line.

There are so much variations of design, with small or big size depends on consumer needs. The sample of product design is like threaded lamp but there is a LED circuit with small battery inside. This lamp can be operated as normal lamp connected to AC power line, but if the failure of power line electricity occurs, the lamp is still on for hours depends on battery capacity. Uniquely, this lamp can be switched off by disconnecting its contact to power line installation. It is not like conventional emergency lamps, when the electricity failure, the lamp is switched on, but if the electricity establishes, the lamp would be switched off and the battery is being charged [1, 2].

It is important to analyze the unique system inside these lamps and to know how the system works. So, this article would like to explain the analysis done by exploring main circuit of the lamp with its operational steps and characteristics.

2. System design

The diagram in figure 1 shows that LEDs circuit is driven by main circuit with two sub circuits inside, they are voltage activated sensor circuit and resistance activated sensor circuit. Due it has two sub circuits, so it also has two inputs, voltage and resistance. As the emergency lamp, this system must have the energy source from battery unit to supply voltage to the circuits [3, 4].

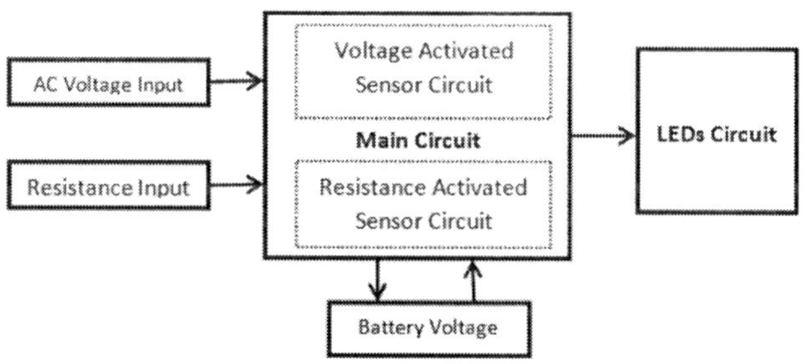

Figure 1. LEDs circuit system design.

3. Research method

The research is conducted by observing physical shape, functional characteristic and operational performance of the lamp. Physical shape contains it design and operation manual procedures, including power dissipation in watt and voltage optimal range in volt listed on its body name plate. Functional characteristic and operational performance is done by implementing the lamp on actual line installation and testing it in dummy circuit load, as behalf of voltage or resistance. This research is different from previous research that conducted to analyze its characteristic only without its operational performance and implementation test.

4. Data and analysis

There are about five components in LED intelligent emergency lamp unit. The 3,7V battery as emergency power supply, main board with sub circuits included that also called smart boards, the wick as LEDs circuit board, light body as the cover of lamp unit, and screw/threaded electrode connector as voltage and resistance input line are shown in figure 2.

Figure 3 shows the circuit with AC voltage input rectified by rectifier circuit block which supplies LED circuit to be on. There is two kind of LED circuits, serial one and parallel one. The serial one is functioned to be light and to be line rectified voltage to charge the battery at once when the voltage power supply is established. And the parallel one is just being light, and to be indicator that the voltage is enough to supply battery. The important one to note here is the battery circuit; it has a control system connected to resistance input circuit showed in figure 4 [4, 5].

IOP Conf. Series: Materials Science and Engineering **403** (2018) 012030 doi:10.1088/1757-899X/403/1/012030

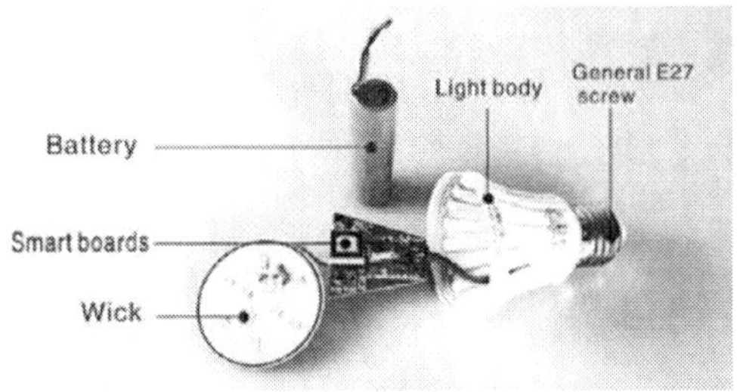

Figure 2. LED emergency lamp components.

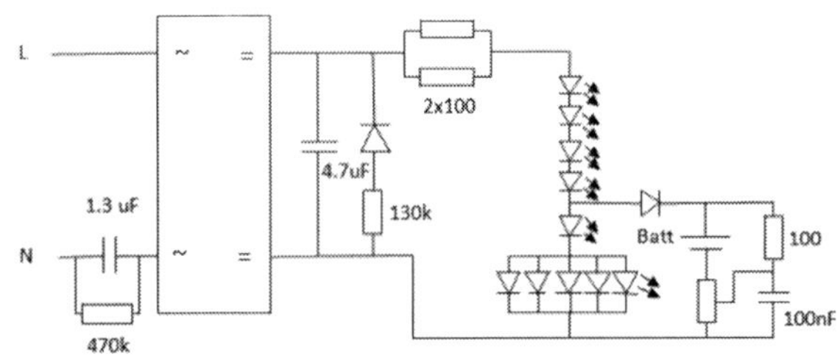

Figure 3. Voltage input circuit.

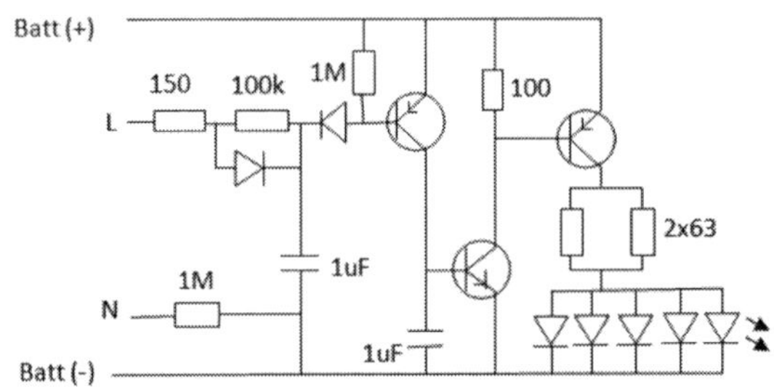

Figure 4. Resistance input circuit.

The resistance input circuit will work when energy failed and no voltage detected in L and N inputs. Then the circuit will check the inputs, if there is detected value resistance threshold or short circuited, the battery discharged to supply LEDs lamp circuit, otherwise the LEDs circuit will be shut down and the battery is in standby mode.

Three tests are conducted to know operational characteristic and performance of the LED lamp unit. Every test is using 3 lamps with difference watt. Figure 5 is scenario diagram to test lamp using a transformer unit as behalf of electricity network in real installation line. By estimating that the battery in the lamp unit is fully charged, S1 and S2 switched to open and close. First step, all switches S1-S2 are closed, the current flows and the lamp lights on. When S1 is opened that denotes failed energy, the lamp is still on, because inductor coil in the transformer is still connected to the lamp and gives the impedance input as resistance to the lamp. But, if S2 is opened the lamp will be switched off, because there is no voltage and no resistance input.

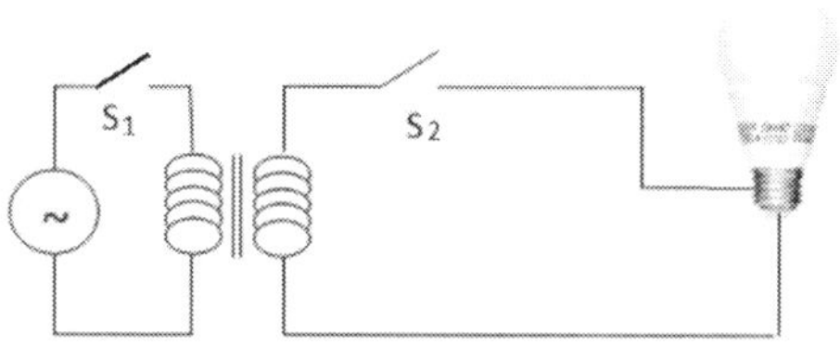

Figure 5. Voltage activation input test diagram.

In figure 6, there is no external energy operated. When S2 is closed, it means that the lamp electrode is short-circuited, the lamp will on. And when S2 is opened and S1 is closed, the electrode will meet variable resistor VR unit that able to change the value of its resistance. It is about 10 Mega Ohm (MΩ) variable resistors, considering of dry skin resistance in about 1 MΩ. When the VR is set to skin resistance range (about 0 – 1M), the lamp stays to keep light on. Otherwise, the lamp will be off [6, 7].

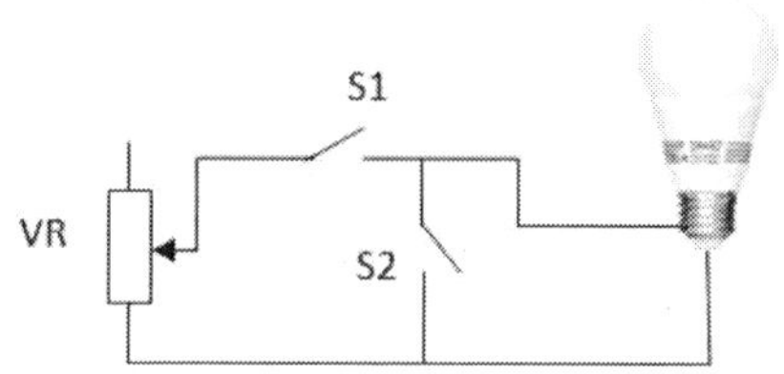

Figure 6. Resistance input test diagram.

If resistive or inductive load added on the diagram of figure 5 become to diagram in figure 7, the lamp will always on although the switch S is opened. It is because off the resistance value detected by resistance sensor in the lamp unit. Resistive or inductive load can be other lamp, electric fan, wash machine, radio or television, computer unit, charger, or the other electric/electronics devices. Capacitive load takes no effect to this system, because if switch S2 is opened, there is only DC power supply of lamp internal battery unit operated.

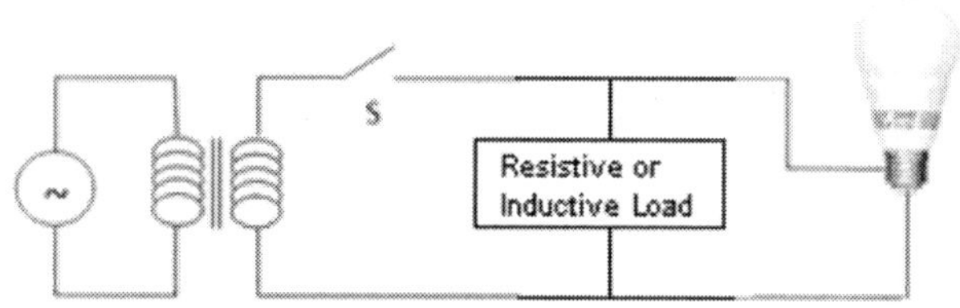

Figure 7. Inductive or resistive parallel load diagram.

5. Conclusion and suggestion

An intelligent LED emergency lamp has main circuit with two sub circuits, voltage activated circuit and resistance activated circuit, to control the light on and the battery charging time. The lamp can operate correctly when it becomes a single load in a single line, no other resistive or inductive load. Range of

IOP Conf. Series: Materials Science and Engineering **403** (2018) 012030 doi:10.1088/1757-899X/403/1/012030

resistance to activate the lamp is about 0 Ω (short circuit) to 1 MΩ (dry skin resistance). To operate this lamp well, it is strongly recommended not to use it in parallel connection with other electric device unit although with the same lamp.

References

[1] Budisusila E N and Arifin B 2017 Joule-thief circuit performance for electricity energy saving of emergency lamps *Int. Conf. Electr. Eng. Comput. Sci. Informatics* vol 190 no 1 p 012017

[2] Harjunowibowo D 2015 Simple blocking oscillator performance analysis for battery voltage enhancement *J. Mob. Multimed.* vol 11 no 3-4 pp 321–29

[3] Li H, Wakimoto T, Murase K and Tomiyoshi K 2012 0.25 μm 1.2 MHz Boost-PFM-Continuous triple mode LED driver integrated in analog front end IC for portable application *Analog Integr. Circuits Signal Process. Int. J.* vol 72 no 3 pp 595–603

[4] Liu Z and Lee H 2015 Design of high-performance integrated dimmable LED driver for high-brightness solid-state lighting applications *Analog Integr. Circuits Signal Process. Int. J.* vol 82 no 3 pp 519–32

[5] Pope C A 2013 Testing the efficiency of a joule thief LED light circuit (California: University of Southern California)

[6] Boucsein W 2013 *Electrodermal Activity* 2nd ed (US: Springer US)

[7] Reilly J P 1998 *Applied Bioelectricity: From Electrical Stimulation to Electropathology* 2nd ed. (New York: Springer-Verlag New York)

Prototype of online expert system for electrical energy audit with benchmarking method

A Solichan and L Assaffat

Electrical Engineering, University Muhammadiyah Semarang, Semarang, Central Java, Indonesia

solichan@unimus.ac.id

Abstract. An energy audit as the primary process of an energy conservation program should be the responsibility of all parties. This should be done not only because of the government's advice but also driven by the energy crisis while alternative energy is not enough to meet energy needs. In fact, the task of energy audit has not been widely applied in the community due to the lack of popularity of this activity and the difficulty of its implementation. The aim of this research is to make efforts to succeed energy conservation by making computer applications as a tool in the implementation of the electrical energy audit. The method of drawing conclusions and recommendations of the electrical energy audit is done by benchmarking process. This application is made in two modes that are web-based and mobile. The original application is web-based to improve data storage and processing capabilities. While mobile applications as navigation and guides are made to improve the mobility of energy audit implementation. The results showed that with the development of this application the energy audit is increasingly known to the public and the audit process can be carried out more easily.

1. Introduction

Increased energy consumption due to technological and industrial developments that have continued to increase since the industrial revolution up to now and the development of population caused the crisis of energy and global warming [1]. Some research explores for new and renewable energy as an alternative [2]. At now the supply of renewable energy has not fully become alternative energy, therefore it needs the best management and energy efficiency [3]. Energy management and efficiency are believed to be a major step towards for solution the energy crisis and global warming [4, 5]. Energy conservation is the responsibility of government and society [6]. Unfortunately, the responsibility of energy management is not yet popular in Indonesia. According to official ESDM web data for new and renewable energy and energy conservation, by 2017 the number of energy managers across Indonesia will be 275 persons [7]. This shows the level of implementation of management and energy audit in Indonesia is still very low. Factors causing energy audits have not been widely implemented namely the assumption that Indonesia is endowed with abundant natural resources and lack of widespread training on energy audits [8].

Advancement of information technology is very useful as a tool and support activities because of its superiority in the collection, access, processing and data analysis [9]. Information technology has an important role in delivering the right information to the right target [10]. Good Search Engine Optimization is very influential on the success of information a web [11]. Community difficulties in the energy audit process can be assisted by a system of energy audit experts [12]. Expert systems will guide the energy audit process until an energy audit report is generated [13].

The development of mobile computing using gadget equipment gives a powerful impact on information systems. Mobile equipment becomes the solution for jobs that require high flexibility [14]. Energy audits are conducted in two stages: initial audit and detailed audit [15, 16]. Field trips are conducted both in the initial audit and detailed audit stages and are in dire need of flexibility and mobility. This problem is solved with a mobile-based application. EBTKE Directorate General of ESDM Ministry has made a web-based application online report on energy management but has not yet covered its energy audit process [17].

The aim and state of the art of this research are to make online application server web and mobile to facilitate the sustainability energy conservation program. The audit process is slightly different from SNI 03-6196-2000 and SNI 6196-2011 that is by conducting all audits to study the relationship of the initial audit result with the detailed audit.

Hypothesis: with the development of electrical energy audit applications, the energy management can be applied more widely and users can run the audit process with more ease and more precise results.

2. Process of electrical energy audit

Energy audits as the backbone of energy management, aimed at calculating the amount of energy consumption in buildings, recognizing the savings opportunities and ways of saving. Energy audits are conducted by a mini-audit process (walk through audit) followed by preliminary and detailed audits [15, 16, 18]. Audit data is contained in a printed worksheet or using a computer application. Analysis of energy-saving opportunities can be done with the help of computer programs (computer aided) that has been recognized profession society [19].

The walk-through audit is a site visit of the building utility to visually inspect each of the energy using systems. This includes an evaluation of energy consumption data to analyze energy use quantities and patterns as well as provide comparisons for industry averages or benchmarks for similar facilities. It is a preliminary estimate and provide a list of low cost savings opportunities through improvements in operational and maintenance practices. This audit process into three components, pre-site work, site visit and post-site work. The pre-site work is collect and review at least two years of uses energy data and accuracy of the billings. The site visit will be spent inspecting actual systems and answering specific questions from pre-site review, fill out the audit data sheets, and take pictures during walk through the building [20].

Detailed energy audit for the electrical system typically includes gathering the data lighting system survey, power factor and demand, motor inventory and loads. Lighting accounts for a significant portion of electrical energy consumed in a building. Energy is saved in the lighting system by reducing illumination levels if needed, improving lighting system efficiency, curtailing operating hours and by taking advantage of available daylighting. Further, the Heating, Ventilation and Air-Conditioning (HVAC) Systems is a very important portion of the overall energy audits program [21].

3. Prototype and simulation

The energy audit flow developed on the prototype of the system is in accordance with SNI 03-6196-2000 and SNI 6196-2011. A little difference is made by performing a detailed audit both when the initial audit results are below or above the IKE benchmark standard. The purpose of this step is to know the subsequent relationship of the initial audit result with detailed audit result.

Electrical energy audit system is made gradually with waterfall model [22]. Energy audits on this prototype consist of an assessment of electrical energy consumption (IKE), illumination, watt intensity of lighting, room temperature, humidity chamber and watt intensity of AC.

The simulation is done on building Rajawali RSUP Dr. Karyadi, Semarang, Indonesia. The lighting conditions, temperature and humidity based on KEMENKES 2011 standard as benchmark. The conditions measured that out of the standard tolerances are recommended to be changed according to the standard. The intensity of power in lighting and air conditioners that exceeds the standard becomes an energy saving opportunity. The results are made from the assessment of energy audit as table 1.

Table 1. Relationship measurement of air condition and lighting room with energy intensity used.

Room condition	Energy Intensity	Results	Recommendation
Under standard	Under standard	Probably Normal	Room condition and energy must be standardized
	Standard	Probably Waste	Energy-saving opportunities
	Exceed standard	Probably Waste	Energy-saving opportunities
Standard	Under standard	Efficient	Maintain
	Standard	Normal	Maintain
	Exceed standard	Waste	Energy-saving opportunities
Exceed standard	Under standard	Efficient	Energy-saving opportunities
	Standard	Probably Efficient	Energy-saving opportunities
	Exceed standard	Probably Normal	Energy-saving opportunities

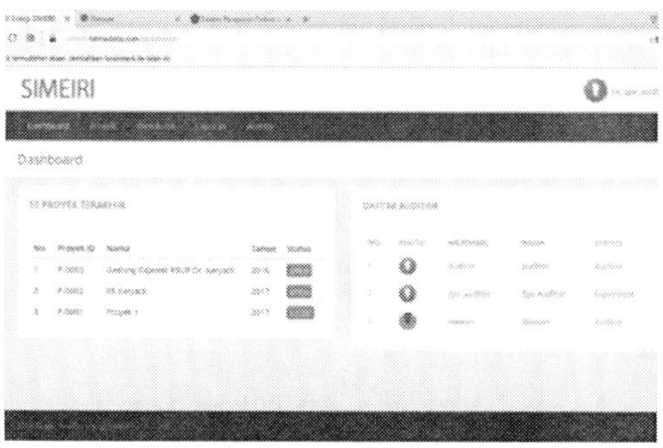

(a) Page of "Dashboard".

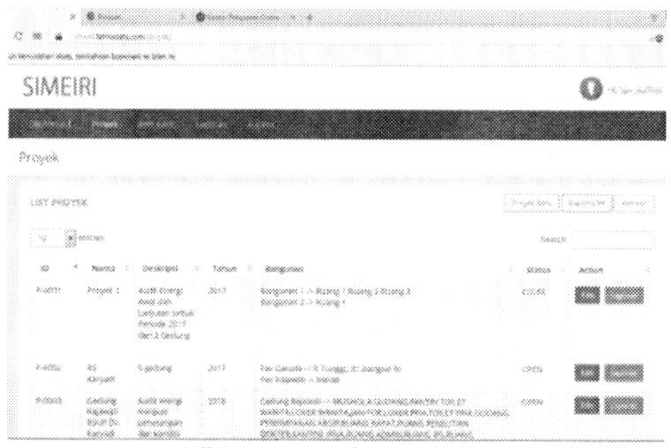

(b) Page of "Project".

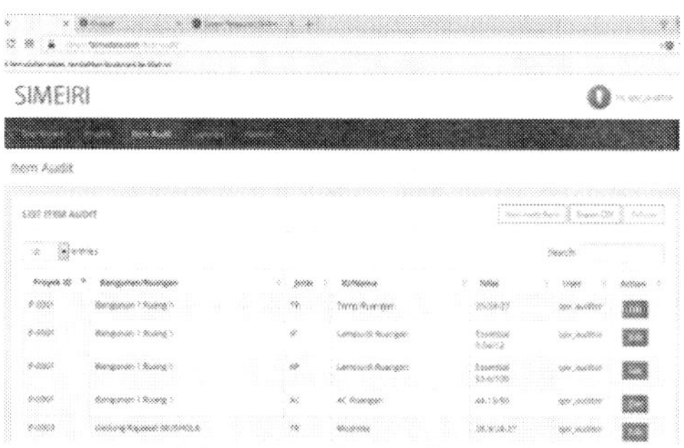

(c) Page of "Audit Item".

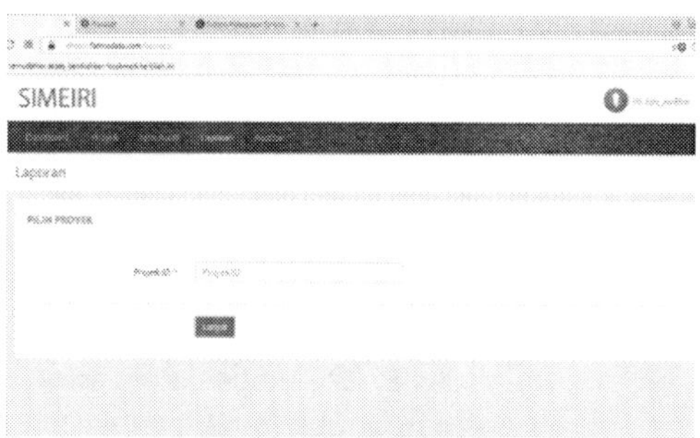

(d) Page of "Report".

Figure 1. The user interface of the application.

Web-based parent application so that the data input on the client side and data storage is on the server. This system model relieves user-side storage media and facilitates teamwork. Users consist of supervisors and auditor members. To ensure the security of user data, the protection is done by using username and password login procedure and used https-protected internet communication protocol.

After logging into the system, the start page displays "Dashboards" showing the list of existing projects and auditors. The audit process is done by following the next page flow from the "Project" which contains the project identity and building data, the "Item Audit" page contains space and fills the audit data as well as the benchmark results of each item. The "Auditor" page contains auditor data. Page views are shown in figures 1(a) - (d). The last page is used to generate electrical energy audit reports.

4. Conclusions

An important note of this research is that the application of information technology is very helpful and support the energy conservation program. The applications created in this study are designed to facilitate managers and auditor members in assessing, processing and producing collaborative teamwork reports. Application alpha testing stage is done by a case study of building project of Rajawali RSUP Karyadi Semarang and got result according to audit process done manually. Further development of the website not only as an audit tool but also contains all information related to energy conservation to attract website visitors and help spread the science of energy conservation.

Acknowledgments

We acknowledge valuable contributions by Luqman Assaffat, Wawan Kurniawan, Catur Novianto, Rizal Fajar Rahmawan. This work is part of the research program of the *'Penelitian Produk Terapan'*, the first of two years program, which is financially supported by the 'Ditjen Penguatan Riset dan Pengembangan Kemenristekdikti'.

References

[1] Energy Coservation 2013 Policies and programs on energy efficiency and conservation in Indonesia *Ecid* 11 October
[2] Dartanto T 2005 BBM, kebijakan energi, subsidi, dan kemiskinan Indonesia *Inovasi* 5/XVII pp 61–4
[3] Roslizar A, Alghoul M A, Bakhtyar B, Asim N and Sopian K 2014 Annual energy usage reduction and cost savings of a school: End-use energy analysis *Sci. World J.*
[4] Blomqvist E, Tholander P, Keskisarkka R and Paramonova S 2014 Energy efficiency measures as linked open data *IOS Press* vol 1 no 1 pp 1–5
[5] Solichan A 2013 Optimasi jadwal operasi dan pemeliharaan pembangkit tenaga listrik menggunakan metode algoritma genetika *Media Elektr.* vol 6 no 2 pp 25–31
[6] Republik Indonesia UU NOMOR 30 TAHUN 2007 TENTANG ENERGI 2007 pp 1–21
[7] Direktorat Konservasi Energi Kementrian ESDM 2017 Daftar Auditor Energi Bersertifikat pp 1–11
[8] ESDM 2012 Energi Bersih : Indonesia mau, Indonesia mampu *ESDM Mag* p 2
[9] Mawarsari V D and Solichan 2015 A Keefektifan Penerapan Perangkat Pembelajaran Berkarakter Dengan Pendekatan Inquiry Padamatakuliah Geometri Ruang Berbasis ICT *JKPM* vol 2 pp 39–46
[10] Deb S 2014 Information Technology , Its Impact on Society and Its Future *Adv. Comput.* Vol 4 no 1 pp 25–29
[11] Saefullah A 2015 Analisa Pengaruh Penggunaan Search Engine Optimization (SEO) Pada Website E-Commerce Open Access *J. Inf. Syst.* November pp 2–3
[12] Mayadevi N, Ss V and Ushakumari S 2014 A Review on Expert System Applications in Power Plants *Int. J. Electr. Comput. Eng.* vol 4 no 1 pp 116–126
[13] Borthic A F and West O D 1987 Expert Sistem Account *Horizons* March pp 9–16
[14] Begley J and Howell D 2004 An overview of Gadget, the globally applicable area-disaggregated general ecosystem toolbox *ICES* C. pp 1–15
[15] Badan Standardisasi Nasional 2010 Prosedur Audit Energi Pada Bangunan Gedung *SNI 03-6196-2000*

[16] Badan Standardisasi Nasional 2011 SNI. 6196 : 2011. Prosedur Audit Energi Bangunan Gedung Badan Standardisasi Nasional
[17] Dirjen EBTKE Kementrian ESDM 2017 Sistem Pelaporan Online Manajemen Energi - Dirjen EBTKE - Kementerian ESDM [Online]. Available: http://aplikasi.ebtke.esdm.go.id/pome/l. [Accessed: 16-Aug-2017].
[18] Solichan A 2010 Audit Dan Konservasi Energi Sebagai Upaya Pengoptimalan Pemakaian Energi Listrik Di Kampus Kasipah Unimus Pros. Semin. Nas. UNIMUS 1 1 p 309–13
[19] Rizkani T and Ciptomulyono U 2012 Audit Energi dengan Pendekatan Metode MCDM-PROMETHEE untuk Konservasi J. Tek. ITS 1 September p 465–70
[20] Thumann A 2007 Handbook of Energy Audits Seventh ed. (London: The Fairmont Press, Inc.)
[21] Andarini R 2014 The role of building thermal simulation for energy efficient building design Energy Procedia 47 p 217–26
[22] Petersen K, Wohlin C and Baca D 2009 The waterfall model in large-scale development Lect. Notes B*us. Inf. Process.* 32 LNBIP p 386–400

Flex sensors and MPU6050 sensors responses on smart glove for sign language translation

A Yudhana, J Rahmawan and C U P Negara

Electrical Engineering Department, Faculty of Industrial Technology, Universitas Ahmad Dahlan, Indonesia

eyudhana@ee.uad.ac.id, jihadrahmawan20@gmail.com, cahya.utama63@gmail.com

Abstract. Flex sensor is a transducer that measures the level of curvature of the sensor. One type of flex sensor is a resistive flex sensor. This type of sensor has output resistance value that proportionally to the level of flexibility of the sensor. The gyroscope and accelerometer sensor used in the research is the MPU6050 that has measurements on earth gravity. The glove was attached by the flex sensor and gyroscope sensors for sensing fingers and hand movement. This paper presented data on flex and MPU6050 sensor response. The microcontroller is used in the research was Arduino Nano controller to read flex sensors and MPU6050 sensors.

1. Introduction

Sign language has an important role for people with speech disorders. In Indonesia alone persons with disabilities at 18-30 years of age are 1.15% of the population, even in underfives with a disability of 0.26% [1]. Species of speech impaired at the age of 24-59 months in Indonesia in 2010 reached 0.14% and in 2013 reached 0.15% increase from the previous year [1]. In the data the person with speech impairment increases in the next year, it takes innovation to help our brother with those special needs. People with speech impaired use of sign language in daily communication. In figure 1, a simple gesture of American sign language. Each letter in the sign language system has its own gesture, the different finger density of the fingers of different fingers.

In the processing of sign language based electronic systems have many techniques and sensors are applied. Such as flex sensor, metal material, image processing [2]. Each technique has its own advantages and disadvantages. The use of gloves in an electronic system has practicality and convenience for its users [3]. Image processing techniques have high levels of computing and system complexity, while the use of flex sensors and applied to a lower level complexity glove, and easy to use for users [3].

Sensing gloves with flex sensors are of many applicative variants in assisting human tasks, the use of flex sensor gloves to drive DC motors designed to drive one wheelchair [4]. In addition, flex sensor gloves can be used on the robot arm control, even inserted accelerometer sensor to increase the sensation of hand movement, not only the finger curve that is read movement but also the slope of the wrist [5].

The research done by [6] shows data on the sensor response to each sign language gesture, which presented data from flex and accelero sensor in forming a gesture alphabet sign language. Data glove is can be applied in various aspects, such as virtual reality, robotics, tele electric, biomechanics.

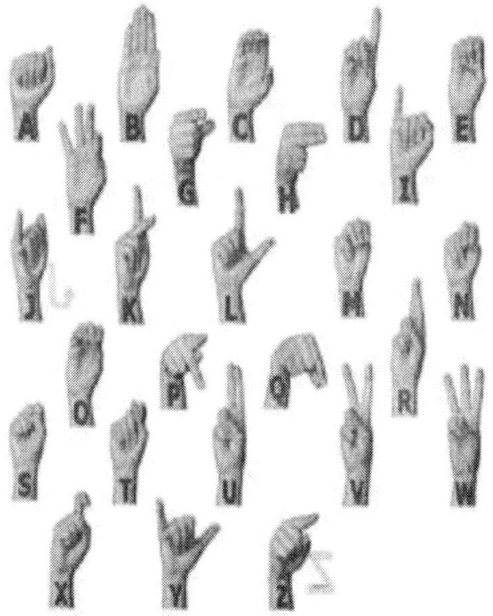

Figure 1. Gesture of American sign language.

2. Overview of smart glove

To perform gesture readings on sign language with gloves the gloves require electronic sensor to perform the gesture reading of the sign language. The characteristics of the sign language gesture in figure 1 are the fingertip plays and the slopes on the hands.

There are two keywords: flexibility and slope. One sensor for flexibility reading is flex sensor. To measure the tilt can use accelerometer sensor and one of the series is MPU6050. In processing the sensor can use a microcontroller in this glove using Arduino as a sensor reader and sensor processor. The component material is packed and sewn into a glove with a flexible fabric and is designed to be easy to use, and can read gestures well.

3. Working and data smart glove

Gloves are made with sensors and electronics to perform the task of reading gesture sign language and can process the sensor output and converted into sign language in the form of text or sound to facilitate communion. With the data can then be made algorithm to make the sign language cues from gesture to the form of sound or text.

3.1 Smart glove hardware

The design of a hardware glove system using sensors and sensors MPU6050 with Arduino microcontroller for processing and a Personal computer to monitor the value of sensor output value. The block diagram of hardware groove of smart glove is shown in figure 2.

In accordance with figure 2 Arduino as sensor signal processor from sensor flex through ADC (Analog to Digital Converter) and MPU6050 gyroscope & accelerometer sensor through I2C (Inter-Integrated Circuit) line. In the flex sensor there is a circuit for dividing the voltage with a 10k ohm resistor according to figure 3.

The flex sensor is placed on top of each finger and the MPU6050 gyroscope & accelerometer sensor is placed on top of the hand, using a flexible Fabric both sensors closed and sewn over the fabric. While the Arduino placed on a cloth that can be worn on the wrist. In figure 4 detail the layout of the components of the smart glove.

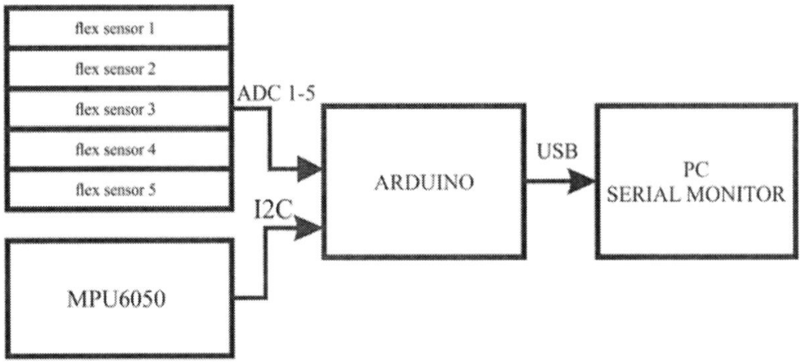

Figure 2. Diagram block of smart glove.

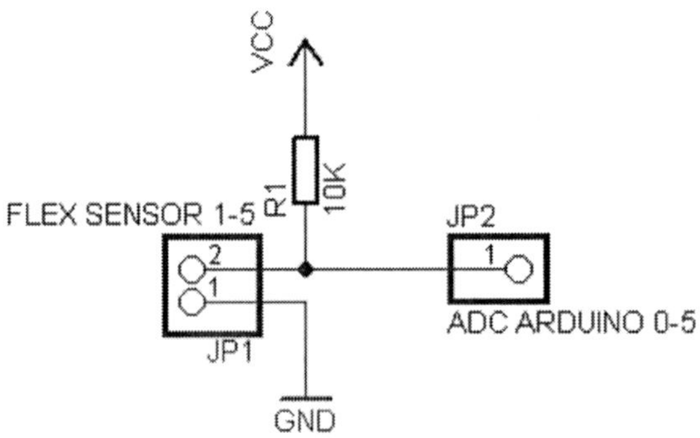

Figure 3. Flex sensor schematic.

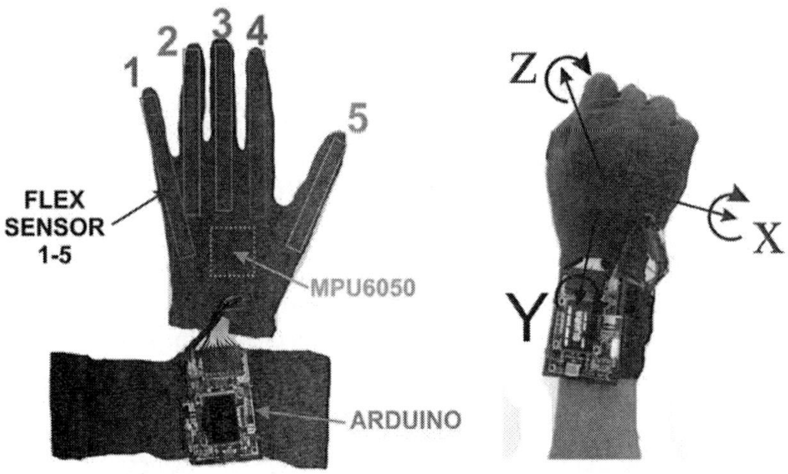

Figure 4. Attachment sensors on smart glove.

In figure 4, seen laying accelerometer sensor with 3 axis. In the above smart gloves on the picture 4 using a 4.5" sensors flex that is on the sensors 1-4 and on the sensors 5 flex sensor using size 2.2".

3.2 Response sensor data.

In sign language each finger letters form a certain pattern so that every pattern of gesture to form letter sign has its own character. In Figure 1 the gesture pattern of each letter has its own gesture. From A to Z has a different gesture. Due to the difference, that the sensor MPU6050 sensor and sensor output has different output values also on each gestures. Tested every sign language gesture from A to Z is taken 100 data on each gesture and data is taken at 1 second interval in each gesture data. The observed data are the ADC flex sensor 1-5 and the accelerometer value of MPU6050 axis x and y only and then take of mean on each gesture. Obtained data in table 1.

Table 1. flex and accelero sensors data.

Alphabet	Flex sensor 1	Flex sensor 2	Flex sensor 3	Flex sensor 4	Flex sensor 5	Acc X	Acc Y
A	754.58	747.51	744.64	719.41	858.10	-15,557.58	-1,840.04
B	511.52	510.90	508.09	511.18	903.05	-15,816.73	-672.97
C	582.22	628.07	614.12	623.98	884.06	-16,345.37	-214.06
D	636.70	660.67	647.79	539.22	879.64	-16,344.69	700.44
E	735.55	743.28	738.26	737.46	915.11	-16,111.43	1,526.06
F	513.90	516.21	516.24	690.08	902.97	-16,142.18	1,360.61
G	748.31	730.81	715.81	554.66	865.23	-5,215.11	-14,692.16
H	719.56	690.13	521.63	548.09	899.10	-7,971.96	-12,637.82
I	518.04	694.56	724.84	713.03	910.89	-14,670.14	2,956.93
J	517.34	693.31	719.17	714.79	902.89	-13,427.64	-8,908.12
K	733.44	692.25	541.05	515.78	861.43	-16,198.18	-1,150.46
L	747.43	728.96	709.44	526.73	883.46	-15,605.86	-643.03
M	740.67	696.80	713.58	719.70	863.32	-16,118.46	-2,071.88
N	746.54	737.91	721.14	704.46	845.14	-16,332.00	1,118.06
O	686.04	703.77	726.02	706.33	899.67	-16,210.18	1,050.26
P	730.39	685.46	600.62	537.13	858.93	-7,027.80	-2,663.96
Q	735.96	720.07	721.28	626.74	870.78	-6,910.79	-1,540.73
R	716.81	685.86	514.83	528.53	895.44	-15,701.58	487.88
S	733.17	728.11	753.30	747.41	907.92	-16,101.21	1,346.71
T	730.64	691.82	731.19	709.17	839.01	-16,238.75	-503.07
U	713.02	675.46	514.44	527.93	876.55	-16,302.91	-358.59
V	712.46	682.93	519.15	541.91	903.59	-16,202.99	-1,172.08
W	686.46	512.00	518.26	538.97	884.65	-16,290.91	387.68
X	749.36	732.85	754.41	639.26	910.17	-16,333.58	-552.81
Y	511.11	690.39	708.58	713.37	861.89	-15,940.85	-894.34
Z	728.59	721.08	726.21	569.25	897.17	-15,278.87	-296.89

3.3 Result

Obtained results in table 1, flex sensor response with unit bit of ADC Output. The results are taken from the flex sensor readings as they are attached to glove and form a specific curve at each gesture formation of sign language. The value is derived from the ADC conversion of the Arduino Nano with input according to figure 3 and the ADC value reference voltage in the Arduino is 5V. While the accelerometer value of x and y is the value of hand slope in implementing hand movement in formatting gesture. The reading of that value with the figure 4. Figure 5 and figure 6 are graphs obtained from table 1 to clarify the difference in values on each gesture.

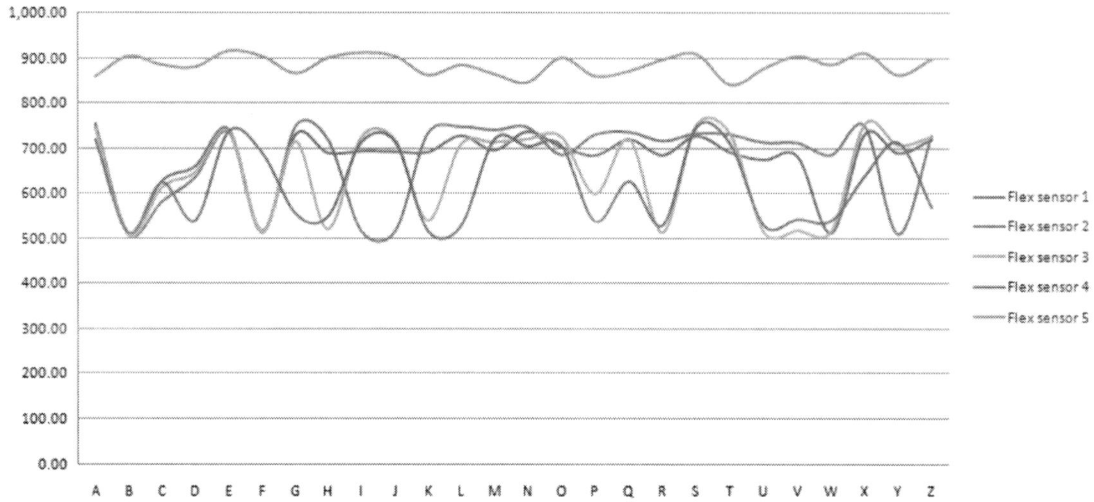

Figure 5. Flex sensors response for alphabet.

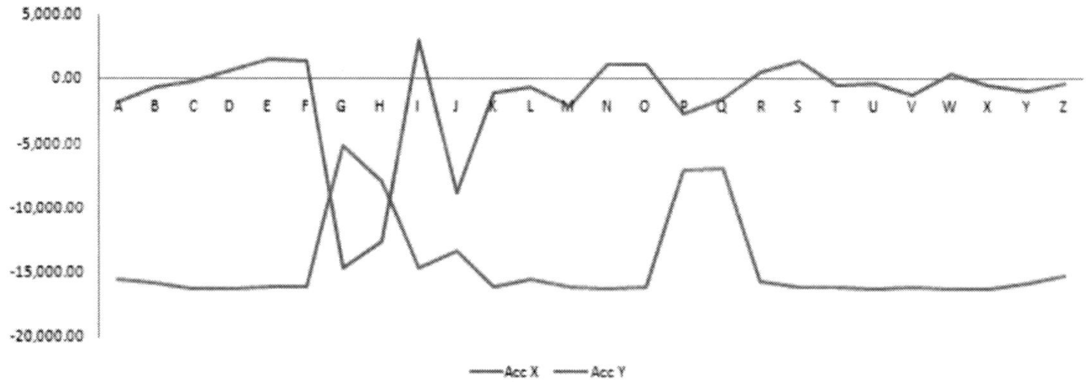

Figure 6. Accelerometer response for alphabet.

4. Conclusion

This paper has presented data flex and accelerometer sensor in forming gesture sign language. Flex sensor placed on each finger and MPU6050 placed on the hand. Then the sensor is read using the Arduino Nano controller, and is displayed on the PC using the serial path. Hand finger forms a certain curve and angle according to the gesture of sign language and the accelerometer's flex sensor reads each flex and slope and obtained data according to the paper presented.

Using additional accelerometer sensors can detect gestures that have a different slope than other gestures, such as the letters G, H, J, P and Q. The gesture accelerometer value has a significant difference value than the other letters

References

[1] Kementrian Kesehatan RI 2014 Situasi Penyandang Disabilitas *Bul. Jendela Data Inf. Kesehat.* vol 2 no 1 pp 1–5

[2] Rastogi K 2016 A Review Paper on Smart Glove - Converts Gestures into Speech and Text *International Journal on Recent and Innovation Trends in Computing and Communication* vol 4 no 5 pp 92–94

[3] Jadhav A J and Joshi M P 2016 Hand Gesture recognition System for Speech Impaired People : A Review *International Research Journal of Engineering and Technology (IRJET)* vol 3 issue 7 p 1171–75

[4] Tameemsultana S and Saranya N K 2011 Implementation of Head and Finger Movement Based Automatic Wheel Chair *Bonfring International Journal of Power Systems and Integrated Circuits* vol 1 December pp 48–51

[5] Syed A, Taj Z, Agasbal H, Melligeri T and Gudur B 2012 Flex Sensor Based Robotic Arm Controller Using Micro Controller *J. Softw. Eng. Appl.* vol 5 May pp 364–66

[6] Tarchanidis K N and Lygouras J N 2003 Data glove with a force sensor *IEEE Trans. Instrum. Meas.* vol 52 no 3 pp 984–89

Comparative study of magnetic fields measurements with logger lite and Arduino on electronic devices

S D Fatmaryanti, Y Al Hakim and Ashari

Department of Physics Education, Universitas Muhammadiyah Purworejo, Jl. KH. Ahmad Dahlan 3 Purworejo, 54111, Indonesia

sd_fatmaryanti@yahoo.com

Abstract. A study of magnetic field measurements of electronics devices has been conducted. Magnetic field measurement is done by two methods, that is Arduino based measurement by using magnetic field sensor with C + language. Other measurements use magnetic sensors and Logger Lite software with Labquest Mini interface. In Arduino measurements, the measurement results are made in units of Tesla. ADC Arduino has 10 bits. The resolution of this magnetic sensor is 1.6 gauss /volt. Display data in the form of sensor read voltage is changed in milli Tesla (mT) by dividing the data reading voltage sensor value with a value of slop 0.160 mT and intercept -0.32 mT. Measurement results are compared with measurements using magnetic field sensors and free Logger Lite software. The results found that there was no significant difference on measurement data between Arduino and logger lite. The larger difference is 0.032, a result of Servo measurement and the lower is 0.002 mT, result of Coil with 1 Ampere measurement. The conclusion of these two methods is the measurement of the magnetic field using Arduino is recommended to be used.

1. Introduction

The magnetic field is a physical phenomenon related to electrical quantities. An electric current creates a magnetic field, and vice versa electric current can be generated from a magnet driven near a wire coil, Faraday's induction law. To know the magnitude and direction, the magnetic field needs to be measured. The magnetic field is expressed in units of Tesla, for example, a normal earth magnetic field near the equator of about 45,000 nano Tesla, with north-south direction.

Such measurements have already been published [1-3]. The magnitude of the magnetic field is one of the most important physics is the magnetic field and to measure it required magnetic sensors. At this moment magnetic sensors are growing rapidly along with the advance of science and technology. One of them is the hall's magnetic sensor effect. This sensor is designed to sense the presence of magnetic objects with changes in position. Continuous magnetic field changes cause the occurrence of pulses which can then be determined frequency, this type of sensor commonly used as a speed gauge.

The development of science and technology is very rapid and closely related to human needs. In this case, human needs are much related to electronic devices. Every electrical equipment from household appliances up to communication tools reacts to the electromagnetic field. For this purpose, the process of easy reading of magnetic sensor data becomes a necessity.

2. Literature Review

2.1 Sensor of magnetic field

Hall effect occurs when a current carrying conductor is held in a magnetic field, the field gives a lateral force to the charge flowing on the conductor. All Hall Effect appliances are activated by the presence of a magnetic field [4]. The magnetic field has two important characteristics, flux density, and polarity. Most of the Hall Effect digital switches are designed to die if there is no magnetic field (open circuit at the output). The switch will be active if only subjected to a magnetic field that has sufficient density and proper direction. Magnetic field sensors are used instead of Gaussmeter for data to be automatically displayed on the computer and easy to analyze, and better results can also be obtained. The magnetic field sensor in this study used the Vernier manufacturer's output.

2.2 Sketch for Arduino

Fortunately, the vast majority of Vernier analog (BTA) sensors have linear calibrations. For these sensors, we only have to add a few lines of code to Vernier Analog Voltage sketch to get meaningful readings. Specifically, we need to convert from a raw voltage to sensor reading using equation (1).

$$\text{Sensor reading} = \text{intercept} + \text{voltage} * \text{slope} \tag{1}$$

For any Vernier sensor with a linear calibration, we can get the slope and intercept from the user manual from any sensor's web page or in the Vernier support section. The Arduino comes with a built-in, 10-bit analog-to digital (A/D) converter to read voltages. The voltage is read as a count in the range of 0 volts and 1023 representing 5.0 volts [5].

2.3. Logger lite

Logger lite is software with support tool which consists of Labquest and two sensors. Labquest as the link between the Current Probe Sensor, Voltage Probe Sensor and Magnetic Field Sensor to the computer, the lab quest used is Vernier's mini lab quest. Two sensors are Current Probe Sensor and Voltage Probe Sensors that exist in energy sensor devices. Function as a substitute Amperemeter and volt meter, so that data directly read by the computer and also possible to get more accurate results [6].

2.3 Vernier Arduino Interface Shield

The SparkFun Vernier Interface Shield, a simple add-on to give Arduino the capability to interface with Vernier sensors. Attached to this interface shield are four British Telecom sockets (2 Analog, 2 Digital) which all Vernier sensors use as connectors. Simply attach it to an Arduino development platform. This shield was developed to be used with the Arduino Uno platform [7].

3. Method

In this study used magnetic field sensor based on Vernier production hall effect with 3.2 mT measuring interval and 1.6 Gauss/volt sensor output. The sensor is useful for detecting the magnetic induction of electronic devices.

This observation was using 7 sample. There were handphone, motor stepper, magnet U with four times various distance, coil 500 with 1 A DC. The magnetic field detected by the sensor will be displayed and processed data in 2 ways. First by using Vernier production interface before reading on the laptop that has been installed Logger Lite +Labquest Mini. Secondly by using Arduino + Vernier Arduino Interface Shield with the same sensor. The results of the data processing displayed then compared to get whether there is a difference between the two applications.

IOP Conf. Series: Materials Science and Engineering **403** (2018) 012033 doi:10.1088/1757-899X/403/1/012033

4. Results and Discussion

The comparison of two measurement data was the aim of this research. The result of magnetic field measurement can be seen in table 1.

Table 1. Comparison of measurement data from 7 samples.

Sample	Measurement data with Logger Lite (mT)	Measurement data with Arduino (mT)	Difference between measurement with Logger lite and Arduino (mT)
1	0.023 ± 0.001	0.03 ± 0.00	0.007
2	-0.318 ± 0.000	-0.32 ± 0.00	0.002
3	0.011 ± 0.001	0.02 ± 0.00	0.009
4	0.032 ± 0.001	0.04 ± 0.00	0.008
5	0.104 ± 0.010	0.11 ± 0.00	0.006
6	-0.061 ± 0.001	-0.053 ± 0.001	0.008
7	0.396 ± 0.001	0.364 ± 0.0001	0.032

Data obtained from mini Labquest and lite logger at every second and data retrieval can be obtained as many as 5 data, 10 data even 100 data. The measurement result using lite logic is very precise, as evidenced by the standard deviation value of each sample is very small. The average standard deviation value of 7 sample size is 0.002, with the largest standard deviation value in the 5th sample (0.104 ± 0.010) mT. For magnetic field measurements using Arduino, a sketch of voltage data reading by Analog to Digital Converter (ADC) from Arduino is built-in with float command as shown in equation (2).

$$\text{Voltage} = \text{Count} / 1023.0 * 5.0 \tag{2}$$

The value 1023 refers to the maximum value of digital data that indicated by the Arduino, while the value 5, and indicates the magnitude of the reference voltage used by the Arduino of 5 volts. Next, the value of the Arduino is converted to the magnitude of mille Tesla with the float equation Sensor Reading. For a 3.2 mT range, the slop value of the magnetic field change to the voltage by the Vernier magnetic sensor is 0.160 mT / V, and the intercept value is -0.32 mT. The Arduino measurement results have a good accuracy level with an average deviation standard score of 7 samples measured 0.001 with the highest deviation standard in sample 6 (-0.053 ± 0.001) mT.

From this result, we saw that there was no significant difference between two measurement data. It means that the using of Arduino with magnetic field sensor from Vernier was recommended. The validation level can be increased by multiplying the numbers behind the comma from the measurement results. This is done by adding a sketch after the conversion process into mille Tesla in the form of `Serial.println(SensorReading, X);` where X denotes the number of numbers to be displayed behind the comma. The more of numbers behind the comma the greater the level of accuracy.

Several studies on magnetic field measurements have also been performed. Design instrumentation measurement instruments for strong magnetic fields using Atmega8535 microcontroller [8]. The use of magnetic sensors in magneto static experiments has also been done with smartphones [9]. Our result in measurement by Arduino and Logger lite device was able to provide an alternative in experimental measurement of a more economical and easy magnetic field.

5. Acknowledgments

The author would like to thank the Laboratory of Physics, Universitas Muhammadiyah Purworejo, for the use of the laboratory equipment for the preparation of this work.

References

[1] Arribas E, Escobar I and Suárez C P 2015 Measurement of the magnetic field of small magnets with a smartphone: a very economical laboratory practice for introductory physics courses *Eur. J. Phys.* vol 36 no 6 p 11
[2] Andris P and Frollo I 2012 Measurement of magnetic field with background using a low-field NMR scanner *Meas. Sci. Technol.* vol 23 no 6 p 65006
[3] Atkin K 2016 Construction of a simple low-cost teslameter and its use with Arduino and MakerPlot software *Phys. Educ.* vol 51 no 2 p 24001
[4] Serway and Jewett 2006 *Physics for Scientists and Engineers* vol 36 no 3
[5] Vernier *Vernier Arduino Interface Shield* [Online]. Available: https://www.vernier.com/products/interfaces/bt-ard/?search=ardui&category=autosuggest
[6] Vernier *Logger lite* [Online]. Available: https://www.vernier.com/products/software/logger-lite/
[7] SparkFun *SparkFun Vernier Interface Shield - DEV-12858 - SparkFun Electronics* [Online]. Available: https://www.sparkfun.com/products/12858.
[8] Premono P, Soedjarwanto N, Alam S and Belakang A L 2015 Rancang Bangun Alat Instrumentasi Pengukuran Digital Kuat Medan Magnetik dengan Menggunakan Mikrokontroler Atmega8535 *Electrican* vol 9 no 3 pp 160–70
[9] Septianto R D, Suhendra D and Iskandar F 2017 Utilisation of the magnetic sensor in a smartphone for facile magnetostatics experiment: magnetic field due to electrical current in straight and loop wires *Phys. Educ.* vol 52 pp 15015

Simplified design of low speed electric permanent magnet generator for small wind power plant

M Irfan, E A Hakim, D Suhardi, N Kasan, M Effendy, A Faruq and M C Anam
Faculty of Engineering, Electrical Engineering Department, University of Muhammadiyah Malang,
246 Raya Tlogomas road, 65144 Malang, Indonesia

irfan_tsd@yahoo.com

Abstract. Wind power plant is one of the renewable energy where energy utilization is ± 3-5% from potential existing of wind power. Wind power plant requires a generator to convert mechanical energy into electrical energy, mostly the generators available in the market is a kind of high speed induction generator which requires high rotational speed and an electricity to generate a magnetic field. While in wind power plant needs low speed generator and without initial electricity. Hence, a radial flux generator is designed in this project to have low speed rotation using permanent magnet type Neodymium Iron Boron (NdFeB). Finite Element Method (FEM) Magnet software is used for designing the generator model. Several modifications are applied to get optimum result by changing generator diameter, number of coils, the copper wire diameter, number of poles, and used slots. The simulation results obtained generator speed 500 rpm, the average series voltage is 185.44 Volt. Generator requires 18 cm diameter, number of turn for each coil is 55, diameter of the copper wire used is 0.6 mm, and number of poles is 8 pairs and 12 unit slots.

1. Introduction

The utilization of wind energy is the most renewable energy developed nowadays. With abounding wind potential in coastal areas of Indonesia, total capacity installed in wind energy conversion system is currently less than 800 kilowatts (kW). In the whole of Indonesia, five windmill units generator already built with each capacity of 80 kW. In 2007, followed by seven units with the same capacity has been built in four locations, each of three units in Selayar Island, North Sulawesi two units, and Nusa Penida, Bali, and Bangka Belitung one unit respectively. Referring to the national energy policy, the wind power plants expected to reach 250 megawatts (MW) by 2025 [1]. At the end of 2008, the electrification ratio in Indonesia is still 65.1%, while the lowest ratio (20% - 40%) in NTB, NTT, Southeast Sulawesi, Papua and West Irian Jaya areas. So that the government continues to strive increase the electrification ratio, especially in remote areas (areas that have not been electricity by the government) with utilizing renewable energy such as wind energy.

One of the important components in wind power station is a power generator, a device that converts wind turbine into electrical energy. The problem is, mostly electrical generators available in the market is high speed induction generator, that is requires a high speed between 1000 rpm – 1500 rpm, and also requires the initial electrical energy to create the magnetic field. While the thermal power station, which has a generator needs low rotation and without an initial electricity, this was due to a wind speed average in Indonesia is 3.47 m/s and in Malang is 2.7 m/s [2]. With wind speeds are very small it is difficult to drive a high speed generator. Therefore, the objective in this project is to design an electric permanent magnet NdFeB generator that able to generate a large voltage and current with low-speed turning.

Application of wind power station has been successfully carried out previously by [2], precisely in the southern coastal district of Malang. However, the prototype made still using a modified car generator (dynamo), so that the voltage and the current generated is still very limited. The use of generator low speed has been successfully performed by [3] for lighting a street. In the axial generator comprising 9 coils was able to work at 560 rpm and a maximum output voltage of 30V.

Suhardi [4] has designed and built an electric generator low speed with 180 rpm. However, the electrical power generator is still very small; it is 25 Watt, with efficiency only 25%. Budiman *et al.* [5], has been created and tested an axial permanent magnet generator. Maximum electrical voltage produced at that time is 12V with a load current of 0.14A. However, the rotation speed of the generator is still quite high i.e. 1200 rpm. With permanent magnet material available on the market, Asy'ari *et al.* [6] has designed a low speed permanent magnet generator for wind power station. It was produced a permanent magnet generator design with a speed of 1000 rpm and a maximum voltage generated at 38V and 114 mA electric current.

An electric permanent magnet Neodymium Iron Boron permanent magnets ($Nd_2Fe_{14}B$) generator has designed in this work. The NdFeB permanent magnet is categorized as strong and rare magnet. Characteristics of the magnetic owned NdFeB better when compared with the other permanent magnet, such as Ferrite, Alnico and Samarium Cobalt. The maximum Energy Product (BHmax) can be produced between a 30 MGOe up to 52 MGOe, which reached 440 KJ/m^3 [7]. Because it has a high magnetic characteristic, then the NdFeB magnet application has small dimensions and volume. In the other hand, these magnets can also replace the use of Samarium Cobalt magnets, in particular the use at a temperature is less than 80^0 C [8].

Dilev *et al.* [9], has designed and built with the induction generator excitation system itself. Magnets used are NdFeB permanent magnets. However, the test has not been shown using certain electrical load, so that it was unknown how much the electric power can be generated. The use of NdFeB permanent magnet type generator for low speed has been successfully done by Alqodri *et al.* [10] in 2015 with a speed of 500 rpm, and 2V voltage generated. Thus it is critical to use the gear since it was built through hardware directly without design strategy. The researchers use CAD software to simplify designing of electrical generator. This tool unable to simulate and getting more result. Therefore, the challenges toward designing a high-performance and efficient electrical generator are necessary to examine with wind energy especially in rural area.

In this paper, the design and simulation of low speed permanent magnet generator has been demonstrated. Using Finite Element Method (FEM), the design of electrical permanent magnet NdFeB in this project referring the work by [11] and examined several steps they are determine the specifications of generator, followed by design of geometry that consists of components of stator and rotor, designing the air gap of stator and rotor, design of permanent magnets and coils used and simulate the model by running and plotting.

2. Initial design and geometry

This section describes an analytical design method of electric permanent magnet generator. The target is to determine an initial rotor and stator design, coil model, air gap model, while magnet type used is NdFeB. The generator designed in this project using FEM Magnet software basis.

Initialization and design geometry is a step to provide naming and determine of the material used on components contained in the generator and provides a measure of the thickness in each parts. By Initialization and geometry design, the generator can be simulated as well; it covers all component parts of the generator. The important components are consisting of geometry design of stator, rotor, magnet, coil, stator and rotor air gap, air box and shaft. Figure 1 shows initialized components of electric generator.

FEM software-based is a method used to solve the problem of complex electromagnetic field, so it can be solved by analytical models, especially in the part related to the nonlinear nature of the material. This method is basically a part of a cross-sectional discredited engine into areas or small volume called finite element or mesh. An initial mesh for stator, bottom rotor and magnet are set to 2 mm. While an initial mesh for top rotor and air gap are set to 1 mm and 0.5 mm respectively. Figure 2 shows a mesh determined for ¼ designs.

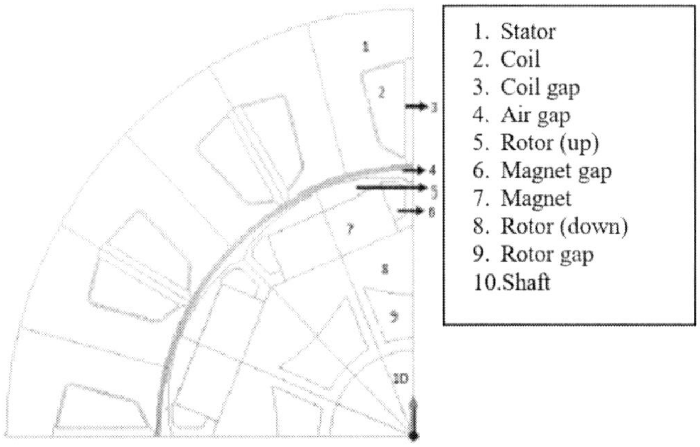

Figure 1. Initial components of electric generator.

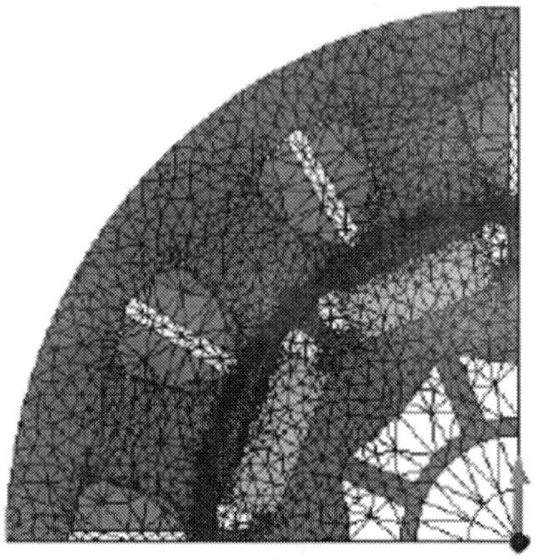

Figure 2. Mesh determined for ¼ designs.

To determine a mechanical degree and electrical degree of generator can be calculated by equation (1) and (2) respectively. While equation (3) to calculate a frequency.

$$\theta_M = \frac{360}{\text{Stator}} \text{ x pole} \tag{1}$$

$$\theta_E = \frac{\text{Pole}}{2} \text{ x } \theta_M \tag{2}$$

$$f = \frac{1}{T} \tag{3}$$

Where θ_M, θ_E, F and T are mechanical degree, electrical degree, frequency and amplitude of the generator respectively. In this work, ¼ design rotor rotated every 1 degree with time 0.1667 ms needed, so that to rotate 360 degrees it has 0.06 sec. Time rotation, rotation speed and energy coefficient can be formulated by equation (4), (5) and (6) respectively. Results of these analyses can be seen at table 1.

$$T_\theta = \frac{\theta_M}{\Sigma \text{ data sampling}} \tag{4}$$

$$\omega = 2.\pi.f \tag{5}$$

$$K_E = \frac{V_{dc}}{\omega} \tag{6}$$

In this work, the electric generator has designed with detailed specifications. Table 1 shows the design variables of electrical permanent magnet generator.

Table 1. The variable designs of electrical permanent magnet generator.

Design variable	Initial value
Magnet poles	8 units
Magnet type	Neodymium Iron Boron Cylinder
Generator diameter	18 cm
Copper wire diameter	0.6 mm
Number of turns	25 pcs
Stator and rotor materials	USS Transformers stator 52-29 materials
Air gap (distance stator and rotor)	40 mm
Number of rotor	16 units
Number of stator	12 units
Stator air gap and rotor air gap	0.5 mm
Coil material	pure copper coil 5.77e7 Siemens/meter
Number of coils	55 units
Number of loops (full model)	660 windings
Total data sampling for ¼ designs	91 times

3. Simulation result and discussion

Figure 3 shows the result of flux linkage and voltage based on simulation respectively. Flux linkage is a flux that is connected or flowing from the rotor to the stator or reverse in. Flux linkage value is to determine the voltage produced by the generator. While each coil voltage is the voltage generated by each coil, in other words it is the voltage of each phase. The graph in figure 4 shows the result of voltage for each coil.

Series voltage is the voltage produced by the generator that is connected in series with load electric current. Thus the series voltage for each coil can be obtained from flux linkage each coil by multiply 4 for full 3600 model. From these data, inter-coil voltage can be obtained. Inter-coil voltage is the voltage between one phase with each other, in the other words it is the differencing voltage between phases. That is coil voltage-one deducted by coil voltage-two, and coil voltage-two deducted by coil voltage-three and then coil voltage-three deducted by coil voltage-one. Output voltage Vdc can be calculated by determining the average of inter-coil voltage. These graphs can be seen in figure 4 respectively. While simulation result with load-generator addressed in table 2.

Table 2. Load generator simulation result.

Parameters	Value	Units
Output voltage	185.44	Volt dc
Output electric-current	5	Ampere
Rotation speed	52.3	Rad/s
Frequency	8.33	Hz
Period/Time rotation ¼ design	0.12	Second
Energy coefficient	1.77	-
Torque	0.23	Nm
Mechanical degree	15	degree
Electrical degree	60	degree

Table 2 indicates load-generator simulation result with an output current 5 Amperes produced. The electric generator simulates with closed-loop circuit, electric current able to flow through the coil windings of generator. If 1 rad/s equal to 9.55 rpm, it can be seen that electric generator design with low speed 500 rpm. The electric generator design has been simulated to get maximum electric current and time needed when it reaches maximum speeds, with customized load electric current. Figure 3 shows (a) the flux linkage of each coil and (b) the voltage of each coil. Figure 4 shows (a) the series voltage of each coil and (b) the inter-coil voltage and Vdc.

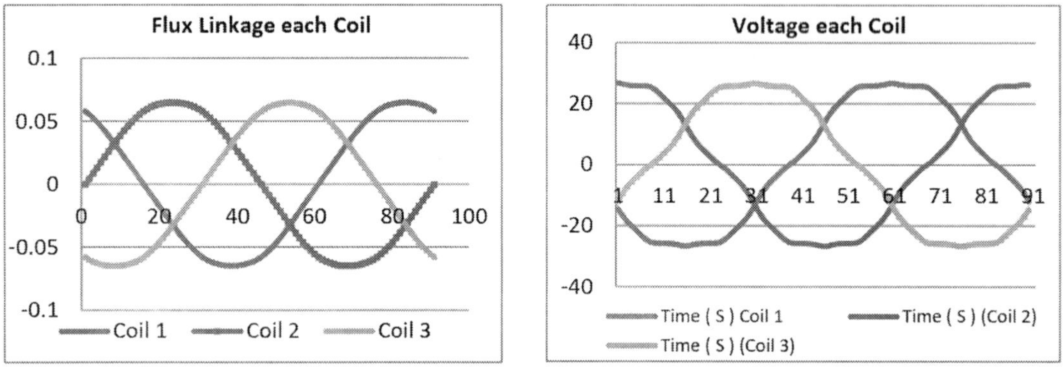

(a) (b)

Figure 3. (a) Flux Linkage each Coil, (b) Voltage each Coil.

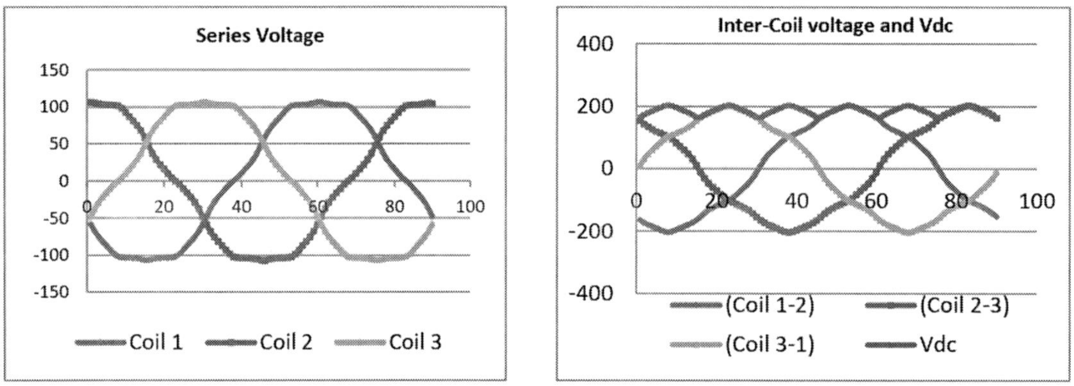

(a) (b)

Figure 4. (a) Series Voltage each Coil, (b) Inter-Coil voltage and Vdc.

4. Conclusion

An analytical design and simulation presented in this work provided a low speed electric generator that can be played for small wind power plant. In terms of accuracy for simulation and generator construction, magnet type Neodymium Iron Boron (NdFeB) and other variables as mentioned (generator diameter, number of turns, poles of magnet) are considered. It is also observed that based on experimental design the air gap between rotor and motor is 1.5 mm. In this work, flux linkage each coil can also be simulated. These form can generate average voltage Vdc. Experimental design of magnet type materials and changing value of variables stated in this work can be as a future work in term of performance for electric generator. The effectiveness of different sample software is also of interest.

Acknowledgements

This project is supporting by Indonesian higher educational and research institutions (DIKTI). Special thanks to Electrical Department University of Muhammadiyah Malang (UMM) for giving full support and cooperation. Also warmest thanks to research and development department of UMM (DPPM). Their supports are gratefully acknowledged.

References

[1] ESDM 2010 Renewable energy development and energy conservation *Energy and Human Resources Development*

[2] Irfan M, Effendy M, Mokhtar A and Suwignyo 2015 Penerapan teknologi pengoperasian, perawatan dan pelatihan mikro hidro di Desa Karangsuko Kecamatan Pagelaran Kabupaten Malang *Dedikasi* vol 12 no 5 pp 39–42

[3] Farhan M, Ramdhan D H and Mahdi S 2013 Design of prototype permanent magnetic generator axial flux on wind power for light street lighting *National Conference of Electrical Engineering*

[4] Suhardi D 2012 Generator listrik 100 Watt putaran rendah untuk pembangkit listrik tenaga air dan angin mikro: disain, perencanaan dan pembuatan *Gamma Natl. J.* vol 5 no 2 pp 98–104

[5] Budiman A, Aji D Y and Asy'ari H 2013 Pembuatan dan pengujian awal generator axial magnet permanen kecepatan rendah *Seminar Nasional Aplikasi Teknologi*

[6] Asy'ari H, Jatmiko and Ardiyatmoko A 2012 Desain generator magnet permanen kecepatan rendah untuk pembangkit listrik tenaga angin atau bayu (PLTB) *Seminar Nasional Aplikasi Teknologi Informasi (SNATI)*

[7] Hendershot J and Miller T 1994 *Design of Brushless Permanent-Magnet Motors* (Oxford: Magna Physic Publishing and Clarendon Press)

[8] Idayanti N, Irasari P, Sudrajat N, Muliani L and Kristiantoro T 2009 Pembuatan magnet permanen bonded hybrid untuk aplikasi generator kecepatan rendah *Jurnal Sains Materi Indonesia* vol 536 pp 141–45

[9] Diļev G, Ose-Zaļā B and Jakobson E 2012 Self-Excitation of low-speed inductor generator *Latv. J. Phys. Tech. Sci.* vol 49 no 4 pp 21–8

[10] Alqodry F, Rustana C and Nasbey H 2015 Rancang bangun generator fluks aksial putaran rendah magnet permanen jenis Neodymium (NdFeB) untuk turbin angin sumbu vertikal tipe double-stage savonius *Prosiding seminar nasional fisika (e-journal)* vol 4 pp SNF2015-VII

[11] Anam M C, Nurhadi and Irfan M 2017 Perancangan Generator 100 Watt Menggunakan Software Elektromagnetik Infolityca *KINETIK* vol 2 no 1 pp 27–36

Irigation distribution automatization based on scheduling system

Sunardi[1], S A Akbar[2], F Noviyanto[3], E Wibowo[4], R Naufal[5]
[1,2,5]Department of Electrical Engineering, Universitas Ahmad Dahlan, Jalan Prof. Dr. Soepomo, SH, Yogyakarta
[3,4]Department of Informatics Engineering, Universitas Ahmad Dahlan, Jalan Prof. Dr. Soepomo, SH, Yogyakarta

sunargm@gmail.com[1], sonali@ee.uad.ac.id[2], fiftin.noviyanto@tif.uad.ac.id[3], ekowb@tif.uad.ac.id[4], raivan.naufal20@gmail.com[5]

Abstract. Paddy fields become areas of rice cultivation, which considered as one of the staple foods for Indonesians. Limited availability of water for irrigating requires regulation of water distribution for each field area. The area of irrigated rice fields that are not proportional to the available of water discharge triggers conflict potential among farmers. Currently, the water distribution is supervised by P3A (Association of Water User Farmers) officers. However, several problems are still around, for example the injustice of the distribution process and water corruption committed by farmers who violate the distribution rules. These matters cause the farmers to have some extra guards at night, and may contribute to social conflict and crop failure in the future. This study designed a system as an instrument to distribute water on a scheduled basis in accordance with the agreement of farmer groups. The system is developed by using Arduino Mega 2560 Microcontroller which is equipped with water level sensors (height) and water moisture sensor (humidity). The system was successfully designed by using two valves as the irrigation door on two fields. In the scheduling process in each of four conditions where the instrument was tested, the tool was proven that able to work within the schedule with 100% success rate, as well as the experiment result of water level stability.

1. Introduction

Forty percent regions in Indonesia are still rely on rainwater as the irrigation sources [1]. The rain-fed type of paddy fields is always depending on the weather that is typically unpredictable. In addition, paddy fields also have limitation in the frequency of planting. Efforts in water utilization through irrigation require improvements so that it can be implemented effectively and efficiently for a better management system [2]. Currently, In Indonesia, the water distribution is supervised by P3A (Association of Water User Farmers) officers. Irrigation is aimed to facilitate the farmers in the process of plant watering especially for type of plants that needs plenty of water supply. If the water shortages during the process of planting, crop failure or decreasing stock of food are potentially happened. By the time it raises the market price, consumers' purchasing and food security may be badly affected. Limited availability of water for irrigating requires regulation of water distribution for each field area. The area of irrigated rice fields that are not proportional to the available of water discharge triggers conflict potential among farmers within the irrigation unions [2][3][4]. Social conflicts among farmers are very likely to happen due to the clash of irrigation shifts [5]. The seizure of water also occurs due to the difference in demand between pool owners and rice farmers as well as triggering the conflict. One solution to be proposed for the root of this problem is the allocation of irrigation. This research exploits Arduino Mega 2560-based scheduling method which is combined with water level sensor and soil moisture sensor. The scheduling system works under a setting schedules and protected by the water level sensor.

2. Research Methods

The first step to do is designing a flowchart as a guide for operating the system, as seen in Figure 1.

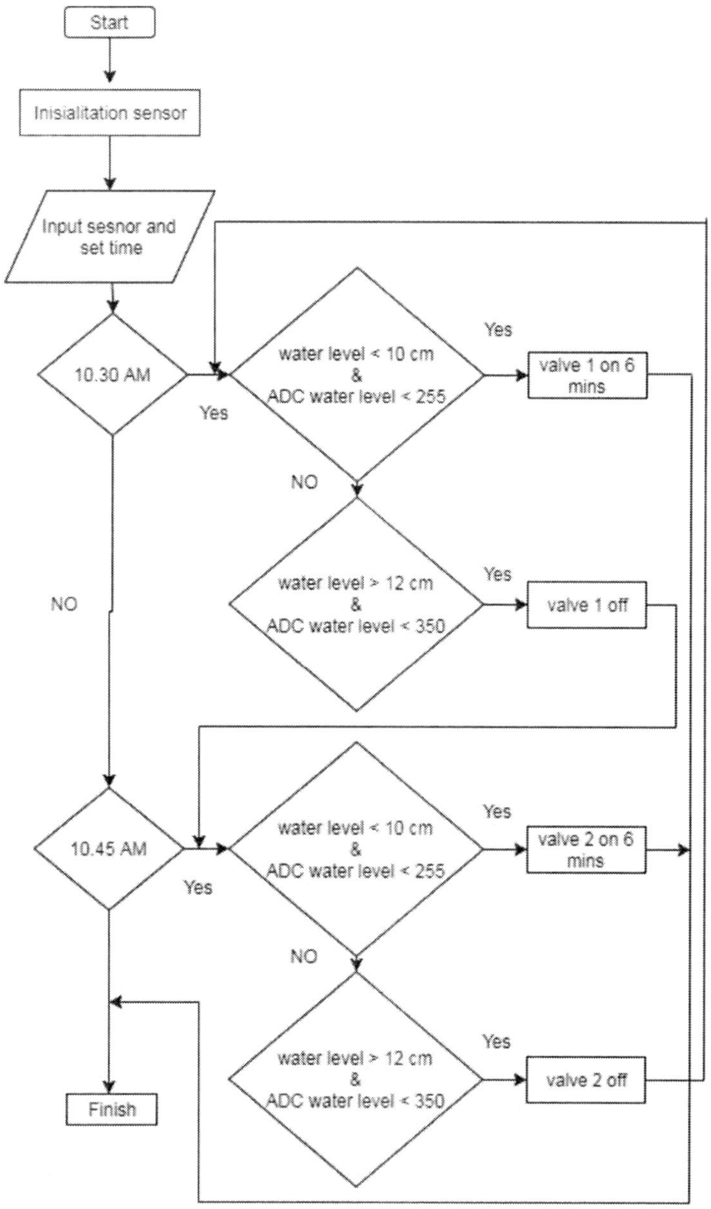

Figure 1. Flowchart of Irrigation Distribution Automatization Based on Scheduling System

As depicted in the flowchart above, the process starts from initialize the sensor to the execution process using a 12V DC motor actuator. This tool works using a scheduling system that has been agreed by the users over two paddy fields with different schedules. In the evening at 10:30 am until 10:45 am, system has ready to check water volume on paddy fields. The System can be actived less then 10 cm water level or ADC values of 255 byte on motor DC that process to open valve among 6 minute. Futhermore, the valves close and system turn-off are sensors detected too high (more than 12 cm) . That system process work is cycled in everyday by default schedules.

3. Result of Research Data

Results obtained from 8 (eight) experimental studies by observing ADC values on water level sensors and digital multimeters are listed in Table 1.

Table 1. Data of voltage and ADC value of water level sensor

Experiment	Plot Number	Water Level ADC	Water Level Volt	Result (volt)	Difference
1	1	425	0,67	0,78	0,11
2	2	451	0,52	0,64	0,12
3	1	429	0,63	0,74	0,11
4	2	393	1,07	1,13	0,06
5	1	397	1,03	1,09	0,06
6	2	389	1,10	1,16	0,06
7	1	362	1,16	1,21	0,05
8	2	374	1,15	1,19	0,04

Table 1 shows the readings of ADC values and voltage on the water level sensor after 8 times of experiments have successfully conducted into 2 plots of rice fields; plot number 1 and plot number 2. The result of voltage used as a comparison with the measurement results to obtain the value of difference. These experiments from all plots resulted in several assessments, such as the percentage of error value and accuracy on time settings and water level sensors, which can be seen in Table 2.

Table 2. The overall data

NO	Swath	Condition Biginning setting	Condition Biginning real	Setting (Time)	Start (Time)	Under Limit (cm/mins)	Upper Limit (cm/mins)	Duration Irrigation Swath (mins)	Finish	Error %	Accuracy %
1	1	0	0	10:30	10:30	10/5	12,5/6	10 mins	10:40:00	0	100
2	2	0	0	10:41	10:41	10/5	12,5/6	10 mins	10:51:00	0	100
3	1	1	1	10:52	10:52	0	0	10 mins	11:02:00	0	100
4	2	1	1	11:03	11:03	0	0	10 mins	11:13:00	0	100
5	1	0	0	11:14	11:14	10/5	12,5/6	10 mins	11:24:00	0	100
6	2	1	1	11:25	11:25	0	0	10 mins	11.35:00	0	100
7	1	1	1	11:36	11:36	0	0	10 mins	15:46:00	0	100
8	2	0	0	11:47	11:47	10/5	12,5/6	10 mins	15.47:00	0	100

From the observation result reported in Table 2, it is found that the error percentage value at the initial condition of setting and real during 8 experiments on each plot obtained by the following error rate:

$$Error\ (\%) = \frac{setting-error}{setting} x100\%$$

$$= \frac{8-8}{8} x100\%$$

$$= 0\% \qquad (1)$$

Error value is the total of similarities between the setting and the real throughout 8 experiments. The following calculation from Equation (1) obtained error rate of 0%, or in other words, the level of accuracy of water level in all plots despite of all conditions reached 100% rate of success. The accuracy level of irrigation distribution scheduling on all plots in all conditions gained perfect score, since the setting time and start time have no differences.

3.1. Graphical Data

Figure 2 visualizes the difference points between the measurement and calculation of voltages as listed in Table 1.

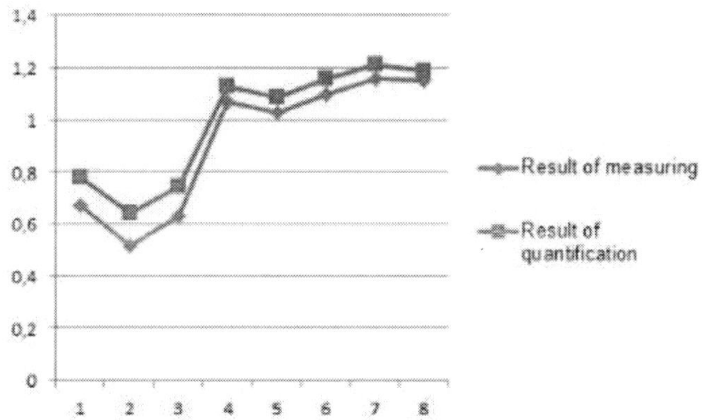

Figure 2. Comparison between result of voltage measurement and calculation

In figure 2 is showed consist test of 8 measurment, red line and blue line that describes of different testing methods both lower and highest voltage by multimeter. It is explain deviation of value each testing. By the graphic that used for check peformance of system which is made a robust system.

4. Results and Analysis

The process of data retrieval in this study starts from the installation of all tools and sensors to be used, as portrayed in Figure 3 and Figure 4.

Figure 3. Installation of tools

Figure 4. Installation of sensors

The system can be executed after all the tools and sensors are installed. The process of opening the valve towards the paddy fields based on the scheduled system can be represented in Figure 5.

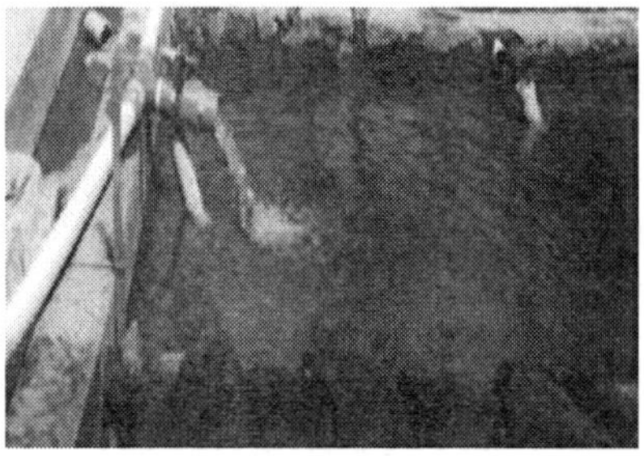

Figure 5. The process of opening the valve on the rice field (1) on schedule

After the process of valve opening is started, it then continued with the data collection process of each sensor, i.e. water level sensor. The ADC current value condition shown on the monitor serial was 425, while the output voltage signal was 0.78 VDC by using (2) and (3) methods [10].

$$ADC\ data = \frac{5V}{521} = 0.009\ V \qquad (2)$$

The 5V value is the output voltage of Arduino Mega 2560 which is divided by the maximal value of ADC at the water level sensor, that is 512. The result is 0,009 V / 1 ADC.

$$Vout = 512 - 425 = 83\ x\ 0.009 = 0.78\ VDC \qquad (3)$$

The value of ADC equals 512, which is the maximum value of the water level sensor. While the value of ADC equals 425, that is obtained from the actual condition of the water level sensor multiplied by the value per 1 ADC which equals 0.009. Therefore, the output voltage signal obtained 0.78 VDC, which means that the water flows in the paddy field is in a good condition, not less than 10 cm and no more than 12 cm above the soil surface. The actual condition results show the difference 0.11 VDC of the measurement result if using a multimeter. Soil moisture sensor test used to compare the measurement results read on Arduino Mega 2560 with the measurement result of the digital multimeter. The results can be seen in Figure 6.

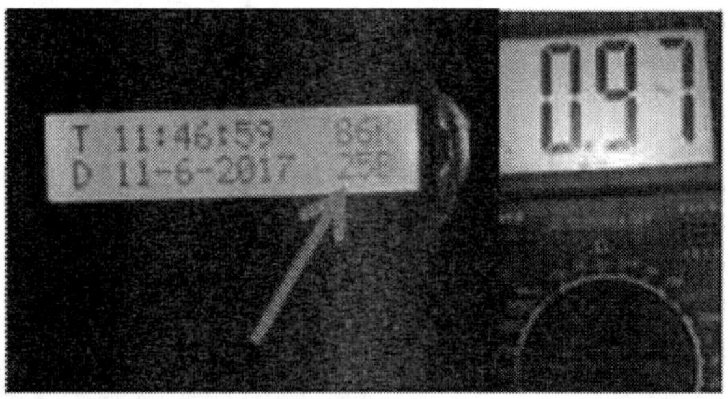

Figure 6. Display of moisture sensor on 16x2cm screen of LCD and multimeter

The ADC value condition shown on the serial monitor is 25, then the output voltage signal is obtained as 1.01 VDC by using methods (4), (5) and (6) [11].

$$ADC\ data = \frac{4{,}12}{1023} = 0.0040273705\ V \tag{4}$$

The value of 4.12 is the output voltage of Arduino Mega 2560, which is divided by the maximum value of ADC towards soil moisture sensor, that is 1023. It then resulted in 0,0040273705 VDC / 1 ADC.

This study uses the condition of ADC value where it is divided by 10 points, so that the scale of ADC value will be transformed into the range of 0-102, causes the reading is written as follows:

$$Condition = \frac{1023}{10} \tag{5}$$

The output voltage signal from ADC value as follows:

$$Vout = 1023 - (77x10) = 253\ x\ 0.00402 = 1.017\ VDC \tag{6}$$

Value of 77 is the result of the maximum scale of the ADC, which is subtracted over the actual conditions score of the rice field. The value of 10 is the multiplier factor so that the ADC value scale is in range of 0-1023. The value of 1023 is the ADC value in 10 binary bits, which resulted in 253 points. The ADC value equals 25 is obtained from the output voltage signal of 1.01 VDC which means that the soil condition is wet as the ADC value < 50. Scores of actual conditions show differences from measurement results by using a digital multimeter which is 0.04 VDC.

5. Conclusion

Based on the observations, the average accuracy of the water distribution scheduling by applying this tool was 100% rate of success, so it can be concluded that the proposed tool works optimally. It is hoped that this automation tool can assist the Water User Farmers Empowerment (P3A) officers in distributing water through the irrigation door. Moreover, this tool hopefully can reduce conflict between farmers about water distribution issues. The accuracy of the water level in the field reached 87.5%. While mistakes previously occurred due to several factors such as inconsistent water discharge, less optimal valve openings, which caused the water level was inaccurate.

6. References

[1] S. M. Ade Supriyatna, S. Dhanang Susatyo, Am. Mokh. Subehi, S. Hety Sulistiyowati, S. S. Aulia Azhar Abdurachman, and S. S. Uliyah, "Statistics of Agricultural Land 2009-2013," p. 216, 2014.

[2] A. Pardossi *et al.*, "Root Zone Sensors for Irrigation Management in Intensive Agriculture," *Sensors*, vol. 9, no. 4, pp. 2809–2835, 2009.

[3] R. Muñoz-Carpena and M. D. Dukes, "Automatic Irrigation Based on Soil Moisture for Vegetable Crops," *Dep. Agric. Biol. Eng. UF/IFAS Ext.*, pp. 1–5, 2014.

[4] M. Dursun and S. Ozden, "A wireless application of drip irrigation automation supported by soil moisture sensors," *Sci. Res. Essays*, vol. 6, no. 7, pp. 1573–1582, 2011.

[5] S. Chandra, "Agriculture Irrigation System for Rice Crops using Neural Network (AIS)," *Int. J. Eng. Technol. Manag. Appl. Sci. www.ijetmas.com*, vol. 2, no. 4, pp. 79–83, 2014.

Preliminary study plate capacitor as a Plethysmometer sensor

W S Aji

Electrical Engineering, Ahmad Dahlan University, Yogyakarta, Indonesia

wahyusaji@gmail.com

Abstract. In this research, we successfully designed a capacitor plate as a Plethysmometer sensor with a measured volume capacity of 1.5 ml, this capacity is obtained from a capacitor plate with a thickness of 0.5 mm width of 15 mm and a height of 200 mm. The sensitivity obtained is 0.36 / 0.1ml with a standard deviation of 0.012. The above results show that the linear capacitor plate sensor for the tested area has sufficient sensitivity that can measure the change of 0.1 ml to be used as a transducer in the Plethysmometer.

1. Introduction

The application of capacitive sensors for distance measurement has been one of the studies of the researchers. This cannot be separated from the ease of manufacture [1] and the ability to detect a wide range including the water [11]. The application of capacitive sensors for water level is also examined [3, 11], in this research, capacitive sensor is used for measure water level height in order cm. Capacitive sensor application for small displacement also observed [2, 5]. Capacitive sensor implementation also shows high precision for long distance measurement [8]

2. Plethysmometer and Plethysmometer sensor

Plethysmometer is a rat paw edema gauge. The working principle of Plethysmometer generally works based on Archimedes Law. Based on the displacement of the measuring fluid when the rat's paw is immersed in the measuring vessel, it will be known the volume of rat's paw swelling.

Most design work areas range from 0-7 ml for rats and 0 to 0.5 ml for mice as in Plethysmometer 37140 made by Ugo basile which is widely used by researchers [6]. Observation of volume change can be done visually it is plural used for manual plethysmometer, while for pletysmometer used sensor. The sensors used are generally conductive sensors such as plethysmometer 37140 Ugo basile. However, with this method, a special sensor fluid is required that the barrier of the type is known and has the potential to change due to contamination of the dirt from the mouse's feet when immersed.

In addition to conductivity based, the volume measurement can be based on the specific gravity of the sensor fluid [6]. Microscale scales to measure the weight of the moving sensor fluid, this method is accurate but less practical as it is still manual. In this research, capacitive sensors will be developed as sensors reader volume changes which will be applied to Plethysmometer.

3. Material and methods

The Plethysmometer will be designed based on Archimedes law, so that the outline contains the following components:

a. Water vessel

b. Tube Drop legs rat
c. The measuring tube contains a capacitive sensor

Water vessels are used to hold water. Water is used as an Archimedes fluid [6] whose volumes change will be read. Water from this vessel will be inserted into the capacitive sensor portion using the valve. Generally, for fluids a special substance is used, such as mercury in a manual plethysmometer, but water can be used as a fluid in a plethysmometer [6]

The immersing vessel is where the operator inserts the mouse's legs into it, while the measuring vessel is the vessel in which the capacitive sensor is installed. Between the water vessel and the dye is connected through the valve in order to adjust the amount of fluid volume present in the rat foot vessel and also equipped with a valve for the discharge of the unused fluid. The capacitive sensor portion is a U-tube. The full image of the capacitive sensor section is shown in figure 1.

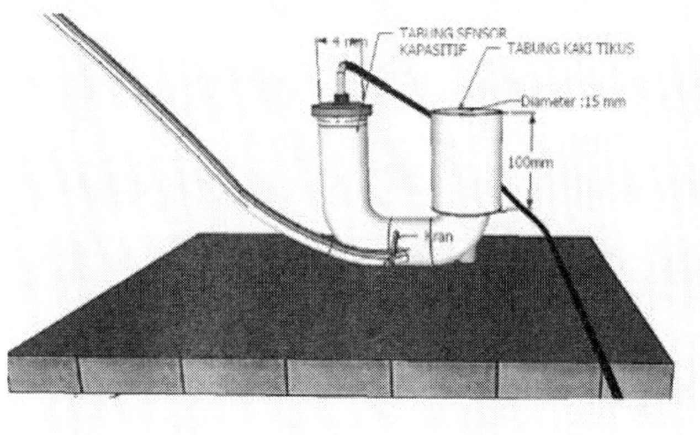

Figure 1. Plethysmometer sensor from front view.

Capacitive sensor with the shape of this U pipe has two tubes of tube sensor and tube measuring or tube rat feet, sensor tubes containing capacitive sensors are composed of two plates as an electrode. The length of the plate is 200, the distance between the plate is 0.5 mm and the width is 15mm. The sensor tube construction is shown in figure 2. In figure 2 the electrode plate will be placed in the tube. The electrode plate will form a capacitor as shown in figure 3.

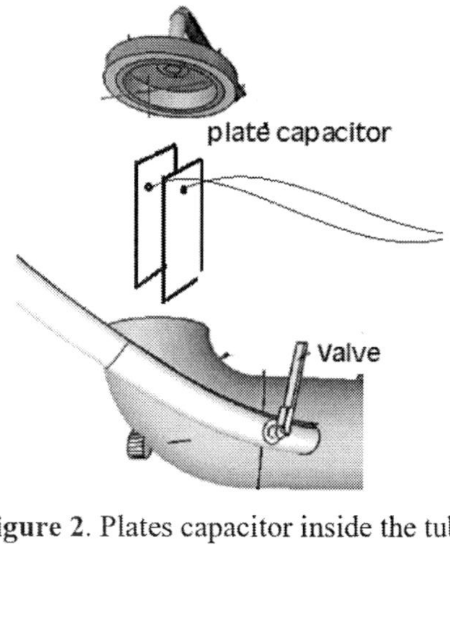

Figure 2. Plates capacitor inside the tube.

IOP Conf. Series: Materials Science and Engineering **403** (2018) 012036 doi:10.1088/1757-899X/403/1/012036

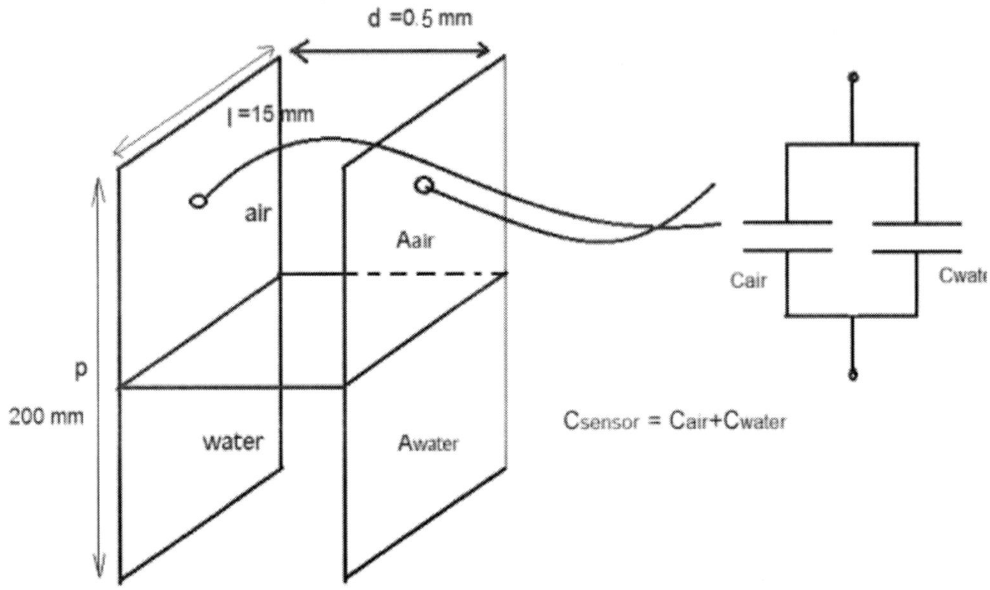

Figure 3. Modeling the capacitor of Plethysmometer sensor.

Referring to the figure 3, the total plate area, A is 3000 mm^2, d = 0.5 mm is the distance between plates, so the volume is 1500 m^3 or 1.5 ml. The capacitance total of the sensor is the sum of capacitance capacitor with area of A_{air} and the capacitance of the sensor with the width of A_{water} or can be written in equation (1).

$$C_{sensor} = C_{water} + C_{air} \tag{1}$$

$$C_{sensor} = \varepsilon_{water}\varepsilon_o \frac{A_{water}}{d} + \varepsilon_o \frac{A_{air}}{d}$$

$$\varepsilon_{water} = 80$$

$$A_{air} = A - A_{water}$$

$$C_{sensor} = \varepsilon_o \frac{79xA_{water} + A}{d}$$

$$\Delta C_{sensor} \approx 79\Delta A_{water} \tag{2}$$

Seen in the equation (2), the change in the volume of the filled sensor of water will be proportional to the magnitude of the capacitance change of the sensor.

4. Result and Comprehension

Sensor testing is done by inserting fluids into the sensor tube in sequence of 0.1ml. The readings of capacitance values are performed using LCR-9073. Table 1 shows the intended test results.

Based on table 1, the mean sensitivity was obtained at 0.37 uF / 0.1 ml and 0.36 uF / 0.1 ml with standard deviation of 0.008 for the first data and 0.013 for the second data. These valuesindicate good repeat ability and resolution.

Table 1. The test results.

First data sampling				Second data sampling			
No.	Volume, ml	C, uF	ΔC, uF	No.	Volume, ml	C, uF	ΔC , uF
1	0.1	5.34	0.37	1	0.1	5.26	0.37
2	0.2	5.71	0.37	2	0.2	5.63	0.37
3	0.3	6.09	0.38	3	0.3	6.01	0.38
4	0.4	6.46	0.37	4	0.4	6.34	0.33
5	0.5	6.82	0.36	5	0.5	6.70	0.36
6	0.6	7.20	0.38	6	0.6	7.06	0.36
7	0.7	7.57	0.37	7	0.7	7.43	0.37
8	0.8	7.93	0.36	8	0.8	7.80	0.37
9	0.9	8.31	0.38	9	0.9	8.15	0.35
10	1.0	8.68	0.37	10	1.0	8.51	0.36
11	1.1	9.04	0.36	11	1.1	8.88	0.37
12	1.2	9.40	0.36	12	1.2	9.23	0.35
13	1.3	9.77	0.37	13	1.3	9.59	0.36
14	1.4	10.13	0.36	14	1.4	9.96	0.37
Mean :0.37				Mean: 0.36			
Standard deviation=0.008				Standard deviation=0.013			

The results of the first and second data experiments also show a linear nature, as shown in the first graph and the graph into figure 4 and figure 5.

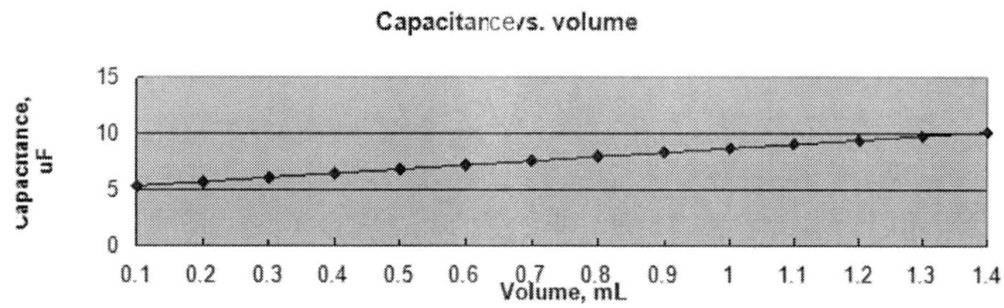

Figure 4. Data plotting from first sampled data.

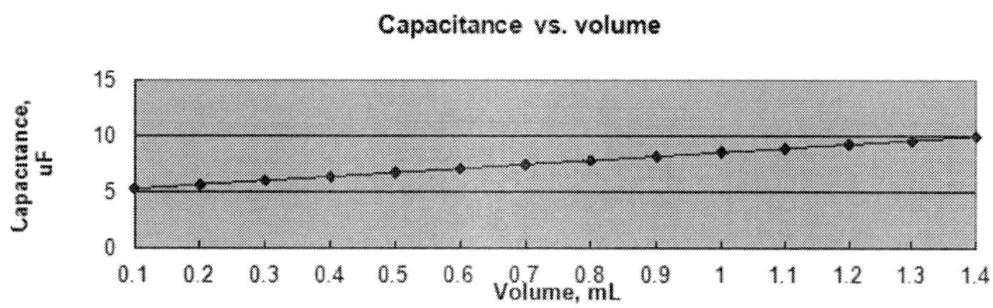

Figure 5. Data plotting from second sampled data.

Modeling with linear regression yields an equation Y = 3.687*X + 4.981, for the first graph and Y = 3.607*X + 4.906 for the second graph. These results indicate that the capacitive sensor being tested has a linear character.

5. Conclusion

In this study successfully demonstrated the feasibility of plate capacitors to be used as sensors in the Plethysmometer measuring instrument. Sensor made has a capacity of 1.5 ml, this capacity is obtained from the plate capacitor with a distance between plate 0.5 mm, width of the plate 15 mm and height 200 mm.

The results show that the sensor has a good sensitivity of 0.36 / 0.1ml, and has a good repeatability is shown by a small standard deviation value of 0.012. The sensor also shows a linear character that can be approximated by a linear equation $y = 98998\ x + 333$.

References

[1] Aezinia, Fatemeh 2014 "Design of Interface Circuits for Capacitive Sensing Application", DoctoralThesis, Simon Fraser University, Canada

[2] Avramovic Svetlana 2010 "Embedded Capacitive Displacement Sensor for Nanopositioning Applications",Conference on IEEE Transactions on Instrumentation and Measurement

[3] Canbolat Huseyin 2009 "A Novel Level Measurement Technique Using Three Capacitive Sensors for Liquids",IEEE Transactions on Instrumentation and Measurement Vol.58

[4] Dingxin Yang, Hu Zheng May 2011, "Research on capacitive sensor for online oil monitoring", 2011 Prognostics and System Health Management Conference

[5] Guohong Gao, Yun Wang, Shitao Yan, Yafeng Han 2014 "Design of capacitive displacement sensor and measuring algorithm based on modulated differential pulse width", Journal of Chemical and Pharmaceutical Research,

[6] Shejawal 2013 "A simple, sensitive and accurate method for rat paw volume measurement and its expediency in preclinical animal studies", SAGE Journal

[7] Toth,Ferry N,Meijer 1992 "A Low-Cost, Smart Capacitive Position Sensor", IEEE TRANSACTIONS ON INSTRUMENTATION AND MEASUREMENT, VOL 41

[8] Kim, Moojin, Moon, et al 2005 "A NEW CAPACITIVE DISPLACEMENT SENSOR WITH HIGH ACCURACY AND LONG RANGE", Digest of Technical Paper International Conference on Solid-State Sensors, Actuators and Microsystems

[9] Smith,J Ralp, Dorf ,C Richard 1991 "Circuits, Devices and Systems: A First Course in Electrical Engineering, 5th Edition", John Wiley and Sons

[10] Wang, David, Application Report SNOA927–December 2014 FDC1004: Basics of Capacitive Sensing and Applications, texas instrument page 1

[11] Qurthobi, A, R. F. Iskandar, A. Krisnatal, and Weldzikarvina 2016 Department of Engineering Physics, Telkom University (8th International Conference on Physics and its Applications (ICOPIA) IOP Publishing Journal of Physics: Conference Series 776 (2016) 012118 doi:10.1088/1742-6596/776/1/012118).

IOP Conf. Series: Materials Science and Engineering **403** (2018) 012037 doi:10.1088/1757-899X/403/1/012037

An excellent system in palmprint recognition

Muhammad Kusban*, **Aris Budiman**, **Bambang Hari P.**
Teknik Elektro, Universitas Muhammadiyah Surakarta 57169 Indonesia
E-mail: *Muhammad.Kusban@ums.ac.id

Abstract. According to Federal Bureau of Investigation (FBI) of United States, more than 30% of record of evidence in criminal cases are derived from the palmprint, instead of fingerprint. For that reason, the research on palmprint biometric continues to be developed. One of the effort to improve the system is to implement the right algorithms for all stages of biometric research. An excellent system in this paper consists of four powerful algorithms to identify and verify a person through palm of the hand particularly increasing the ROI display image with wavelet, the selection of Gabor parameters, the use of appropriate dimension reduction, and the choice of proper matching method. According to research findings, the method is considered reliable to estimate the error rate with value 0.273% and success level of verification 99.727%.

1. Introduction

Palmprint recognition is a human identification system that is carried out through palm of either right or left hand that works exactly the same as using the fingerprint. Since the size of the palms is much larger than the fingerprint, the researchers in the biometric field are interested in obtaining the important information easily, fast, inexpensive, and flexible in a way that is suitable to use at the airport to of the hand in moving state by simply waving a palm of the hand. Despite the palmprint recognition is relatively new compared to face recognition or fingerprint, but the system has its own advantages compared to any other method particularly due to the process of data matching that can be carried out by using a low-resolution image [1]. On the other hand, the system has several weaknesses such as position of hands is different from one another and hand grips tend to shrink when data input is being acquired.

As an identification medium, the principal lines, branch lines, and wrinkles of the palms is employed to match feature information. This is possible because no human has the same line pattern, even for twins, where the outline of palms began to form in the embryo since week 13 and reached the perfect shape in week 18 [2]. In the medical field, the palm pattern recognition are used to diagnose genetic disorders such as down syndrome, Aarskog syndrome, Cohen syndrome, and fetal alcohol syndrome [3]. Researchers aim at obtaining the best algorithms that meet three main objectives, namely lower EER rate, higher verification, and faster processing time. Some researchers [5, 6, 28] classify the palmprint recognition in several stages, specifically data acquisition, pre-processing, dimension reduction, and the matching process. Besides, the general review declares that the best way to increase the performance of system is to improve the algorithms at each stage.

Badrinath in [6] states that the data images from the acquisition process do not have the same level of brightness so need to be filtered to enhance the image clarity, Kang [7] strongly recommends that the filtering is necessary to overcome the grayscale imperfections as well as the shear angle. After all the images have same contrast level, then it is necessary to normalize the image positions as a point reference. Sung in [8] employs the Gabor method to normalize the position and scale for all the data. Furthermore, Wang [9] also uses the Gabor method to strengthen the feature information that weakened due to the influence of the low local value. The impact of the method is the emergence of the huge data as much as multiplication between orientation variable and scale. To reduce the immense number, the researchers use the dimension reduction method - the algorithm that can decrease the number of feature information as much as possible without compromising the quality of the system. Generally, the method is divided into two groups according to the processed data namely linear and non-linear. Connie in [10] uses three types of dimension reduction to analyze the palmprint images: principal component analysis (PCA), fisher discriminant analysis (FDA) and independent component analysis (ICA). The dimension reduction method is also used for the 3D data such as the one conducted by Cui [11] using LDA method to obtain feature information of palms. Despite the low number for the input data, the dimension reduction method remains in use as practiced by Yang [12] by using an unsupervised discriminant projection (UDP). After the dimension reduction is completed, the process continued with the method of matching between the sample data and the overall data. According to Zheng [13] using different matching methods will gain distinct performance achievement in the biometric system. Therefore, Chen in [14] uses the hierarchical-based similarity or matching process to achieve a high degree of accuracy in the palmprint recognition. Furthermore, Jia in [15] combines two matching score of principal lines and locality preserving projections features.

Although the biometric algorithms are promising, but invention and innovation is still an open issue [16]. In order to gain the reliable performance system - detection with low level EER and high percentage verification - then the authors offer four algorithms, starting from using a wavelet filter in pre-processing, selecting the Gabor's parameters of orientation and scale, setting value of the KPCA-based in the dimension reduction method, and using cosine technique for matching process, abbreviated as warkac (Wavelet, Gabor, KPCA, and Cosine). First, the wavelet filter, the proposed improvements combine the Wiener method [17] and the weighting method [18] accompanied by searching normalization form and providing histogram equalization. To enable Gabor method determine the amount of orientation and scale variable, the four values of entropy, variance, energy, and the dissimilarity from the researcher [9] are used and then the power value $\sqrt{2}$ from the researcher [19] is applied. To improve dimension reduction method the midpoint equation from the researcher [20] is developed then amplified by the Wiener filter [17] and weighting method [18] as in the wavelet process earlier. A significant reduction of feature information is carried out by using the SVD method by simply taking the diagonal value. Finally, the last proposed method is to use the cosine similarity from the researcher [21] with modification in normalized value and covariance function.

2. Proposed Methods

With the amount of input data continues to grow from different databases then homogeneous illumination for all the palmprint image is required. A filter method is a general system to get an image enhancement through a series of pixel operations for the purposes of the next application process. in the biometric field, some researchers used filter, there are the Gaussian smoothing by Deepika [22], the two-dimensional (2-D) masking by Doi [23], and the sine method by Jain [24]. Huang in [25] states that the complex wavelet filter can be used to improve the appearance of ROI image which will generate low error rate and high acceptance verification. To attain contrast uniformity, the authors modify the wavelet algorithms belonging to Ercelebi [17] and Kim [18]. In the field of image and signal processing, the wavelet is a mathematical expression that decomposes the signal into the varying waveform. The method can be used to reinforce certain values and reduce the other value in addition to eliminating noise. The system principle is similar to the Fourier transformation process. The wavelet function can be defined by multiplying operation between a function $f(t)$ and a mother wavelet $\psi(t)$ as follows [18].

$$\Psi(\kappa,\tau) = \int_{-\infty}^{\infty} f(t)\psi_{(\kappa,\tau)}(t)dt, \tag{1}$$

with $\Psi(\kappa,\tau)$ is wavelet function, κ for scale and τ for duration of time. In general, the Haar 2D DWT (discrete wavelet transform) is widely used in the image processing application by using the lowpass and highpass filters where each the filter produces two outputs. The output is known as a decomposition value which comprises four coefficients: approximation ξ_A, detail in horizontal ξ_H, detail in vertical ξ_V, and detail in diagonal ξ_D. Equation 1 can be rewritten as follows.

$$\begin{aligned}\Psi(\kappa,\tau) &= \frac{1}{\sqrt{2}}\begin{bmatrix}2 & 0\\ 0 & -2\end{bmatrix}\psi_{(\kappa,\tau)}(t)\\ &= [\xi_A, \xi_H, \xi_V, \xi_D]\end{aligned} \tag{2}$$

Authors offer algorithms for ROI image enhancement by developing an Ercelebi algorithm [17] which is multiplied by the Wiener filter ($\mathcal{W}$) for all wavelet coefficients and then with the weighing value (δ) from Kim algorithms [18] which has the value of $\delta = 1.5$ for the three coefficients detail. The coefficient approximation value is multiplied by a normalization ($\mathcal{N}$) in order to avoid an anomalous value which can be expressed as follows.

$$\mathcal{N} = I(x,y) - \frac{A}{B} - \min\left[I(x,y)\right] \times I(\hat{x},\hat{y}), \tag{3}$$

with image I in pixel of (x,y), $A = \min\left[I(x,y)\right] \times I(\hat{x},\hat{y})$ and $B = \max\left[I(x,y)\right] \times I(\hat{x},\hat{y})$, whereas the notation of $(\hat{x},\hat{y})$ is pixel with matrix of ones. The final output of the wavelet process $I_{\mathcal{W}}(x,y)$ from equation 1 until 3 is obtained by using inverse wavelet that can be expressed as follows.

$$I_W = I[\xi, \text{haar}], \xi \subseteq \begin{cases}\hat{\xi}_A = \xi_A \mathcal{W} \wedge \mathcal{N}\\ \hat{\xi}_H = 1.5\mathcal{W}\xi_H\\ \hat{\xi}_V = 1.5\mathcal{W}\xi_V\\ \hat{\xi}_D = 1.5\mathcal{W}\xi_D\end{cases} \tag{4}$$

with ξ is the four value of decomposition wavelet method and $\wedge$ is process histogram equalization.

After getting the uniformity of brightness the next process is finding a reference point for the beginning of the process. The Gabor method is a means to get the reference point that some researchers [8, 9, 22] have been using in the palmprint recognition.

The Gabor method is a linear filter that is commonly used for the edge detection process. The result is obtained through convolution operation between a sine wave and a Gaussian function. In addition to obtaining a reference point, another advantage of using the Gabor method is the ability to obtain important information in different time spans. Although other methods such as the windowed short-time Fourier transform method or WSTFT can do the same thing, but it is difficult to obtain the desired information for different periods of time, while in many applications, researchers need the important information of frequency at different duration time. To address weaknesses the WSTFT, the Gabor method is used. The working principle of Gabor filter is similar to the wavelet function mainly the Morlet-wavelet with the intention of reducing the impact of standard deviation values that accompany it within the time and frequency.

IOP Conf. Series: Materials Science and Engineering **403** (2018) 012037 doi:10.1088/1757-899X/403/1/012037

General expression of the Gabor function ($G_{(\theta,s)}$) is the multiplication between a sinusoidal wave (S) and an exponential Gaussian function (Γ) that can be expressed as follows [19],

$$G(\phi, \kappa) = S\Gamma = \frac{||\xi_{\phi,\kappa}||^2}{\sigma^2} e^{(-\frac{\xi^2+\gamma^2\xi^2}{2\sigma^2})} e^{i(2\pi\frac{\xi}{\lambda}+\delta)}, \tag{5}$$

with ϕ for an orientation, κ for scale, S for frequency, λ for a wavelength, σ for a phase, and δ for standard deviation. Of all the variables that exist in the Gabor function, only two important variables that are always considered, there are orientation and scale. From equation 5 with the image of wavelet process is as follows.

$$I_G(x, y) = I_w(x, y)G(\phi, \kappa), \tag{6}$$

where $I_G(x, y)$ is image result of Gabor method, $I_w(x, y)$ is wavelet image, and $G(\phi, \kappa)$ is Gabor process. Figure 2. Shows the image selection 40 variations Gabor with the value of orientation is 8 and the scale is 5.

The authors modify Wang's algorithms [9]: entropy, variance, energy, and dissimilarity and then replaced by variable of : $f_{\max} = 0.25$, $n_i = \sqrt{2}$, $\gamma = \sqrt{2}$, $\Delta = \sqrt{2}$ successively. Next, the authors also adopt Perez's algorithm [19] in scale series with the number of $\sqrt{2}$. The scale with seven value in series order can be presented as $\kappa = 2, 2\sqrt{2}, 4, 4\sqrt{2}, 8, 8\sqrt{2}, 16$ meanwhile the eight series value in the orientation can be presented as $\phi = 0, 1, 2, 3, 4, 5, 6, 7$. If the $\phi = 8$ and $\kappa = 7$ then the image is increased into 56 of a new image. With the large number of images and ever growing input data, the dimension reduction method ($\mathfrak{DR}$) is required to eliminate most of the insignificant information data.

The main purpose using the $\mathfrak{DR}$ method is to get the output system with the same result both in untouched and touched the input image data. The analogy is similar to image compression process as in the JPG image. Three advantages of using this $\mathfrak{DR}$ method are reduction of process time thus speeding up the computation time, reduction of complex data space thereby reducing the parameters used, and saving the observation of feature information hence making it more rigorous in supervision. There are two type data for the dimension reduction, linear and non-linear. Using linear type has drawback in process when involving big data and also hard to to represent the object data visually in chart. Researchers believe manifold method is the best solution to change linear form to non-linear either in the form of 2D or in 3D. Several authors [13, 24, 25] have been using the linear dimension reduction method for their research in the palmprint recognition. Generally, the linear type of the PCA is the most used in biometric. However, according to Jaswal [26] the PCA method has a drawback in recognition process when the data input continues to grow. Instead of using the method, the other researchers use the KPCA (*kernel principle component analysis*) method. A kernel is a technique that maps the distribution of data in multiple arrangements of axes (dimensional manifold).

The KPCA process is the development and expansion of the PCA method by using the kernel support. In order to map the data point of x_i in advance in the space $\phi(x_i)$ of the feature information. According to Jaswal [26] to get a kernel can be obtained from multiplication matrix with its transpose that can be expressed as follows.

$$\mathcal{K}(x_i, x_j) = \phi(x_i)^T \phi(x_j), \tag{7}$$

with $\mathcal{K}(x_i, x_j)$ for the KPCA process and T for transpose operation.

In equation (7), the value of $\phi(x_i)$ is too big, thus solving this problem can be carried out by making a midpoint of centered features $\hat{\phi}(x_i)$. Then, Equation (7) can be changed in the following form.

$$\mathcal{K}(x_i, x_j) = \hat{\phi}(x_i)^T \hat{\phi}(x_j) \tag{8}$$

IOP Conf. Series: Materials Science and Engineering **403** (2018) 012037 doi:10.1088/1757-899X/403/1/012037

Translation and simplification Equation (8) can be defined in another form as a follows.

$$\mathcal{K}_c = \mathcal{K} - 2\ell\mathcal{K} + \ell\mathcal{K}\ell, \tag{9}$$

with ℓ is a $[N \times N]$ matrix with the value of $1/N$ for all elements. Varon [20] states that the determination of midpoint in Equation (8) is critical value in KPCA system. In this paper, authors novelty research is modifying Jaswal's method [26] by adding several point value that can be defined as follows.

$$\mathcal{K}(x_i, x_j) = \left[\hat{\phi}(x_i)^T \hat{\phi}(x_j)\right]^{\sqrt{2}}, \tag{10}$$

and also the new form of center kernel in Equation (9) is can be defined as a follows.

$$\mathcal{K}_c = \mathcal{K} - \ell\mathcal{K} - \mathcal{K}\ell + \ell\mathcal{K}\ell. \tag{11}$$

Last process in dimension reduction is reducing the amount data of $\mathcal{K}$ significantly without sacrificing the important feature then the process should be resumed with the diagonal value of SVD (singular value decomposition).

After $\mathcal{DR}$ process is finish then the next step is matching method. Matching or similarity method is a technique of intentionally inserting some test images into the palmprint system to recognize it. Generally, to recognize how good or bad the system performance in biometric is by using indicator value of FAR and FRR. The false accepted rate is the ratio system to recognize the test images as part of a reference database. Whereas, in fact, it does not belong to the reference. The false rejection rate is the ratio system to recognize database, while in fact, they do not to the reference. Whereas in fact, it does belong to the reference. A critical point of intersection line between FAR and FRR known as a EER value. The error equal rate is the easy way how to look at the system performance. The smaller value of EER, means the better the system. According to Senoussaoui [21], the matching method for uncertain condition with forecast noise in big volume of data can be resolved by using a cosine method. In general, the cosine method can be expressed as follows.

$$d(x, y) = \frac{x^T y}{|x| \cdot |y|}, \tag{12}$$

with $x = \left(\sum_{i=1}^{n} x_i^2\right)^{1/2}$ and $y = \left(\sum_{i=1}^{n} y_i^2\right)^{1/2}$. In this paper, author propose to use a normalization $[x_N, y_N]$ and covariance Cov value to anticipate the random changing that may occur in all value data then, the cosine method can be changed as follows.

$$d(x, y) = \frac{x^T \cdot \mathrm{Cov}(y)}{x_N \cdot y_N}. \tag{13}$$

Finally, this paper uses four curves for performance identification: ROC (receiver operating characteristic), CMC (cumulative match curve), EPC (expected performance curve), and DET (detection error trade-off), in addition to the table showing the value of the research results.

3. Results and Discussion

The study aims to determining the reliability of the system by using the PolyU image database of ROI palm hand. The number of ROI is 550 subjects with 10 data variations for each subjects, so total data input is 5500 images with the size of [128 × 128]. Especially for the testing of dimension reduction methods, input research is added with database the Casia and the IITD-India with the number of 650 and 450. Each has a variety of images as much as 5 and 6. To

Table 1. Various methods of image enhancement filter to select the most reliable in palmprint recognition

Filter	Time	EER	Ver.
Original	0.8769	0.00362	99.636
Wavelet	0.68244	0.00272	99.727
Multiple	0.90415	0.00274	99.727
Shock	0.71127	0.00729	99.273
Skeleton	0.66315	0.00835	99.182
Anisotropic	0.69574	0.00363	99.636
Histogram	0.72881	0.00254	99.727

show the reliability filter options, seven different method is used. This research employs PolyU database which the results is shown in Table 1.

The value in Table 1 involves the use of filters wavelet, multiple, shock, skeleton, anisotropic, and histogram equalization. It shows that the wavelet method has a value of 0.00272 which is the lowest EER value. The value is equivalent to 99.727% rate of success in verification. The wavelet filter method reliability can be proved with a curve of DET and CMC in Fig. 1 and 2.

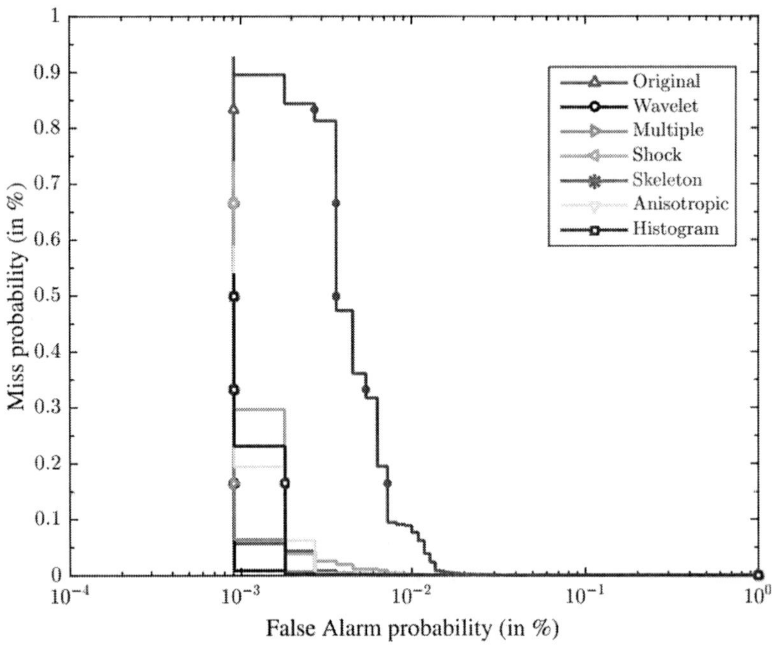

Figure 1. The DET curve in selection filter method

After completing the filter process then continuing with the selection Gabor parameters, the results is shown in Table 2. The option parameters of 8×7 is the best choice compared others with the process time of 1.92639 seconds. In addition, the EER is only 5.233% and verification accuracy degree is 94.769%. Although in terms of process time, another selection of 8×5 is better with the value of 1.69758 seconds. To support the research with evidence, Fig. 3 presents ROC curve which clearly shows that with $\phi = 8$ and $\kappa = 7$ has an image display with the best performance.

After Gabor process is completed, dimension reduction method is started. However, this

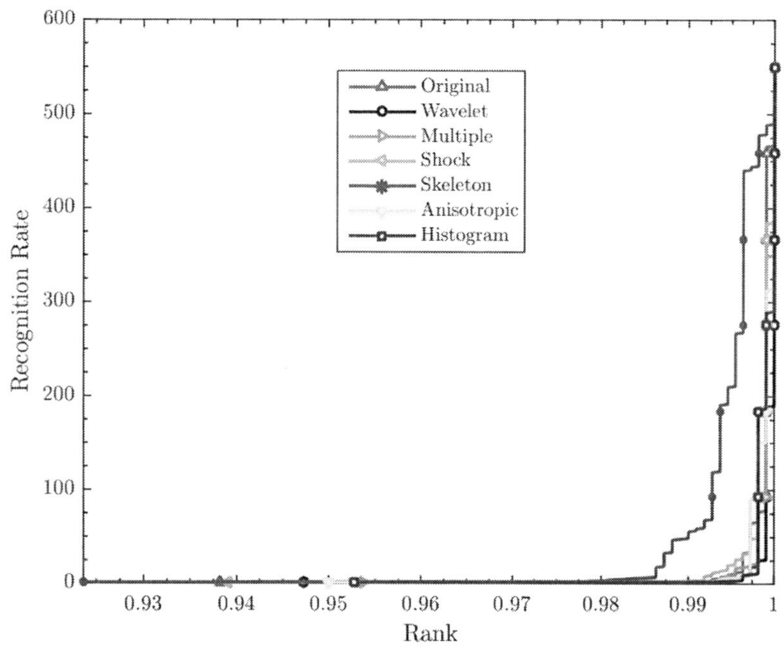

Figure 2. The CMC curve in selection filter method

Table 2. The results of examination of seven pairs of Gabor parameters for different of orientation and scale

$[\phi \times \kappa]$	Time	EER	Ver.
5×8	1.93044	0.40207	0.59846
8×7	1.92639	0.05233	0.94769
8×4	1.71137	0.35799	0.64231
8×5	1.69758	0.42606	0.57462
8×6	1.80697	0.31604	0.68385
5×8	1.77491	0.40207	0.59846
15×8	2.44745	0.29442	0.70538

paper puts the matching process earlier than dimension reduction. The reason is to emphasize the importance of selecting dimension reduction appropriately. Table 3 shows the results of 7 different matching methods on palmprint. As seen from the Table, there are similarities in terms of the processing time of 2.5 seconds for all the methods. But in terms of EER value, the cosine algorithm is superior compared with another method with the value of 0.543%. That value is equivalent 99.455% in system verification. Evidence of cosine method superiority can be seen in the EPC curve in Fig. 4. In the curve, the green curve line represents cosine method that is displayed in the lowest position axis of error rate.

Finally, to find the best method in reduction dimension, four algorithms: KFA, KPCA, LDA, and PCA and three database: Casia, IITD-India, and PolyU are used in this research with the results displayed in Table 4 and curve in Figure 8.

The best EER value occurs in the database PolyU using KPCA method to the value of 0.00273 for the EER, 0.84399 seconds for processing time, and 99.727% for successful verification. With other databases, LDA method is sometimes better than KPCA, but LDA has drawback in processing time is longer than KPCA.

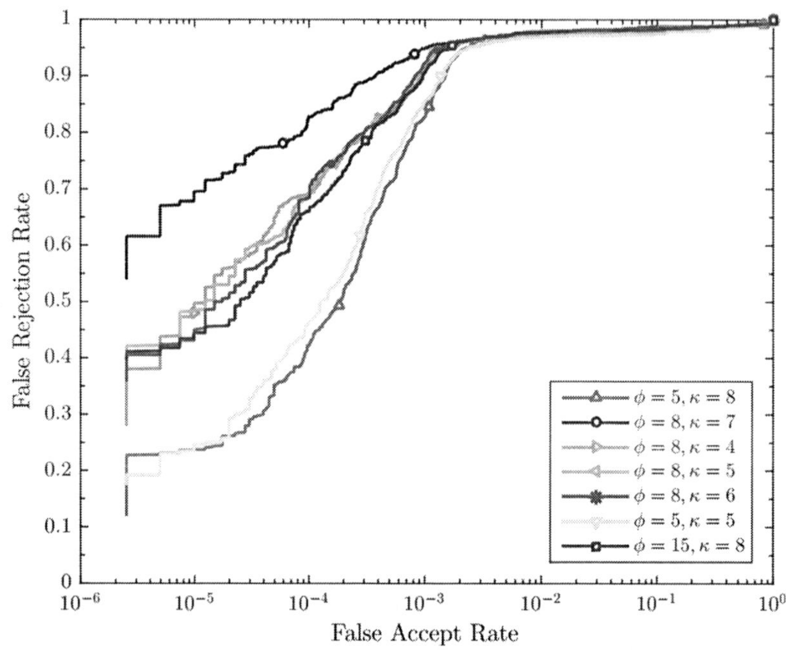

Figure 3. Selection the Gabor parameter of [8 × 7] is the best choice for palmprint system

Table 3. Outcomes of research involving some kind of similarity method

Matching	Time	EER	Ver.
Euclidean	2.40994	0.00638	0.99364
CTB	2.55865	0.01089	0.98909
Cosine	2.52237	0.00543	0.99455
MahCos	2.53797	0.01180	0.98818
ModEuc	2.47730	0.00638	0.99364
Hausdorff	2.52343	0.50955	0.49000
Ndistance	2.51170	0.47964	0.52000

4. Conclusion

From a series of trials that have been conducted in order to obtain the optimum palmprint recognition system, the wavelet filter is a good option for image enhancement, the value of [8 × 7] is the best selection of the Gabor method, the use of KPCA is the superior for the dimension reduction, and the cosine algorithm is the right choice for similarity method. When all of these methods are combined, or we believe that warkac would be a powerful system for verification and identification of palmprint, which is proven by attainment of EER value 0.272%, processing time of 0.84399 seconds, and success verification of 99.727%.

Acknowledgment

The author would like to thank Casia, IITD-India, and PolyU who provided forensic palmprint database used in our experiments. The research leading to these results has received funding from *the Direktorat Riset dan Pengabdian Masyarakat Direktorat Jenderal Penguatan Riset dan Pengembangan Kementerian Riset, Teknologi, dan Pendidikan Tinggi* under grant agreement number *211.35/A.3-III/LPPM/V/2017*.

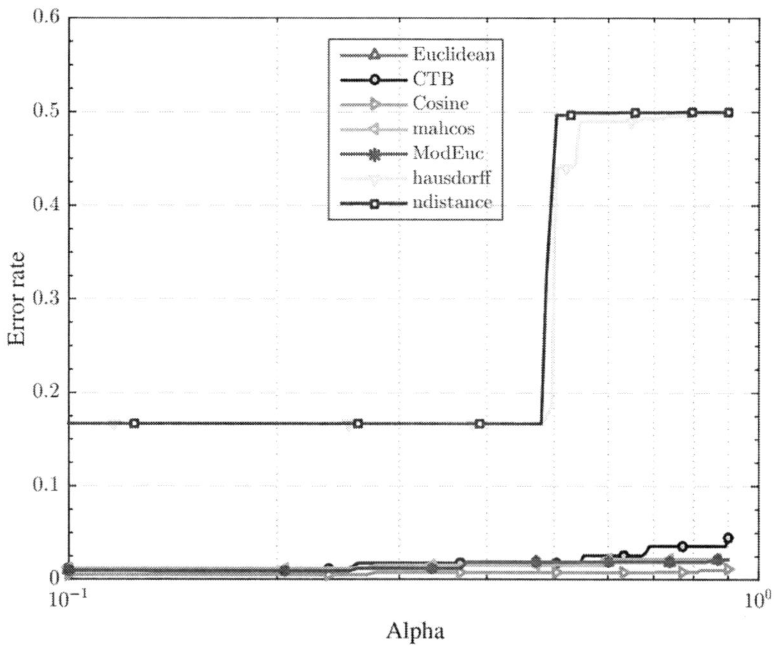

Figure 4. View of the EPC curve showing the use of cosine similarity for the matching method has the best performance characteristics showing the smallest value of error rate with green curve line

Table 4. The results of the research to find the reliability of the four method of dimension reduction in palmprint recognition

Database	D	Time	EER	Ver.
Casia	KFA	2.24663	0.42126	0.57846
	KPCA	1.48114	0.00998	0.99
	LDA	5.89667	0.00774	0.99231
	PCA	4.59728	0.0146	0.98538
IITD	KFA	1.38616	0.3364	0.66444
	KPCA	0.45817	0.02002	0.98
	LDA	2.99767	0.01885	0.98111
	PCA	2.23623	0.02088	0.97889
PolyU	KFA	1.41342	0.03701	0.96273
	KPCA	0.84399	0.00273	0.99727
	LDA	4.32426	0.00358	0.99636
	PCA	3.04235	0.00364	0.99636

References

[1] A. Kumar and D. Zhang, "Personal recognition using hand shape and texture," *IEEE Trans. Image Process.*, vol. 15, no. 8, pp. 24542461, 2006.

[2] G. K. Ong Michael, T. Connie, and A. B. Jin Teoh, "Touch-less palm print biometrics: Novel design and implementation," *Image Vis. Comput.*, vol. 26, no. 12, pp. 15511560, Dec. 2008.

[3] A. Kong, D. Zhang, and M. Kamel, "A survey of palmprint recognition," *Pattern Recognit.*, vol. 42, no. 7, pp. 14081418, Jul. 2009.

[4] B. Zhang, W. Li, P. Qing, and D. Zhang, "Palm-print classification by global features," *IEEE Trans. Syst. Man, Cybern. Part A Systems Humans*, vol. 43, no. 2, pp. 370378, Mar. 2013.

[5] D. Zhang, Z. Guo, G. Lu, L. Zhang, Y. Liu, and W. Zuo, "Online joint palmprint and palmvein verification," *Expert Syst. Appl.,* vol. 38, no. 3, pp. 26212631, 2011.

[6] G. S. Badrinath and P. Gupta, "Palmprint based recognition system using phase-difference information," *Futur. Gener. Comput. Syst.,* vol. 28, no. 1, pp. 287305, 2012.

[7] W. Kang and Q. Wu, "Contactless palm vein recognition using a mutual foreground-based local binary pattern," *IEEE Trans. Inf. Forensics Secur.,* vol. 9, no. 11, pp. 19741985, 2014.

[8] J. Sung, S. Y. Bang, and S. Choi, "A Bayesian network classifier and hierarchical Gabor features for handwritten numeral recognition," *Pattern Recognit. Lett.,* vol. 27, no. 1, pp. 6675, 2006.

[9] H. Jun Wang, H. Nian Qi, and X. F. Wang, "A new Gabor based approach for wood recognition," *Neurocomputing,* vol. 116, pp. 192200, 2013.

[10] T. Connie, T. B. Jin, M. G. K. Ong, and D. N. C. Ling, An automated palmprint recognition system, Image Vis. Comput., vol. 23, no. 5, pp. 501515, May 2005.

[11] J. Cui and Y. Xu, "Three dimensional palmprint recognition using linear discriminant analysis method," *Innovations in Bio-inspired Computing and Applications, International Conference,* pp. 107-111, 2011.

[12] J. Yang, D. Zhang, J. Y. Yang, and B. Niu, "Globally maximizing, locally minimizing: Unsupervised discriminant projection with applications to face and palm biometrics," *IEEE Trans. Pattern Anal. Mach. Intell.,* vol. 29, no. 4, pp. 650664, 2007.

[13] Q. Zheng, A. Kumar, and G. Pan, "Suspecting Less and Doing Better: New Insights on Palmprint Identification for Faster and More Accurate Matching," *IEEE Trans. Inf. Forensics Secur.,* vol. 11, no. 3, pp. 633641, 2016.

[14] F. Chen, X. Huang, and J. Zhou, "Hierarchical minutiae matching for fingerprint and palmprint identification., *IEEE Trans. Image Process.,* vol. 22, no. 12, pp. 496471, Dec. 2013.

[15] W. Jia, B. Ling, K. W. Chau, and L. Heutte, "Palmprint identification using restricted fusion," *Appl. Math. Comput.,* vol. 205, no. 2, pp. 927934, 2008.

[16] A. Kumar, and A. Kumar, "A Cell-Array-Based Multibiometric Cryptosystem," *IEEE Access* vol. 4, pp. 1525, 2016.

[17] E. Ercelebi and S. Koc, "Lifting-based wavelet domain adaptive Wiener filter for image enhancement," *Chinese J. Electron.,* vol. 14, no. 1, pp. 115118, 2005.

[18] S. Kim, W. Kang, E. Lee, and J. Paik, "Wavelet-domain color image enhancement using filtered directional bases and frequency-adaptive shrinkage," *IEEE Trans. Consum. Electron.,* vol. 56, no. 2, pp. 10631070, 2010.

[19] C. A. Perez, L. A. Cament, and L. E. Castillo, "Methodological improvement on local Gabor face recognition based on feature selection and enhanced Borda count," *Pattern Recognit.,* vol. 44, no. 4, pp. 951963, 2011.

[20] C. Varon, C. Alzate, and J. A. K. Suykens, "Noise Level Estimation for Model Selection in Kernel PCA Denoising," *IEEE Trans. Neural Networks Learn. Syst.,* vol. 26, no. 11, pp. 26502663, 2015.

[21] M. Senoussaoui, P. Kenny, T. Stafylakis, and P. Dumouchel, "A study of the cosine distance-based mean shift for telephone speech diarization," *IEEE Trans. Audio, Speech Lang. Process.,* vol. 22, no. 1, pp. 217227, 2014.

[22] C. L. Deepika, A. Kandaswamy, C. Vimal, and B. Satish, "Palmprint authentication using modified legendre moments," *Procedia Comput. Sci.,* vol. 2, no. 2009, pp. 164172, 2010.

[23] J. Doi and M. Yamanaka, "Discrete finger and palmar feature extraction for personal authentication," *IEEE Trans. Instrum. Meas.,* vol. 54, no. 6, pp. 22132219, 2005.

[24] A. K. Jain and J. Feng, "Latent palmprint matching," *IEEE Trans. Pattern Anal. Mach. Intell.,* vol. 31, no. 6, pp. 10321047, 2009.

[25] D. Huang, W. Jia, and D. Zhang, "Palmprint verification based on principal lines," *Pattern Recogn.,* vol. 41, pp. 13161328, 2008.

[26] G. Jaswal, R. Nath, and A. Kaul, "Multiple resolution based palm print recognition using 2D-DWT and Kernel PCA," *Int. Conf. Signal Process. Commun. ICSC 2015,* pp. 210215, 2015.

[27] W. Deng, J. Hu, J. Guo, H. Zhang, and C. Zhang, "globally maximizing, locally minimizing: Unsupervised discriminant projection with application to face and palm biometrics," *IEEE Trans. Pattern Anal. Mach. Intell.,* vol. 30, no. 8, pp. 15031504, 2008.

[28] B. Zhang, W. Li, P. Qing, and D. Zhang, "Palm-print classification by global features," *IEEE Trans. Syst. Man, Cybern. Part ASystems Humans,* vol. 43, no. 2, pp. 370378, 2013.

IOP Conf. Series: Materials Science and Engineering **403** (2018) 012038 doi:10.1088/1757-899X/403/1/012038

Design a system of measurement of heart rate, oxygen saturation in blood and body temperature with non-invasive method

M T Tamam[1], A J Taufiq[1] and A Kusumawati[2]
[1]Electrical Engineering Department, [2]Medical Faculty
Universitas Muhammadiyah Purwokerto
Jl. Raya Dukuhwaluh Kembaran, Banyumas, Jawa Tengah, Indonesia 53182

tamam@ump.ac.id

Abstract. Heart disease is the number one cause of death in the world. The heart functions to pump blood flowing through the veins or arteries to all parts of the body. In the blood contains many levels of molecules or constituents, such as oxygen in blood HbO_2 (oxyhemoglobin). This study aims to design a system used to measure changes in voltage fluctuations to the transducer (photoplethysmography), oxygen saturation and changes in body heat. From the test result got error value at measurement of SpO_2 equal to 0.89% and for measurement of BPM (heart rate) equal to 3.095%, difference or difference measurement of body temperature sensor got error value equal to 0.78%.

1. Introduction

The human body consists of various vital organs that are interconnected and have a unique and complex function. The heart as a vital organ that serves to provide oxygen throughout the body and cleanse the body of the metabolism (carbon dioxide). Heart disease is the number one cause of death in the world, so the need for early detection of this disease. The pulse can be measured at any point on the body where the vibrations in the arteries are sent to the surface frequently when it is pressed against a structure that lies beneath such a bone [1].

Pulse oximetry is a non-invasive method for measuring the percentage rate of hemoglobin saturation. This method uses the different wavelengths of red light (660 nm) and infrared light (940 nm) captured by the detection sensors after passing through the veins and capillaries at the tip of the index finger [2]. Pulse oximetry serves to manage long-term patients with oxygen therapy and to identify patients with congenital heart disease. Pulse oximetry has a high accuracy for detecting congenital heart disease in newborns [3].

Photoplethysmography is an optical technique of detecting cardiovascular pulse waves (heart) from the fingertips. By exploiting optical sensor reflection then warming the area around the skin, this is evidenced by the increasing beats component of photoplethysmography [4]. The design of monitoring of integrated signal monitoring system on a platform of system on chip (SoC) and wireless technology in hospital with monitored parameter is ECG and body temperature then processed using microcontroller. Results from data processing are sent to the web server using wireless communications [5].

Detection of beat pressure (pulse) is an integration of most analytical techniques for arterial blood pressure (ABP), intracranial pressure (ICP) and pulse oximetry signals (SpO2). Pulse detection is used to estimate heart rate in ABP signals, to classify ICP morphology and to estimate blood pressure using pulse oximeter wave [6]. Therefore, medical record media of heartbeat, oxygen saturation in blood and body temperature are computationally computed and study of observation approach of health condition of body without any surgery or physical or non-invasive, and is expected to prevent human death due to abnormalities heart early because of the less than optimal medical record and responsive.

2. Materials and methods
This section explains the materials and methods used in this research.

2.1 Materials
Heart rate sensor and oxygen saturation in blood, body temperature sensor, Arduino Board Due, MULTISIM® software, and BASCOM® SIMULATOR software.

2.2 Method
The selection of sensor types used for the measurement of heart rate parameters, oxygen saturation and body temperature with the measuring unit of each monitoring parameters placed on the fingers, then by analyzing the work characteristics and suitability of each component supporting the idea of design obtained the initial design is steady and optimal both in terms of mechanical, electronic and ergonomic. The block diagram of the proposed method is shown in figure 1 and the design of the pulse oximetry sensor is shown in figure 2.

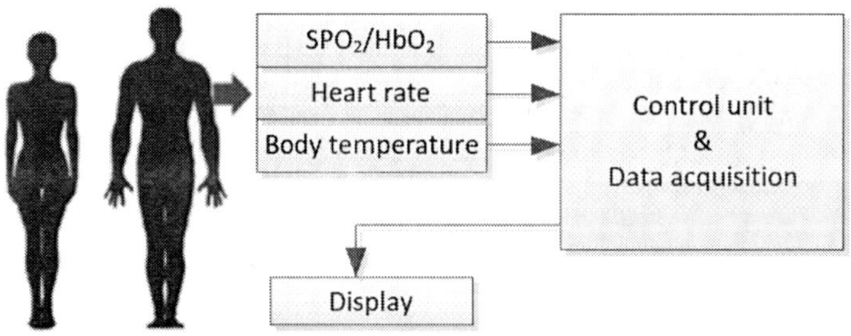

Figure 1. Block diagram of the proposed method.

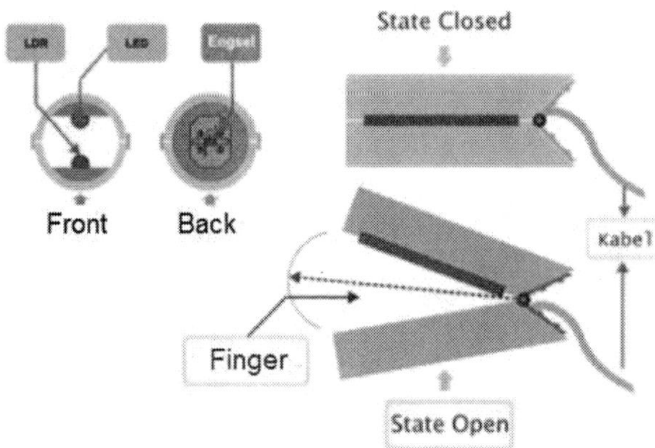

Figure 2. Design of pulse oximetry sensor.

3. Results and discussions

The system is a form of heart rate detection application by sensing the change in blood volume in the finger arteries as the heart pumps blood. It consists of an infrared LED that transmits an IR signal through the subject's fingertip, from the part reflected by the blood cells. The reflected signal is detected by the photo diode sensor. The change in blood volume with the heart rate at the output of the photodiode is not too large to be detected directly by the microcontroller.

Therefore, two stages of high gain and active low pass filter are required, the design uses two operational amplifiers (Op-Amp) to filter and amplify the signal at the appropriate voltage level so that the analog signal can perform computation by the microcontroller. Table 1 shows the results of SPO_2 testing of ten respondents with an average error of 0.89%. Table 2 shows the test results from BPM (heart rate) to ten respondents with an average error of 3.095%.

Table 1. Results of SPO_2 testing.

No	Respondent	R			SPO_2
		Led	IR	Difference	%
1	Respondent 1	2.0	2.0	1.0	85.0
2	Respondent 2	1.0	2.0	0.5	97.5
3	Respondent 3	1.8	2.0	0.9	87.5
4	Respondent 4	1.8	1.8	1.0	85.0
5	Respondent 5	1.4	2.0	0.7	92.5
6	Respondent 6	1.0	1.0	1.0	85.0
7	Respondent 7	2.0	2.0	1.0	85.0
8	Respondent 8	1.4	1.4	1.0	85.0
9	Respondent 9	1.4	2.0	0.7	92.5
10	Respondent 10	1.0	2.0	0.5	97.5

Table 2. Heart rate test results.

No.	*Heart Rate* (bpm)	ECG BIOPAC (bpm)	*Error* (%)
1	74	73.35	0.886
2	68	67.80	0.295
3	78	75.00	4.000
4	82	77.00	6.494
5	70	72.90	3.978
6	64	69.40	7.781
7	86	83.00	3.614
8	80	81.00	1.235
9	78	77.00	1.299
10	74	75.00	1.333

Table 3 shows the value of human body temperature test results against ten respondents with an average error of 0.78% The results of the measurement of heart rate parameters, oxygen saturation in blood and body temperature are represented in units of measured values displayed on 16x2 LCD. The measurement value is fluctuating, because the measurement uses non-invasive method (without surgery).

Table 3. Body temperature test results.

No	Respondent	Body Temperature (°C)		*Error* (%)
		Digital Thermometer	Measuring Instrument	
1	Respondent 1	37.0	37.12	0.32
2	Respondent 2	36.5	36.76	0.71
3	Respondent 3	36.6	36.71	0.30
4	Respondent 4	36.3	36.76	1.26
5	Respondent 5	35.6	36.14	1.51
6	Respondent 6	36.6	37.02	1.14
7	Respondent 7	36.7	36.92	0.59
8	Respondent 8	36.8	37.12	0.86
9	Respondent 9	36.9	36.98	0.22
10	Respondent 10	36.7	37.04	0.92

4. Summary

The system created in this study has been able to work in accordance with the planning, which can measure heart rate, oxygen saturation in blood and body temperature. From the test results can be concluded that the difference or different pulse oximetry sensor measurement obtained error value on measurement of SPO_2 of 0.89% and for the measurement of BPM (heart rate) of 3.095%. Differences or differences in body temperature sensor measurement obtained error value of 0.78%. Difference or difference between measurable and measured measurements caused by several errors, including finger shape output error for sensor readings. The sensor response time is average when the tool is turned on for approximately 20 seconds.

Acknowledgments

The authors are grateful to Directorate of Research and Community Service (DRPM) Ministry of Research, Technology and Higher Education Republic of Indonesia in providing a grant to conduct this research.

References

[1] Smeltzer S C and Bare B G 2005 *Buku Ajar Keperawatan Medisal Bedah Edisi 8* (Jakarta: ECG. Harold S. Koplewich)

[2] Tavakoli M, Turicchia L and Sarpeshkar R 2010 An ultra-low-power pulse oximeter implemented with an energy-efficient transimpedance amplifier *Biomedical Circuits and Systems, IEEE Transactions On* vol 4 no 1 pp 27-38

[3] Bimbaum S 2009 Pulse oximetry: identifying its applications, coding, and reimbursement *American College of Chest Physicians* vol 135 pp 838 - 41

[4] Mendelson Y 2002 Non-invasive pulse oximetry utilizing skin reflectance photoplethysmography *Biomedical Engineering, IEEE Engineering in Medicine and Biology Society* vol 35 no 10 pp 798 - 805

[5] Lin J S, Huang S Y, Pan K W and Liu S H 2009 A physiological signal monitoring system based on an SoC platform and wireless network technologies in homecare technology *Journal of Medical and Biological Engineering* vol 29 pp 47 - 51

[6] Beattie Z T 2008 Algorithm for automatic beat detection of cardiovascular pressure signals *Engineering in Medicine and Biology Society, 30th Annual International Conference of the IEEE* pp 2594 – 7

IOP Conf. Series: Materials Science and Engineering **403** (2018) 012039 doi:10.1088/1757-899X/403/1/012039

Analysis of integration flow control valve and electronic load controller for micro hydro power plant frequency regulation

R Mulyadi, M Effendy, I Pakaya and Zulfatman

Department of Electrical Engineering, University of Muhammadiyah Malang
Jl. Raya Tlogomas No. 246, Malang 65144, Indonesia

zulfatman@umm.ac.id

Abstract. Frequency regulation is very important for microhydro power system. Flow control valve (FCV) and electronic load controller (ELC) are common control systems developed for micro hydro power plant (MHPP) with synchronous generator. However, these control systems each other has a drawback in term of frequency response or harmonics issues. This paper proposes an integrate control system, in which FCV and ELC system work together for MHPP in order to obtain a better performance in frequency response and total harmonic distortion (THD). PID and PI controllers were used for FCV and ELC, respectively. The integrated control method influences on the MHPP performance was verified by MATLAB simulation, when the load reduced 10KW in two sequences time. Results of the simulation confirm that the integration of FCV and ELC offered a better trade-off between frequency response and THD issues for MHPP, which cannot be solved by standalone FCV or ELC. By integrating both control systems, they gave the best values for frequency response and THD in around 1.4071 second and 8.57%, respectively.

1. Introduction

The power system should be able to provide power to consumers load with a constant frequency [1]. Swinging frequency should be within the tolerable range of power plants [2]. Frequency control systems in MHPP are basically two kinds, such: governor or flow control valve (FCV) as water flow control system [3] and electronic load controller (ELC) as power electronic control system [4].

Governor or flow control valve (FCV) is a mechanical control equipment, where in the process of control frequency is more emphasis on the regulation of the amount primary energy into the turbine [5]. But this power system has a drawback due to inability when sudden load changes occur. So, it requires a regulation to result a faster response of frequency on the load changes such as ELC [6].

Electronic Load Controller (ELC) is a frequency control equipment that focused on dumping the active power to the dump load (resistive load). If frequency control with ELC run with MHPP, the generator will be work in full rating which is the ELC can drive the frequency return to its normal frequency very fast when the load changing. However, in terms of power savings this equipment is inefficient due to the excess of power is expelled easily to the dump load and mostly run into overheat on dump load [7] and also ELC with trial component has a high level of harmonics on MHPP [8].

Objective of this paper is to integrate FCV and ELC frequency control systems and uses PID for controller for each frequency control system to obtain a better trade-off between frequency response and harmonics. The system layout and control scheme integrates of FCV and ELC are developed in MATLAB® Simulink 2014b environment.

2. Methodology

2.1. Synchronous generator

The generator used in this analysis is a 3 phase of synchronous generator. The model shown in figure 1 with the parameters as shown in table 1. Nomenclature: Q = turbine flow (m^3/s), G = Gate Opening (rad), H = Net Head (m), Pm = Power Mechanic (pu), Tw = Time Constant Water (s).

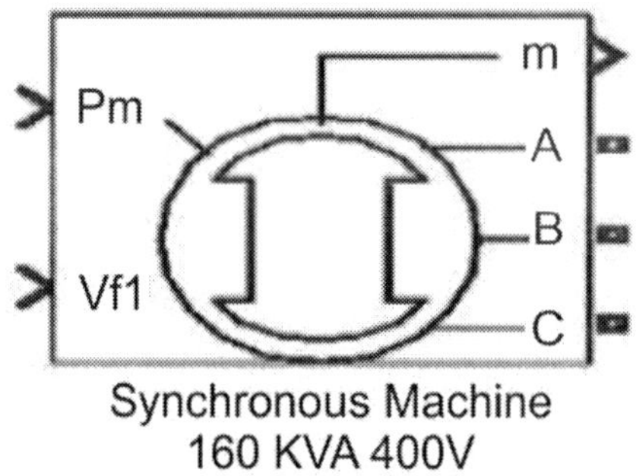

Figure 1. Synchronous generator model.

Table 1. Parameter of synchronous generator.

Parameter	Value
Nominal Power (VA), Line to line voltage (V)	160e3, 400
Frequency	50
Reactance (Xd, Xd', Xd'', Xq', Xq'', Xl) (pu)	[2.24, 0.19, 0.13, 1.38, 0.17, 0.07]
Time Constant [Td',Td'',Tq''] (s)	[0.035, 0.011, 0.011]
Stator Resistance (pu)	0.024
Inertia Coefficient (s), friction factor (pu), pole pairs	[4 0 2]
Rotor Type	Salient-Pole

2.2. Excitation system

The excitation modelling designed by IEEE [9], and available in Matlab Simulink inside MathWorks website with AC4A excitation type. The excitation model system is shown in figure 2.

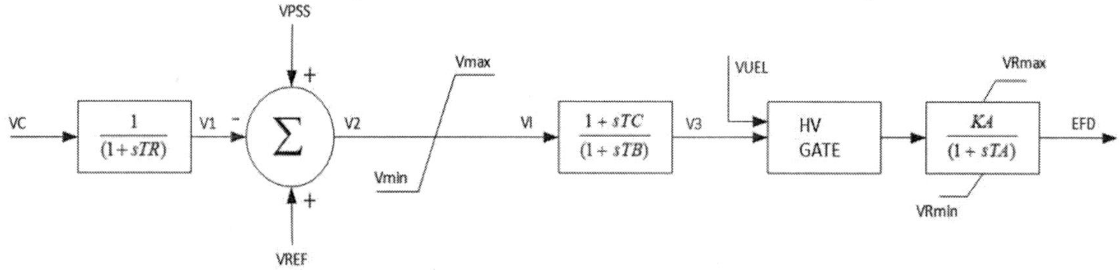

Figure 2. Excitation model system.

The sample data for the AC4A type excitation system is shown in table 2.

Table 2. Excitation system parameter.

Parameter	Value
Tr	0 s
[Ka. Ta]	[200 pu , 0.015 s]
[Kc, Tc]	[0 pu, 1 s]
Tb	10 s
[Vi_{max}, Vi_{min}]	[10, -10] (pu)
[Vr_{max}, Vr_{min}]	[5.64, -4.53] (pu)

2.3. Modelling of Flow Control Valve (FCV) System

There is some of the models needed to create and simulate FCV according to hydraulic turbine and governor modeling designed by the IEEE work group [10], include:

2.3.1. Hydraulic turbine model. Turbine is an important part of microhydro power system that receives potential energy from water and converts it into rotational (mechanical) energy, then this mechanical energy will rotate the turbine axis on the generator. The hydraulic turbine model is depicted as in figure 3. The parameters of hydraulic turbine show in table 3.

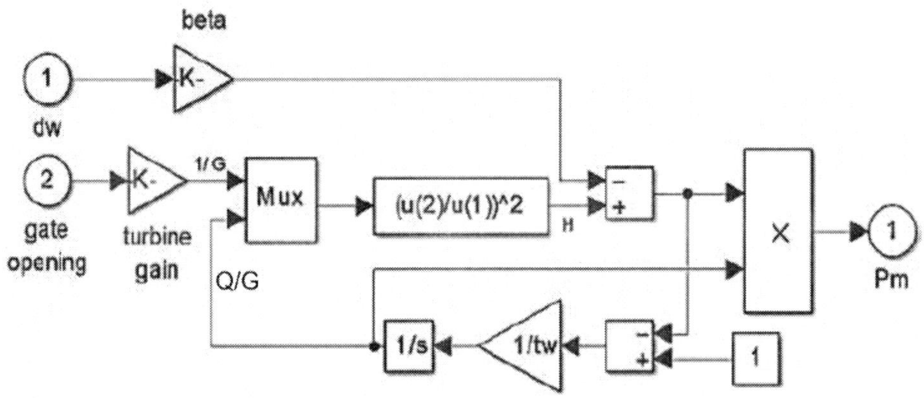

Figure 3. Hydraulic turbine model.

Table 3. Hydraulic turbine parameters.

Parameter	Value
Turbine Flow	0.95 m^3/s
Head base	16.74 m
Penstock Area	0.384 m^2
Length Penstock	60 m
Gravitation	9.8 m/s^2
[G_{max}, G_{min}, beta]	[1,0,0] (pu)

2.3.2. Hydro-electric servo system model as actuator. The motor servo in the FCV used to gate valve according signal from controller to regulate velocity of water flow to stand in set-point value. Hydro-electric servo system shows in figure 4. Hydro-electric servo system parameter shows in table 4.

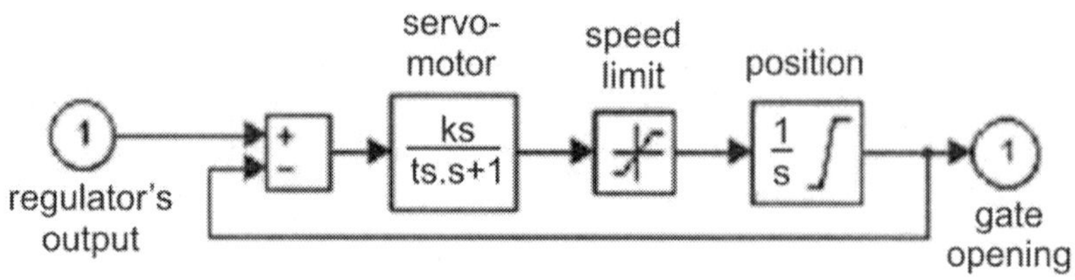

Figure 4. Hydro-electric servo system model.

Table 4. Hydro-electric servo system parameter.

Parameter	Value
Gain constant motor servo	10
Time constant motor servo	0.001
[V_{gmin}, V_{gmax}]	[-0.1, 0.1]

2.3.3. Modelling of PID controller. The PID controller is inserted in the controller system when desired. Improved transitional response as well as steady state responses. PID controller is shown in equation (1). The PID controller has two zeros and one pole. One of the methods used to design PID controllers. It is used to design the PI portion to provide a steady state response, then the PI controller is considered part of the process / plant. In the PD section is designed to improve the transition response.

$$G_o(s) = K_p + K_p s + \frac{K_i}{s} = \frac{K_p S^2 + K_p S + K_i}{s} \tag{1}$$

2.4. Modelling of Electronic Load Controller (ELC) system

Designing the ELC model also uses hydraulic turbine model and hydro-electric servo system model which already been explained on the previous, but both systems are not functioned for controlling the frequency. The model that used for controlling the frequency is needed to simulate ELC is:

2.4.1. AC Regulator Model. AC regulator model uses three phase bidirectioal thyristor is shown on the figure 5. The parameters data for the thrystor component are shown in table 5.

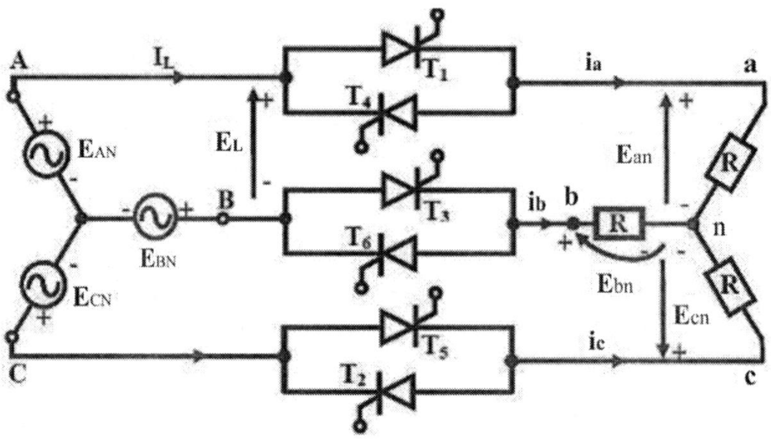

Figure 5. Three phase bidirectional AC regulator circuit.

:

Table 5. Thyristor parameters.

Parameter	Value
Resistance Ron (ohms)	0.001
Inductance Lon (H)	0
Forward voltage Vf (V)	0.8
Initial Current Ic (A)	0
Snubber resistance Rs (ohms)	1e6
Snubber capacitance Cs (F)	inf

2.4.2. Complementary load model. The complementary loads used are resistive loads that is connected in star (Y) connection with the ability to absorb 180 KW of power that is connected with output. The complementary load model is shown in figure 6. The parameters data for the complementary load model are shown in table 6.

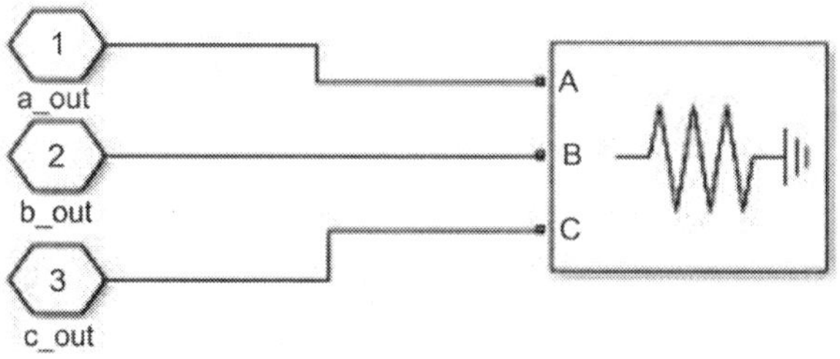

Figure 6. Complementary load model.

Table 6. Complementary load parameters.

Parameter	Value
Configuration	Y (grounded)
Nominal Phase-to-phase voltage Vn (Vrms)	400
Nominal frequency fn (Hz)	50
Active power (W)	180e3
Inductive reactive power Ql (var)	0
Capacitive reactive power Qc (var)	0

2.4.3. PI controller design. PI (Proportional Integral) controller has a transfer function and can be viewed in equation (2) and (3). This controller has a pole at the center point and zero at –Ki/Kp. Due to a pole is very close to the center point compared with zero, this controller is belonging to lagging-phase compensator and it increases the negative angles to angular TKA criteria. Therefore, the PI controller is used to correct the response of steady state of the system.

$$G_c(s) = K_p + \frac{K_i}{s} \quad (2)$$

or

$$G_c(s) = \frac{K_p s + K_i}{s} = \frac{K_p\left(s + \frac{K_i}{K_p}\right)}{s} \quad (3)$$

2.5. FCV-ELC integration system design

After develop the above frequency control systems in standalone form, both of the frequency control can be combined into an integrated control system. The combination of FCV and ELC is proposed to meet the best trade-off between frequency response and THD-V. The integrated scheme for the FCV and ELC can be seen in figure 7.

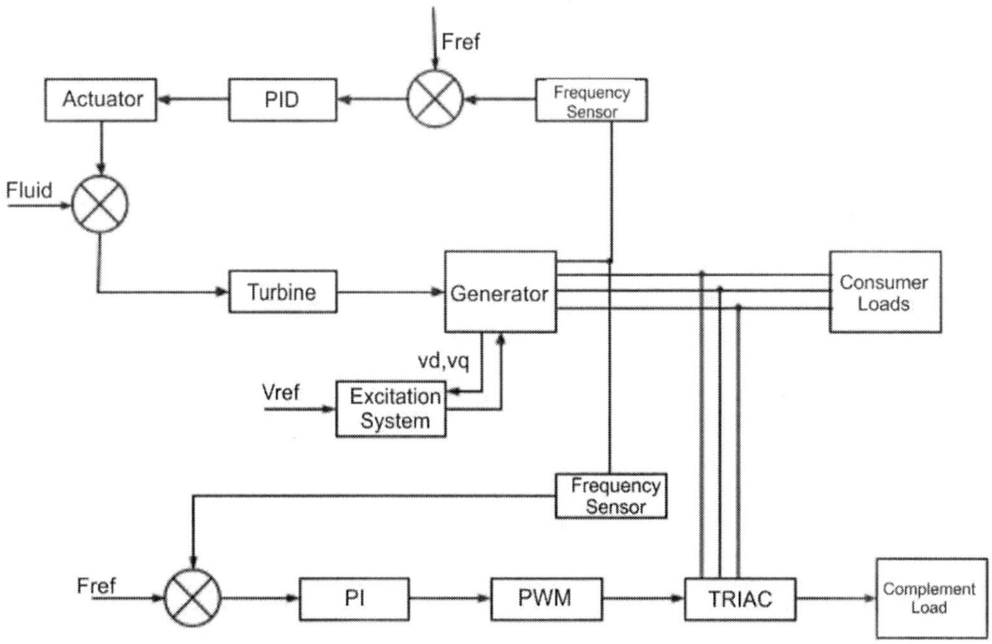

Figure 7. Complement load model.

In figure 7, FCV is developed to control the actuator which regulates the fluid flow rate based on the reference frequency of generator. PID controller is employed to control the performance of the actuator. Accuracy of the actuator will guarantee the turbine receives the fluid amount to rotate in the required frequency. In another side, ELC system control the compliment load based on the change of consumer loads to maintain the generator frequency in its bounded frequency. In the task the ELC system performance is improved by employing PI controller. Both of the control systems performance are investigated under the change of consumer loads.

3. Results and Discussion

In order to verify the proposed control system, each type of the control system was simulated and verified in details to answer the objectives of the study. The consumer loads were assumed to reduce two times in 10KW at 10th and 20th seconds as seen in figure 8 for all types of simulation.

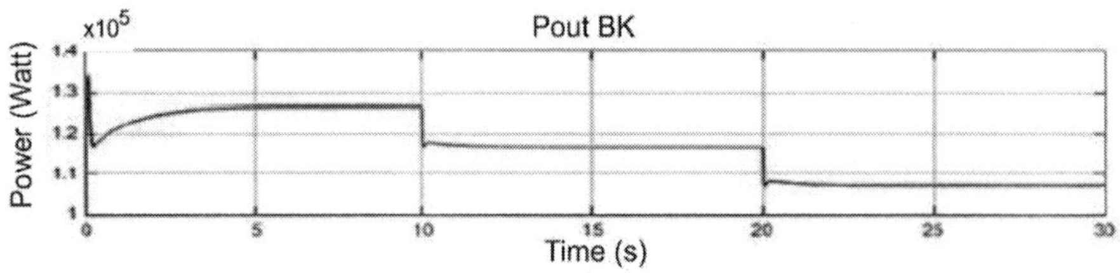

Figure 8. Consumer power load FCV system.

The first verification with the load was done for FCV system with the results as shown in figure 9. Figure 9 shows that there was a change in the current of the phase to ground when the consumer loads reduce in 10 KW at the 10th and 20th seconds. However, for the phase to ground voltage remains stable in $230V_{rms}$.

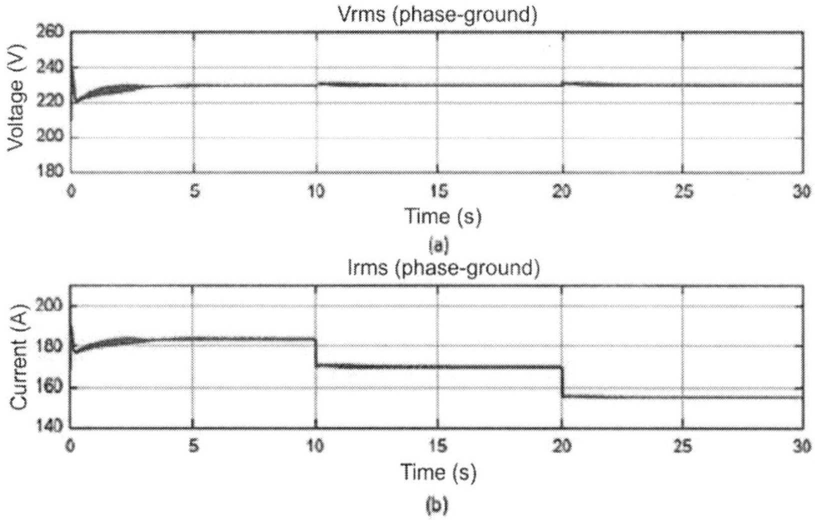

Figure 9. (a) V_{rms} and (b) I_{rms} of FCV system.

The reduction of the load may cause the change of power mechanic and the increase of frequency to around 50.51 Hz in figure 10 and figure 11, respectively. From those figures, it can be seen that, the FCV result a higher change of frequency, but it can ensure the frequency to return to the nominal value of 50 Hz by reducing the valve opening of the turbine, and decrease the turbine generated power. This type of control system was very capable to result a very low THD, but it was very slow to return to its normal frequency.

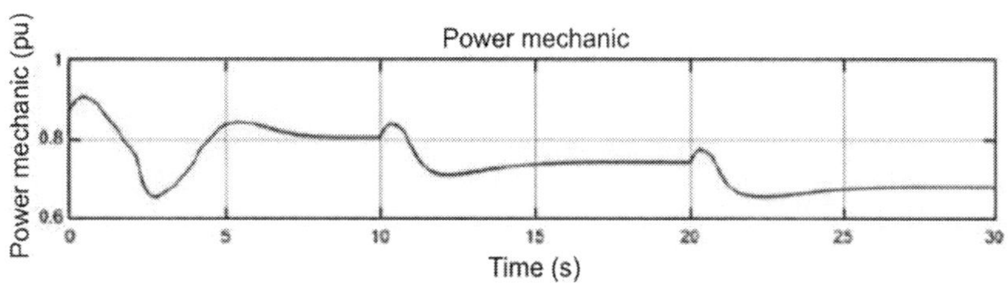

Figure 10. Power mechanic of FCV system.

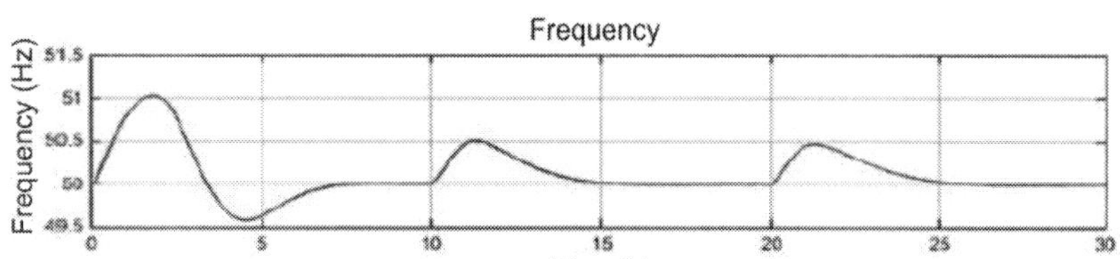

Figure 11. Frequency of FCV system.

While, simulation results of verification for ELC system in the MHPP system model can be seen in figure 12 to figure 14. In figure 12, there was a reduction in the consumer load of 10 KW that also affected on V_{rms} and I_{rms}. However, the ELC responded the change of load by enlarging the aperture angle of the regulator and the power that can be applied to the complement load. Such that, the total power generated by the generator remains constant. Consequently, impact of the load change on I_{rms} were not significant as the MHPP with FCV system. In the contrary, the ELC system result THD higher than the FCV.

Power mechanic and frequency of the MHPP with the ELC system can be seen in figure 13. Figure 13 shows that a load reduction in the MHPP with the ELC system did not cause any change in power mechanic. Because there is no frequency feedback from the ELC system to governor system to adjust the turbine rotation. The ELC system works to adjust the frequency change to return to the normal frequency of 50 Hz. As shown in figure 14, with the ELC system capability, the generator capable to return to its normal frequency much faster, around 11 times compared to the FCV systems.

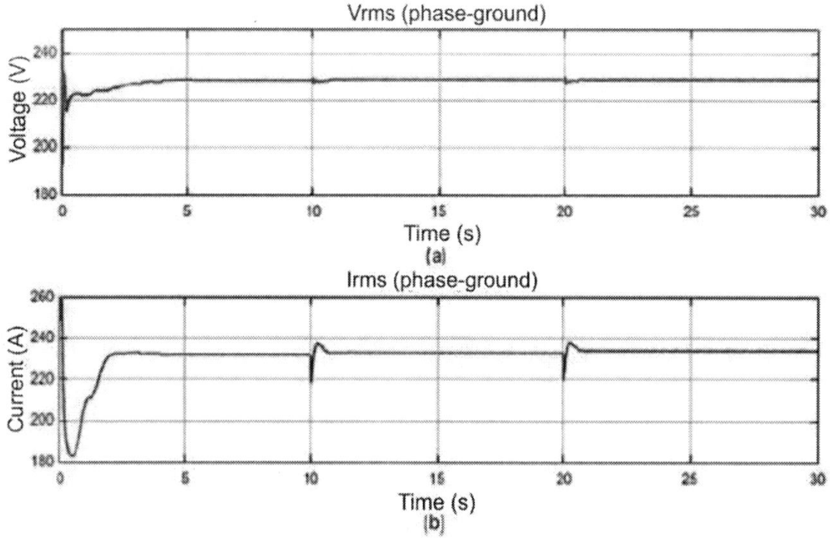

Figure 12. (a) V_{rms}, (b) I_{rms} of ELC.

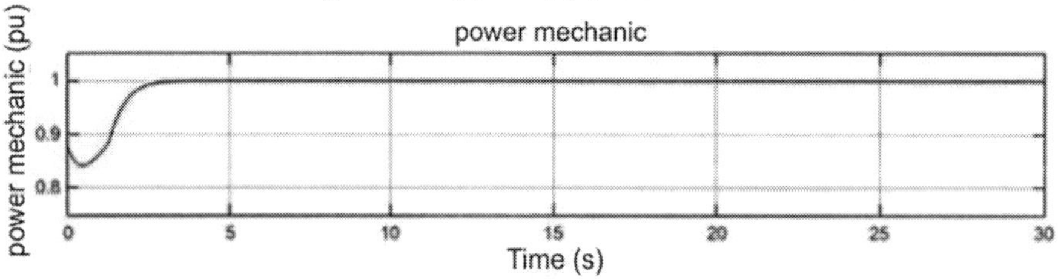

Figure 13. Power Mechanic of ELC system.

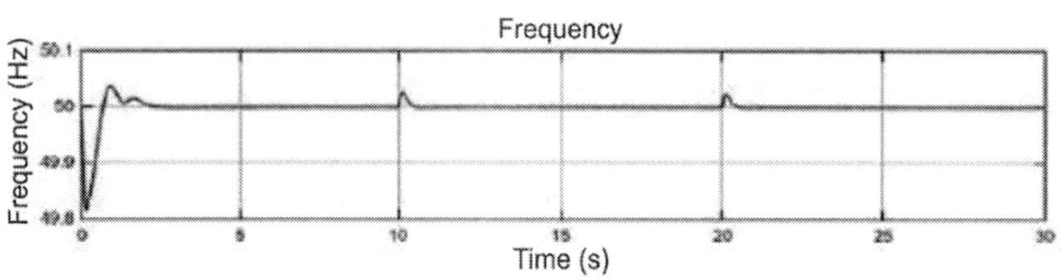

Figure 14. Frequency of ELC system.

Then, results of the integration between the FCV and ELC systems on MHPP system are shown as in figure 15 to figure 18. Figure 15 indicates that no significant change in the phase to ground current when a consumer loads reduction occurred in 10 KW at the 10th and 20th seconds, and for the phase to ground voltage remains stable at 230 V_{rms}. As shown in figure 16, when consumer load reduced in 10KW, the ELC worked by enlarging the aperture of the ac regulator angle and the power can be adjusted to the complement load so that the total power generated by the generator remains constant. However, the existence of FCV caused the power mechanic influencing significantly with lower THD.

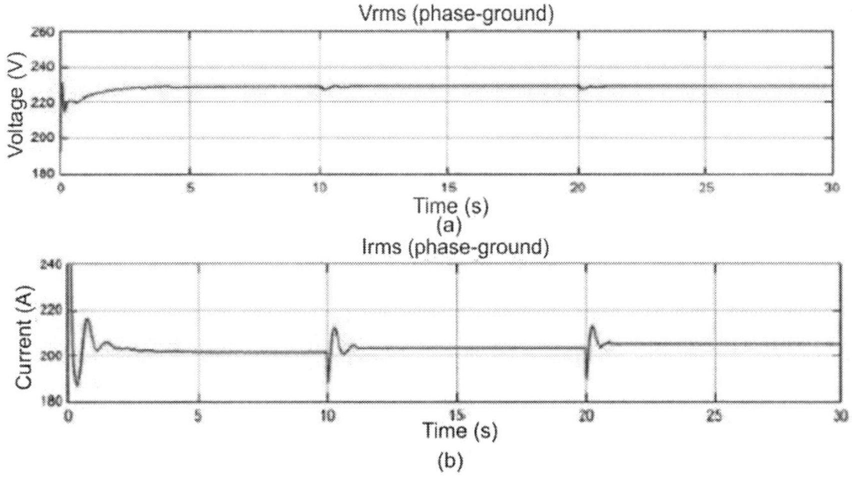

Figure 15. (a) V_{rms}, and (b) I_{rms} of FCV-ELC integration system.

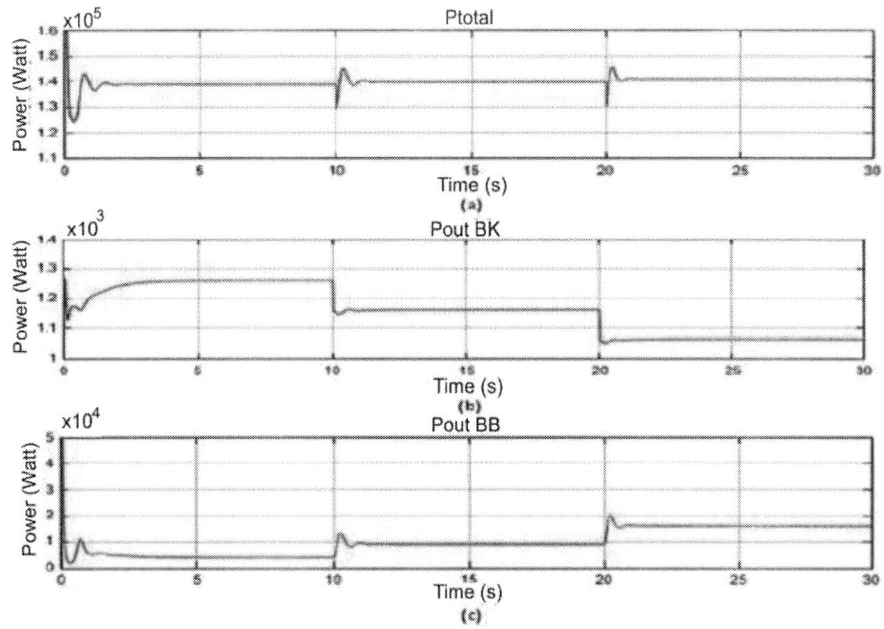

Figure 16. (a) Power Consumer Loads (b) Power Complement Load (c) Total Power of Integrated FCV-ELC system.

Following a significant change in its power mechanics, it also shows the performance of the FCV systems to manage changes in frequency to return to the normal value by reducing the valve opening on the turbine. But, the integrated cannot manage the frequency response faster as the ELC response.

Figure 18 shows that the frequency response capability to return to the nominal value takes 1.3575 second.

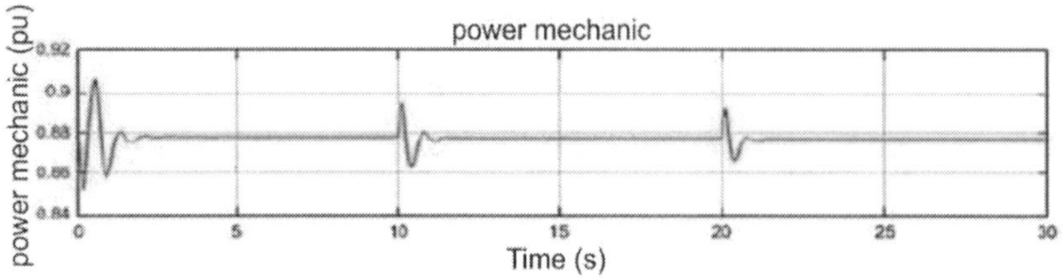

Figure 17. Power mechanic FCV-ELC integration system.

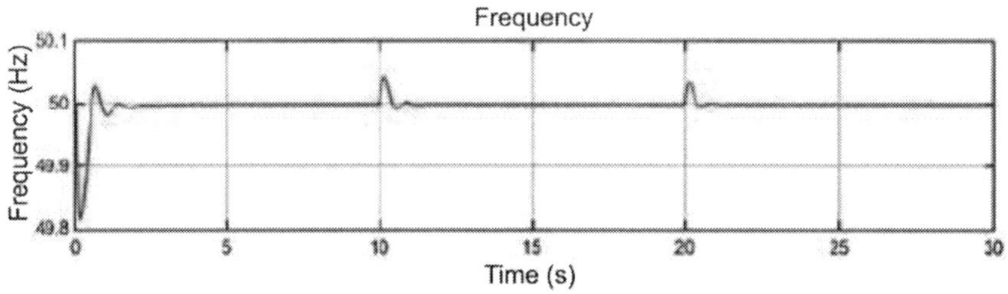

Figure 18. Frequency of FCV-ELC integration system.

From figures 8-18, frequency performance of the integrated control using FCV-ELC compared to the standalone FCV and ELC is summarized in table 7. While, summary of THD-V of the integrated FCV and ELC compared to the standalone controller system can be seen in table 8.

Table 7. Frequency performance of the integrated FCV-ELC on FCV and ELC.

Performance	FCV	ELC	FCV-ELC
Settling Time (s)	5.4548	0.4046	1.3573
Settling Min (Hz)	49.99	50	49.993
Settling Max (Hz)	50.53	50.03	50.042
Overshoot (%)	1.06	0.06	0.084
Peak (Hz)	50.53	50.03	50.042
Peak Time (s)	1.3071	0.0944	0.1519

Table 8. Summary of the THD-V of the three.

No	Consumer Load (KW)	THD-V		
		FCV	ELC	FCV+ELC
1	128	0.17%	10.65%	8.4%
2	118	0.07%	12.24%	9.75%
3	108	0.10%	13.67%	11.62%

THD-V in the FCV-ELC integration system decreased after both systems were integrated for both the first load reduction and for the second load reduction, for the first load reduction the THD-V value reached 7.75% and for the second load reduction the THD-V value reached 11.62%, with 5 cycle sample.

4. Conclusion

Integration of FCV and ELC to improve the performance of the MHPP in term of frequency and THD-V has been done successfully. Simulation on the MHPP model test shows that the generator is capable to generate electrical power more than 128 KW with 50 Hz of frequency and 230 V of V_{rms}. Results of the simulation confirm that the integration of FCV and ELC offered a better trade-off between frequency response and THD-V issues for MHPP, which cannot be solved by standalone FCV or ELC. By integrating both control systems, they gave the best values in term of frequency response and THD-Vin around 1.4071 second and 8.57%, respectively.

Acknowledgment

This research is funded by Ministry of Higher Education, Research and Technology (Kemenristekdikti) Indonesia and supported by University of Muhammadiyah Malang (UMM) through *Penelitian Terapan Unggulan Perguruan Tinggi* (PTUPT) Grant 2017. Authors are grateful to the Ministry and UMM for their support on the present work.

References

[1] Schener L G and Tambara R V 2016 Voltage and Frequency Regulation of Standalone Self-excited Induction Generator for Micro-hydro Power Generation using Discrete-time Adaptive Control *IET Renewable Power Generation* 10 4 p 531-540

[2] Marinescu C and Serban I 2009 Analysis of Frequency Stability in A Residential Autonomous Microgrid Based on A Wind Turbine and A Microhydro Power Plant IEEE Power Electronics and Machines in Wind Applications Conference Lincoln NE, USA, 24-26 June

[3] Blegojevic V and Stojiljković M 2013 Guide Vane Position Control of the Mini Hydro Power Plant with the Reduction of the Impact of Disturbances on the System FactaUniversitatis Series: Mechanical Engineering 11 1 p 55-64

[4] Gyawali N P 2016 Universal Electronic Load Controller for Microhydro Power Plant 12th IEEE Int. Conf. on Control and Automation (ICCA) Katmandhu, Nepal, 1-3 June.

[5] Borkowski D and Wegiel T 2013 Small Hydropower Plant with Integrated Turbine-Generators Working at Variable Speed IEEE Trans. on Energy Conversion 28 2 p 452-459

[6] Sandy A 2010 Studi Pengontrolan Beban Elektronik Pada Pembangkit Listrik Tenaga Mikrohidro Seloliman, Trawas Kabupaten Mojokerto Electrical Engineering Department, ITS, Indonesia Undergraduated Thesis

[7] Singh A K 2013 Modeling and Simulation of Micro Hydro Diesel Hybrid Power Systemfor Localized Power Requirement Using MATLAB/Simulink Master of Power Engineering, Jadavpur University Thesis

[8] Mashar A and Sodiq D 2011 Analisis Hamornisa Elektronic Load Control (ELC) PadaPembangkit Listrik Tenaga Mikro Hidro (PLTMH) Poli-Teknologi Politeknik Negeri Bandung 10 2

[9] IEEE 2005 IEEE Recommended Practice for Excitation System Models for Power System Stability Studies IEEE Standard 421 5 (Revision of IEEE 521.5-1992)

[10] IEEE 1992 Hydraulic Turbine and Turbine Control Models for Dynamic Studies IEEE Trans. on Power Systems 7 1 p 167-179

A study on electrical characteristics of Pentacene based MOS diodes and transistors

Fadliondi, Budiyanto and H Muchtar

Department of Electrical Engineering Faculty of Engineering, Universitas Muhammadiyah Jakarta, Indonesia

fadliondi@ftumj.ac.id

Abstract. Recently, organic electronics has become an object of research in field of electronic devices because organic semiconductor materials have many advantages such as flexibility, low-cost production and low-temperature process. The purpose of this research is to fabricate metal-oxide-semiconductor (MOS) diodes and transistors using organic semiconductor materials called pentacene which is commonly used p-type organic semiconductor to fabricate organic devices such as MOS diodes and transistors because of its availability and stability. The lowest minimum capacitance in depletion region and the highest maximum capacitance in accumulation region obtained in this research were 0.1μF/cm^2 at 5 V and 0.9μF/cm^2 at -5 V, respectively. The highest magnitude of drain current measured in this research was around 1.4 μA at drain and gate voltage of -5 V.

1. Introduction

Organic thin film transistors have many potentials to be used in low cost, large area, light and flexible electronics. Moreover, organic thin film transistors can be fabricated by low-temperature process like thermal evaporation and spin coating. There was a laboratory group that fabricated inverter using organic thin film transistors [1]. Pentacene has become an attention because of its stability, performance and availability. In many publications, the mobility of pentacene based organic thin film transistors varies from 0.01 – 1 cm^2/(Vs) which is close to that of amorphous silicon. Pentacene based devices are expected to be used as humidity sensors because, it is very sensitive to humidity [2, 3]. The surface energy of pentacene is around 38 mJ/m^2 [4]. Organic electronics have become so far from being just specific research [5]. The development of organic semiconductor material offers the probability to fabricate organic devices on a large surface [6]. Organic electronics also has many advantages like being light, flexible, and environmentally friendly so it attracts many researchers over the world [7]. Organic devices have many applications like RFID, sensor and displays [8]. Even though organic devices have lower mobility and larger subthreshold swing value than silicon based conventional devices, organic devices can be applied for flexible electronics. There is a laboratory who have fabricated analogue and digital circuit using organic material [9]. Devices fabricated from organic semiconductor pentacene can be in forms of MOS diodes, bottom contact and top contact transistors [10]. Pentacene is deposited using thermal evaporation and its performance depends on what it is deposited, surface treatment and process condition like deposition rate, substrate temperature and annealing after deposition [11]. Organic devices fabricated from pentacene can also be applied for humidity sensor because it is very sensitive towards water vapor and gases [12]. The purpose of this research was to investigate the electrical characteristics of pantacene based MOS diodes and transistors.

2. Methods

Figure 1(a) and (b) illustrate how pentacene MOS diodes and transistors were fabricated and figure 1(c) and (d) describe how they were measured. Overall, there were 4 steps to fabricate pentacene MOS diodes and transistors. The pentacene MOS diodes and transistors were fabricated at the same time using different masks. First of all, the n$^+$-Si (100) wafer was cleaned to remove the organic material from its surface. Secondly, SiO_2 gate insulator was grown using thermal oxidation method. Then, the organic semiconductor material, pentacene was deposited on the SiO_2 using thermal evaporation method. Lastly, the Au electrodes were formed using shadow masks. The C-V characteristics of pentacene MOS diodes and the I_D-V_D characteristics of pentacene transistors were measured using C-V meter and semiconductor parameter analyzer.

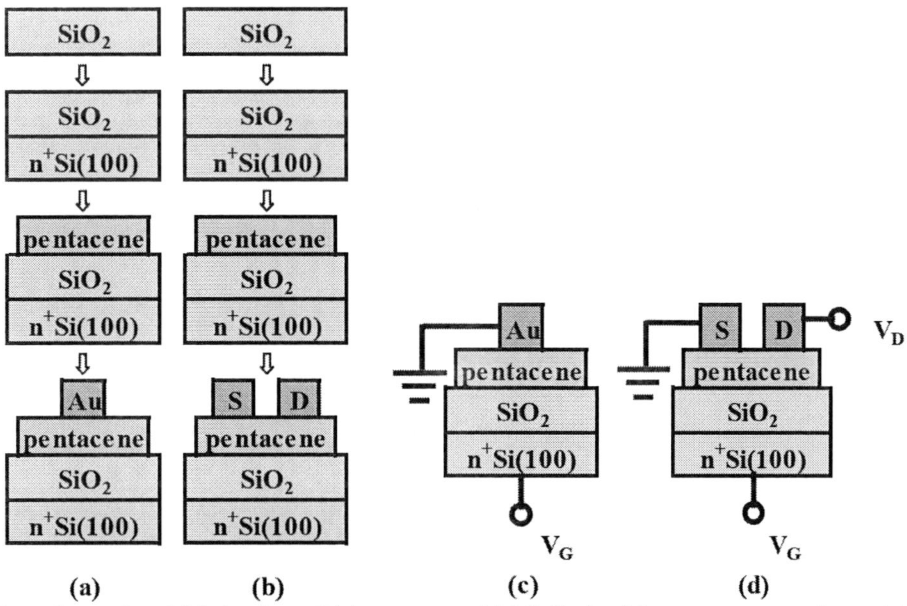

Figure 1. The schematic of fabrication of (a) pentacene MOS diode, (b) pentacene transistor, (c) MOS diode measurement points and (d) transistors measurement points.

3. Result and discussion

The experiment results are shown from figure 2 to figure 4. Figure 2 shows how the capacitance changed when the applied voltage was changed and how C-V characteristics were influenced by different frequencies and by different pentacene thickness. The SiO_2 grown by thermal oxidation thickness was around 9 nm. Overall, the capacitance increased when the applied gate voltage was decreased from 5 V to -5 V and when the frequency was increased from 15 kHz to 1 MHz. The capacitance was higher when the pentacene thickness was around 15 nm. The horizontal axis and vertical axis represents applied gate voltage measured in volt and capacitance measured in $\mu F/cm^2$, respectively.

In figure 2 (a), with regards to frequency of 15 kHz, the capacitance began at 0.2 $\mu F/cm^2$ at 5 V and then increased rapidly to 0.5 $\mu F/cm^2$ at around 0 V before remaining almost constant at 0.5 μF/cm2 when the voltage was decreased to -5 V. In terms of frequency of 110 kHz, the capacitance was around 0.2 $\mu F/cm^2$ at 5 V and then rose significantly to 0.4 $\mu F/cm^2$ at around 1.5 V before staying the same at 0.4 $\mu F/cm^2$ until the voltage was reduced to -5 V. Likewise, the capacitance started at 0.2 $\mu F/cm^2$ at 5 V and rose rapidly to around 0.35 $\mu F/cm^2$ at 1 V before remaining the same. In figure 2 (b), in terms of frequency of 15 kHz, the capacitance began at just 0.1 μF/cm² at 5 V and remained almost constant until the voltage was 2 V before increasing rapidly to around 0.45 μF/cm². When the frequency was increased to 110 kHz, the maximum capacitance decreased by around half to 0.2 μF/cm². The maximum and the minimum capacitance differed slightly at around 0.1 – 0.15 μF/cm² when the frequency was further increased.

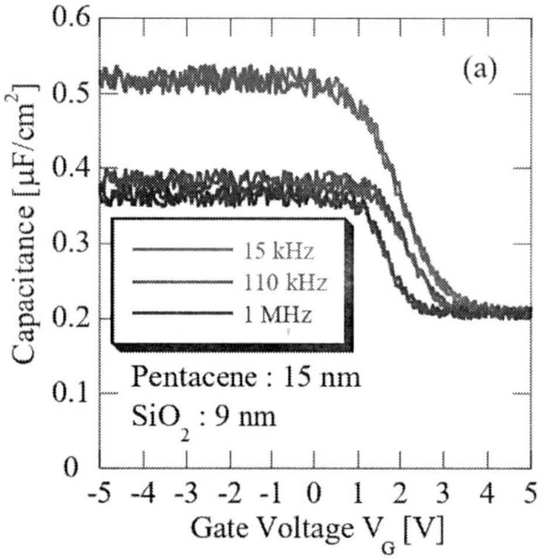

Figure 2. The frequency dependence of C-V characteristics of Au/Pentacene/SiO_2(9 nm)/n^+-Si(100) with pentacene of (a) 15 nm (b) 120 nm.

Figure 3 illustrates how parameters like applied gate voltage, pentacene thickness and SiO_2 thickness affected the capacitance of pentacene MOS diodes. Overall, the capacitance of diodes whose pentacene thickness was 15 nm was higher for compared to that of diodes with pentacene of 120 nm. Moreover, when the SiO_2 thickness increased, the total capacitance in the accumulation region decreased.

In figure 3(a), when the SiO_2 thickness was around 5 nm, the minimum capacitance at the depletion region was around 0.25 μF/cm² at 5 V and the maximum capacitance at accumulation region was around 0.9 μF/cm² at -5 V. As the SiO_2 thickness was increased to 9 nm, the minimum capacitance decreased slightly from 0.25 μF/cm² to 0.2 μF/cm² but the maximum capacitance decreased significantly from 0.9 μF/cm² to around 0.5 μF/cm². When the SiO2 thickness was further increased to 13 nm, the minimum capacitance remained constant at 0.2 μF/cm² but the maximum capacitance further decreased to 0.35 μF/cm².

In figure 3(b) where the pentacene thickness was 120 nm, the minimum capacitance at depletion region did not vary even though the SiO_2 thickness was changed from 5 nm to 13 nm but the maximum capacitance at accumulation region decreased from 0.7 μF/cm² to 0.2 μF/cm² when the SiO_2 thickness was increased from 5 nm to 13 nm.

Figure 4 shows how the drain current of the transistors changed when the applied drain voltage V_D and gate voltage V_G were changed. Overall the magnitude of the drain current in the linear region increased when the drain and gate voltage were changed to the more negative voltage. When the gate voltage was changed from 0 V to – 5 V, the magnitude of drain current increased to around 1.4 μA in case of the transistors with pentacene of 15 nm and 0.75 μA for the transistors with pentacene of 120 nm.

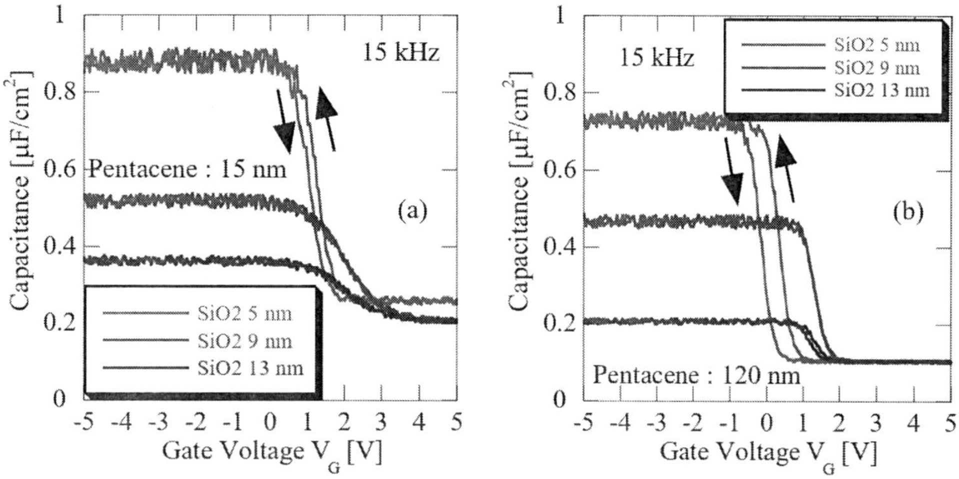

Figure 3. The SiO_2 thickness dependence of C-V characteristics of Au/Pentacene/SiO_2/n^+-Si(100) with pentacene of (a) 15 nm (b) 120 nm.

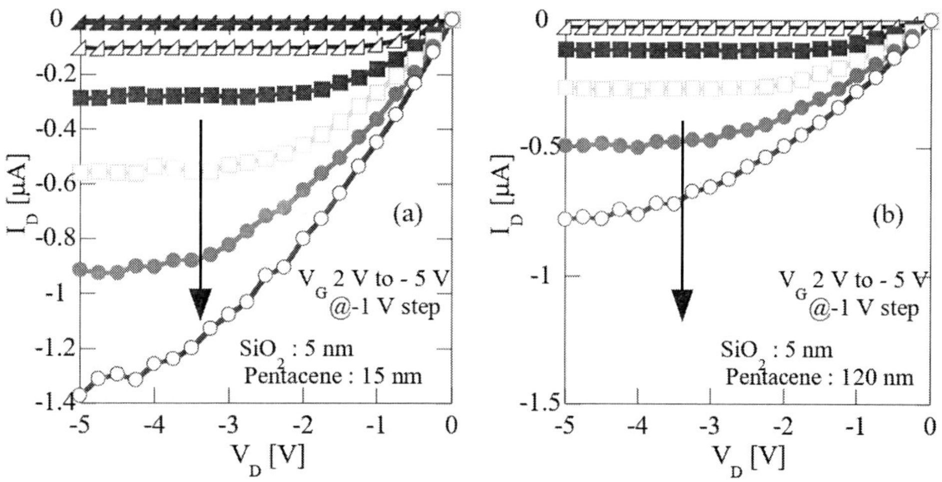

Figure 4. The pentacene thickness dependence of transistor ID-VD characteristics. (a) pentacene 15 nm (b) pentacene 120 nm.

4. Conclusion

The metal-oxide-semiconductor diodes and transistors were fabricated using organic semiconductor material, pentacene. The results show that in the MOS diodes, the capacitance increased when the pentacene thickness was decreased. However, the maximum capacitance in the accumulation region also depended on the measurement frequency. The maximum capacitance in the accumulation region decreased when the measurement frequency was increased. Moreover, in case of equal pentacene thickness, the maximum capacitance in the accumulation region increased when the SiO_2 thickness was decreased. In transistors, the magnitude of drain current increased when the pentacene thickness was decreased.

References

[1] Wei C-Y Huang W-C, Yang C-K, Chang Y-Y and Wang Y-H 2011 Low-Operating-Voltage Pentacene-Based Transistors and Inverters with Solution-Processed Barium Zirconate Titanate Insulators *IEEE Electron Device Letters* vol 32 pp 1755-57

[2] Zhu Z-T, Mason J T, Dieckmann R and Malliaras G G 2002 Humidity sensors based on pentacene thin-film transistors *Applied Physics Letters* vol 81 pp 4643-5

[3] Jacques T and Mohsen E 2013 Stability of pentacene transistors under concomitant influence of water vapor and bias stress *Microelectronics Reliability* vol 53 pp 274-8

[4] Zan H-W, Yen K-H, Liu P-K, Ku K-H, Chen C-H and Hwang J 2007 Low-voltage organic thin film transistors with hydrophobic aluminum nitride film as gate insulator *Organic Electronics* vol 8 pp 450-4

[5] Nakamura M and Matsubara R 2014 Carrier Mobility in Organic Thin-film Transistors: Limiting Factors and Countermeasures *Journal of Photopolymer Science and Technology* vol 27 pp 307-16

[6] Lassnig R, Hollerer M, Striedinger B, Fian A, Stadlober B and Winkler A 2015 Optimizing pentacene thin-film transistor performance: Temperature and surface condition induced layer growth modification *Organic Electronics* vol 26 pp 420–8

[7] Sun H, Yin Y, Wang Q, Jun Q, Wang Y, Tsukagoshi K, Wang X, Hu Z, Pan L, Zheng Y, Shi Y and Li Y 2015 Reducing contact resistance in ferroelectric organic transistors by buffering the semiconductor/dielectric interface *Applied Physics Letters* vol 107 p 053304

[8] Faraji S, Hashimoto T, Turner M L and Majewski L A 2015 Solution-processed nanocomposite dielectrics for low voltage operated OFETs *Organic Electronics* vol 17 pp 178-83

[9] Fukuda K, Takeda Y, Yoshimura Y, Shiwaku R, Tran L T, Sekine T, Mizukami M, Kumaki D and Tokito S 2014 Fully-printed high-performance organic thin-film transistors and circuitry on one-micron-thick polymer films *Nature Communications* vol 5 pp 1-8

[10] Benacer I and Dibi Z 2014 Modeling and Simulation of Organic Field Effect Transistor (OFET) Using Artificial Neural Networks *International Journal of Advanced Science and Technology* vol 66 pp 79-88

[11] Taylor D M, Patchett E R, Williams A, Neto N J, Ding Z, Assender H E, Morrison J J and Yeates S G 2014 Organic Digital Logic and Analog Circuits Fabricated in a Roll-to-Roll Compatible Vacuum-Evaporation Process *IEEE Transactions On Electron Devices* vol 61 pp 2950-6

[12] Yu J, Yu X, Zhang L and Zeng H 2012 Ammonia gas sensor based on pentacene organic field-effect transistor *Sensors and Actuators B: Chemical* vol 173 pp 133–8

IOP Conf. Series: Materials Science and Engineering **403** (2018) 012041 doi:10.1088/1757-899X/403/1/012041

Robust ball tracking scheme for soccer robot under various illuminations and occlusions

R D Puriyanto[1] and A Prahara[2]

[1]Electrical Engineering Department, [2]Informatics Department
Universitas Ahmad Dahlan, Yogyakarta, Indonesia

[1]rikydp@ee.uad.ac.id, [2]adhi.prahara@tif.uad.ac.id

Abstract. The ball tracking method for soccer robot usually done using simple color thresholding and shape detection. The method is designed solely to track specific type of ball determined by the soccer robot regulation. However, the method should be robust and not to be limited to the regulation. In this research, a robust ball tracking method for soccer robot that can afford to track various types of ball under various illuminations and occlusions is developed. The proposed method uses combination of Histogram of Oriented Gradients (HOG) and Linear Support Vector Machine (Linear-SVM) to detect the ball and Optical Flow to track the ball. In order to reduce the computation, the method only scans the whole scene when the ball is not detected. Once the ball is detected, the method only scans the area around previously detected ball. HOG itself has its own mechanism to handle various illuminations and SVM as machine learning method is tolerance to occlusion. The experiment shows that the proposed scheme is accurate and can perform real time.

1. Introduction

Robotics has been widely used not only in industrial but also in educational field. Various competitions are held every year to increase the interest for robotics. One of them is Indonesian Robotics Contest which held on national scale. There are some categories that are included in the competition. One of them is the new category, wheeled soccer robot that first held in 2017.

The wheeled soccer robot uses ball tracking and goal detection to perform actions such as dribbling and kicking the ball towards the opposing goal. The movement of the ball influences the calculation of motor rotation. The motor rotation will determine the movement of wheeled soccer robot while following the ball. In other words, the ball tracking method determines how quick the wheeled soccer robot can find and follow the direction of the ball.

Most ball tracking systems are limited to the soccer robot regulations, for example a regulation that set the ball color to be orange, the goal to be white, and so on. This regulation causes most of the ball tracking systems for wheeled soccer robot use color detection and shape detection to detect the ball. If the regulation changes, the method for ball tracking also need to be adjusted. The adjustment will take time to do research, experiment, and performance test in order to get the suitable method for the competition. In this research, authors want to keep the adjustment to minimum and save the time to do improvement.

This research proposes a robust ball tracking scheme for soccer robot under various illuminations and occlusions. In summary, the contribution of this work is given as follow:

- The proposed scheme uses texture information to detect the ball instead of color because texture information will need a little adjustment to fit the regulation.
- The ball detection uses machine learning method to handle the occlusion and only scans the whole scene if the ball is not tracked in the previous scene otherwise only scans the surrounding area of previously tracked ball. This scheme can localize the detection area and reduce the computation for real time application.
- Finally, the ball is tracked using optical flow to estimate the ball direction and movement.

The rest of this paper is organized as Section 2 presents the related works, Section 3 presents the proposed ball tracking scheme, Section 4 presents the results and discussion, and the conclusion of this work is described in Section 5.

2. Related works

Ball tracking method for soccer robot has been developed by many researches. Most of them [1-4] use combination of color detection and shape detection. As in [1], the authors aim to implement a real time object detection and tracking for soccer robot. The method uses color based segmentation in HSV (Hue-Saturation-Value) color space to differentiate the ball and the goal. Hough transform is used to detect the goal and ellipse detection is used to detect the ball. The result shows that the method can perform well although there are some errors in the detection.

Color based detection method also used in [2]. The authors use omnidirectional camera which has 360 degrees of view in order to increase robot field of vision. The benefit using omnidirectional camera is the distance calculation become accurate. However, the computation become more complex and sensitive to the surrounding illumination.

Detection of ball color and ball position have been done by [3]. The method is implemented for goalkeeper humanoid soccer robot. Their research proposes a method to track ball position using HSL (Hue-Saturation-Lightness) color space, perform blob detection, and using nine cells coordinate model to define the ball position. The result shows that the system is able to detect the ball although the accuracy of ball detection needs to be improved, especially under different illumination.

The environment of the robot tournament is unpredictable where the ball color is not guaranteed to be the same every year. Most of the participants use color detection to detect the ball. The research of aerial ball detection has been done by [4] to predict the direction of ball while in the air. The method is implemented in goalkeeper robot for real-time application. It uses blob detection with Kinect sensor and color filtering using depth information to discard the color of the object that found farther than certain distance.

In this research, a robust ball tracking scheme for soccer robot is proposed. The ball tracking scheme uses combination of Histogram of Oriented Gradients (HOG) and Support Vector Machine (SVM) to detect the ball and Optical Flow to track the ball. In order to reduce the computation, the method only perform tracking on the whole scene if the ball is not detected otherwise only on surrounding area of the previously tracked ball. The proposed scheme uses ball texture information rather than color and the detector uses various types of ball in the training process. This scheme can track the ball with good accuracy and perform real time. Moreover, the proposed scheme will adapt the regulation easily.

3. Methodology

The general procedure of robust ball tracking for soccer robot under various illuminations and occlusions is shown in figure 1. The procedure of the proposed ball tracking scheme consists of three steps: training the ball detector, perform ball detection, and tracking the ball.

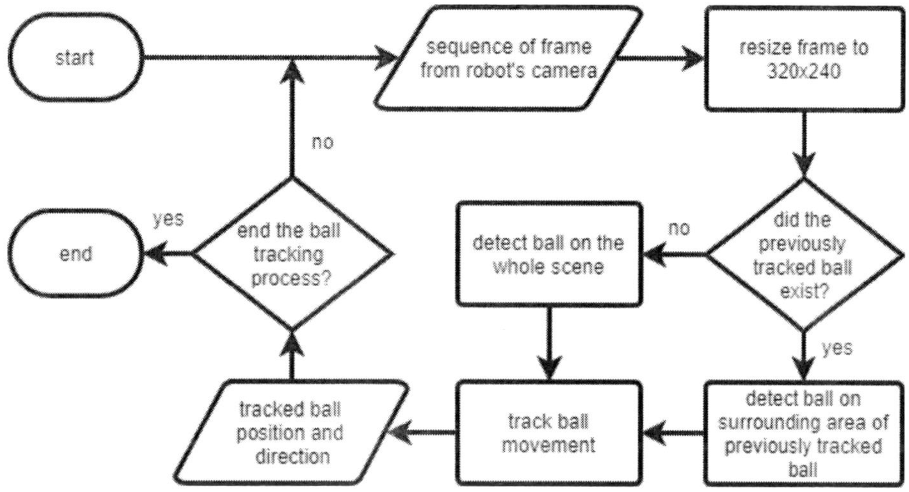

Figure 1. General procedure of the proposed ball tracking scheme.

3.1. Training the ball detector

At the first step, before performing ball detection, the ball detector needs to be trained. The detector is trained using SVM from extracted HOG descriptors. HOG descriptors can be extracted using the following steps [5]:

- Convert image to grayscale.
- Normalize gamma from input image then calculate the gradient.
- Divide image into small spatial regions called cells.
- For each cell, accumulates a local 1-D histogram of gradient directions or edge orientations over the pixel of the cell. The orientation bins are evenly spaced over 0-180 degree (unsigned gradient) with 20 degrees for each space. The nine orientation bins are filled with the weighted magnitude for each orientation.
- The cells are organized into overlapping blocks and perform contrast normalization on the overlapping blocks for better invariance to illumination and shadowing.
- A set of features is obtained by collecting features on all blocks.

SVM is a classifier that determined by separation which called hyperplane. Hyperplane can be calculated by maximizing margin or distance from two set of object from two different class. Classification with SVM consist of training and classification. In the training step [6], if there are given training data $x_i \in R^n, i = 1, \ldots l$ in two classes and label $y \in R^l$ such that $y_i \in \{1, -1\}$, it can be solved using equation (1).

$$\min_{w,b,\xi} \frac{1}{2} w^T w + C \sum_{i=1}^{l} \xi_i \tag{1}$$

where w is weight vector, b is bias, ξ_i is slack variables, $\phi(x_i)$ maps x_i into higher dimensional space and $C > 0$ is the regularization parameter and the decision function is shown in equation (2).

$$sgn(w^T \phi(x) + b) = sgn\left(\sum_{i=1}^{l} y_i \alpha_i K(x_i, x) + b\right) \tag{2}$$

where $K(x_i, x_j)$ is the kernel function. After training process, parameter $y_i \alpha_i \forall i$, b, label names, support vectors, and kernel parameter saved as output model from training SVM.

Optimal parameter is selected using k-fold cross validation which is a method to do cross validation by dividing training data into k set which has (k-1) as training data and the rest will be the test data. After training process, we will get variable w, x, and b for each class.

Authors collect some samples of various types of ball on the internet and rendered ball image from model 3D as positive training data. The negative training data are collected from soccer robot arena. The samples of positive and negative training data are shown in figure 2 (a) and (b) respectively. HOG descriptors are extracted from training data then linear SVM will classify the features into ball or non-ball category.

(a) positive training data

(b) negative training data

Figure 2. Samples of positive and negative training data

3.2. Ball detection

At the second step, the ball is detected using previously trained ball detector. HOG has own mechanism to handle various illumination by applying contrast normalization. SVM is a machine learning method that tolerance to occlusion if proper trained. In the classification process, voting strategy is performed for each data which will be designated to be in a class with the maximum votes [6]. The classification process can be done with the following steps.

- Calculate linear SVM kernel.
- Calculate decision function with (2).
- Repeat the first step and second step for other classes.
- Determine the class by function which gives the most maximum result.

Because the heavy computation of HOG and SVM, authors set a scheme for the detection as shown in figure 3. The method only detects the ball in the whole scene if the ball is not tracked from the previous iteration otherwise the method will detect the ball only in the surrounding area of previously tracked ball. This scheme can minimize the area of detection and reduce the computation time.

3.3. Ball tracking

At the final step, the detected ball is tracked using Optical Flow. Optical Flow assumes that the intensity from the same object is not changing from sequences of frame. The neighborhood pixel and object pixel will move with the same movement.

If a pixel in the first frame $I(x, y, t)$ and moves by distance (dx, dy) in the next frame after dt time, we will have equation (3).

$$I(x, y, t) = I(x + dx, y + dy, t + dt) \tag{3}$$

According to Taylor series approximation of right-hand side then remove the common terms and divide by dt, we will have equation (4).

$$f_x u + f_y v + f_t = 0 \tag{4}$$

where $f_x = \frac{\partial f}{\partial x}$ and $f_y = \frac{\partial f}{\partial y}$ are the image gradients, f_t is the gradient along time and u, v can be solved using Lucas-Kanade method [7] as shown in equation (5).

$$\begin{bmatrix} u \\ v \end{bmatrix} = \begin{bmatrix} \sum_i f_{x_i}^2 & \sum_i f_{x_i} f_{y_i} \\ \sum_i f_{x_i} f_{y_i} & \sum_i f_{y_i}^2 \end{bmatrix}^{-1} \begin{bmatrix} -\sum_i f_{x_i} f_{t_i} \\ -\sum_i f_{y_i} f_{t_i} \end{bmatrix} \quad (5)$$

Optical Flow will match the features from current and previous ball area to find the distance and direction of ball. The distance and direction are used to predict the movement of the ball.

3.4. Performance evaluation

The performance evaluation procedure uses confusion matrix then calculates the accuracy, specificity, sensitivity, and precision using equation (6), (7), (8), and (9) respectively where TP is true positive, FP is false positive, TN is true negative, and FN is false negative.

$$Accuracy = \frac{TP+TN}{TP+FP+TN+FN} \quad (6)$$

$$Specificity = \frac{TN}{(TN+FP)} \quad (7)$$

$$Sensitivity = \frac{TP}{(TP+FN)} \quad (8)$$

$$Precision = \frac{TP}{(TP+FP)} \quad (9)$$

4. Result and discussion

The proposed ball tracking scheme runs in a simulation written in C++ and additional OpenCV library on Laptop with Intel i5 processor and 8GB of RAM. Authors analyze the fps generated in the result. The test data are two videos with ten seconds duration collected from soccer robot point of view. The test data contain change of illumination and occlusion. Authors use approximately 1,000 positive data and 20,000 negative data to train the ball detector.

4.1. Ball detection result

The proposed ball detection is tested under sudden change of illumination and occlusion. The performance evaluation of proposed ball detection is shown in table 1. From table 1, the accuracy and the precision of proposed ball detection is high which has average 92.42% and 98.13% respectively. The high accuracy and precision mean the proposed method can detect ball under sudden illumination change and occlusion well. The low sensitivity means that the method is less sensitive when detecting the ball. The large number of negative data when training the ball detector affect in less false positive but more false negative.

Table 1. Performance evaluation of proposed ball tacking scheme.

Data	TP	TN	FP	FN	Accuracy	Precision	Specificity	Sensitivity
Video 1	146	130	5	24	0.9049	0.9669	0.9630	0.8588
Video 2	239	45	1	16	0.9435	0.9958	0.9783	0.9373

Figure 3 shows the result of ball detection under sudden illumination change on sequences of frame. Figure 4 shows the result of ball detection under occlusion on sequences of frame. The detected ball is marked with blue rectangle. From figure 3, the proposed ball detection can handle sudden illumination. The ball remains detected although the scene suddenly become bright and darker. From figure 4, the proposed ball detection method can handle occlusion. The ball also remains detected if the large portion of ball area is shown in the camera and not detected if only small portion of ball area is shown in the camera.

The proposed scheme produces approximately 12 fps when scanning the whole scene to find the ball and approximately 30 fps when the ball is already tracked in the previous scene. This result is

IOP Conf. Series: Materials Science and Engineering **403** (2018) 012041 doi:10.1088/1757-899X/403/1/012041

accepted because once the ball is tracked the robot can quickly follow the ball because the detection is real time. The red rectangles on figure 3 and figure 4 show the surrounding area of previously tracked ball.

The drawback using combination of HOG and linear SVM is the detector need to be large enough to produce better feature descriptors. The yellow rectangles on figure 3 and figure 4 show the size of the detector. Large size of detector means more computation to perform that is why author minimize the detection area by using the mentioned scheme. The range of detection also depend on the scale and pyramid level used in the detector.

Figure 3. The result of ball detection under sudden illumination change on sequences of frame

Figure 4. The result of ball detection under occlusion on sequences of frame.

4.2. Ball tracking result

The ball is tracked using optical flow. Figure 5 and figure 6 show the result of proposed ball tracking scheme on sequences of frame. From figure 5 and figure 6, blue rectangle is the detected ball, red rectangle is the surrounding area of previously tracked ball, yellow rectangle is the detector size, yellow snapshot with arrowed line is the distance and direction from robot's center to the tracked ball, green line is the ball trails, and the magenta text is the information including the direction of ball movement.

Figure 5 and figure 6 show that the proposed method can tracked the ball and predict the movement. Optical flow will produce arrows for each matched features from previous and current detected ball that represent the direction and distance. The average of distance and angle are used to estimate the speed and direction. By using the ball trails which marked by green line, the proposed method can track the history of ball's movement if going out of scene.

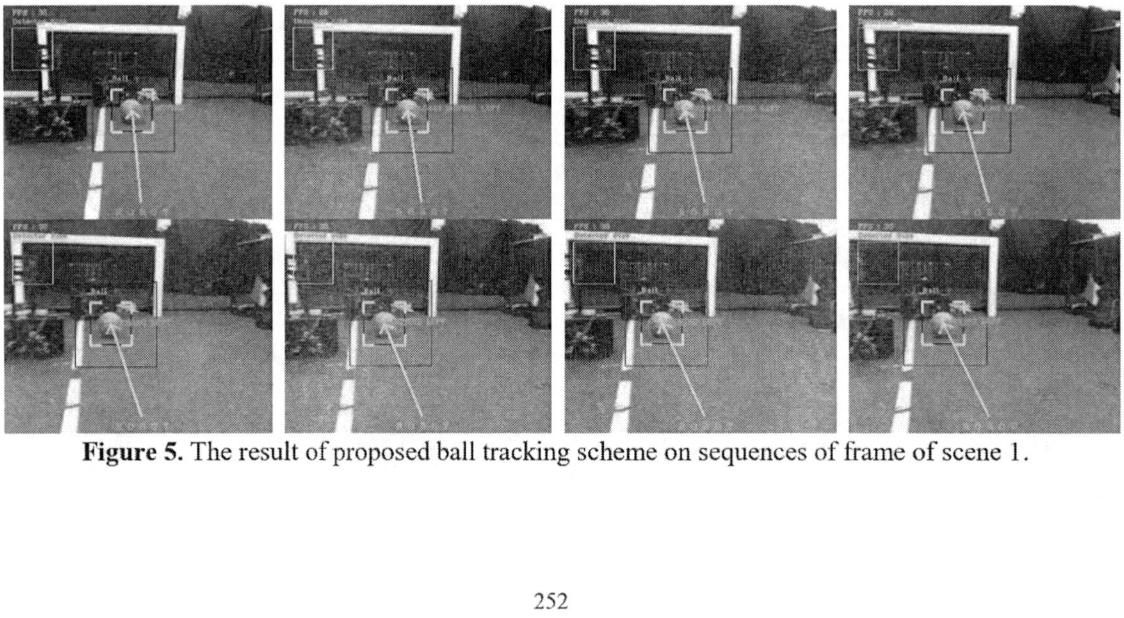

Figure 5. The result of proposed ball tracking scheme on sequences of frame of scene 1.

Figure 6. The result of proposed ball tracking scheme on sequences of frame of scene 2.

5. Conclusion

In this research, authors proposed robust ball tracking scheme for soccer robot under various illuminations and occlusion. The proposed ball tracking scheme can perform well which proven by the high accuracy and precision from the result of performance evaluation. The proposed scheme also can handle sudden illumination change and occlusion in the sequences of frame. The fps is acceptable despite of the heavy computation of combination of HOG and SVM and with the proposed scheme, the fps can be further increase to real time if the ball is already tracked in the previous scene. By using this scheme, the adjustment for robot soccer competition is only adding more training data and size of the detector. For future work, the proposed scheme can be improve using more training data and applied in the real hardware of robot soccer.

References

[1] Fitriana A N, Mutijarsa K and Adiprawita W 2016 Color-based segmentation and feature detection for ball and goal post on mobile soccer robot game field *2016 International Conference on Information Technology Systems and Innovation (ICITSI). IEEE* pp 1–4

[2] Mulya A K, Ardilla F and Pramadihanto D 2016 Ball tracking and goal detection for middle size soccer robot using omnidirectional camera *2016 International Electronics Symposium (IES). IEEE* pp 432–7

[3] Rahman A and Widodo N S 2013 Colored Ball Position Tracking Method for Goalkeeper Humanoid Robot Soccer *TELKOMNIKA (Telecommunication Computing Electronics and Control)* vol 11 no 1 pp 11 - 6

[4] Neves A J R, Trifan A, Dias P and Azevedo J L 2015 Detection of Aerial Balls in Robotic Soccer Using a Mixture of Color and Depth Information *2015 IEEE International Conference on Autonomous Robot Systems and Competitions. IEEE* pp 227–32

[5] Dalal N and Triggs B 2005 Histograms of Oriented Gradients for Human Detection *2005 IEEE Computer Society Conference on Computer Vision and Pattern Recognition (CVPR'05). IEEE* pp 886–93

[6] Chang C-C and Lin C-J 2011 LIBSVM *ACM Transactions on Intelligent Systems and Technology. ACM* vol 2 no 3 pp 1–27

[7] Lucas B D and Kanade T 1981 An iterative image registration technique with an application to stereo vision *Proceedings of the 7th international joint conference on Artificial intelligence - Volume 2. Morgan Kaufmann Publishers Inc.* pp 674–9

Water quality monitoring at paddies farming based on android

A Yudhana and A C Kusuma
Electrical Engineering Department, Universitas Ahmad Dahlan Yogyakarta
Jln. Prof. Dr. Supomo Yogyakarta. Tlp. 0274-379418, Fax. 0274-381523

eyudhana@ee.uad.ac.id, kusumaadichandra@gmail.com

Abstract. Continuous rice cultivation results in changes in the environment due to poor water quality. House hold waste pollution and industrial waste have been made the degradation of water quality at the paddies farming. The changing of temperature and pH of rice fields influenced on rice crops. The process of rice water monitoring is needed by farmers, to know the quality of water in the rice fields. So in the process of cultivation of rice plants can be monitored the quality of temperature and water pH in rice fields using android. This research basically using DS18B20 Waterproof sensor and pH meter analog kit sensor to know the condition of water quality of paddy field, then can be known the measurement value in response by sensor then read and sent to android with distance up to 10 meter by using Bluetooth module which in with Arduino Uno. The data testing conducted at three difference time i.e. morning, afternoon and at evening. Test results show that if the daytime test, afternoon da early morning the tool can respond to the state of the quality of the rice water. The prototype has been show that parameters of water quality successfully monitored via android.

1. Introduction

Water has a very important role for human life; even water becomes the foundation of society for daily activities and become a source of community livelihoods such as irrigation to farmers and farmers needed as the needs of society in general. A monitoring system that can monitor current water quality has become an urgent need for rice fields spread across Indonesia [1]. This is due to the adverse record caused by underwater activity and the quality of the water did not monitored in real time, which comes suddenly resulting in losses to rice farmers.

The development of increasingly sophisticated technology at this time gave birth to a product that performed resembles a desktop computer device. Smartphone is a minicomputer model product as the development of mobile phone devices. Different applications similar to computer devices could been installed on it. The application of water quality monitoring of rice fields using a receiver inventor on android worked accurately [2] and some monitoring works using technology also have been developed by many researcher [3-7].

2. Research methods

2.1. System design

System design is presented in the form of block diagram. Blocks of this diagram that will help the manufacture of water quality monitoring tools based on android rice. The design block diagram and work system of the tool can be seen in figure 1.

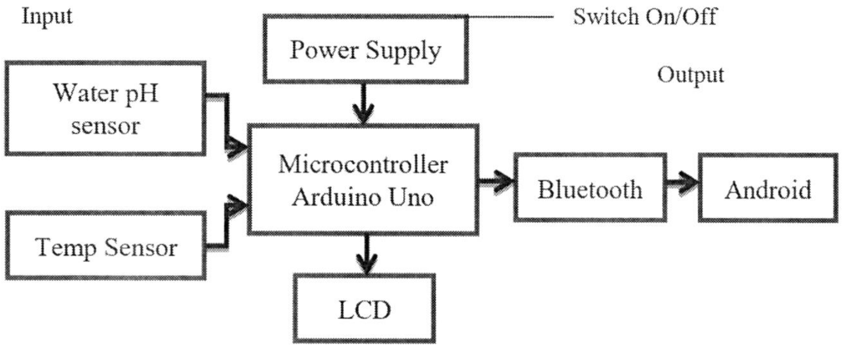

Figure 1. Block system design diagram.

2.2. Hardware Design

The design of hardware consists of a control system Arduino system that served as a sensor controller and perform data processing. The whole system set can be seen in figure 2.

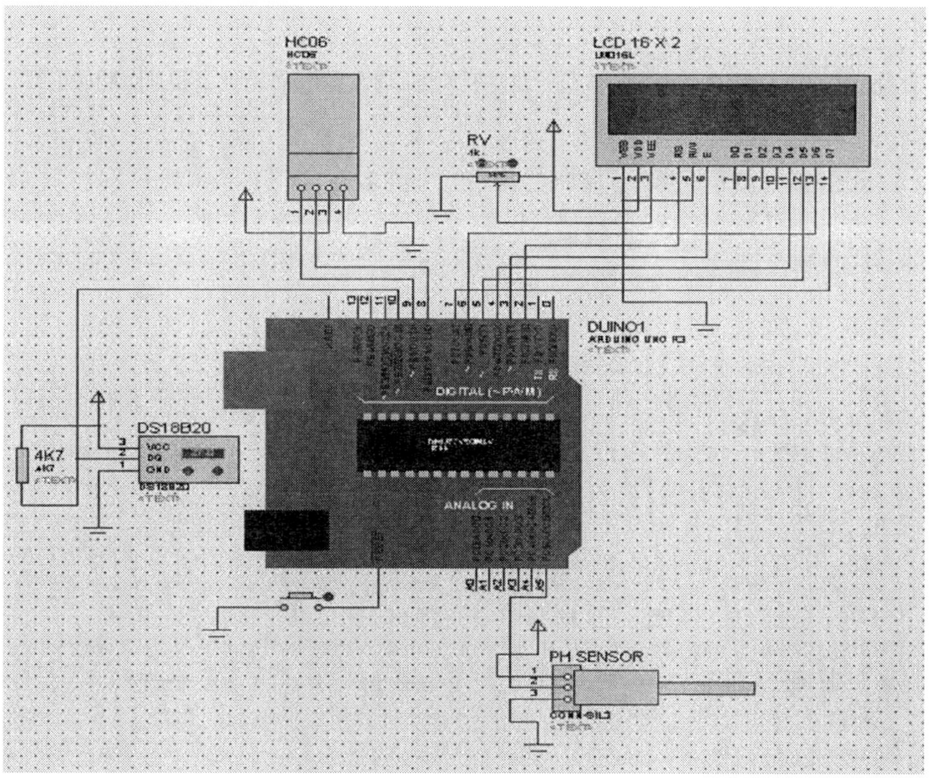

Figure 2. Overall system set.

2.3. Software Design

Software design using Arduino software which is then downloaded on Arduino Uno. Software design is displayed in the following flowchart diagram shown in figure 3.

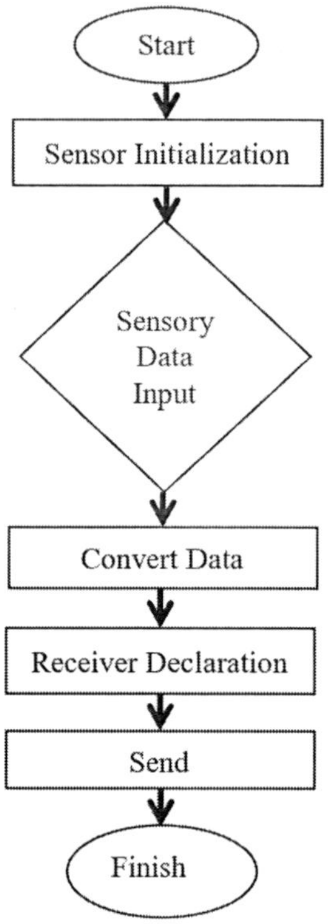

Figure 3. Flowchart diagram.

2.4. Android Device Design

The design of the android device is designed as a receiver. The design of the android device using MIT programming Inventor online mode. App Inventor is a tool for creating android apps. Interestingly this tool is based on visual block programming [5].

3. Results and discussion

3.1 Temperature sensor test DS18B20 waterproof

Waterproof DS18B20 sensor testing is done to determine the response provided by the sensor. Sensor testing is performed to be able to display the sensor measurements that present a temperature value in the water of paddy fields. Water temperature value test is done by using pure water sample, hot water and cold water which is assumed as wetland water to know the expected state of sensor reading. Waterproof DS18B20 temperature sensor testing can be seen in figure 4. Searched for average on sensors and measuring instruments and searched for average error then there will be comparable final results of both are calculated using equation (1) – (3).

$$\overline{x} = \frac{\sum_{i=1}^{n}}{n} = \frac{x1 + x1 + x1 ...... Xn}{n} \quad (1)$$

$$\text{E} = \left[\frac{x - \overline{x}}{x}\right] \text{x } 100\ \% \quad (2)$$

$$\overline{E} = \frac{\sum_{i=1}^{n} Ei}{n} \quad (3)$$

Information:

$\overline{x}$: Average measurement results with the sensor
E : *Error measurement with sensor*
x : Measurement with sensor
x_i : Number of measurement results with the sensor
n : Number of measurements
$\overline{E}$: Average measurement error with the sensor
Ei : Number of errors

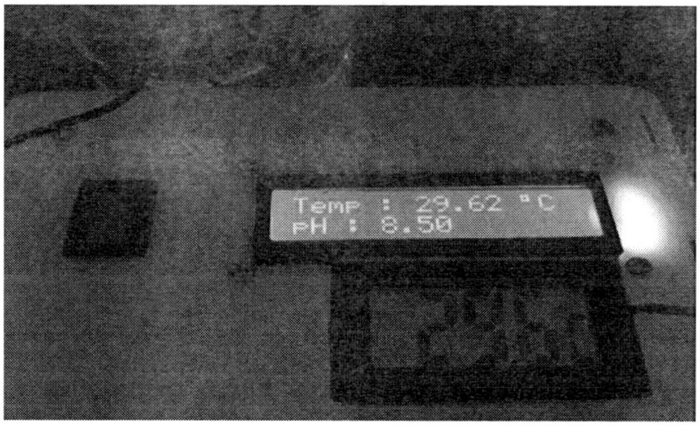

Figure 4. Testing temperature sensor DS18B20 waterproof.

3.2. Sensor analyzer pH meter testing kit
Basically testing the pH Meter Analog Kit sensor is the same as the previous temperature sensor test, done to find out a value indicated on the sensor. Sensor testing is performed to be able to display pH measurement of rice field water. The test was performed three times as in figure 5 the first test was performed using a well water test sample, the second test was carried out using soap water and the third test was performed using orange juice samples.

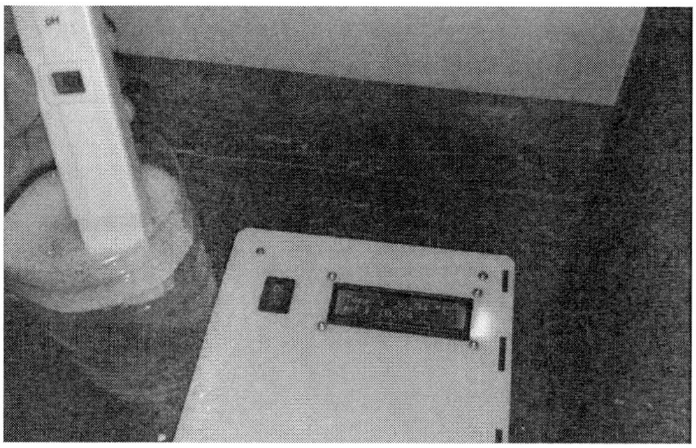

Figure 5. Testing pH meter analog kit sensor.

In the test of figure 5 it can be explained the sensor reading value will be in accordance with the water condition at that time, because basically soap water has a high or alkaline pH. The test is done three times, the first test, second and third then the results obtained as table 1.

Table 1. Test analyzer pH Meter kit sensor.

No	Test Material	Digital pH Meter	pH Meter Analog Kit	Info	Accuracy (%)
1	Orange Water	3.50	2.88	Acid	82.28
2	Pure Water	6.80	6.39	Normal	93.98
3	Soap water	10.60	10.84	Bases	97.79
		Average accuracy			**91.35**

3.3. Android testing

In testing Bluetooth connection to android is a communication between hardware with android smartphone with the aim to know the transmission distance between Bluetooth found on android smartphone with Bluetooth module HC-06 which has been integrated with Arduino Uno circuit. The test of Bluetooth connection is shown in table 2.

Table 2. Testing Bluetooth connection

No	Distance (m)	Information
1	1	Current receive orders
2	5	Current receive orders
3	6	Current receive orders
4	7	Current receive orders
5	8	Current receive orders
6	9	Current receive orders
7	10	Current receive orders

3.4. Comparison of DS18B20 waterproof temperature value with digital thermostat

The temperature sensor and pH testing is done at day time. The results (table 3) obtained is the average DS18B20 Waterproof temperature of 29.44°C and the measuring instrument is a temperature of 29.55 °C. For the average of the measurement error generated from the two temperature sensors is 0.37, while for the average accuracy of the temperature measurement is 98.77% which means the accuracy of the measurement is approaching a significant result.

Table 3. Results of data collection on the temperature sensor.

No	Time	Temperature(°C)		Difference	Measurement Accuracy (%)
		DS18B20 *Waterproof*	***Digital Thermostat***		
1	13.33	29.30	29.7	0.40	98.65
2	13.36	29.32	29.7	0.38	98.72
3	13.39	29.32	29.7	0.38	98.72
4	13.42	29.40	29.4	0.00	100.00
5	13.45	28.99	29.1	0.11	99.62
6	13.48	29.56	29.2	0.30	98.98
7	13.51	29.67	29.2	0.47	98.41
8	13.54	29.88	29.3	0.58	98.05
9	13.57	30.01	29.4	0.61	97.96
10	14.00	30.11	29.7	0.41	98.63
Average		**29.55**	**29.44**	**0.36**	**98.77**

3.5. Comparison of pH Meter Analog Kit value with Digital pH Meter

The results (table 4) obtained are the average measurement of Digital pH Meter with a pH of 6.9°C and the average result on pH Meter Analog Kit sensor is 6.95. For the average of measurement error resulting from both pH sensors is 0.57, whereas for the average accuracy of pH measurement is 97.25%.

Table 4. Test results of data on the pH sensor.

No	**Time**	**pH**		**Difference**	**Measurement Accuracy (%)**
		pH Meter Analog Kit	***Digital pH Meter***		
1	13.33	7.11	7.0	0.11	98.45
2	13.36	7.09	7.0	0.09	98.73
3	13.39	7.09	7.0	0.09	98.73
4	13.42	6.92	6.9	0.02	99.71
5	13.45	6.80	6.9	0.10	98.26
6	13.48	6.68	7.0	0.40	94.28
7	13.51	6.76	6.9	0.14	96.84
8	13.54	6.79	6.8	0.01	99.85
9	13.57	7.23	6.8	0.43	94.60
10	14.00	7.16	6.7	0.46	93.57
Average		**6.95**	**6.9**	**0.18**	**97.25**

3.6. Comparison pH Meter Analog Kit and Digital pH Meter

The result of observed data on pH sensor is shown in table 5. Measurements using pH meter analog kit sensors obtained an average measuring value of pH 6.60 and measurements using digital measuring instruments obtained an average pH value of 6.99.

Table 5. The observed data on the pH sensor

No	**Time**	**pH**		**Difference**	**Measurement Accuracy (%)**
		pH Meter Analog Kit	**Digital pH Meter**		
1	06.33	6.55	6.8	0.25	96.32
2	06.36	6.55	6.8	0.27	96.04
3	06.39	6.60	6.9	0.30	95.65
4	06.42	6.70	6.9	0.20	96.82
5	06.45	6.72	6.9	0.20	96.82
6	06.48	6.65	7.0	0.35	94.72
7	06.51	6.50	7.1	0.60	91.42
8	06.54	6.50	7.0	0.60	91.54
9	06.57	6.51	7.2	0.69	90.41
10	07.00	6.68	7.1	0.42	93.42
Average		**6.60**	**6.98**	**0.38**	**94.51**

3.7. Android application tests

Testing this system is done by installing applications that have been created by using MIT Inventor with step barcode on android smartphone, application named "Rice Field". The display to open the app is shown in figure 6.

Figure 6. App display on android smartphone.

Display android smartphone applications in general just like any other application. Testing of this application is done to determine the response given by the application on android for monitoring the quality of rice water with temperature parameters and water pH. Once the application can already be installed on android then the next step is to activate the existing Bluetooth on android smartphone as wireless on the smartphone.

In the monitoring application android step to connect to the hardware is to click the Bluetooth button on the application, then connect with Bluetooth HC-06 with a note between the android with hardware not too far when connecting because it can interfere with the response given to the application and cause errors. After being able to connect with Bluetooth eat the results of monitoring will be displayed measuring value on android.

Following on the results of research and testing obtained evidence when the tool started on (connected to the battery) and both sensors dipped into wet water then the LCD will display the results of measuring. Then Bluetooth will send digital data to android which means monitoring tool can run as expected. The data will be displayed and always update if on the hardware change then the android application also changed with the same value. Next is when finished using the app on android then the next thing is logout on the app by pressing the back button on android then press "Yes". Here is the view when logout. Quality of connection is less or unstable.

4. Conclusion

In the results obtained in all tests, analog pH meter kit sensor and DS18B20 sensor performed water quality. The comparison of data between temperature and pH water readings using sensors and standard measuring instruments in can be temperature and pH of water on android can run and according to display data on LCD. If the temperature changes then the android display also changes and runs as where expected.

Acknowledgement

This research was funded by ministry of Research and Higher Education Indonesia by scheme PTUPT with contract number PUPT/-085/SP3/LPP- UAD/IV/2017.

References

[1] Warlina N 2004 Pencemaran Air, Sumber dan Dampak Mara Menanggulanginya Institut Pertanian Bogor, Indonesia *Thesis*

IOP Conf. Series: Materials Science and Engineering **403** (2018) 012042 doi:10.1088/1757-899X/403/1/012042

[2] Pradipta B S 2017 Rancang Bangun Sistem Pemantauan Suhu dan Kelembapan Udara Pada Tanaman Hidroponik Dalam Rumah Kaca, Electrical Engineering, Universutas Ahmad Dahlan, Indonesia *Undergraduate Thesis*
[3] Ardiansyah N 2015 Rancang Bangun pH Meter Air di Utilities Refinery Unit IV PT PERTAMINA (PERSERO) Berbasis Arduino Uno R3 Electrical Engineering, Universitas Muhamadiyah Purwokerto, Indonesia *Thesis*
[4] Hidayat M 2008 Perancangan Kendali Suhu Dan Kelembaban Untuk Pertanian Rumah Kaca Berbasis Mikrokontroler ATMega 8535 Electrical Engineering, Universitas Ahmad Dahlan, Indonesia *Undergraduate Thesis*
[5] Ramadhan, A W 2016 Purwarupa Monitoring Ph Air Pada PDAM Berbasis Arduino Uno R3, Vocational School, Universitas Gadjah Mada, Indonesia *Final Project*
[6] Iswahyudi N 2017 Pengendalian Sirkulasi pH Air Pada Tambak Udang Berbasis Arduino Uno, Informatics, Universitas Islam Negeri Alauddin Makasar, Indonesia *Undergraduate Thesis*
[7] Taofik J Rancang Bangun Alat Ukur Suhu Pada Aquarium Air Tawar Berbasis Arduino Terintegrasi WEB, Program Keahlian Teknik Komputer, Institut Pertanian Bogor, Indonesia

Neural network simulation for obstacle avoidance and wall follower robot as a helping tool for teaching-learning process in classroom

T Dewi, P Risma, Y Oktarina and M Nawawi

Department of Electrical Engineering, Politeknik Negeri Sriwijaya
Jalan Srijaya Negara Palembang, Indonesia, 30139

tresna_dewi@polsri.ac.id

Abstract. One of the most applied technologies is robotics, and one the most applied type of robots is a mobile robot. To provide a tool for the interactive teaching-learning process in the classroom, a simulation program is needed. Low price and easy to use software is preferable considering not every polytechnic can provide expensive simulation software for its students. Robotics related curriculum should include artificial intelligence since it is functioning as a brain to the robot. One type of artificial intelligences is neural networks. This study presents the application of neural networks by using low price simulation software. The simulation results show the feasibility of utilizing the software for the students to learn neural networks by variating the scenarios of mobile robots application, adding more robots, more sensors and writing the programs to control the robots. The contribution of this paper is to show and inspire teachers to create a more interactive teaching-learning process in the classroom by using low price, user-friendly software, and to encourage them to search more alternatives of low price or even free software to be the teaching tool in the classroom.

1. Introduction

Polytechnic as a vocational education has to focus its curriculum and educational process to the application of current technology rather than theoretical analysis [1]. The vocational education insists the teachers to leave the conventional learning method into a more interactive and applicative learning method. The curriculum has to be always updated with the technology in the industry [2]. Robotics is the rapidly increasing technology in the industry, therefore polytechnic also has to provide the basic knowledge for its students to understand robots and how to design it [3-6]. To provide robotics course means to integrate several subjects that support it, such as mathematical analysis, computing, electronics, mechanics, etc. The integration of this several subjects can be a subject discussing the main controller or the brain of the robot.

In order to function well, a robot has to be equipped with a brain, therefore one of the classes should be provided in robotics curriculum is the one that discussed the brain of the robot. This brain is called artificial intelligence [7]. The installed artificial intelligence in a robot will provide a robot a way to think and decide the best options to finish the assigned task. Artificial intelligence class should be included in the polytechnic curriculum if that polytechnic decides to include robot's related topics [3-6, 8].

Before applying to the real robot, simulation is the best method to get the feel of the environment and avoid dealing with the uncertainties for the beginner when interacting with the robot for the first time [9-11]. Simulation is bringing the textbook to live (animation) with much less risk of damaging real system with trial and error. The method of applying the right simulation software in vocational education has made the class more interesting rather than focus within textbook theory. Students can observe the effects of creating a certain system justification and prepare them to work with the real system.

Neural networks are a method for artificial intelligence [7]. It imitates the function of neural in a human brain. This type of artificial intelligence is very suitable for a system with a series of sensors and not easy to be modelled into certain mathematical equations. Data and information received from sensors are utilized to be the input for training the neurons or synaptic weights. The learning process will create output that will be fed back to the controller system. The basics applications of neural networks in robotics are obstacle avoidance and wall follower robot [12-22]. From this basic, students can improve and develop the examples to a more complicated system.

The current method of teaching-learning in the classroom is still in a conventional way, where the teacher teaches and students only listen. This conventional way can be improved by creating an interactive class and involving students more, this way not only can improve students understanding but also their interest to participate more in class [3, 4, 6]. Providing simulation program and involving students actively in working on simulation to show the theoretical and application of the taught subject is one of the excellent ways to create an interactive class. The problem with providing simulation is to find the right software program. In many cases of universities and polytechnics in developing countries cannot provide expensive software program such as MATLAB [9-11]. Therefore, choosing the right simulation software is a crucial decision. Product price becomes one of the considerations in purchasing the software.

Mobotsim is low price software and offers 1-month trial to simulate a robot. Obstacle avoidance and wall follower robot can be easily simulated in this software [21]. Students can easily develop the robot system from the examples to any systems. This paper presents the application of low price software to create teaching module for learning neural networks in the classroom. The procedure to develop the program and program examples of obstacle avoidance and wall follower robots will be provided in simulation results to show the feasibility of using this software (and any software alike) in creating a helping tool for teaching-learning process of artificial intelligence class. The contribution of this paper is to show and inspire teachers to create more interactive teaching-learning process in the classroom by using low price, user-friendly software, and to encourage them to search more alternatives of low price or even free software to be the teaching tool in the classroom.

2. Introduction to Neural Networks

The Neural Networks was first introduced by McCulloch and Pitts in 1943, as the result of a brain study, and in 1949 Hebb presented the concept of synaptic weight [6]. Neural networks are the type of artificial intelligence that use a large number of nonlinear processors called neurons. This neuron imitates human brain and has a number of internal parameters called synaptic weights. The neurons are characterized by a state, weighted inputs defined by a learning process and a state equation for their dynamic operation. Neural networks get the input from the sensory data information, process it and learn from the data information.

The most popular type of neural networks application is multilayer perceptron where the feedback control can approximate any functions such as continuous and integrated function. The structure of layers design consists of input layers, hidden layers, and output layers. Hidden layers are the part where the learning process is happening and adjusting online.

In robotics, neural networks get input data from sensory information and/or robot modelling; kinematics and/or dynamics modelling. The inputs are processed in hidden layers, and the results are the output of the system. The processes are updated online based on data received by the sensors and/or the updated mathematical equations in Jacobian derived from kinematics and dynamics of the system. The generic output equations for neural networks is given by equation (1).

$$y_i = \sum_{j=1}^{l}\left[v_{ij}\sigma\left(\sum_{k=0}^{n} w_{jk}x_k\right)\right], \qquad i = 1,2,\dots,m, \tag{1}$$

where $y_i$is the output, w_{jk} are weight function as the input to hidden layers, v_{ij} are the output from hidden layers, $x_k(k = 0,1,2,\dots,n)$ is the control input, l is numbers of neurons in the hidden layers, m is the output neurons, l is a number of neurons in the hidden layers, m is the output neurons, and σ is the nonlinear hyperbolic tangent activation function, given by equation (2).

$$y = V^T\sigma(W^T x), \tag{2}$$

where$N_1, N_2, N_3, \dots, N_n$ is the vector function; related to neurons, and W and V are weight of neurons.

3. Simulating mobile robot with neural networks

For students learning in the classroom, the most applied type of mobile robots is considered, the two-differential driven mobile robot, shown in the figure 1, where v and ωare the translational and rotational velocity of the robot, ϕ is the orientation of the robot, L is the distance from driving wheels to the robot axis in y-direction, θ_L and θ_R are left and right tires' angles, and S_1 to S_4 are the installed proximity sensors. This type of robot is suitable to show the application of neural networks in robotics.

In order to control the robot, modelling of the robot is necessary. There are two types of modelling, kinematics modelling; describing robot movement without considering the input to the robot, and dynamics modelling where the input forces are considered.

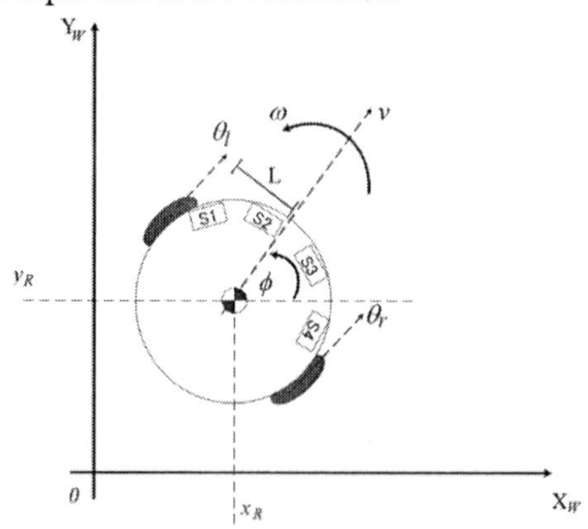

Figure 1. Two-differential driven mobile robot.

3.1. Kinematics modeling

In kinematics modelling, the robot movement in its coordinate frame is derived relative to world coordinate frame. In this modelling, the pose, position and orientation, of the robot is defined as $q = [x \quad y \quad \phi]^T$where x and y are robot's position in x and y-axis, and $\phi$is the orientation of the robot. From this pose, the translational and rotational velocities are derived as in equation (3).

$$\dot{q} = \begin{bmatrix}\dot{x}\\ \dot{y}\\ \dot{\phi}\end{bmatrix} = f(\dot{\theta}_R, \dot{\theta}_L, L, r, \phi), \tag{3}$$

where $\dot{x}$ and $\dot{y}$ are the translational velocities in x and y axis respectively, and $\omega = \dot{\phi}$ is the rotational velocity.

In order to get the value of tires orientation and velocity, the inverse kinematics of the robot is derived as in equation (4).

$$\begin{bmatrix}\dot{\theta}_R \\ \dot{\theta}_L\end{bmatrix} = f(\dot{x}, \dot{y}, \dot{\phi}). \tag{4}$$

From equation (4), the relation between robot's translational v and rotational ω velocities and both tires velocities are shown in equation (5).

$$v = r\frac{\dot{\theta}_R - \dot{\theta}_L}{2}, \quad \text{and } \omega = \frac{r}{2L}(\dot{\theta}_R - \dot{\theta}_L). \tag{5}$$

3.2. Dynamics modelling

Dynamics modelling considers the total forces applied to the system, including disturbance. The velocities and accelerations are derived from the pose (positions and orientations) given in Kinematics. The complete generic dynamics modelling of two differential driven mobile robots that suffers the non-holonomic constraints is shown in equation (6).

$$D(q)\ddot{q} + C(q, \dot{q}) + G(q) + M^T\lambda = E\tau, \tag{6}$$

where $D(q)$is the inertia matrix, $C(q, \dot{q})$ is the centripetal and coriolis matrix, $G(q)$ is the gravitational vector, M^T is the $m \times n$ matrix of the m non-holonomic constraints, E is a non-singular transformation matrix, and τ is total force and/or torque applied to the system, including frictions occurs due to actuators motions.

The constraints matrix M^T can be eliminated by a constraint-free model of $n \times (n - m)$matrix $B(q)$. By eliminating M^T, the dynamics equations of two differential driven mobile robot can be written using equation (7).

$$\bar{D}(q)\ddot{q} + \bar{C}(q, \dot{q}) + \bar{G}(q) = \bar{E}\tau, \tag{7}$$

where $\bar{D}$, $\bar{C}, \bar{G}$and $\bar{E}$show the justification of the original representation in equation (6).
Jacobian matrix is derived from kinematics and dynamics of the robot and defined as in equation (8).

$$J = \frac{\partial q}{\partial v_c} = \frac{\partial q}{\partial v}\frac{\partial v}{\partial \tau}\frac{\partial \tau}{\partial e_c}\frac{\partial e_c}{\partial v_c} \tag{8}$$

where v_c is the control law and $e_c$is the error dynamics.The Jacobian is needed to be the input for hidden layers of neural networks and updated online.

The application of neural networks to control a mobile robot can be applied by deriving kinematics and dynamics model, as shown in figure 2 or directly using the sensory data information as the input, shown in figure 3. Figure 2(a) and 3(a) show the diagram block representation of both methods, and figure 2(b) and 3(b) show the layer presentation, where v and ω are the translational and rotational velocity, S_1 to S_4 are sensor 1 to 4, and m_1 and m_2 are motor left and right velocity.

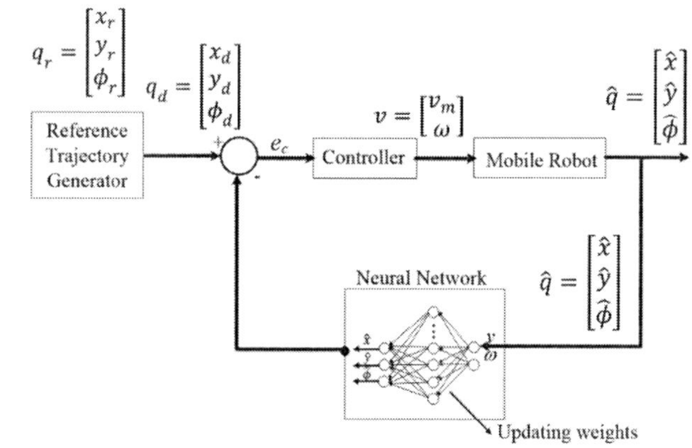

(a) Diagram block

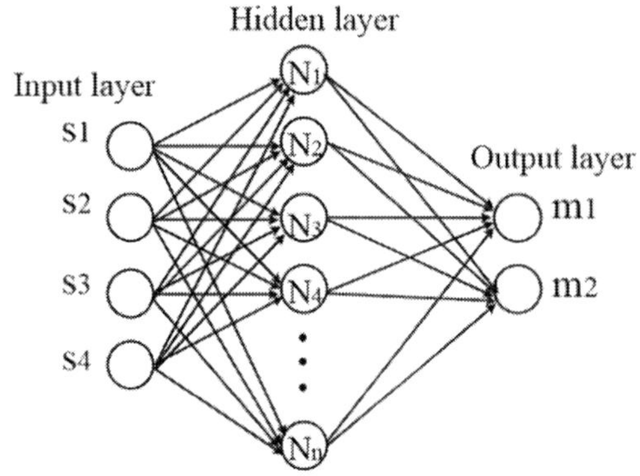

(b) Layer presentation

Figure 2. Neural networks controller design without considering robot's kinematics and dynamics.

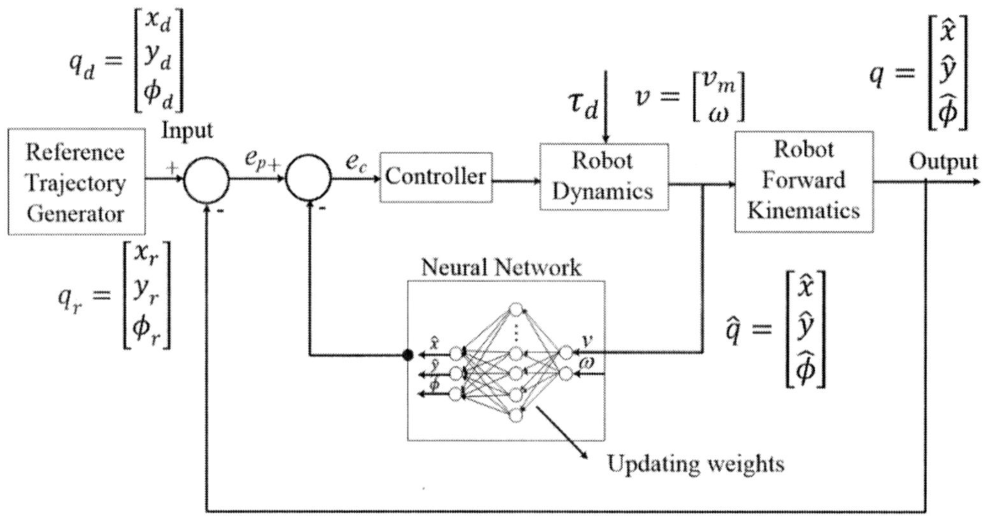

(a) Diagram block

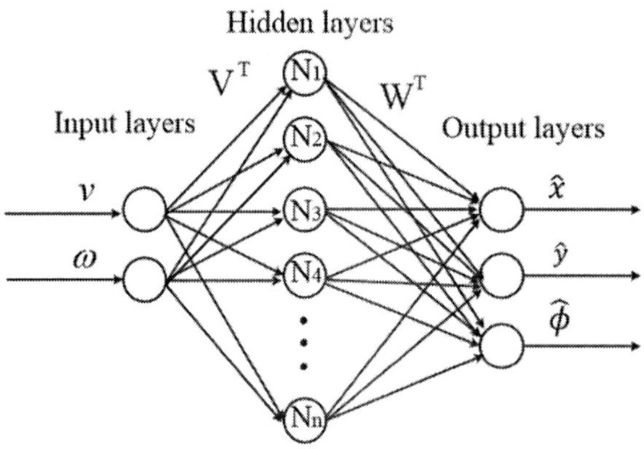

(b) Layer presentation

Figure 3. Kinematics and dynamics modelling application in neural networks controlled design.

4. Simulation results

A simulation of a system is operating the model of the system by reconfiguring and experimenting using the software. Teaching-learning process in vocational education insists the real application to understand the topic more. Learning and understanding neural networks will be easier and deeper by simulating its application using mobile robot simulation software, such as Mobotsim from Mobotsoft.

Mobotsoft is low price software that is possible to purchase by students. The students can use the trial version for 30 days or for 100 times usage, which one comes first. The simulation is written with Basic programming language and the environment can be customized according to the scenarios assigned by the programmer. Mobotsim is already giving examples of a mobile robot applying neural networks controller that can be developed for more sensors, more complicated environment and/or for more robots. This study shows four cases of neural networks application for two-wheel differential driven mobile robot. Robot(s) has (have) 4 to 8 proximity sensors for obstacle avoidance and/or wall follower.

4.1. Case 1: a wall follower robot with 4 stopping points

Case 1 in figure 4 has the scenario of a patrol robot. It comes to 3 checking points for 3 rounds, and in the 4th round, the robot goes to its base. Robot follows the wall and stops at the stopping points. One stopping point for one round, after four rounds, the robot will stop at the final position.

Mobile robot applied in figure 4 is equipped with 8 sensors assigned to follow the wall. The neural network controller is used making the robot follows the wall, and this wall-following scheme is combined with the virtual force application. The stopping points considered to have an attractive force for one time after the robot stops one time at a point, the point has no attractive force anymore, therefore for the next round the robot will skip it and stop at the next stopping point. Figure 4(a) shows the robot at the starting point and figure 4(b) shows robot at the finish point.

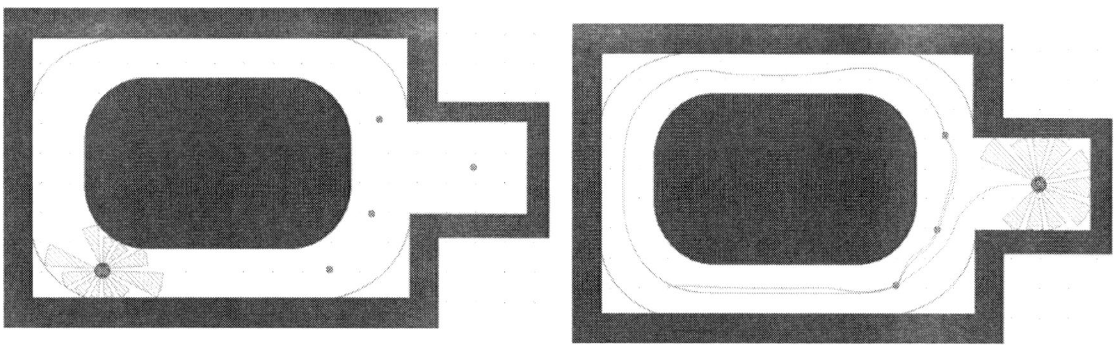

(a) Robot at starting point (b) Robot at final point

Figure 4. Case 1: a robot follows the wall, and stop at one point at one round, and on the 4th round, the robot reaches the final point.

4.2. Case 2: two wall follower robots with many stopping points for one robot

Figure 5 shows the environment of the simulation program. The programmer can develop the program according to any scenarios, sensors applied to the robot can be reduced, robots and obstacles can be added, and the syntax can be developed as required.

Case 2 shows in figure 5 is two robots in one environment. The first robot (the green robot) is a robot that applied in a cookies manufacture company. The robot will work from supply and ingredient (suing) post to stir post, to bake post and finally to packing post. After the first robot reaches the packing post, robot 2 will move bringing the cookies to the market.

The robot is also equipped with 8 proximity sensors to follow the wall. The points are applying virtual attractive force to make the robot stop at some time, using a time delay. Figure 5(a) shows robot 1 on starting point, figure 5(b) and 5(c) show the robot is on the move from point to another. Figure 5(d) shows robot 1 reaches the final position and robot 2 starts to move. Figure 5(e) shows robot 2 reaches its final position. Figure 6 shows the Basic programming language for setting the simulation.

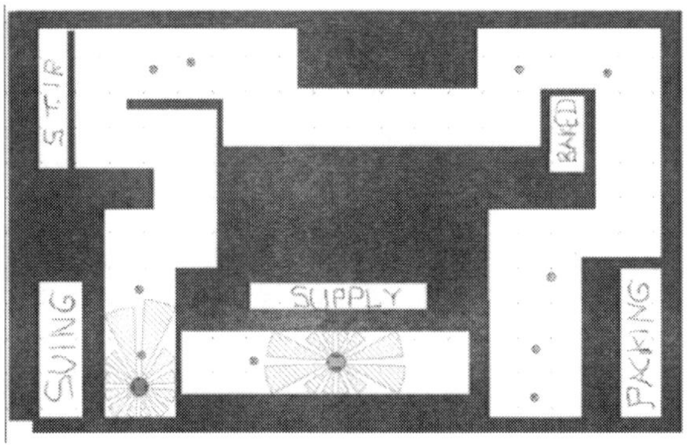

(a) Robot 1 on the starting point

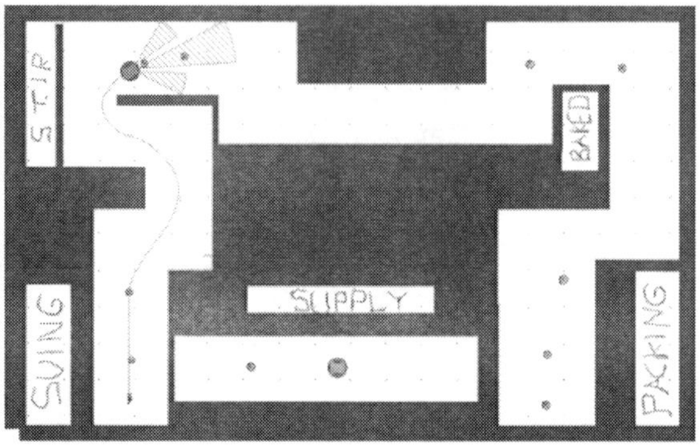

(b) Robot 1 is on the move from point to point

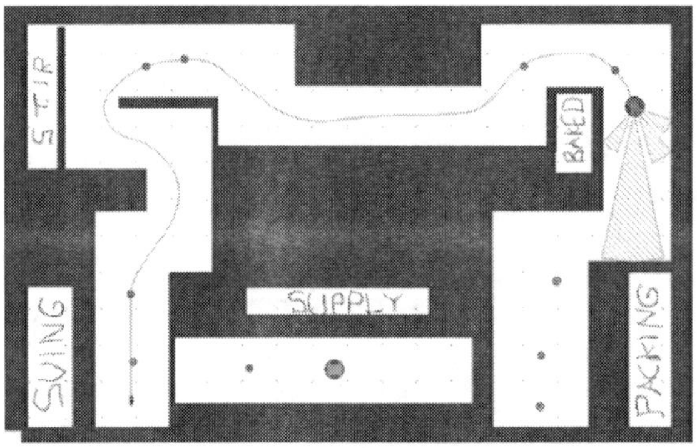

(c) Robot 1 is on the move from point to point

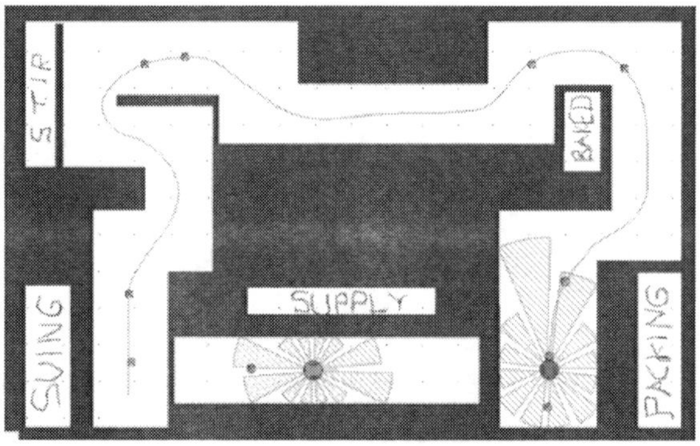

(d) Robot 1 reaches the final point and robot 2 starts to move

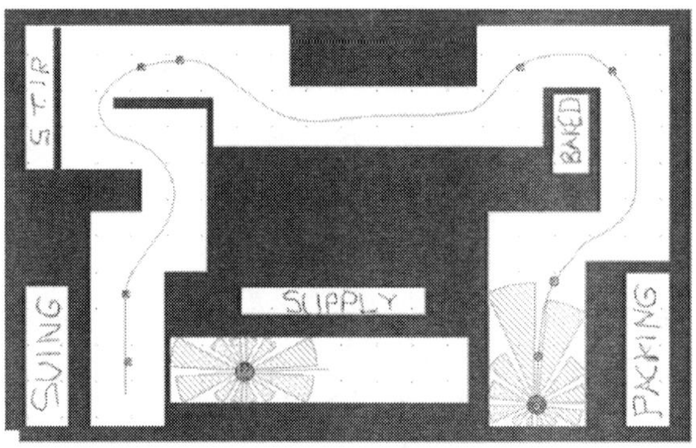

(e) Robot 2 reaches the final position

Figure 5. Case 2: 2 robots, one robot stops in many points and robot 2 moves after robot 1 reach the final position.

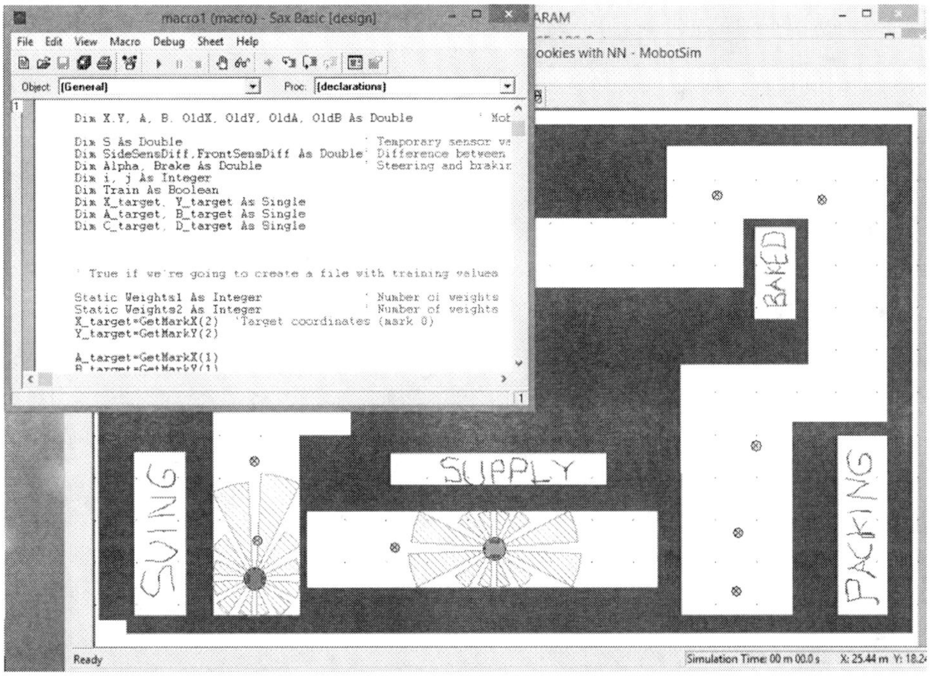

Figure 6. Basic programming language for setting up the simulation.

4.3. Case 3: a delivery robot

Case 3 shown in figure 7 is a delivery robot. The robot delivers food to the final position. The robot will reach the target by choosing the path automatically and avoid obstacles. This scenario is not applying the optimum path; it chooses the path with fewer obstacles. The robot is also equipped with 8 proximity sensors and attractive virtual forces on points.

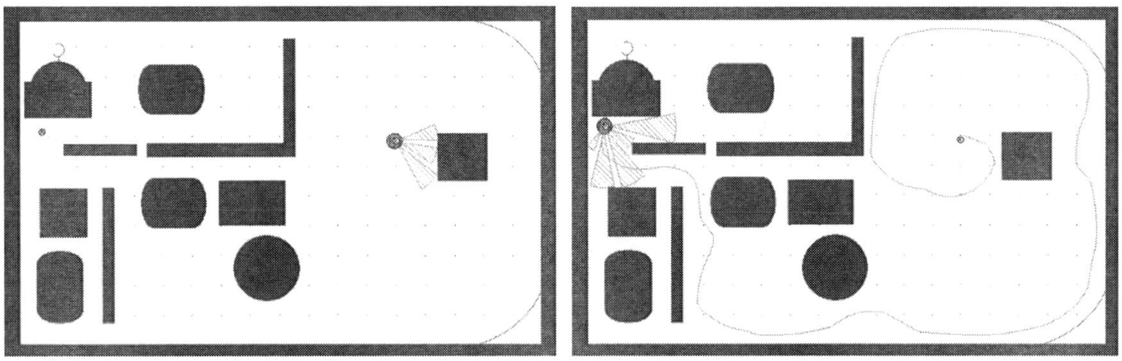

(a) Robot at starting point (b) Robot at final point

Figure 7. Case 3: a food delivery robot, considering obstacles avoidance

4.4. Case 4: 2 robots in two parallel environments

Case 4 in figure 8 is the same with case 2 in figure 5. Case 4 just to show that more robots can be added and more environments can be set up. Case 4 is to show the flexibility of simulating neural networks application in a two-wheel differential driven mobile robot and more.

(a) Robot at starting point (b) Robot at final point

Figure 8. Case 4: 2 parallel environment setting.

Figure 4 to 8 show the examples of simulating neural networks application in a mobile robot. The flexibility of the simulation software program helps the students to understand the topic more and deeper by improving and developing it. The environments and robots addition can be done in a fun way and the controller can be set in Basic programming language. This simulation helps the teacher to reach the objective of vocational education where more applications are better rather that theoretical study.

5. Conclusion

Simulation for learning artificial intelligence is necessary to show the application of robotics without having to deal with the uncertainties of the real system. This paper has presented the utilization of low price software namely Mobotsim to learn neural networks interactively. The basic applications of neural networks in mobile robot are chosen, the obstacle avoidance and wall follower. Students can learn the application of Neural networks in mobile robot by setting the scenario, choosing the robot or robots, and programming it (them) with Basic programming language. The simulation results of 4 scenarios show that teaching-learning process in the classroom for understanding neural networks application for obstacle avoidance and wall follower robots can be conducted by using a simulation software as a helping tool. This paper has shown that application of low price software can be an alternative to create an interactive teaching-learning process in the classroom, and authors would like to inspire and encourage teachers to search more alternatives of low price or free software (open source) for teaching tool in the classroom.

References

[1] Jianyu W, Xi L and Chen X 2012 Vocational Ability Oriented Modularized Curriculum for Advanced Vocational School *2012 International Conference on Future Computer Supported Education, IERI Procedia* vol 2 pp 897-900
[2] Kuangdi X 2008 Engineering Education and Technology in a Fast-Developing China *Technology in Society* vol 30 pp 265-74
[3] Assaf D, Larsen J C and Reichardt M 2012 Extending Mechanical Construction Kits to Incorporate Passive and Compliant Elements for Educational Robotics *3rd International Conference on* Robotics *in Education* pp 33-40
[4] Ospennikova E, Ershov M and Iljin I 2015 Educational Robotics as an Innovative Educational Technology *Worldwide trends in the development of education and academic research, Procedia – Social and Behavioral Sciences* vol 214 pp 18-26
[5] Baros T T T and Lages W F 2012 Development of a Firefighting Robot for Educational Competition *3rd International Conference on Robotics in Education* pp 47-54
[6] Petrovic P Having Fun with Learning Robots *3rd International Conference on Robotics in Education* pp 105-12
[7] Tzafestas S G 2014 *Introduction to Mobile Robot Control* Elsevier
[8] Alemany J and Cervera E 2012 Appealing Robots as a Means to Increase Enrolment Rate: a Case Study *3rd International Conference on Robotics in Education*, pp 15-9
[9] Arlegui J, Moro M and Pina A 2012 Simulation of Robotic Sensors in BYOB *3rd International Conference on Robotics in Education* pp 25-32
[10] Pandey A and Parhi D R 2014 MATLAB Simulation for Mobile Robot Navigation with Hurdles in Cluttered Environment Using Minimum Rule Based Fuzzy Logic Controller *2nd International Conference on Innovations in Automation and Mechatronics Engineering ICIAME 2014, Procedia Technology* vol 14 pp 28-34
[11] Casini M and Garulli A 2016 MARS: a Matlab simulator for mobile robot experiments *IFAC-PaperOnline* vol 49 no 6 pp 069-074
[12] Farooq U, Amar M, Asad M U, Hanif A and Saleh S O 2014 Design and Implementation of Neural Network Based Controller for Mobile Robot Navigation in Unknown Environment *International Journal of Computer and Electrical Engineering* vol 6 no 2 pp 83-9
[13] Engedy and Horváth G 2009 Artificial Neural Network Based Mobile Robot Navigation *2009 IEEE* International *Symposium on Intelligent Signal Processing*, Budapest pp 241-6
[14] Capi G, Kaneko S and Hua B 2015 Neural Network based Guide Robot Navigation: An Evolutionary Approach *2015 IEEE International Symposium on Robotics and Intelligent Sensors (IRIS2015)* pp 74-79
[15] Chi K H and Lee M F R 2011 Obstacle avoidance in mobile robot using Neural Network *2011 International Conference on Consumer Electronics, Communications and Networks (CECNet)* pp 5082-5
[16] Seyr M, Jakubek S and Novak G 2005 Neural Network Predictive Trajectory Tracking of an Autonomous Two-wheeled Mobile Robot *16 Triennial World Congress, Prague, Elsevier IFAC Publication* pp 385-90
[17] Boukens M and Boukabou A 2017 Design of an Intelligent Optimal Neural Network-Based Tracking Controller for Non-holonomic Mobile Robot System *Neurocomputing* 226 p 46-57
[18] Janglová D 2004 Neural Network in Mobile Robot Motion *International Journal of Advanced Robotics Systems* vol 1 no 1 pp 15-22
[19] Shamsfakhr F and Bigham B S 2016 A Neural Network Approach to Navigation of a Mobile Robot and Obstacle Avoidance in Dynamic and Unknown Environments *Turkish Journal of Electrical Engineering & Computer Science*
[20] Larasati N, Dewi T and Oktarina Y 2017 Object Following Design for a Mobile Robot Using Neural Network *Computer Engineering and Application Journal* vol 6 no 1
[21] Fierro R and Lewis F L 1998 Control of a Nonholonomic Mobile Robot Using Neural Network

IOP Conf. Series: Materials Science and Engineering **403** (2018) 012043 doi:10.1088/1757-899X/403/1/012043

IEEE Transactions on Neural Networks vol 9 no 4 pp 589-600
[22] Mohareri M, Dhaouadi R and Rad A B 2012 Indirect Adaptive Tracking Control of a nonholonomic Mobile Robot Via Neural Network *Neurocomputing* vol 88 pp 54-66

Market risk assessment on poultry industry using Monte Carlo simulation

R Purwaningsih[1], M Arief[2], N U Handayani[3], D Rahmawati[4] and A Mustikasari[5]

[1,3,4, 5] Industrial Engineering Department, Engineering Faculty, Diponegoro University
Jl. Prof. Soedarto, SH, Kampus Undip Tembalang, Semarang, Indonesia 50275
[2]PT. Ciomas Adisatwa east java region
ratna.tiundip@gmail.com

ratna.tiundip@gmail.com; ariefpkpmjkt@gmail.com; naniekh@gmail.com; rahmawatidaru@gmail.com;anita_mustikasari@yahoo.com

Abstract. One of risk in poultry industry is fluctuations in the selling price of live bird. This study aims to (1) calculate the cost of production and cost structure, (2) calculate the profit and loss, (3) measure the market risk of the farms. The study was done on 3000 population broilers farm at some district on East Java province. Value at risk and Monte Carlo simulation methods used to measure business risk. Research result define that the production cost of live bird production from farm is Rp 14338/kg and the selling price is Rp 15476/kg. The highest percentage of cost structure is feed cost. Feed cost reach 74% of total production cost. Simulation results show that broiler farm have a probability of loss risk of 54, 27%, categorize as high risk.

1. Introduction

Indonesian chicken meat consumption reached 1.6 million tons per year in 2015 [1]. Indonesians spent 1, 85 percent of their income on meat and 62% of which is poultry. Indonesians consumption of poultry meat was reach 3650 kg per-capita on 2013[2]. The broilers are grown in company farms owned by the large integrators (10 percent), contract farms (70 percent), and independent farms (20 percent). The average farm size is small, with a capacity for 5000-20000 birds [3]. The broiler farms have various risks. The business risks include social, political, economic, technological and environmental aspects [4]. Risk from the economic aspect is caused by several factors: (1) fluctuation of live bird selling price that is difficult to predict, (2) supply chain actor activity on the downstream sector, and (3) fluctuation of production input price. In a certain period of time there is a quite significant difference between the prices of live bird selling price at farmers with selling price at end consumers. Selling price formed by market price. If the market price is below cost of production then there will be a loss for farmer.

The purpose of this research is to calculate the production cost per kilogram, profit and loss for a 3000 population farm, and estimate the business risk of independent farm. The location of the research is broiler farms located in East Java (Mojokerto, Gresik, Jombang, and Lamongan). Data collection was done by direct observation, deep interview, and collected historical data. Risk assessment done by VaR method. Value at Risk (VaR) simulation is a statistical risk measurement method that estimates the maximum possible loss of a portfolio at a certain level of confidence [5]. VaR is a method of estimating potential maximum losses in certain periods with certain confidence levels and in a normal market [6].

Currently the Indonesia government has been finalizing the regulation to stabilize the price of live bird and reduce the fluctuations by define a minimum price policy or the lowest selling price policy of live bird. This policy try to prevent farmers from lose. However, broiler is a livestock model. While farmers are prohibited from selling chickens at low prices, the farmer must continue to keep the chickens and need more cost for feed, then farmer losses will be bigger. Measurement of business risks for broiler farm needs to be done to know the amount of risk and estimate the probability of losses.

2. Literature review

Risk is defined as an event that has an unfavorable impact, difficult to accept or even unacceptable [7]. Currently, risk assessment is an important research theme because risk is always present as an important issue for the industry [8]. The role of risk management in financial firms has evolved into the use of complex econometrics and uncertainty models in finance [9]. The interaction between all the interconnected risks in the logistics chain is known as the Supply Chain Risk Management (SCRM) concept. The SCRM definition was first provided by Juttner [10]. SCRM is an effort to identify and manage supply chain risk, through coordination among supply chain members, to reduce overall supply chain vulnerability. Supply chain risk management is defined as a risk mitigation process through collaboration, coordination and implementation of risk management tools among partners, to ensure long-term benefits from supply chain actors.

Recent years, the industry experienced violent incidents, health crises and natural disasters. This incident has a major impact on the industrial economy. Risk management becomes an important topic in the world of logistics chain. The risk is related to uncertainty in the supply chain industry [11]. Then the method of financial risk analysis method developed rapidly. Especially for agribusiness, the risk dependence on natural factors is high due to uncertainty.

The market price is determined by the balance of supply and demand. Therefore, the accuracy of supply and demand data is required to maintain the balance of live bird selling price. The market price of live bird from farmer is determined from the amount of supply and the amount of market demand. Government role is needed on the regulation of the supply and demand of live birds to avoid price fluctuation. Government policies on import permits for chicken breeders should be reviewed to regulate the supply of domestic stock. The number of GGPS (Great Grand Parent Stock) imported determines the amount of DOC production and live bird. The imbalance on supply and demand on broiler meat happens in Indonesia in 2014. Over supply on live bird give impact on live bird selling price. Problem of falling selling price in 2014 occurs because the number of imports of the GGPS is more than it should be. Oversupply live chicken in 2010 is still 31.27% of national consumption, but in 2013 reached 55.08%, then in 2014 reached 65.75% [12].

3. Research method

This study attempts to calculate the cost of production per kg of chicken meat from farmers, determine the level and the probability of loss of farmers. The research object is an independent farmer (has no partnership with feed mill industry) in several districts in East Java. The research methods are:

- Production cost estimation done by deep interview to 20 independent farmers of around 3000 population broiler farm. The interview questions are related to the production cost on growth of 3000 doc (day old chick) to average body weight of 2.05 kg when harvested.
- Profit and loss probability value was done by compare the production cost (from step one) to the adjusted selling price of live bird. The selling price data was obtained from year 2014 to 2015. The probability value is based on selling price real condition and constant production cost.
- Business risk calculation done by Value at Risk method to estimate the probability of loss from simulation on selling price for long periods. It is different from step 2, the probability

value is estimate from a simulation process. Value of market price and production cost on 2014 and 2015 was inputted into the software and simulated by Monte Carlo.

VaR can measure maximum losses that may occur on the desired time period. The advantage of VaR is that this method focuses on the downside risk and independent assumption of the returns distribution. The figures obtained from the measurements represent aggregate or comprehensive accounting results of business risks. Monte Carlo simulation is known as simulation sampling or Monte Carlo Sampling Technique. This simulation uses the existing data sampling (historical data) to identified distribution pattern of data. Simulation of Monte Carlo will generate random numbers based on the distribution of input data and calculate the probability of the event to be known [13]. This method can be divided into several stages: (a) Define probability distributions from past data, (b) Convert the distribution into cumulative frequency, (c) Create a random number or a random number, (d) Conducting simulations with random numbers analyzing the simulation output.

4. Result and Discussion

4.1. Production cost estimation

The structure of production cost consists of (1) investment cost (costs to build farm and purchase of production facilities), (2) direct raw materials cost (cost of purchasing DOC, chicken feed, and vitamins), (3) labor costs (salaries of workers), (4) other fees/overhead cost (cost of purchasing chaff, heating for the cage, electricity, water medicine). The value of each component of broiler farm production cost is given in table 1. Food consumption ratio (FCR) assumed in 1.6.

Table 1. Estimation of production cost in broiler farm

Cost item	Value	Price /unit (Rp)	Total (Rp)
Investment cost (farm and equipment)	370 m^2		85000000
Investment cost per growing period	30 days		4250000
Direct material cost			
DOC	Rp 3000	4500	13500000
Feed *			
— **Pre-Starter (1 - 18 days old)**	Rp 1500/kg	6700	10050000
— **Starter (19 – 25 days old)**	Rp 4500/kg	6600	29700000
— **Finisher (25 – harvest time)**	Rp 3500/kg	6400	22400000
Medicine and vitamin			568763
Total direct material cost			76218763
Total of overhead cost			1452500
Total of labor cost			1850000
Total production cost/growth period			83771263
Chick in population	Chick in		3000
Percentage of Mortality	5 %		150
Population harvest (f - g)			
Average Body Weight	Kg/chick		2,05
Total harvest/growth period (h x i)	kg		5843
Margin of Production cost (e / j)	Rp/kg		14338
Selling price *	Rp/kg		15476
Revenue per period (j x l)	Rp/period		90418530
Profit per period (m - e)	Rp/period		6647267
Profit per kg (n / j)	Rp/kg		1138

Table 1 explains that the estimation of production cost for broiler farm 3000 population is Rp 83771263. The investment cost use assumption of live time 5 year and 4 times period of growth per year. The farm gain profit per period about 6.6 million in a certain selling price (Rp 15476). Selling price is average of selling price in 2014 and 2015. In fact, the selling price is highly fluctuated per month during 2014 and 2015. In 2014 lowest selling price was Rp 12263, happened on March 2014, while the highest selling price was Rp 17514, happened on August. The gap between the highest and the lowest selling price was Rp 4776. The gap is more than 30 % of the lowest price. Table1 also show the cost structure of production cost on broiler farm. The composition of broiler farm production cost is given in the figure 1.

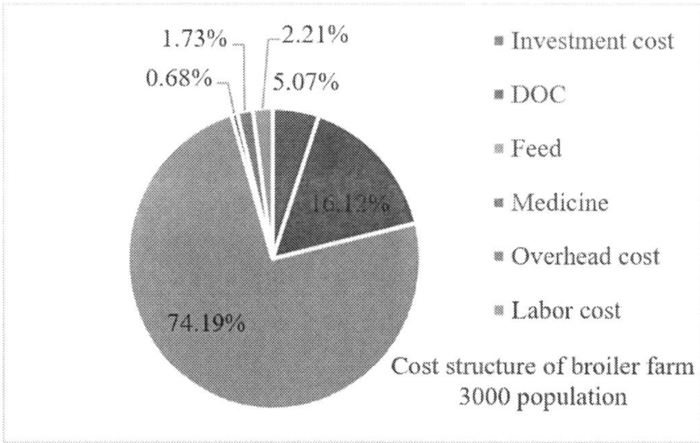

Figure 1. Cost structure of 3000 population broiler farm

Figure 1 shows that the cost of feed has highest percentage of all production cost, it reach 74%, then, followed by doc cost 16% and labor cost 5%. These three components already reach 95% of total production cost on broiler farm 3000 population.

4.2. Profit – Loss Estimation

The risk of broiler farm came from highly fluctuated on live bird selling price. Figure 2 show the fluctuation of selling price on 2014 and 2015 (in Rp per kg). Determine the market risk of farm done by a simple simulation based on historical data and based on simulation by *crystal ball* software (Monte Carlo simulation) which generates random value according to distribution pattern of historical data inputted.

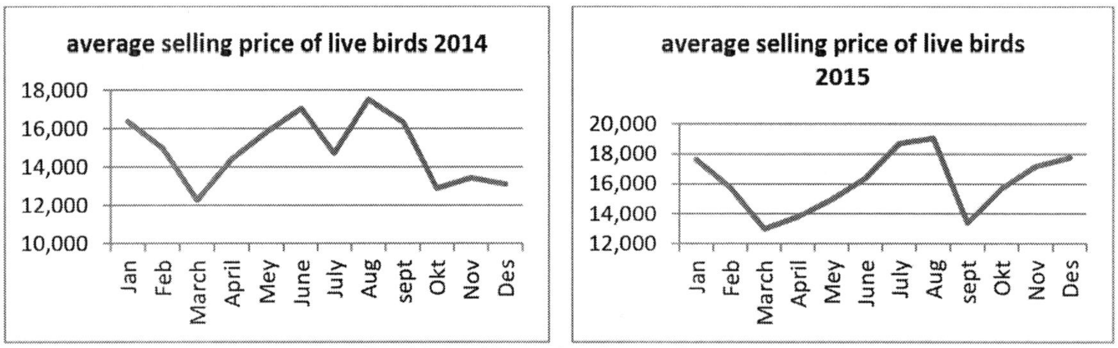

Figure 2. The live bird selling price fluctuation in 2014 and 2015

We make a simulation based on production cost data on table 1, and assumed that all of production component cost are constant, then we adjust the selling price of live bird using the real selling price of live bird on figure 2. After the simulation, some indicator of business risk was identified and shown on table 2. Table 2 shows that the probability to loss on 2014 is 33% and 25% on 2015. Total losses on 2015 also lower compare to total losses on 2014.

Then, the simulation was continued with Monte Carlo method. Before the simulation, we determine the distribution pattern assumption of input variables. These input variables are data revenue and total production cost per month of 2014 to 2015. Already identified that the distribution of data from revenue and production cost are *uniform* distribution. The result of simulation is given on figure 3. It can be seen that the probability of loses was 54.27%. The value quite difference with the simple simulation result (33% on 2014 and 25% on 2015). The forecast value is given on table 3.

Table 2. Simple simulation based on historical data of selling price

Month	Selling price 2014	Revenue	profit/lose 2014	Selling price 2015	Revenue	profit/lose 2015
Jan	16362	95604668	11833405	17610	102895074	19123811
Feb	14974	87491059	3719796	15724	91872761	8101498
March	12263	71652470	(12118793)	12986	75874553	(7896710)
April	14407	84181196	409933	13794	80600491	(3170772)
May	15813	92395026	8623763	14964	87435562	3664299
June	17039	99560831	15789568	16389	95762694	11991431
July	14712	85962241	2190978	18694	109228867	25457604
August	17514	10233218	18560922	19041	111259078	27487815
Sept	16328	95403497	11632234	13376	78153109	(5618154)
Oct	12874	75222594	(8548669)	15682	91628457	7857194
Nov	13431	78476950	(5294313)	17171	100332191	16560928
Dec	13093	76504560	(7266703)	17759	103766864	19995601
Total loss			(33228478)			(16685636)
Total Profit			72760599			140240181
Average profit/period			6063383			11686681
Probability			0.33			0.25

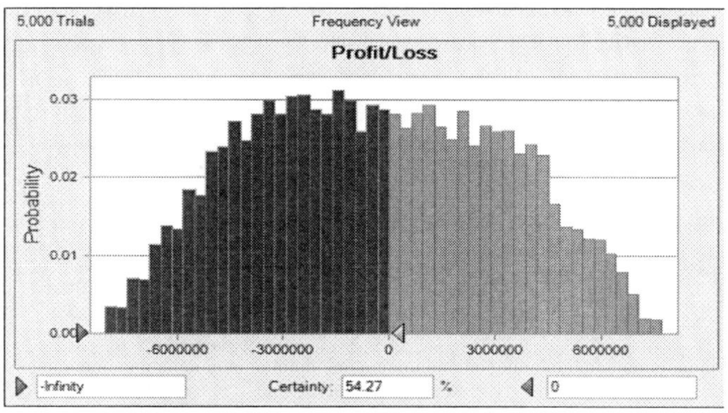

Figure 3. Diagram of lose probability

Table 3. Forecast value

Statistics	Forecast values
Trials	5,000
Mean	-382,98983
Standard	3,565,39050
Variance	12,712,009,450,73
Minimum	-8,061,20532
Maximum	7,726,63771

The simulation result on table 3 shows the minimum range is Rp 8061205 in negative value, it means there will be a possibility of breeder suffering loss at that value. But there is also the possibility of making a profit of Rp 7726637, that is the maximum range. The average amounts of farm will loss of Rp 382989. However, the standard deviation value is quite large because the duplication of data per generate random (5000 data) is only based on 24 data. Simulation does not guarantee optimal results but can guarantee that the results will be close to optimal.

Poultry agribusiness system is a series of activities that unite natural resources, human resources, financial resources and technology resources to process poultry through a biological and industrial process to become a product that can meet human needs and wants [14]. Agribusiness system can be mapped into several subsystems that are: Upstream off-farm subsystem, cultivation subsystem (On-farm), processing subsystem (downstream off-farm), marketing subsystem (downstream off-farm) and supporting services subsystem (supporting institution). The five sub-systems each consist of several supply chain groups that are interconnected and have a role in the poultry agribusiness system [15]. All of sub-system on poultry agribusiness gives influence to market risk. The systemic and holistic study is needed to form a brief understanding on poultry business supply risk management.

Poultry farm face some risk such as biological risk, safety risk for worker and economic risk. Farm has risk profile on introduction of pathogens, subsequent development of disease, and spread of pathogens to other farms. Farm intensive apply in biosecurity measures appropriate to the level of risk of virus incursion. Many research on that field and became the focus of some organization [15]. Similarly, the study on risk of security and safety also has done a lot [16]. On the other hand, the economic risk for poultry farm received less attention. The risk caused by fluctuation on population which hard to controlled.

In developed countries, price stability is well maintained because the information system and population data collection already good enough. Farms in developed countries are farms with large populations. Developed countries in the world such as Australia, the European Union, Canada and the United States have used Livestock Identification and Traceability System (LITS) as Livestock Management System to monitor the national stock of livestock supply. LITS is a livestock management system that has the basic components of animal identification system and the system tracks the existence or movement of animals along the supply chain to the final destination. All supply chain actors, from government to industry players, must work together in implementing and running the system. The LITS system is managed directly by the government and is run by all stakeholders or actors in the livestock industry chain. The Government has the authority to ensure that all stakeholders can comply, support the implementation of the system and help to run the system [17].

Currently the Indonesian government has been analyzing a regulation to stabilize chicken price with minimum price policy or lowest selling price policy of chicken. This policy is intended to keep farmers from losing. But what should be observed further is the nature of broiler products as a live stock. If the breeder is prohibited from selling at a low price, while the selling price is formed by the market mechanism, the farmer must continue to keep the chicken and need more money to feed. Thus, the losses at the farmer level will be greater. If the broiler should be stored as stock frozen meat, cold storage technology is not owned at the farmer level. The lowest price policy can't protect the people's farms. Need more concrete regulatory efforts such as the study of broiler chains distribution due to the high value-added process, occurred in distribution and trading activities. The length of the distribution chain also formed the price of broiler.

5. Conclusion

The margin of production cost per kg of broiler farm population 3000 in one growth period is Rp14338/kilogram of live bird. The biggest percentage on cost structure is feed cost and day old chick cost. The probability of loses by simple simulation based on historical data of live bird selling price is 33% in 2014 and 25% in 2015, categorize at high risk. While, the probability of losses based on Monte Carlo simulation is 54.27%. The average profit per growth period from simple simulation was Rp 6.06 million and from Monte Carlo was Rp 7726637. Consider the finding of the result, can be conclude that the value from Monte Carlo simulation not yet give a good prediction on indicators of broiler farm risk assessment. Market risk can be reduced by price stability, it need good information system on farm population data collection.

References

[1] Muliany, H.P (2015) Agriculture commodity outlook. Sub-sector broiler, agriculture centre of data and information system, secretariat general of agriculture ministry

[2] BPS, 2014, Indonesia National Socio-Economic Survey, 2009-2013, centre of catalogue data micro SUSENAS

[3] USAID. (2013). Indonesia's Poultry Value Chain Costs, Margins, Prices, and Other Issues. Nathan associates Inc., United States Agency for International Development

[4] Leat, P., & Giha, C. R. (2013). Risk and resilience in agri-food supply chains : the case of the ASDA Porklink supply chain in scotland. *Supply Chain Management: An International Journal Vol.18*, 219-231

[5] Best, P (1998) *Implementing Value at Risk* West Sussex: John Wiley & Sons Inc

[6] Butler, C (1999) *Mastering Value at Risk : A Step-by-step Guide to Understanding and Applying VaR* Great Britain: Prentice Hall

[7] Tsai M.C., Lai K.H., Lloyd A.E., Lin H.J. (2012). The dark side of logistics outsourcing-Unravelling the potential risks leading to failed relationships, *Transportation Research Part* E (48), 178-189

[8] Olsson R. (2007). In search of opportunity management: Is the risk management process enough? *International Journal of Project Management* (25), 745–752, 2007

[9] Alexander, C. (2005). The Present and Future of Financial Risk Management, *Journal of Financial Econometrics*, 2005, Vol. 3, No. 1, 3–25

[10] Jüttner, U., Peck, H., Christopher, M. (2003). Supply chain risk management: Outlining an agenda for future research. *International Journal of Logistics: Research & Applications*, Vol. 6, No. 4, 2003, pp 197-210

[11] Tang O., Musa S.W. (2011). Identifying risk issues and research advancements in supply chain risk management, *International Journal of Production Economics* (133), 25–34

[12] Khudori. (2016). Kartel dan Industri Perunggasan (Cartel and poultry industries). *Media Indonesia*, p. 6

[13] Kakiay, T J. (2003) *Pengantar Sistem Simulasi* (Introduction to simulation system) Yogyakarta: Andi Offset

[14] Purwaningsih, R. (2001). Identifikasi Kesesuaian Karakteristik Peternakan Ayam Pedaging Sebagai Plasma Kemitraan Pola Inti – Plasma, *Thesis,* ITB. Bandung

[15] Purwaningsih, R., Arief, M., & Rahmawati, D. (2016). Analisis Rantai Pasok dan Distribusi Ayam Pedaging. SENTI - Seminar Nasional Teknik Industri UGM 2016 (pp. 176-183)

[16] Sims, L.D. (2007). Risks associated with poultry production systems, *poultry in the 21st century,* FAO

[17] HSA. (2006). Farm safety code of practice risk assessment document, Health and Safety Authority

IOP Conf. Series: Materials Science and Engineering **403** (2018) 012045 doi:10.1088/1757-899X/403/1/012045

Solving CCVRPTW using biased random key genetic algorithm (BRKGA) with multiple parent

H Prasetyo[1], M Qoyyiimah[1] and G Fauza[2]

[1] Dept. of Industrial Engineering, Universitas Muhammadiyah Surakarta, Jl. A Yani Tromol Pos I Pabelan Surakarta Indonesia
[2] Dept. of Food Science and Technology, Universitas Sebelas Maret, Jl. Ir. Sutami 36A Kentingan Surakarta Indonesia

hari.prasetyo@ums.ac.id

Abstract. This research addresses a Capacitated Vehicle Routing Problem with Time Windows (CCVRPTW) which is a Vehicle Routing Problem (VRP) with fleet capacity and delivery time constraints. In the research, a Biased Random Key Genetic Algorithm (BRKGA) is proposed to solve the problem. The standard BRKGA is improved by modifying the crossover process accommodating Multiple-Parent. The proposed BRKGA is coded using the Matlab programming language. A case of soft drink distribution is used to evaluate the performance of the algorithm. From the research, it can be concluded that (1) The proposed BRKGA is better than the heuristic method proposed by previous research for dealing with the same case; (2) the Multiple-Parent BRKGA can further improve the general BRKGA; (3) The Multiple-Parent BRKGA with third parent selected randomly from NON-ELITE group results in a better solution than that of selected randomly from the whole population.

1. Introduction

Vehicle Routing Problem (VRP) aims to identify routes using some vehicles for delivering products or services to consumers in different locations [1]. Generally, VRP consists of one depot where one vehicle functions to fulfill the demands or needs of consumers in particular geographical areas [2]. One important aspect is determining delivery routes so that all outlet or consumer demands can be fulfilled with minimal delivery distance or cost. VRP has been commonly implemented, such as for delivering products, waste collection or removal, and newspaper pickup and delivery.

In VRP, vehicles generally depart from a depot and return to the depot commonly known as *Closed Vehicle Routing Problem* [3]. In general, solutions for VRP case do not adequately consider the capacity of vehicles used [4]. The limited capacity of vehicles used to serve consumers is the main factor in VRP case, which is called *Capacitated Vehicle Routing Problem* (CVRP) [5]. Besides capacity limitation, there are other limitations which require all vehicles serving consumers to be in a particular time frame. This means all vehicles serving consumers should not exceed the time specified by consumers or the company. The problem of VRP due to service time limitation is called *Vehicle Routing Problem with Time Windows* (VRPTW) [6]. Reference [7] conducted a study on VRPTW using *Nearest Neighbor* method. However, the method needs much time if the scope of the problem is broad. Further, reference [2] examined CVRPTW in a soft drink company using *Evolution Strategies*.

In practices, limitations of both capacity and time are often simultaneously occurred in VRP also known as *Capacitated Closed Vehicle Routing Problem with Time Windows* (CCVRPTW). The problems of CCVRPTW have been examined by [8] for the distribution of bottled water. This study compared two solution methods, *Branch & Bound* and *Clarke & Wright Savings*. The results of the comparison indicate that *Branch and Bound* method is better in solving the problem of CCVRPTW; however, solutions for the problem are difficult to find when the scope becomes bigger.

Generally, VRP is a problem of *Non-Deterministic Polynomial-Time Hard* (NP-*Hard*). When the number of nodes increases, both the solution space and the computation time will explode exponentially [3]. Research on VRP should consider the quality of the solutions in terms of optimal solution and computation time [9]. Analytic and exact approaches are often difficult to apply due to either impractical to find derivative formula or excessive computation time [10]. Therefore, some researchers use heuristic method as it can generate solutions more quickly, for example [11] who solved the distribution problem of soft drink and [12] who solved the problem of waste disposal. Yet, the solution generated by heuristic method is not optimal, although it requires less time. Thus, many researchers use metaheuristic method to generate optimal solutions in reasonable computation time, such as [13] who solved the VRP using *Ant Colony*. Furthermore, [14] used genetic algorithm to solve CVRP problems.

This study uses metaheuristic method by designing BRKGA through an efficient modification of multiple parents to solve CCVRPTW in the case of soft drink distribution investigated by [11]. The BRKGA with multiple-parent modification uses three parents to produce an offspring. The performance of the proposed BRKGA is then compared to the heuristic procedure which has been used to deal with CCVRPTW in [11]. In addition, a comparison with the standard BRKGA is also conducted to show the performance improvement by the proposed BRKGA.

2. Problem and solution approach

2.1. Problem definition

The problem of soft drink distribution investigated by [11] is addressed in this research. The distribution company serves the demand of 45 outlets geographically dispersed consisting of supermarkets, minimarkets, canteens, cooperatives, business entities, and restaurants. A truck with limited capacity is used to serve all outlets for 7 days in a week. The demand per week of all outlets is known and constant. The truck must depart from the depot and return to the depot within the working hours prescribed by the company. Thus the problem is to find the best set of sub-routes passed by the truck for one week. The data required for the problem includes truck capacity, demand of each outlet, service time of each outlet, setup time of the vehicles, company's working hours, and transportation time and cost matrices among depot and outlets.

2.2. Solution approach

The proposed solution for dealing with CCVRPTW is Biased Random Key Genetic Algorithm (BRKGA) which has been extensively used for other combinatorial optimization problems [15] due to its flexibility in representing various problems [16]. This algorithm consists of initial p individual population each of which is represented by x-vector length randomly generated from 0 to 1 [3]. Each chromosome/individual comprises a set of genes representing outlets (as seen in figure 1).

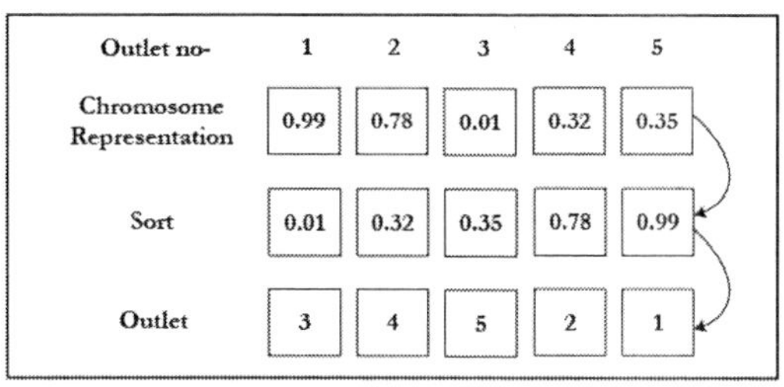

Figure 1. Coding process.

IOP Conf. Series: Materials Science and Engineering **403** (2018) 012045 doi:10.1088/1757-899X/403/1/012045

Figure 1 illustrates the encoding process of a chromosome comprising 5 genes. Genes 1-2-3-4-5 have their corresponding gene values which are 0.99-0.78-0.01-0.32-0.35. By sorting the gene values in ascending order, one has a sub-route of 3-4-5-2-1. Note that for this study this sub-route must meet the truck capacity and distribution time constraints.

In BRKGA, as explained by [17], and [3], the p initial population is divided into two groups; small group called ELITE (pe) individuals approximately 10-20% of most fit solutions, and NON-ELITE individuals with the rest of individual from p-pe. This population evolves to obtain the next generation (figure 2) until stopping criteria is met. The ELITE individual will be copied to the next generation, called TOP chromosomes, while pm individuals are randomly generated and placed in the next generation, also called BOT chromosomes. The number of population in the next generation is kept to be p chromosomes. Thus, the remaining members of the next generation (p-(pe+pm)) are filled by offspring obtained from mating between ELITE individual and NON-ELITE individual. This crossover process is conducted by selecting multiple parents, one ELITE chromosome and two NON-ELITE chromosomes and combining them to produce one offspring. The crossover process is illustrated in figure 3. Chromosome 1 refers to ELITE and chromosomes 2 & 3 to NON-ELITE. In this example the probability of allele inheritance is 0.5 for chromosome 1, 0.25 for chromosome 2 and 0.25 for chromosome 3 to produce offspring chromosome.

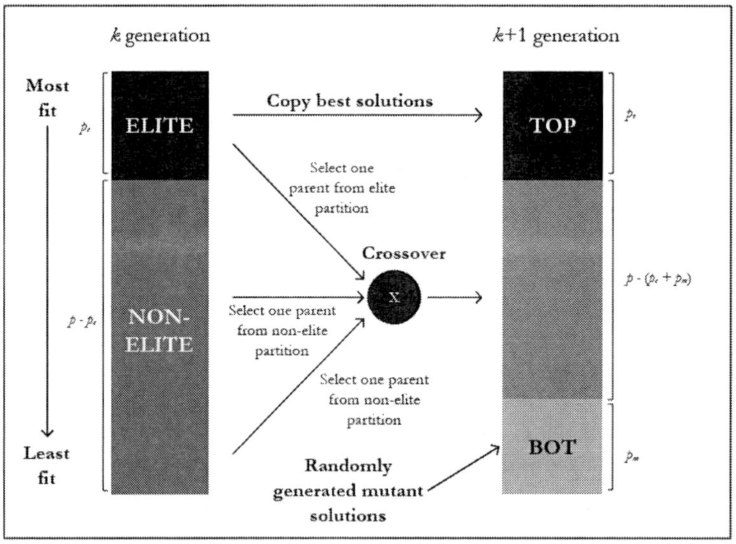

Figure 2. New generation transition.

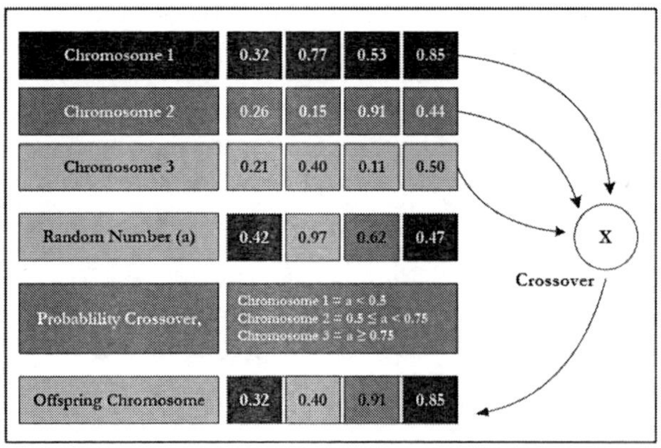

Figure 3. Crossover process.

The modification of crossover process using more than two parents or called multiple-parent has previously been done by [18], [19], and [20]. In BRKGA, there are some parameters which need to be set in order to obtain the best solutions. Based on research conducted by [16], and a survey done by [15] on the implementation of BRKGA in various fields, the recommended parameter setting of BRKGA is presented in table 1.

Table 1. BRKGA parameters [15].

Parameter (a)	Description (b)	Recommended value (c)
p	size of population	$p = ax$, where $1 \leq a \in \mathbb{R}$ is a constant and x is the length of the chromosome
p_e	size of elite population	$0.10p \leq p_e \leq 0.25p$
p_m	size of mutant population	$0.10p \leq p_m \leq 0.30p$
ρ_e	elite allele inheritance probability	$0.5 \leq \rho_e \leq 0.8^*$

*for two-parent crossover

3. Results and discussion

3.1. BRKGA parameter setting

BRKGA algorithm is coded in Matlab version 7.11.0.584 (R2010b) 64-bit (win64) and run using a notebook with the following specifications: processor Intel® Core™ i5-2450M, processor speed 2.50 GHz and RAM 4GB. The program is run to obtain a parameter with solutions which generate the lowest average cost. The best parameter was obtained through 12 parameter combinations, consisting of number of population, ELITE percentage, and mutation percentage. The parameter setting is evaluated based on the survey conducted by [15]. The number of population is 45, the elite percentages are 0.1, 0.15, and 0.25, and the percentages of mutation are 0.1, 0.2, and 0.3. From each combination, 50 samples are collected; to identify the best parameters, each sample is iterated 700 times. Based on the experiment, the best parameter combination is population size 45, elite percentage 0.25, and mutation percentage 0.1.

Figure 4 shows how the BRKGA with the selected parameter is able to converge to the optimal solution faster and better in comparison with another set of parameter. It reaches the convergence point at iteration 2000. This result demonstrates that the selection or setting of appropriate parameters plays an important role in implementing the proposed BRKGA.

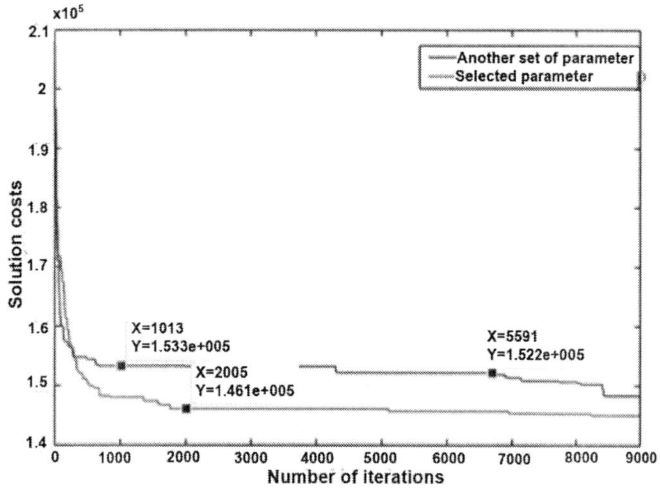

Figure 4. Convergence point of BRKGA.

3.2. Performance of multiple-parent BRKGA and standard BRKGA

The performance of multiple-parent BRKGA designed in this research is evaluated by comparing this algorithm with the standard BRKGA. The aim of the comparison is to indicate that the modification of multiple-parent BRKGA is able to improve the results of standard BRKGA in terms of quality of generated solution. Based on the application of BRKGA algorithm conducted by [15], the average allele percentage or crossover probability often used in previous research is 0.7. Therefore, standard BRKGA in this study uses crossover probability of 0.7, meaning that descendant opportunity of ELITE chromosome is 70%.

In this study, the modification of multiple-parent BRKGA algorithm is done not only by selecting the third parent from NON-ELITE group, but also from the whole population (both from ELITE and NON-ELITE groups). The total costs generated by multiple-parent BRKGA and standard BRKGA are presented in table 2. As table 2 column b indicates, the cost generated by multiple-parent BRKGA using the third parent from NON-ELITE group is the best solution cost, that is IDR 144,948. This BRKGA is better than the standard BRKGA with a lower cost, which is IDR 11,672. The BRKGA using the third parent from the whole population is not better than the NON-ELITE; however, it can improve the standard BRKGA with a cost difference of IDR 5,710 as shown in table 3 column c. Therefore, with insignificant time difference, the use of multiple-parent in the crossover process is able to improve the standard BRKGA. However, in the research, the selection of third parent from NON-ELITE group results in a lower cost than that of from the whole population.

Table 2. Cost of BRKGA results.

Methods (a)	Cost (IDR/week) (b)	Gap with standard BRKGA (IDR/week) (c)
General BRKGA	156620	-
Mutiple-Parent BRKGA with all population	150910	5710
Mutiple-Parent BRKGA with NON-ELITE	144948	11672

The performance comparison of standard BRKGA and BRKGA with multiple-parent (MP BRKGA) is presented in figure 5.

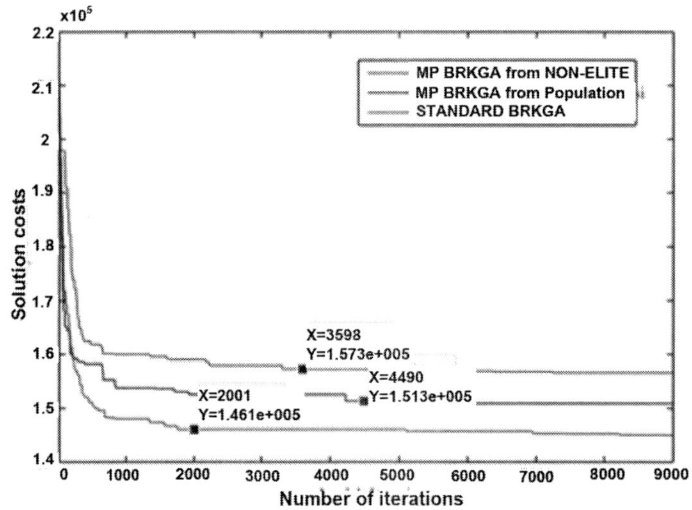

Figure 5. Convergence points of BRKGA and *Multiple-Parent* BRKGA.

Figure 5 indicates the number of iterations for obtaining solutions toward convergence. Multiple-parent BRKGA with the selection of the third parent from NON-ELITE group is faster in generating solutions toward convergence, that is at iteration 2000, followed by standard BRKGA at point 3500 and multiple-parent BRKGA with the selection of third parent from the whole population at point 4400. After many iterations, the three algorithms still experience the decrease of solution cost. However, the cost decrease is not significant compared to the number of iterations needed. Although standard BRKGA reaches convergence faster than the multiple-parent BRKGA using parent from the whole population, multiple-parent BRKGA is better in generating low-cost solutions.

3.3. Performance of the proposed BRKGA and Heuristic

Vehicle routing problem in a soft drink company in Medan has previously been solved by [11] using heuristic method comprising two phases: divide and conqueror. The problems of CCVRPTW in the soft drink company should be solved so that consumer demands can be fulfilled effectively with low distribution cost. Heuristic method used by [11] can decrease the distribution cost of the company.

In this study, multiple-parent BRKGA algorithm for solving CCVRPTW problems is designed and evaluated using CCVRPTW case which has been investigated by [11]. The results indicate that multiple-parent BRKGA is able to generate low-cost solutions. The costs generated by the heuristic method and multiple-parent BRKGA can be seen in table 3. As table 3 indicates, the solution obtained using multiple-parent BRKGA is able to decrease the distribution cost, that is IDR 91,552 or 38.7%. The main advantage of the implementation of multiple-parent BRKGA in this case is that it can improve service quality, which is the fulfillment of consumer demand without exceeding working hours with minimal cost.

Table 3. Cost of heuristic method and multiple-parent BRKGA.

Method (a)	Cost (IDR/week) (b)	Gap (IDR/week) (c)
Heuristic	236500	-
Multiple-Parent BRKGA	144498	91552

4. Conclusions and recommendations

This research has produced the design of an efficient multiple-parent BRKGA algorithm program for solving optimization problems in the distribution of soft drink in Medan. The proposed BRKGA has identified subroutes set reducing distribution cost for fulfilling the demand of outlets. The set of subroutes was determined by taking into account the working hours and capacity of vehicles used, which become constraints in the CCVRPTW of the soft drink company. Not only does multiple-parent BRKGA provide better solutions than those generated by heuristic method (18.7% saving), it also improves the results of standard BRKGA. Multiple-parent BRKGA generates solution result of 7.4% better than the standard BRKGA. The designed multiple-parent BRKGA program is easy to apply to various CCVRPTW problems with similar characteristics. It can also be adapted for solving other distribution problems by modifying its obstacles and objective function. This study has some limitations. The demand is deterministic; thus, probabilistic cases should be investigated. Also, further research using more than one type of vehicles should be conducted.

References

[1] Gunawan G, Maryati I and Wibowo H K 2012 Optimasi penentuan rute kendaraan pada sistem distribusi barang dengan ant colony optimization *Semantik* vol 2 no 1
[2] Harun I A, Mahmudy W F and Yudistira N 2014 Implementasi evolution strategies untuk penyelesaian vehicle routing problem with time windows pada distribusi minuman soda doro *J. PTIIK Universitas Brawijaya* vol 4 no 1
[3] Grasas A, Ramalhinho H, Pessoa L S, Resende, M G C, Caballé I and Barba N 2014 On the improvement of blood sample collection at clinical laboratories *BMC Health Services Research* vol 14 no 1

[4] Arvianto A, Setiawan A H and Saptadi S 2014 Model vehicle routing problem dengan karakteristik rute majemuk, multiple time windows, multiple products dan heterogeneous fleet untuk depot tunggal *J. Teknik Industri* vol 16 pp 83–94.

[5] Awansari S A and Abusini S 2013 Implementasi model capacitated vehicle routing problem pada pengiriman pupuk urea bersubsidi *J. Mahasiswa Matematika* vol 5 p 372

[6] Nugraha D C A and Mahmudy W F 2015 Optimasi vehicle routing problem with time windows pada distribusi katering menggunakan algoritma genetika *SESINDO*

[7] Amri M, Rahman A and Yuniarti R 2014 Penyelesaian vehicle routing problem dengan menggunakan metode nearest neighbor; studi kasus di MTP Nganjuk Distributor PT. Coca cola *J. Rekayasa Dan Manajemen Sistem Industri* vol 2 pp 36–45

[8] Nurhayati S 2013 Perbandingan metode branch and bound dengan metode clarke and wright savings untuk penyelesaian masalah distribusi aqua galon di PT. Tirta Investama Yogyakarta *UNY*

[9] Gonçalves J F and Resende M G C 2011 A biased random-key genetic algorithm for job-shop scheduling *AT&T Labs Res. Tech. Report* vol 46 pp 253–71

[10] Gonçalves J F and Resende M G C 2011 A parallel multi-population genetic algorithm for a constrained two-dimensional orthogonal packing problem *J. of Comb. Opt.* vol 22 pp 180-201

[11] Sembiring C A 2008 *Penentuan rute distribusi produk yang optimal dengan menggunakan algoritma heuristik pada pt. coca cola bottling indonesia medan* Medan

[12] Al'Aziz M R, Utomo B and Nurhadi K 2013 Analisis sistem pengangkutan sampah kota surakarta dengan metode penyelesaian vehicle routing problem (VRP) *J. Matriks Teknik Sipil* vol 1

[13] Sari I N and Widodo A 2014 Penentuan rute yang optimal pada distribusi kacang menggunakan ant colony system: studi kasus di PT Qlauworks Indonesia *J. Mahasiswa Matematika* vol 2 pp 271

[14] Shahab M L and Irawan M I 2016 Algoritma genetika ganda untuk capacitated vehicle routing problem *J. Sains dan Seni ITS* vol 4 no 2

[15] Prasetyo H, Fauza G, Amer Y and Lee S H 2015 Survey on applications of biased-random key genetic algorithms for solving optimization problems *IEEE International Conference on Industrial Engineering and Engineering Management* (IEEM) pp 863–70

[16] Gonçalves J F and Resende M G C 2011 Biased random-key genetic algorithms for combinatorial optimization *J. of Heuristics* vol 17 pp 487–525

[17] Gonçalves J F and Resende M G C 2015 A biased random-key genetic algorithm for the unequal area facility layout problem *Euro. J. of Op. Res.* vol 246 pp 86–107

[18] Gonçalves J F and Resende M G C 2015 A parallel multi-population biased random-key genetic algorithm for a container loading problem *Comp. & Op. Res.* vol 39 pp 179–90

[19] Fontes D B M M and Gonçalves J F 2013 A multi-population hybrid biased random key genetic algorithm for hop-constrained trees in nonlinear cost flow networks *Opt. Letters* vol **7** pp 1303-24

[20] Lucena M L, Andrade C E, Resende M G C and Miyazawa F K Some extensions of biased random-key genetic algorithms 2014 *In Simposio Brasileiro de Pesquisa Operacional* XLVI

IOP Conf. Series: Materials Science and Engineering **403** (2018) 012046 doi:10.1088/1757-899X/403/1/012046

Key performance indicators (KPIs) on Vannamei shrimp supply chain performance (a preliminary research)

D Widyaningrum

Department of Industrial Engineering, Muhammadiyah Gresik University, Gresik, Indonesia

dzakiyah@umg.ac.id

Abstract. Higher competition makes the company should be able to fulfill consumer needs quickly, precisely and low cost. It can achieve by designing good supply chain. To know supply chain's success, we must measure its performance with proper KPIs (Key Performance Indicators). But, to get proper KPIs is not easy because of each supply chain has unique characteristics, one of them is vannamei shrimp supply chain. Vannamei shrimp is perishable products, seasonal in production, and highly dependence on nature. This preliminary work will propose a set of proper KPIs by considering its uniqueness to measuring its performance. This work done by mapping vannamei shrimp supply chain network first, then determines the KPIs from six literatures. The stakeholders mostly consist of small and medium enterprises (SMEs). We get 34 KPIs from this study, which are 31 KPIs can be use in general supply chain and three special KPIs which only use for vannamei shrimp supply chain. Furthermore, this research will be done to complete the KPIs by considering stakeholders suggestions, weighting the KPIs by AHP and develop vannamei shrimp supply chain performance measurement system (PMS). The PMS will be used to measure vannamei shrimp supply chain and compared with another PMS.

1. Introduction

Indonesia is the second largest producer of aquaculture fish [1]. According to [2], the aquaculture fish production reaches almost 2.5 million tons since 2015, its increase almost three times over the last ten years. The fish supply will rely on aqua culture because the supply of marine fish is reduced due to over fishing [3] so the aquaculture fish production must be increased in line with the demand in fish consumption.

Aquaculture fish in Indonesia is dominated by shrimp, i.e. prawns and vannamei [3]. Vannamei shrimp more desirable than prawns. It makes the production in Indonesia is growing rapidly. In 2013, vannamei shrimp production is 390.278 ton, and increased to 411.729 ton in 2014 [3].

But in the other side, as the second largest shrimp producer after China [1], in numbers, Indonesia is only 50% of China. If the vannamei shrimp is not properly considered, then Indonesia will be more left behind than China, and the worst thing is Indonesia's position may go down.

The current competition is a global competition. Its competition is not only between company, but also between supply chain networks. Companies are required to meet consumer demand accurately, quickly and cheaply [4]. To get success in globally market we must build the right supply chain network [5, 6]. It should be designing appropriately [7] because it is most crucial planning problems in supply chain management [8]. To assess the success of supply chain can be done by measuring its performance with the proper instrument [9]. To get the right performance measurement results, we must use proper tool in accordance with the supply chain characteristics [10].

There are several studies to establish instrument on supply chain (examples are [11, 12]. Besides, we can use SCOR (Supply Chain Operation Reference) to establish supply chain KPIs based on its process, i.e. Plan, Source, Make, Deliver and Return. Balanced Scorecard (BSC) also can be used for measuring supply chain performance [13]. Another method is DEMATEL (Decision Making Trial and Evaluation Laboratory) [14] and combine DEMATEL and AHP (Analytical Hierarchy Process) [15]. Dynamics system model used for choosing KPIs which are influence on performance measurement significantly [10].

Those studies did not consider supply chain characteristic, whereas every supply chain has unique characteristic and should be measure using proper indicators based on its characteristics. Example is vannamei shrimp supply chain. There was no supply chain performance measurement tool in accordance with the characteristics of vannamei shrimp.

Vannamei shrimp is perishable product. It makes vannamei shrimp has special handling. Vannamei shrimp also high dependence on climate and season, high dependence on nature, especially for aquaculture vannamei shrimp. This condition makes gap on production quantity. Sometimes high in production, but sometimes low in production. Such unique characteristics should be considered, when designing performance measurement system, one of which is indicator determination.

This study aims to develop vannamei shrimp supply chain performance measurement system (instrument) based on its characteristics. Designing the proper instrument can be done by determining its KPIs first. The biggest aquaculture fish producer in Indonesia is East Java, i.e. Banyuwangi and Gresik [2]. Because Banyuwangi is far, so this study held in Gresik.

By using this KPIs, it is expected that the performance of aquaculture fish supply chain can be measured properly. Then, it can be known KPIs or things that need to be prioritized for improvement. With the improvement of the supply chain according to the priority, it is expected supply chain will be built better and can meet the needs of aquaculture fish consumers. In long-term use, hopefully it can make Indonesia is the largest aquaculture fish producer in the world and big on its export number.

2. Literature review

One of key elements in effective supply chain management is performance measurement [16]. The purpose of performance measurement is to evaluate and to determine the best strategy to improve the supply chain [4]. To make performance measurement system, we must determine its KPIs first.

There are some previous studies, either determining the KPIs alone or creating a model or framework of performance measurement. Current models are Balanced Scorecard (BSC), hybrid (DEMATEL - Decision Making Trial and Evaluation Laboratory and AHP – Analytical Hierarchy Process), dynamics system model, and Supply Chain Operation Reference (SCOR).

BSC measures supply chain performance by involving four perspectives. They are financial perspective, customer perspective, business process perspective and learning and growth perspective. It is balance because this framework involves financial and non-financial aspects [17]. BSC used by approximately 59% researchers [13]. But, BSC still has shortages, they are difficult to use and need many data.

Another researcher using DEMATEL [14] or combinations between DEMATEL and AHP methods [15]. Relationship between KPIs is described in this research using DEMATEL. While AHP is used to integrate the indicator of the entire supply chain. This is mathematical model, so it needs more understanding.

Research for the development of KPIs has also been done with dynamic systems [10]. System dynamic used to describe the relationship between the KPIs. But, if we want to use this model we must know and define the relationship between KPIs well.

SCOR was developed in 1996 by the Supply Chain Council. This framework is hierarchical with KPIs (Key Performance Indicators) on each level [18], SCOR measures performance based on process-oriented of the supply chain, i.e. Plan, Source, Make, Deliver and Return. This framework involves the process from start to finish, so it does not provide specific results, but by building models based on supply chain processes, it is expected to cover all processes in the supply chain from upstream to downstream without any process being missed. It needs additional tool to measure its KPIs [13]. SCOR combining business process reengineering (BPR), performance measurement, and logistics management to establish the supply chain performance measurement framework [19].

Previously built models, including SCOR, have not considered the characteristics of the vannamei shrimp supply chain. To apply the supply chain performance measurement model should be adjusted to the characteristics of the supply chain. Vannamei shrimp characteristic is different from other supply chain. So far, no one has done research on supply chain performance with special characteristics such as vannamei shrimp [4] have developed a framework for measuring the performance of sea fishery supply chain. Sea fishery and vannamei shrimp have similar characteristics. They are having differences, such as the season in the vannamei shrimp influenced by rainy and dry season, while the sea fishery is influenced by wind, and storm. In addition, vannamei shrimp are cultivated, so can be more controlled whereas sea fishery is not. The location of the two were different, shrimp vannamei in the pond while the sea fish in the ocean. Therefore, this study will be based on sea fishery supply chain performance measurement research [4].

Framework is built based on [9] framework and SCOR framework. The KPIs in [4] study is collected from several literatures and stakeholders. Metric set is done by researchers based on agriculture food supply chain framework [9], and verified by stakeholders (especially Department of Marine and Fisheries). Each indicator tested by the validity and reliability. The framework is already used to measure its supply chain in Sadeng. Gunung Kidul, Yogyakarta.

3. Methodology

This preliminary study focuses on determine KPIs on vannamei shrimp supply chain. This study conducted by several steps. First step is mapping the vannamei shrimp supply chain network to know its stakeholders and business processes. The vannamei shrimp supply chain to be mapped is located in Cerme, Gresik, East Java. Second step is collecting KPIs from literatures. This study use sea fishery supply chain performance measurement system [4] as a primary literature because of its similarity. Another literature also used in this research [5, 6, 14, 16, 19]. KPIs from literatures then sorted. The unsuitable KPIs are eliminated. Then, if there are similar KPIs with different name, they will be merged or combined.

4. Results and discussion

4.1. Vannamei shrimp supply chain network

There are two types of supply chain network (figure 1), namely traditional and intensive. Intensive supply chain networks are medium and large scales networks, while traditional networks are small and medium scale.

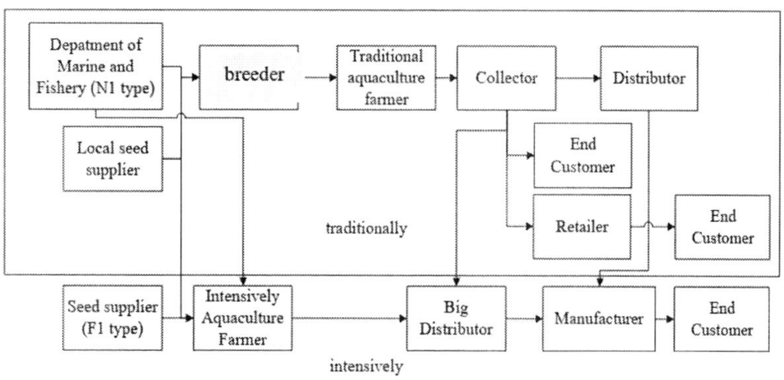

Figure 1. Vannamei shrimp supply chain network.

Intensive vannamei shrimp supply chain produce high quality vannamei shrimp. Its seeds are high quality product from outside Gresik. Intensively aquaculture farmer breeds the seeds in intensive aquaculture with high level of control. They try to control many dangerous factors so that they will get high quality product. It farmers have big distributors as their customers. This big distributor will supply vannamei shrimp to manufacturers. If intensive is cultivated with high control, then traditional is cultivated in open ponds with less control.

The upstream of traditional vannamei shrimp supply chain are local breeders and Fisheries and Marines Department. Then seed nurseries, aquaculture farmers, collectors, distributors, end customers, and manufacturer. The samples are seed nursery, aquaculture farmers, small distributor, big distributor, and end customers.

Fisheries and Marine Department is a provider of N1 (the best vannamei shrimp seed) vannamei shrimp seed. There are also local seed come from Tuban and surroundings in East Java. Both seeds are very small. It will usually die, if it directly cultivated by aquaculture farmers. It needs treatment by breeder. Breeder will take care of the seeds for 7 – 12 days. The treatment is intended to give time to the vannamei shrimp seeds to adapt from sea water to fresh water. During the treatment, the water in the seed pool that was originally the sea water was reduced the salt level slowly to become fresh water. The best time treatment duration 12 days. But the seeds can be removed within 7 days, if there is an urgent request from farmers.

Farmers are vannamei shrimp producers. Farmers buy seeds from breeder. Seeds are not directly spread to pond, but quarantined first for one until two nights. It aims to minimize the risk of vannamei shrimp death. Quarantine is done by placing a net on the edge of the pond and then the seeds are inserted in the net.

The vannamei shrimp treatment does not cost a lot. For example, vannamei shrimp does not need to be fed, the vannamei shrimp will eat plankton in the pond. However, it takes patience in shrimp vannamei farming. For example, once or twice a week, pond should be checked. Pond checking is done by diving to ensure the condition of the shrimp, because the die shrimp will drown. If the diseased and dead shrimp are left unchecked, it will infect the other shrimp and may result in crop failure or non-optimal production. Another way that can be done to minimize the risk is to cultivate other fishes in the same pond, so they can eat the dead shrimp.

Another problem is weather. The best time for vannamei shrimp farming is when the rainy season because vannamei shrimp live in the cold waters. However, if the rainy season, the water in the pond will be high and flood or overflow. This can make crop failure, because the vannamei shrimp will be lost in the flood. Meanwhile, if the dry season arrives, the water will recede, so that the hot sun will reach the bottom of the pond. The water temperature gets too hot for the vannamei shrimp. The heat can be slightly reduced by sowing inorganic urea fertilizers, but if it too much will cause pH levels become acidic and vannamei shrimp may die. Typically, pond farmers will avoid vannamei shrimp farming in the dry season, so the pond will be used for other fish cultivation.

The best time to harvest vannamei shrimp is three months for a high selling price. Vannamei shrimp are harvested by netting, not by emptying the water, so that shrimp that are not ready to be harvested can be released again.

Once harvested, the shrimp will be sold to collectors. Shrimp collected from pond farmers, will be sent to distributor, then distributor will distribute it to the consumer (big class consumers/companies). While household consumers or stalls and traders in the market usually buy shrimp from collector.

Stakeholders in this study are small and medium enterprises from vannamei shrimp supply chain network. Object of this research is supply chain of small and medium enterprises or traditional on vannamei shrimp supply chain in Gresik, East Java, Indonesia (figure 1 grey area).

4.2. Collecting KPIs from literatures

We got 48 KPIs from six literatures. Table 1 shows the list of KPIs. There are unsuitable KPIs to be used in vannamei shrimp supply chain, also similar KPIs in different name. So the KPIs would be sorted. List of sorted KPIs shown in table 2.

Table 1. List of KPIs from literatures.

No	KPIs	Author(s)					
		D. Widyaningrum and N.A. Masruroh	Lee, Tzong-Ru, Shiu Yi Siang, P. Sivakumar	R. Tarasewicz	F. Uysal	I. B. Bukhori, K. H. Widodo, D. Ismoyowati	M. A. Wibowo, M. N. Sholeha
1	Total Cost	√	√		√	√	√
2	Asset Turn	√					
3	Profit	√					
4	Inventory Levels	√					√
5	Return on Investment (ROI)	√					
6	Cash-to-cash Cycle	√		√			√
7	Delivery Flexibility	√				√	
8	Customer satisfaction	√					√
9	Productivity value-added/ innovations	√			√		
10	Volume flexibility	√					
11	Product Lateness	√					
12	Fill rate	√					
13	Lead time	√				√	√
14	Appearance	√					
15	Shelf life	√					
16	Salubrity	√					
17	Energy use	√			√		
18	The Number of Employee	√			√		
19	Computerized system	√			√		
20	Equipment	√					
21	Ice availibility	√					
22	Cold storage	√					
23	Training	√			√		
24	Organization	√					
25	Accuracy season forecast	√					
26	Price stability	√					
27	total sales				√		
28	Turnover rate of Employee				√		
29	forecast accuracy					√	
30	Customer complaints					√	
31	Delivery Cycle Time		√			√	
32	Perfect order fulfilment		√			√	√
33	Product cycle time					√	
34	Production flexibility					√	√
35	Delivery order fulfilment					√	
36	Transportation cost		√			√	
37	Supply chain management cost		√				√
38	Inventory days of supply		√				
39	Quality						√
40	Customer Satisfaction						√
41	Cost						√
42	Environmental cost						√
43	Business waste						√
44	order fulfilment cycle time		√				
45	Source cycle time		√				
46	make cycle time		√				
47	Dock	√					
48	Fish Auction	√					

Table 2. List of sorted KPIs.

No	KPIs	Author(s)					
		D. Widyaningrum and N.A. Masruroh	Lee, Tzong-Ru, Shiu Yi Siang, P. Sivakumar	R. Tarasewicz	F. Uysal	I. B. Bukhori, K. H. Widodo, D. Ismoyowati	M. A. Wibowo, M. N. Sholeha
1	Asset Turn	√					
2	Profit (Total Sales - Total Cost)	√	√		√	√	√
3	Inventory Levels	√					√
4	Return on Investment (ROI)	√					
5	Cash-to-cash Cycle	√		√			√
6	Delivery Flexibility	√				√	
7	Customer satisfaction/ complaints	√				√	√
8	Productivity value-added/ innovations	√			√		
9	Volume flexibility	√					
10	Product Lateness	√					
11	Fill rate	√					
12	Lead time	√				√	√
13	Energy use	√			√		
14	The Number of Employee	√			√		
15	Equipment	√					
16	Ice availibility	√					
17	Cold storage	√					
18	Training	√			√		
19	Organization	√					
20	Season forecast Accuracy	√					
21	Price stability	√					
22	Turnover rate of Employee				√		
23	forecast accuracy					√	
24	Delivery Cycle Time		√			√	
25	Perfect order fulfilment		√			√	√
26	Product cycle time					√	
27	Production flexibility					√	√
28	Delivery order fulfilment					√	
29	Inventory days of supply		√				
30	Quality	√					√
31	Business waste						√
32	order fulfilment cycle time		√				
33	Source cycle time		√				
34	make cycle time		√				

After being shorted, the number of KPIs is only 34 items. Three KPIs are eliminated, they are dock, fish auction, and computerized system. Dock is not suitable for vannamei shrimp supply chain. Dock is used for marine fishery products or offshore ponds, while vannamei shrimp is aquaculture product. Vannamei shrimp does not need fish auction because it sold directly to distributors without auction process. Computerized system is also eliminated because object of this research is small and medium vannamei shrimp enterprises, where they have not yet used a computer system. If this KPI persists, its performance score is bad and will affect the overall performance result.

Sorting process is also done by combines similar KPIs. KPI related to cost (cost, environmental cost, supply chain management cost, and transportation cost), merged to be total cost. Then, total sales and total cost joined with profit. They are used to count profit (no. 2 on table 2). Customer satisfaction and customer complaints are combined (no. 7 on table 2). Users can choose which one is easier to use.

Productivity value-added is similar with innovations, so it can be merged. Appearance, shelf life and salubrity are metric to show quality.

There are three special KPIs from 34 KPIs. They are ice availability, cold storage, and season forecast accuracy. Ice availability and cold storage are used for preserve vannamei shrimp. Vannamei shrimp is perishable product which needs special material handling (kept in frozen state to keep its shelf life from expired quickly). While season forecast accuracy is needed to maintain best time to spread the seeds.

5. Conclusions

This is a preliminary research. From this research, there are 34 KPIs suitable for small and medium enterprise vannamei shrimp supply chain collected from six literatures. Some of 31items are general KPIs that can be used in general supply chain. They are Asset Turn, Profit (Total Sales - Total Cost), Inventory Levels, Return on Investment (ROI), Cash-to-cash Cycle, Delivery Flexibility, Customer satisfaction/ complaints, Productivity value-added/ innovations, Volume flexibility, Product Lateness, Fill rate, Lead time, Energy use, The Number of Employee, Equipment, Training, Organization, Price stability, Turnover rate of Employee, forecast accuracy, Delivery Cycle Time, Perfect order fulfilment, Product cycle time, Production flexibility, Delivery order fulfilment, Inventory days of supply, Quality, Business waste, order fulfilment cycle time, Source cycle time, make cycle time. Another three KPIs are special used on vannamei shrimp supply chain. They are ice availability, cold storage, and season forecast accuracy. In further research, more in-depth research can be done to determine the KPIs on the vannamei shrimp supply chain and establish a performance measurement system. First step is giving explanation at each KPIs including how to measure it. Then, stakeholders are asked to propose KPIs by interview method and questionnaire. AHP will be used for weighting the KPIs. The results of the measurement of supply chain performance with the lowest score and the highest weight is the priority to be improved. It would be better if those KPIs are tested to measure vannamei shrimp supply chain. Then compared it with other measuring instruments. For long-term research, it can be done for general aquaculture supply chain not just vannamei shrimp.

References

[1] Shrimpnews 2015 *The World Estimated Farmed Shrimp Production in 2013 and 2014* [Online]. Available: shrimpnews.com/FreeReportsFolder/NewsReportFolder/WorldProductionIn2013And2014.html [Accessed 11 June 2017].

[2] Badan Pusat Statistik 2016 *Statistik Indonesia 2016* [Online] [Accessed 11 June 2017].

[3] Ministry of Marine Affairs and Fisheries Republic of Indonesia 2015 Ministry of Marine Affairs and Fisheries Republic of Indonesia [Online]. Available: http://www.djpb.kkp.go.id. [Accessed 11 June 2017]

[4] Widyaningrum D and Masruroh N A 2012 Development of the Sea Fishery Supply Chain Performance Measurement System: A Case Study *International Journal of Supply Chain Management (IJSCM)* vol 1 No 3 pp 20-32

[5] Tarasewicz R 2016 Integrated approach to supply chain performance measurement – results of the study on Polish market *Transportation Research Procedia* vol 14 pp 1433–42

[6] Wibowo M A and Sholeh M N 2015 The Analysis of Supply Chain Performance Measurement at Construction Project *Procedia Engineering* vol 125 pp 25–31

[7] Butzera S, Schötzb S, Petroschkea M and Steinhilper R 2017 Development of a Performance Measurement System for International Reverse Supply Chains *Procedia CIRP* 61 pp 251–6

[8] Chopra S and Meindl P 2007 *Supply Chain Management Strategy, Planning and Operation, 3rd ed.* (Upper Saddle River: NJ) pp 38-59

[9] Aramyan L H, Lansink A G J M O, Vorst J G A J and Kooten O 2007 Performance Measurement in Agri-food Supply Chain: A Case Study *Supply Chain Manangement: An International Journal* vol 12 no 4 pp 304-15

[10] Jamil C M and Mohamed R 2011 Performance Measurement System (PMS) In Small Medium

Enterprises (SMEs): A Practical Modified Framework *World Journal of Social Sciences* Vol 1 No 3 pp 200-12

[11] Fauske H, Kollberg M, Dreyer H C and Bolseth S 2006 Criteria for Supply Chain Performance Measurement Systems *14th International Annual EurOMA conference* Ankara, Turkey

[12] Cai J, Liu X, Xiao Z and Liu J 2009 Improving Supply Chain Performance Management: A Systematic Approach to Analyzing Iterative KPI Accomplishment *Decision Support Systems* Vol 46 No 2 pp 512 –21

[13] Neely 2005 The Evolution of Performance Measurement Research: Developments in the Last Decade and a Research Agenda for the Next *International Journal of Operations & Production Management* Vol 25 No 12 pp 1264-77

[14] Uysal F 2012 An integrated model for sustainable performance measurement in supply chain *Procedia - Social and Behavioral Sciences* vol 62 pp 689-94

[15] Najmi and Makui A 2010 Providing Hierarchical Approach for Measuring Supply Chain Performance Using AHP and DEMATEL Methodologies *International Journal of Industrial Engineering Computations* Vol 1 No 2 pp 199-212

[16] Lee, Tzong-Ru, Shiu Yi-Siang and Sivakumar P 2012 The Application SCOR in Manufacturing: Two Cases in Taiwan *Procedia Engineering* vol 38 pp 2548-63

[17] Kurien G P and Qureshi M N 2011 Study of Performance Measurement Practices in Supply Chain Management *International Journal of Business, Management and Social Sciences* Vol 2 No 4 pp 19-34

[18] Delipinar G E and Kocaoglu B 2016 Using SCOR model to gain competitive advantage: A Literature Review *Procedia - Social and Behavioral Sciences* vol 229 pp 398-406

[19] Bukhori B, Widodo K H and Ismoyowati D Ev2aluation of Poultry Supply Chain Performance in XYZ Slaughtering House Yogyakarta using SCOR and AHP Method *Agriculture and Agricultural Science Procedia* vol 3 pp 221-5

Theorizing of spirit 212 in predicting Muslim behavioral intentions: A hypothesized model

W S Jatiningrum[1] and A Y Astuti[2]

[1,2]Industrial Engineering, Faculty of Industrial Technology, Ahmad Dahlan University, Indonesia

wandhansari.sekar@ie.uad.ac.id

Abstract. Total number of Muslim dominates the population in Indonesia. However, Muslim entrepreneurs only contribute quite small to the Indonesian economy. As a result, these economy benefits couldn't be perceived directly by dominated Muslims. Action of defending Islam 212 by Muslims is able to evoke spirit of 212 which has positive energy to build solidarity and solidity among Muslims. One of the results from spirit of 212 is campaign to shop to neighborhood store that aimed to develop economic aspect. This is because the owner of these store is dominated by Muslim. This paper investigates the relationship between spirit of 212 and behavioral intentions to shop to neighborhood store. Three factors in Theory of Planned Behavior, they are subjective norm, perceived behavioral control, and attitude toward behavior are proposed to explain this relationship. Based on the literature review, religiosity is also added as mediating variable which is considered to explain. Therefore, a conceptual model is built to determine the implications of spirit 212 on Muslim behavioral intentions using these four mediating variables.

1. Introduction

Islam is considered to the fast growing religion in the world. This follower is about 1,8 billion or roughly 24% of world's population as of 2015 [1]. However, Islamic society only contributed for 5% to the world economy [2]. It's also happened in Indonesia which is one of the largest number Muslim in the world. According to data from The Central Bureau of Statistics [3], total number of Muslim in Indonesia is about 87,18% of total population in Indonesia. Nevertheless, the total number of successful entrepreneurs in Indonesia was not dominated by Muslim. Total number of the richest Muslim entrepreneurs in Indonesia was only 16% [4]. As a result, the benefit of the economic activities couldn't be used by Muslims evenly.

Action for defending Islam 212 that was held on December 2, 2016, was become an Islamic revival. This would be potential to unify Muslims in Indonesia to develop many aspects in life of nation, such as aspects in economics, politics, social, and others. This action also has gotten big attention from many parties, both from domestic and abroad. The spirit of 212 is a result from action for defending Islam 212. After action for defending Islam 212 was held, things related to spirit 212 always get positive responses from Muslims, including to strengthen economics aspect. This is emphasized by statement from the chairman of Indonesian Ulema Council (MUI), K.H. Ma'ruf Amin who invited Muslims to divert the positive energy for strengthening nation's economy [5].

The strengthening of the economics aspect is motivated by Muslims in Indonesia who prefer to shop in conventional retails owned by richest entrepreneurs than small stores around their home. This causes many small stores are losing their markets so that economic inequality was occurred. The economic empowerment of Muslim by guiding store owners and consumers is proposed to overcome these problems. There is a guidance for store owners to raise their small stores for increasing their competitiveness and store image. To the consumers, there is a campaign to shop in their neighborhood store which usually owned by Muslims [2].

Spirit of 212 is estimated become a factor related religion which influence Muslim consumer behavior to shop to neighborhood store. This is because the effect of the spirit 212 looks quite great to remind Muslims to build up the kinship among them. Moreover, Muslim consumer always used Islamic principle for purchasing decision [6, 7]. Islamic perspective stated that the focus of consumer behavior is the welfare of individual and social surroundings [8]. Some previous researches has been conducted to explain about behavioral intentions to use Islamic principle adopting Theory of Planned Behavior. However, there are no researches has been conducted taking into account spirit of 212. Therefore, it would be needed to have deeply review about how the implications of spirit 212 on Muslim behavioral intentions to shop to neighborhood store which is one of many ways to strengthen nation's economy through Islamic principle.

2. Review of Literature

Theory of Reasoned Action (TRA) and Theory of Planned Behavior (TPB) whose providing conceptual framework are used to explained the complexities of social human behavior. Theory of Planned Behavior is the result of developing Theory of Reasoned Action proposed by Ajzen and Fishbein at 1975. According to TPB model, most of all actions are managed by intentions. Behavioral intentions are motivational factor that influence an action or a behavior. They are related the indications about the planning or willingness of a person to perform an action [9]. Some previous researches have proven that subjective norm, perceived behavioral control, and attitude toward behavior could be used to predict behavioral intentions correctly. Some previous researches have been conducted by adopting Theory of Planned Behavior model. Echchabi and Aziz [10] used TPB model to investigate behavioral intentions of applying Islamic banking services for Moroccan customers. Variables on TPB which consist of subjective norm, perceived behavioral control, and attitude toward behavior are analyzed. These three variables proved to have a significant effect on behavioral intentions. While, Khalek and Ismail [11] also adopted TPB model to investigate the determinant factors of urban Generation Y in related to consume halal food in Malaysia. TPB model is proved to be used effectively to predict behavioral intention to consume halal food. Similar researches by taking other cases are conducted by [12] and [13]. Their research aims to investigate the factors that allegedly influencing consumers in adopting of Islamic home financing. Respondents in this study are clients of Islamic banks. The empirical study shows that three variables in TPB have significant impact to consumer acceptance in Islamic home financing. Therefore, this model has proved empirically to be applied in other related cases.

Khan [14] stated that consumer behavior is decision making process and consumer activity related in seeking, choosing, purchasing and evaluating goods and services to meet their needs and desires. Consumer behavior underlies consumers to make purchasing decisions. Religious aspect is not considered by conventional consumer, while norms and religious value always be used to Muslim consumer to behave [6, 7]. This corresponds to research conducted by [15] that Muslim consumers feel more secure to buy products made by Muslim or to buy products in retails owned by Muslim. This is due to their expectation that those products have been guaranteed to be halal and allowed to be consumed by Muslims. Muslim dominated market is an opportunity for Islamic retailer stores. They will be attracted to the Islamic atmosphere that carry out culture, traditions or values related to Islam. Nevertheless, it is not accordance with the phenomenon occurred in Indonesia, Muslim prefer to shop in conventional retail owned by rich entrepreneurs.

Religion is considered as one of important cultural factors. This due to the fact that religion has significant impact on values, attitudes, and behaviors of a person, both at the individual and social level

[16]. While religiosity or also called religious commitment indicates an obedience of a person toward belief and practice in his or her religion [17].

Kotler [18] stated that religion as one of cultural factor, can shape behavior of a person. Religion make people having their decisions and actions based on certain value in their religion. Khalek and Ismail [11] suggested to consider religiosity factor to predict intention for the next research. In their prior research, three factors in TPB have been proved as independent variables to predict behavioral intention in consuming halal food by generation Y. While research conducted by [19] stated that religiosity acts as mediating variable which explained relationship between relative and contextual variables, and purchase behavior of Muslim consumers. Religiosity is also considered as independent variable that influence behavioral intention of consumer to choose Islamic home financing product [13]. The result of this research revealed that religiosity acts as important role in determining behavioral intention of consumer to choose Islamic home financing product. Therefore, religiosity is also needs to be considered as variable that effect consumer behavioral intentions to shop to neighborhood store.

Spirit of 212 is the result from action for defending Islam 212, which is a series of peaceful activities as a defense from Muslims toward their religion, through dhikr, tausyiah, and pray together. This aimed to tighten the kinship among Muslims. One of the results from spirit of 212 is campaign to shop to neighborhood store that aimed to developed economic aspect. Therefore, it would be needed to investigate about spirit of 212 related to Muslim behavioral intentions to shop to neighborhood store whose majority owners are Muslims.

3. Constructs of the hypothesized model for behavior intention of shopping to neighborhood store

Spirit of 212 is considered to affect three variables in Theory of Planned Behavior, which predict behavioral intentions of consumer. They are subjective norm, perceived behavioral control, and attitude toward behavior. Religiosity or the degree to which one practices their beliefs is also predicted by spirit of 212. A person who has higher support into spirit of 212 is considered to have higher religiosity. These forth variables are considered as mediating variables which affect behavioral intentions of shopping to neighborhood store. The relationship among these variables can be seen in figure 1.

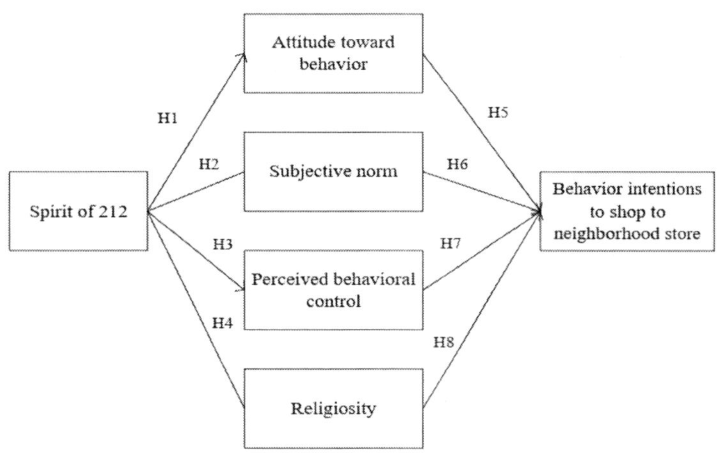

Figure 1. Conceptual model of spirit of 212 and its implications for Muslim behavioral intentions.

H1: Spirit of 212 has positive effect on attitude toward behavior.
H2: Spirit of 212 has positive effect on subjective norm.
H3: Spirit of 212 has positive effect on perceived behavioral control.
H4: Spirit of 212 has positive effect on religiosity.
H5: Attitude toward behavior has positive effect on behavioral intentions to shop to neighborhood store.
H6: Subjective norm has positive effect on behavioral intentions to shop to neighborhood store.

H7: Perceived behavioral control has positive effect on behavioral intentions to shop to neighborhood store.
H8: Religiosity has positive effect on behavioral intentions to shop to neighborhood store.

3.1 Spirit of 212
Spirit of 212 is the result of the action of defending Islam 212 that unify many Muslims to evoke the power of Islam to build the nation, including in economic aspect. Spirit of 212 is indicated by the extent to which a Muslim has a passion to evoke the power of Islam in various sectors. Spirit of 212 is considered has a positive effect on religiosity and three other variables which are predictors of consumer behavior intention in Theory of Planned Behavior.

3.2 Religiosity
Religion is a system of unity that regulates beliefs and practices of a person toward the sacred things. While religiosity is still related to religion, is also termed as religious commitment. Religiosity is the degree of person's commitment to hold and practice values in his or her religion [20]. Some other prior researches has explained that religiosity affected towards purchase behavior [19, 20]. Therefore, religiosity is considered has positive effect towards consumer behavior intention.

3.3 Attitude toward behavior
Attitude toward behavior is one of three variables in TPB to predict behavioral intention of consumer. This is a person's evaluation to conduct an assessment related to whether a behavior is favorable or unfavorable. In other words, someone intends to conduct a behavior if they think this is positively and vice versa. According to [21], attitude of the Muslim consumers is always justified by religion.

3.4 Subjective norm
Subjective norm is also considered to be predictor toward consumer behavioral intentions in Theory of Planned Behavior. It refers to individual belief about expectation from influential people around, either from individual or group for doing or not doing a particular behavior. Subjective norm is measured directly by respondents' sense of willingness to follow their trusted people [22]. Generally, people will get social pressure for doing a particular behavior if they believe that most of referent will allow their selves to conduct that behavior and they are motivated to follow that behavior.

3.5 Perceived behavioral control
Perceived behavioral control represents the degree of ease or difficulty in performing particular behavior. This is reflected by past experience as well as anticipated impediments and obstacles. This determine by individuals past experiences and also individual estimated about how ease or difficulty for doing a certain behavior. Ajzen [9] explained that a person's behavior not only controlled by himself, but also requires control from others factors, such as availability of resources, opportunities, and certain skill. Perceived behavioral control is considered to mediate between spirit of 212 and behavioral intentions of shopping to neighborhood store.

3.6 Behavioral intentions of shopping to neighborhood store
Intentions is predicted as variable that affect a person to perform behavior. Nevertheless, behavioral intentions are not necessarily representing the actual performance. The measurement of behavioral intentions might be failed to predict. This can be due to intentions changed after the measurement. According to TPB model, behavioral intentions is affected significantly by subjective norm, perceived behavioral control, and attitude toward behavior. Behavioral intentions are motivational factor which an action or a behavior. They are related the indications about the planning or willingness of a person to perform an action. Generally, the stronger effect of behavioral intention to predict behavior, the more likely should be its performance [9]. This could be applied to predict behavioral intentions of shopping to neighborhood store. Variables in TPB model (subjective norm, attitude toward behavior, and

perceived behavioral control) and religiosity are considered have positive effect toward behavioral intentions of shopping to neighborhood.

4. Conclusions

This research aims to determine and analyze the implications of the spirit of 212 toward one of economic program initiated by Muslim, called campaign of shopping to neighborhood store. The literature review of related matters has been conducted. A conceptual model is developed that spirit of 212 influences three variables in TPB model, such as subjective norm, attitude toward behavior, and perceived behavioral control. Religiosity factor is also considered influenced by spirit of 212. These forth variables are considered as mediating variables that relate between spirit of 212 and behavioral intentions to shop to neighborhood store.

References

[1] Centre P R 2017 World's Muslim population more widespread than you might think [Online] Available: http://www.pewresearch.org/fact-tank/2017/01/31/worlds-muslim-population-more-widespread -than-you-might-think/ [Accessed: 25-Jul-2017]

[2] Nurfatoni 2017 Di Sambiroto Surabaya, gerakan belanja di warung tetangga telah dimulai [Online] Available: https://www.pwmu.co/22571/2017/01/di-sambiroto-surabaya-gerakan-belanja-di-warung-tetangga-telah-dimulai/ [Accessed: 25-Jul-2017]

[3] T. C. B. of Statistics 2010 Kewarganegaraan, suku bangsa, agama, dan bahasa sehari-hari penduduk Indonesia : Hasil Sensus Penduduk 2010 Jakarta

[4] Simanjuntak D A 2017 Roh jihad ekonomi umat [Online] Available: http://www.republika.co.id/berita/koran/opini-koran/17/01/17/ojwws619-roh-jihad-ekonomi-umat [Accessed: 04-Jun-2017]

[5] Farhan S 2017 Ketua MUI: Alihkan energi 212 untuk perkuat gerakan ekonomi [Online] Available: http://food.detik.com/read/2017/04/06/055219/3466417/294/ketua-mui-alihkan-energi-212-untuk-perkuat-gerakan-ekonomi [Accessed: 19-Jun-2017]

[6] Hamid M A 2009 *Islamic economics: An introductory analysis* (Bangladesh: Md.Iqbal Aziz Khan, CSE, Rajshahi University)

[7] Hossain B 2014 Application of Islamic consumer theory : An empirical analysis in the context of Bangladesh *Glob. Rev. Islam. Econ. Bus.* vol 2 no 1 pp 69–83

[8] Wani T A and Islamia J M 2013 Buying behaviour –An Islamic perspective : An analysis of an ideal muslim buying behaviour *Int. J. Res. Commer. Manag.* vol 4 no 10 pp 152–5

[9] Ajzen I 1991 The Theory of Planned Behavior *Organ. Behav. Hum. Decis. Process* vol 50 no. 2 pp 179–211

[10] Echchabi A and Aziz H A 2012 Empirical investigation of customers' Perception and Adoption Towards Islamic Banking Services in Morocco *Middle-East J. Sci. Res.* vol 12 no 6 pp 849–58

[11] Khalek A A and Ismail S H S 2015 Why are we eating halal – Using the Theory of Planned Behavior in predicting halal food consumption among generation Y in Malaysia *Int. J. Soc. Sci. Humanit.* vol 5 no 7 pp 608–12

[12] Amin H, Abdul-Rahman A-R and Razak D A 2014 An empirical study of consumer behaviour of Islamic mortgage in Malaysia *J. Islam. Mark.* vol 5 no 2 pp 273–301

[13] Ibrahim M A, Fisol W N M and Haji-Othman Y 2017 Customer intention on Islamic home financing products: An application of Theory of Planned Behavior (TPB) *Mediterr. J. Soc. Sci.* vol 8 no 2 pp 77–86

[14] Khan M A 2007 *Consumer behaviour and advertising management* (New Delhi: New Age International Limited Publishers)

[15] Fauzi W I M, Muhammad N, Mokhtar S S M and Yusoff R Z 2016 What motivate muslim consumer to patronage islamic based–retail store? *Int. Rev. Manag. Mark.* vol 6 pp 199–203

[16] Mokhlis S 2009 Relevancy and Measurement of Religiosity in Consumer Behavior Research *Int.*

Bus. Res. vol 2 no 3 pp 75–84

[17] Delener N 1990 The effects of religious factors on perceived risk in durable goods purchase decision *J. Consum. Mark.* vol 7 no 3 pp 27–38

[18] Kotler P 2000 *Consumer market and consumer behavior, Principles of Marketing*, 8th ed (NJ: Prentice-Hall, Upper Saddle River)

[19] Alam S S and Sayuti N M 2011 Applying the Theory of Planned Behavior (TPB) in *halal* food purchasing *Int. J. Commer. Manag.* vol 21 no 1 pp 8–20

[20] Essoo N and Dibb S 2004 Religious Influences on Shopping Behaviour: An Exploratory Study *J. Mark. Manag.* vol 20 no 7–8 pp 683–712

[21] Baker A A 2011 *The Principles of Islamic Marketing* (Surrey: Gower Publishing Limited)

[22] Tjahjono H and Ardi H 2008 Kajian niat mahasiswa manajemen Universitas Muhammadiyah Yogyakarta untuk menjadi wirausaha *Util. J. Manaj. dan Bisnis* vol 16 pp 46–63

IOP Conf. Series: Materials Science and Engineering **403** (2018) 012048 doi:10.1088/1757-899X/403/1/012048

Evaluation of loading and hauling technology for improving andesite mine performance

J S Adiansyah, K Payadi, Alpiana and D Rahmawati

Mining Department, Engineering Faculty, Universitas Muhammadiyah Mataram, Indonesia

joni.adiansyah@ummat.ac.id; joni.adiansyah@gmail.com

Abstract. This paper evaluates the utilization of loading and hauling technology at andesite mine using cycle time and match factor methods. One mine site operated by PT. Anugerah Karya Agra Sentosa was taken as a case study. PT. Anugerah Karya Agra Sentosa located in Mujahiddin Village, Plampang Sub-district, West Sumbawa Regency is an andesite mining using open pit system with quarry type method. Mechanical equipment that operates is one unit Excavator (Kobelco SK 200) capacity 0.9 m^3 and five units Dump Truck (Toyota Dyna 130 HT) with bucket capacity 3.6 m^3. The analysis results found that the average cycle time for loading equipment was 0.54 minutes while the average cycle time for hauling equipment was 14.58 minutes. Recalculation for match factor (MF=1) resulted in additional of two hauling equipments from five units to seven units (Dump Truck).

1. Introduction

Mining industry contributes significantly not only to the revenue of a region but also to the communities. On the other side, the mining operation could create an environmental disaster [1, 2]. Therefore, proper management of mine operation is required to minimize the negative impact and increase the positive outcomes. One of mine management that essential to be considered is the fleet management. Heavy equipment plays crucial role in a project including construction and mining projects. The availability of heavy equipment in project site would contribute in accelerating project implementation process and increase project effectivity and efficiency. Therefore, the utilization of heavy equipment should be managed properly to ensure that the project production target is achieved [3, 4].

Two major problems in terms of loading and hauling for mining operations are energy consumption and Greenhouse gases (GHG) emissions [5, 6]. The largest contribution for GHG emissions in iron and ore mining was loading and hauling was noted in [6]. Furthermore, increasing for energy consumption would automatically elevate mine operational cost. The proper design for loading and hauling fleets will also avoid the mine site from the problems associated with operational and investment cost [4].

Therefore, the aims of this study are evaluating cycle time of loading and hauling equipment and calculating match factor of the current equipment. This study also finds the optimum requirement for loading and hauling fleets to increase the project efficiency and sustainability.

2. Methods

This study is using a quantitative method that combines four study stages: literature review, field survey, data analysis, and conclusion. Literature review is aimed to obtain current studies general description on cycle time and match factor of andesite mine. One month field survey was conducted for observing actual cycle time of loading and hauling equipment. The movement of these equipments was recorded by stopwatch. Data obtained from field survey are then analyzed using cycle time and match factor methods. An andesite mine was taken as a case study (see section 2.1).

2.1. Case study

PT. Anugrah Karya Agra Sentosa (PT. AKAS) is a private company with two final products: stone aggregate and Hot Mixture Asphalt. In order to generate these products, PT. AKAS operates Andesite mine located in Brang Ene District, West Sumbawa Regency, West Nusa Tenggara Province and has two main heavy equipment: excavator and dump truck. These two equipments play a significant role in loading and hauling in mining and processing area. Generally, Andesite stone excavated using an excavator and transporting with dump truck to the stockpile prior delivered into crushing (processing) plant.

2.2. Cycle time

Cycle time is total time required by mechanical equipment i.e. excavator and dump truck for finishing one working cycle (start to end task) and ready for the next cycle. Cycle time for loading equipment is described in equation (1) [7].

$$CT_m = B_t + S_{tf} + D_t + S_{te} \quad (1)$$

Where CT_m is cycle time for loading equipment (second), B_t is time required for digging (second), S_{tf} is swing time with full bucket (second), D_t is dumping time (second), S_{te} is swing time with empty bucket (second). Cycle time for hauling equipment is described in equation (2) [7].

$$CT_a = ST_l + LT + TT_f + ST_d + DT + TT_e \quad (2)$$

Where CTa is cycle time for hauling (second), ST_l is positioning time for loaded (second), LT is loading time (second), TT_f is transporting time with material (second), ST_d is positioning time for dumping (second), DT is dumping time (second), TT_e is transporting time with empty truck (second).

2.3. Match Factor (MF)

In order to obtain a suitable fleet composition number between loading and hauling equipment, the calculation of match factor is required. According to [8] that MF value equal with one indicated an efficiency operation between loading and hauling equipment. While, If MF value less than one (MF < 1) indicated that hauling equipment works 100% effective and loading equipment allocates more time for waiting. The last scenario is when MF value greater than one (MF > 1), this indicated that loading equipment works effectively but waiting time is generated by hauling equipment. MF value can be calculated by using equation (3) [8].

$$MF = \frac{N \; x \; CTm \; x \; Na}{CTa \; x \; Nm} \quad (3)$$

Where: MF is match factor, N is the number of filling for each hauling equipment, Na is the number of hauling equipment (unit), N_m is number of loading equipment (unit), C_{tm} is cycle time for loading equipment (minute), C_{ta} is cycle time for hauling equipment (minute).

3. Result and discussion

There are four mining activities associated with material handling which occurred at PT. AKAS: loosening, collecting, loading, and hauling. These processes involved one unit excavator for loading (Kobelco SK 200 – see figure 1) and five unit dump trucks with 3.6 m^3 bucket capacity (see figure 2).

Figure 1. Loading equipment.

Figure 2. Hauling equipment.

The cycle time data compilation for loading and hauling activities has been generated after one month field survey at the mine site. Average cycle time data are presented in table 1 and table 2.

Table 1. The average cycle time for loading.

Variable	Digging (B_t)	Swing full (S_{tf})	Dumping (D_t)	Swing empty (S_{te})	Total
Cycle time - CT_m (second)	10.20	9.00	7.20	6.00	32.40
				Cycle time (minute)	0.54

Excavator with 0.9 m^3 capacities required 0.54 minute to conduct one cycle for filling in dump truck bucket. Therefore, four cycles (2.16 minutes) were required by excavator to complete filling a dump truck with 3.6 m^3 bucket capacity.

Table 2. The average cycle time for hauling.

Variable	Loading position (ST_l)	Loading (LT)	Transporting full (TT_f)	Dumping position (ST_d)	Dumping (DT)	Transporting empty (TT_e)	Total
Cycle time - CTa (second)	16.80	126.60	360.60	18.00	15.60	337.20	874.80
						Cycle time (minute)	14.58

The average cycle time for a dump truck to haulage the andesite stone from mining area to stockpile area was 14.58 minutes. Therefore, total average cycle time for five dump trucks that currently operated by PT. AKAS was 72.9 minutes.

Cycle time required by both heavy equipments and number of equipment operated are two main parameters for calculating MF value. Calculation of MF value using equation (3) resulted that MF value less than one (0.74). MF value shows that dump truck works 100% effective and excavator has waiting time for each loading cycle. Improving MF value into one (MF = 1) is then required to increase mine performance. A strategy that can be taken is by adding more dump truck to reduce the excavator waiting time. By using equation (3), the total number of dump truck required to generate MF value equal with one are seven units.

4. Conclusion

This study found that the excavator needs 2.16 minutes to fill in a dump truck (DT) and DT requires 14.58 minutes for transporting material from mining area to stockpile area. The current cycle time which required by both loading and hauling equipment generates MF value less than one. Therefore, in order to improve mine performance, the additional of two dump trucks is required for balancing work load between excavator (loading) and dump truck (hauling).

References

[1] Adiansyah J S, Rosano M, Vink S and Keir G 2015 A framework for a sustainable approach to mine tailings management: Disposal strategies *J. Clean. Prod.* vol 108 part A pp 1050-62
[2] Adiansyah J S, Haque N, Rosano M and Biswas W 2017 Application of a life cycle assessment to compare environmental performance in coal mine tailings management *J. Environ. Manage.* vol 199 pp 181-91
[3] Salama A, Greberg J and Schunnesson H 2014 The use of discrete event simulation for underground haulage mining equipment selection *Int. J. Min. Miner. Eng.* vol 5 issue 3 p 256
[4] Santelices G, Pascual R, Lüer-Villagra A, Mac Cawley A and Galar D 2015 Integrating mining loading and hauling equipment selection and replacement decisions using stochastic linear programming *Int. J. Mining, Reclam. Environ.* vol 30 issue 1 pp 52–65
[5] Awuah-Offei K 2015 Energy efficiency in mining: A review with emphasis on the role of operators in loading and hauling operations *J. Clean. Prod.* vol 117 pp 89-97
[6] Norgate T and Haque N 2010 Energy and greenhouse gas impacts of mining and mineral processing operations *J. Clean. Prod.* vol 18 issue 3 pp 266–74
[7] Yanto I 2014 *Pemindahan Tanah Mekanis* (Yogyakarta: Fakultas Teknologi Mineral UPN)
[8] Prodjosumarto P 1995 *Pemindahan Tanah Mekanis* (Bandung: Teknik Pertambangan ITB)

IOP Conf. Series: Materials Science and Engineering **403** (2018) 012049 doi:10.1088/1757-899X/403/1/012049

Coordinated stock pre-positioning model to support emergency relief response

P F Opit

Industrial Engineering Department, Faculty of Engineering, Universitas Katolik De La Salle, Kombos Kairagi I, Manado, Indonesia

prudensy_f@yahoo.com

Abstract. This paper aims to develop a model that coordinates the pre-positioning of critical supplies in multiple distribution centers owned by different parties or organizations. This model determines the maximum proportion of relief demand covered in multiple distribution centers and it is built to support emergency relief response in the event of an earthquake. A linear programming approach is used to solve this model. First, we assign the service area of each distribution center (Stage I). Next, by using the result of the assigned service area, we determine the minimum value of the lower bound of proportion of unsatisfied relief demand (Stage II). Afterwards, we use the output of stage II as an input to determine the maximum amount of critical relief supplies to be stocked in each distribution center (Stage III). This model is applied to the eastern part of Indonesia with 10 disaster areas and 8 existing distribution centers, of which 4 are owned by the National Agency for Disaster Management (BNPB), 2 are owned by Ministry of Health and the other 2 are owned by Indonesian Red Cross (PMI).

1. Introduction

Natural disaster such as earthquake is notorious for its uncertainty factors. Even with all the new technologies possessed in this modern age, it is still nearly impossible to predict the exact time of occurrence and the extent of the area hits by a disaster. The decision makers in disaster relief do not know when, where, what, how much, where from and how many times demand is required in the early days of the post disaster response [1]. Rather than waiting passively to launch humanitarian operations, it is better to show pro-activity by mobilizing supplies or other material and non-material resources in anticipation [2]. In this case, stock pre-positioning that involves preparing critical relief supplies in strategic locations and determining the amount of demand to be released in disaster areas is needed.

In reality, the case related to the planning of disaster operations involves many parties. These parties include the government, military, Non-Governmental Organizations (NGOs), etc. To support a quick emergency response, it is highly recommended that these parties discuss and plan the pre-positioning of critical relief supplies together in advance. Although in reality, this coordination—especially in Indonesia, is still difficult to do [3].

In the previous paper, the authors develop a stock pre-positioning model that simultaneously determines the decision of DCs to cover a single disaster area and the proportion of critical items to be stocked in each DC [4]. Their model is applied to Indonesia with 33 disaster areas and 16 DCs.

IOP Conf. Series: Materials Science and Engineering **403** (2018) 012049 doi:10.1088/1757-899X/403/1/012049

Although the tested result is satisfying, the model only considers the 16 DCs which are belong to one party: The National Agency for Disaster Management (BNPB). The other parties who also owned some DCs located in certain disaster areas and normally take part in disaster operations such as Ministry of Health (KEMKES), Indonesian Red Cross (PMI), Ministry of Social Welfare, and so on, are not considered into the model.

The later research proposes a more complex model that integrates the decisions of the maximum proportion of relief demand covered in DCs and the maximum amount of relief supplies delivered to a single disaster area within a certain period of time [5]. This research is extended to a new research that considers all potential scenarios of facility disruptions [6]. Nonetheless, these two models also do not consider the coordination of disaster operations into account. Another authors provide some crucial information about the relief logistics network structures in Indonesia [3]. They mention that the government of Indonesia has established BNPB as the policy maker and the main coordinator in the event of major disaster. They also propose two models of facility location. Their model however, do not consider the coordination of disaster operations. The next authors aim to map and analyze humanitarian logistic in Indonesia [7]. Their research focuses on the coordination of humanitarian logistic system using RASIC method. There is no mathematical model related to disaster operations is discussed in their research.

The above issues motivate us to extend the previous model developed [4] to a new stock pre-positioning model that coordinates the pre-positioning of critical relief supplies in multiple DCs owned by different parties or organizations. This new stock pre-positioning model is divided into three stages, where each stage is solved by using the Linear Programming approach. The objective of this paper is to maximize the expected relief demand covered by DCs owned by different parties.

2. Model formulation

In this model, each distribution center (DC) owned by one party is located in several different disaster areas. Each DC is assigned to provide services to one or more disaster areas that located inside the range of a given maximum response time limit (refer to Stage I). Let I = $\{1, 2, 3,\ldots,\bar{\imath}\}$ be the set of disaster areas, J = $\{1, 2, 3,\ldots, \bar{\jmath}\}$ be the set of DCs owned by BNPB, M = $\{1, 2, 3,\ldots, \bar{m}\}$ be the set of DCs owned by KEMKES and N = $\{1, 2, 3,\ldots, \bar{n}\}$ be the set of DCs owned by PMI. Let K = $\{1, 2, 3\ldots, \bar{k}\}$ be the set of item types. Table 1 depicts the parameters and decision variables used in this model.

Stage I: Grouping Service Area

Obj. function: Maximize $Z = XA_{ij} + XB_{im} + XC_{in}$ (1)

s/t:

$$T_{ij}XA_{ij} \leq \theta \qquad \forall\, i \in I, j \in J \qquad (2)$$

$$T_{im}XB_{im} \leq \theta \qquad \forall\, i \in I, m \in M \qquad (3)$$

$$T_{in}XC_{in} \leq \theta \qquad \forall\, i \in I, n \in N \qquad (4)$$

$$\textstyle\sum_{j\in J}\sum_{m\in M}\sum_{n\in N} XA_{ij} + XB_{im} + XC_{in} \geq 2 \qquad \forall\, i \in I \qquad (5)$$

$$XA_{ij} \in \{0,1\} \qquad \forall\, i \in I, j \in J \qquad (6)$$

$$XB_{im} \in \{0,1\} \qquad \forall\, i \in I, m \in M \qquad (7)$$

$$XC_{in} \in \{0,1\} \qquad \forall\, i \in I, n \in N \qquad (8)$$

The objective function in equation (1) maximizes the number of DCs to serve disaster areas. Constraint sets in equation (2), (3) and (4) ensure that the existing DC owned by BNPB, KEMKES and PMI, respectively, can provide service in specific disaster area if the expected time to satisfy relief demand is smaller than the maximum response time limit. Constraint set in equation (5) assures that at least two DCs will provide services in a single disaster area. Constraint sets in equation (6), (7) and (8) define the binary variables.

Table 1. Parameters and decision variables

Parameters:	
T_{ij}, T_{im}, T_{in}	expected time to satisfy relief demand in disaster area i from DC j (hr), $i \in I, j \in J$; expected time to satisfy relief demand in disaster area i from DC m (hr), $i \in I, m \in M$; expected time to satisfy relief demand in disaster area i from DC n (hr), $i \in I, n \in N$, respectively,
θ	maximum response time limit to perform emergency response (hr),
u_k	upper bound of the proportion of unsatisfied relief demand of item type k, $k \in K$,
m_k	degree of importance of item type k; where $m_k = m,\ k \in K$,
P_i	probability of occurrence of earthquake in each disaster area, $i \in I$,
d_{ik}	expected demand for item type k in disaster area i (unit), $i \in I$, $k \in K$,
UA_j, UB_m, UC_n	capacity of DC j (m^3), $j \in J$; capacity of DC m (m^3), $m \in M$; capacity of DC n (m^3), $n \in N$, respectively,
γ_k	unit volume of item type k (m^3), $k \in K$,
BA_0, BB_0, BC_0	pre-disaster budget of BNPB ($), pre-disaster budget of KEMKES ($), pre-disaster budget of PMI ($), respectively,
BA_1, BB_1, BC_1	post-disaster budget of BNPB($), post-disaster budget of KEMKES ($), post-disaster budget of PMI ($), respectively,
GA_{jk}, GB_{mk}, GA_{jk}	unit cost of acquiring item type k at DC j ($/unit), $j \in J$, $k \in K$; unit cost of acquiring item type k at DC m ($/unit), $m \in M$, $k \in K$; unit cost of acquiring item type k at DC n ($/unit), $n \in N$, $k \in K$, respectively,
CA_{ijk}, CB_{imk}, CC_{ink}	unit cost of shipping item type k from DC j to demand point i ($/unit), $i \in I$, $j \in J$, $k \in K$; unit cost of shipping item type k from DC m to demand point i ($/unit), $i \in I$, $m \in M$, $k \in K$; unit cost of shipping item type k from DC n to demand point i ($/unit), $i \in I$, $n \in N$, $k \in K$, respectively,
α_k	criticality weight for item type k; $\sum_k \alpha_k = 1$ and $\alpha_k \geq 0$, $k \in K$,
WA_j, WB_m, WC_n	cost of operating a single DC j, $j \in J$; cost of operating a single DC m, $m \in M$, cost of operating a single DC n, $n \in N$, respectively,
Decision variables:	
XA_{ij}, XB_{im}, XC_{in}	set of potential DC j to provide service in disaster area i (X_{ij}= 1, if DC j provides service in disaster area i, 0 otherwise); set of potential DC m to provide service in disaster area i (X_{im}= 1, if DC m provides service in disaster area i, 0 otherwise); set of potential DC n to provide service in disaster area i (X_{in}= 1, if DC n provides service in disaster area i, 0 otherwise), respectively,
fA_{ijk}, fB_{imk}, fC_{ink},	proportion of item type k relief demand satisfied by DC j that provides service in disaster area i ; proportion of item type k relief demand satisfied by DC m that provides service in disaster area i ; proportion of item type k relief demand satisfied by DC n that provides service in disaster area i, respectively,
N_{ik}	proportion of unsatisfied relief demand of item type k in disaster area i for each scenario s,
Z_k	the lower bound of the proportion of unsatisfied relief demand of item type k,
QA_{jk}, QB_{mk}, QC_{nk}	units of item type k stored at DC j ; units of item type k stored at DC m ; units of item type k stored at DC n, respectively.

Stage II: Generating lower bounds of the proportion of unsatisfied demand

This stage is developed to prevent zero results of the proportions of relief demand satisfied. In this stage, the values of XA_{ij}, XB_{im} and XC_{in} have been determined in Stage I.

Obj. function: Minimize $Z = Z_k$ (9)

s/t:

$$\sum_{k\in K} fA_{ijk} \leq M\, XA_{ij} \qquad \forall\, i \in I,\ j \in J \qquad (10)$$

$$\sum_{k\in K} fB_{imk} \leq M\, XB_{im} \qquad \forall\, i \in I,\ m \in M \qquad (11)$$

$$\sum_{k\in K} fC_{ink} \leq M\, XC_{in} \qquad \forall\, i \in I,\ n \in N \qquad (12)$$

$$\sum_{j\in J} \sum_{m\in M} \sum_{n\in N} (fA_{ijk} + fB_{imk} + fC_{ink}) = 1 - N_{ik} \qquad \forall\, i \in I,\ k \in K \qquad (13)$$

$$N_{ik} \leq Z_k \qquad \forall\, i \in I,\ k \in K \qquad (14)$$

$$fA_{ijk}\, d_{ik} \leq QA_{jk} \qquad \forall\, i \in I,\ j \in J,\ k \in K \qquad (15)$$

$$fB_{imk}\, d_{ik} \leq QB_{mk} \qquad \forall\, i \in I,\ m \in M,\ k \in K \qquad (16)$$

$fC_{ink}\ d_{ik} \leq QC_{nk}$	$\forall\, i \in I,\ n \in N,\ k \in K$	(17)
$\sum_{k\in K} \gamma_k\, QA_{jk} \leq UA_j$	$\forall\, j \in J$	(18)
$\sum_{k\in K} \gamma_k\, QB_{mk} \leq UB_m$	$\forall\, m \in M$	(19)
$\sum_{k\in K} \gamma_k\, QC_{nk} \leq UC_n$	$\forall\, n \in M$	(20)
$\sum_{j\in J} \sum_{k\in K} \left(WA_j + \left(QA_{jk}\, GA_{jk}\right)\right) \leq BA_0$		(21)
$\sum_{m\in M} \sum_{k\in K} (WB_m + (QB_{mk}\, GB_{mk})) \leq BB_0$		(22)
$\sum_{n\in N} \sum_{k\in K} (WC_n + (QC_{nk}\, GC_{nk})) \leq BC_0$		(23)
$\sum_{j\in J} \sum_{k\in K} CA_{ijk}\, fA_{ijk}\, d_{ik} \leq BA_1$	$\forall\, i \in I$	(24)
$\sum_{m\in M} \sum_{k\in K} CB_{imk}\, fB_{imk}\, d_{ik} \leq BB_1$	$\forall\, i \in I$	(25)
$\sum_{n\in N} \sum_{k\in K} CC_{ink}\, fC_{ink}\, d_{ik} \leq BC_1$	$\forall\, i \in I$	(26)
$fA_{ijk} \geq 0$	$\forall\, i \in I, j \in J, k \in K$	(27)
$fB_{imk} \geq 0$	$\forall\, i \in I, m \in M, k \in K$	(28)
$fC_{ink} \geq 0$	$\forall\, i \in I, n \in N, k \in K$	(29)
$N_{ik} \geq 0$	$\forall\, i \in I, k \in K$	(30)
$Z_k \geq 0$	$\forall\, k \in K$	(31)

The objective function in equation (9) minimizes lower bound of the proportion of unsatisfied relief demand of each item type. Constraint sets in equation (10), (11) and (12) ensure the amount of supplies is stocked in a DC owned by BNPB, KEMKES and PMI, respectively, only when the DC provides service in the designated disaster area. Constraint set in equation (13) means that the actual demand is equal to the amount of satisfied relief demand summed with the amount of unsatisfied relief demand. Constraint set in equation (14) guarantees that the proportion of unsatisfied relief demand does not exceed the desired lower bound limit. Constrain sets in equation (15), (16) and (17) assure that the amount of demand is smaller than the inventory level on DC owned by BNPB, KEMKES and PMI, respectively. Constraint sets in equation (18), (19) and (20) impose the capacity restrictions on each DC. Constraint sets in equation (21) to (26) state the maximum budgets available of pre- and post-disasters programs. Constraint sets in equation (27) to (31) describe the non-negativity constraints.

Stage III: Maximizing Expected Relief Demand Covered by Distribution Centers
In this stage, the value of the upper bound of the proportion of unsatisfied relief demand is determined, where $u_k = Z_k + (1 - Z_k) * m_k$, $Z_k < u_k < 1$, and $0 < m_k < 1$, $\forall k \epsilon K$. Value of Z_k has been generated in Stage II.
Obj. function: Maximize $Z = \sum_i \sum_j \sum_m \sum_n \sum_k P_i\, \alpha_k\, d_{ik}\, fA_{ijk}\, fB_{imk}\, fC_{ink}$ (32)
s/t:

$$N_{ik} \leq u_k \qquad \forall\, i \in I,\ k \in K \qquad (33)$$

The objective function in equation (32) is now maximizing the total expected relief demand by each DC. Constraint set in equation (14) is now replaced by constraint set in equation (33) that guarantees the proportion of unsatisfied relief demand in each disaster area is smaller than the desired upper bound limit. The rest of the constraints are remaining the same (refer to Stage II).

3. Data construction

For the sake of simplicity, in this paper we apply our model to the eastern part of Indonesia with 10 disaster areas. According to the data provided by [3], there are 8 existing distribution centers located in eastern Indonesia, of which 4 are owned by BNPB, 2 are owned by KEMKES and the other 2 are owned by PMI. The disaster areas are: (1) South Sulawesi—with one existing DC, (2) West Sulawesi,

(3) South East Sulawesi, (4) Central Sulawesi, (5) Gorontalo, (6) North Sulawesi—with one existing DC, (7) North Maluku, (8) Maluku—with one existing DC, (9) West Papua, (10) Papua—with one existing DC. The 4 DCs owned by BNPB are located in: (1) Makassar, (2) Manado, (3) Ambon, (4) Jayapura. The 2 DCs owned by KEMKES and 2 DCs owned by PMI are located in the same cities: (1) Makassar, (2) Manado. Table 2 describes the data estimation. The critical item types are: (A) Medicine (box), (B) Instant food (box), (C) Drinking water (box) and (D) Tent (unit). In this case, DCs owned by KEMKES can only be used to stock medical supplies (item type 1). Helicopters are used to transport each item to disaster area.

Table 2. Data estimation

Exp. resp. time (hr)	(distance from DC to the affected area (km)) / vehicle speed (km/hr)) + (expected loading time (hr)).
Max. resp. time (hr)	Expected to be 8 hours (the same for each disaster area).
Probability of earthquake	Calculated based on the frequency of earthquakes hit each disaster area during 2005-2013. The earthquakes magnitude varies between 1.0 to 9.0 Mw.
Demand	Assumed to be 50% of the total population of each province in year 2010.
Criticality weight	Weight of item type A to D = 0.286, 0.286, 0.286 and 0.143, respectively.
Volume (m^3)	Unit volume of item type A to I = 0.018, 0.054, 0.054, and 0.200, respectively.
DC capacity (m^3)	Dimension of each DC is assumed to be 100 x 100 x 12 = 120,000 m^3. Normally, only 70% space is used for the storage. The capacity of DCs owned by BNPB is 84,000 m^3 (the same for each DC). While the capacities of DCs owned by KEMKES and PMI are assumed to be 25,200 m^3 and 42,000 m^3, respectively (the same for each DC).
Cost of operating a DC ($)	For each DC owned by BNPB, KEMKES and PMI, the costs are assumed to be $6,100/5 years, $1,830/5 years and $3,050/5 years (the same for each DC). (1 USD = 13,300 IDR).
Unit cost of acq. ($)	Purchase cost of item type A to D = $236.842, $2.406, $488.722 and $751.445, respectively.
Unit cost of ship. ($)	(Expected response time (hr)) x (fuel needed (liter/hr)) x 2 (round trip).
Maximum budgets available ($)	Pre-disaster budgets of BNPB, KEMKES, PMI assumed to be $857,317,919.075 (100%); $257,195,375.722 (30%) and 428,658,959.537 (50%); respectively.
	Post-disaster budgets of BNPB, KEMKES, PMI assumed to be $116,589,595.375 (100%); $34,976,878.612 (30%) and $58,294,797.687 (50%); respectively.

4. Computational results and analysis

LINGO 8.0 is used for finding the optimal solutions. All experiments solving each problem are tested on a personal computer with an Intel® Core™ 2 Duo CPU 2.93GHz and 2.00 GB of RAM. The computation time of all the test problems is less than 3 minutes. The result of stage I, which is grouping service area for each distribution center (DC), stated that for all DCs located in Makassar (owned by BNPB, KEMKES and PMI) can provide service to all disaster areas except West Papua and Papua, while all DCs located in Manado (owned by BNPB, KEMKES and PMI) can provide service to all disaster areas except Papua, and DC located in Ambon (owned by BNPB) can provide service to all disaster areas. Meanwhile, DC located in Papua (owned by BNPB) can only provide services to Gorontalo, Maluku, West Papua and Papua. From the results also can be stated that Papua receives the least services compared to the other disaster areas due to its remote location.

The results of stage II, which is the lower bound of proportion of unsatisfied relief demand (for item type A to D) are all 0.000. These results are used as inputs in stage III. Furthermore, in stage III, we set the value of degree of importance of each item type (denoted by m_k) to be 0.85. Thus, the result of total expected relief demand covered by distribution centers is 109,577.900 (objective function). Table 3 shows the result of the proportion of satisfied relief demand in disaster area 9 (West Papua) for all item types. In this case, DCs 3 and 4 (owned by BNPB) are covering half of the items type B (instant food), type C (drinking water) and type D (tent) to be sent to West Papua, while the other half is covered by DC 2 owned by PMI. Meanwhile, DC 2 owned by KEMKES is covering 100% of item type A (medicine). Table 4 shows the result of the total proportion of satisfied relief demand in each disaster area. Given the assumption of budgets available and 50% of the total population in each disaster area that will be affected by the earthquake and needed to be treated immediately, disaster area 1 to 9 (South Sulawesi to West Papua) can be covered 100% (1.000), while disaster area 10 (Papua) can be covered only as big as 99.2% (0.992) for item type A, 15% (0.150) for item type B and type C, and 23.7% (0.237) for item type D.

Table 3. The proportion of satisfied relief demand in disaster area 9 (West Papua).

DC		Item type			
		A	B	C	D
BNPB	1	0.0	0.0	0.0	0.0
	2	0.0	0.0	0.0	0.0
	3	0.0	0.5	0.0	0.0
	4	0.0	0.0	0.5	0.5
KEMKES	1	0.0	0.0	0.0	0.0
	2	1.0	0.0	0.0	0.0
PMI	1	0.0	0.0	0.0	0.0
	2	0.0	0.5	0.5	0.5

Table 4. The total proportion of satisfied relief demand in each disaster area.

Disaster area	Item type			
	A	B	C	D
1	1.000	1.000	1.000	1.000
2	1.000	1.000	1.000	1.000
3	1.000	1.000	1.000	1.000
4	1.000	1.000	1.000	1.000
5	1.000	1.000	1.000	1.000
6	1.000	1.000	1.000	1.000
7	1.000	1.000	1.000	1.000
8	1.000	1.000	1.000	1.000
9	1.000	1.000	1.000	1.000
10	0.992	0.150	0.150	0.237

Compared to the previous papers as discussed in the Introduction part, especially one written by [4], this paper considered multiple DCs owned by different parties, in this case BNPB, KEMKES, and PMI. This paper gives a solution to determine the optimum number of critical items to be stocked in each DC owned by different parties based on the available budgets and capacity restrictions. This coordinated stock pre-positioning model can minimize the oversupply and/or undersupply of critical items at the affected areas which will lead to the improvement of the emergency relief response.

5. Conclusion and future work

This paper proposes a new model that coordinates the pre-positioning of critical relief supplies in multiple distribution centers (DCs) owned by different parties. This model applied to eastern Indonesia with 10 disaster areas and 8 existing distribution centers, of which 4 are owned by the National Agency for Disaster Management (BNPB), 2 are owned by Ministry of Health (KEMKES) and the other 2 are owned by Indonesian Red Cross (PMI). The result of Stage I (grouping service area) shows that West Papua and Papua receive the least services from DCs compared to other disaster areas due to its remote location. Based on the results of the final stage, the proportions of satisfied relief demand in most disaster areas, except in Papua, can be fully satisfied. This coordinated stock pre-positioning model can minimize the oversupply and/or undersupply of critical items at the affected areas. As for the future work, we will consider the distribution or transportation planning problem into the model.

References

[1] Decker M 2013 *Last Mile Logistics for Disaster Relief Supply Chain Management: Challenges and Opportunities for Humanitarian Aid and Emergency Relief* (Hamburg: Anchor Academic Publishing)

[2] Hale T and Moberg C R 2005 Improving supply chain disaster preparedness *Int. J. Phys. Distrib. Logist. Manag.* vol 35 pp 195–207

[3] Kusumastuti R D, Wibowo S S and Insanita R 2013 Modeling facility locations for relief logistics in indonesia *Humanitarian and Relief Logistics: Research Issues, Case Studies and Future Trends* ed V Zeimpekis, S Ichoua and I Minis (New York, NY: Springer New York) pp 183–205

[4] Opit P F, Lee W-S, Kim B S and Nakade K 2013 Stock pre-positioning model with unsatisfied relief demand constraint to support emergency response *Oper. Supply Chain Manag. An Int. J.* vol 6 no 2 pp 103–10

[5] Opit P F and Nakade K 2015 Emergency response model of stock-prepositioning with transportation constraints *2015 IEEE International Conference on Industrial Engineering and Engineering Management (IEEM)* (IEEE) pp 239–43
[6] Opit P F and Nakade K 2015 Pre-positioning of emergency relief supplies under facility disruptions *Proc. of the 2nd East Asia Works. on Industrial Engineering 2015* pp 164-169
[7] Octavia T, Widyadana I G A and Palit H C 2016 Analyzing humanitarian logistic coordination for disaster relief in indonesia *Proceedings of Second International Conference on Electrical Systems, Technology and Information 2015 (ICESTI 2015)* ed F Pasila, Y Tanoto, R Lim, M Santoso and N D Pah (Singapore: Springer Singapore) pp 583–8

IOP Conf. Series: Materials Science and Engineering **403** (2018) 012050 doi:10.1088/1757-899X/403/1/012050

Development of an integrated production-inventory model for food products considering exponential perceived value loss

G Fauza[1], H Prasetyo[2] and B S Amanto[1]

[1] Department of Food Technology, Universitas Sebelas Maret, Surakarta, Indonesia
[2] Department of Industrial Engineering, Universitas Muhammadiyah Surakarta, Indonesia

gustifauza@staff.uns.ac.id

Abstract. Studies on deteriorating inventory models have been done extensively from the simple EOQ to more complex joint economic lot size (JELS). The majority of studies in this area assume the quantity depletion of deteriorating inventories. In most food products, however, it is quality or value that losses over time while the quantity remains the same during a certain period. Regarding packaged food products whose expiration dates are stamped on their label, the products' value is perceived by customers by examining their remaining life time. The customers' willingness to pay for a product declines when they realize that the product approaches its expiration dates. This phenomenon is called perceived value loss. The losses can be linear or exponential depending on products' characteristics. In this study, the exponential perceived value loss is considered when developing an integrated production-inventory model in a single vendor single buyer supply chain system. The numerical test shows that the performance of the proposed model in terms of the total profit of the joint system is 2.08% higher compared to the benchmark model.

1. Introduction

Managing perishable inventories across a supply chain is challenging since items that are deteriorating over time are forced to wait until they are being consumed. Studies on the integrated production-inventory system for deteriorating items in a multi echelon system have been done. Most of the studies assume that some fractions of inventories vanish during storage time (some of them can be seen in [1-4]). In food products, however, it is quality which decreases over time while the quantity remains the same during a certain period. Therefore, current models need to be modified in order to develop a more representative model for managing food inventories through a multi stage system.

Some inventory studies of food products have addressed quality decrease and lifetime constraint in their models [5]. An optimization approach for managing quality degradation in bell peppers across a supply chain was introduced in [6]. As mentioned in [6], firmness and color are the quality factors observed in the peppers. The value loses over time following a kinetic model where the storage temperature affects the rate of quality loss. A supply chain strategy for fresh produce (e.g. melons and sweet corn) was proposed by [7]. In their model, three variables, namely, the picking rate, the size of transfer batch, and the transportation mode, are determined to minimize the total cost of the supply chain system. Both approaches are suitable for fresh food where physical degradation can easily be examined by customers.

In some food products, quality changes (such as microbial changes) are not easy to be detected. Therefore, shelf-life is used to justify whether the food products are still acceptable to consume. Food shelf-life in planning and scheduling of yogurt production was considered in [8]. In their studies, the benefit of selling an item depends on its remaining lifetime. The longer the remaining life time of food sold to customers, the more benefits will be obtained. The model is suitable for packaged food products where customers examine the quality of the products from its remaining lifetime. In a multi echelon system, [9] developed an integrated production-inventory model accommodating perceived value loss in food products. Customers perceived the value of finished goods as degrading once the products approach their expiration dates. Model in [9] only focuses on value loss in finished goods, while model in an integrated system also accommodates the quality loss of raw materials, which can be beneficial for reducing food waste. The model of [9] then advanced in [10] by considering quality loss in raw materials. As mentioned in [9] and [10] the perceived value loss is assumed to be linear, which is suitable for some typical food (e.g. lettuce or carrots) but not for some others (such as beef or chicken). Therefore, in this research, an integrated production-inventory model considering exponential perceived value loss in food product is proposed.

The remaining sections of this paper are organized as follows. Section 2 presents the characterization of the observed system as well as the mathematical models which represent the system. Further, the experiment and its result are discussed in Section 3. Finally, Section 4 summarizes this study.

2. System characterization and mathematical model development

In this section, the system of interest is defined and the mathematical models related to the system are developed and discussed.

2.1. System characterization

The system observed in this study is similar to [10], i.e. a single vendor-single buyer (SVSB) inventory system as presented in figure 1. In manufacturer (vendor) system, raw material is procured m times during the production period of T. At the same period, the finished goods are produced and delivered n times using single setup multi delivery strategy (SSMD) introduced by [11].

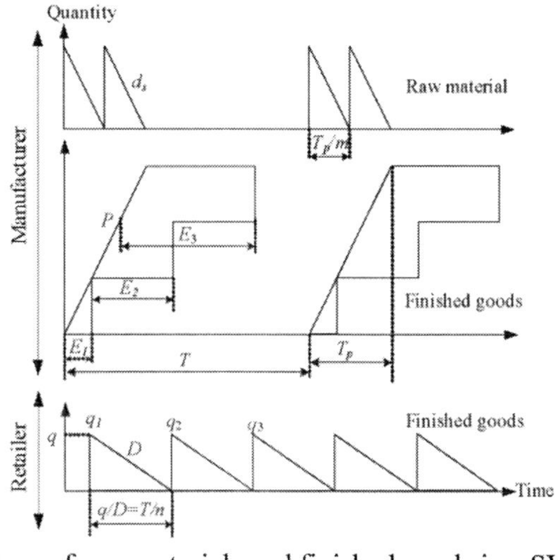

Figure 1. Inventory of raw materials and finished goods in a SVSB system [10].

The raw material is a deteriorating item whose quality loses over time with deteriorating rate k following the kinetic model. Meanwhile, the finished goods are perishable packaged products with expiration dates stamped on their label. The price of the product is not constant during the selling period or Ct. It is because the customer's willingness to pay (WTP) for a product decreases when the product approaching its expiration dates or during the period of T_{start}-T_{SL} as seen in figure 2. T_{SL} is product shelf-life time while T_{start} is the time when a customer perceives that the value of the aging product has decreased and feels reluctant to buy it at the same price. Consequently, the product demand rate becomes slower (as seen in figure 2, below section). Without any action, such as discounting product price, the perceived value loss (PVL) potentially causes more outdated items. In [10] the PVL follows the linear pattern (figure 1, line A). However, in this study the exponential decrease (referring to line B in figure 1) is used. This exponential pattern is more suitable for representing perceived value loss in some food products such as beef and chicken [12].

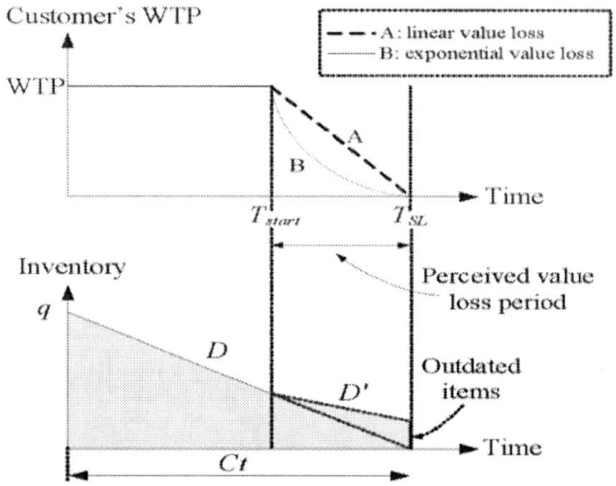

Figure 2. Decreasing customer's WTP of finished goods in retailer.

Further, as the customer's WTP decreases during the PVL period, retailer will offer discounted products to maintain demand and avoid outdated items. There will be three regions of price offered during Ct (as seen in figure 3). During the first region, the products will be sold with normal price or p_{max}, while during the second region, a PVL occurs; hence, the price of the products is discounted. Unlike (10), the decrease in product price will be exponential. Finally, in the last region, i.e. the region in which the products have been expired, the price of expired products is set as p_{min}. Selling products with p_{min} occurs if *CompTime* is longer than T_{SL} (*CompTime* is time for the last product in the batch sold to customers or $CompTime = E_i + Ct$). Different from [10], the age of batch entering retailer or E_i is set to be younger than T_{start}. This is to ensure that retailer will have fresh products with a good price (p_{max}).

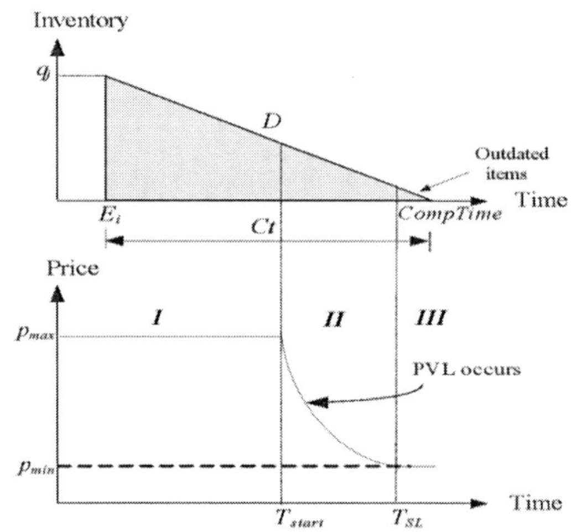

Figure 3. Shelf-life based price function.

IOP Conf. Series: Materials Science and Engineering **403** (2018) 012050 doi:10.1088/1757-899X/403/1/012050

2.2. Mathematical model development

The proposed model is developed based on the following assumptions:

a. The production rate and demand rate are constant.
b. Once the customer's WTP starts decreasing (i.e. the product approaches its expiration dates), its price will decline exponentially.
c. All delivering batches enter the retailer's warehouse before the customer's WTP declines.
d. Lead time is neglected and no shortage is allowed.

Further, the following notations are used in the mathematical model.

For the manufacturer:

P	:	production rate (unit product/unit time)
D	:	demand rate from retailer (unit product /unit time)
K	:	deterioration rate (unit quality/ unit time)
c_{raw}	:	purchasing cost of raw material (\$/unit product)
A_{raw}	:	delivering cost of raw material (\$/shipment)
q_{raw}	:	ordering size of raw material (unit product)
H_{raw}	:	holding cost of raw material (\$/unit product/ unit time)
c_{loss}	:	cost of quality loss (\$/unit quality/unit time)
$TC_{\text{raw}}\ (m,T)$	:	total cost of raw material system (\$/unit time)
c_{mfc}	:	manufacturing cost (\$/unit product)
A_{mfc}	:	production setup cost (\$/setup)
H_{mfc}	:	holding cost of finished goods (\$/unit product/unit time)
$TC_{\text{mfc}}(m,T)$	:	total cost of finished goods system (\$/unit time)
$TP_{\text{mfc}}(m,n,T)$	:	total profit of the manufacturer (\$/unit time)

For the retailer:

c_{ret}	:	product selling price from manufacturer to retailer (\$/unit product)
H_{ret}	:	holding cost of finished goods at retailer (\$/unit product/ unit time)
A_{ret}	:	transportation cost for delivering product to retailer (\$/shipment)
l	:	shape parameter of the exponential distribution
T_{SL}	:	product shelf-life (unit time)
T_{start}	:	time when customer's WTP starts decreasing (unit time)
$p_{\max}$	:	maximum product price (\$/unit product)
$p(t)$	:	price of product at age t (\$/unit product) or the shelf-life based price function
E_i	:	age of batch i when entering retailer (unit time)
Ct	:	finished goods delivery cycle (unit time)
$CompTime$	:	completion time of the last product in the transferred batch (unit time)
TC_{ret}	:	total cost of the retailer (\$/unit time)
R_{ret}	:	revenue of the batch (\$/unit time)
$TP_{\text{ret}}(n,T)$	:	total profit of the retailer (\$/unit time)
$TP(m,n,T)$	:	total profit of the integrated supply chain system (\$/unit time)

Firstly, the proposed model is developed by formulating the joint total cost (*JTC*) as can be seen in equation (1). Detailed steps in constructing equation (1) can be found in [10].

$$JTC(m,T,n) = C_m D + \frac{A_m}{T} + H_m \frac{DT}{2n}\left(\frac{D}{P}(2-n) + (n-1)\right) + C_{\text{raw}} D + A_{\text{raw}} \frac{m}{T} + H_{\text{raw}} \frac{D^2 T}{2mP} + L(m,T) + \frac{O_{\text{ret}}}{T} + A_{ret} \frac{n}{T} + H_{\text{ret}} \frac{DT}{2n} \tag{1}$$

The next step is formulating joint total revenue (JTR) which is obtained by adding revenue in manufacturer (R_{manu}) with revenue in retailer (R_{ret}). Equation 2 represents the total revenue in manufacturer system where C_{ret} denotes selling price from manufacturer to retailer and D is demand.

$$R_{manu} = C_{ret} D \tag{2}$$

As previously mentioned, the price of products from retailer to end customers is varied during the selling period because of the PVL. For representing this, an exponential shelf-life based price function is developed as represented by equation (3) where $p(t)$ is the price of product at time t.

$$p(t) = \begin{cases} p_{\max} & 0 \le t < T_{\text{start}} \\ p_{\max} - \dfrac{(p_{\max} - p_{\min})}{(e^{-\lambda(T_{SL} - T_{\text{start}})} - 1)}(e^{-\lambda(t - T_{\text{start}})} - 1) & T_{\text{start}} \le t < T_{sl} \\ p_{\min} & t \ge T_{sl} \end{cases} \tag{3}$$

Further, equation (4) represents a formula for calculating the age of batch entering retailer (E_i) where i denotes i^{th} batch or the transferred batch sequence (i = 1… n). The detailed procedure for modelling the age of batch can be seen in [10].

$$E_i = (i-1)\frac{T}{n} - (i-2)\frac{DT}{nP} \tag{4}$$

The information of Ei will be used to compute the completion time of the batch ($CompTime_i$) in retailer i.e. $CompTime_i = E_i + Ct$. Hence, the general procedure to calculate revenue in retailer (R_{ret}) is represented by equations (5), (6) and (7).

Case 1: If $CompTime < T_{\text{start}}$ then the revenue of the batch (R_{ret}) is maximized as all products are sold with p_{max} as defined by:

$$R_{\text{ret}}(n,T) = \frac{d\,p_{\max} Ct}{T} \tag{5}$$

Case 2: If $T_{\text{start}} \le CompTime$ then R_{ret} is defined by:

$$R_{\text{ret}}(n,T) = \frac{d}{T}\left[p_{\max}(T_{\text{start}} - E_i) + \int_{T_{\text{start}}}^{E_i + Ct} p(t)dt \right] \tag{6}$$

Case 3: If $T_{sl} < CompTime$ then R_{ret} is represented by:

$$R_{\text{ret}}(n,T) = \frac{d}{T}\left[p_{\max}(T_{\text{start}} - E_i) + \int_{T_{\text{start}}}^{T_{SL}} p(t)dt + p_{\min}(E_i + Ct - T_{SL}) \right] \tag{7}$$

Finally, the joint total revenue (JTR) in the SVSB system is represented by equation (8).

$$JTR(n,T) = \sum_{i=1}^{n} R_{\text{ret}\,i}(n,T) + R_{\text{manu}} \tag{8}$$

Hence, the joint total profit of the system (JTP) is shown by equations (9), (10), (11) and (12), where equation (11) is a constraint to ensure that the age of batch received by retailer is always younger than T_{start}.

Maximise:

$$JTP(m,T,n) = JTR(n,T) - JTC(m,T,n) \tag{9}$$

Subject to:

$$P \,\varepsilon\, D; \tag{10}$$

$$E_i < T_{\text{start}};\ \text{for } i = 1, 2, \ldots, n \tag{11}$$

$$T > 0; \tag{12}$$

The proposed model will be evaluated and compared to the benchmark model in the numerical test section.

3. Numerical test

In this experiment, parameters in [9] are adopted with some adjustment, as shown in table 1.

Table 1. Parameters in numerical test

Manufacturer	P	D	k	c_{raw}	A_{raw}	H_{raw}	c_{loss}	c_{mfc}	A_{mfc}	H_{mfc}
Value	19,200	4,800	0.01	5	1000	1	10	5	600	6
Retailer	c_{ret}	H_{ret}	A_{ret}	O_{ret}	T_{sl}	T_{start}	λ			
Value	30	7	25	50	0.208	0.167	200			

To determine the decision variables m, T and n, the genetic algorithm (GA) tool in Matlab 2009 is utilized. The GA's parameters used in this test are the same with those in [10]. The result can be seen in table 2.

Table 2. Total cost, revenue and total profit of the SVSB system

Benchmark Model [10]	Decision Variables		
	m=2	n=5	T=0.1486
	Manufacturer	**Retailer**	**SVSB**
Total Cost	132,672.81	146,350.29	279,023.10
Revenue	144,000.00	308,133.90	452,133.90
Total Profit	11,327.19	161,783.61	173,110.80
The Proposed Model	Decision Variables		
	m=2	n=5	T=0.1448
	Manufacturer	**Retailer**	**SVSB**
Total Cost	132,901.64	146,385.84	279,287.48
Revenue	144,000.00	311,997.10	455,997.10
Total Profit	11,098.36	165,611.26	176,709.62

As seen in table 2, the linear perceived value loss model leads to m = 2, n = 5 and T = 0.1486. These decision variables result in the total profit of $173,110.80. Meanwhile, the values of the decision variables in the exponential model are m=2, n=5 and T=0.1449 leading to $176,709.62 total profit (±$3,598 higher than the linear model). It means that using linear approach while managing product with exponential perceived value loss characteristic may lead to profit reduction (losing a chance to improve the profit until 2.08%).

Further, still from table 2 in the manufacturer column, it can be seen that the total cost of manufacturer increases around $228. Reducing the length of production cycle or T (±2.53%) can be one of the causes since it affects the setup cost. On the other hand, although the total cost in retailer also increases (around $35), the shorter T gives a significant benefit to retailer i.e. the total revenue increases up to $3,863.21. It seems that, in this case, using the shorter T when dealing with exponential perceived value products gives more advantages to retailer.

4. Conclusion

In this research, an integrated production-inventory model considering an exponential perceived value loss of finished goods has been addressed. The numerical test shows that the exponential model is potential to increase the profit if it is used to manage a product where the customer's WTP for the product nonlinearly decreases once the product approaches its expiration dates. For the given parameters, the performance of the exponential model is better than the benchmark model i.e. the joint profit of the proposed model significantly increases. However, the proposed strategy seems to give more benefit to retailer as the total cost of manufacturer slightly increases. Therefore, a benefit sharing scenario is worth to be proposed to encourage the manufacturer which may be resistant to adopt this strategy since it disadvantages them.

References

[1] Wee H-M 1998 Optimal buyer-seller discount pricing and ordering policy for deteriorating items *Eng. Econ.* vol 43 pp 151–68.

[2] Rau H, Wu M-Y and Wee H-M 2003 Integrated inventory model for deteriorating items under a multi-echelon supply chain environment *Int. J. Prod. Econ.* vol 86 pp 155–68.

[3] Wee H M, Chung C J and Yang P C 2008 An integrated vendor-buyer deteriorating item ordering policy *2008 International Conference on Machine Learning and Cybernetics* (IEEE) pp 3927–32.

[4] Chen T-H and Chang H-M 2010 Optimal ordering and pricing policies for deteriorating items in one-vendor multi-retailer supply chain *Int. J. Adv. Manuf. Technol.* vol 49 pp 341–55.

[5] Pahl J and Voß S 2014 Integrating deterioration and lifetime constraints in production and supply chain planning: A survey *Eur. J. Oper. Res.* vol 238 pp 654–74

[6] Rong A, Akkerman R and Grunow M 2011 An optimization approach for managing fresh food quality throughout the supply chain *Int. J. Prod. Econ.* vol 131 pp 421–9

[7] Blackburn J and Scudder G 2009 Supply chain strategies for perishable products: the case of fresh produce *Prod. Oper. Manag.* vol 18 pp 129–37

[8] Lütke entrup M, Günther H-O, Van Beek P, Grunow M and Seiler T 2005 Mixed-Integer Linear Programming approaches to shelf-life-integrated planning and scheduling in yoghurt production *Int. J. Prod. Res.* vol 43 pp 5071–100

[9] Fauza G, Amer Y, Lee S-H and Prasetyo H 2015 A vendor-buyer inventory model for food products based on shelf-life pricing *Oper. Supply Chain Manag. An Int. J.* vol 8 pp 67–73

[10] Fauza G, Amer Y, Lee S H and Prasetyo H 2015 An inventory model of production-inventory policy for food products considering quality loss in raw materials *2015 IEEE International Conference on Industrial Engineering and Engineering Management (IEEM)* (IEEE) pp 910–4

[11] Kim S-L and Ha D 2003 A JIT lot-splitting model for supply chain management: Enhancing buyer–supplier linkage *Int. J. Prod. Econ.* vol 86 pp 1–10

[12] Tsiros M and Heilman C M 2005 The effect of expiration dates and perceived risk on purchasing behavior in grocery store perishable categories *J. Mark.* vol 69 pp 114–29

IOP Conf. Series: Materials Science and Engineering **403** (2018) 012051 doi:10.1088/1757-899X/403/1/012051

Optimization of storage temperatures to maintain Lycopene content of tomato from moderate water stress irrigated greenhouse

A N Khairi[1], M A F Falah[2], A P Pamungkas[2] and N Takahashi[3]
[1]Departement of Food Technology, Faculty of Industrial Technology, Universitas Ahmad Dahlan, Yogyakarta, Indonesia
[2]Departement of Agroindustrial Technology, Faculty of Agricultural Technology, Universitas Gadjah Mada, Yogyakarta, Indonesia
[3]Departement of Biomechanical System, Faculty of Agriculture, Ehime University, Matsuyama, Japan

amalya.khairi@tp.uad.ac.id

Abstract. This research objective is to analyze the storage temperatures that optimize lycopene of tomato (*Lycopersiconesculentum*, cv.Taiankitijitsu) from moderate water stress irrigation and to compare it with normal irrigated tomato in hydroponic greenhouse system. Harvested tomatoes were stored in various temperatures (10, 15, 25 and 30 °C) until mature stage and lycopene content was observed with destructive method using spectrophotometer. Optimum temperature was analyzed by Minitab 15 Statistical Software using *surface response methodology*. After 16 days of storage, lycopene of water stress tomato attained 8,152 mg/100 g (10 °C), 11,636 mg/100 g (15 °C), 11,747 mg/100 g (25 °C), and 11,947 mg/100 g (30 °C) respectively. Storaging tomato at 10 °C remained the lycopene of water stress tomato in stable trend. Temperatures above 15°C increased lycopene of both water stress and normal irrigated tomato. Lycopene of normal irrigated tomato increased only for first 8 days with temperature 15 °C (slightly increase), 25 °C, and 30 °C (highest lycopene value). After 8 days of storage, the trends tend to drop. Analysis of multivariate multiple regressions showed low R^2 value of lycopene during storage. Optimization using Minitab 15 Statistical Software showed temperature of 24.52°C with 11 days storage and might reached 10.71 mg/100g.

1. Introduction

Plants suffer from water shortages when supplies to the roots are limited or when the transpiration process is excessive. Water stress is largely occurred due to water shortages from drought or salinity of soil is high. In the case of high soil salinity or other conditions such as flooding or low temperatures of soil, soil water is available however plants cannot absorb it, or often called ‘physiological drought’. Water stress is caused by the retain of water affects physical changes on the environment and physiological crop [1].

One of the most popular tomato was a sweet tasting tomato called “tomato fruit” or “high brix tomato”, “high sugar content tomato”, and “high lycopene tomato”. High lycopene tomato was developed to answer the needs on functional quality of fresh product with a high market demand. However, the selling price was very high; reached ¥10.000/kg [2]. This tomato was produced by imposing abiotic stress. Ehime University Greenhouses evolved a mid-level controlled irrigation called moderate water stress treatment for tomato implantation [3]. The lycopene content of tomato is well known for its impact on inhibiting cancer cell developments. Therefore, the method to preserve its quality is necessary to develop.

Tomato is climacteric and easily rotten fruit that requires technology to retard the decay process, preserve qualities, and prolong postharvest shelf life [4]. The methods to restrain the decay of tomatoes were widely developed. However, the study of effectiveness of those methods to preserve quality especially water stress tomato was very limit. The context of this study is to find the simply best method to store and preserve the lycopene content of tomato from water stress irrigation. The lycopene content of water stress tomato and no stress tomato after storage will be compared after adjusting some storaging methods.

Generally, fresh products require low temperature and high relative humidity to reduce physiological and biochemical decay, also microbiological activity that are major causes of spoiling process occurred [5]. As temperatures play significant role in preserving the postharvest qualities of normal irrigated tomato, the objectives of this study was to find the effects of temperatures to lycopene content of water stress irrigated tomato. Optimization of storage temperatures will be conducted to obtain the best storage condition for water stress tomato.

2. Material and method

Tomato (*Lycopersicon esculentum*, cv.Taiankitijitsu) was sowed hydroponically with rockwool slabs in greenhouse of Ehime University, Japan. Average daily air temperature during the day was 22 °C and relative humidity was 70%, while at night was 13 °C and 83%. Tomatoes were irrigated with moderate water stress and normal irrigation. moderate water stress irrigation began at week 14 after sowing with 83% of nutrient water subtracted from normal irrigation which is 100%. After the 15th week reduced again 63% of normal irrigation be diminished. Tomatoes were harvested at the 21st week at ripening levels (bright reddish color). After harvesting, tomatoes were put at 10 °C, 15 °C, 25 °C, and 30 °C chambers for 16 days storage [3].

2.1. Lycopene content analysis

The content of lycopene was observed with absorbance value determined from spectrophotometer. First step was extracting juice tomato to obtain the standard of lycopene solution. The determination of solution absorbance value was done with spectrophotometer (U-1900 Hitachi spectrophotometer, Japan) clarity check at 505 nm wavelength. The measurement of absorbance value used blank solvent, acetone, and used equation (1) to obtain the value of lycopene content.

$$Lycopene\ content = (10 \times absorbance\ value)/(0.315 \times sample\ volume\ (gr)) \quad (1)$$

2.2. 3-D plot, multiple regression, and optimum storage condition

The (x,y,z) plot was done by Matlab Simulink computational software and multiple regression by Ms. Excel to make estimation of each quality change caused by temperature and time. Temperature as x axis, storage time as y axis, and quality parameters as z axis.

a) Test of linearity and curve estimation was done by SPSS 16 that analyzed the value of R^2 of the model. The highest value of R^2 have been considered as the best model of quality parameters changes during storage.
b) Goodness of fit was done by test of significancy of SPSS 16.
c) Optimization of Storage Temperature was done by Minitab 15 Statistical Software using *surface response methodology* to select the best temperature to store the tomato.

3. Results and discussion

3.1. Lycopene content

Changes in lycopene content during storage could be seen in figure 1. The initial conditions of lycopene of both water stress and normal irrigated tomato were relatively similar. Initial lycopene content of water stress tomato was 7.024 mg/100g and 6.459 mg/100g of normal irrigated tomato.

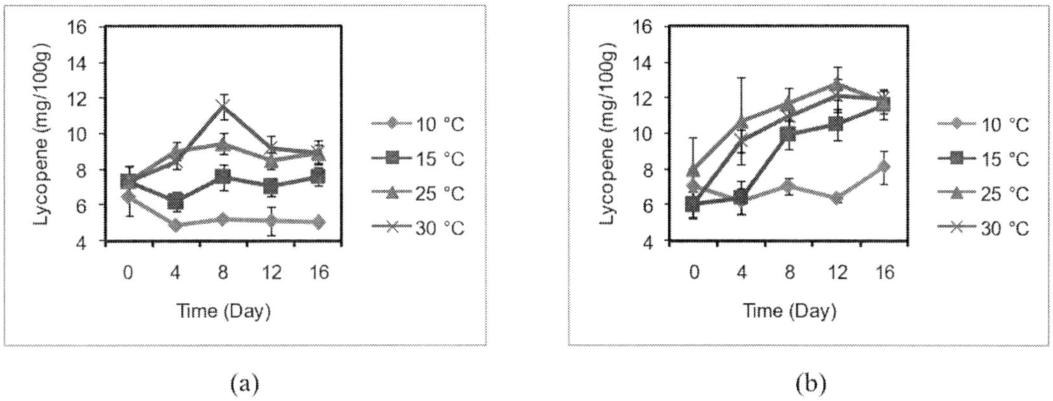

Figure 1. Lycopene value during storage (a) normal irrigated tomato (b) water stress tomato.

Lycopene of water stress tomatoes showed increase trend at more than 10 °C storage temperature, while at 10 °C increased slightly. At day 16, the lycopene of water stress tomato reached 8.152 mg/100 g, 11.636 mg/100 g, 11.782 mg/100 g, and 11.947 mg/100 g for 10 °C, 15 °C, 25 °C, and 30 °C. The highest value obtained at day 12 of 25 °C storage temperature for water stress tomato. Normal irrigated tomato with 10 °C storaging slightly increasing of lycopene content occurred in first 4 days and stable at day 4 until day 16. Storaging at 10 °C and 15 °C did not significantly affect lycopene of the normal irrigated tomato. Normal irrigated tomato increased the lycopene content only for 8 days storage with temperature 15 °C, 25 °C, and 30 °C. More than 8 days storage showed the value of lycopene tended to decrease. Meanwhile in water stress tomato only on 10 °C storage temperature the lycopene content value was stable, relatively. Higher temperatures (25-30°C) increase the lycopene content of water stress and normal irrigated tomato. In this study, temperatures of 15 °C, 25 °C and 30 °C induce an increase trend in lycopene of water stress tomato during storage.

Storage temperature is significantly affect the lycopene of tomato harvested at mature and light red level. Low temperature induces low lycopene content of tomato during storage. An increased of lycopene content occurs as a correspond of tomato ripening process [6]. The development of lycopene structure will be optimized at temperatures of 12 °C and 32 °C [7]. Some research results also suggest that the optimal temperature for increasing tomato lycopene content is 16-18 °C and 26 °C [8]. Temperatures above 25 °C will increase respiration rate, red color pigment formation, and develop more ethylene than fruit in chilled storage [9]. Treatment with hot water (60 °C for 3 minutes) before storage prolong the shelf life of tomatoes up to 29 days [10]. In fresh cut tomatoes, storage with temperatures above 8 °C might increase lycopene content [11].

3.2. Multiple Regression and 3-D Plot of Lycopene Content

Lycopene parameter during storage of water stress tomatoes were higher than no water stress tomatoes in all temperatures respectively. Significant increase occurred on 30 °C storaging temperature for water stress tomato.

In this study, temperature and storage time become two independent variables that caused the changes of lycopene content of no water stress and water stress tomato. Multiple regression model of quality parameters calculated to estimate the changed of each parameter caused by storage temperature and time. The mathematical models then were tested with R^2 value to confirm the applicability of the models. The result of this analysis shown in figure 2 that explain the 3D-plot graphics using Matlab made to see the surface of quality changing during storage.

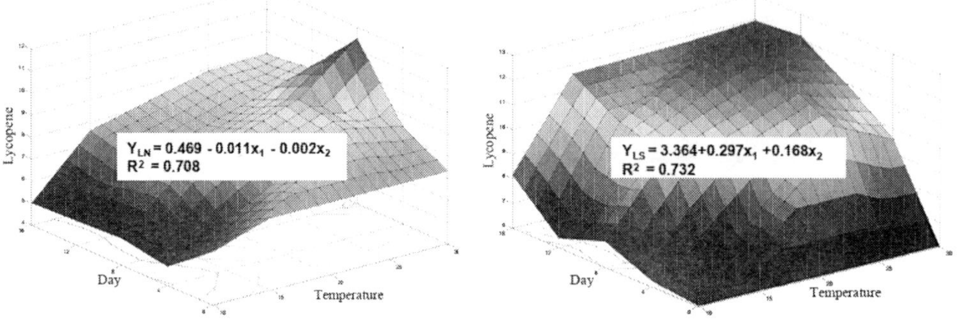

Figure 2. 3-D plot of lycopene content of tomato.

3.3. Optimization of storage condition of water stress tomato

Temperatures that good for water stress tomato were different according to the quality parameter that want to be preserved. Physical qualities (firmness and weight) preservation better to be stored with low temperature (10°C) and chemical qualities parameters (lycopene content) was better to be stored with room temperature. For next analysis, optimum storage condition to store water stress tomato will be analyzed further. The step for optimization were:

3.3.1. Goodness of fit to be optimize. This step completed with test of linearity, significant test, and curve estimation using SPSS. Curve that has high coefficient correlation (R^2) and significant to error degree α, will be considered as the curve that fit the reality. The result of this analysis shown in table 1. The table below showed that the lycopene fit to be optimized (linear, quadratic, and cubic curve).

Table 1. Test of significancy (α=0.05).

Quality Parameters	Regression Function	Significancy	Fittess To Be Optimized
	Linear	0.001	√
Lycopene	Quadratic	0.004	√
	Cubic	0.015	√

- √ = fit to be optimized

Lycopene content increased 100.91% compared with initial value in storaging at 30 °C. Using this result, lycopene content of water stress tomato will be optimized to get the optimum storage condition. This result also supported many previous study that water stress tomato is 'high lycopene content' tomato thus lycopene was the most important quality parameter inside the tomato.

3.3.2. Test of surface response methodology. This analysis was done by the test of steepest ascent method and supported by Minitab 15 statistic software and was finalized with surface and contour analysis. This test started with optimizing model analysis to get fitted model. The x_1 was temperature, x_2 was time, and x_3 was lycopene factor estimated by Minitab. With degree of significancy, α=0.05, the result of this test shown in Fig 4. below. The optimization model of lycopene content of water stress tomato was shown in equation (2).

$$Y=10.53+0.21x_1+0.82x_2+0.4951x_3-0.51x_1^2-1.46x_2^2-0.096x_3^2+0.595x_1x_2+0.24x_1x_3+0.13x_2x_3 \quad (2)$$

a) Model goodness of fit test

Significancy or model fit will be obtained if the P_{value} and *lack of fit* value was higher than degree of significancy. From figure 2, we can see that the P_{value} of all regression coefficients and *lack of fit* value were higher than 0.05. This result can be considered that the model fit the optimization model of lycopene content of water stress tomato.

b) Optimization test

Test of optimum storage condition was done by surface and contour plot analysis. Lack of fit value from the analysis was 0.27. The result that was done by Minitab 15 shown in table 2 and figure 3.

Table 2. Regression analysis test for optimizing model.

	Constant	Temperatures	Times	Lycopene	Temperatures vs temperatures	Times vs times	Lycopene vs Lycopene	Temperature vs times	Temperatures vs lycopene	Times vs lycopene
F	0.00	0.785	0.298	0.522	0.496	0.071	0.898	0.555	0.813	0.899

.

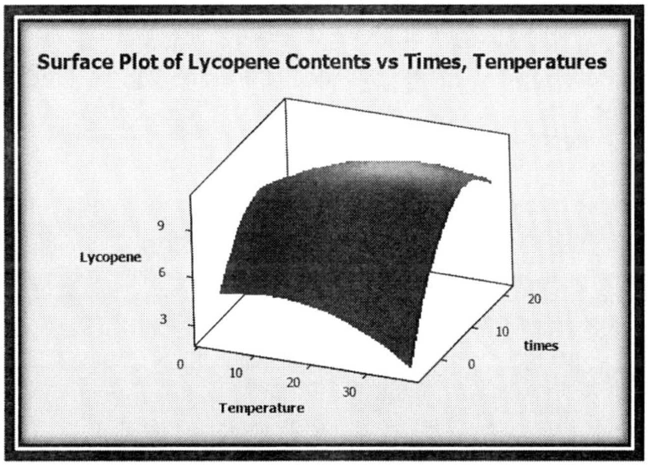

Figure 3. Surface plot of lycopene vs times and temperatures.

From figure 3, optimum value of lycopene was in range of 20-30 (temperatures axis) and range of 10-20 (times axis). To know the exact value of optimum storage condition, contour plot was conducted, shown in figure 4.

Optimum value of lycopene content was reached at:
Temperature: 24.52°C.
Time: 11 days.
Lycopene: 10.71 mg/100g.

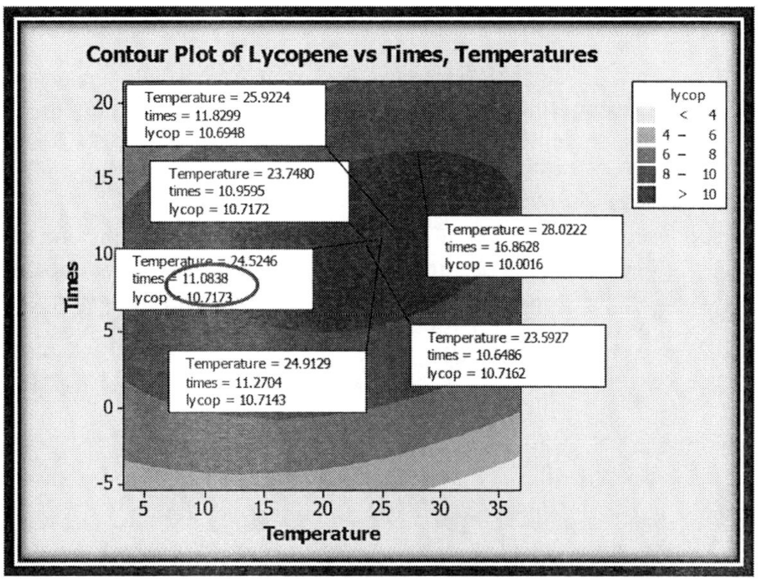

Figure 4. Contour plot of lycopene vs temperatures and times (*lycop= lycopene content).

From figure 5, the darker the green color, the higher the value of lycopene. To get the optimum value, plant flag analysis was conducted with randomly shot on the darkest green area. From the value in the boxes of the figure, highest value of lycopene was considered as optimum storage condition.

4. Conclusion

Temperature above 10 °C increase lycopene of water stress irrigated tomato. On 12nd day of storaging in 25 °C, lycopene reached the highest value. After 16 days of storage with temperatures of 10 °C, 15 °C, 25 °C, and 30 °C, lycopene of water stress tomato attained 8,152 mg/100 g, 11,636 mg/100 g, 11,747 mg/100 g, and 11,947 mg/100 g respectively. The optimal storage temperature of water stress tomato was 24.52 °C during 11 days storage with lycopene content potentially reach 10.71 mg/100g, analyzed by Minitab 15 software for optimization. Tomatoes with normal irrigation showed nearly stable trends of lycopene content on 10 °C and 15 °C storage temperatures. Lycopene of normal irrigated tomato increased only for first 8 days with temperature 15 °C (slightly increase), 25 °C, and 30 °C (highest lycopene value of normal irrigated tomato). After 8 days of storage, lycopene contents showed the trends that tend to drop.

References

[1] Kebede E Processing of horticultural produces in ethiopia 1991 *Acta Hortic.* vol 270 no 1 pp 270–301
[2] Takayama K and Nishina H 2007 Early detection of water stress in tomato plants based on projected plant area *Environ. Control Biol.* vol 45 no 4 pp 241-9
[3] Khairi, AN Falah M A F, Suyantohadi A Takahashi N Nishina H 2015 Effect of storage temperatures on color of tomato fruit (*solanum lycopersicum* mill.) cultivated under moderate water stress treatment *Agriculture and Agricultural Procedia* vol 3 pp 178-83
[4] FAO (Food and Agriculture Organization, United Nations) 2014 Tomato production *www.fao.org*. Accessed 8th May 2014
[5] Kader AA 1986 Effects of postharvest handling procedures on tomato quality *Acta Hortic* vol 190 pp 209–21
[6] Javanmardi J Chieri K 2006 Variation of lycopene, antioxidant activity, total soluble solids and weight loss of tomato during postharvest storage *Postharvest Biology and Technology*; pp 41151–5
[7] Leoni C 1992Industrial quality as influenced by crop management *Acta Hortic.* vol 301 pp 177–84.
[8] Turk R Seniz V Ozdemir N Suzen MA 1994 Changes in the chlorophyll carotenoid and lycopene contents of tomatoes in relation to temperature *Acta Hortic*. vol 368 pp 856–62
[9] McDonald RE McCollum TG Baldwin EA 1999 Temperature of water heat treatments influences tomato fruit quality following low-temperature storage *Postharvest Biology and Technology* vol 16 pp 147–55
[10] Islam MP Morimoto T Hatou K 2012 Storage behavior of tomato inside a zero energy cool chamber *Agric Eng Int: CIGR Journal Open access* vol 14 no 4 p 209
[11] Lana MM Kooten OV 2005 Effects of cutting and maturity on lycopene concentration of fresh-cut tomatoes during storage at different temperatures *Proc. 5th Int. Postharvest Symp Acta Hortic. ISHS* p 682

IOP Conf. Series: Materials Science and Engineering **403** (2018) 012052 doi:10.1088/1757-899X/403/1/012052

Stability analysis of trawls type traditional fishing boat with modification of eco-friendly fishing-gear on the north coast of Central Java

Kiryanto, M Ridwan, B A Adietya and D Chrismianto

Department of Naval Architecture, Faculty of Engineering, Diponegoro University, Semarang, Indonesia

deddychrismianto@yahoo.co.id

Abstract. The prohibition of the use of fishing gear trawls in Indonesia's fishing territory requires fishermen to make replacements of fishing gear against their boats. The replacement of fishing gear on the boat affects the performance of the boat. This study was conducted to determine the extent to which the replacement can be applied to traditional fishing boat Trawls North Coast of Central Java. The study focused on the ring net and gill net as a catch tool. Based on the results of stability and seakeeping analysis, it can be concluded that the replacement of fishing gear is influenced by the loading factor of the used fishing gear equipment. The vessel with the replacement of the ring net and gill net has a stability below the value of stabilities owned by the traditional trawler of Juwana. For the results of motion analysis, the vessel has met the Tello criteria of acceptance of fish ship characteristics for vessels with full velocity and wave as high as 2m. The vessel has the highest rolling value of 3.31 deg, and slamming probability with the greatest value of 2.10% for speed 6 knots and 180 deg heading on head sea condition.

1. Introduction

Regulation of Ministry of Marine Affairs and Fisheries No. 2/Permen-KP/2015 has a big influence that must be resolved soon. Basically a significant change is very difficult to quickly apply. Fishermen need time and solutions of any changes or new rules that affect their lives. One solution that can be done is to impose a transition period and the transfer of fishing gear with other fishing gear according to the characteristics of the ship and of course the permissible fishing gear. In accordance with Ministerial Decree No. KP.KEP.06/MEN/2010 and Ministerial Regulation No. 2/2015 there are 8 (eight) fishing gear that is allowed in Indonesia.

The replacement of fishing gear on the boat affects the performance of the boat itself. The change of fishing gear must prioritize several factors on the vessel which will be very influential when the ship is in sailing condition. Ships must maintain good stability and seakeeping for safety and operational convenience.

In this study it is important to know the extent to which fish gear replacement can be effectively applied to Juwana's traditional Trawl fishing boats. The study focused on gear nets and ring nets as a catch tool. The selection of the gill net for gill net and purse seine for the ring net is based on the catch target, the operational ease of the net, as well as the habits of the origin of the traditional ship. Researchers will analyze the stability and movement as a result of replacement fishing gear.

The purpose of this research are: to get a good general arrangement and equipment layout to modify the trawler with gill net and ring nets, and getting the changes in stability and seakeeping of the boat due to the replacement of fishing gear type Trawls with ring nets and gill nets.

2. Literature review

Ring net, according to National Oceanic and Atmospheric is a large net that forms walls deployed to catch clumped fish. According to the book Fishing Boat [1], the fish that are the catch of the ring net are "pelagic shoaling species" fish which means the fish must form a gang, close to the surface of the water and it is desirable that the fish gang is high, which means the distance of fish with other fish should be as close as possible.

Gill nets are a type of fishing tool that has long been used but still able to survive until now. The gill net is a rectangular net with the same mesh size. On the top side of the net is equipped with several floats and the bottom is equipped with sinker. The size of fishing vessels of this type of gill nets is very varied from small to large vessels operated in the open sea. Species of fish caught include *tetengkek* (*Megalacpis cordyla*), flying fish (*Cypselurus sp*), mullets (*Mugil sp*), *kuro* fish (*Polynemus sp*), *alu-alu* (*Sphyraena sp*), and mackerel (*Scromberomorus commersoni*).

The mandatory stability requirements of ships shall refer to the standards established by the local Indonesian Bureau of Classification or the Marine Authority such as International Maritime Organization (IMO). Thus the process of stability analysis carried out must be in accordance with IMO (International Maritime Organization) standards [2]. Code A.749 (18) Ch 3 - design criteria applicable to all ships Section A.749 (18), Chapter 3.1.2.3: The area under the curve GZ at oscillation angle 0°- $30^o \geq 3{,}101$ m.deg, area 0°- 40° (deg) $\geq 5{,}157$ m.deg and area 30°- $40^o \geq 1{,}719$ m.deg. Chapter 3.1.2.2: Maximum GZ value occurring at an angle of 30°-$180^o \geq 0.2$ m. Chapter 3.1.2.2: angle at the maximum GZ value $\geq 25\,^o$. Chapter 3.1.2.4: Early GM at $0^o \geq 0.15$ m.

In the study of ship movement, the movement under consideration is a movement that can only be responded by the ship, namely rolling, heaving, pitching [3]. The seakeeping fruits are evaluated by adjusting the seakeeping criteria standards depending on the type of vessel. In this study, the standard seakeeping used is the standard criterion [4] contained in table 1.

Table 1. Seakeeping criteria for fishing boat.

No	Criteria	Value
1	Green water deck	5% (prob)
2	Slamming	3% (prob)
3	Propeller emergence	15% (prob)
4	VA at Bridge	0.2 g (rms)
5	LA at Bridge	0.1 g (rms)
6	VA at work deck	0.2 g (rms)
7	LA at Work deck	0.1 g (rms)
8	Roll	6° (rms)
9	Pitch	3° (rms)

The ship movement response to regular waves is expressed in RAO (Operator Amplitude Response), where RAO is the ratio between ship's amplitude of movement (either translation or rotation) to wave amplitude at certain frequencies. The RAO movement's response to translational motion is a direct comparison between the amplitude of the ship's movement (Z0) with wave amplitude ($\zeta 0$) (both in unit length) as shown in equation (1).

$$RAO = \frac{z_0}{\zeta_0} \qquad \text{(m/m)} \qquad (1)$$

While the rotational motion is the ratio of the amplitude of the rotational motion (in radians) with the wave inclination which is the multiplication of wave numbers, kw = ω2 / g with wave amplitude as shown in equation (2).

$$RAO = \frac{\theta_0}{kw\,\zeta_0} = \frac{\theta_0}{(w^2/g)\zeta_0} \qquad \text{(rad/ rad)} \qquad (2)$$

In fact, the waves at sea are random waves so the ship's response to the regular waves expressed in the RAO cannot describe the ship's response to the real state of the ocean. To obtain a motion response of ships against random waves can be described by the response spectrum [5] and [6]. Probability of slamming can be calculated using equation (3).

$$Nw = \frac{1}{2\Pi}\sqrt{\frac{2m_{OR}}{2m_{2R}}} \text{ x } Pr\ (slamming) \qquad (3)$$

where:

$Pr\ (slamming) = exp^{-y}$

$y = \frac{T^2}{2m_{0s}} + \frac{V_{cr}{}^2}{2m_{2s}}$

T = draught of ship

V_{cr} = velocity threshold 0,093 $(gl)^{1/2}$

Nw = intensity of *slamming* per second

Mor = area under curve spectrum response moment in zero 0

M_2r = area under curve spectrum response moment 2nd

Probability of deck wetness can be calculated using equation (4).

$$Pr\ (deck\ wetness) = exp^{-f} \qquad (4)$$

where:

$f = \frac{T^2}{2m_{0s}}$ = effective Freeboard

T = draught of ship

Mor = area under curve spectrum respon moment in zero 0

3. Results and discussion

This study used four (4) variations or changes to the general plan layout of each type of fishing gear used in accordance with the loading of the fishing gear equipment [7, 8]. Given the replacement of fishing gear on a fishing boat will cause a change of ship's center of gravity which will affect stability.

Laying of equipment above deck or change of layout of general plan baracu at Indonesian Fish Shipowner from BPPI. The following is a drawing of the ship's general plan layout plan after reconstruction of the fishing gear and auxiliary equipment on the deck. Figure 1(a) shows the general arrangement of trawl at Juwana area, figure 1(b) shows the arrangement after changed using ring net and figure 1(c) shows the arrangement after changed using gill net. Figure 2(a) shows the general arrangement of trawl at Tegal area, figure 2(b) shows the arrangement after changed using ring net and figure 2(c) shows the arrangement after changed using gill net. Figure 3(a) shows the general arrangement of trawl at Batang area, figure 3(b) shows the arrangement after changed using ring net and figure 3(c) shows the arrangement after changed using gill net.

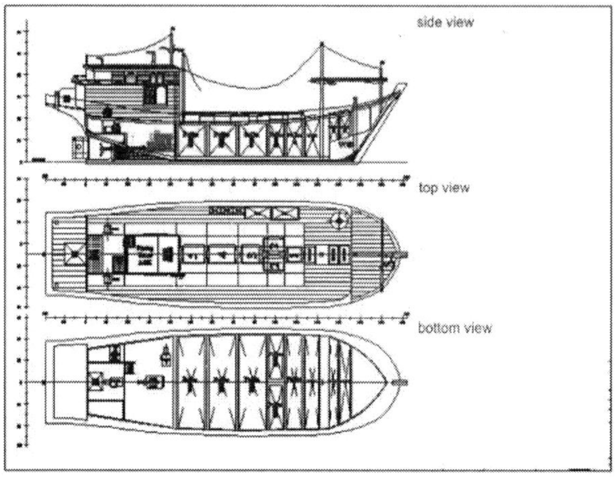

(a) General arrangement of trawl at Juwana area

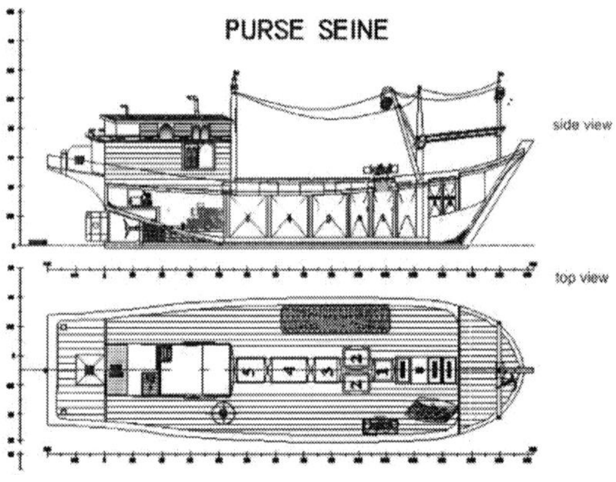

(b) The arrangement after changed using ring net

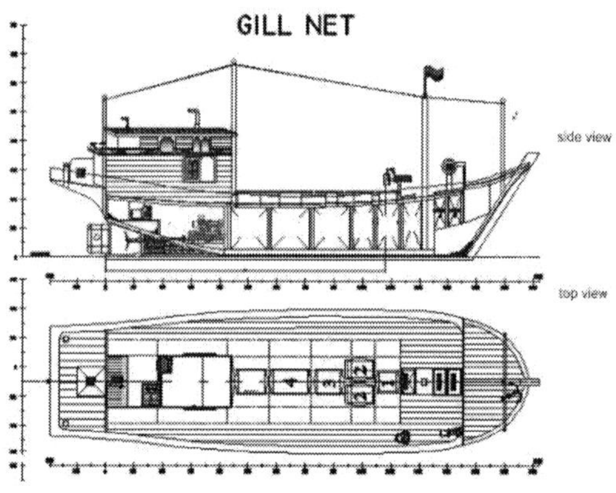

(c) The arrangement after changed using gill net

Figure 1. The ship's general plan layout plan after reconstruction at Juwana area.

The calculation of ship stability performed in this study analyzes the 3 conditions of each model that have been made, in accordance with IMO A.749 (18) Chapter 3.5. As for the 3 conditions are:

1) Condition I: Represents a ship condition before departing from port to fishing ground. Under these conditions, the vessel in a consumable charge state is empty.
2) Condition II: The ship is already on the fishing ground. In this condition, the ship has been given additional cargo ship (mackerel) 50%.
3) Condition III: The ship is still on the fishing ground. In this condition, the boatload (mackerel) is 100%.

IOP Conf. Series: Materials Science and Engineering 403 (2018) 012052 doi:10.1088/1757-899X/403/1/012052

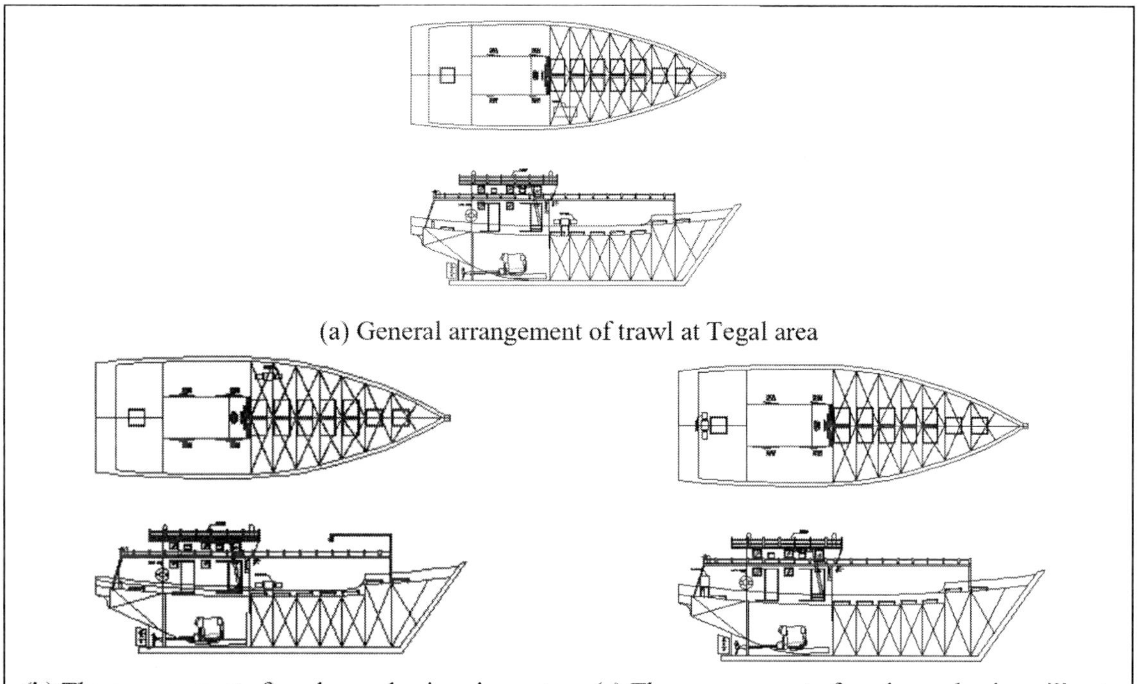

(a) General arrangement of trawl at Tegal area

(b) The arrangement after changed using ring net (c) The arrangement after changed using gill net

Figure 2. The ship's general plan layout plan after reconstruction at Tegal area.

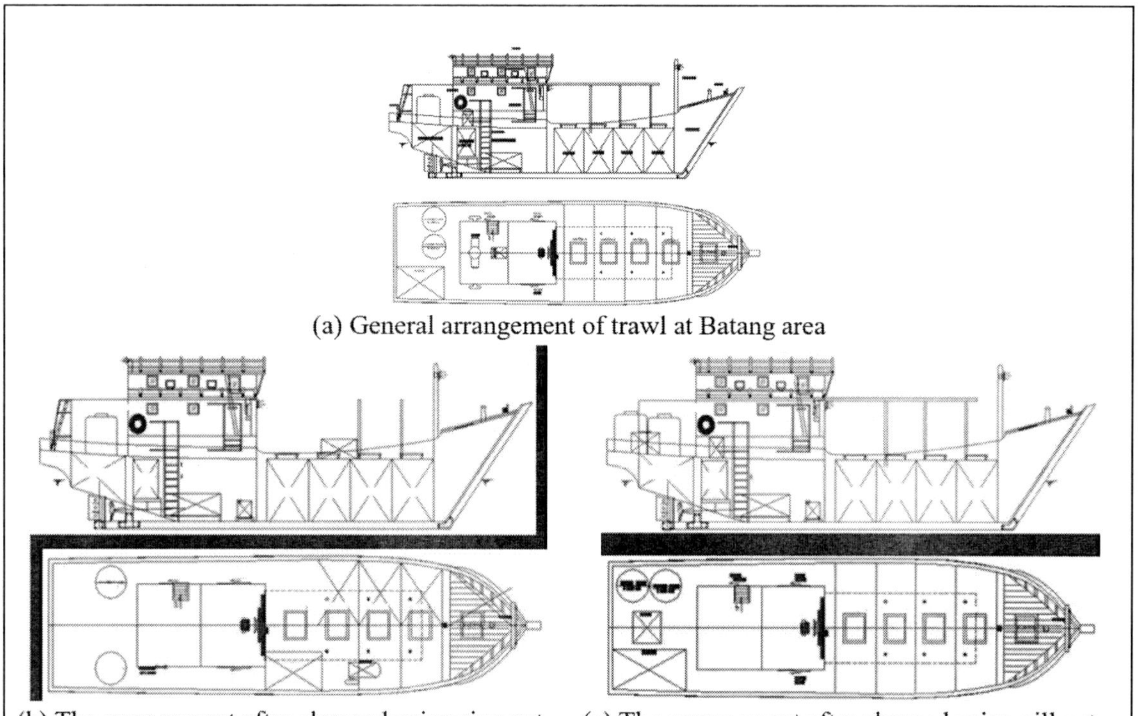

(a) General arrangement of trawl at Batang area

(b) The arrangement after changed using ring net (c) The arrangement after changed using gill net

Figure 3. The ship's general plan layout plan after reconstruction at Batang area.

The stability of the ship can be determined by looking at the ship's static stability curve. The stability curve shows the value of the control arm (righting arm) at different angle values. A comparison of the GZ value of the vessel with each capture device and loading conditions is carried out to determine the best loading conditions owned by the ship and the layout model that best meets the criteria of stability acceptance based on IMO. The comparison of stability analysis results for trawling boats in Juwana, Tegal, and Batang areas and all layouts with change of fishing gear in each condition are shown in figure 4, figure 5 and figure 6.

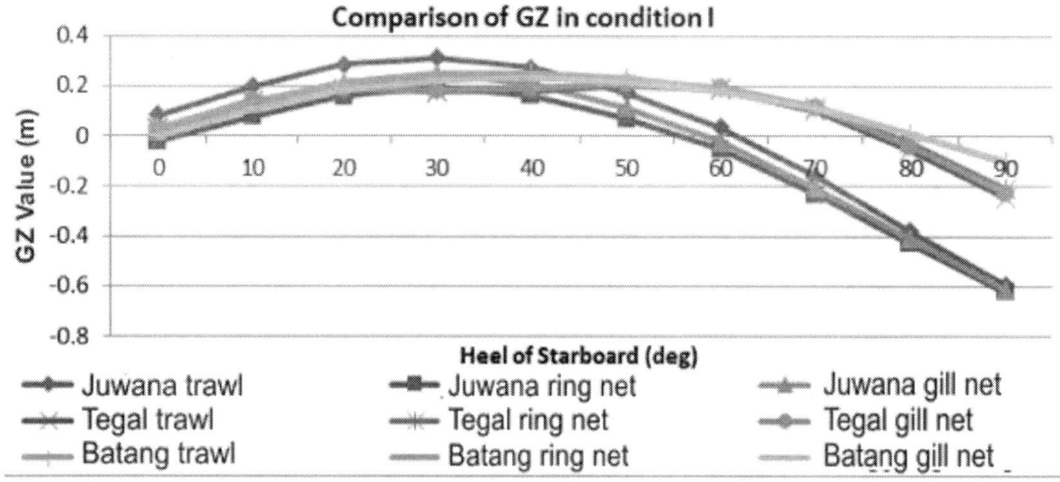

Figure 4. Comparison of GZ value in condition I.

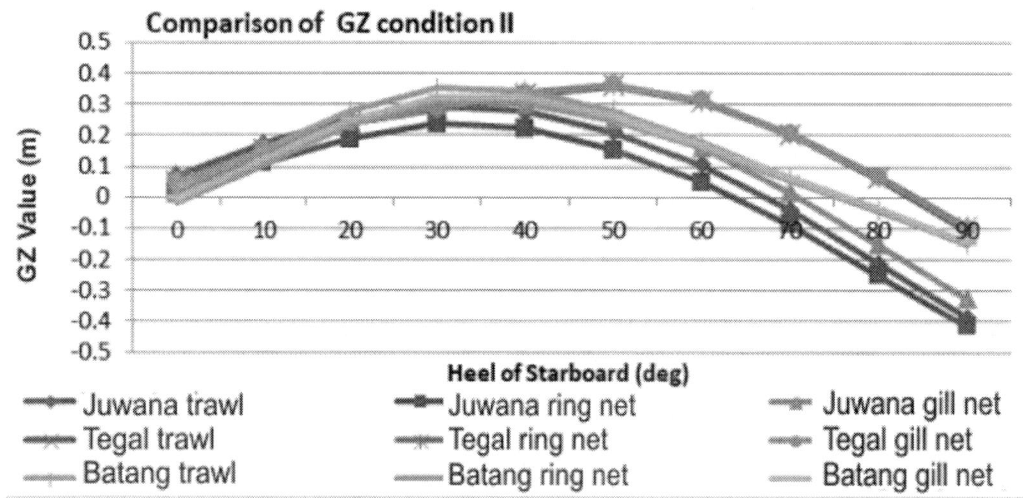

Figure 5. Comparison of GZ value in condition II.

Comparison of some GZ arm graphs to the three conditions presented above indicates that the best stability or value of the highest IMO acceptance criteria for each ship loading condition as well as the modeling of the layout of the respective equipment occurs in the third condition. Condition when the ship is in the fishing ground area. In this condition, the boatload (mackerel) is 100%.

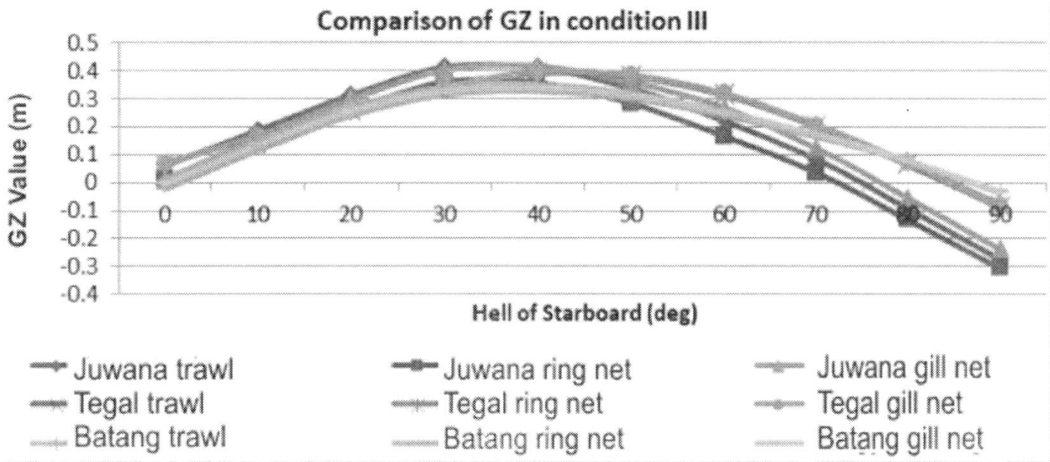

Figure 6. Comparison of GZ value in condition III.

The stability of the vessel with the replacement of fishing gear has a stability value that is still below the stability value that owned by the Juwana's traditional trawl boat of KM *Sumber Mino Mulyo* and Batang MV *Barokah Rezeki*. For the largest GZ arm values on boats with the addition of fishing gears are boats with gill net gear for Juwana boat, and ring net for Tegal and Batang boat.

4. Conclusions

A general plan of good equipment layouts for modifying trawlers with gill nets and ring nets can be done taking into account the weight of the apparatus, the weight of the apparatus is adjusted to keep the vessel's point of weight balanced transversely and without extreme trim. Of the 8 layouts that have been analyzed then obtained the best results of the layout to 4 of each substitute fishing gear. The 4th layout has an arrangement with equipment and load points located backward to the rear of the stern with a LCG nets below 5.2 m, and LCG from a total of 5.47 m vessels for ring net and 5.74 m for gill nets, and a TCG valued at 0.02 m to the left of the girder ship's deck center.

Stability analysis of vessel with replacement of fishing gear equipment with three conditions has met IMO acceptance standard. The value of area under GZ curve at oscillation angle 0°- 30° > 3.101 m.deg, area 0° - 40° > 5.15 m.deg and an area of 30° - 40° > 1.719 m.deg. As well as the maximum GZ value occurring at an angle of 30° - 180° > 0.2 m, the angle at the maximum GZ value > 25°, and the initial GM score at 0° > 0.15 m. However, there is a Layout that does not meet the IMO stability acceptance parameter, which is the first Layout plan of ring net. GZ arm values at 30 deg or more less than 0.2 m or only have a value of 0.194 m. Analysis of ship movement has fulfilled the Criteria of Admission of fish ship characteristics for vessels with full speed and wave as high as 2 m vessel has the highest rolling value of 3.31°, and probability slamming with the greatest value 2.10% for vessel velocity 6 knot and corner 180 deg heading head sea.

Acknowledgement

Direktorat Riset dan Pengabdian Masyarakat Direktorat Jenderal Penguatan Riset dan Pengembangan Kementerian Riset, Teknologi, dan Pendidikan Tinggi according with grant No: 007/SP2H/LT/DRPM/V/2017.

References

[1] Ayodhyoa A U 1985 Fishing Boat Corespondence Course Centre IPB Bogor
[2] Hind J A 1982 *Stability and Trim Of Fishing Vesels And Other Small Ships,* Second Edition (Famham, Surrey, England: Fishing News Book Ltd.)

IOP Conf. Series: Materials Science and Engineering **403** (2018) 012052 doi:10.1088/1757-899X/403/1/012052

[3] Robert F B 1998 *Motion In Waves and Controllability, Principles of Naval Architecture Volume III* (USA: The Society of Naval Architecture and Marine Engineers)

[4] Tello M, Ribeiro e Silva S and Guedes Soares C 2010 *Seakeeping Performance of Fishing Vassels in Irregular Wave* (Elsevier)

[5] Iqbal M and Rindo G 2016 Pengaruh anti Slamming Bulbous Bow Terhadap Gerakan Slamming Pada Kapal Perintis 200 DWT *KAPAL* vol 13 no 1

[6] Iqbal M and Rindo G 2015 Optimasi Bentuk Demihull Kapal Katamaran Untuk Meningkatkan Kualitas Seakeeping *KAPAL* vol 11 no 1 pp 19-24

[7] Chrismianto D and Dong J K 2014 Parametric bulbous bow design using the cubic Bezier curve and curve-plane intersection method for the minimization of ship resistance in CFD *J Mar Sci Technol* vol 19 pp 479–92

[8] Chrismianto D, Zakki A F, Arswendo B and Dong J K 2015 Development of Cubic Bezier Curve and Curve-Plane Intersection Method for Parametric Submarine Hullform Design in order to Optimize the Hull Resistance by Using CFD *Journal of Marine Science and Application* vol 14 issue 4 pp 399–405

The use of importance-performance analysis for Indonesian smoked fish production strategy

H Mastrisiswadi, D N Izzhati and T Talitha

Industrial Engineering Department, Universitas Dian Nuswantoro

mastrisiswadi@dsn.dinus.ac.id

Abstract. Food and beverage industry is currently facing tight competition, especially due to ASEAN free market (MEA). Local products in Indonesia including smoked fish products must compete with imported products in ASEAN region. In order to be competitive in global market, Indonesian smoked fish industry must understand the current product quality position to formulate the right strategy. One of the best methods is the Importance Performance Analysis (IPA) that aims to measure the relationship between consumer perceptions and the priority of product quality improvement. This research result that the Indonesian smoked fish production strategy has to improve two variables which are hygiene and safety for consumption of the product. These two variables are categorized as action and located in concentrate quadrant. It means that these two variables have high priority of importance to the customer but less performance. In order to make Indonesian smoked fish competitive in ASEAN market competition, these two variables need to be improved.

1. Introduction

Indonesia is a maritime country where most of its territory is oceans. As a maritime country, Indonesia has a wealth of abundant seafood, as well as its processing product. One of the fish processing product that has a good prospect in the future is smoked fish [1].

Smoked fish is the result of fish processing product that combines salting and smoking process to reduce the water content in the fish meat that aims to prevent the bacteria live. It will make the rotting process become longer. The smoking process is easy, it does not require large power, and it is cheap so these industries significantly emerge.

The food and beverage industry is currently facing tight competition [2] especially due to ASEAN free market (MEA). Local products in Indonesia including smoked fish products must compete with imported products in ASEAN region. In order to be competitive in global market, Indonesian smoked fish industry must understand the current product quality position so that it can formulate the right production strategy. One of the best methods to do is the Importance Performance Analysis (IPA) that aims to measure the relationship between consumer perceptions and the priority of product quality improvement [3].

2. Smoked fish

Smoked fish is one of traditional fish processing product that using the smoke to preserve, colorize and add the distinctive flavor of the fish [4]. Principally, this fumigation process has several stages that are:

salting, drying, heating, and fumigation. But in general, the fumigation is carried out with the following steps:

2.1. *Preliminary treatment*
Preliminary treatment is the process of preparation for the next process. In this process the grouping is done based on the type, size and quality of freshness to be cleaned (removed the contents of the stomach and gills) before it can be processed further.

2.2. *Salting*
After the preliminary treatment, the next step is salting process. The salting process has a purpose to attract water and coagulate protein in fish meat so that the texture of the fish becomes denser. At a rather high concentration, salt can inhibit the growth of bacteria and discoloration and can give the typical flavor.

2.3. *Drying*
This process aims to remove the water content in the fish so the quality of the fish is stable. If the fish meat is smoked directly when it still has the water content high enough, the fish meat will be dark brown and ugly. Through proper drying, the fish surface on the inside will becomes drier. Many moisture contents evaporate from the intercellular portion of the fish and leave the gaps between the cells in the surface layer. This can cause the fish to absorb the color and good smell of smoke at the time of smoking.

2.4. *Arrangement*
The arrangement of the fish is arranged in the smoking chamber in such a way that aim to obtain a uniform flow of smoke and heat which greatly determines the quality of the final product. To get the flow of smoke and heat evenly, the distance between the fish on the smoker rack and the distance between each of the curing racks in the smoking chamber should not be too tight.

2.5. *Smoking*
The purpose of smoking in the preservation of fish is to preserve and give the color and flavor of a typical smoke in fish. In fact, the long-lasting power generated by smoke is very limited, so that the fish can preserve must be followed or preceded by other preservation methods. Smoking also aims to remove the vapors from phenol or aldehyde elements that contained in wood species and then attach it to the fish meat. Delicious flavors that characterize smoked fish products, especially from phenol and aldehyde element.

2.6. *Cooling and packaging*
The smoking process generally ends with a cooling and packaging stage. After completion of the smoking stage, the product is kept in a clean room and left so it can reach the room temperature, after that the packaging is carried out. Packaging can use the polyethylene plastic and if it wants to extend the storage life of the product, it can be done with vacuum packaging.

3. Importance performance analysis
Importance Performance Analysis (IPA) measure the satisfaction level of product or service into four quadrants in Importance-Performance matrix (figure 1). This method firstly introduced by [5]. The interpretation of the IPA is graphically presented on a grid divided into four quadrants [6]. It helps to analyze and compare the performance that perceived by the customers. The quadrants are:

a. Concentrate here quadrant
   In this quadrant, the variables have the good importance for the customers but less performance on the product. We have to concentrate to increase the performance of the variable to satisfy the customer.

b. Keep up the good work quadrant
In this quadrant, the variables have the good importance for the customer and the product has a good performance. We have to keep the performance stable.
c. Low priority quadrant
In this quadrant, the variables are less importance to the customer and less performance on the product. These variables that include in this quadrant have to studied further to attract the importance of customer.
d. Mostly overkill quadrant
In this quadrant, the variables are less important to the customer but it has good performance on the product. We have not to do anything for these variables.

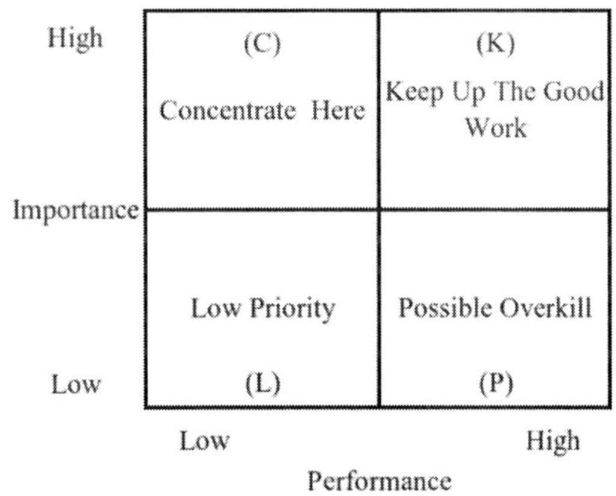

Figure 1. Importance-Performance Matrix [7].

4. Methodology

This research is divided in to three stages which will be explained in the following subsection.

4.1. Preparation stages

This stage aims to find out the needs and desires of smoke fish products by costumers. We interview some people who often consume it to get it done. The result of the interview is a variable that will be used to make the importance and performance questionnaire, that are:

1. Freshness
2. Color
3. Flavors
4. Typical smell
5. Can be stored in a long period of time
6. Cleanliness
7. Prices
8. Security for consumption
9. Undamaged fish meat
10. Fish dryness level
11. Organic processing (without chemicals)
12. Uniqueness shape

4.2. Data collection, testing and processing stages

After the questionnaire based on the variables for the importance and performance products was made, the questionnaire was distributed to respondents in Semarang. However, there was only 40% of 100 questionnaires could be used. The data from the questionnaire were then tested for validity and reliability analysis. The variables that not valid and reliable were eliminated and then the other was re-tested until all variables are valid and reliable. After that, the data was processed using IPA method.

4.3. Data analysis stages

The data analysis is performed based on IPA method, where the variables have been divided into four quadrants so it is known where the quality position and what strategy should be done for each variable.

5. Results and discussions

The discussion of the results is explained in the following subsection.

5.1. Validity and reliability test

The data from the data collection is tested using validity test and reliability test. The validity test result is shown in table 1.

Table 1. Validity test result

Variables	Importance			Performance		
	r_{xy}	r_{table}	Results	r_{xy}	r_{table}	Results
1	0.754	0.320	Valid	0.470	0.320	Valid
2	0.755	0.320	Valid	0.513	0.320	Valid
3	0.717	0.320	Valid	0.519	0.320	Valid
4	0.455	0.320	Valid	0.374	0.320	Valid
5	0.756	0.320	Valid	0.280	0.320	Not Valid
6	0.724	0.320	Valid	0.456	0.320	Valid
7	0.708	0.320	Valid	0.505	0.320	Valid
8	0.709	0.320	Valid	0.335	0.320	Valid
9	0.729	0.320	Valid	0.580	0.320	Valid
10	0.557	0.320	Valid	0.246	0.320	Not Valid
11	0.625	0.320	Valid	0.284	0.320	Not Valid
12	-0.143	0.320	Not Valid	0.161	0.320	Not Valid

From table 1, we know that variables 5, 10, 11, and 12 are not valid for the performance data, and variable 12 is not valid for importance data. These variables are eliminated and the rest is re-tested again. The validity re-test result is shown in table 2.

Table 2. Validity re-test result

Variables	Importance			Performance		
	r_{xy}	r_{table}	Results	r_{xy}	r_{table}	Results
1	0.782	0.320	Valid	0.473	0.320	Valid
2	0.729	0.320	Valid	0.581	0.320	Valid
3	0.712	0.320	Valid	0.446	0.320	Valid
4	0.420	0.320	Valid	0.473	0.320	Valid
5	0.759	0.320	Valid	-	-	-
6	0.746	0.320	Valid	0.441	0.320	Valid
7	0.713	0.320	Valid	0.470	0.320	Valid
8	0.752	0.320	Valid	0.396	0.320	Valid
9	0.739	0.320	Valid	0.489	0.320	Valid
10	0.586	0.320	Valid	-	-	-
11	0.624	0.320	Valid	-	-	-
12	-	-	-	-	-	-

From table 2, we have that all variables are valid so we can process to reliability test that can be seen in table 3. From table 3, we know that the results for importance and reliability test is reliable for both of them so we can use the data for the IPA process. For the next processing, we just use 8 variabels that valid for both importance and performance.

Table 3. Reliability test results

Importance		Performance	
Cronbach's Alpha	N of Items	Cronbach's Alpha	N of Items
0.920	11	0.764	8

5.2. IPA process

The first step in IPA process is determine the Hold or action for each variables by comparing the suitability scores with decision scores that can be seen in table 4.

Table 4. IPA-Hold or Action

No	Variables	Importance scores	Performance scores	Suitability scores	Decision scores	Hold or Action
1	Freshness	125	109	87.20%	86.99%	H
2	Color	121	105	86.78%	86.99%	A
3	Flavors	127	116	91.34%	86.99%	H
4	Typical smell	124	108	87.10%	86.99%	H
6	Hygiene	129	105	81.40%	86.99%	A
7	Price	120	112	93.33%	86.99%	H
8	Safety for consumption	129	104	80.62%	86.99%	A
9	Undamaged fish meat	127	112	88.19%	86.99%	H
	Total	1002	871	86.99%		

There are 3 variables that categorized as Action (see table 4) that are: Color, Hygiene, and Safety for consumption. These variables need to be improved to get the customer satisfaction. The rest can be Hold to keep the customer satisfaction. After IPA-Hold or Action table is made, the next step in IPA is to make the IPA matrix that devide the variables into 4 quadrants. That diagram is based on the importance and performance scores for each variables (figure 2).

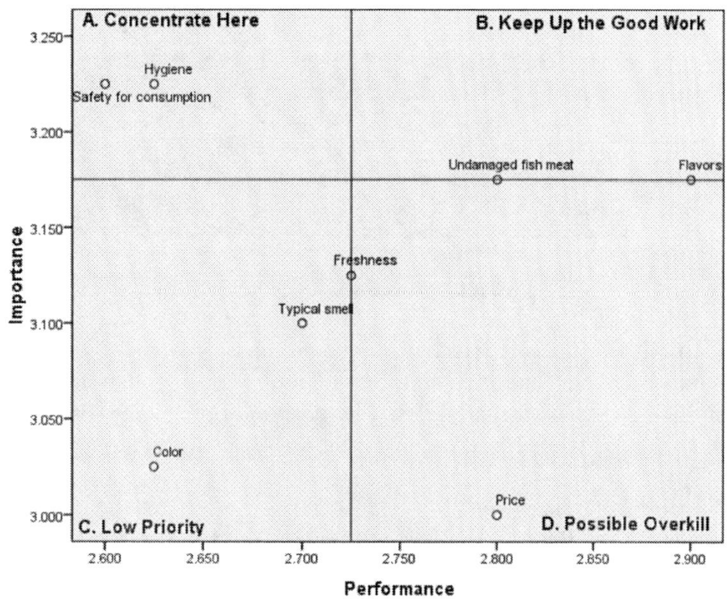

Figure 2. IPA matrix

a. Concentrate here quadrant

The variables that conclude on this quadrant are Hygiene and Safety for Consumption. Both of them are listed in the action categorized that need to be repaired soon to increase the satisfaction of customer. Most of the Indonesian smoked fish is less hygiene because it processed traditionally with less hygiene tools and place (figure 3). The Indonesian traditional smoked fish industry is identical with low hygiene levels and poor sanitation aspects so that smoked fish products have a bad image as food for consumers. In addition, smoked fish products are also considered less qualified and nutritious, its functional properties are inconsistent and there is no guarantee of quality and safety [1].

Figure 3. Indonesian traditional smoked fish production process

The hygiene and the safety for consumption are two variables that strongly related. There is no guarantee for the product safety if the product is less hygiene. Not only the process, but the Indonesian smoked fish packaging is also less hygiene (figure 4).

Figure 4. Indonesian smoked fish

The strategy to solve this problem is to make the production center that has good hygiene production and quality assurance systems, modern facilities and tools, good sanitation, hygiene packaging that can keep the quality and safety of the products. The research on smoked fish packaging has been done previously, that concludes the best packaging method is by using vacuum packaging [8, 9].

b. Keep up the good work quadrant

Undamaged fish meats and flavor is in the Keep Up the Good Work Quadrant. It means these variables has a good priority to the customers and the industry has the good performance on it. Although the production process is less hygiene, but the flavor is very delicious. Indonesian smoked fish has delicious flavor due to its recipe and the smoking process. The fish meats that undamaged is thanks to the sticking process that carried out carefully. Although its process is very easy but the process is risky.

c. Low priority quadrant
In this quadrant, the variables have low priority for the consumer and it has low performance too. It can be something that customers want in the future so we have to pay attention to these variables too. The variables that conclude in this quadrant are freshness, typical smell, and color.

The typical smell is reached by using some Indonesian wood that has different aromas each other. It can be the attractiveness to the customer if it explores further. The color is something special. It has to be studied because the perception of one to another is different. There is someone think that the darker the color is better and the other think it is worse.

d. Possibly overkill quadrant
Price is the only variable that concludes in this quadrant. This variable is less important for the customer but the performance is too high or even exaggerate. It means that the price is too cheap so everyone can buy. There is no improvement for this variables and we can concentrate on the others.

6. Conclusion
The variables that have to be improved are hygiene and safety for consumption. These two variables are categorized as Action and located in Concentrate here quadrant. It means that these two variables have the high priority of importance to the customer but less performance. In order to make Indonesian smoked fish survive the competition in ASEAN market, these two variables need to be improved. The solution for this problem is to make the production center that has good hygiene production systems, modern facilities and tools, good sanitation, hygiene packaging and the quality assurance system that can keep the quality and safety of the products.

References
[1] Sulistijowati R, Djunaedi O S, Nurhajati J, Afrianto E and Udin Z 2011 *Metode Pengasapan Ikan* (Bandung: UNPAD Press)
[2] Chen S-H and Pai C-K 2014 Using the QFD Technical to improve Service Quality in Vegetarian Foods Industry *Int. J. Acad. Res. Bus. Soc. Sci.* vol 4 pp 162–8
[3] Ortinau D J, Bush A J, Bush R P and Twible J L 1989 The Use of Importance-Performance Analysis for Improving the Quality of Marketing Education: Interpreting Faculty-Course Evaluations *J. Mark. Educ.* vol 11 pp 78–86
[4] Horner W F A 1997 Preservation of fish by curing (drying, salting and smoking) *Fish Processing Technology* ed G M Hall (Boston, MA: Springer US) pp 32–73
[5] Martilla J A and James J C 1977 Importance-Performance Analysis *J. Mark.* vol 41 p 77
[6] Pangow V L, Massie J D and Tielung M V 2017 Evaluating in-flight service of batik air using importance and performance analysis *Jurnal Riset Ekonomi, Manajemen, Bisnis Dan Akuntansi* vol 5
[7] Ho L-H, Feng S-Y and Yen T-M 2014 Using Modified IPA to Improve Service Quality of Standard Hotel in Taiwan *J. Serv. Sci. Manag.* vol 7 pp 222–34
[8] Nur M 2012 Pengaruh cara pengemasan, jenis bahan pengemas, dan lama penyimpanan terhadap sifat kimia, mikrobiologi, dan organoleptik sate bandeng (Chanos chanos) *J. Teknol. Ind. Has. Pertan.* vol 14
[9] Paputungan T S, Wonggo D and Damongilala L J 2015 Kajian mutu ikan cakalang (katsuwonus pelamis l.) asap utuh yang dikemas vakum dan non vakaum selama proses penyimpanan pada suhu ruang *J. Media Teknol. Has. Perikan.* vol 3

Muslim and non-muslim perceptions on Halal Meat Logistics (HML)

I Masudin[1], F W Fernanda[2], F Jie[3], H Djajadikerta[3] and Widayat[2]

[1]Department of Industrial Engineering, [2]Department of Management
Universitas Muhammadiyah Malang
Jl. Raya Tlogomas 26 Malang, Indonesia
[3]School of Business & Law, Edith Cowan University
270 Joondalup Dr, Joondalup WA 6027, Australia

masudin@umm.ac.id

Abstract. The study explores customer preferences of willingness to pay halal meat products based on logistics activities for Indonesian consumers in muslim and non-muslim areas. Two different Indonesian places were chosen to explore and compare the difference preferences about halal meat products. This paper investigates the factors influencing muslim consumers to pay halal meat products especially in term of logistics activity factors. The respondents of East Java and Bali provinces are collected to be compared to explore consumers' preferences in purchasing halal meat products in term logistics activities. Statistics descriptive is used to answer the research question using Cut off Point (COP) or average ration to figure out the distribution data of respondents in two different areas. Two variables considered in this study are halal logistic performance and customer's satisfaction.

1. Introduction

The increasing demand on halal products in most countries in the world tends to widen the study about halal logistics (H-logistics) recently as the result of fast growing population of muslims around the world that enlarged size of halal market. A study about halal logistic by [1] identified shows that the number of muslim is more than one quarter of the world's population that spreading over 112 countries. In addition, the fast demographic movement across the world and some regions regulations that has reduced the nations' borderless and the principles of halal food for muslims for daily consumption, it is important to determine determining the level of certification of halal food products in areas of both muslim and non muslim countries.

As most other Muslim countries, Indonesia and Malaysia require halal product that needs halal logistic to keep halal status, particularly for meat product that needs more complex halal logistic than other products. It is because halal meat status was started from giving halal food to the cattles into halal meat display in traditional and modern market. Good halal logistic performance will improve halal meat product quality and create greater Muslim customer's satisfaction. This paper attempts to investigates the factors influencing muslim consumers to pay halal meat products in term of logistics activity factors. The objective of this study could be met by finding the answers of these questions: Are the willingness to pay halal meat of muslim consumers in muslim and non muslim areas influenced by halal logistics performance? And in what are the factors affecting the decisions of muslim people who are living in muslim and non muslim areas when purchasing halal meat products?

2. Literature review

Meat is one of major sources of proteins and has social meaning in mostly community [2]. For Muslim, halal status is major requirement before making decision to purchase meat [3]. Halal is term which means of permitted and allowed according Islamic syariah law [4]. Halal is not only applied in the physical product, but also on the process of production, distribution, and sale [5]. It means that halal meat is meat that permitted to be consumption if the manufacturing process and meat quality is in accordance to Islamic syariah law. The process starts from feeding halal food to cattle, halal slaughter process, halal meat process until the display in market. Because the process is very complex, halal meat is vulnerable to be non halal meat.

Because of its vulnerable, demand of halal meat is high especially for Muslim customer in non Muslim area. Jusmaliani and Nasution [6] have investigated Muslim customer's behavior in Jakarta and Melbourne toward halal meat product. The result is halal meat availability is the only one significant determinant factor in customer's behavior between Jakarta and Melbourne Muslim customer. In non muslim area, halal process (halal logistic) mostly does not use as a standard and make it more difficult for Muslim customer to purchase a product. Stitou and Rezgui [7] make research in Muslim community in France, and the result showed 56% of 223 Muslim respondent will not purchase a product if they are doubt about halal status of the product. Otherwise, 87% respondent is willing to pay for a product that has guaranteed the halal status. It indicates a requirement of halal logistic for a Muslim customer, especially in non Muslim area, even if it is more expensive than conventional logistic.

Willingness to pay for good performance halal logistic has been investigated in research by [8-10]. Customer's perception on halal meat logistics is investigated by [8]. His research compares Muslim consumer's behavior between Malaysia (Muslim country) and Netherland (non Muslim country). He found Muslim consumer in Malaysia has higher willingness to pay and request for higher segregation level than Muslim consumer in Netherland. But the research had similar result with [7] and [9] that found Muslim consumer cares and willing to pay for halal logistic, in Muslim and non Muslim country. Research by [10] investigated about driver of consumers' willingness to pay for halal logistics. According to the result, halal logistic certification of 3PL provider is key drivers of consumers' willingness to pay for halal logistic. Service capability and image of 3PL are insignificant factors to create intention to pay for halal logistic. It indicates that companies should educate and emphasize the importance of the halal logistic to current/potential customers [10].

2.1. Halal logistic performance

Definition of halal logistic stages of logistic activities that applying halal concept start from point of origin to point of consumption [4]. To ensure halal concept is applied, halal certifiers have to monitor all halal logistic activities. Example of halal certifiers such as Majelis Ulama Indonesia (MUI) and Jabatan Kemajuan Islam Malaysia (JAKIM) that are operated by government. However, there are some countries using private halal laws and halal certifiers from private institutions, such as Netherland and United States [11]. They will monitor all halal logistic activities start from breeding farm, abattoir, and then halal meat will be distributed and sold to domestic customer, halal meat importer, and food industry [4].

To build customer's satisfaction, logistic provider has to create good logistic performance [12]. According to Davis [13], competitive advantage which is built by stronger customer relationship can be generated from good logistic performance. Logistic provider's ability to handle order process will be measured by logistic performance [14-19]. Logistic performance is measured with nine parameter logistic services quality (LSQ) by [20]. The concepts are personnel contact quality; (2) order release quantities; (3) information quality (4) ordering procedures (5) order accuracy; (6) order condition; (7) order quality; (8) order discrepancy handling and (9) timeliness [13-16, 21].

However, respondent of this research ends halal meat customer. That's why the concepts should be adapted by concept that can be measured by customer. Items such as ordering procedures, timeliness, etc cannot be measured because customer can't see that process. To adapt the concepts, research from [22] is used. So that the concept that used for halal logistic performance is personnel contact quality, information quality, order quality, order condition, and order discrepancy. Personnel contact quality refers to customer valuation of the treatment that is given by supplier's logistic contact people [23]. Information quality refers to customer's perception toward information of product which customer may choose [23]. Information quality must be accurate, reliable, and complete to support good logistic performance [24]. Order quality refers to how well the supplier conform customers' needs in a product and how well the products work [23, 25]. Order condition refers to how well condition of product is (damage level in product) [26]. Order discrepancy refers to how well logistic provider handles any discrepancies after the order arrived in customer [27].

2.2. Customer satisfaction

Customer satisfaction is an essential and a fundamental element in marketing [28, 29]. It is essential because it is very important for a marketer to fulfil customer's need and desire in a product [30]. Paul *et al.* [31] defined customer's satisfaction as mental state where the customer compares expectation before purchase with customer's perception about product performance after purchase. This valuation depends on product availability, information, and demand. Then customer will compare expectation and perception based on the three factors. If customer expectation fulfilled with product performance, there will be the positive perception as customer's satisfaction indicator [32, 33]. Customer's satisfaction is not a guarantee of repeat business, but most unsatisfied customer will not buy again [34]. Therefore, many industries put customer satisfaction as criteria to assess product performance [35]. Customer satisfaction can also as exit barrier to protect customer from other competitor product [36]. Satisfied customer tends to use product from current producer, because if they switch to other product there will be a risk if the other product doesn't fulfil their expectation [37].

3. Methodology

This study will explore the factors influencing muslim consumers to pay halal meat products especially in term of logistics activity factors. Statistics descriptive is used to answer the research question using Cut off Point (COP) or average ration to figure out the distribution data of respondents in two different areas. The aspects considered in this study is halal logistic performance. In this study, 2 different places are determined as the location of research where these two areas are chosen based on the muslim and non muslim population. In this case, the respondents of East Java and Bali provinces are collected to be compared to explore consumers' behavior in purchasing halal meat in term logistics activities. The explanatory single cross sectional design is used to explore the comparisons of respondents or variables in the two different places.

4. Data collection and results

In this section will be discussed about the data processing of questionnaires distributed in Muslim majority areas (Malang) and the majority of non-Muslims (Bali). The data were taken as many as 24 respondents in each of these areas. This data is divided into 3 types, first data contains brief question about social demography data, and second data is descriptive statistic data when buying halal meat product. The third is the result of hypothesis test from the respondent's statement using likert scale. Likert scale data will be processed by using statistical software PLS. The results of processing will be the basis for the proving hypothesis that has been prepared.

4.1. Non muslim majority area (Denpasar)

Sample profile from 24 respondents in Denpasar divided of man (54.17%) and woman (45.83%). Percentage of sample profile between man and woman almost equal and that is mean distribution of data is pro rata. The largest age group of respondents is group 17-25 (58.34%). Occupation group that has biggest number is student/scholar (54.17%). High school (SMA/SMK) is biggest education level group (66.66%). The number of meat purchased was collected in questionnaire, but it is not demography data. The number of meat purchased was collected to check the respondents is meat frequent buyer. The result is 62.5% of respondents buy meat in 2-3 times a month. 20.83% of respondents buy meat 4-5 times a month, and 16.67% buy meat more than 5 times a month. Table of data distribution for demography and the number of meat purchased data can be seen in table 1 and table 2.

Table 1. Demography information of respondents in Denpasar.

Demography informations	Frequency	Percentage
Gender		
Man	13	54.17%
Woman	11	45.83%
Missing Value	0	0.00%
Age		
17-25	14	58.34%
26-35	5	20.83%
36-45	2	8.33%
46-55	2	8.33%
Missing Value	1	4.17%
Occupation		
Government	1	4.17%
Private	5	20.83%
Entrepreneur	3	12.50%
Student/scholar	13	54.17%
Housewife	2	8.33%
Missing Value	0	0.00%
Education Level		
Elementary School	0	0.00%
Junior High School	0	0.00%
Senior High School	16	66.66%
Diploma Degree	1	4.17%
Bachelor Degree	6	25.00%
Master Degree	0	0.00%
Doctorate Degree	1	4.17%
Missing Value	0	0.00%

Table 2. The number of meat purchased.

The Number of Meat Purchased	Frequency	Percentage
2-3 times	15	62.50%
4-5 times	5	20.83%
>5 times	4	16.67%
Missing Value	0	0.00%

4.2. Muslim majority area (Malang)

Sample of respondents in muslim majority area (Malang) as many as 24 people. Male respondents in Malang as many as 13 people (54,17 %) and female respondents as many as 11 people (45,83 %). Percentage in gender group almost equal, it means distribution of data is pro rata. Second item in demography data is age group, which is divided in 4 group, 17-25, 26-35, 36-45, and 46-55. 17-25 group has percentage as many as 95,83% (23 people). 26-35 group has percentage as many as 4,17% (1 people). Other items in demography data is occupation and education level. In occupation group, government section is the biggest group that has 87,5% from sample. For education level group, senior high school has 62,5% from sample. The result of number of meat purchased is 87,5% of respondents buy meat in 2-3 times a month. Other respondents as many as 12,5% buy meat 4-5 times a month. Table of data distribution for demography and the number of meat purchased data can be seen in table 3 and table 4.

Table 3. Demography information of respondents in Denpasar.

Demography informations	Frequency	Percentage
Gender		
Man	13	54.17%
Woman	11	45.83%
Missing Value	0	0.00%
Age		
17-25	23	95.83%
26-35	1	4.17%
36-45	0	0.00%
46-55	0	0.00%
Missing Value	0	0.00%
Occupation		
Government	21	87.50%
Private	1	4.17%
Entrepreneur	2	8.33%
Student/scholar	0	0.00%
Housewife	0	0.00%
Missing Value	0	0.00%
Education Level		
Elementary School	0	0.00%
Junior High School	0	0.00%
Senior High School	15	62.50%
Diploma Degree	0	0.00%
Bachelor Degree	7	29.17%
Master Degree	0	0.00%
Doctorate Degree	1	4.17%
Missing Value	1	4.17%

Table 4. Demography information of respondents in Denpasar.

The Number of Meat Purchased	Frequency	Percentage
2-3 times	21	87.5%
4-5 times	3	12.5%
>5 times	0	0.0%
Missing Value	0	0.0%

4.3. Descriptive statistical test

This section divided into 2 sub chapters, that is sub chapters halal logistic performance and sub chapters customer satisfaction. Result of descriptive statistical test provide customer's preference about halal meat product. Each dimension of variable will be explained for more detailed test result.

5. Discussion

5.1 Non muslim majority area (Denpasar): Halal logistic performance

Range average value for halal logistic performance start from 5.00 to 5.38. This result indicate halal logistic performance for halal meat product in Denpasar is good. Halal logistic performance has 5 dimensions, that are order quality, personnel contact quality, information quality, order condition, and order discrepancy. Order quality has an average value of 5.38. These results prove that consumers agree that the quality of halal meat in the market is good enough. Personnel contact quality tests the service quality of staff in the market that sells halal meat in Bali. The average value of this dimension is 5.08 which is indicating that market staff are willing and able to provide good information regarding halal meat products.

The third dimension is information quality that has an average value of 5.04. This value indicates that consumers are confident that the halal certification and logos present on the product are from a trusted source. The order condition tests the condition of the product, if there is no damage then the condition of the product is considered good. The average value in this dimension is 5.17. This shows the condition of the product in good condition when it is in the end consumer. Order discrepancy has an average value of 5.00 which is the lowest average among other halal logistic performance dimensions. However, the value already indicates consumer approval that halal meat products in Denpasar have a product return warranty if there is damage in the product.

5.2 Non muslim majority area (Denpasar): Customer satisfaction

This variable has 3 indicators in the questionnaire that examines how the satisfaction level of halal meat consumer. The "good service" indicator has an average grade of 5.25. An "appropriate services and products" indicator has an average of 5.63. This value is equal to the average value of "service and product that make consumers happy" indicators. These results indicate that halal meat consumers are already satisfied with the products and services they get. These results are relevant with previous study by [38] who found that service quality and appropriate service of product are the most elements considered by customers at the beginning of purchase.

5.3 Muslim majority area (Malang): Halal logistic performance

Descriptive statistical results in Malang area showed varying results with the lowest average ranging from 3.92 and the highest of 5.13. In general, these results show that halal logistic performance in Malang is quite good, although the value is lower than in Denpasar. The dimension that has the lowest average value is order discrepancy. A value of 3.92 means the consumer gives a neutral value for this dimension. The dimension of personnel contact quality has an average value of 4.67. This value means that respondents are quite agree that the personnel in the market is good enough in providing information related to halal meat products to consumers.

The information quality dimension has an average value of 5.02. This dimension serves to measure whether the logos and halal certifications that are provided come from trusted sources. The resulting value of the questionnaire indicates that the respondent agrees that the halal logo and certification on the product come from a trusted source. Order quality dimensions show an average value of 5.13 which means respondents agree that product quality is good. The highest average value of 5.21 is on the order condition dimension. In general, respondents agree that halal meat products have good condition without any damage.

5.4 Muslim majority area (Malang): Customer satisfaction

Customer satisfaction measures whether consumers feel satisfaction after using the product. The 3 indicators in this variable have an average value of 4.99. The highest value is a "good service" indicator with an average value of 5.13. The "service and product that make consumers happy" indicators has an average value of 4.96. The "appropriate service and product" indicator has the least average value of 4.88. It can be concluded that consumers are quite satisfied in using halal meat products in Malang. These results reflected the relevance of the arguments stated by [32, 33] that customer satisfaction depends on the quality of product purchased by customers in fulfilling their expected demand.

6. Conclusion

In conclusion, the factors affecting the decisions of muslim people who are living in muslim and non muslim areas when purchasing halal meat products are customer satisfaction on the product purchased and the performance of halal logistics. The outcome of the tests indicate that halal logistic performance has positive and significant effect on customer satisfaction. Good halal status of products obtained from halal logistic performance will improve customer satisfaction and willingness to pay. Results of the tests in non muslim majority area (Denpasar) show higher score than result test in muslim majority area. In Denpasar, average of halal logistic performance score is ranging from 5.00–5.38. Also, customer satisfaction score is ranging from 5.25-5.63. Score of halal logistic performance in Malang is ranging from 3.92-5.13, and customer satisfaction score is ranging from 4.88–5.13. This gap can be taken as future studies of halal product marketing research.

References

[1] Alam S S and Sayuti N M 2011 Applying the Theory of Planned Behavior (TPB) in halal food purchasing *International Journal of Commerce and Management* vol 21 pp 8-20
[2] Fiddes N 1992 Meat : A Natural Symbol (New York: Routledge)
[3] Nakyinsige K, Che Man Y, Sazili A, Zulkifli I and Fatimah A 2012 Halal meat: A niche product in the food market 2nd International Conference on Economics, Trade and Development IPEDR 2012 pp 167-173
[4] Omar E N and Jaafar H S 2011 Halal supply chain in the food industry-A conceptual model Business, Engineering and Industrial Applications (ISBEIA), 2011 IEEE Symposium on pp 384-389
[5] Kirwan J 2006 The interpersonal world of direct marketing: examining conventions of quality at UK farmers' markets Journal of Rural Studies vol 22 pp 301-12
[6] Jusmaliani J and Nasution H 2009 Religiosity aspect in consumer behaviour: Determinants of halal meat consumption ASEAN Marketing Journal vol 1 pp 1-11
[7] Stitou N and Rezgui H 2012 The Muslim Consumer as the Key Player in Halal ASIDCOM 2012
[8] Tieman M, Ghazali M C and Van Der Vorst J G 2013 Consumer perception on halal meat logistics British Food Journal
[9] Kamaruddin R, Iberahim H and Shabudin A 2012 Willingness to pay for halal logistics: The lifestyle choice Procedia-Social and Behavioral Sciences vol 50 pp 722-9
[10] Fathi E, Zailani S, Iranmanesh M and Kanapathy K 2016 Drivers of consumers' willingness to pay for halal logistics British Food Journal vol 118 pp 464-79
[11] Dahlan H A and Sani N A 2016 "Halal Laws": From Conception to Current Challenges
[12] Liang H-C 2008 Impact of logistics service performance on tourist satisfaction and loyalty RMIT University
[13] Davis E R 2006 The role of logistics service quality in creating customer loyalty
[14] Mentzer J T, Flint D J and Hult G T M 2001 Logistics service quality as a segment-customized process *Journal of marketing* vol 65 pp 82-104
[15] Mentzer J T, Myers M B and Cheung M-S 2004 Global market segmentation for logistics services Industrial Marketing Management vol 33 pp 15-20
[16] Stank T P, Goldsby T J, Vickery S K and Savitskie K 2003 Logistics service performance: estimating its influence on market share Journal of Business Logistics vol 24 pp 27-55
[17] Stank T P, Goldsby T J and Vickery S K 1999 Effect of service supplier performance on satisfaction and loyalty of store managers in the fast food industry Journal of Operations Management vol 17 pp 429-47
[18] Man Y S 2006 Performance Measurement and Management of Third Party Logistics: An

Organizational Theory Approach Hong Kong Baptist University
[19] Cheung C F, Chan Y L, Kwok S K, Lee W B and Wang W M 2006 A knowledge-based service automation system for service logistics Journal of Manufacturing Technology Management vol 17 No 6 pp 750-71
[20] Mentzer J T and Williams L R 2001 The role of logistics leverage in marketing strategy," Journal of Marketing Channels vol 8 pp 29-47
[21] Bienstock C C, Royne M B, Sherrell D and Stafford T F 2008 An expanded model of logistics service quality: Incorporating logistics information technology International Journal of Production Economics vol 113 pp 205-22
[22] Siu N Y and Tak-Hing Cheung J 2001 A measure of retail service quality Marketing Intelligence & Planning vol 19 pp 88-96
[23] Mentzer J T, Flint D J and Kent J L 1999 Developing a logistics service quality scale Journal of Business logistics, vol 20 p 9
[24] Jie F, Parton K A and Cox R J 2012 An agile supply chain analysis of Australian beef wholesalers and retailers International Journal of Agile Systems and Management vol 5 pp 297-318
[25] Novack R A, Rinehart L M and Langley Jr C J 1994 An internal assessment of logistics value Journal of Business Logistics vol 15 p 113
[26] Mentzer J T, Rutner S M and Matsuno K 1997 Application of the means-end value hierarchy model to understanding logistics service value International Journal of Physical Distribution & Logistics Management vol 27 pp 630-43
[27] Rinehart L M, Cooper M B and Wagenheim G D 1989 Furthering the integration of marketing and logistics through customer service in the channel Journal of the Academy of Marketing Science vol 17 pp 63-71
[28] Brady M K and Robertson C J 2001 Searching for a consensus on the antecedent role of service quality and satisfaction: an exploratory cross-national study Journal of Business research vol 51 pp 53-60
[29] Kotler P and Armstrong G 2004 Principles of Marketing (Upper Saddle River: NJ Prentice Hall)
[30] Han H and Ryu K 2009 The roles of the physical environment, price perception, and customer satisfaction in determining customer loyalty in the restaurant industry Journal of Hospitality & Tourism Research vol 33 pp 487-510
[31] Paul J, Mittal A and Srivastav G 2016 Impact of service quality on customer satisfaction in private and public sector banks International Journal of Bank Marketing vol 34
[32] Jhandir S U 2012 Customer satisfaction, perceived service quality and mediating role of perceived value International Journal of Marketing Studies vol 4
[33] Lenka U, Suar D and Mohapatra P K 2009 Service quality, customer satisfaction, and customer loyalty in Indian commercial banks Journal of Entrepreneurship vol 18 pp 47-64
[34] La K V 2005 Customer Loyalty in Web-based Retailing Doctor of Philosophy, School of Management, Royal Melbourne Institute of Technology, Melbourne
[35] Anderson E W and Sullivan M W 1993 The antecedents and consequences of customer satisfaction for firms Marketing science vol 12 pp 125-143
[36] Gundersen M G, Heide M and Olsson U H 1996 Hotel guest satisfaction among business travelers Cornell *Hospitality Quarterly* vol 37 p 72
[37] Selnes F 1998 Antecedents and consequences of trust and satisfaction in buyer-seller relationships European journal of marketing vol 32 pp 305-22
[38] Jahanshani A A, Hajizadeh G M A, Mirdhamadi S A, Nawaser K and Khaksar S M S 2014 Study the effects of customer service and product quality on customer satisfaction and loyalty

IOP Conf. Series: Materials Science and Engineering **403** (2018) 012055 doi:10.1088/1757-899X/403/1/012055

Remanufacturing sustainability indicators: An Indonesian small and medium enterprise case study

Y A Fatimah and M Aman

Engineering Faculty, Universitas Muhammadiyah Magelang, Jl. Mayjen Bambang Sugeng, KM 5, Mertoyudan, Magelang, Central Java, Indonesia

yun.fatimah@ummgl.ac.id

Abstract. Remanufacturing which transforms end of life products to similar to a new product with a warranty to match is growing unstoppably toward a circular economy. Sustainability is an important issue for remanufacturing SMEs to stay competitive in global market. However, appropriate indicators that help remanufacturing SMEs understand their sustainability performance is very few discussed. This research develops Remanufacturing Sustainability Indicators (RSI) with the case of Indonesian small and medium enterprises as a guide for achieving sustainability in developing countries. The RSI integrates economic, social and environmental indicators, values and capacity that was determined based on national and international literature and case studies. The purposed RSI was then endorsed to local remanufacturing SMEs. The purposed remanufacturing sustainability indicators is expected to provide a guide/tool that will help remanufacturing SMEs in Indonesia and developing countries in achieving sustainability.

1. Introduction

Remanufacturing is a series manufacturing steps transforming an end of life (EoL) product or component to be similar as new product or better performance with a better or equivalent warranty [1]. Sustainability, which is meeting the needs of the present without compromising the ability of future generations to meet their own needs, has become an important issue to achieve remanufacturing competiveness, as they promote economic, social and environmental balancing [2]. The essential considering sustainability in remanufacturing process is to maintain the need of long-term solutions that secure economic values and social equity and wellbeing, and to conserve the needs for natural resource, energy and environment biodiversity. Improvement of the remanufacturing sustainability performance should be initiated from the first stage of remanufacturing process (e.g. core collections and material elections) to the final stage of the remanufacturing process (e.g. EoL management), as the potential economic, social and environmental impacts could be evaluated and monitored continuously at low risks.

Remanufacturing offers numerous sustainable values. Remanufacturing intuitively brings potential economic values not only in developing countries, but also in developed countries, together with the creation of a million-job opportunity. Along with these economic and social benefits, remanufacturing has great environmental benefit through resource and energy conservation, and GHG emission, waste and pollution reduction [3-6]. Decision made to improve the sustainability performance during remanufacturing process will have great consequences on whole sustainability impacts.

However, the application of sustainability during remanufacturing process especially for SMEs is still problematic. There is lack of sustainability evaluation applied in the remanufacturing operation. In addition, the sustainability of the remanufacturing process cannot be evaluated because there are no indicators or methods to measure them correctly and inclusively. While there are many global sustainability manufacturing indicators adopted by large industry, very few sustainability indicators are able to be adopted by SMEs. Therefore, there is a need to develop a sustainability indicators framework which is applicable, understood and relevant with SMEs.

The objective of this research is to determine a set of sustainability indicators in the scope of remanufacturing process especially for SMEs. The framework of this paper consists of three main ideas that understand the research background to find the key factors for developing remanufacturing sustainability indicators, taking into account the methods and measurement scope to determine the sustainability indicators, and discussing the remanufacturing sustainability indicators measurement and framework. The purposed remanufacturing sustainability indicators framework is expected to provide a guide/tool that will help remanufacturing SMEs in Indonesia and developing countries to monitor and evaluate the consequences of their remanufacturing activities at different phase, and eventually to generate a sustainable and high performance of their remanufactured products through its life cycle.

2. Remanufacturing process and business

This section explains the remanufacturing process and business into several subsections as follow:

2.1. Remanufacturing process

A sustainable remanufacturing requires an integration process and approach. Ensuring that the remanufacturing process are technically, economically, socially and environmentally sounds needs an integrated and comprehensive strategies, as the remanufacturing process consist of a sequence of phase. Generally, remanufacturing process begin with core collection, dismantling, cleaning, reconditioning, reassembling and testing [7] as presented in figure 1.

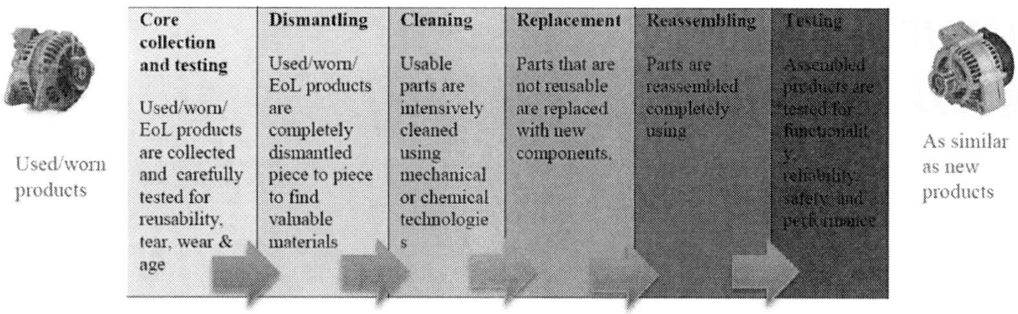

Figure 1. Remanufacturing process.

Figure 1 shows the steps of common remanufacturing process. Through this process, there are material, energy and resources, waste, GHG emission, pollution flowing in and out of the remanufacturing process. Adoption of innovative and appropriate technology, effective and efficient equipment often significantly influence the consumption of the material, energy and resource and the pollution, emission and waste generation.

2.2. Key benefits from remanufacturing

Remanufacturing has undoubtedly become engine of economy, social and environmental sustainability worldwide. Remanufacturing not only offers significant benefits for industry itself but also provide significant contribution to the improvement of society and environment. A number of research found that remanufacturing. Figure 2 summaries the economic, social and environmental advantages from remanufacturing industry gathered from literatures [3, 4, 6, 8-12].

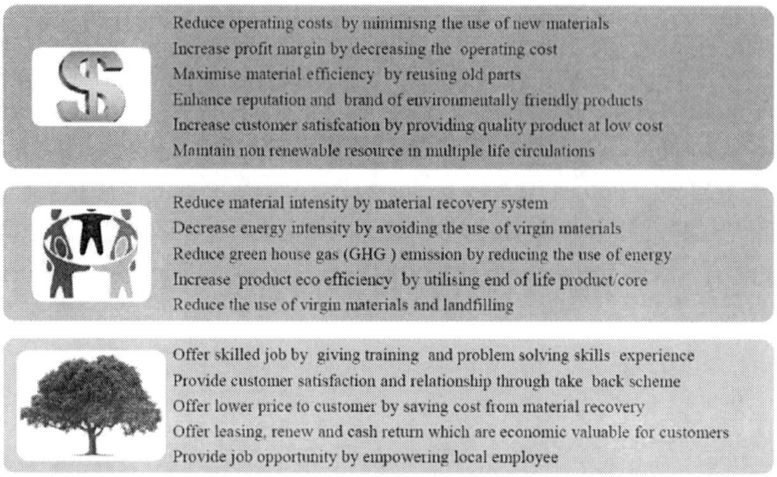

Figure 2. Key benefits from remanufacturing.

2.3. Indicators measurement

Indicators are critical key of defining, evaluating, monitoring and improving sustainability performance of an industry. Using this indicator, every industry can track their business performance sales, customer complaint, employee productivity) everyday which can help the industry to maximize their sustainability performance every year. Therefore, the indicators should be able to illustrate the interaction of industry not only within the industry itself (i.e. labor, management, equipment, resource, energy) but also with external stakeholders (i.e. society, suppliers, consumers). In the case of remanufacturing industry, the interaction between the remanufacturing industry and its environment should consider all input, process, impacts and products created from remanufacturing process as presented in figure 3.

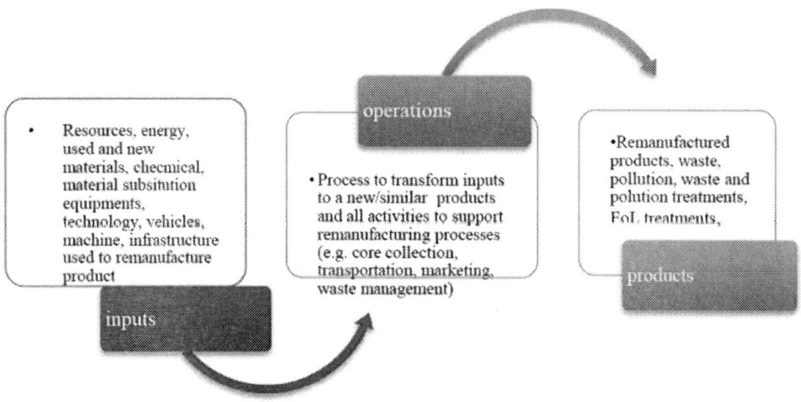

Figure 3. Relationship between remanufacturing and its environment.

3. Study methods

This section presents methods used in this research. An intensive study in accordance with research background and objectives is conducted to develop sustainability-remanufacturing indicator (SRI). A systematic approach based on literature review (e.g. sustainability framework review), case studies in developing countries and expert judgments from industry and academic is used to identify and select the appropriate indicators. An intensive brainstorming with local SMEs and Community is also presented to determine relevant indicators. Figure 4 presents a systematic approach implemented in this research.

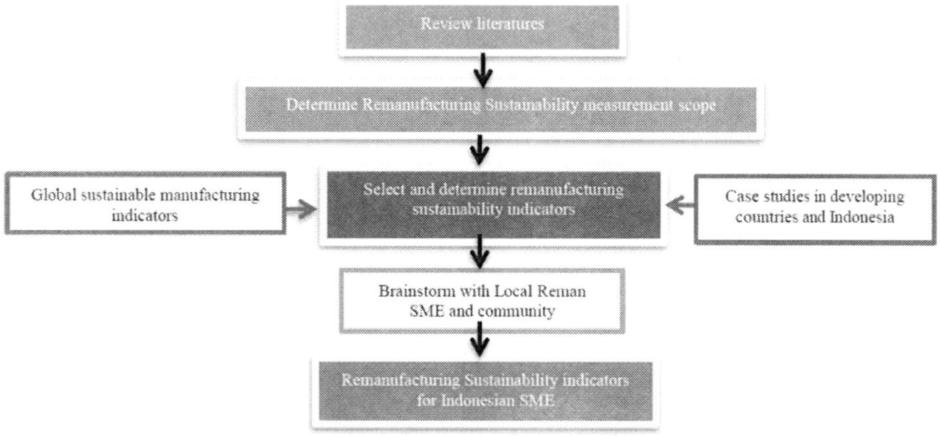

Figure 4. A systematic approach of study.

The remanufacturing sustainability indicators cover completely remanufacturing process, including all activities from core collection stage up to EoL management. The sustainability indicators are categorized into triple bottom sustainability.

4. Results and discussion

The results are explained and discussed in the following subsections.

4.1. Key success factors

Key success factors to achieve remanufacturing sustainability are the implementation of all remanufacturing process requirements holistically. The ability of resource associated with the main source for conducting remanufacturing process (i.e. cores, materials, energy) and enablers (i.e. labor, management, equipment and technology, information communication technology (ICT)) are very important to remanufacturing industry. Many remanufacturing gets trap on the process and neglects the enablers, thus the objective of the remanufacturing is difficult to be achieved. For example, an enterprise has already had energy efficient technology, but inefficiency is still high due to unavailability efforts to change the behavior of the labor and lack of knowledge in saving the energy.

Therefore, appropriate sustainability indicators in related to the remanufacturing standards in related to remanufacturing process and its enablers are urgently required by remanufacturing enterprises. Sustainability indicators open possibility to monitor whether the standards meet or exceed the requirements. These sustainability indicators could also help in monitoring the remanufacturing process over time. Sustainability indicator can help access the economic, social and environmental values and impacts at any stage of remanufacturing processes.

4.2. Indicator selection principles

There are three general criteria including understandable, applicable and relevant need to be concerned while selecting sustainability indicators. Indicator should be understandable which means that the indicator should be clear and not ambiguous for remanufacturing to understand [13]. To make the indicators easily applied by SMEs, four element of indicators including quantification technic (i.e. calculation formula), measurement unit (i.e. kg, kWh), development goal (i.e. improvement direction) and measurement period (i.e. yearly) are used. The Indicators are also considered to meet the remanufacturing purposes - raise economic values, improve social equity and increase environmental conscious.

4.3. Global sustainable manufacturing indicators

A review on sustainability manufacturing indicators has been conducted to identify the appropriate sustainability indicators for remanufacturing. In 2006, Ford Europe established a product sustainability index (PSI) consisting of eight indicators involving economic, societal and environmental aspects [14]. In this PSI, sustainability of a product considers all aspects of raw material extraction, production, use and product recovery. The Organization for Economic Co-operation and Development (OECD) published a big toolkit consisting of 18 main indicators to help measure the environmental problems associated with sustainable manufacturing [10], as presented in figure 5. In this framework, OECD stated that the main goal of sustainability in manufacturing is to minimize the use of materials while improving the benefits of product sent to society and to industry.

Walmart has delivered sustainability index of environmental products accordance to the marketability with the goals to meet 100% renewable energy, zero waste design and sustainable products sales. The EU Eco-Management and Audit Scheme (EMAS) developed by European Commission provide organization environmental performance information. The international Organization for Standardization (EPE) ISO 14031 presented a set of environmental key indicator to measure, assess and communicate the environmental performance of organization. General motors set metrics for sustainable manufacturing using fifty metrics of 6 major aspects. Global reporting initiative (GRI) provides 91 measures of company sustainability indicators, which covers economic, social and environmental aspects [15, 16]. This guideline helps industry in making good decisions toward economic, social and environmental sustainability for their industry, customer and stakeholders. Most studies often only concerned on environmental sustainability alone. However, this study considers economic, social and environmental sustainability in the same equal level.

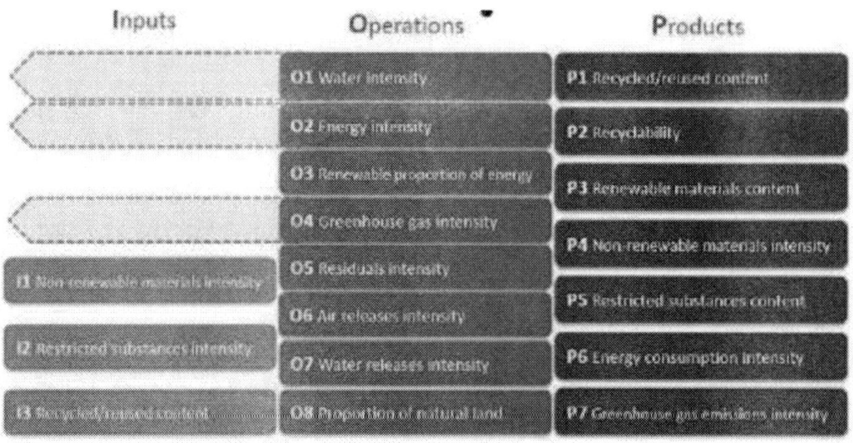

Figure 5. OECD sustainability indicators [17]

4.4. Lessons learn from developing countries

A review on the development of sustainability indicators from developing countries was carried out [4, 15, 17-20]. In developing countries, sustainable manufacturing has become a critical issue due to rapid development of population and industrialization. Achieving sustainable manufacturing through resource, energy and water efficiency have become popular among the industries, and environmental sustainability has become main consideration when the study discussed about sustainability. However, most of the study agreed that sustainable manufacturing is always about economic, social and environmental integration.

4.5. SMEs Remanufacturing Sustainability Indicators Selection

According to a comprehensive literature review and case studies analyses of the current and worldwide manufacturing and remanufacturing sustainability indicators, triple helix sustainability indicators of economy, social and environment are constructed. The economic sustainability indicators concern on the economic value created from the remanufacturing process including job creation, salary improvement etc. The social sustainability indicators focus on all social aspects created from the remanufacturing SMEs including health and safety, customer satisfaction etc. The environmental sustainability indicators consider all environment issues created from remanufacturing activities in SMEs including material efficiency, GHG emission etc. The detail result of the selected indicators is gathered from [21-24], and presented as follows.

4.5.1. Economic sustainability indicators. Economic sustainability of remanufacturing relates to all economic values associated with the use/utilization/management of resource, energy, labor, technology, method, waste, supply chain and Information Communication technology during remanufacturing process, which not only ascertains benefits for current generation but also provide opportunities for future generation. In sustainability assessment, it is urgently required to understand the level of economic benefits gathered from remanufacturing enterprise. Based on literature review and case study conducted from previous research, a number of specific economic indicators are considered to be adopted to measure the economic sustainability of remanufacturing enterprise, as presented in table 1.

Table 1. Economic indicators of remanufacturing sustainability.

Elements	Indicators	Unit Measurement	Formula
Job creation	Job creation	% (percentage)	$Job\ creation = \frac{total\ reman\ job}{total\ formal\ job}$
Employee	Employment remuneration	US Dollar	$ER = \sum man\ days \cdot average\ wage\ per\ man\ day$
	Salary improvement	% (percentage)	$SI = \frac{improvement\ salary}{total\ salary}$
Process	Tax revenue	US Dollar	Total tax paid
	Production cost	US Dollar	$PC = \frac{total\ cost}{total\ revenue}$
	Net profit	US Dollar	$NP = total\ revenue - total\ cost$
	Material efficiency cost	US Dollar	$ME = \frac{total\ material\ cost}{total\ production\ cost}$
	Energy efficiency cost	US Dollar	$EF = \frac{total\ energy\ cost}{total\ production\ cost}$
	Productivity	% (percentage)	$P = \frac{all\ remanufacturing\ inputs}{all\ remanufacturing\ products}$
Technology	Technology investment (i.e. Rate of Return – ROI,	US Dollar	$ROI = \frac{revenue\ gain\ from\ investment - cost\ of\ investment}{cost\ of\ investment}$
	Net present value – NPV)	US Dollar	$NPV = \sum_{t-1}^{T} \frac{cash\ flow}{(1-i)^t} - \frac{initial\ cash}{investment}$
Market	Market development	% (percentage)	$MD = \frac{total\ market\ improvement}{total\ market}$
	Brand image recognition	% (percentage)	$BI = \frac{total\ customer\ recognition}{total\ customer}$
	Enterprise competiveness	% (percentage)	$CL = \frac{sales}{total\ market\ sales}$
Waste and pollution	Waste treatment cost	US Dollar	$WsTC = total\ cost\ spent\ for\ waste\ treatment$
	Water treatment cost	US Dollar	$WrTC = total\ cost\ spent\ for\ water\ treatment$
	Pollution treatment cost	US Dollar	$PTC = total\ cost\ spent\ for\ pollution\ treatment$

4.5.2. Social sustainability indicators. Social issues in the rapid development of remanufacturing industry are predicted to be prominent. Main social advantages of remanufacturing include job opportunity and labor productivity. However, possible social disadvantages of remanufacturing are also need to be considered. For example, while there could be a great potential of employment opportunity, unprotected and unhealthy environment during remanufacturing process especially in developing country, may create social disadvantages impacts on the labor. Accordingly, measurement of the social impacts in this research considers employee, customer and community, as presented in table 2.

Table 2. Social indicators of remanufacturing sustainability.

Elements	Indicators	Unit Measurement	Formula
Employee	Health & safety	% (percentage)	$HS = \frac{total\ employee\ sick}{total\ labor}$
	Work injury rate	% (percentage)	$WIR = \frac{total\ injuries}{hour\ worked}$
	Labor productivity	% (percentage)	$LP = \frac{total\ revenue\ cost}{total\ labor\ cost}$
	Remanufacturing training	% (percentage)	$RT = \frac{employee\ received\ reman\ training}{total\ number\ of\ employee}$
	Education level	% (percentage)	$EL = \frac{total\ educated\ employee}{total\ number\ of\ employee}$
	Skill level	% (percentage)	$SL = \frac{total\ skilled\ labor}{total\ number\ of\ employee}$
	Gender equity	% (percentage)	$GE = \frac{total\ female\ labor}{total\ number\ of\ employee}$
Customer	Customer satisfaction	% (percentage)	$CS = \frac{satisfied\ customers}{total\ customers}$
Community	Community complaints	% (percentage)	$CC = \frac{total\ people\ complaint}{total\ community}$
	Public acceptability	% (percentage)	$PC = \frac{total\ people\ accepting\ product}{total\ community}$

Remanufacturing is complex system, which needs a certain level of knowledge and understanding. Level of employee education has strong linkages to the ability of employee to understand complex system. It is believed that good education will provide better knowledge and skill in associated with the remanufacturing process.

4.5.3. Environmental sustainability indicators. Environmental sustainability of remanufacturing is associated with all environmental consequences created from remanufacturing process and received by current generations without neglecting future generation. The consequences could be both positive and negative which could affect human and environmental performance directly and indirectly in different manners. The system boundary of remanufacturing process considers supply of cores and material, remanufacturing process, marketing activity and waste management. Taking this system boundary into consideration, the environmental sustainability adopts all material, energy, waste and pollution flowing through remanufacturing process, as presented in the following table 3.

Table 3. Environment indicators of remanufacturing sustainability.

Elements	Indicators	Unit Measurement	Formula
Materials	Used material acquisition	kg used material/kg reman product	$MA = \frac{total\ reused\ materials}{kg\ reman\ product}$
	Material efficiency	kg raw material/kg reman product	
Energy	Energy intensity	kWh/reman product	$EI = \frac{energy\ consumption}{kg\ reman\ product}$
Wastes	Solid waste intensity	kg waste/kg reman product	$SW = \frac{mass\ of\ solid\ waste\ produced}{kg\ reman\ product}$
	Water waste intensity	m3 waste water/kg reman product	$MA = \frac{water\ consumption}{kg\ reman\ product}$
	Residual intensity	m3 residual/kg reman product	

Elements	Indicators	Unit Measurement	Formula
Emission	GHG emission intensity	$kgCO_2$/kg reman product	$GHG = \frac{kgCO_2}{kg\ reman\ product}$
	Hazardous gas intensity	m3 gas/kg reman product	$HG = \frac{m3\ gas}{kg\ reman\ product}$
	Dusk	kg/kg reman product	$D = \frac{kg\ dusk}{kg\ reman\ product}$
Restricted substance	Hazardous chemical	kg/kg reman product	$HC = \frac{kg\ chemical}{kg\ reman\ product}$
	Acidification substance	kg/kg reman product	$AS = \frac{kg\ adictive\ substance}{kg\ reman\ product}$
Water	Water intensity	m3 water used/kg reman product	$WI = \frac{m3\ water\ used}{kg\ reman\ product}$
Land use	Landfilling	m3 solid waste sent to landfill/kg reman product	$L = \frac{m3\ solid\ waste}{m3\ landfill\ area}$
Ecosystem	Biodiversity impacts	% (percentage)	$BI = \frac{biodiversity\ impacts}{total\ biodiversity}$
	Water treatment	% (percentage)	$WT = \frac{water\ treatment}{total\ water\ used}$
	Waste treatment	% (percentage)	$WsT = \frac{waste\ treated}{total\ waste}$
	Pollution treatment	% (percentage)	$PT = \frac{pollution\ treated}{total\ pollution}$

4.5.4. Local SME and community brainstorming. Since the remanufacturing sustainability indicator will be applied for SMEs, the final selection process was delivered to the SMEs. The purposed RSI was endorsed to remanufacturing SMEs (i.e. photocopy, alternator, computer products) located in Java, Indonesia to receive some important feedback. The enterprises were chosen based on their willingness to participate in this pilot study. Direct interview and electronic mail communications were conducted to the three enterprise managers. A number of questions were structured to identify key indicators (i.e. core indicators, additional indicators) based on SMEs perceptions. Table 4 presents the selected sustainability indicators for remanufacturing SMEs.

Table 4. SME remanufacturing sustainability indicators.

Economy indicators			
Core indicators	Job creation	1	Job creation
	Employee	2	Salary improvement
	Process	3	Net profit
		4	Production cost
		5	Material efficiency cost
		6	Energy efficiency cost
		7	Productivity
	Market	8	Market development
	Company	9	Brand image recognition
		10	Enterprise competitiveness
		11	Market development
Additional indicators	Employee	1	Employment remuneration
	Process	2	Tax revenue
	Technology	3	Green/eco technology
	Waste& pollution	4	Environment tax
		5	Waste treatment cost
		6	Pollution treatment cost

IOP Conf. Series: Materials Science and Engineering **403** (2018) 012055 doi:10.1088/1757-899X/403/1/012055

Social indicators			
Core indicators	Employee	1	Health and safety
		2	Work injury rate
		3	Labor productivity
		4	Skilled labor
	Customer	5	Customer satisfaction
	Community	6	Community complaint
Additional indicators	Employee	1	Education level
		2	Remanufacturing training
		3	Gender equity
	Community	4	Public acceptability
Environmental indicators			
Core indicators	Material	1	Material acquisition
		2	Material efficiency
	Water	3	Water intensity

4.5.5. Future study and remanufacturing sustainability level. The objectives of this study is to determine Indonesian SME remanufacturing sustainability indicators which has finally could be achieved based on comprehensive literature review, case studies of developing country and brainstorming with local remanufacturing SMEs. However, due to limited sample of the pilot study, further research involving more SMEs is urgently required. Future research on how to measure the level of the sustainability is also essential to be conducted at enterprise level. A purposed concept on the measurement of the SMEs remanufacturing sustainability level is briefly discussed below. The remanufacturing sustainability levels are determined accordance to a technical standard of ISO/IEC 15504 Information technology – Process assessment. The remanufacturing sustainability level is estimated using three attributes (i.e. economic, social and environment) which consist of general practice indicators as stated in the previous section. Figure 6 presents the remanufacturing sustainability level, which can be achieved by SMEs.

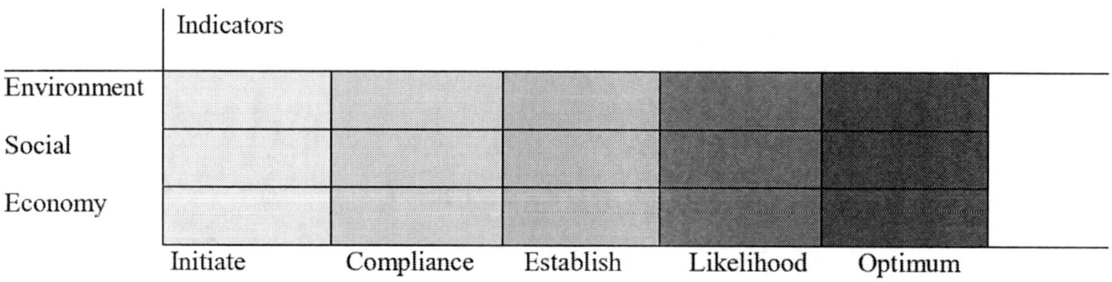

Figure 6. The SMEs remanufacturing sustainability level.

The initiate level is the existing level in which both indicators (i.e. core indicators and additional indicators) met the threshold values/standard. The threshold value is estimated based on intensive and comprehensive research on the remanufacturing SMEs case study and literature review. The compliance level is the sufficient level in which at least core indicators meet the threshold value/standard. The establish level is the ideal/standard level in which both indicators meet the threshold value. The likelihood level is managed level in which core indicators are exceeded and additional indicators meet threshold value. The optimum level is the highest level in which all indicators exceed the threshold value, showing significant improvements. The further research will consider the involvement of more SMEs to apply the sustainability indicator framework, and to assess the sustainability performance of Indonesian SMEs level.

5. Conclusion

Remanufacturing implementation is greatly recommended to increase resource and energy efficient, to reduce waste and pollution, to improve economic and social equity especially for developing countries. This research has successfully developed Remanufacturing Sustainability Indicators (RSI) as a guideline for remanufacturing SMEs to understand their sustainability performance. The indicators can measure economy, environmental and social sustainability separately of remanufacturing enterprise. The economic indicators cover net profit, rate of return, salary improvement, tax revenue, material and resource cost efficient, material recovery, investment in technology, productivity etc. The environmental indicators consist of waste production, energy and water consumption. The social indicators involve employment rate, gender involvement, professional occupations, level of education and skill of labor, consumer satisfaction rate etc. This RSI guideline can be applied to any other remanufacturing SMEs in developing countries with small modifications. However, further future research is urgently required to more comprehensively know the remanufacturing sustainability in the SMEs.

References

[1] Macarthur E 2006 Towards the circular economy *J. Ind. Ecol.* vol 10 pp 4–8
[2] King S 2011 What does sustainable manufacturing mean to Australia? *Csiro*
[3] Gutowski T G, Sahni S, Boustani A and Graves S C 2011 Remanufacturing and energy savings *Environ. Sci. Technol.* vol 45 pp 4540–7
[4] Rathore P, Kota S and Chakrabarti A 2011 Sustainability through remanufacturing in India: a case study on mobile handsets *J. Clean. Prod.* vol 19 pp 1709–22
[5] Ferrer G and Ayres R U 2000 The impact of remanufacturing in the economy *Ecol. Econ.* vol 32 pp 413–29
[6] Fatimah Y, Biswas W, Mazhar I and Islam M 2013 Sustainable manufacturing for Indonesian small- and medium-sized enterprises (SMEs): the case of remanufactured alternators *J. Remanufacturing* vol 3 no 6
[7] Jawahir I S and Bradley R 2016 Technological elements of circular economy and the principles of 6r-based closed-loop material flow in sustainable manufacturing *Procedia CIRP* vol 40 pp 103–8
[8] Fatimah Y A and Biswas W K 2016 Remanufacturing as a means for achieving low-carbon SMEs in Indonesia *Clean Technol. Environ. Policy* vol 18 pp 2363–79
[9] Ullwer J, Campos J K and Straube F 2016 Waste and pollution management practices by german companies *IFAC-Papers On Line* vol 49 pp 102–7
[10] Nardo M, Saisana M, Saltelli A, Tarantola S, Hoffman A and Giovannini E 2005 *Handbook on constructing composite indicators*
[11] Kerr W and Ryan C 2001 Eco-efficiency gains from remanufacturing: A case study of photocopier remanufacturing at Fuji Xerox Australia *J. Clean. Prod.* vol 9 pp 75–81
[12] Zhang J-H and Chen M 2015 Assessing the impact of China's vehicle emission standards on diesel engine remanufacturing *J. Clean. Prod.* vol 107 pp 177–84
[13] Sustainable Measures 2009 *Sustainable Measures*
[14] Schmidt W-P and Taylor A 2006 Ford of Europe's product sustainability index *13th CIRP Int. Conf. LIFE CYCLE Eng.* pp 5–10
[15] Amrina E and Yusof S M 2011 Key performance indicators for sustainable manufacturing evaluation in automotive companies *2011 IEEE International Conference on Industrial Engineering and Engineering Management* (IEEE) pp 1093–7
[16] Niemeijer D and de Groot R S 2008 A conceptual framework for selecting environmental indicator sets *Ecol. Indic.* vol 8 pp 14–25
[17] Seadon J K 2010 Sustainable waste management systems *J. Clean. Prod.* vol 18 pp 1639–51
[18] Boret Cordoba D and Veshagh A 2013 Managing eco design and sustainable manufacturing *Re-engineering Manufacturing for Sustainability* (Singapore: Springer Singapore) pp 59–67

[19] Fatimah Y A and Biswas W K 2016 Sustainability assessment of remanufactured computers *Procedia CIRP* vol 40 pp 150–5

[20] Tan H X, Yeo Z, Ng R, Tjandra T B and Song B 2015 A sustainability indicator framework for singapore small and medium-sized manufacturing enterprises *Procedia CIRP* vol 29 pp 132–7

[21] Joung C B, Carrell J, Sarkar P and Feng S C 2013 Categorization of indicators for sustainable manufacturing *Ecol. Indic.* vol 24 pp 148–57

[22] Feng S C and Joung C B 2011 A measurement infrastructure for sustainable manufacturing *Int. J. Sustain. Manuf.* vol 2 p 204

[23] Feng S C, Joung C-B and Li G 2011 Development overview of sustainable manufacturing Metrics *Proc. 17th CIRP Int. Conf. life cycle Eng.* pp 6–12

[24] Chengcheng Fan, Carrell J D and Hong-Chao Zhang 2010 An investigation of indicators for measuring sustainable manufacturing *Proceedings of the 2010 IEEE International Symposium on Sustainable Systems and Technology* (IEEE) pp 1–5

Collaboration of multi agent systems and web service in temperature data collecting process

E S J Atmadji[1], H Y Riskiawan[2] and D P S Setyohadi[3]

[1,2,3]Politeknik Negeri Jember, Indonesia
[1]ery@polije.ac.id, [2]yufit@polije.ac.id, [3]dwi.putro@polije.ac.id

Abstract. Web service is one of the emerging technologies later which one example of SOA (Service-Oriented Architecture). Therefore Web Service widely used in cross platform communication applications. In addition, in the retrieval of data on the Web Service generally still using the model of synchronization or request at a certain time. This paper will explain how merging multi agent system will help in data retrieval in Web Service autonomous. The intelligent agent-based software was designed using the Prometheus methodology, and developed through the JADE platform using the Java programming language. Model development focuses on how agents can retrieve data on the Web Service, declare data based on metadata and deliver the results of the data retrieval. The concept applied to the agent in the data retrieval is based on the WSDL (Web Service Definition Language), with the type of data is temperature, humadity that is in many rooms at a certain time. Evaluations were performed on 10, 100 and 200 data obtained from the Web Service and sensors.

1. Introduction

The development of technology, especially on the use of the internet has brought many developments in various sectors, one sector that has a very fast development is IoT (Internet of Things). In IoT the thing that often becomes a lot of attention is on the use of data and communication protocol between machines or commonly called M2M Communication [1]. Due to the difference between one machine with another machine then required a global language that is able to bridge the difference therefore proposed Web Service as the bridge [2]. But in implementing it there are various problems that often occur, one of them is kind of order of Language or which is usually called semantic on Web Service that will be used. Web Service itself consists of many types ranging from SOAP to REST, each of which has different characteristics [3].

Due to the difference between one type of web service with another web service then the process of taking data based on meta tagging was finally become one of its own challenges. In addition, the challenge in developing Web Service is the development and retrieval of data that must be fast [4], because the use of Web Service is usually used for small memory applications [5].

To overcome this then required an intelligent system capable of managing data based on Meta tagging contained on the Web Service. Smart and Autonomous agent system is a technique in computer program that can be utilized in terms of Web Service data retrieval automatically. Agent can be used in assisting the system in taking data based on fields or components that have been made to the database that has been provided, in addition agent can work autonomous and can solve problems quickly in accordance with the needs of the system [6] In order to recognize data retrieval, or what has been learned, how much data can be retrieved [2]. To help the agent system do the data retrieval and map the data, it needs a more optimal development approach, there are many approaches that can be applied in this problem. In taking data JSON-shaped websevice using a library that will help the agent in doing the translation so it will be easy to read by the agent, one of the libraries that will be used is JSONViewer. This research will analyze the utilization of Autonomous Multi Agent System for retrieving data from Web Service that has form of tagging which have been conditioned.

2. Research Method

Web Service is a mechanism that used as a bridge between applications or usually called multiplatform integration. Web services are triggered by a stateless process mechanism, which is executed based on requests from other applications or objects. There are several types of web services that are often used in the world such as SOAP and REST, SOAP is an XML-based data exchange protocol by encoding http headers with XML so that it will produce an output that can be read by other programs, while REST or usually called degan RESTfull is a data exchange mechanism that allows the exchange of data between clients and servers simultaneously without any knowledge of the client about the process, the output REST is often generated is JSON.

JADE (Java Agent Development Framework) is a framework for developing software based on multi-agent systems and intelligent agent applications that comply with FIPA standards. JADE is a software development framework for developing multi-agent systems as stated in [4]. The JADE platform is made up of various containers on a single computer, or may be scattered on a computer network. This container is the place of life for agents. Container provides JADE run-time and all services required for agent placement and execution.

JADE provides a platform service called AMS (Agent Mobility Service) so that it can implement intra-platform mobility. This gives agents the ability to move over two or more container in the same platform. Therefore, JADE supports the development of systems that require mobile agents. An agent can have a specific task called behavior [5]. A behavior will present a task to be performed by an agent. An agent can have more than one behavior and an agent can run more than one behavior at a time. JADE architecture is shown at Figure 1.

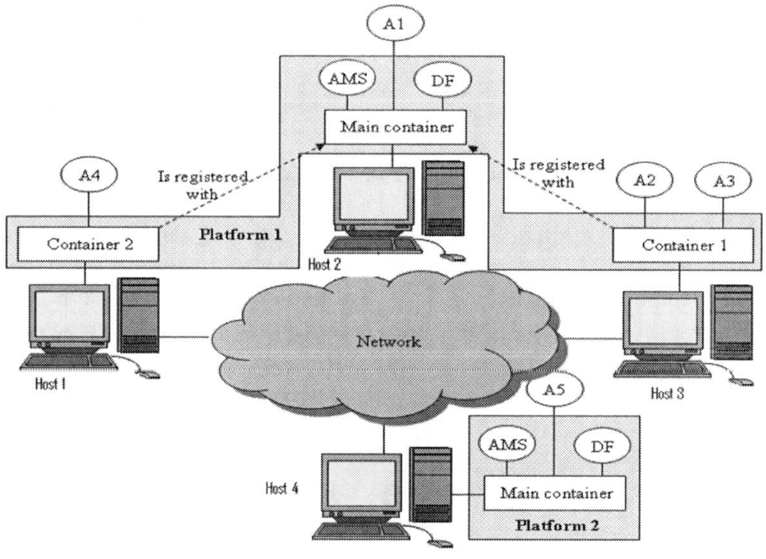

Figure 1. JADE Architecture [4]

Interoperability in a system can be reduced by the presence of a Web Service. Web Services can be used as one of the bridges between one application with another application, both on embeded systems and computer conventional systems [6]. From the results of the analysis that has been done, during this system interoperability is one of the problems that often occur and difficult to overcome. Using Web Service, the interoperability problems can be solved. Data retrieval and data transmition will also be more optimal.

Based on the explanation above, to solve the problem, the researcher proposes a system capable on bridging interoperability by combining with autonomous mechanism owned by agent. The system is used to help overcome interoperability and autonomous delegation of a Web Service with the REST approach, resulting in data retrieval, while the agent itself acts in accordance with the environment created by the user itself [7]. This system works in real time that the system will monitor the activities of the server and will display and perform the mechanism of data parsing from the web server. Data parsing results can later be utilized for many things in accordance with the needs and needs utilization. The agent work system diagram is shown in Figure 2.

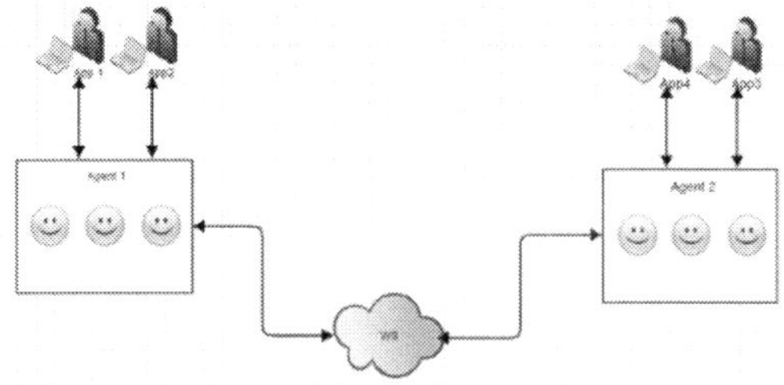

Figure 2. Agent Work System Diagram

Autonomous smart agent is used to solve the problem of how to find out data and parsing data using Agent. The term Agent is well known in many fields, both in the field of informatics and computer science, such as software engineering, artificial intelligence (AI), and distributed systems [8]. As for the proposed method of sending data on the server can be seen in Figure 3.

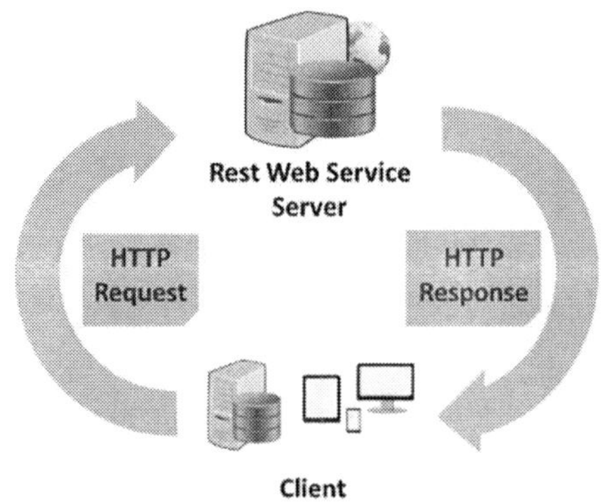

Figure 3. Web Service with REST method

3. Results and Discussion

Modeling an agent system must also do an environmental modeling where the agent will respond. Any changes that occur in the environment will result in changes also to the way of the movement of the agent, the environment used is the activity that occurs in the server environment, in this case the server will be used is a server that will accommodate data from the temperature sensor so it will provide impact on the system Agent. This environment will be modeled into a database that will be updated based on updates that occur on the system that has been made before.

Prior to the process of data retrieval by agents, the first process to do is mapping data. Data to be taken is data stored on the server in the form of humidity temperature data, and other supporting data. Once obtained the data, then the next process is taking metadata in accordance with the metadata that issued by the Web Service as exemplified in Figure 4. After obtained the data then the next process is to process the data retrieval done by agentRetieval through the mechanism of sending messages by AgentRequester to AgentRetrieval as shown in Figure 5.

After getting the message from agentRequester then agentRetrieval will perform the process of taking metadata generated by Web Service. The generated data will be sent through the mechanism of sending the message with FIPA-ACL to agentRequester. That output will be produced in accordance with the agent needs. After the data obtained then the next process is the process of extraction of information based on metadata. This is obtained by agentRetrieval to be developed into a prediction mechanism for the temperature needs of a room at a certain time interval. Thus the use of electricity will be more efficient as data resulted. Learning process on multi agent at the time of data retrieval through restful mechanism. The process is done by taking the metadata which will then be done data-solving process. With the structure of the Web Service obtained, then the agent will more easily take the distribution of data and map data based on the distribution of data.

```
{
- channel: {
    id: 84008,
    name: "Temperature & Humidity",
    description: "Temperature & Humidity of
    latitude: "50.94498",
    longitude: "-1.425",
    field1: "Temperature",
    field2: "Humidity",
    created_at: "2016-02-03T14:08:38Z",
    updated_at: "2016-12-01T22:22:59Z",
    elevation: "49",
    last_entry_id: 303352
  },
- feeds: (100) [
  - {
      created_at: "2016-12-01T21:46:10Z",
      entry_id: 303253,
      field1: "21.80",
      field2: "24.40

      "
    },
  - {
      created_at: "2016-12-01T21:46:33Z",
      entry_id: 303254,
      field1: "21.90",
      field2: "24.50

      "
    },
```

Figure 4. Web Service example

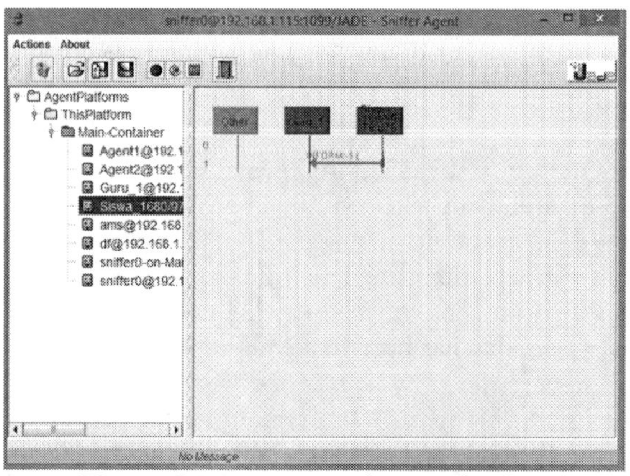

Figure 5. Message passing mechanicm

After the process of data retrieval and solving of the data then the next process is to store the clustering results into the database and perform the process of filtering data in accordance with the metadata and characteristics of each data. The results of the filtering data will be sent back through FIPA-ACL to the agent Requester so that Data can be done next process that is predicted temperature at interval time and also certain room. After the message received by AgentRequester, then next is to display the data retrieval and forecasting process to the data obtained to the next temperature. Delivery questions via ACl Message mechanism is shown in Figure 6.

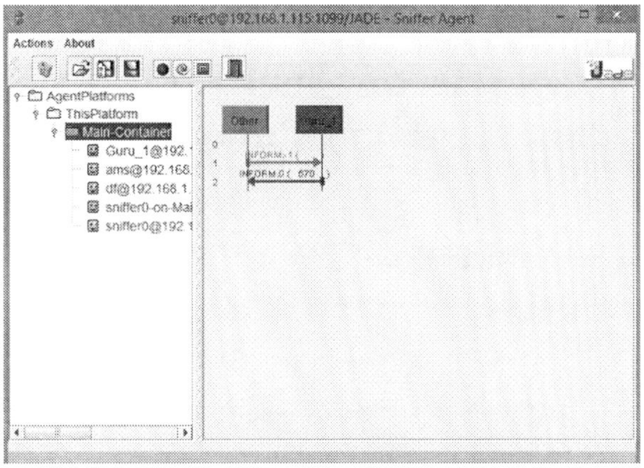

Figure 6. Delivery questions via ACl Message mechanism

4. Conclusion

Based on the architectural model proposed in this research there are two agents that are implemented, namely AgentRequester and AgentRetrieval. Each agent is able to perform tasks in data retrieval and data split in accordance with the given behavior. AgentRequester has successfully retrieved the data through the Web Service and sends it to AgentRetrieval which is tasked with breaking the data using string parser and JSONParser to get the desired data by type and time. By using Multi-agent in data retrieval on Web Service makes data retrieval more flexible, this is because the agent is able to adapt by utilizing data sent by Web Service as object learning from agent.

In addition, agents are able to collaborate with each other so that in subsequent research the results of the retrieval will be used as a basis for forecasting and other research.

References

[1] P. Leong and L. Lu, "Multiagent web for the internet of things," ICISA 2014 - 2014 5th Int. Conf. Inf. Sci. Appl., pp. 1–4, 2014.
[2] O. Kwon, G. P. Im, and K. C. Lee, "An agent-based web service approach for supply chain collaboration," Sci. Iran., vol. 18, no. 6, pp. 1545–1552, 2011.
[3] A. Micsik, P. Pallinger, and A. Klein, "Soap based message transport for the jade multiagent platform," … Auton. Agents Multiagent …, pp. 101–104, 2009.
[4] I. Špeh and I. Heđ, "A Web - Based IoT Solution for Monitoring Data Using MQTT Protocol," pp. 249–253, 2016.
[5] X. Bai, G. Dai, D. Xu, and W.-T. Tsai, "A multi-agent based framework for collaborative testing on Web services," Softw. Technol. Futur. Embed. Ubiquitous Syst. 2006 2006 Second Int. Work. Collab. Comput. Integr. Assur. SEUS 2006/WCCIA 2006. Fourth IEEE Work., p. 6 pp., 2006.
[6] B. M. Susanto et al., "Analisa Dan Penerapan Metode Single Exponentian Smoothing Dan Multi Agent Sistem Pada Prediksi," J. Ilm. Inov., vol. 1, no. 2, pp. 172–177, 2016.
[7] J. Song, W. Chen, and D. Gao, "The Adaptive On-line Exam System Based on Agent," 2011 Int. Conf. Futur. Comput. Sci. Educ., pp. 576–579, Aug. 2011.
[8] F. Bellifemine, G. Caire, and D. Greenwood, Developing Multi-Agent Systems with JADE. 2007.
[9] A. Mokhamad Azwar, E. S. Jullev A, T. Baratha adji, and I. ; Hidayah, "Penggunaan Multi-Agent Sistem Untuk," SEMNASTEKNOMEDIA ONLINE, pp. 6–8, 2015.

Application of fuzzy selection method of accompany computer laboratory assistant recommendation

Cholifia and Yulmaini

Institute Darmajaya Bandar Lampung, Indonesia

choco.1311010127@mail.darmajaya.ac.id, yulmaini@darmajaya.ac.id

Abstract. The admission process computer lab assistant candidates are a routine activity to get a candidate assistant in accordance with the desired criteria. In the selection process is often difficult to determine the candidate who passed the selection because there are some candidates who have the ability not much different. Therefore, to optimize the selection process for the admission of computer lab assistant candidates, a system is needed in decision making based on the weighting of each predetermined criteria, such as Semester, GPA, Practical Course Value, Written Test, Competency Test, and Test Interview. The method used in decision making is by applying the TOPSIS fuzzy problem-solving method, which is one of the multi-attribute fuzzy methods for decision-making based on the concept where the chosen alternative not only has the shortest distance from the ideal solution but also has the longest distance from the ideal solution negative. In the application, this method will generate an alternate ranking where the selected alternative is the top ranking alternative that is tailored to the needs/quota on the acceptance selection of the assistant and will produce a competent candidate according to the criteria which has been specified.

1. Introduction

Computer laboratory is one of the facilities to support teaching activity. This facility has a function as a mean for conducting a practice or a theoretic implementation, research, and scientific development. To support the teaching and learning process in a computer laboratory, a laboratory assistant is needed. The laboratory assistant is a student who has been selected through several admission phases. This assistant's duty is to assist other students in learning, supervising and assisting processes and to carry out other tasks related to the practicum class conducted in the computer laboratory.

Registration process for the IBI Darmajaya laboratory assistant candidate was still use manual process. The candidate submitted the requirement to the laboratory administrative division. Moreover, the assessment process for the assistant candidate was also still subjective due to a lack of the consideration on the weight criteria. Furthermore, the assessment process also had different assessment standards and difficulties to determine the selected candidate due to the candidates' similar skills. To optimize the assessment process, a decision making system had to be built and it was based on the weight criteria which had been prepared through TOPSIS fuzzy method. Previously done research [1] build decision support system computer laboratory assistant recommendation using fuzzy Simple Additive Weight based offline. The research entitled the implementation of TOPSIS fuzzy method for the selection of employee acceptance builds a system that can provide recommendations as a consideration for proper decision making in selection of employee [2]. Previous research [3] implemented fuzzy inference system Mamdani to help the management decision in appropriateness financing to apply for goods credit. While the research conducted [4] builds a multi-dimensional data-driven Decision Support System using Fuzzy TOPSIS as a recommendation against evaluation of supply on Electronic Government Procurement System (SPSE).

Based on the background of the problem, the formulation of the problem was build the decision making system through the TOPSIS fuzzy method for the admission of the laboratory assistant candidate based on each the weighting of any predefined criteria web-based.

2. Theoretical review

2.1. Admission of assistant candidate

Admission of laboratory assistant candidate is the routine activity conducted in every academic year. The admission of laboratory assistant candidate is done by selecting the candidate based on the predetermined criteria e.g., semesters, practicum course scores, written tests, competency tests, and interview tests. They are used to optimize the actual learning process. In the admission of the laboratory assistant candidate, there is a work instruction which is used as a reference for the laboratory manager to implement the admission of a new laboratory assistant involving all IBI Darmajaya academic communities.

2.2. Decision support system

The decision support system concept uses Management Decision System [5]. Decision-making is basically a form of selection from a variety of alternative actions which have been chosen through a certain mechanism for creating the best decision. The decision support process begins with the intelligence, design, selection, and implementation.

2.3. Fuzzy Multiple Attribute Decision Making (FMADM)

Fuzzy Multiple Attribute Decision Making (FMADM) is the method used to find the optimal alternative of a number of alternatives with the certain criteria. The essence of FMADM is to determine the weight score for each attribute and to determine the ranking process used to select the provided alternatives. Several methods which can be used to solve FMADM problems are Simple Additive Weighting Method (SAW), Weighted Product (WP), ELECTRE, Technique for Order Preference by Similarity to Ideal Solution (TOPSIS), and Analytic Hierarchy Process (AHP). These methods are used to take a conversion from fuzzy to crisp data by determine the ranking process.

2.4. Fuzzy TOPSIS (*Technique for Order Preference by Similarity to Ideal Solution*)

Fuzzy TOPSIS is the method that can assist the systematic evaluation chosen from several alternatives based on the certain criteria by which each criterion has a fuzzy score. The TOPSIS fuzzy method is simple and easy to understand. Moreover, the computation is efficient, and has the ability to measure a relative performance of the decision alternatives in simple mathematical form [6]. The TOPSIS fuzzy method procedures are as follows [6]:

a) To determine the criteria related to the case study.
b) To determine the weight and important rate of each criterion.
c) To determine the membership degree of each criterion on each alternative.
d) To convert fuzzy numbers of a decision matrix into crisp numbers.
e) To determine the normalized decision matrix (R).
f) TOPSIS requires a performance rating on each prospective employee at any normalized criteria or sub criteria. The normalized matrix is formed from the equation (1).

$$Rij = \frac{xij}{\sqrt{\sum_{i=1}^{n} x_{ij}^2}} \tag{1}$$

g) To calculate a weighted normalized decision matrix (Y)

To calculate the normalized matrix by (Y_{ij}), the W weight is firstly determined to represent the absolute preferences of the decision maker. The weight value preference indicates the relative important rate of each criterion or sub criteria in the equation (2):

$$Yij = Wij\,.Rij \qquad \text{where } W = \{W1, W2, \ldots, Wn\} \tag{2}$$

h) To calculate the positive and negative ideal solution matrix (A- and A +)
The positive and negative ideal solution can be determined by a normalized weight rating. The requirement to calculate the ideal solution score which is used to determine whether it is beneficial or costly is needed using equation (3) and (4).

$$A^+ = \{Y_1^+, Y_2^+, \ldots, Y_n^+\} \tag{3}$$
$$A^- = \{Y_1^-, Y_2^-, \ldots, Y_n^-\} \tag{4}$$

With the requirement that: $y_j^+ = \begin{cases} \max y_{ij}; & If\ j\ is\ benefit\ attribute \\ \min y_{ij}; & If\ j\ is\ cost\ attribute \end{cases}$

$y_j^- = \begin{cases} \min y_{ij}; & If\ j\ is\ benefit\ attribute \\ \max y_{ij}; & If\ j\ is\ cost\ attribute \end{cases}$

i) To calculate the distance between each alternative score and the ideal positive and negative solution matrix (Di^+ and Di^-). Through an equation (5) and (6).

$$D_i^+ = \sqrt{\sum_{i=1}^{n} (y_i^+ - y_{ij})^2} \tag{5}$$

$$D_i^- = \sqrt{\sum_{j=1}^{n} (y_{ij} - y_i^-)^2} \tag{6}$$

j) To calculate the preference score for each alternative **(Vi)**
The preference value **(Vi)**for each alternative is formulated using equation (7).

$$V_i = \frac{D_i^-}{D_i^- + D_i^+} \tag{7}$$

3. Result and discussion

3.1. Data Source Analysis

The data used in this research was the data of registrants as IBI Darmajaya computer laboratory assistant candidate in the academic year of 2015/2016. Moreover, the assistant candidate came from Informatics Engineering Study Program seen on table 1.

Table 1. Registrant data.

No	Student Number	Candidates	SEMESTER	GPA
1	1311010046	Indra Sahari	5	2.80
2	1311010059	Ahmad Farhan Afifi	5	3.44
3	1311010062	Yogi Maulana	5	1.93
4	1311010102	Irvan Rimhot Sijabat	5	3.15
5	1311010126	Nina Mutmainah	5	3.33
6	1111010017	Robby Muksin	9	3.12
7	1311010134	Mandaleka Saputri	5	2.93

No	Student Number	Candidates	SEMESTER	GPA
8	1411010024	Teddy Pratama	3	3.39
9	1411010048	Firdaus	3	3.72
10	1411010073	Ananda Dharmayana Rani	3	3.22
11	1311010081	Paksia Margono, S.Pd	5	3.63
12	1411010029	Muhammad Harisuddin	3	2.79

3.2. Fuzzy TOPSIS analysis

The calculation process of this research used TOPSIS fuzzy method. This method was used to rank the assistant candidates based on predetermined criteria. The alternative candidate data could be seen on table 2.

Table 2. Alternative candidate.

Alternative Candidates	Criteria					
	Semester	GPA	Practicum Course Score	Written Test	Competency Test	Interview Test
Farhan	5	3.44	81.2	65	40	90
Irvan	5	3.15	79.2	25	40	55
Nina	5	3.33	76.1	70	42	65
Paksia	5	3.39	85.4	30	41	65
Teddy	3	3.39	80.0	70	64	95
Firdaus	3	3.72	80.0	80	65	90
Ananda	3	3.22	80.0	75	66	90

Criteria used as a reference in the admission process can be seen on the table 3. Each criterion had the important weight rate. In addition, all criteria had the benefit attribute. This was because the highest score of each criterion was the best one so it was beneficial.

Table 3. Important rate criteria.

No	Name	Important Rate	Weight	Attribute
1	Semester (C_1)	Important	10%	*Benefit*
2	GPA (C_2)	Important	10%	*Benefit*
3	Practicum Course Score (C_3)	Important	10%	*Benefit*
4	Written Test (C_4)	Very Important	30%	*Benefit*
5	Competency Test (C_5)	Very Important	30%	*Benefit*
6	Interview (C_6)	Important	10%	*Benefit*

The score given for each criterion based on the specified fuzzy numbers could be seen in table 4 to table 9.

Table 4. Semester Criteria (C1).

Semester	Fuzzy Numbers	Crisp Numbers
Semester $<$ 3 or Semester $>$ 8	Limited (K)	0.00
3 – 4	Sufficient (S)	0.25
5 – 6	Good (G)	0.50
7 – 8	Very Good (VG)	0.75

IOP Conf. Series: Materials Science and Engineering **403** (2018) 012057 doi:10.1088/1757-899X/403/1/012057

Table 5. GPA Criteria (C2).

IPK	Fuzzy Numbers	Crisp Numbers
GPA< 3	Limited (K)	0.00
3	Sufficient (S)	0.25
3 >GPA< 3,5	Good (G)	0.50
GPA> 3,5	Very Good (VG)	0.75

Table 6. Practicum Course Score Criteria (C3).

Practicum Course Score	Fuzzy Numbers	Crisp Numbers
< 56	Limited (K)	0.00
56 – 69	Sufficient (S)	0.25
70 – 79	Good (G)	0.50
≥ 80	Very Good (VG)	0.75

Table 7. Written Test Criteria (C4).

Written Test Score	Fuzzy Numbers	Crisp Numbers
< 50	Limited (K)	0.00
50 – 59	Sufficient (S)	0.25
60 – 69	Good (G)	0.50
≥ 70	Very Good (VG)	0.75

Table 8. Competency Test (C5).

Competency Test Score	Fuzzy Numbers	Crisp Numbers
< 50	Limited (K)	0.00
50 – 59	Sufficient (S)	0.25
60 – 69	Good (G)	0.50
≥ 70	Very Good (VG)	0.75

Table 9. Interview Criteria (C6).

Interview Test Score	Fuzzy Numbers	Crisp Numbers
< 50	Limited (K)	0.00
50 – 59	Sufficient (S)	0.25
60 – 69	Good (G)	0.50
≥ 70	Very Good (VG)	0.75

Score given to each alternative on each criterion could be seen on table 10.

Table 10. Alternative score on each criteria.

Alternative (A)	Variable					
	C1	C2	C3	C4	C5	C6
A1	0.50	0.50	0.75	0.50	0.0	0.75
A2	0.50	0.50	0.50	0.00	0.0	0.25
A3	0.50	0.50	0.50	0.75	0.0	0.50
A4	0.50	0.50	0.75	0.00	0.0	0.50

Alternative (A)	Variable					
	C1	C2	C3	C4	C5	C6
A5	0.25	0.50	0.75	0.75	0.5	0.75
A6	0.50	0.75	0.75	0.75	0.5	0.75
A7	0.25	0.50	0.75	0.75	0.5	0.75

According to the fuzzy TOPSIS procedure and equation (1) – (7), Alternative Preference Score was obtained on the table 11.

Table 11. Alternative preference score.

Alternative	Preference (V)	Graduation Result
Firdauss	0.9083	Passed
Teddy	0.8872	Passed
Ananda	0.8872	Passed
Nina	0.4272	Unpassed
Farhan	0.3634	Unpassed
Paksia	0.1196	Unpassed
Irvan	0.0917	Unpassed

A result of the calculation on the rank process by using TOPSIS fuzzy method showed that there were 3 candidates who were passed with the highest score. It meant that there was a feasibility score on the laboratory assistant candidate.

3.3. Discussion

The advantages of TOPSIS fuzzy system is to facilitate the computer laboratory in recruitment process prospective assistant because registration is online. Disadvantages of this system is the absence of login system for the registrant, so it is less secure, the system has not been able to directly manage the value of the assistant candidate so that the recapitulation process of the assistant candidate data value is still calculated manually.

4. Conclusion

TOPSIS fuzzy method can be used in the admission process for assistant candidate, so that it can produce the ranking about computer laboratory assistant candidate. Thus, the admission process for this assistant candidate is more objective and related to the specified criteria. Based on the data of candidate assistant (alternative) that has been calculated using TOPSIS fuzzy method there are 3 candidates who pass the selection and occupy the top ranking is Firdaus (with value = 0.9899) followed by Teddy and Ananda (with value = 0.9840).

References

[1] Ivanjelita L A, Utami E and Luthfi T 2015 Sistem Pendukung Keputusan Seleksi Penerimaan Calon Asisten Praktikum *Jurnal Ilmiah DASI* vol 16 no 4 pp 37-46
[2] Lestari S and Priyodiprodjo W 2011. Implementasi Metode Fuzzy Topsis Untuk Seleksi Penerimaan Karyawan *IJCCS* vol 5 no 2 pp 20–26
[3] Arfida S 2012 Implementasi Fuzzy Terhadap Sistem Pendukung Keputusan Menentukan Kelayakan Pembiayaan Pengujian Kredit Barang *Jurnal Informatika* vol 12
[4] Enre H A, Mardi S.N. S and Purnomo M H 2015 Sistem Pendukung Keputusan Berbasis Data Multidimensi Menggunakan Fuzzy Topsis pp 1–10
[5] Informatics & Business Institute Darmajaya 2016 *Panduan Penyusunan Karya Ilmiah* (Bandar Lampung, Indonesia: IBI Darmajaya)
[6] Kusrini M K 2007 *Konsep dan Aplikasi Sistem Pendukung Keputusan 1st Ed.* Editor: F. S. Suyantoro (Yogyakarta: Andi)

IOP Conf. Series: Materials Science and Engineering **403** (2018) 012058 doi:10.1088/1757-899X/403/1/012058

Evaluation QOS and energy consumption for DSDV and DSR in MANET (Mobile ad hoc network)

H A Sidharta, Sidharta and C Huda

Computer Science Study Program, Bina Nusantara Institute of Creative Technology Malang, East Java

hanugra.sidharta@binus.ac.id, sidharta@binus.ac.id, ChoirulHuda@binus.edu

Abstract. A mobile ad hoc network (MANET) is an infrastructure less, autonomous, and standalone network. Routing protocol mechanism on MANET can be classified proactive, reactive and hybrid. On this paper researcher comparing QOS with reactive and proactive, reactive represented by DSR and proactive by DSDV. Performance of these routing protocols is evaluated with performance metrics such as Dropped Packets, Packet Delivery Ratio, Throughput, End to End Delay, Jitter, and Residual energy. Based on experiment shows that DSDV is better than DSR, performance DSDV more stable than DSR, DSR on little node better than DSDV, but in line with increment number of node performance is degrade.

1. Introduction

MANET (Mobile ad hoc network) is multi-hop wireless network that are composed of mobile hosts communicating with each other through wireless links. MANET is likely to be use in many practical applications, including personal area networks, home area networking, military environments, and so on. The wide range of potential applications has led to a recent rise in research and development activities [1]. MANET also can use for NFC-based mobile payment services [2]. In MANET, nodes move arbitrarily, therefore the network may experience rapidly and unpredictably topology changes. Additionally, because nodes in MANET normally have limited transmission ranges, some nodes cannot communicate directly with each other. Hence, routing paths MANETs potentially contain multiple hops, and every node in mobile ad hoc net-works has the respond to act as a router.

There is a classification for MANET protocols as table-driven (proactive) and source initiated (routing) protocols. In proactive routing protocols, routing tables are used to keep route information from each source to every destination in network before this route is needed. On the other hand, in reactive routing protocols, a source sends a route discovery through the network, only when the route is required. The limited battery resources consumed by nodes in a MANET must be considered as a limited resource in using a routing protocol [3].

Efficient energy conservation plays an important role in the performance of MANET because mobile host in such networks are usually battery-operated [1]. Most existing network protocols do not consider power consumption an issue since they assume the presence of static hosts and routers, which are powered by mains. However, mobile devices today are mostly operated by batteries. Battery technology is still lagging behind microprocessor technology. The lifetime of Li-ion battery today is only 2-3 hours. Such a limitation in the operating hours of devices implies the need for power conservation. In particular, for ad hoc mobile networks, mobile devices must perform both the role of an end system and that of intermediate system. Hence, forwarding packets on the behalf of others will consume power, and this can be quite significant for nodes in an ad hoc wireless network [4]. In this paper, researcher investigates evaluation of energy consumption of two routing protocols between DSDV and DSR.

2. Literature review

This section reviews the literature used in this research.

2.1 Routing in MANET

The term routing refers to the process of selecting paths in a computer network along which to send data. This process is used to exchange information about topology and link weights, and a routing algorithm, that actually computes paths between nodes. Mobile Ad-hoc networks are self-organizing and self-configuring multi hop wireless networks where, the structure of the network changes dynamically. This is mainly due to the mobility of the nodes. Nodes in these networks utilize the same random access wireless channel, cooperating in a friendly manner to engaging themselves in multi hop forwarding. The nodes in the network not only act as hosts but also as routers that route data to/from other nodes in network. In mobile ad-hoc networks where there is no infrastructure support as is the case with wireless networks, and since a destination node might be out of range of a source node transmitting packets; a routing procedure is always needed to find a path so as to forward the packets appropriately between the source and the destination [5, 6].

Within a cell, a base station can reach all mobile nodes without routing via broadcast in common wireless networks. In the case of ad-hoc networks, each node must be able to forward data for other nodes. This creates additional problems along with the problems of dynamic topology which is unpredictable connectivity changes. Routing is difficult since mobility causes frequent network topology changes and requires robust and flexible mechanism to search and maintain routes. Routing protocols must also deal with other constraints such as low bandwidth, limited energy consumption, and high error rates [6].

2.2 DSR dynamic source routing protocol

DSR Dynamic Source Routing Protocol uses specific packet header to carry information. This particular packet header can be placed in any existing IP packet. A packet of DSR options header by four bytes of the length of the fixed part and the immediately subsequent sequence of DSR options, DSR option sequence composed by a number of DSR options, DSR options carrying optional information. DSR Options header sequence of DSR options, that is the end of the end of the DSR Options header, so the total length of the DSR Options header is equal to 4 bytes and the byte length of the sequence of DSR options combined [7].

2.3 DSDV

Destination Sequenced Distance Vector [7], is hop-by-hop distance vector routing protocol which requires every node at regular interval transmit routing updates which should be based on principles of classical Bellman-Ford Routing algorithm. Every routing node maintains a routing table which consists of record of the "next hop" for every reachable destination, also the number of hops attainable to destination and the sequence number assign by sink node. The sequence number can be used to make difference between stale routes and the fresh ones and thus avoiding the loop formation. It will also transmit its routing table if any significant change that has occur in their table from the final update sent [8].

1) Advanced uses of DSDV:
- DSDV protocol guarantees loop free paths.
- We can avoid extra traffic with incremental updates instead of full dump updates.
2) Limitation of DSDV:

It is difficult to maintain the routing table's advertisement for larger network. Each and every host in the network should maintain a routing table for advertising. But for larger network this would lead to overhead, which consumes more bandwidth. DSDV doesn't support Multi path Routing [9, 10].

2.4 Quality of service
The performance metrics helps to characterize the network that is substantially affected by the routing algorithm to achieve the required Quality of Service (QoS). In this work, the following metrics are considered [9]. End-to-End Delay (EED) is the time taken for an entire message to completely arrive at the destination from the source. Evaluation of end-to-end delay mostly depends on the following components:

- Propagation time (PT).
- Transmission time (TT).
- Queuing time (QT).
- Processing delay (PD).

EED is evaluated as PT + TT + QT + PD. The throughput measures how fast a node can actually send the data through a network. So throughput is the average rate of successful message delivery over a communication channel. Control overhead is the ratio of the control information sent to the actual data received at each node. Packet Delivery Ratio (PDR) is the ratio of the total data bits received to total data bits sent from source to destination. Energy Consumption per Delivered Packet measure the energy expended per delivered data packet [10].

3. Materials and method
This section explains the materials and method used in this research.

3.1 Simulation model
We have used Network Simulator (NS)-2 in our evaluation. The NS-2 is a discrete event driven simulator. NS-2 is suitable for designing new protocols, comparing different protocols and traffic evaluations. It is an object oriented simulation written in C++, with an OTcl interpreter as a frontend. NS uses two languages because simulator got to deal with two things:

- detailed simulation of protocols which require a system programming language which can efficiently manipulate bytes, packet headers and implement algorithms.
- research involving slightly varying parameters or quickly exploring a number of scenarios [11].

3.2 Simulation parameter
In our work, we used routing protocol DSDV and DSR which evaluated by varying the network size (number of mobile nodes). The simulation parameter is shown in table 1.

Table 1. Simulation parameter.

Parameter	Value
Simulator	NS-2 Simulator 2.1b9
Channel Type	Channel/Wireless Channel
Radio Propagation Model	Propagation/Two Ray Ground
Network Interface Type	Phy/Wireless Phy
MAC Type	Mac/802.11
Interface Queue Type	Queue/DropTail/CMUPriQueue
Link Layer Type	LL

Parameter	Value
Antenna	Antenna/Omni Antenna
Maximum Packet in ifq	50
Area(M*M)	500 * 500
Number of Nodes	5, 10, 15, 20, 25, 30
Traffic Type	TCP
Simulation Time	150
Routing Protocols	DSDV & DSR
Network QOS	*Jitter, Packet Delivery Ratio, Delay, Throughput, Energy*

4. Result and discussion

The simulation results are shown in the following section in the form of line graphs. Graphs show comparison between the DSDV and DSR by varying different numbers of nodes on the basis of different performance metrics.

4.1 Dropped packets

As shown in figure 1 with increase in number of nodes, Dropped Packets of all the protocols increased, performance of DSDV is better than DSR on node number 15 through 30, but on 5 to 10 node DSR better than DSDV.

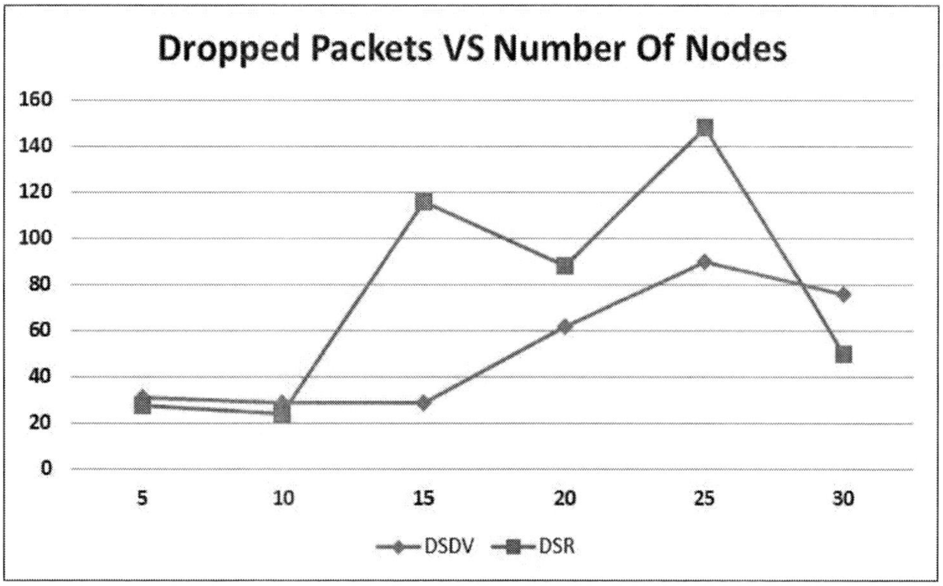

Figure 1. Dropped packets.

4.2 Packet delivery ratio

As shown in figure 2 with increase in number of nodes, Packet Delivery Ratio of all the protocols increased, performance of DSDV is better than DSR on node number 5 through 25, but anomalies on node 30 DSR better than DSDV.

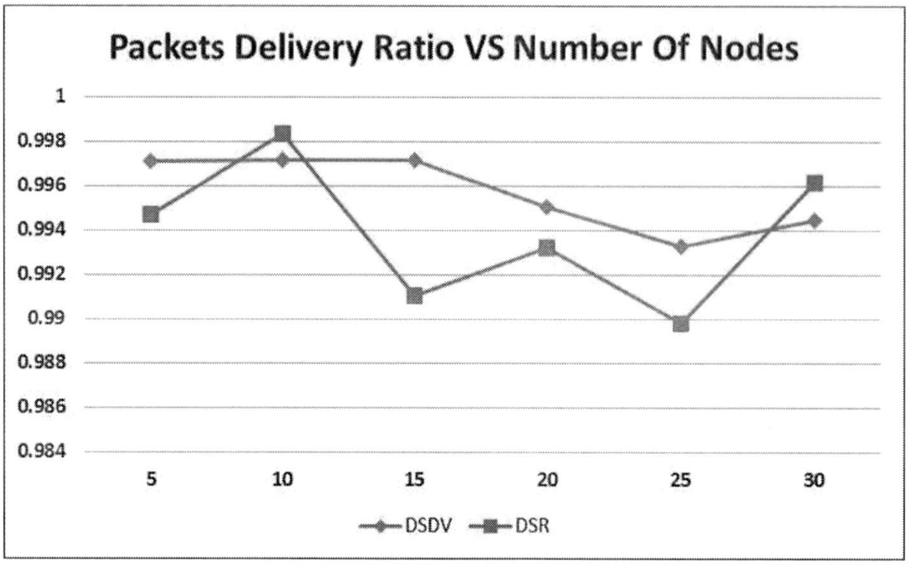

Figure 2. Packet delivery ratio.

4.3 Throughput

As shown in figure 3 with increase in number of nodes, Throughput of all the protocols increased, performance of DSDV stable from node 5 through node 30 with average throughput is 409,72. Meanwhile DSR performance on node 5 is astonishing with 662.89 but from node 10 to 30 performances degrade with average on all nodes are 454.93.

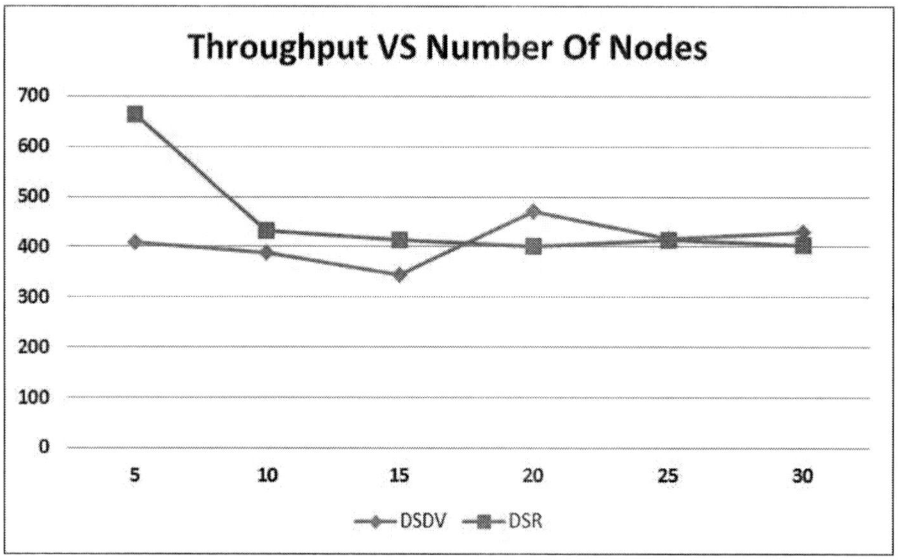

Figure 3. Throughput

4.4 End to end delay and jitter

Based on figure 4 between end to end delay and jitter have similarity symptom, DSDV is better from DSR on end to end delay also on jitter, also DSDV have more stable than DSR.

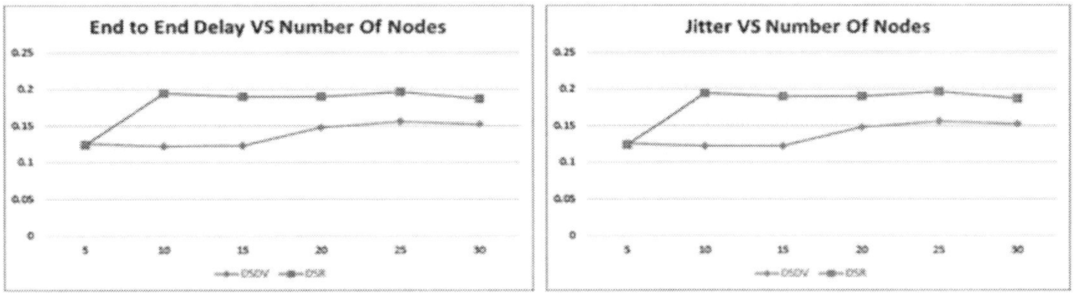

Figure 4. End to end delay and jitter.

4.5 Residual energy

As shown in figure 5 with increase in number of nodes, performance of DSDV is better than DSR. With DSDV on all node energy that have been used is more stable than DSR with average DSDV is 909.708 and DSR 887.05 from initial value 1000.

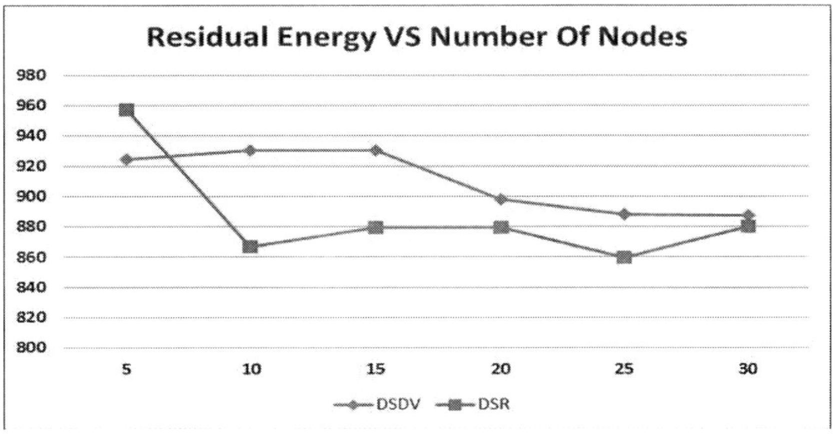

Figure 5. Residual energy.

5. Conclusion

In this paper performance comparison of DSR and DSDV routing protocols for Mobile Ad hoc Networks is presented as a function of number of nodes (network size). Performance of these routing protocols is evaluated with performance metrics such as Dropped Packets, Packet Delivery Ratio, Throughput, End to End Delay, Jitter, and Residual energy. As per statistic based on our trial shows that DSDV is better than DSR, performance DSDV more stable than DSR, DSR on little node better than DSDV, but in line with increment number of node performance is degrade. In future work, need to compare performance with other protocol such as geological, multi path, hierarchical, and multicast. Due vulnerability of routing protocol from attacker, need have proposed secure routing protocols for ad hoc networks.

References

[1] Li X, Zi-Wen W and Bao-Yu Z (n.d.) 2003 TPBDSR: a new DSR-based energy saving routing in MANET 2003 *International Conference on Computer Networks and Mobile Computing (ICCNMC 2003)*

[2] Wu W and Liaw H 2015 A Study on High Secure and Efficient MANET Routing Scheme *Journal of* Sensors pp 1-10

[3] Barati M, Atefi K, Khosravi F and Daftari Y A 2012 Performance evaluation of energy

IOP Conf. Series: Materials Science and Engineering **403** (2018) 012058 doi:10.1088/1757-899X/403/1/012058

consumption for AODV and DSR routing protocols in MANET *2012 International Conference on Computer & Information Science (ICCIS)*

[4] Elavarasipriyanka S and Latha V 2015 Design of Routing Protocol for MANET Through Energy Optimization Technique *International Journal of Innovations & Advancement in Computer Science* vol 4 pp 173-80.

[5] Radwan A, Mahmoud T and Houssein E 2011 Evaluation comparison of some ad hoc networks routing protocols *Egyptian Informatics Journal* vol 12 no 2 pp 95-106

[6] Jadhav S S, Kulkarni A V and Menon R 2014 Mobile Ad-Hoc Network (MANET) for disaster management *2014 Eleventh International Conference on Wireless and Optical Communications Networks (WOCN)*

[7] Ya-Qin F, Wen-Yong F and Lin-Zhu W 2010 OPNET-based Network of MANET Routing Protocols DSR Computer Simulation *2010 WASE International Conference on Information* Engineering

[8] Gupta M and Kumar S 2015 Performance Evaluation of DSR, AODV and DSDV Routing Protocol for Wireless Adhoc Network *2015 IEEE International Conference on Computational Intelligence & Communication Technology*

[9] Mohapatra S and Kanungo P 2012 Performance analysis of AODV, DSR, OLSR and DSDV Routing Protocols using NS2 Simulator *Procedia Engineering* vol 30 pp 69-76

[10] Khediri S E, Nasri N, Benfradj A, Kachouri A and Wei A 2014 Routing protocols in MANET: Performance comparison of AODV, DSR and DSDV protocols using NS2 *The 2014* International *Symposium on Networks, Computers and Communications*

[11] Gaharwar P and Gupta S R 2013 Performance Comparison of AODV, DSDV & DSR Routing Protocols in MANET *International Journal of Advanced Research in Computer Science and Software Engineering* vol 3 no 4 pp 1059-63

IOP Conf. Series: Materials Science and Engineering **403** (2018) 012059 doi:10.1088/1757-899X/403/1/012059

Discriminant independent component analysis for hyperspectral image classification

Murinto and N R D P Astuti

Departement of Informatics Engineering, Universitas Ahmad Dahlan, Yogyakarta, Indonesia

murintokusno@tif.uad.ac.id, rochmahdyah@tif.uad.ac.id

Abstract. A novel approach to discriminative extraction and independence of remote sensing hyperspectral imagery features using Discriminant Independent Component Analysis (DICA) is introduced in this paper. This approach maximizes the variance and negentropy of a given feature. The experimental results show that the proposed method has better classification accuracy when compared to using one of the most commonly used classifiers in the hyperspectral image of SVM. Classification accuracy obtained from DICA dimensionality reduction on Indian Pines dataset, value of average accuracy (AA) is 92.87 %, overall accuracy (OA) is 92.94 %, Kappa index is 0.82, whereas on Washington DC mall dataset, classification result obtained by using DICA dimensionality reduction is average accuracy amounted to 84.89 %, overall accuracy 84.69 % OA, Kappa index is 0.82. From experiment we can see that classification accuracy increases when using DICA approach compared than technique classical SVM.

1. Introduction

Hyperspectral imagery consists of more hundreds bands and have high spatial resolution, range from visible to infrared of areas. Spectral resolution increases with wide spectral range. It will further clarify characterizing the materials. This makes the hyperspectral image very suitable for source separation and classification processes. On the other hand, with very large data dimensions, the computational load will be heavier and can also decrease the result of the classification processes, because of the curse of dimensionality i.e. Huge effect [1]. In many studies have been developed techniques that overcome this problem, but the results still need to be improved again.

The linear transformation of the original data is often used to display data in order to facilitate its computation. Some of linear transformation methods include Principal Component Analysis (PCA) [2], Linear Discriminant Analysis (LDA) [3] and Factor Analysis (FA) [4]. The linear transformation technique that proposed after that is Independent Component Analysis (ICA) in which the components are statistically independent in the displayed data [5, 6]. Although widely used in hyperspectral image processing, the ICA has the disadvantage that independent components extracted on the ICA may be appropriate for representation but may not suitable to classification. Instead, it could use an optimization change on the problem [7]. To solve this problem, one of the solutions offered is to use a single, semi-supervised Discriminant independent component analysis (DICA) linear. In DICA, Fisher linear Discriminant and Negentropy are simultaneously used to process and hidden independent factors are extracted from multivariate data.

Some classifiers such as Support Vector Machine (SVM) [8] and Neural Network (NN) [9] are widely used in hyperspectral image classification, because of their ability to handle high dimensional data problems. However, they have disadvantages that may influent to result the classification process. The disadvantage of NN techniques is the number of hidden layers that are numerous, as well as the number of iterations that sometimes take too long on the computation process. Although SVM produces relatively good classification accuracy in limited training data sets, SVM also has a major disadvantage of training time, depending on the training sample data. In kernel usage, it is often necessary to find the optimal parameter, usually using a cross-validation approach [10].

In this study we propose a method for Discriminant analysis based on the extraction of independent features obtained by using entropy maximization i.e. Discriminant Independent Component Analysis (DICA) [7, 11]. The value of the fisher criterion and the sum of negentropy from features that independently extracted is maximized simultaneously. The results obtained using the proposed method was then compared with one of the most widely used classifiers in hyperspectral image processing i.e. the support vector machine (SVM).

2. Independent Component Analysis (ICA)

ICA is a linear transformation that minimizes dependencies statistically among the components. It was first introduced in 1994 by Pierre Common and has been used in a variety of applications including Blind Source Separation (BSS) [5]. ICA has been used to reduce high dimensions through the projection of the original hyperspectral image to a lower dimensional space. In this projection, the number of observed signals n is the original dimension of the data and the number of source m signals is a low dimensional space corresponding to material present in the hyperspectral image

ICA is based on the problem BSS recovering unknown source signal by simplifying the assumption of n independent signal that can be write as $s(t) = s_1(t), \dots, s_n(t)$ and observed mixture of the signals are denoted as $x(t) = x_1(t), \dots, x_n(t)$, Suppose mixture of signal as linear mixing equation can be given as equation (1).

$$x(t) = As(t) \tag{1}$$

Where t is time, A is unknown mixing matrix. S (t) is a source signal for transformation into a mixed signal. If mixing matrix, A is invertible there is no source M less than or equal to number of sensors P, then source can be split directly and output from y(t) can be write as equation (2).

$$y(t) = Bx(t) = BAs(t) \tag{2}$$

Where B is mixing matrix obtained from matrix A.

ICA is a linear approximation that maximizes a Gaussian measure or minimizes objective function. In general, the objective function used in ICA algorithm is mutual information of a y (t), which is written in equation (3).

$$I(y_t, B) = \sum_i H(y_{t_i}) - H(y_t) \tag{3}$$

3. Discriminant Independent Component Analysis

Multivariate data with lower dimensions and independent features in Discriminant Independent Component Analysis (DICA) were obtained by using entropy maximization [7]. Here the Fisher criterion and marginal summation of negentropy extracted independent features are maximized simultaneously. Optimization function that maximize the incorporation of negentropy and the function of classification performance measurement at constrains normalize vector v_i given through equations (4).

$$\hat{L}(W) = \check{L}(W) + k\emptyset(W, Z, C) \tag{4}$$

Where k is constant and $\emptyset(W, Z, C)$ is measurement function classification performance from features Y given C, while $\breve{L}(W)$ is a constraint of Lagrangian formula as given in [7].

Adaptation of w_i can be improved by using *gradient ascent algorithm* along the function in equation (4), then followed by symmetric orthonormalization of W. Learning rule as given in equation (5).

$$\Delta w_i = \eta\left((\gamma_i(E\left(Zg^1(w_i^T Z)\right)\right) + \gamma_i kE(Zg^2(W^T Z))\frac{\partial\emptyset(w_i,Z,C)}{\partial w_i} + 2\beta_i w_i\Big) + \eta k\frac{\partial\emptyset(W,Z,C)}{\partial w_i} \quad (5)$$

$$W \leftarrow (WW^T)^{-\frac{1}{2}}W \quad (6)$$

Value of β_i in equation (5) can be obtained from completion of stationary point of equation (4) and is given as:

$$\beta_i = -\frac{1}{2}\left((\gamma_{1i}(k_1 E(y_i g^1(y_i)) + \gamma_{2i} k_2 E(y_i g^2(y_i) + kw_i^T \frac{\partial\emptyset(w_i,Z,C)}{\partial w_i}\right) \quad (7)$$

Function measuring classification perform $\phi(W, Z, C)$, Which is corresponds to demixing vector W given by the equation (8).

$$\phi(W, Z, C) = \sum_{i=1}^{R} log \sum_{c=1}^{C} N_c(\mu_{ic} - \mu_i)^2 \quad (8)$$

Where $\mu_{ic} = E(z_i|c)$ is feature mean i class c, and N_c is the number of samples corresponding to class c, while $\mu_i = E(z_i)$ is feature mean i-th class. While optimal linear transformation can be write as equation (9).

$$\breve{W} = arg \max_W \phi(W, Z, C) = arg \max_W \sum_{i=1}^{R} log \sum_{c=1}^{C} N_c(\mu_{ic} - \mu_i)^2 \quad (9)$$

Partial derivatives from $\phi(W, Z, C)$ Which corresponds to the vector basis w_i can be written as an equation (10).

$$\frac{\partial\phi(W,Z,C)}{\partial w_i} = 2\sum_{c=1}^{C}\sum_{n\in class\ c}\frac{(\mu_{ic}-\mu_i)}{\sum_{c'=1}^{C} N_{c'}(\mu_{ic'}-\mu_i)^2} z_n \quad (10)$$

Where z_n is sample n[th] from whitened observation. By substituting the equation (10) into equation (5), it produces a discriminative feature extraction algorithm that maximizes sum of marginal negentropy among classes plotted on each feature. The algorithm for extraction of DICA features on asymmetric distribution is given as follows [7, 12]:

1. Center the data, $X: X \leftarrow X - E(X)$.
2. Whitening the data center X to get orthonormal features, Z using equation (4).
3. Initialize $W = W_0: ||W_0|| = I_R$.
4. Obtain DICA features, $Y = W^T Z, Y \in R^R$.
5. Update W using equations (10) and (5).
6. Symmetric orthogonalization $W: W \leftarrow (WW^T)^{1/2}W$.
7. If the sum negentropy y and discriminant function $\emptyset$ is convergent then iteration stopped. If other, back to step 4.

4. Experiment results

This experiment uses two hyper spectral image datasets obtained from Aviris and Hydice sensors. Data from Aviris sensors used in this experiment is Indian pines hyperspectral image, while Hydice sensor is hyperspectral image of Washington DC Mall. First image is Indian Pines hyperspectral image obtained from Aviris sensor that flew over the north western Indiana area in 1992. This image has a size of its spectral resolution of 145 x 145 pixels. Where in this image consists of 220 spectral bands (Noise already omitted) [12]. Ground truth Indian pines hyper spectral image consists of 16 classes. Second image is hyper spectral image obtained from the Hydice sensor. This sensor gets the image of a flight over the Washington DC Mall. The hyperspectral image of Washington DC Mall consists of 210 spectral bands. Band noise is eliminated and only 191 spectral bands are used. This hyperspectral image was taken in August 1995, in which each band consists of 1280 lines, with each band consisting of 307 pixels. Ground truth image Washington DC Mall consists of 6 classes. In the experiment, 35 pixels were taken randomly from each class and used as training samples [8]. The test data in this experiment are each taken from each class on the image data of Indian pines and Washington DC Mall. Complete data of training and testing are shown in table 1. In this experiment training sample was taken 20 pixels randomly from each class.

Table 1. Samples of train-test for two hyper spectral image datasets.

	Indian Pines			Washington DC Mall		
No. Class	Class	Train	Test	Class	Train	Test
#1	Alfalfa	20	54	Roof	35	3794
#2	Corn no-till	20	1434	Road	35	376
#3	Corn-min till	20	834	Trail	35	135
#4	Corn	20	234	Grass	35	1888
#5	Grass-pasture	20	497	Tree	35	365
#6	Grass-Trees	20	747	Water	35	1184
#7	Grass-Mowed	20	26	Shadow	35	57
#8	Hay-windowed	20	489			
#9	Oats	20	20			
#10	Soybean-no till	20	968			
#11	Soybean-min till	20	2468			
#12	Soybean-clean	20	614			
#13	Wheat	20	212			
#14	Woods	20	1294			
#15	Bldg-Trees-Drive	20	380			
#16	Stone-Steel-Tower	20	95			

Results of experiments conducted in this study using DICA then compared with results obtained using SVM, with polynomial kernels. Compared performance is in terms of Overall Accuracy (OA), Kappa index (K) and Average Accuracy (AA) of both techniques used. OA represents percentage of sample correctly classified. AA represents the average of classification accuracy for each class in the data set, while Kappa index (K) is a correct estimate of the percentage estimation parameters without the amount due chance alone. Table 2 shows the classification results using DICA and SVM original methods. From table 2 it shows that in AVIRIS Indian Pines data set when using SVM obtained an OA value of 86.18%, index of K is 0.85 and AA is 85.95%, while with same data set increase in the value of OA, K and AA are 92.94%, 0.92 and 92.98%, respectively. Likewise, in Washington DC Mall data set, there is an increase in classification accuracy when using DICA approach. The values obtained are OA is 75.59%, K is 0.73 and AA is 80.86%, using SVM method. And when we were using DICA the value obtained by OA is 84.69 %, K is 0.82, and AA is 84.89%. From this result so can be concluded that by using DICA method it will be increase the accuracy of hyperspectral image classification compared with SVM original methods.

Table 2. Comparison of classification accuracy resulting from DICA and SVM methods of hyperspectral image datasets of Indian Pines and Washington DC Mall.

	Indian Pines Image		Washington DC Mall Image	
Methods	SVM	DICA	SVM	DICA
OA	86.18%	92.94%	75.59%	84.69%
K	0.85	0.92	0.73	0.82
AA	85.95	92.87	80.86	84.89
Class #1	47.62	80.95	60.10	70.31
Class #2	80.00	90.00	75.00	80.48
Class #3	90.91	100.00	90.91	92.50
Class #4	85.00	95.00	75.00	80.00
Class #5	85.00	100.00	75.00	80.50
Class #6	100.00	100.00	100.00	100.00
Class #7	100.00	100.00	90.00	91.00
Class #8	100.00	100.00	-	-
Class #9	100.00	95.00	-	-
Class #10	95.24	95.24	-	-
Class #11	90.48	95.24	-	-
Class #12	65.00	90.00	-	-
Class #13	75.00	75.00	-	-
Class #14	100.00	100.00	-	-
Class #15	73.91	78.26	-	-
Class #16	86.96	91.30	-	-

5. Conclusions

In this research we have proposed an approach for hyperspectral image classification using Discriminant Independent Component Analysis (DICA). The proposed approach is based on the application of Independent Component Analysis (ICA) to obtain independent components of the processed data, using entropy maximization and Fisher criterion. The study used two different data sets: Aviris Indian Pines and Hydice Washington DC Mall data sets to implement the proposed approach. The experimental results show that the proposed method has better classification accuracy when compared to using one of the most commonly used classifiers in the hyperspectral image of SVM. Classification accuracy obtained by using DICA dimensionality reduction on Indian Pines Aviris dataset, the value of average accuracy (AA) is 92.87%, overall accuracy (OA) is 92.94%, Kappa index is 0.82, whereas on Washington DC mall dataset, classification result obtained by using DICA dimensionality reduction is average accuracy amounted to 84.89%, overall accuracy 84.69 % OA, Kappa index is 0.82. This classification accuracy is higher when compared to using the technique classical SVM.

Acknowledgment

This research is funded by DP2M-RISTEKDIKTI with applied products research (PPT) scheme as outlined in contract No: PPT-071/SP3/LPP-UAD/IV/2017.

References

[1] Hughes G F 1968 On the mean accuracy of statistical pattern recognizers *IEEE Trans. Inform. Theory* vol IT-14 pp 55–63
[2] Rodarmel C and Shan J 2002 Principal component analysis for hyperspectral image classification *Surveying and Land Information Science* vol 62 no 2 p 115
[3] Bandos T V, Bruzzone L and Camps-Valls G 2009 Classification of hyperspectral images with regularized linear discriminant analysis *IEEE Transactions on Geoscience and Remote*

Sensing vol 47 issue 3 pp 862-873

[4] Hans G and Geladi P ed. 2007 *Techniques and applications of hyperspectral image analysis* (John Wiley & Sons)

[5] Hyvärinen A, Karhunen J and Oja E 2002 *Independent component analysis* (John Wiley & Sons, Inc.)

[6] Villa A, Chanussot J, Jutten C, Benediktsson J A and Moussaoui 2009 On the use of ICA for hyperspectral image analysis *Geoscience and Remote Sensing Symposium, 2009 IEEE International, IGARSS 2009* Cape Town, South Africa

[7] Dhir C S and Lee S-Y 2011 Discriminant independent component analysis *IEEE transactions on neural networks* vol 22 issue 6 pp 845-57

[8] Melgani and Bruzzone L 2004 Classification of hyperspectral remote sensing images with support vector machine *IEEE Transaction on Geoscience and Remote Sensing* vol 42 issue 8

[9] Ratle F, Camps-Valls G and Weston J 2010 Semisupervised neural networks for efficient hyperspectral image classification *IEEE Transactions on Geoscience and Remote Sensing* vol 48 issue 5 pp 2271-82

[10] Camps-Valls G and Bruzzone L 2005 Kernel-based methods for hyperspectral image classification *IEEE Transactions on Geoscience and Remote Sensing* vol 43 issue 6 pp 1351-62

[11] Mollaee M, Moattar M H and Mahdavi S J S 2014 Feature extraction based on modified discriminant independent component analysis via particle swarm optimization *International Journal of Software Engineering and Its Applications* vol 8 no 12 p 89-100

[12] Murinto and Harjoko A 2016 Dataset feature reduction using independent component analysis with contrast function of particle swarm optimization on hyperspectral image classification *2016 2nd International Conference on Science in Information Technology (ICSITech)*

IOP Conf. Series: Materials Science and Engineering **403** (2018) 012060 doi:10.1088/1757-899X/403/1/012060

Development of monitoring and hospital patient alert systems using smartwatch application

Yuantoro K[1], M Rosmiati[2] and Y Siradj [3]

[1,2,3]Fakultas Ilmu Terapan Universitas Telkom, Bandung, Indonesia

[2](corresponding author to provide phone: +6281223228338; e-mail: mia@tass.telkomuniversity.ac.id).

Abstract. Health is particularly important family. When a person is sick and hospitalized, the number of family will be limited to just one or two from the patient's family. The family often responds slowly to the critical situation due to the distance traveled from home to hospital far enough. It would be easier if they able to monitor the condition by using a smart watch. it sends the current condition through the API displayed in the mobile application. The Android mobile app will process any data retrieved from the watch then decide the real condition of the patient. This research could support the Internet of Things in the field of health, may improve the chances of life of patients by providing a quicker and accurate information.

1. Introduction

Hospitals provide inpatient services for patients who should receive intensive care with a doctor's monitoring. Most hospitals only allow some family members to assist patients during the treatment period. During the treatment time, if family members want to know the condition of the patient they should ask the doctor or nurse who can understand how to read the health monitor device. This method is impractical because not every time the doctor or nurse is willing to serve the family questions An intelligent device that connected to the internet, would notify the family about the patient's health condition in real time. A smart watch as a potentially wearable device may become a solution.

As one of the wearable devices, smartwatch holds great potential to improve the health of human life. Currently, smartwatch has been supported by various sensors that can display the user's body condition in real time. Sensors are like accelerometer, gyro meter, motion tracking, temperature, humidity and more. Integration sensors supported by Big Data technology, and AI (Artificial Intelligent) will further bring smartwatch usability to a higher level, especially in health. At least in the future smartwatch can contribute in the field of preventive health monitoring, disease therapy, aids for patients with specific diseases and digital medical record technology. With the ever-evolving sensor technology, smartwatch has a great hope of revolutionizing a healthy human lifestyle in the future.

The initial phase of research is making mobile applications that will retrieve health data from the API provided by smartwatch vendors. The purpose of this research is to develop a mobile application which can real-time present the patient's health condition as seen from the parameters of heart rate and blood oxygen level.

This research is expected to generate a new perspective in monitoring the development of patient recovery through mobile applications. The smart watch used in this study should have at least the features of heart rate reading, blood oxygen levels, GPS, Recovery advisors, and provide APIs to be developed by third party applications. The data retrieval technique is done by integrating vendor smartwatch's API into Android app. Further Android Applications will translate raw data taken from smartwatch in real time into a description of health conditions and divided into several alerts from normal to emergency levels. In addition to being a tool to monitor the patient's health, the application will also tell the patient's family if the patient's condition worsens through the notification mechanism. This paper is divided into chapters. The introductory chapter contains the background of the study. In the second chapter contains the method used in the research of the Requirement Prototyping model, while in the third part is the installation stage and system configuration to get the exp The fourth section contains the analysis of the research data, and in the last section is th the research results along with the suggestions for further research to get another better

2. Research Methods

The research method used is a method of Design Science Research Methodology (DSI method, there are six stages to be performed, i.e., problem identification, goal setti development, case studies, evaluation, and communications. On this research will onl up to the stage of the evaluation. It can be shown by the DSRM methods Figure 1.

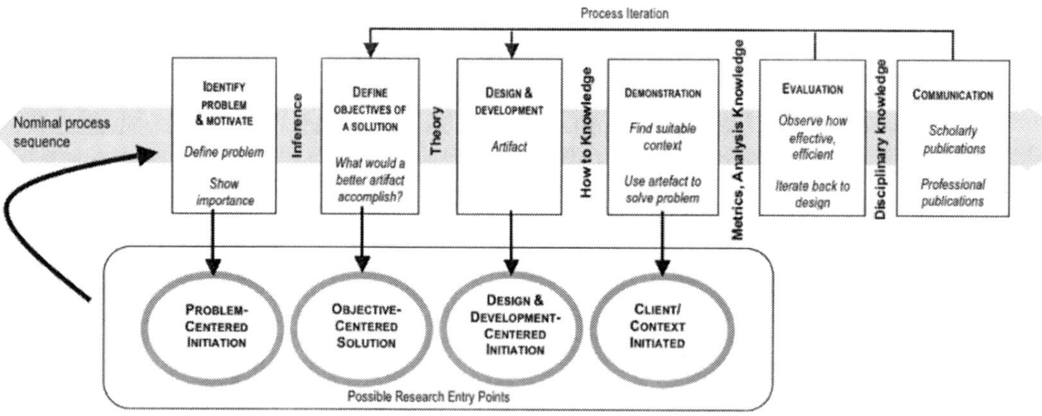

Figure 1. Design Science Research Methodology

In Figure 1 above, the identification of problems is made by monitoring the healt hospitals, while the development of smartwatch technology using heart rate sensors, oxy and heart health monitoring modeling is done through a literature study. Regarding solution, the smartwatch API is used to build an alert system application from the patie it changes to an abnormal condition. Moreover, with the integration of Android techı provided by smartwatch vendors, the heartbeat condition of smartwatch users (the patie via smartphone. The process of measuring the success rate of the application functionali built will be tested directly through the measurement of the heartbeat of patients who l pressure disorder in Banjaran Health Center.

2.1. *Smartwatch*

The saturation conditions of smartphone usage and sales are influenced, for one reason, t of wearable devices market - especially smartwatches - across the Smartwatch offers users new interaction experience in the digital world. One of the things is its ability to monitor and display data related to user health. Smartwatches o resemble watches have been embedded with powerful processors, memory subsyste sensor variations such as Accelerometer, gyroscope and optical sensors [2]. Sen smartwatch has opened up opportunities for smart devices to improve the quality of Accelerometer and sensor Gyroscopes, for example, are ideal for gait-based biometric [3] with the ability to send data and calculations via Bluetooth, or with the help of smaı internet [3]. making smartwatch more reliable in monitoring human health than smartp below shows the parts of the smartwatch.

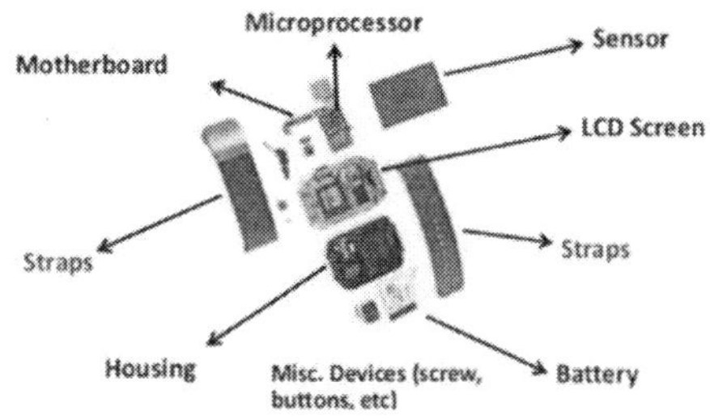

Figure 2. Parts of Smartwatch

In Figure 2 above GodsTale [4] created a smartwatch consisting of Arduino Microcontroller with ATmega328 chipboard and 2Kbytes RAM, Bluetooth module HC-06 equipped with status, low power display with 1-inch size supporting I2C, SPI compatible with board Arduino and Lithium Polymer Battery. This Smartwatch has a minimum specification that can run the Android v4.3 operating system and supports notification services.

2.2. Health data and heart rate

Health is one of the most important factors in life, and one of the benchmarks is the heart rate condition. Heart rate is the number of heartbeats per unit of time, usually minutes. The pulse rate is the propagation of the heart rate calculated per minute by repetition count (times/min), generally represented as bpm (beats per minute). The condition of a person's heart rate will affect the volume of blood and oxygen that flows throughout his body.This condition is critical to be observed at any time, especially someone who has a history of heart disease. Significant changes in the status of heart rate at any given time can cause death. The monitoring of heart rate at all times is crucial both for the sufferer and his family. Table 1 below shows the heart rate and associated conditions.

Table 1. Human Heart Rate Data

Category	Heart Rate (bpm)
Low	< 60 bpm
Normal	60 – 100 bpm
High	> 100 bpm

2.3. Android SDK

Android Studio SDK is an IDE (Integrated Development Environment) for Android application development. In addition to being a powerful code editor and developer tool, Android Studio offers more features to improve productivity when creating Android apps [5].

2.4. ANT+

ANT + is a wireless technology that allows other electronic devices to communicate with each other on condition that the device supports ANT + communication or embedded ANT + technology. ANT + technology is widely used by smartwatch manufacturers to detect heart rate to retrieve data for health monitoring applications easily. ANT + also provides APIs for developers to create similar applications.

3. Data and Analysis

Testing aims to see the functionality of the system built. The system uses two devices that are smartwatch and smartphones that have mobile applications installed and both support ANT + technology. The ANT + protocol can communicate without further adjustment. Mobile apps can take heartbeat data from the smartwatch. Through the process of parsing API smartwatch, it can display patient's heartbeat data. The built-in flowchart system is shown in Figure 5 below.

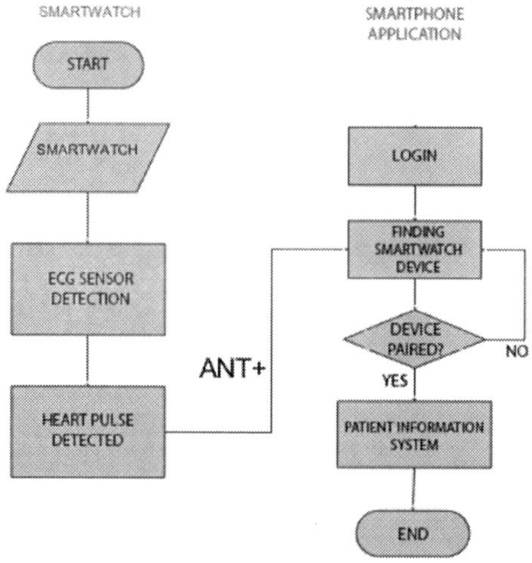

Figure 3. Flowchart System

Figure 3 illustrates the process of integrating smartwatch data into Android apps, which the smartwatch owner data can be displayed on Android smartphone. First, the login view to know the patient's name and number. The second is the smartwatch device search field. Thirdly, mobile apps display health information. Health information features such as patient identity, time, device status, real-time value and average heart rate. By leveraging the existing ANT + APIs in the smartwatch, the mobile app allows taking heart rate data.

The testing process is done through 3 stages:

3.1. Test of heartbeat data retrieval from a smartwatch.

Intake of heartbeat data from smartwatch can be done by using ANT + communication where previously have added library from ANT + API so that smartwatch heartbeat data can be processed using a smartphone. The results of this test are shown in Figure 5.

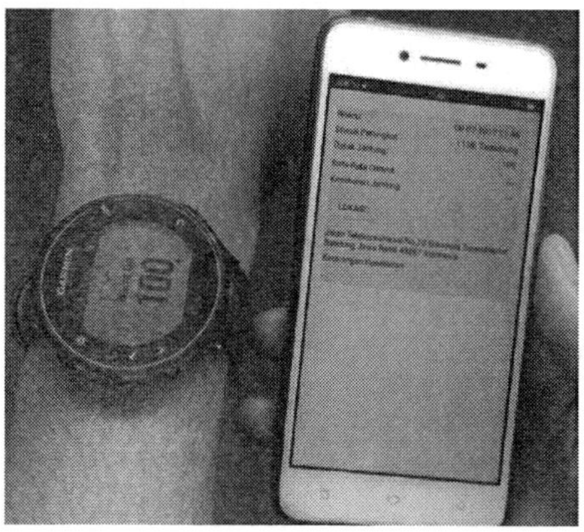

Figure 4. Test of heartbeat data retrieval from smartwatch

IOP Conf. Series: Materials Science and Engineering **403** (2018) 012060 doi:10.1088/1757-899X/403/1/012060

Figure 5 shows that smartphone managed to take heartbeat data from smartwatch through ANT + technology.

3.2. Heartbeat data processing.

Heart rate data processing aims to determine the average heart rate of patients. Average is obtained by using the formula (1).

$$\bar{x} = \frac{x_1 + x_2 + x_3 + \cdots + x_n}{n} \quad \ldots\ldots (1)$$

The mean value of the patient's heartbeat test results repeatedly performed in a minute will be searched according to equation one above. Furthermore, this average value will be displayed and categorized into the patient's heartbeat health status consisting of Low, Normal, or High. This health categorization of the heartbeat is done through heartbeat testing after the patient performs walking, jogging, and running activities. Also, the category testing process carried out on patients who have a history of high blood disease. Table 2 below shows the patient's heartbeat category.

Table 2. Categorization of patient's heartbeat

No.	Time	Data	Heartbeat	Condition	Status
1	16:43	90,91,92,96,95,96, 95,94,93,94,95,96, 97,95,97,96,95,92, 91,90 = 1880	Amount of Data = 20 X=1880/20 = **94**	Walking	Normal.
2	20:29	128,129,130,129, 130,131,132,133, 132,129,124,114, 107,106,105,106, 108,109,110,111, 112,113,114 2742	Amount of Data = 23 X=2742/23 = **119,21**	Jogging	High
3	20:39	142,144,143,141, 139,138,141,143, 149,152,150,149, 151,150,152,153, 150,152,151,150, 153,152,149,146, 147,148,149,150, 151,150,149,148, 147,148,145 5178	Amount of Data = 35 X=5178/35 = **147,94**	Running	High

Table 2 above shows the variation in the patient's heart rate repeatedly measured to determine the patient's heartbeat category, while Table 3 below lists the results of system and application testing with different patient activity conditions. This test is intended to find out whether mobile applications can work if user activity varies by displaying status changes from Normal to High and vice versa according to the average heart rate generated based on data from the smartwatch.

Table 3. Testing Mobile Application

No.	TIme	Average Heartbeat	Condition
1.	16:43	94	Walking
2.	20:29	119,21	Jogging
3.	20:39	147,94	Running

After doing the activity test, then the next test will be done to the patient at Cikalong Public Health Center, Banjaran. This test is intended to determine the difference in heart rate of patients who have different diseases, ages, and sexes. Table 4 shows direct application testing to the patients of Community Health Centre Cikalong Banjaran, Kab. Bandung.

Table 4. Tests Against Patients

No.	Identity of Patient	Test Data		
		Time	Heartbeat	Status
1.	Name : Ase Somdani.	10:30	96	Normal.
	Age: 75	10:31	93	Normal.
	Male	10:32	93	Normal.
	Disease: Maag and Hypertension	10:33	92	Normal.
		10:34	90	Normal.
2.	Name : Edi Bartas.	10:37	79	Normal.
	Age: 70	10:38	77	Normal.
	Male	10:39	77	Normal.
	Disease: Cholesterol and Hypertension	10:40	75	Normal.
		10:41	75	Normal.
3.	Name: Wiwik Suciati.	10:50	72	Normal.
	Age: 50	10:51	61	Normal.
	Female	10:52	63	Normal.
	Disease: Heart Failure and Hypertension	10:53	66	Normal.
		10:54	65	Normal.

Based on table 4 above, the process of measuring heart rate data and determining the health category of patients based on smartwatch heart rate data is successfully displayed via smartphone. The data shows the success rate of the system built to reach 100%.

In addition to displaying the user's heartbeat smartwatch data, this application can also display the Geolocation of smartwatch users because the built application integrates the GPS API for user location data retrieval. Figure 8 below shows the smartwatch user position shown through the built application.

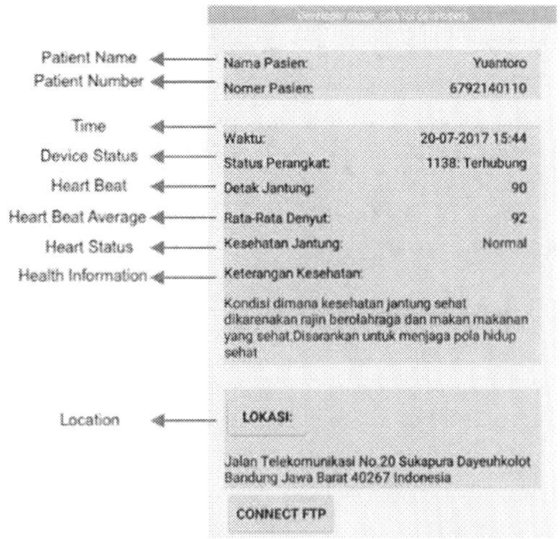

Figure 5. Smartwatch application test for location

Figure 6 indicates that the built application successfully displays the location of the smartwatch user, i.e., at Jalan Telekomunikasi No.20 Sukapura Dayeuh Kolot Bandung, West Java 40267, Indonesia. Smartwatch user location retrieval process can be done if the smartphone is connected to the internet.

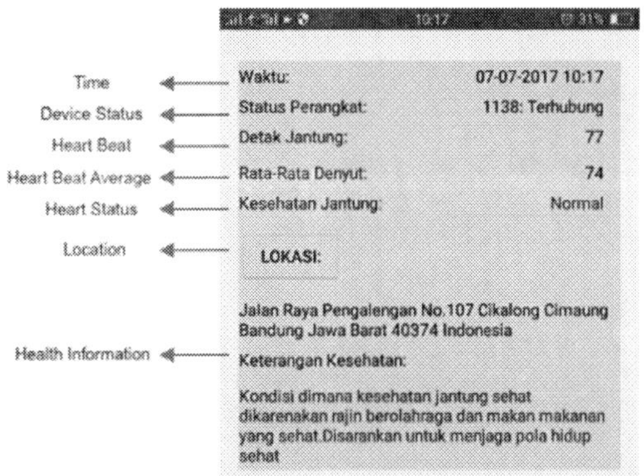

Figure 6. Information of Testing Location Centre

Figure 7 shows the location information when the built system is tested at the Cikalong Public Health Center, Banjaran.

4. Conclusion

Some of the tests performed indicate that the Smartwatch integration process with Android has been successfully built. The application can display the patient's heart rate in real timeBy integrating smartwatch with GPS, the user position can be known directly.

For further research, it is necessary to build a SMS Gateway to inform the patient's health condition to his family. It is crucial that the family can get notified as soon as possible if something terrible happens to the patient.

References

[1] "Healthcare and Smartwatches Were Meant To Be Together in Taiwan, BuzzOrange." [Online]. Available: http://buzzorange.com/global/2015/06/11/healthcare-and-smartwatches-were-meant-tobe-together-in-taiwan/. [Accessed: 21-Mar-2016].

[2] D. Phan, L. Y. Siong, P. N. Pathirana, and A. Seneviratne, "Smartwatch: Performance evaluation for long-term heart rate monitoring," 2015 Int. Symp. Bioelectron. Bioinforma., pp. 144–147, 2015.

[3] A. H. Johnston and G. M. Weiss, "Smartwatch-Based Biometric Gait Recognition," 2014.
[4] "Make your own smartwatch - 2." [Online]. Available: http://www.instructables.com/id/Make-your-own-smart-watch/step2/Preparing forRetroWatch/. [Accessed: 22-Mar-2016].
[5] J. Altmann, and J. Mathew, "Multiple band-pass autoregressive demodulation for rolling element bearing fault diagnosis," Mech. Syst. Signal Process., Vol. 15, no. 5, pp. 963_77, Sep. 2001.
[6] Y. Cai, Y. Zhao, X. Ding, and J. Fennelly, "Magnetometer basics for mobile phone applications," *Electron. Prod. (Garden City, New York)*, vol. 54, no. 2, 2012.
[7] J. M. Azzarelli, K. A. Mirica, J. B. Ravnsbæk, and T. M. Swager, "Wireless gas detection with a *smartphone* via rf communication.," *Proc. Natl. Acad. Sci. U. S. A.*, vol. 111, no. 51, pp. 18162–6, 2014.
[8] Siradj, yahdi, "Smartwatch Potential For healthcare", Telekontran, vol. 4 no. 1, April 2016
[9] Peffers, Ken and Tuunanen, Tuure and Rothenberger, Marcus and Chatterjee, Samir, "A Design Science Research Methodology for Information Systems Research", Journal of Management Information Systems, vol.24, no. 3, M. E. Sharpe, Inc., Dec. 2007.

Comparative analysis of three MCMC methods for estimating GARCH models

D B Nugroho

Department of Mathematics, Satya Wacana Christian University,
Jln. Diponegoro 52–60 Salatiga 50711 Central Java, Indonesia

didit.budinugroho@staff.uksw.edu

Abstract. GARCH model have been considered as an important and widely employed tool to analyse and forecast variance of the financial market. This study develops three MCMC methods, namely adaptive random walk Metropolis, Hamiltonian Monte Carlo, and Independence Chain Metropolis–Hastings algorithms. It is used to estimate GARCH (1,1) under Normal and Student-t distributions for conditional return distribution. Results on real financial market data indicate that the best method is the approach based on the Independence Chain Metropolis–Hastings algorithm.

1. Introduction

Financial markets have played an important role in global economic growth . On this subject one of the important aspects of the markets is to model and estimate the fluctuative financial market It is caused by the volatility estimation,is often relied by policy makers as a barometer for the sensitivity of financial markets [1]. Meanwhile, in stock exchange markets, the volatility of stock price is an important indicator of the stock price fluctuations [2]. Volatility can be determined by using statistical measures of standard deviation (square root of variance) between returns from that same asset. Commonly, the high volatility (the wide range of fluctuations) reflects high trading risk. In a rising market, however, high volatility can favor the return potential of an investment [3].

The popular class of volatility model for the financial time series that exhibit heteroscedasticity (refers to the volatility fluctuation) is represented by the Autoregressive Conditional Heteroskedasticity (ARCH) model of [4] and the Generalized ARCH (GARCH) of [5]. On the empirical level, for example, [6,7,8,9,10,11,12,13] provided several applications of GARCH-type models for estimating and forecasting volatility in the stock and foreign exchange markets.

In order to infer the model parameters from GARCH models, the numerical procedures based on a maximum likelihood (ML) are often employed. For example, [14,15,16] developed the ML estimation for the GARCH-type models with normal and Student-*t* distributions and [17] proposed the non-Gaussian Quasi-Maximum Likelihood estimation for the GARCH model with Student-*t* distribution.

Although the ML estimation is conceptually simple, in practice it may be possible to encounter difficulties in applying the ML-based methods for the GARCH-type models, see the detail in [18]. To overcome the computational difficulties of ML-based methods, [18] proposed Bayesian method via Markov Chain Monte Carlo (MCMC) estimation method. A variety of algorithms which might implement MCMC methods effectively for the GARCH-type models have been proposed and applied. For example, the Gibbs sampler is used in [19], the Metropolis–Hastings-based algorithms are used in

[15,18,20,21,22,23,24,25,26], and the Hamiltonian Monte Carlo method is used in [27,28]. In particular, [15] reported that the GARCH(1,1) parameters obtained by the ML and Metropolis–Hastings methods are close to each other. Furthermore, [20,29] showed that the Bayesian approach via the MCMC methods may lead to more reliable and suitable than the ML-based methods for estimating the GARCH model based on the use of real financial data. Hence this studi focuses on applying MCMC algorithms to GARCH models.

Since the implementation of the MCMC algorithm is not unique and may involve high-dimensional integration, it is important to consider how rapidly the convergence of the Markov chain produced by MCMC algorithms to its stationary distribution. It is therefore necessary to investigate the convergence performance of MCMC methods in estimating the GARCH parameters. The convergence property of these methods is revealed by measure of the sampling efficiency and Markov chain mixing, which are defined by integrated autocorrelation time (IACT) and numerical standard error (NSE), respectively. The literature on comparison of the MCMC methods in estimating the GARCH models is much limited. In the GARCH context, only [23] numerically compared some of MCMC methods, including Griddy Gibbs, Metropolis–Hastings, adaptive rejection Metropolis, and acceptance-rejection Metropolis–Hastings.

In the context of seeking a method that draws samples efficiently from any univariate conditional density function, this study has two main contributions. The first is a novel performance comparison of three MCMC methods, including adaptive random walk Metropolis (ARWM), independence chain Metropolis–Hastings (ICMH), and Hamiltonian Monte Carlo (HMC), for the estimation of the GARCH parameters. Our survey indicate that none of research apply those methods separately to estimate the Student-*t* GARCH model in particular and even compare their performance. The ARWM method is proposed by [30] for improving the efficiency of the random walk Metropolis, which is the simplest and the most widely used MCMC method [31]. The ICMH method is proposed by [32] and among the most popular alternative Metropolis–Hastings (MH) methods. This method works best when there is another probability measure which is easy to sample from. The HMC method is proposed by [33] and has better convergence properties than MH and Gibbs sampler [34]. The second one is to contribute to extend the empirical study on GARCH estimation in emerging financial markets.

Although the extension of the standard GARCH is numerous, this study chooses the GARCH(1,1) models with normal and Student-*t* distributions, without limiting the methodological potential of estimation procedures, because of their application to many financial market data. An empirical application that compares the MCMC methods is illustrated by fitting the models to the daily volatility of currency exchange rate returns on the Indonesian Currency Exchange Market. This is to provide a performance comparison of the proposed algorithms in terms of autocorrelation time and standard error. This study is structured into four sections as follows. Immediately preceding Introduction in Section 1 is Section 2, which outlines the methodology and real data. Section 3 presents the empirical results, and Section 4 provides the concluding remarks.

2. Methodology and Real Data

2.1. Methodology

The Bayesian framework was chosen because this allows the application of MCMC methods which are known as very powerful techniques for calculating the numerical integration of the joint density. The key steps of our proposed framework are as follows. First, within a particular GARCH model, a Markov chain having an identical stationary distribution to the posterior density of parameter is constructed by using an MCMC method. After discarding early samples and sufficient iterations, the Markov chain provides samples from the posterior distribution of parameter. Second, for a given set of competing MCMC methods, we employ each method separately and base their comparison in terms of autocorrelation time and numerical standard error.

Let R_t be a serially uncorrelated return whose mean equation is given by $R_t = \sigma_t\varepsilon_t$, where $\{\varepsilon_t\}$ is a sequence of iid random innovations (errors) with mean 0 and variance 1. The GARCH(1,1) approach estimates the current variance, σ_t^2, as a linear function of past squared innovation and past variance, i.e.,

$$\sigma_t^2 = \omega + \alpha R_{t-1}^2 + \beta\sigma_{t-1}^2,$$

where $\omega, \alpha, \beta \geq 0$ to ensure the positivity of the conditional variance and $\alpha + \beta < 1$ to ensure stationarity; see [5] for details. The above model is then called the normal GARCH (GARCHn hereafter) model. By considering $\mathbf{R} = (R_1, R_2, ..., R_T)$ and let $p(\omega,\alpha,\beta)$ and $f(\mathbf{R}|\omega,\alpha,\beta)$ be a joint prior density and a likelihood function, respectively, the logarithm of the joint posterior density of parameters can be obtained by combining the log joint prior density with the log likelihood function via Bayes' Theorem, i.e.,

$$L = \log p(\omega, \alpha, \beta|\mathbf{R}) = \log p(\omega, \alpha, \beta) - \frac{1}{2}T\log(2\pi) - \frac{1}{2}\sum_{t=1}^{T}\log(\sigma_t^2) - \frac{1}{2}\sum_{t=1}^{T}\frac{R_t^2}{\sigma_t^2}. \quad (1)$$

If ε_t has the Student-t distribution with degrees of freedom ν, the model is called the Student-t GARCH (GARCHt hereafter) model. In this case, the logarithm of the joint posterior density of parameters $(\omega,\alpha,\beta,\nu)$ is similar to that given by [35],

$$L = \log p(\omega, \alpha, \beta, \nu|\mathbf{R}) = \log p(\omega, \alpha, \beta, \nu) + T\log\Gamma\left(\frac{\nu+1}{2}\right) - T\log\Gamma\left(\frac{\nu}{2}\right) - T\log\Gamma\left(\frac{1}{2}\right)$$
$$-\frac{1}{2}\sum_{t=1}^{T}\log(\sigma_t^2(\nu-2)) - \frac{\nu+1}{2}\sum_{t=1}^{T}\log\left(1 + \frac{R_t^2}{\sigma_t^2(\nu-2)}\right). \quad (2)$$

This study develops the procedure to sample the parameter vector $\theta = (\omega,\alpha,\beta,\nu)$ in the GARCHt model from the log joint posterior density (2) at each sampling stage in the MCMC algorithm as follows.

1. Draw ω from $p(\omega|\alpha,\beta,\nu,\mathbf{R})$.
2. Draw α from $p(\alpha|\omega,\beta,\nu,\mathbf{R})$.
3. Draw β from $p(\beta|\omega,\alpha,\nu,\mathbf{R})$.
4. Draw ν from $p(\nu|\omega,\alpha,\beta,\mathbf{R})$.

In the GARCHn case, the sampling procedure is similar to the above but ignoring parameter ν and on the basis of the joint posterior (1). Since the above conditional posteriors are not of standard form, we draw each parameter using the same method in a MCMC simulation.

Two popular MH chains are the random walk chain and the independence chain. The random walk chain centers the proposal density over the current value of the chain. Improving the performance of Random Walk Metropolis algorithm has been proposed by [30] through ARWM algorithm which automatically handles the the proposal value. In contrast to the random walk chain, "independence" of ICMH refers to the fact that the proposal value is not necessary depend on the current value. For this method to work well, this study adapts the proposal value using the method which is based on the behavior of the density about its mode (see [36]). In the case of the sampling of degrees of freedom ν, the first and second derivatives of $L(\nu)$ are respectively given as in [35].

An alternative MH chain is the HMC method which was proposed in the statistical physics literature. This method avoids the random walk behavior by combining Gibbs updates with Metropolis updates. The proposal value is computed based on a trajectory which is guided by first-order derivative of the log of the posterior density. In particular, this study applies the similar algorithm which was employed by [37,38,39] for estimating the degrees of freedom ν.

For the unknown parameters in the GARCH models, following standard practice, we complete the models by the truncated normal priors with mean 0 and variance 1000 on (ω,α,β) and the exponential prior with parameter $\lambda = 0.01$ on ν. For initial parameter values, we use $\omega_0 = 0.01$, $\alpha_0 = 0.2$, $\beta_0 = 0.7$, and $\nu_0 = 0.01$. For the HMC updates, we apply 100 leapfrog iterations with a step size of 0.01.

Desirable properties for method in the MCMC algorithm are high efficiency and well chain mixing, which yield fast convergence. In this study, the efficiency is measured by the integrated autocorrelation time (IACT) which is estimated using the Sokal's adaptive truncated periodogram estimator [40]. The quantity can be roughly interpreted as the number of iterations required to produce one independent draws. A lower value indicates a higher sampling efficiency in a set of draws while large values indicate a poor sampling efficiency (little movement between successive draws). Meanwhile, the mixing performance is checked by assesing the Monte Carlo error that can be accomplished by estimating the variance which is also known as the numerical standard error (NSE); see [41] for details.

2.2 Data for analysis

For illustrative and comparative purposes, this study uses the series of daily closing prices $\{P_t\}$ of the EUR, JPY, and USD selling exchange rates to IDR over a period from January 2010 to December 2016. The series of return are daily percentage returns given by

$$R_t = 100 \times (\log P_t - \log P_{t-1}).$$

Notice that our daily returns are neither serially correlated and normally distributed on the basis of Ljung–Box correlation and Jarque–Bera normality tests.

3. Empirical Results

For each sampling method, we ran two MCMC simulations, i.e., for 6000 iterations (first 1000 iterations discarded as burn-in) and 15000 iterations (first 5000 iterations discarded as burn-in). This runs are long enough to obtain stationary distribution, and so follow realistic statistical practice. The remaining iterations are then used for calculating the posterior mean, IACT, and NSE. The MCMC methods were implemented in Matlab R2013a by writing own code and the simulations were ran on a personal machine with Intel Core i-7-4790 CPU 3.60GHz x 16.0 GB of memory on a 64-bit Windows 10 Pro operating system.

Table 1. Parameter estimates and convergence speeds in the GARCHn and *GARCHt* models adopting IDR/EUR return series. *Note*: 'Time' denotes CPU time.

Stat.	Method	*N*= 6000, *Burn-in* = 1000					*N*= 15000, *Burn-in* = 5000				
		ω	α	β	ν	*Time (sec.)*	ω	α	β	ν	*Time (sec.)*
	ARWM	0.0243	0.0621	0.8833		2.6	0.0184	0.0557	0.9029		6.9
		0.0373	*0.0781*	*0.8380*	*8.8*	*4.9*	*0.0583*	*0.0960*	*0.7735*	*8.7*	*13.6*
Mean	HMC	0.0221	0.0585	0.8918		256	0.0233	0.0612	0.8865		410
		0.0452	*0.0840*	*0.8142*	*8.7*	*507*	*0.0428*	*0.0800*	*0.8238*	*8.7*	*1157*
	ICMH	0.0130	0.0474	0.9231		66	0.0146	0.0504	0.9166		178
		0.0280	*0.0687*	*0.8683*	*8.4*	*152*	*0.0346*	*0.0750*	*0.8472*	*8.3*	*253*
	ARWM	219.9	130.3	216.7			245.3	197.2	259.1		
		195.3	*140.1*	*191.7*	*6.4*		*221.4*	*173.1*	*223.2*	*6.3*	
IACT	HMC	116.2	96.5	129.7			131.2	83.7	133.0		
		150.9	*113.4*	*161.0*	*15.8*		*198.2*	*170.7*	*202.3*	*17.2*	
	ICMH	72.3	55.1	75.1			99.3	71.0	111.7		
		94.3	*70.2*	*95.2*	*1.8*		*94.1*	*64.0*	*90.0*	*1.9*	
	ARWM	1.55	1.82	5.06			1.00	1.37	3.47		
		3.02	*2.90*	*9.30*	*67.2*		*3.96*	*3.15*	*11.9*	*36.9*	
NSE ($\times 10^{-3}$)	HMC	1.39	1.52	4.58			1.48	1.35	4.63		
		3.00	*2.50*	*9.00*	*83.4*		*4.06*	*3.18*	*12.10*	*80.7*	
	ICMH	0.49	0.75	1.80			0.41	0.66	1.56		
		1.37	*1.58*	*4.59*	*22.3*		*2.13*	*1.86*	*6.58*	*20.8*	

Tables 1–3 show estimation results for the GARCHn(1, 1) and GARCH*t*(1,1) models adopting IDR/EUR, IDR/JPY and IDR/USD return series, respectively. Although the methods produce similar parameter estimates, the ICMH method is most efficient and gives the best mixing in terms of IACT and NSE, followed by the HMC method and the ARWM method in the case of sampling (ω,α,β). On estimating the parameter degrees of freedom ν, all methods experimentally show to have high efficiency, where the ARWM and HMC methods are very competitive. Regarding computational time, the HMC

method runs much slower than the others caused by the use of leapfrog method. Among three competing methods, the ARWM method is least time-demanding.

Table 2. Parameter estimates and convergence speeds in the GARCHn and *GARCHt* models adopting IDR/JPY return series.

		N= 6000, *Burn-in* = 1000					*N*= 15000, *Burn-in* = 5000				
Stat.	Method	ω	α	β	v	*Time (sec.)*	ω	α	β	v	*Time (sec.)*
Mean	ARWM	0.0377	0.0984	0.8428		2.6	0.0333	0.0925	0.8554		6.8
		0.0399	*0.1117*	*0.8285*	*5.1*	*4.8*	*0.0426*	*0.1152*	*0.8212*	*5.1*	*12.9*
	HMC	0.0300	0.0880	0.8653		246.6	0.0335	0.0928	0.8548		401.8
		0.0379	*0.1065*	*0.8374*	*5.1*	*506.5*	*0.0416*	*0.1115*	*0.8267*	*5.1*	*1149.9*
	ICMH	0.0255	0.0818	0.8783		65.7	0.0281	0.0855	0.8704		157.6
		0.0280	*0.0920*	*0.8668*	*5.0*	*152.0*	*0.0299*	*0.0958*	*0.8603*	*5.0*	*425.2*
IACT	ARWM	223.8	166.2	206.9			236.0	264.3	295.7		
		284.6	*258.5*	*287.6*	*9.4*		*154.2*	*182.1*	*184.2*	*10.1*	
	HMC	67.9	56.9	75.7			94.2	81.0	99.5		
		146.9	*107.0*	*143.7*	*5.0*		*103.3*	*95.3*	*110.2*	*5.9*	
	ICMH	48.2	47.7	53.0			58.5	51.6	60.8		
		62.5	*56.0*	*67.1*	*2.9*		*83.2*	*72.5*	*93.2*	*3.3*	
NSE ($x10^{-3}$)	ARWM	2.42	3.60	7.10			1.72	2.44	5.03		
		2.95	*4.57*	*8.85*	*27.5*		*2.19*	*3.55*	*6.56*	*17.4*	
	HMC	1.03	1.64	3.16			1.50	2.32	4.67		
		2.37	*3.58*	*7.06*	*15.2*		*1.76*	*2.70*	*5.21*	*16.4*	
	ICMH	0.57	1.13	1.98			0.91	1.43	2.78		
		0.83	*1.54*	*2.77*	*10.9*		*0.80*	*1.42*	*2.58*	*7.82*	

Table 3. Parameter estimates and convergence speeds in the GARCHn and *GARCHt* models adopting IDR/USD return series.

		N= 6000, *Burn-in* = 1000					*N*= 15000, *Burn-in* = 5000				
Stat.	Method	ω	α	β	v	*Time (sec.)*	ω	α	β	v	*Time (sec.)*
Mean	ARWM	0.0092	0.2065	0.7719		2.4	0.0098	0.2138	0.7631		6.2
		0.0040	*0.2005*	*0.7939*	*4.2*	*4.2*	*0.0045*	*0.2137*	*0.7802*	*4.2*	*12.0*
	HMC	0.0098	0.2145	0.7628		243.8	0.0097	0.2113	0.7657		416.1
		0.0039	*0.1970*	*0.7972*	*4.2*	*504.7*	*0.0040*	*0.1961*	*0.7979*	*4.2*	*1111*
	ICMH	0.0089	0.2063	0.7743		65.7	0.0093	0.2135	0.7670		155.6
		0.0038	*0.2015*	*0.7946*	*4.1*	*148*	*0.0036*	*0.1936*	*0.8026*	*4.1*	*246*
IACT	ARWM	109.4	144.0	152.9			105.2	142.3	160.6		
		211.2	*267.8*	*266.5*	*10.3*		*213.7*	*277.6*	*279.3*	*8.6*	
	HMC	46.0	43.8	52.4			33.0	36.0	38.5		
		80.0	*125.9*	*124.4*	*5.2*		*90.7*	*176.6*	*176.2*	*4.2*	
	ICMH	24.2	37.2	37.7			23.5	31.0	34.7		
		68.4	*95.1*	*95.8*	*2.7*		*51.8*	*96.9*	*97.3*	*2.8*	
NSE ($x10^{-3}$)	ARWM	0.23	3.92	4.19			0.17	2.84	3.00		
		0.15	*4.65*	*4.77*	*13.6*		*0.16*	*4.30*	*4.39*	*8.9*	
	HMC	0.18	2.56	2.86			0.15	2.33	2.55		
		0.12	*4.03*	*4.06*	*12.0*		*0.09*	*2.61*	*2.65*	*7.5*	
	ICMH	0.08	1.55	1.55			0.09	1.360	1.451		
		0.08	*2.67*	*2.71*	*4.5*		*0.06*	*2.08*	*2.12*	*5.1*	

4. Conclusions

This study adopted three popular MCMC methods for estimating the GARCH models with normal and Student-t distributions for the return errors. For each method, we describe the difference in proposing a candidate value. The comparison of the proposed methods performance was grounded on the real data All proposed methods work well and provide similar accuracy. The results declare that the ICMH is the best method in term of the efficiency, mixing, and computational time of the MCMC algorithm.

References

[1] Poon S-H and Granger C W J 2003 *Journal of Economic Literature* **41** 478–539
[2] Raja M and Selvam M 2011 *International Journal of Applied Economic and Finance* **5** 18–29
[3] Keppler M 1990 *Die Bank* **11** 610–14
[4] Engle R F 1982 *Journal of Econometrica* **50** 987–1007
[5] Bollerslev T 1986 *Journal of Econometrics* **31** 307–27
[6] Abdalla S Z S 2012 *International Journal of Economics and Finance* **4** 216–29
[7] Emenike K O and Aleke S F 2012 *European Journal of Business and Management* **4** 52–60
[8] Kamal Y, Hammad-Ul-Haq, Ghani U, and Khan M M 2012 *African Journal of Business Management* **6** 2830–38
[9] Wei S-Y, Yang J J W, Chen J-T, and Hong W-C 2013 The Volatility for Pre and Post Global Financial Crisis: An application of Computational Finance *Modeling Applications and Theoretical Innovations in Interdisciplinary Evolutionary Computation* ed W-C S Hong (USA: Information Science Reference) pp 134–48
[10] Nortey E N N, Ngoh D D, Doku-Amponsah K, and Ofori-Boateng K 2015 *SpringerPlus* **4** 1–10
[11] AL-Najjar D 2016 *Asian Journal of Finance & Accounting* **8** 152–67
[12] Kutu A A and Ngalawa H 2016 Investment Management and Financial Innovations **13** 203–11
[13] Epaphra 2017 *Journal of Mathematical Finance* **7** 121–43
[14] Bucevska V 2013 *Business Systems Research* **4** 49–64
[15] Malik V, Kumar R, and Rather A H 2016 *International Journal of Finance* **7** 79–118
[16] Augustyniak M, Boudreault M, and Morales M 2017 *Methodology and Computing in Applied Probability* **19** 1–24
[17] Fan J, Qi L, and Xiu D 2014 *Journal of Business & Economic Statistics* **32** 178–91
[18] Ardia D 2008 *Financial Risk Management with Bayesian Estimation of GARCH Models: Theory and Applications* (Berlin: Springer) pp 1–38
[19] Bauwens L and Lubrano M 1998 *Econometrics Journal* **1** c23–46
[20] Nakatsuma T 2000 *Journal of Econometrics* **95** 57–69
[21] Vrontos I D, Dellaportas P, and Politis D N 2000 *Journal of Business and Economic Statistics* **18** 187–98
[22] Mitsui H and Watanabe T 2003 *Journal of the Japan Statistical Society* **33** 307–24
[23] Asai M 2006 *Journal of the Japan Statistical Society* **36** 199–212
[24] Takaishi T 2009 *New Advan. in Intel. Decision Techno. (SCI 199)* ed K. Nakamatsu et al. (Berlin: Springer) p 635
[25] Takaishi 2013 *Procedia Computer Science* **22** 1056–64
[26] Salim F C, Nugroho D B, and Susanto B 2016 *Jurnal MIPA Unnes* **39** 78–84
[27] Takaishi T 2006 *Proc. of the 9th Joint Conference on Information Sciences* (Amsterdam: Atlantis Press) p 214
[28] Onyeka-Ubaka J N, Abass O, and Okafor R O 2014 *International Journal of Probability and Statistics* **3** 1–7
[29] Haitao P 2010 *Eighth Annual Conference Proceedings of Uncertain Systems in China* (Beijing: Electronic Industry Press) p 12
[30] Atchade Y F and Rosenthal J S 2005 *Bernoulli* **11** 815–28
[31] Sharma S 2017 *Annual Review of Astronomy and Astrophysics* **55** 1–49
[32] Tierney L 1994 *Annals of Statistics* **22** 1701–62

[33] Duane S, Kennedy A D, Pendleton B J, and Roweth D 1987 *Phys. Lett.* B**195** 216–22
[34] Neal R M 2003 *Annals of statistics* **31** 705–67
[35] Nugroho D B and Susanto B 2017 *AIP Conference Proceedings of the 4th ICRIEMS* **1868** 040005-1–8
[36] Albert J 2009 *Bayesian Computation with R* (Springer: New York) p 94.
[37] Nugroho D B and Morimoto T 2014 *Journal of the Japan Statistical Society* **44** 83–118
[38] Nugroho D B and Morimoto T 2015 *Computational Statistics* **30** 491–516
[39] Nugroho D B and Morimoto T 2016 *Journal of Applied Statistics* **43** 1906–27
[40] Sokal A D 1997 Monte Carlo Methods in Statistical Mechanics: Foundations and New Algorithms *Functional Integration: Basics and Applications* ed C DeWitt-Morette, P Cartier, A Folacci (New York: Springer) pp 131–92.
[41] Geweke J 2005 *Contemporary Bayesian Econometrics and Statistics* (New Jersey: John Wiley & Sons) p 106

IOP Conf. Series: Materials Science and Engineering **403** (2018) 012062 doi:10.1088/1757-899X/403/1/012062

The development of road databases and analysis of traffic accident-prone section (blacksite) based on Geographic Information System (GIS)

M E Bolla and B S Djahi

Nusa Cendana University, NTT, Indonesia

E-mail: margiebolla@staf.undana.ac.id

Abstract. Historical data on traffic accidents can provide information on blacksite areas. This can aid in the approach for reducing the number of traffic accidents. Therefore, this research aims to develop database of traffic accidents by the implementation of Quantum Geographic Information System (QGIS) and its interface uses ArcView GIS. We found that there are 69 roads which are identified as the location of traffic accidents. Of these, there are 15 roads as blacksite. The most vulnerable road is the Timor Raya road which has an accident rate (APW) of 813.2 and a UCL value of 71.55. This database provides comprehensive information on geometric conditions of the road, accident data and blacksite, and also features road photographs and cross-sectional road drawings.

1. Introduction

East Nusa Tenggara with the city of Kupang as the provincial capital is one of the provinces in Indonesia which has the highest traffic accident rate in Indonesia. Based on data from the Police Traffic Directorate, the level of traffic accidents in the province is the third largest in Indonesia. The number of traffic accidents in Kupang City has increased from 2011-2015 where there were 940 traffic accidents with 1558 victims[1]. The cost for the traffic accidents is around 14,365,572,490 Indonesian Rupiahs[2].

Therefore, Determination of traffic accident prone locations requires systematic historical data, but until now it has not been available in Kupang City. The traffic accident data at the Kupang Kota Resort Police is still in written form that are easily damaged, so it is necessary to create a database to facilitate the management and search of accident data. Database based on Geographic Information System (GIS) has been widely used in facilitating the performance of policy holders, where the GIS-based database is capable of performing data processing as well as advanced spatial analysis. Geographical Information Systems (GIS) stand out for their ability to perform complex spatial analysis. However, sometimes the GIS, has been used only as a geographical database to store and represent data about accidents and road characteristics[3].

There are 2 different GIS software became a star in compiling Gis-based database that are ArcGIS and Quantum GIS (QGIS). Many studies have been developed regarding the use of both software in mapping of traffic accident prone sites such as Reference [4] creates a website page using Quantum GIS 1.8.0 Lisboa as software for make digital maps, and a study conducted by [5] that used arcGis in analyzing sight distance for road safety. QGIS software is also used by [6] to geocode five years (2009-2013) road accidents locations over the digitized map of Varanasi. The both software have their own advantages and disadvantages, but the main advantage of QGIS is that there are no license levels in QGIS. Open source QGIS software does not limit which tools can be used.

Based on the conditions as described above then it is critically important to have comprehensive database in order to develop and find better strategies for better traffic safety. Therefore, this research aims to develop a GIS-based database using Quantum GIS program, analyze road network that are accidents prone, and arrange the interface of accident data with ArcView program. This will provide information in particular for public policy makers and help them to finding better strategies to reduce the number of traffic accidents.

2. Literature Review

It is stated in Reference [7] that traffic accident is an unexpected and unintentional highway event which involve vehicles and other road users. This results in human casualties and/or property losses. Blacksite identification is done by investigating historical data.

2.1. Weighting the accident data

Weighting is a value used to calculate the accident rate. This calculation is based on the death, severe and minor injuries. The weighting of traffic accidents (APW) is performed using accident-equivalent as given by [8], following this ratio shown in equation (1).

$$MD : LB : LR : K = 12 : 3 : 3 : 1 \tag{1}$$

where:

MD = death
LB = severe injuries
LR = minor injuries
K = material loss due to accidents

2.2. Rank method of accident-prone road

According to [8], the determination of accident prone-roads may use quality-control statistics using UCL (Upper Control Limit) Chart, with the following formula:

$$\mathrm{UCL} = \lambda + \Psi \, x \sqrt{\frac{\lambda}{m} + \frac{0{,}829}{m} + \left(\frac{1}{2} \times m\right)} \tag{2}$$

where:

λ = average accident rate for all road
Ψ = probability factor = 2.576 (for probability 99%)
m = accident rate on each road

Roads with accident rates above UCL are defined as accident-prone roads (blacksite).

2.3. Quantum GIS (QGIS)

GIS software that will be used in this research is Quantum GIS Chugiak, which is a tool used to manage (input, management, process and output) spatial data or geographically referenced data. The use of QGIS in the world of spatial information systems has reasons:

a. It is free software.
b. It is frequently updated by enhancing and added new tools.
c. Users can also create their own tool functions.

2.4. ArcView GIS

The development of GIS application requires an effective and efficient interface since this is a bridge between system and users. In this research, user interface is created using ArcView, a popular GIS software for spatial data management. This has been developed by ESRI (Environmental Systems Research Institute, Inc) by the use of Avenue programming language.

3. Methods

3.1. Development of GIS-based spatial data
GIS-based spatial data is used to analyze and sketch road network. Map data is arranged in the form of shapefile (.shp) which consist of vector data maps of points, lines, and areas (polygons). The shapefile format maps come from various sources, such as the Quantum GIS forum, the Open Street Map forum and from the BAKOSURTANAL website. The map sketching begins by the recapitulation of GPS survey results data, geometric survey of roads and recapitulation of road name and road function data.

3.2. Analysis of traffic accident data
The field calculator feature is used to analyze traffic accident data from 2011-2015. The analysis is carried out by multiplication between the attribute fields, where the field is filled with accident data recording, accident weighting (formula 1) and data required to calculate UCL (formula 2), so that the formulas included in field calculator can be executed. The result can also be presented in the form of the attribute field.

3.3. Arrangement of GIS-based database
The data that has been collected and analyzed is then arranged in the database. They are displayed in the form of interface, which is formed by the use of *ArcView* GIS software.

4. Analysis
The following steps are conducted in the analysis and database compilation.

a. The first step is to collect primary and secondary data, which is then arranged in the form of GIS-based spatial data using QGIS software. Data consists of city road network, polygon, and coordinate points of GPS survey. This needs to be re-sketching for the interface arrangement. We then input the coordinate points of GPS survey using GPS tools.
b. We then use Map Composer to create map in the form of JPEG, which aims to depict the entire Kupang city road network. The data used in the depiction is the result of the shapefile image map (point a).
c. Blacksite analysis; Accident weighting and UCL calculations is performed using field calculator features. The obtained results are arranged by adding an attribute column.
d. Results showed that there are 69 road segments in Kupang City classified as traffic accident location. Of these, 15 road segments are classified as blacksite. The most vulnerable road segment is the Timor Raya road that has an accident weight of 813.2 and the UCL value of 71.55.
e. Arrangement of interface; the interface is named 'Database dan Analisis Lokasi Rawan Kecelakaan di Kota Kupang berbasis SIG (Road Database and GIS based Blacksite Analysis'. At the beginning of the interface there are two options: 'Menu Utama (Main Menu)' option and 'Keluar (Exit)' option to end interface usage.

   The entire process of compiling this database is presented in Figure 1.

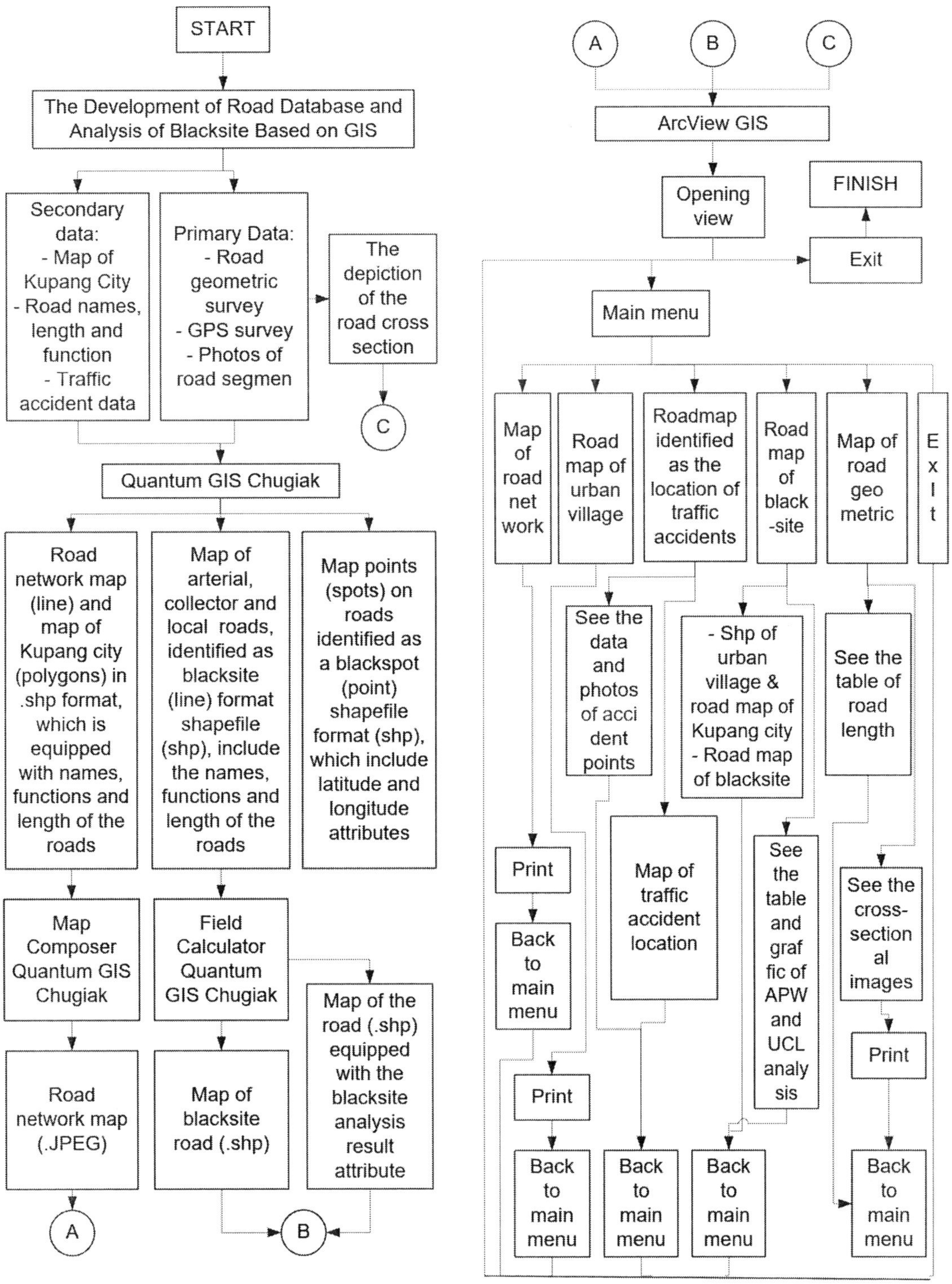

Figure 1 Flowchart of Database Compilation

5. Conclusion and recommendation

5.1. Conclusion

Based on the analysis above, it showed that the road network maps, blacksite map, geometric data and cross-sectional maps of road segments, as well as photographs of roads which are identified in traffic accident data can be arranged into GIS-based databases by using QGIS software and being depicted using ArcView GIS interface. The blacksite analysis can be done in the database which is compiled, where the result of the accident data analysis using QGIS showed that there are 69 roads identified as traffic accident locations, where fifteen of them are classified as blacksite. The top blacksite road is Timor Raya where the accident point weightage (APW) and UCL are is 813.2 and 71.55 respectively.Traffic accident data in the databases are in line with the other information: maps, results of blacksite analysis, road photographs and cross-sectional road drawings where data can be printed as needed. GIS-based data that are neat and informative will be able to accommodate the needs of stakeholders to make quicker and more accurate decisions in minimizing the number of accidents that occur, as well as in preparing the program of handling the location of traffic accident prone (blacksite).

5.2. Recommendation

a. An analysis needs to be conducted often as data always changes overtime due to the changes in the blacksite locations overtime.
b. The database has to be updated to accommodate the other aspects that are required for road safety analysis such as traffic accident modelling and road safety audit.

References

[1] M. E. Bolla and Y. A. Messah, "Collection of Papers Submitted for Presentation in The 19th Int., Symp., of Indonesian Inter-University Transportation Studies Forum," in *The 19th FSTPT Symposium*, Yogyakarta, 2016.
[2] M. E. Bolla, "Analisis Karakteristik Dan Biaya Kecelakaan Lalu Lintas Di Kota Kupang," *M E*, no. 1, p. 12, 2015.
[3] R. Satria and M. Castro, "GIS Tools for Analyzing Accidents and Road Design: A Review," *Transp. Res. Procedia*, vol. 18, pp. 242–247, 2016.
[4] R. Romadoni, "Sistem informasi geografis pemetaan lokasi rawan kecelakaan lalu lintas di kota pangkalpinang berbasis web," p. 8.
[5] D. S. Berbel, "Sight distance for road safety analysis using GIS," p. 10, 2013.
[6] J. Choudhary, A. Ohri, and B. Kumar, "Identification of Road Accidents Hot Spots in Varanasi using QGIS," p. 8.
[7] Ministry of Transportation, "Law of the Republic of Indonesia Number 22 Year 2009 Concerning Road Traffic and Transportation." Ministry of Transportation, Directorate General of Land Transportation, 2010.
[8] Departemen Permukiman dan Prasarana Wilayah, "Indonesia Traffic Accident Prevention Site Handling Manual." Departemen Permukiman dan Prasarana Wilayah, 2004.

E-commerce development using AngularJS framework and RESTful API

A Hidayati and R Nabila

Department of Informatics and Computer Engineering, Jakarta State Polytechnic
University of Indonesia, Depok City, West Java - Indonesia

anita.hidayati@tik.pnj.ac.id, rusyda.nabila.tik13@mhsw.pnj.ac.id

Abstract. There are many social enterprises that use e-commerce in running their business, Emcekaqu is one of them. It wants to change hygiene and sanitation behaviours by building a non-subsidized latrines. This achieved by selling local products to people in the city. E-commerce becomes a solution to maximize the usability of the web as a media transaction. Frontend systems are built using AngularJS framework and backend systems are built using the RESTful API. System test is done using user acceptance test that has two stages, are alpha testing and beta testing. The testing technique is black box testing. In addition, application responsiveness testing is also done to prove that the web is accessible from multiple platforms. Based on the tests, functional system is running in accordance with business processes.

Keywords: AngularJS, RESTful API, E-commerce, Rapid Application Development

1. Introduction

Emcekaqu is a pre-social enterprise that focus on eradicating open defecation in remote area of Pandeglang, Banten. This business was built by students of the Faculty of Nursing University of Indonesia from a sense of caring for the low level of community sanitation in Cinibung village, Pandeglang. Emcekaqu will build a non-subsidized latrines by selling local products, such as rengginang and some other snacks.

Website is one of the most popular promotion media today. The use of websites can help businesses to reach more customers with unlimited reach of space and time. In starting the business, emcekaqu created a website, namely www.emcekaqu.com as a media campaign to the people in the city. In 2016, emcekaqu.com is built with mobirise. Mobirise is a free app for building a mobile-friendly web without coding experience. However, the web is static, so the business process in selling the product can't be achieved on this web.

Based on the explanation, emcekaqu website needs to be maximized by build an e-commerce. AngularJS framework has been selected, it's because AngularJS has the ability to create a Single Page Application (SPA) website.

2. Literature Review

E-commerce is the use of the internet, web, and application to perform business transactions. Formally, the transaction is done digitally between organizations and individuals. E-commerce is divided into six kinds, namely Business-to-Consumer (B2C) E-commerce, Business-to-Business (B2B) e-commerce,

Consumer-to-Consumer (C2C) e-commerce, Social E-commerce, Mobile E-commerce, and Local E-commerce. The project is a B2C e-commerce because of the online transaction is conducted from online businesses (emcekaqu) to the individual consumer [1].

Single Page Application (SPA) is composed of individual component that can be replaced or updated independently, without refreshing whole page so that the entire page does not need to be reloaded on each user action, which saves bandwidth as well as no loading of external files every time when page is loaded, such as images or CSS files, etc. The purpose behind this is to make the subsequent page loads very fast as compared to traditional Request-Response cycle [14].

AngularJS is one of the widely used frameworks for modern single-page web application development which is designed to support dynamic views in the applications [13]. AngularJS is a front-end JavaScript Framework that used to build a Single Page Application uses the Model View Controller (MVC) architecture. The approach used by AngularJS is to extend the functionality of HTML using JavaScript custom attributes tied to the HTML tag. AngularJS ability to expand HTML functionality, enables DOM manipulation becomes easier. It's done through the client-side and two way data binding, when it's synchronizing the model and the view. AngularJS using MVC and dependency injection to facilitate the code structure organizing and testing process [2]. AngularJS has several advantages, that are [3]:

- Provides capability to create Single Page Application in a very clean and maintainable way
- Provides data binding capability to HTML. Thus, it gives user a rich and responsive experience
- Code is unit testable
- Uses dependency injection and make use of separation of concerns
- Provides reusable components
- With AngularJS, the developers can achieve more functionality with short code
- Views are pure html pages, and controllers written in JavaScript do the business processing

Web service is software designed to provide communication and interaction of two or more applications with the nature of the program-to-program [4]. Web service can be divided into two types, namely Simple Object Access Protocol (SOAP) and Representational State Transfer (REST). REST is a kind of API that uses the HTTP protocol as data communication lines. REST web using a standard format, such as HTTP, HTML, XML, JSON, URI, and MIME. REST web service architecture came to be known as a RESTful Web Service [5].

My Structured Query Language (MySQL) is a open source Database Management System (DBMS) program. MySQL is a database program that can be accessed over a network, so it can be used for multiple users. Another advantage of MySQL is using a SQL standard. SQL is a Structured Query Language, SQL has been standardized for all DBMS program [6].

Rapid application development (RAD) is an object-oriented approach to systems development that includes a method of development as well as software tools. RAD aim is to shorten the system development cycle. RAD can be adapted in user requirements quickly. Figure 1 represents the stage in the RAD cycles [7].

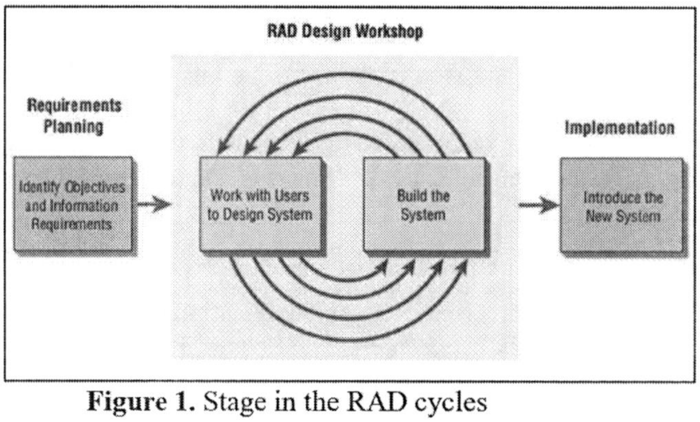

Figure 1. Stage in the RAD cycles

- Requirements Planning Phase
- In the requirements planning phase, users and analysts meet to identify objectives of the application or system and to identify information requirements arising from those objectives.
- RAD Design Workshop
- The RAD design workshop phase is a design-and-refine phase that can best be characterized as a workshop. During the RAD design workshop, users respond to actual working prototypes and analysts refine designed modules based on user responses.
- Implementation Phase
- As soon as previous stage is agreed on and the systems are built and refined, the new systems or part of systems are tested and then introduced to the organization.

User Acceptance Test (UAT) is an information system testing that refers to the perspective of the end-user or system owner. UAT focuses on generating systems that can deliver the expected business benefits when operated by the users [8]. UAT is divided into two stages, that are alpha and beta testing. Alpha testing is a repeated test by the developer to ensure that there is no error in the application. Beta testing is performed by end-users by entering real data and observing displayed errors [9].

Black box testing is one type of testing technique that is also called behavioral testing. Black box testing focuses on software functional testing. It's mean, black box testing allows developers to create some input conditions that will be tested on the program function [10]. Black box testing has four testing design techniques, that are equivalence classes, boundary value analysis, expected result coverage, and intuition & experience. This research uses the expected results coverage design technique, which is a technique that focuses on the expected output results of the various input types entered [11].

3. Result and Analysis

3.1. Requirements Planning Phase

This study aims to generate e-commerce on emcekaqu.com, that can be accessed from various platform, so the process of promotion and ordering products becomes easier.

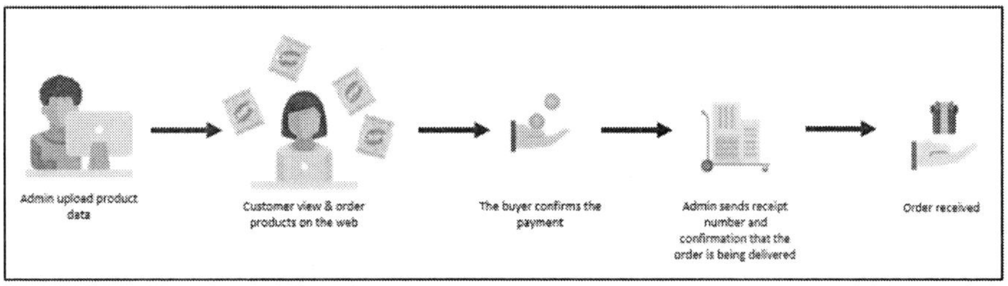

Figure 2. The proposed business process

Figure 2 represents the proposed business process. It begins when admin upload a product and customer view and order products from the web. Buyer need to makes a payment at ATM and send the payment confirmation in the web. Admin will update the order status and send receipt number when the order is being delivered. Buyer can track the package from the shipping provider's web official.
To realize the goal, we need minimum requirements, i.e.
Programming Language:

- AngularJS 1.2.16
- Angular-bootstrap 0.12.0
- Bootstrap 3.1.1
- Code Igniter 3.0.3

Hardware:

- Minimum processor version: Intel Celeron
- Minimum RAM memory: 1 GB
- Minimum hard disk memory: 10 GB

Software:

- Xampp 3.2.2 version (Apache 2.0, PHP 5.6.14, MySQL 5.0.11)
- Sublime
- Chrome 58.0

Based on business process analysis, there are admin, member, and guest as actors. Table 1 shows the activity of each actor.

Table 1. The Activity of Each Actor

Admin	Member	Guest
Update the order status and send it to the buyer's email	Order the product	Send message
Product management	Access the order status and get it via email	Access emcekaqu news
Category management	Payment confirmation	-
News management	Address management	-
Know the donation progress	Login to member area and edit profile	-
Login to admin area and edit profile	-	-

4. RAD Design Workshop

The design and repair process is carried out if there is still a design nonconformity. Outputs from this stage are database design, backend design, frontend design, and interface design.

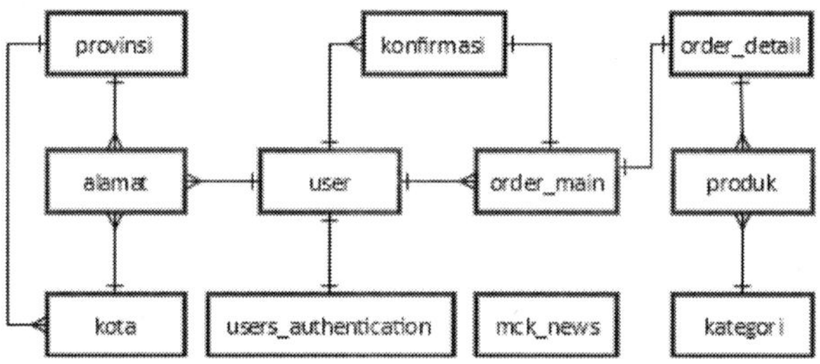

Figure 3. Entity Relational Diagram

Database design. Figure 3 represents the entity relational diagram of the system. Data in the news management menu is stored in the mck_news table. While other tables are used in the process of e-commerce transactions.

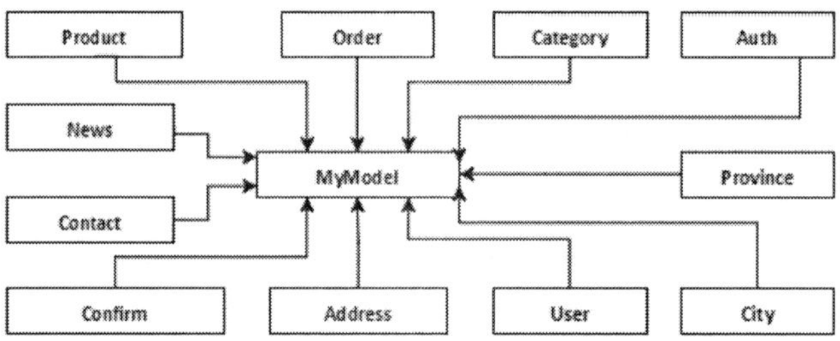

Figure 4. Class Diagram Backend System

Backend design. Backend systems are built from various classes. Figure 4 represents the class diagram of the backend system. The MyModel class is used to authenticate the user. Other classes are connecting to database to display the specific data.

Frontend design. Frontend design is represented by Unified Modelling Language (UML). UML used is use case and activity diagram. Figure 5 represents use case diagram of the system. Scenarios of the actors are based on Table 1.

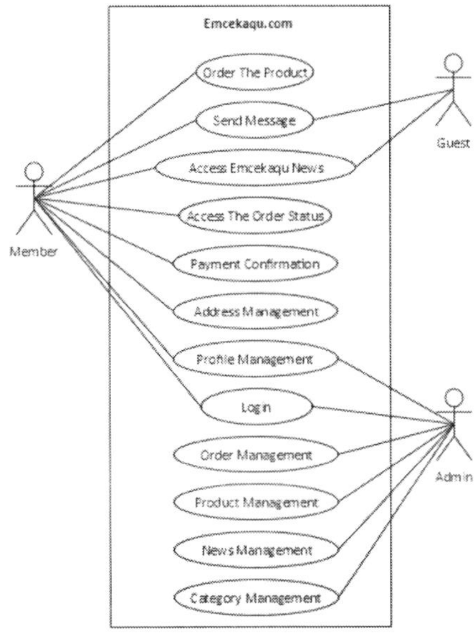

Figure 5. Use Case Diagram

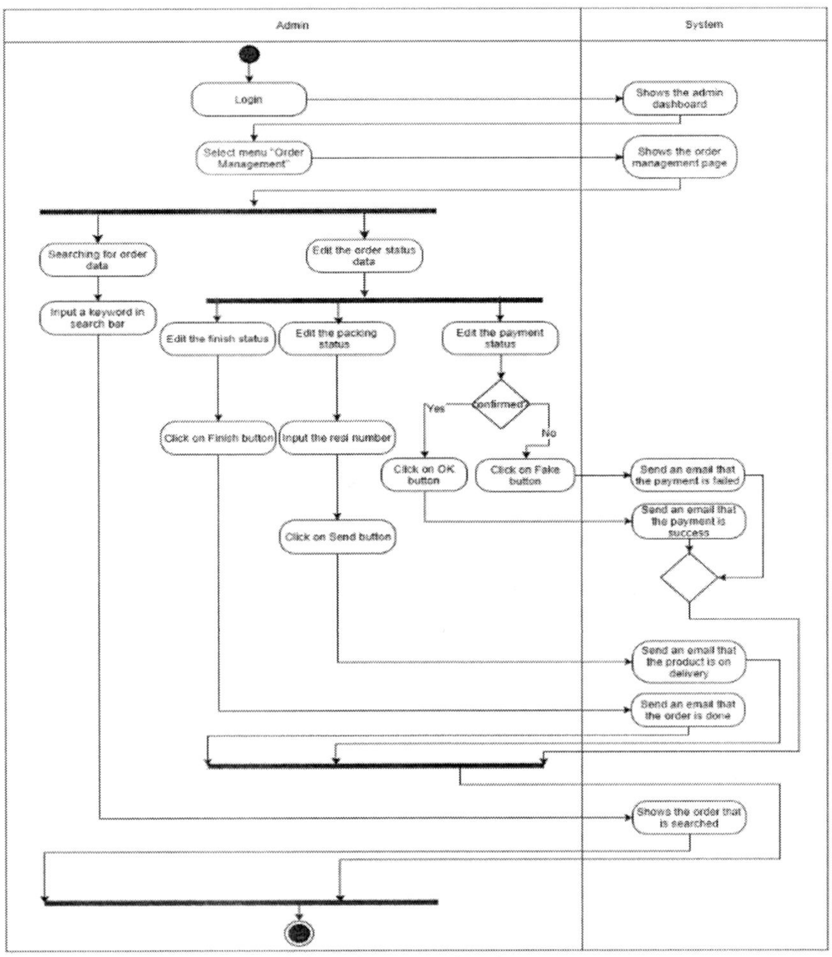

Figure 6. Activity diagram of order management

The activity diagram is representing of activity flow in the system, about how the flow is started until ended. The example of activity diagram shows in Figure 6.

Interface Design. Figure 7 represents the shopping area design. This page is main page of the purchase process. On product page, user can search product by name or category. Users choose the product by pressing the buy button, then the product will enter into the basket. To add or decrease the number of products, the user can press the plus (+) or minus (-) button. To delete the entire contents of the baskets, user can press the clear button. When finished shopping, user can press the checkout button.

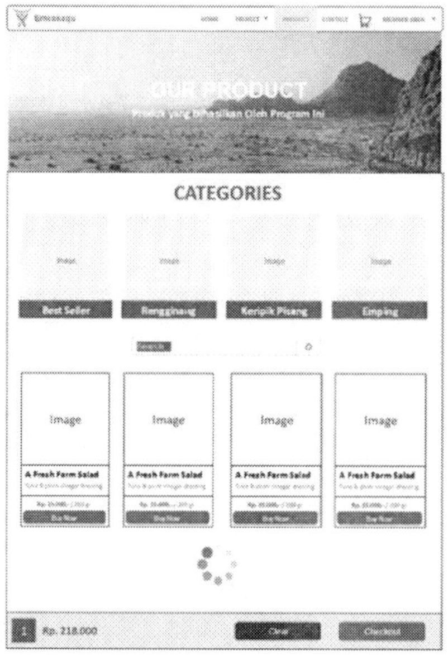

Figure 7. Shopping area design (product page)

5. Implementation Phase

5.1. Web Realization

Shopping area consists of three different pages, are product page, shopping cart, and checkout page. Figure 8 is the realization of product page in web view and Figure 9 is the realization of product page in mobile view.

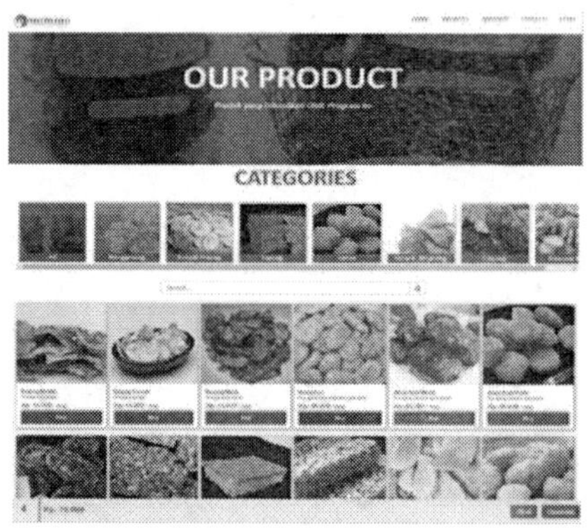

Figure 8. The realization of product page in web view

IOP Conf. Series: Materials Science and Engineering **403** (2018) 012063 doi:10.1088/1757-899X/403/1/012063

Figure 9. The realization of product page in mobile view

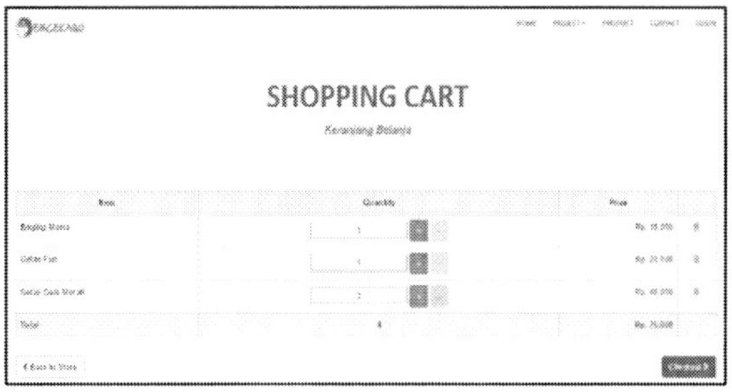

Figure 10. The realization of shopping cart page

Figure 10 is the realization of the shopping cart page. On this page users can edit the number of products to be purchased. Purchased product data will be stored on the local storage browser so it can be loaded by another page. The checkout page has three sections, ie login, shipping form, and bill. Figure 11 is the realization of checkout page.

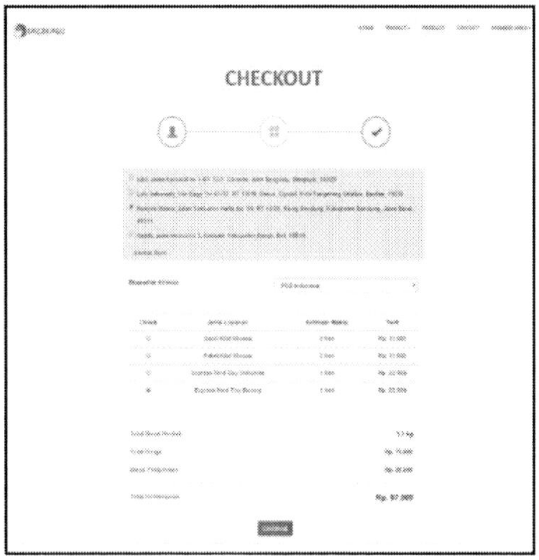

Figure 11. The realization of checkout page

- Testing
  The author performs a user acceptance test that has two stages, are alpha testing and beta testing. The testing technique is black box testing that focuses on application behaviour when receiving input. Table 2 presents the test detail. Alpha and beta testing results earn 100% success percentage. All modules can run according to the functionality.
- In addition, application responsiveness testing is also done to prove that the web is accessible from multiple platforms. It is also important to provide a support system to handle potential issues with products and services that are acquired through the platform [12].

Table 2. Test result

Test Item	Test Detail
Authentication Module	Login
	Logout
User Module	Edit profile
	Change password
	Forgot password
	Add user admin
	Delete user admin
	Add user member (sign up)
Contact module	Send message
Product Module	Read product data
	Add product data
	Edit product data
	Delete product data
Category Module	Read category data
	Add category data
	Edit category data
	Delete category data
Address Module	Read address data
	Add address data
	Edit address data
	Delete address data
News Module	Read news data
	Add news data
	Edit news data
	Delete news data
Order Module	Put the product into shopping cart
	Make an order (checkout)
	Read order detail
	Edit order status
Payment Confirmation Module	Read payment confirmation detail
	Add payment confirmation
	Edit payment confirmation data

- Application Responsiveness Testing
  Testing is done by using toggle device toolbar feature in Google Chrome. This test is performed to prove that the application is responsive and can be accessed on various platforms. Figure 12 represents the application display when loaded on iPad Pro screen with 1024 x 1366 resolution. Figure 13 represents the application display when loaded on Samsung Galaxy S5 with 360 x 640 resolution.

Figure 12. App display on iPad Pro screen

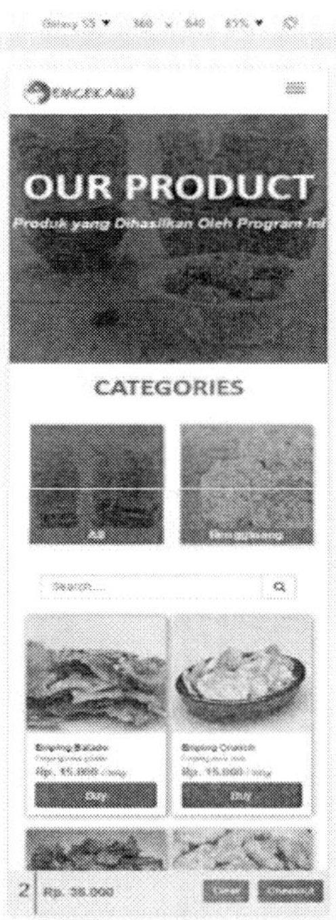

Figure 13. App display on Samsung Galaxy S5 screen

6. Conclusion

Based on the implementation of the research that has been done, can be concluded that this research produces e-commerce on the emcekaqu.com which can be used as media for buying and selling emcekaqu products. The application was developed using AngularJS framework with features: order, payment confirmation, product management, category management, address management, user management, authentication, contact, and news management. Alpha and beta testing results earn 100% success percentage. All modules can run according to the functionality. The results of application responsiveness testing prove that the emcekaqu.com can be accessed from various platforms.

References

[1] K. C. Laudon and C. G. Traver, E-commerce: business, technology, society, 10th ed., Upper Saddle River: Pearson, 2014.

[2] A. Q. Haviv, MEAN Web Development, Birmingham: Packt Publishing, 2014.

[3] V. Waghade and B. V. Chaudhari, "Study Of AngularJS with Other Frameworks", *International Journal of Research in Computer & Information,* vol. 1, no. 2, pp. 151-154, 2016.

[4] H. D. Purnomo, D. A. Saputro, R. Somya and C. Fibriani, 'The Application of Restful Web Service and JSON for Poultry Farm Monitoring System", *Journal of Electrical Engineering and Computer Sciences,* pp. 25-30, 2016.

[5] D. Jacobson, G. Brail and D. Woods, APIs: A Strategy Guide, California: O'Reilly Media, 2011.

[6] B. Nugroho, Latihan Membuat Aplikasi Web PHP dan MySQL dengan Dreamweaver, Yogyakarta: Gava Media, 2009.

[7] K. Kendall and J. Kendall, System Analysis and Design 8th Edition, Upper Saddle River: Prentice Hall, 2011.

[8] B. Hambling and P. Goethem, User Acceptance Testing: a step-by-step guide, Chippenham: BCS, 2013.

[9] A. Dennis, B. H. Wixom and R. M. Roth, System Analysis and Design, New Jersey: John Wiley & Sons, Inc., 2012.

[10] R. Pressman, "Rekayasa Perangkat Lunak Buku 1," in *Pendekatan Praktisi Edisi 7*, Yogyakarta, Andi, 2010, p. 668.

[11] G. D. Everett and R. McLeod, Software Testing: testing across the entire software development life cycle, New Jersey: John Wiley & Sons, Inc., 2007.

[12] L.G.Vandenbosch, M.Nicoletti, "An industry case study: a mobile-based business strategy to improve the customer care service in a major retail company", 5° Simposio Argentino de Informática Industrial, 2016

[13] W.Chansuwath, T.Senivongse, "A Model-Driven Development of Web Applications Using AngularJS Framework", ICIS, June 26-29, 2016, Okayama, Japan

[14] M.A. Jadhav, B.R. Sawant, A.Deshmukh, " Single Page Application using AngularJS", *(IJCSIT) International Journal of Computer Science and Information Technologies*, Vol. 6 (3), 2015, 2876-2879

IOP Conf. Series: Materials Science and Engineering **403** (2018) 012064 doi:10.1088/1757-899X/403/1/012064

An investigation of external factors for technological acceptance model of nurses in Indonesia

D B Setyohadi[1] and N W Purnawati[1]
Magister Teknik Informatika, Universitas Atma Jaya Yogyakarta, Yogyakarta, Jl. Babarsari 44 Yogyakarta, Indonesia 55281.

Abstract. The Indonesian government policy has required the hospitals in Indonesia to implement a health information system. There are still found many barriers in its implementation in the hospitals such as lack of facilities and human resources which lead to the problem complexity. As a result, the e-health system implementation has not been being optimum. This study is purposed to recognize the problem of technology implementation, particularly the adoption of the aplication system by using Technology Acceptance Model (TAM) theory. By reviewing the previous studies and literatures, we detected the need to relate the external variables of PEOU and PEU which are rooted in TAM. The study we conducted shows that TAM model application as the e-Health measurement framework in Indonesia needs to be extended particularly its factors which significantly influence PEOU and PEU. The factors are personal factor and organizational factor. Furthermore, a model is formulated for future empirical testing.

1. Introduction

Hospital information systems become a technology that was instrumental in the development of hospital in Indonesia. It is required since demographic have raised attention regarding the need to escalate the level of life and health. Although e-health is important there are many problems cased form external factor around e-health such as infrastructure, human resource. Therefore only a few of e-Health implementation is successful since the number of user acceptance of e-Heath also rise slowly. To solve that condition, the proper model for analysis of acceptance e-Health is needed.

In 2009 the ministry of health of Indonesia made regulations in article 52 paragraph 1 Number 4/2009 regarding record keeping and reporting of all activities in hospitals to the Ministry of health, it was conducted in a system that is a hospital information system (HIS). It is done to find out the performance on both government and private hospitals. This condition represents that e-Health services are viewed as the best alternative solution of the health quality improvement problem management. Furthermore, it possibly covers both healthcare services and innovation [1] Many countries have created policy with regard improvement of service in health by developing of e-health [2] in order to make ease the process of operations on community service [3][4][5]. Furthermore, a study of e-Health implementation problem in the developing country is also limited. The critical factor should be studied in advanced since the developing countries require a bigger requirement for a better e-Health implementation [6]. It is needed since the demand of health care is increased and service alternatives and health personnel are still becoming scarce. In Indonesia, not all Hospital can be maximized using HIS. This can be seen by the differences between the intention of both private and public hospital management [7]. In Indonesia, the acceptance of HIS is greatly influenced by Human Characteristic, Technology and Organizational [8].

The lack of infrastructure and resources has become the contributory causes the problem of e-Health implementation. In some hospital, many nurses still do the manual hospital report, therefore, the nurses need a sense and knowledge to understand the associated system and the desire to use the system. In short, the use of e-health in developing country is important, and it needs a big effort. This conditons is the reason why the research should be performed. Furthermore, Technology Acceptance Model (TAM) is used since TAM is the frequently used model to assess the acceptance of users to adopt the new technology [9]. TAM has two main concepts. They are usefulness and ease of use perception. This study will be focused to investigate the acceptance factors of the nurses in Indonesia when they adopt e-health. World Health Organization reported that there are some problems or barriers in terms of e-health implementation in hospitals. Some barrier of technology adoption, such as lack of experiences using Informatics Communication Technology (ICT), lack of training, the support of organizations such as the infrastructure problems must be fixed [8]. It is becoming an important issue for management of hospitals in the region of Indonesia. This research will explore the methods with a development approach on problem of TAM that influences the nurse perception of acceptance of adopting e-health. Furthermore, the appropriate model will be developed, and the framework skeleton of acceptance of adopting e-health is proposed. It can hopefully help the hospitals to find the significant factors that influence the acceptance of nurses in e-health in Indonesia.

2. Literature Review

Technology acceptance Model (TAM) was developed as an extension of the Theory of Reaction Action (TRA) [9]. TAM is useful to describe the relationship between attitude and behaviour of human volitional. This relates to a person's motivation to show the reaction based on the intention [10]. TRA aims to predict the intentions of behaviour, including the motives underlying the formation of the behaviour. Using that theory, Davis tries to formulate the relationship with the behaviour of acceptance. Moreover, TAM is focused on the acceptance of users based on 3 factors i.e. ease of perception, perception and attitude toward the use of utilization systems especially on the perception ease of use (PEOU) and Perception of Usefulness (PU). In addition, it is recommended that if the system was already considered to be useful then users will develop the use of these intentions become stronger. With regard to external factor, UTAUT is developed by removing the attitudes for being minor roles against external variables that allows the influence. Furthermore, subjective norm, image, Job Relevance, the Output Quality, and the result of external as demonstrability are connected by a sense of usefulness associated based on voluntary user experience and technology [11].

Based on their stakeholder's point of view, the e-health implementations have many dimensions, consequently it has a more complicated and complex requirement [4]. Due to its complexity, many studies about TAM, with regard to e-Health Care, are performed not only by applying but also extending the required variable. For example, Ahlan, and Ahmad [12], used Output Quality on the original model of TAM and added a factor of Perceived Cost-Effectiveness which can then affect the attitudes towards the use of technology. The same is also done by Kuo, Liu, & Ma [13], which combines the factors of technology readiness index (TRI), 5 these factors such as Optimism, Innovativeness, Insecurity, and the Discomfort became the influence on the two important factors of TAM i.e. PEOU and PU.

In perspective of user, regarding barrier problem of e-health, cultural approach is an important factor which should be considered when e-health is designed [14]. The suitable e-health design was able to figure out the intentions of the user (Intention to Use). Gajanayake, Iannella, NICTA, & Sahama [15], used 3-dimensional technology reporting to individual context, technology and implementation context on the Unified Theory of Acceptance and Use of Technology (UTAUT) which is developed by Venkatesh, Morris, Davis, & Davis [11]. On the other case, Srur & Drew combined factors of several models such as TAM, TAM 2, Diffusion and Innovation (DOI) and the UTAUT to understand the success and influence of the personally controlled Electronic Health Records (PCEHRs) [16].

TAM is also used by Bezboruah, Paulson, & Smith to understand the attitude of the staff and procedures at the care home against use of Hospital Information Technology (HIT) [17]. Holden & Karsh study that the increasing study of the acceptance of e-Health is accompanied by changes and integration of e-Health based on patterns, rules or the concept of the use of e-health [18].

With regard to design, many studies reported that technology is an important factor particularly as the e-health ability to match the local condition [19]. The latest technologies capability made e-health to adapt the changing of the environment in order to make a greater acceptance of the user [20]. Rely on these phenomena; it is in-line that PEOU is the easier design. It could be understood since PEOU will face to the adaptability problem directly [21]. To raise the acceptance, e-Health aplication must be able to provide an interface which supports exchange information [22].

Many studies have been performed to improve the acceptance of e-health. It should be noted that in the implementation of application, the design of interface is focused only to deal with social cultural problem [23, 24]. The previous study related to the acceptance matters reveals that the design aspect is the solution part of social aspect management. Therefore, the most fundamental aspect to explore is the social aspect which should be linked to PEOU. The previous survey research exposed that external factors mainly such as personal and organizational factors have relation with it.

3. Model Development

According to the previous explanation, this study is aimed to develop a model based on TAM ad its extension. Theoretically, institutions, particularly institutions which us e-health, have significant roles that are capable of dealing with the environmental pressure [7]. Many studies report that institution is able to influence its community to adopt a new system when they manage attitudes properly [25]. With regard to institution roles and individual characteristics, the PEOU and PEU of TAM are significantly influenced by organization and its people. Both of them, when combined with technological capability, could be considered as a comprehensive framework for measuring the quality of information system application [26][27][18]. Since the development of technology moves quickly, we believe that technological capability is a must. Furthermore, the factors, which should be considered are two i.e. organization and personal factor. They influence directly to both PEOU and PEU of TAM and it is categorized as an external factor and the detailed is described below:

3.1. Personal factors

Because e-health, particularly on nurses, is limited on the previous literature, it is important to address the question about attitude towards e-health by looking at the literature which related to both TAM e-health. To make it easier, the study is focused on the personal factor and its environment. Personal factors which is the most important in this case is IT knowledge /experience that it may also be affected due to lack of experience and knowledge about information technology [6]. Things that are closely related that can be influential as well are factors of age, gender, and skill, demography factor of staff [25][29]. In addition, another factor which can be influential in adopting e-Health among nurses is the interface usability. Scholtz, Mahmud, & Ramayah [30] understand that user interaction can be associated against acceptance of the adoption of a technology.

Related to PEU problem, in e-health is common when we relate to interface design of e-health [31]. In terms of e-health, interface usability can be a personal factor in maximizing the performance of e-health [23]. Another scholar reported that the efforts of management to optimize the interface design will significantly influence the usability and usefulness particularly when the e-health is implemented [32]. Regarding there is much hospital management in rural area in Indonesia, to summarize, the component should be considered as an important factor in personal factor can be categorised as follow:

- Presentation, e-Health is a design conceived by nurses and organizations.
- Navigation, ease of access is an e-Health that is easily understood by nurses and organizations.
- Learnability as an easier guide for nurses and organizations without having to use the introduction of the system.

3.2. Organizational factors

World Health Organization (WHO) reveals that the Indonesian government had a response to ICT education on a medical professional such as nurse, but the barriers to learn to use e-health is still high. The Organization has an important role in supporting the implementation of e-health in hospitals in

Indonesia. This can be observed on the WHO data regarding the lack response of the government against software, skills training, ongoing support, scholarship and ICT Equipment.

The role of organization in technological acceptance problem is must. Many aspects of user acceptance are build by a variety of organizational factors which could be a main determinant of it [33]. The influence of organizational relationship into positive beliefs about the new technology is proven, and it is performed when the information is well distributed [34].

In information system, the organization significantly influences technological acceptance. When user feels better support the performance is improved [23]. There are many supports regarding acceptance of e-health. As example, the management should provide sufficient guidance regarding lack of knowledge [35], technically support [36]. Regarding familiarity to the system, the management should involve user when system is developed and more over training is needed [23][37]. Training is an important component, and it is suggested to be done during, and after implementation [38].

Shortly, the relation both or Information system user motivation and an external variable are illustrated in Figure 1, and the summary of both two factors are described below:

- Personal factors consist of IT Knowledge/ Experience i.e. knowledge or experience nurses and organizations to find out what activities and information related to e-Health. Personal other factors, interface usability, i.e. are the interaction that must be understood by nurses and organizations regarding e-health.
- Organizational Factors comprising infrastructure, clinical data sharing, and training is a challenge from nurses and organizations to improve the e-Health.

For this purpose, we have proposed and delineated a model to investigate the relationships among user, management roles and attitude towards e-health. To summarize this paper, we focus on the fact that e-health is being used increasingly by hospital management due to regulation in Indonesia and it still has the nurse's acceptance problem. Thus, we recommend well-made studies to be conducted to examine the many relationships which provide valuable knowledge. Hopefully it can be used to design and implement e-health that promotes the benefit of stakeholders in hospital management.

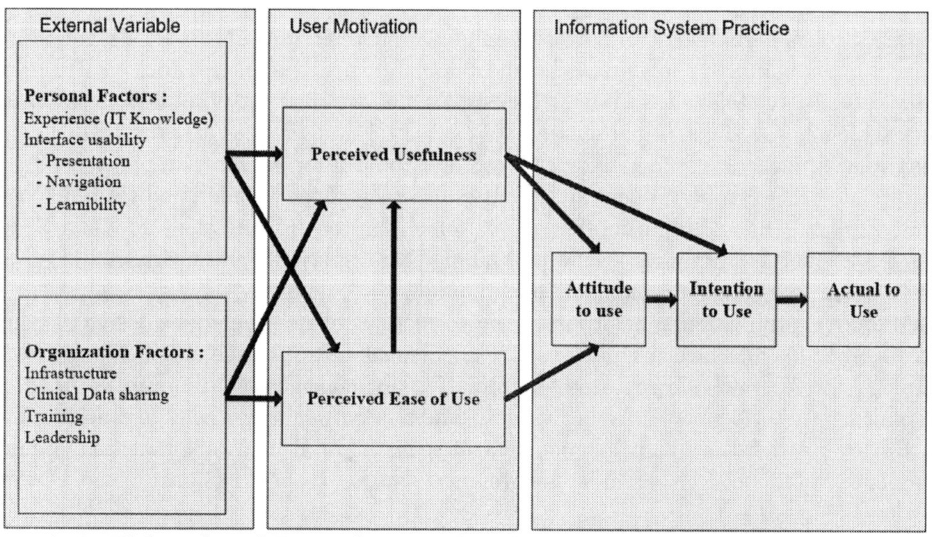

Figure 1. An Elaboration of External Factors for Nurse Adoption in e-health in Indonesia

4. Conclusion

We study an overview of e-health for nurses and TAM. This paper has proposed an external factor which will most significantly influence PEOU and PEU of nurses when they use e-health as an alternative approach analysing acceptance model of e-health by nurses. External variable focuses on personal and organizational factors, and then the motivation factor focuses on the concept of TAM that is PU and PEOU as well as information system practice that focuses on attitude to use, behaviour intention to use and actual system to use e-health.

References

[1] Ross, J., Stevenson, F., Lau, R., & Murray, E. 2016. Factors that influence the implementation of e-health: a systematic review of systematic reviews (an update). Implementation Science : IS, 11, 146. http://doi.org/10.1186/s13012-016-0510-7

[2] Broens TH, Huis in't Veld RM, Vollenbroek-Hutten MM, Hermens HJ, van Halteren AT, Nieuwenhuis LJ., 2007, Determinants of successful telemedicine implementations: a literature study. J Telemed Telecare. 13(6):303–9

[3] N. I. Ismail, N. H. Abdullah and A. Shamsuddin, 2015, Adoption of Hospital Information System (HIS) in Malaysian Public Hospitals, Procedia - Social and Behavioral Sciences, vol. 172, p. 336 – 343

[4] K. Cresswella and A. Sheikh, 2013, Organizational issues in the implementation and adoption of health information technology innovations: An interpretative review, International Journal of medical informatics, vol. 82, p. e73–e86

[5] C. D. Melas, L. A. Zampetakis, A. Dimopoulou and V. S. Moustakis, 2014, An empirical investigation of Technology Readiness among medical staff based in Greek hospitals," European Journal of Information Systems, vol. 23, p. 672–690

[6] Hage, Eveline, et al. 2013, Implementation factors and their effect on e-Health service adoption in rural communities: a systematic literature review." BMC health services research 13.1: 19.

[7] P. Handayani, A. Hidayanto, D. Ayuningtyas and I. Budi, 2016, Hospital Information System Institutionalization Processes in Indonesian Public, Government-owned and Privately Owned Hospitals," International Journal of Medical Informatics, vol. 95, pp. 17-34

[8] P. Handayani, A. Hidayanto, A. Pinem, I. Hapsari, P. Sandhyaduhita and I. Budi, 2016, Acceptance Model of a Hospital Information System, Int. Journal of Medical Informatics, pp. 1-58

[9] F. D. Davis, 1989, Perceived Usefulness, Perceived Ease of Use, and User Acceptance of Information Technology," MIS Quarterly, Vol. 13, No. 3, pp. pp. 319-340

[10] Ajzen and T. J. Madden, 1986, Prediction of goal-directed behavior: Attitudes, intentions, and perceived behavioral control," Journal of Experimental Social Psychology, vol. 22, no. 5, pp. 453-474

[11] V. Venkatesh, M. G. Morris, G. B. Davis and F. D. Davis, 2003, User Acceptance of Information Technology: Toward a Unified View," MIS Quarterly, vol. Vol. 27, no. No. 3 , pp. pp. 425-478

[12] A. R. Ahlan and B. I. Ahmad, 2014, User Acceptance of Health Information Technology (HIT) in Developing Countries: A Conceptual Model," Procedia Technology, vol. 16, p. 1287 – 1296

[13] K.-M. Kuo, C.-F. Liu and C.-C. Ma, 2013, An investigation of the effect of nurses' technology readiness on the acceptance of mobile electronic medical record systems," BMC Medical Informatics and Decision Making, vol. 13, no. 8, pp. 1-14

[14] A. H. H. M. Mohamed, H. Tawfik, L. Norton and D. Al-Jumeily, 2013, A technology acceptance model for a user-centred culturally-aware e-Health design," Sixth International Conference on Developments in eSystems Engineering, pp. 121-125

[15] R. Gajanayake, R. Iannella, NICTA and T. Sahama, 2014, Adoption of Accountable-eHealth Systems by Future Healthcare Professionals : An empirical research model based on the Australian context," IEEE International Conference on e-Health Networking, Applications and Services (Healthcom), vol. 16, pp. 495-499

[16] B. L. Srur and S. Drew, 2012, Challenges in designing a successful e-health system for Australia : Is the increased efficiency created by electronic health records greater than its potential risks?," International Symposium On Information Technology In Medicine And Education, pp. 480-484

[17] K. C. Bezboruah, D. Paulson and J. Smith, "Management attitudes and technology adoption in long-term care facilities, 2014, Journal of Health Organization and Management, vol. Vol. 28, no. Iss 3 , pp. pp. 344 - 365

[18] R. J. Holden and B.-T. Karsh, 2010, The Technology Acceptance Model: Its past and its future in health care," Journal of Biomedical Informatics journal, vol. 43, no. 1, pp. 159-172
[19] Ludwick DA, Doucette J.,2009, Adopting electronic medical records in primary care: lessons learned from health information systems implementation experience in seven countries. Int J Med Inform. 78(1):22–31.
[20] Li, J., Talaei-Khoei, A., Seale, H., Ray, P., & MacIntyre, C. R., 2013, Health care provider adoption of eHealth: systematic literature review. Interactive Journal of Medical Research, 2(1)..
[21] Goldstein DH, Phelan R, Wilson R, Ross-White A, Van Den Kerkhof EG, Penning JP, 2014, Brief review: adoption of electronic medical records to enhance acute pain management. Can J Anesth. 61(2):164–79.
[22] Boonstra A, Broekhuis M., 2010, Barriers to the acceptance of electronic medical records by physicians from systematic review to taxonomy and interventions. BMC Health Serv Res. 10:231.
[23] Djoko Budiyanto Setyohadi, Michael Aristian, Benyamin Langgu Sinaga and Nor Aziati Abdul Hamid, 2017, Social Critical Factors Affecting Intentions and Behaviours to Use E-Learning: An Empirical Investigation Using Technology Acceptance Model. Asian Journal of Scientific Research, 10: 271-280
[24] Lee, T. T., Mills, M. E., Bausell, B., & Lu, M. H. ,2008, Two-stage evaluation of the impact of a nursing information system in Taiwan. Int. Journal of Medical Informatics, 77(10), 698-707
[25] Fitzpatrick L, Melnikas A, Weathers M, Kachnowski S.,2007, Understanding communication capacity. Communication patterns and ICT usage in clinical settings. J Healthc Inf Manag. 22(3):34–41.
[26] Lin, H. C., Hsu, M. H., & Yang, C. W. ,2014, The influences of computer system success and informatics competencies on organizational impact in nursing environments.Computers Informatics Nursing, 32(2), 90-99
[27] Lu, C. H., Hsiao, J. L., & Chen, R. F. ,2012, Factors determining nurse acceptance of hospital information systems.Computers Informatics Nursing, 30(5), 257-264
[28] Pai, F. Y., & Huang, K. I. ,2011, Applying the technology acceptance model to the introduction of healthcare information systems. Technological Forecasting & Social Change, 78(4), 650-660
[29] Oroviogoicoechea C, Elliott B, Watson R. , 2008, Review: evaluating information systems in nursing. J Clin Nurs. 17(5):567–75.
[30] B. Scholtz, I. Mahmud and T. Ramayah, 2016, Does Usability Matter? An Analysis of the Impact of Usability on Technology Acceptance in ERP Settings," interdisciplinary Journal of Information, Knowledge, and Management, vol. 11, pp. 309-330
[31] Fruhling, A. L., & Lee, S. M. ,2006, The influence of user interface usability on rural consumers' trust of e-health services. Int. journal of electronic healthcare, 2(4), 305-321. 10.1504/IJEH.2006.010424
[32] Castillo VH, Martinez-Garcia AI, Pulido JR.,2010, A knowledge-based taxonomy of critical factors for adopting electronic health record systems by physicians: a systematic literature review. BMC Med Inform Decis Mak. 10:60.
[33] W. Lewis, R.Agarwal and V. Sambamurthy, 2003, Sources of influence on beliefs about information technology use: an empirical study of knowledge workers, MIS Quarterly 27(4) , pp.657-678
[34] Noh MJ, Lee KT. , 2015, An analysis of the relationship between quality and user acceptance in smartphone apps. Information Systems and e-Business Management. 1–19.
[35] Kilsdonk E, Peute LW, Knijnenburg SL, Jaspers MW.,2011, Factors known to influence acceptance of clinical decision support systems. Stud Health Technol Inform. 169:150–4.
[36] Jarvis-Selinger S, Chan E, Payne R, Plohman K, Ho K.,2008, Clinical telehealth across the disciplines: lessons learned. Telemed e-health. 14(7):720–5
[37] Stolee P, Steeves B, Glenny C, Filsinger S.,2010, The use of electronic health information systems in home care: facilitators and barriers. Home Healthc Nurse. 28(3):167–79

[38] Gagnon MP, Desmartis M, Labrecque M, Car J, Pagliari C, Pluye P, 2012, Systematic review of factors influencing the adoption of information and communication technologies by healthcare professionals. J Med Syst. 36(1):241–77.

IOP Conf. Series: Materials Science and Engineering **403** (2018) 012065 doi:10.1088/1757-899X/403/1/012065

Designing enterprise architecture based on TOGAF 9.1 framework

H Qurratuaini

Jl. Cemara No. 07, Kelurahan Kayu Putih, Kecamatan Pulo Gadung, Jakarta Timur, DKI Jakarta, Indonesia, 13210

hanqrrtn@gmail.com

Abstract— The development and utilization of information technology in everyday life continues to grow rapidly within the passage of time. As a result, organizations or companies have many choices in determining the information technology to be applied in their organizations or companies. To ensure that information technology solutions implemented by enterprises are able to help companies properly solve their problems, the solution must take into consideration the culture and condition of the enterprises and the capabilities of information technology held at that time. The solution should also be able to integrate the company's data and systems, take advantage of existing enterprises capabilities, and ensure that the enterprises' resources are ready to transform from the current state to the target state. Based on the requirements statement, the right information technology solution to be applied to enterprises is enterprise architecture. From the existing enterprise architecture frameworks, TOGAF framework is chosen as the framework that will be used in this paper by reason of the completeness of the process, integrated processes, and flexibility. Expected results from the design of enterprise architecture in this paper is the target of enterprise architecture to be achieved, gap analysis between the initial architecture and targets, as well as project roadmap to assist enterprises in achieving the desired target and solve their current problems. In this paper, the design of enterprise architecture by applying TOGAF framework will be done on one of state-owned enterprises (BUMN) in Indonesia.

1. Introduction

The development and utilization of information technology in everyday life continues to grow rapidly within the passage of time. As a result, organizations or companies have many choices in determining the information technology to be applied in their organizations or companies. For state-owned enterprises, the information technology solutions implemented should not only in accordance with the needs of the enterprise but also comply with the government's rules and regulations. The chosen solutions need to pay attention to the culture and condition of the organization and the capabilities of information technology owned at the time, as well as able to integrate the data and system of the enterprise. This is because state-owned enterprises tend to have fragmented data and system spread across different branches which can cause several issues such as missing data or system not being able to be implemented across all branches. Enterprises should also ensure that the resources they currently owned is ready to evolve from the current conditions to the expected conditions. On that basis, the right information technology solution to be applied to state-owned enterprises is the enterprise architecture.

Enterprise architecture is a technology and management practice aimed at developing enterprise performance by helping them see themselves in the context of a holistic and integrated view of their technology resources, information flow, business practices, and strategic direction [1]. Enterprise architecture aims to optimize enterprise's fragmented business processes into an integrated and responsive environment towards changes as well as able to support the delivery of business strategies [2]. There are currently many enterprise architecture frameworks that can be used by organizations or companies, but based on Cameron & McMillan's research on the five most commonly used enterprise architecture frameworks, TOGAF frameworks are rated as far superior compared to other frameworks. The advantages of TOGAF include process completion, TOGAF ADM, flexibility in the use of elements, integration / interconnection between layers, vendor neutrality, and alignment with industry standards [3].

The focus of this paper is designing enterprise architecture on multi-branch state-owned enterprise with different business units using the TOGAF 9.1 framework approach, which includes the use of TOGAF ADM (Architecture Development Model), to integrate currently existing and implemented data and systems in the enterprise.

2. Theoretical Basis

2.1. Enterprise Architecture

Enterprise architecture is a technology and management practice aimed at developing enterprise performance by helping them see themselves in the context of a holistic and integrated view of their technology resources, their information flow, business practices, and strategic direction [1]. The key to enterprise architecture is the iterative process, which includes the process of creating, managing, and, ultimately, utilizing enterprise architecture [4]. Enterprise architecture also analyzes the gap between target state and current state, and provides a road map that supports the enterprise's transformation toward the target state to complete and close the existing gap [5].

The purpose of enterprise architecture is to optimize enterprise's business processes that tend to be fragmented into an integrated and responsive environment towards changes as well as able to support the delivery of business strategies. The benefits of implementing enterprise architecture are as follows [2]:

- More efficient business process
- More efficient information technology process
- Increase in return on investment (ROI) and decrease in investment risk
- Faster, simpler, and cheaper procurement

2.2. Enterprise Architecture Framework

There are currently many enterprise architecture frameworks that can be used by organizations or companies according to their needs. The four most widely used frameworks in the industry today are Zachman Framework, Gartner Framework, Federal Enterprise Architecture (FEA), and The Open Group Architecture Framework (TOGAF) [4]. However, based on the results of research conducted by Cameron & McMillan on some of the most commonly used enterprise architecture frameworks, TOGAF framework is considered as far superior compared to other frameworks. The results of this comparison can be seen in Table 1[5].

Table 1. Comparison of Enterprise Architecture Framework

Attributes/Criterias	Zachman	TOGAF	DoDAF	FEA	Gartner
Business alignment with information technology / business focus	1	3	1	1	4
Taxonomy guide	4	2	2	3	1
Reference model	1	3	2	4	1
Completeness of process	1	4	1	2	3
Rating of maturity	1	2	2	3	3
Governance support	1	2	3	3	3
Interoperability / flexibility	2	4	3	3	2
Knowledge repository / information availability	2	4	2	2	1
Standards (architecture, industry, government)	2	4	3	3	1
Best of breed / Best fit	2	4	2	3	1
Integration / Connectivity between layers	3	4	2	3	2
Business alignment with information technology / business focus	2	4	2	3	1

The advantages of TOGAF include process completion, TOGAF ADM, flexibility in the use of elements, integration / interconnection between layers, vendor neutrality, and alignment with industry standards. Based on these reasons, the enterprise architecture framework to be used in this research is TOGAF framework. The core of TOGAF itself is the TOGAF ADM (Architecture Development Method), where TOGAF ADM presents iterative and tested processes to develop the architecture. The processes are divided into different sections and phases as shown in Figure 1.

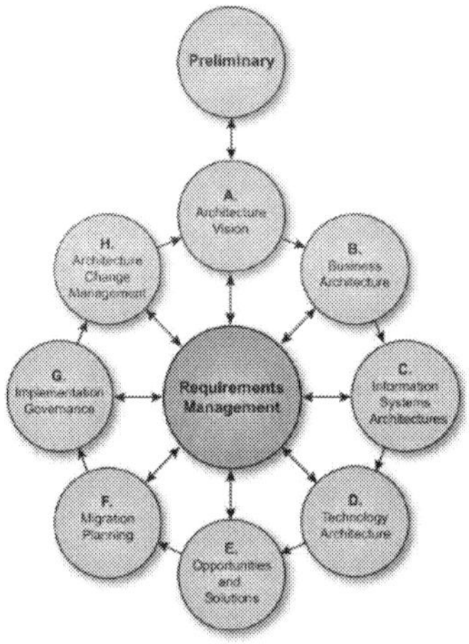

Figure 1. TOGAF ADM's Iterations and Phases

Here are the sections and phases in TOGAF ADM [2].

- Architecture Context
  - Preliminary Phase
  - Phase A: Architecture Vision
- Architecture Development
  - Phase B: Business Architecture
  - Phase C: Information Systems Architecture
  - Phase D: Technology Architecture
- Transition Plan
  - Phase E: Opportunities and Solutions
  - Phase F: Migration Planning
- Architecture Governance
  - Phase G: Implementation Governance
  - Phase H: Architecture Change Management
- Requirements Management

2.3. Related Researches

The explanation of several studies that have been conducted related to the company architecture and its application can be seen in Table 2.

Table 2. Related Researches

Research Title	Author(s)	Research Focus	Lack in Research
Approach in Designing Effective Enterprise Architecture [6]	Glissmann & Sanz, 2011	Propose approach to make effective enterprise architecture. The approach discussed in this study can be combined with other	It has never been tested in real situations where existing enterprise architecturestandards (such as TOGAF or Zachman) have been tailored to the specifications of the companies concerned

IOP Conf. Series: Materials Science and Engineering **403** (2018) 012065 doi:10.1088/1757-899X/403/1/012065

		enterprise architecture standards	
Designing Enterprise Architecture to Ensure Business and IT Strategic Alignment (SAMM Integration with TOGAF 9.1) [7]	Novianto Budi Kurniawan; Suhardi, 2013	Providing solutions for designing enterprise architectures that can provide assurance for business and IT strategic alignment designed through the integration of SAM components with TOGAF 9.1 metamodel	The results are still too generic and testing is not carried out through clear case studies, more about the selection and explanation of each artifact
Enterprise Architecture for Development of Cloud-Based ERP System [8]	Ardian Indra Gunawan, 2014	Application and use of enterprise architecture to manage enterprise IT resources with a major focus on cloud-based ERP adaptation and development with multi-tennant concepts	Evaluation of the application of enterprise architecture implemented without clear or previously-defined samples and references
Designing Enterprise Architecture for Application of Integrated Information System at Multi-Branches Organization [9]	Dilla Anindita, 2016	Designing an enterprise architecture model suitable for multi-branches organizations that can be used to integrate information systems within the organization	The scope of research does not arrive at the implementation stage of enterprise architecture, only to the D (Technology Architecture) phase of the TOGAF 9.1 framework. In addition, no research has been done to verify whether the developed enterprise architecture method can be used for other enterprise scopes other than multi-branches hospitals
Analysis on Recent Trends Related to Enterprise Architecture Frameworks [3]	Hilman Ramadhan, 2016	Studies related to enterprise architecture frameworks that are most often applied to organizations as well as what attributes are required of an organization to ensure that the enterprise architecture designed meet their needs	The scope of the study does not extend to the framework-making stage or methodology that can assist the organization in defining criteria and guidelines for creating and utilizing the hybrid framework of enterprise architecture

3. Analysis of Design Method

3.1. Enterprise Architecture Requirements Analysis

To ensure that the enterprise architecture designed can help enterprises meet their strategic needs and objectives, the enterprise architecture requirements need to be determined. These requirements will be used as references while designing the enterprise architecture. With reference to the definition of enterprise architecture described in the Theoretical Basis section, the enterprise architecture requirements for the targeted enterprise are as follows.

- *Define the enterprise's IT and IS capabilities* [1] [2]
- *Ensure all branch offices have implemented standards and use predefined applications* [1] [2] [5]
- *Integrate the enterprise's data and systems as a whole* [1]
- *Carry out readiness assessment regarding the application of enterprise architecture for the enterprise* [5]
- *Determine the best approach to develop the architecture while considering the current and target conditions* [1] [2]
- *Minimize current constraints* [1] [2]
- *Minimize the infrastructure gap between branch offices* [1] [2]
- *Develop project roadmap to support the company's transformation to meet the target enterprise architecture* [5]

3.2. Determine the Scope of the Enterprise Architecture Framework
As described in sub-chapter 2.2, the design of the enterprise architecture model for the targeted enterprise will utilize the TOGAF 9.1 framework. The method adopted from TOGAF 9.1 and used in the design of enterprise architecture model for this paper is the TOGAF ADM method. While taking the role and needs of enterprise architecture described in sub-chapter 3.1 into account, the scope of TOGAF ADM framework can be determined starting from Preliminary Phase, followed by Phase A: Architecture Vision, Phase B: Business Architecture, Phase C: Information Systems Architecture, and Phase D: Technology Architecture.

3.3. Enterprise Architecture Design Method
Once the scope of the enterprise architecture framework to be used is determined, the next step is to describe the steps of the design method itself. The goal is to make sure that the scope and focus of the development of enterprise architecture has a clear focus and deliverables that are also in accordance with the needs of the company. The steps of the enterprise architecture design method established for this paper can be seen in the diagram in Figure 2.

Figure 2. Enterprise Architecture Design Method

The explanation of the enterprise architecture design method is as follows.

- Architecture Initialization
  The focus of this phase is to define the current state of the enterprise and the principles of enterprise architecture implementation for the enterprise. This phase refers to the Preliminary Phase of TOGAF ADM. Activities undertaken in this phase include identifying organization profile; identifying organization scope, capabilities, and constraints; identifying architecture principles; and identifying business goals and business drivers.
- Architecture Vision
  The focus of this phase is to define the purpose of implementing enterprise architecture for the enterprise and to assess the company's readiness in transforming to achieve the target enterprise architecture. This phase refers to Phase A: Architecture Vision of TOGAF ADM. Activities undertaken in this phase include identifying of architecture vision; assessing current business capabilities; and business transformation readiness assessment.
- Baseline Enterprise Architecture
  The focus of this phase is to define the current state of business, information systems, and technology architecture of the enterprise. This phase refers to Phase B: Business Architecture, Phase C: Information Systems Architecture, and Phase D: Technology Architecture of TOGAF ADM. Activities undertaken in this phase include defining the baseline business architecture; defining the baseline data architecture; defining the baseline application architecture; and defining the baseline technology architecture.

- Target Enterprise Architecture
  The focus of this phase is to define the target state of business, information systems, and technology architecture of the enterprise. This phase refers to Phase B: Business Architecture, Phase C: Information Systems Architecture, and Phase D: Technology Architecture of TOGAF ADM. Activities undertaken in this phase include defining the target business architecture; defining the target data architecture; defining the target application architecture; and defining the target technology architecture.
- Gap Analysis
  The focus of this phase is to analyze the gaps on each architecture layers. This phase refers to Phase B: Business Architecture, Phase C: Information Systems Architecture, and Phase D: Technology Architecture of TOGAF ADM. Activities undertaken in this phase include analyzing gaps between baseline and target business architectures; analyzing gaps between baseline and target data architectures; analyzing gaps between baseline and target application architectures; and analyzing gaps between baseline and target technology architectures.
- Roadmap Components Recommendation
  The focus of this phase is to determine the roadmap components for the development of enterprise architecture based on the gap analysis results. This phase refers to Phase B: Business Architecture, Phase C: Information Systems Architecture, and Phase D: Technology Architecture of TOGAF ADM. Activities undertaken in this phase include determining the projects to be undertaken during the development of enterprise architecture; creating a priority order for the projects; and creating a migration plan.

4. Enterprise Architecture Design Results

4.1. Enterprise Architecture Principles

Based on the TOGAF 9.1 framework, the definition of architecture principles is a set of principles related to architectural work [2]. Architecture principles manage the architecture process, impacting the development, management, and use of enterprise architecture. The architectural principles will be discussed in accordance with the four domains of enterprise architecture, namely business principles, data, applications, and technology. Detailed principles can be seen in Table 3.

Table 3. Enterprise Architecture Principles

Principle Category	Principle Name
Business Principle	Business continuity
	Compliance with the law
	Provide benefits to the company
	IT responsibilities
Data Principle	Data is an asset
	Integrated data between branches
	Data is accessible
	Data security
Application Principle	Application is reliable
	Application is cost-effective
	Application is reliable
Technology Principle	Interoperability
	Responsive change management

4.2. Architecture Vision

The architecture vision can be determined based on the current state of the company as well as the company's mission, business objectives, business drivers, and business constraints. Figure 3 explains the origin of the architectural vision more clearly.

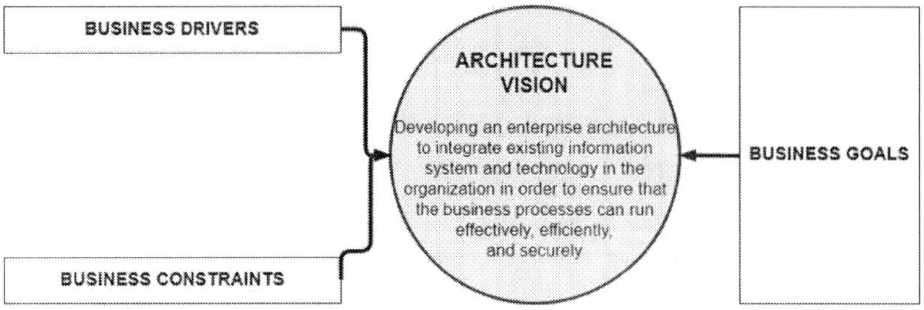

Figure 3. Architecture Vision

4.3. Results of Business Transformation Readiness Assessment

One of the keys to successful architecture transformation is to understand the enterprise's readiness to accept changes, identify issues at hand, and solve them. The goal of the assessment is to evaluate the conformity of the enterprise architecture that has been developed. This activity is called Business Transformation Readiness Assessment. The method suggested by the TOGAF 9.1 framework to assess the organization's readiness is the Business Transformation Enablement Program (BTEP) in accordance with Phase A: Architecture Vision of TOGAF ADM [2].

Readiness factors for the readiness assessment are determined based in the BTEP method. Readiness factors along with the results of the assessment conducted on the targeted enterprise can be viewed in Table 4.

Table 4. Business Transformation Readiness Assessment

Factor	Assessment	Description
Vision	Good	The enterprise has a clear architecture development vision which is in accordance with the needs and capabilities of the enterprise
Desire, Willingness, and Resolve	Good	The enterprise has a strong desire and commitment in executing and completing the enterprise architecture development project and is ready to face the impact of the project development
Need	Good	The enterprise has clear reasons and needs related to the enterprise architecture development project
Business Case	Good	The enterprise architecture development project for the enterprise is supported by the problems, needs, and business opportunities that is currently faced by the enterprise
Funding	Good	The enterprise has a clear source of funding to meet the expenditure and procurement of the enterprise architecture development project
Sponsorship and Leadership	Enough	The enterprise's executives provide the support needed for the project
Governance	Enough	The enterprise involves interested parties to the project with the purpose of ensuring that the company's interests and its objectives are met
Accountability	Enough	The presence of assigned specific and appropriate responsibilities to human resources that have an appropriate role to develop, execute, and implement the results of enterprise architecture development projects across branches
Workable Approach and Execution Model	Bad	The enterprise has a reference in project modeling and uses a reasonable and relative approach for the task
IT Capacity to Execute	Enough	The enterprise have IT resources and can run IT processes smoothly
Enterprise Capacity to Execute	Enough	The enterprise have resources outside of the IT area to do all the work required during the project

Enterprise Ability to Implement and Operate	Enough	The enterprise may accept and adjust the changes that occur as a result of project implementation

Based on the assessment results, the targeted enterprise is considered quite ready in carrying out business transformation.

4.4. Target Enterprise Architecture

The discussion regarding target enterprise architecture can be reviewed by understanding the description of target business, information systems, and target architectures as the realization of the architecture vision

- *Target Business Architecture*
  The realization of architecture vision in the context of business architecture is implemented through improving the effectiveness and efficiency of business processes, the adjustment of business actors and their roles, and the addition of business activities related to decision making on existing business processes and functions. Issues that exist in the baseline business architecture of the targeted enterprise will be solved in the target business architecture. The issue and its solutions are as follows:
  - Data issues that have not been well validated in the baseline business architecture will be solved by the addition of specific business functionality to evaluate data in the target business architecture
  - The issue of the absence of business processes to support the socialization of services provided by the targeted enterprise in the baseline business architecture will be solved by the addition of a marketing division in the target business architecture
  - The issue of the absence of formal-defined activities to manage the organization's data center in the baseline business architecture will be solved with the addition of specific business functionality to maintain in the target business architecture
- *Target Data Architecture*
  Target data architecture can be mapped by understanding the target picture of the data architecture and the target relationship between data architecture and business architecture. Issues that exist in the baseline data architecture of the targeted enterprise will be solved in the target data architecture. The issue and its solutions are as follows:
  - The issue of the absence of specific data entities related to data center maintenance in the baseline data architecture will be solved by the addition of related data entities in the target data architecture
  - The issue of the absence of specific data entities to support the socialization of services in the baseline data architecture will be solved by the addition of related data entities in the target data architecture
- *Target Application Architecture*
  Target application architecture can be mapped by understanding the target application structure and how target applications are connected to each other. Issues that exist in the baseline application architecture of the enterprise will be solved in the target application architecture. The issue and its solutions are as follows:
  - The issue of lack of applications to support business processes in the baseline application architecture will be solved by adding several business applications to support the organization's business processes in the target application architecture
  - The issue that not all applications are integrated in the baseline application architecture will be solved by integrating all business applications so that they can be centrally accessed in the target application architecture
- *Target Technology Architecture*
  Issues that exist in the baseline technology architecture of the targeted enterprise are mostly related to the quality of existing information technology infrastructure as well as the security of information systems. Based on these findings, the target technology architecture for the destination organization has several focuses:
  - Evaluate the quality of information technology infrastructure so as to improve the quality of infrastructure owned by the organization
  - Develop backup system for the enterprise's information system as a whole, unlike current backup system which only focused on some parts of the system

- Improve data center maintenance and governance through the addition of formal-defined business functions related to the data center

4.5. Gap Analysis

Gap analysis between the baseline and target enterprise architectures is done on each architecture layers to ensure that all components are properly mapped. The results of gap analysis that is done on the targeted enterprise can be seen on Table 5. The results of this analysis will be used as a reference in determining the projects for migration plan and project roadmap.

Table 5. Results of Gap Analysis

Gap Title	Related Architecture Layer
Addition of Data Evaluation business functions	Business, Technology
Addition of Data Center Maintenance business function	Business, Data, Technology
Addition of Marketing division	Business Data
Development of business applications	Application
Integration of business applications	Application
Evaluate the security quality of information technology infrastructure	Technology
Development of backup system for whole information system	Technology

4.6 Projects Recommendation as Roadmap Components

Roadmap components for the targeted enterprise are determined based on the main gaps. The results of mapping the main gaps into roadmap components can be seen in Table 6.

Table 6. Mapping Gaps to Projects

Project Title	Related Gap
Business Application Development Project	Development of business applications
Business Application Integration Project	Integration of business applications
Marketing Division Realization Project	Addition of Marketing division
Business Function of Data Center Maintenance Realization Project	Addition of Data Center Maintenance business function
Business Function of Data Evaluation Realization Project	Addition of Data Evaluation business functions
Information Technology Infrastructures Quality Evaluation Project	Evaluate the security quality of information technology infrastructure
Backup System Development for Overall Information System Project	Development of backup system for whole information system

4.7. Migration Plan

Preferred projects are then sorted based on their priorities before being mapped into the roadmap components. The priority mapping can be carried out by taking into account of several factors such as the results of previous business transformation readiness assessment and the dependencies between each project. The list of projects to be worked one based on their priorities can be seen in Table 7.

Table 7. Priority of Projects to be Worked on

Project Title	Priority
Business Application Development Project	High
Business Application Integration Project	High
Marketing Division Realization Project	Medium
Business Function of Data Center Maintenance Realization Project	Medium
Business Function of Data Evaluation Realization Project	High
Information Technology Infrastructures Quality Evaluation Project	Medium
Backup System Development for Overall Information System Project	Medium

Based on the result of priority determination, the sequence of the proposed project work in sequence is as follows

- Business Application Development Project
- Business Application Integration Project
- Business Function of Data Evaluation Realization Project
- Business Function of Data Center Maintenance Realization Project
- Backup System Development for Overall Information System Project
- Information Technology Infrastructures Quality Evaluation Project
- Marketing Division Realization Project

After determining these priorities, a migration plan detailing the implementation of each project including the year of which the implementation will take place. The migration plan of the objective organization in this study can be seen in Figure 4.

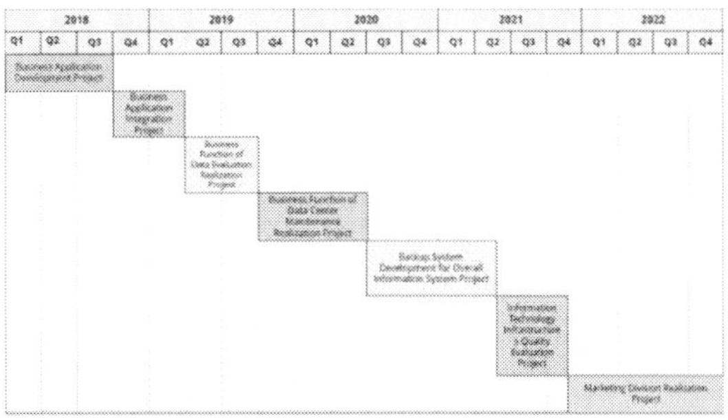

Figure 4. Migration Plan for the Targeted Enterprise

5. Conclusions

In designing and developing enterprise architecture that suits the needs of the state-ownedenterprise, there are several things to be considered including the business objectives, business drivers, business constraints of the company. In addition, the enterprise should also take into account the scope of the enterprise architecture design itself as well as the scope of application of the enterprise architecture framework to be used. Key components required in designing enterprise architecture for enterprises are the components of the four major domains of architecture, ie business, data, application, and technology, with a primary focus on,

- Defining and mapping the enterprise's capabilities,
- Integrating data and system as a whole,
- Assessing the company's readiness to transform, and
- Planning the transition

The application of enterprise architecture will affect the enterprise's business processes, especially in realizing the company's values in the form of well-functioned business processes and functions, data quality improvement, realization of integrated information systems, and evaluation and improvement of information technology infrastructure.

References

[1] S. A. Bernard, An Introduction to Enterprise Architecture, 3rd ed., Bloomington: AuthorHouse, 2012.
[2] The Open Group, TOGAF Version 9.1, The Open Group, 2011.
[3] B. H. Cameron and E. McMillan, "Analyzing the Current Trends in Enterprise Architecture Frameworks," Journal of Enterprise Architecture, 2013.
[4] R. Sessions, "A Comparison of the Top Four Enterprise-Architecture Methodologies," May 2007. [Online]. Available: https://msdn.microsoft.com/en-us/library/bb466232.aspx. [Accessed 14 November 2016].
[5] R. S. Bittler and G. Kreizman, Gartner Enterprise Architecture Process: Evolution 2005, Gartner, Inc., 2005.
[6] S. M. Glissmann and J. Sanz, "An Approach to Building Effective Enterprise Architectures," 2011.
[7] Novianto Budi Kurniawan; Suhardi, "Enterprise Architecture Design for Ensuring Strategic Business IT

Alignment (Integrating SAMM with TOGAF 9.1)," Bandung-Bali, 2013.
[8] Ardian Indra Gunawan, "Enterprise Architecture for Cloud-based ERP System Development," 2014.
[9] Dilla Anindita, "Perancangan Arsitektur Enterprise untuk Penerapan Sistem Informasi Terintegrasi Pada Organisasi Multicabang (Studi Kasus: Rumah Sakit Permata)," 2016.

Design of flood detection system based on velocity and water level sensor in Arduino with SWOD application on Kalimati-Kretek Gantung DAM Banyuwangi

H Yuliandoko[1], Subono[2], V A Wardhany[3], S H Pramono[4] and P Siwindarto[5]

[1,2,3]Informatic Engineering Dept. Politeknik Negeri Banyuwangi,
Banyuwangi – East Java, Indonesia
[4,5]Electrical Engineering Dept. Brawijaya University
Malang – East Java, Indonesia

[1,2,3]{herman.yuliandoko, subono, vivien.wardhany}@poliwangi.ac.id
[4,5]{sholehpramono, ponco}@ub.ac.id

Abstract. As a tropical country Indonesia has the potential to experience rain in several periods due to global warming of the weather becomes erratic resulting in extreme weather. With high rainfall intensity, as well as lack of absorption land can result in water volume in a dam can exceed capacity and overflow. During this observation process of increasing the volume of water using conventional method is considered less effective when viewed in terms of operational by the operator must come directly to the field at a certain period. If the condition of the water volume increases suddenly and the officer is not located then it will cause flood disaster therefore required a proper technology so as to overcome the problem. SWOD (*Safety Water Overflow Detection*) is an application that integrates with web server and sensor device as data taker able to provide real time volume and water speed information from dams through mobile device and comes with warning when reaching emergency condition.

Keywords: flood disaster; real-time; sensory device; SWOD

1. Introduction

Flood disasters always causing harm to humans, both from material and even casualties, and the impact of ecosystem changes, temporary or permanent. Floods that occur due to extreme weather resulting in high rainfall intensity, causing an abundance of water in the land that is marked by rising river water level. Water Monitoring applications utilize fast and accurate information delivery using server computers as containers of information so that handling becomes quick to open the floodgates. In the conventional monitoring method usually Forman Officer of Dam has to standby 24 hours to monitor and detect water condition every day. If there is any dangerous situation and Forman Officer is not in the location, it can be a big problem for disaster information handling.

The results of this water monitoring will be send as information for water management so that the water level is suitable for river capacity. Forman Officer of Dam will open the floodgates when the water level reaches the safe zone. This is one of the efforts to anticipate flooding due to increased water volume. To facilitate the processing of water level monitoring information the authors want to build a water quality information monitoring system that can provide information related to the status of water level in real time and connected with the application of Water Safety Overflow Detection (SWOD) so that anticipation and early warning of flooding can be done more Fast and efficient.

2. Related Works

2.1. Disaster Alert and Notification System Via Android Mobile Phone by Using Google Map

The system consists of a third-party server named Disaster Management Server (DMS), android device on which our application installed and user. Updates of the disaster (tsunami, cyclone or flood) are put on DMS by the local weather office. To get automatic notification of upcoming disaster device user registers on Disaster Management Server (DMS) else user can gets manual notification. The user keeps in touch with DMS to get most renew data obtained by GSM. The proposed application notifies the user located in possible disaster zone with visual and audio disaster warning and evacuation guideline combine with nearest location of shelter or safe zone on the map of the application. DMS also tracked Evacuation progress and national identification of user [1].

2.2. View of Flood Disaster Management in Indonesia and the Key Solutions

Over the years, Indonesia has seen many flood disasters that have brought about great losses. The aim of this study is to address key issues that lead to flooding problems in Indonesia in response to the challenge of recurrent flood events. An overview of the past flood disaster profiles and ongoing flood management are presented. The problems with the current situation are identified and the critical solutions are recommended to manage flooding and mitigate the negative impact in a more sustainable way. This study shows that man-made factors, natural causes, and managerial issues are the factors that have contributed to the problem. The coordination and the public awareness are the challengesin improving the flood management. Efforts have been made to alleviate the problems through legal framework establishment, community participation programs, and flood-control projects. In the post-disaster stage, the authorities and public have been quite responsive. However, prevention and preparedness are still lacking. The overall current flood disaster management may lead to more recurrent events and cause severe impacts.Sustainable actions are needed to solve these problems that include environment-based flood integrated counter measures,improving water retarding function, eradication of deforestation, meteorological and hydrological prediction, and political will and law enforcement[2].

3. The System Overviews

With the progress of increasingly sophisticated technology can be utilized as a means of delivering information more quickly and efficiently. Internet of Things is one of the potential that can be developed which is a new breakthrough that combines the device with a system through the Internet network, no longer just on computers with internet but android devices (Mobile Phone) connected to the internet can be interconnected, the system we made is consist of 3 different part first the sensor device that build by ultra sonic and water flow sensor and sending data using wemos d1-mini, the second is web server and the third is SWOD android aplication that inform the notification report form the sensor device and web server to the officer so they can warn if there is flooding disaster happend earlier.

- The System Design

In this system there are two factors that play the role of managers as decision makers and PPA Officers (water door officer at DAM) From each of these factors have elements that must be fulfilled, among others, from the Officers PPA has 3 elements that determine the success of this warning that is, Knowledge, Attitudes, and Behavior. The managing party as a decision maker provides a decision based on the condition of monitoring data from the field through a Web Server, which then continues into several categories of early warning that can be directly connected to the body of preparedness such as BPBD (Regional Disaster Management Agency) and also the SAR Team & also local community.

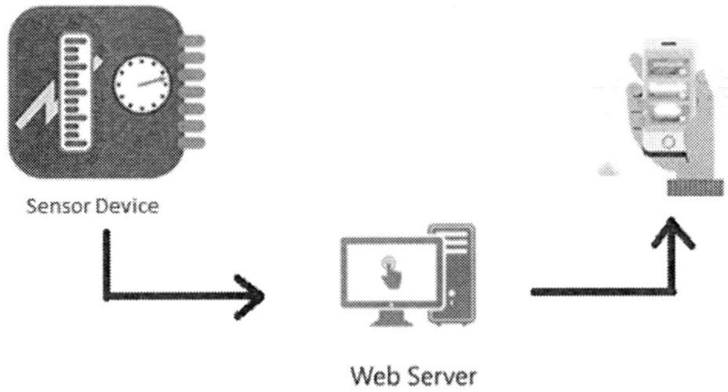

Figure 1. The system Design

As we can see in the Figure 1, the water level detection process uses an additional component of an application via a mobile device. Previously the user must login with an account that has been created by the server. So it can login to the application and receive information provided by the server periodically and quickly. If the water level reaches the safe threshold point, a warning will appear immediately to the point of location. There are details that are accompanied by tips in certain conditions related to the condition when reaching a certain status. In this mobile application added with notification, so without opening the application the user can know the condition of the height and speed of water through the notification received.

- The Rule of Flood Detection

When the officer wants to determining the Status of Flood Disaster Precautions, the Dam Management Party has two criteria such as Water Front and Water Speed. The management of this time is the Public Works Office of Water Resources of Banyuwangi Regency has a scale of height and water velocity with the potential of flood disaster is as in Table 1.

Table 1. The Rules of Flood Preparedness Status

No	Status Category	Water Level	Flow Velocity
1.	Aman	<70 cm	< 61000 L/Sec
2.	Siaga	70 cm s/d 139 cm	61000 L/Sec
3.	Waspada	140 cm s/d 239 cm	122000 L/Sec
4.	Darurat	240 cm s/d > 240 cm	273000 L/Sec

- Ultrasonic sensor

To measure water level in the DAM Kalimati-Kretek Gantung used ultrasonic sensor.

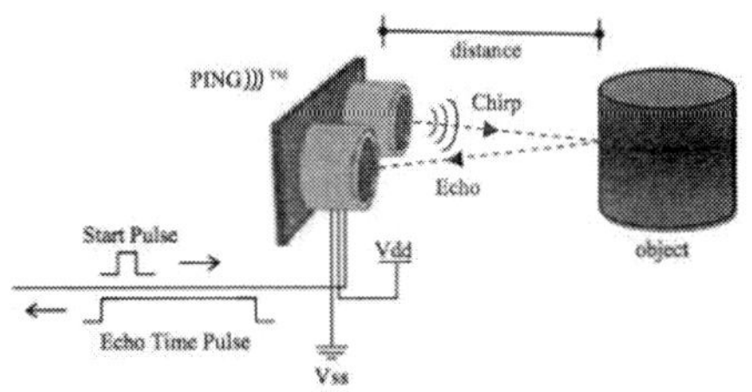

Figure 2. Ultrasonic sensor

As we can see in the Figure 2, the ultrasonic sensor has a transmitter and receiver. This sensor transmits ultrasonic wave for 200 μs and then detect the reflection wave. Ultrasonic wave spread in the air with the velocity about 344 m/s (sound wave velocity), touch the object and reflected back to the sensor [6]. That mechanism is used ultrasonic to measure the distance as equation (1).

$$S = (t_{IN} \times V) \div 2 \quad (1)$$

Where S denotes the distance between ultrasonic with object detected, V indicates sound-wave velocity, and t_{IN} denotes the wave deviation transmiting and receiving time [7].

IOP Conf. Series: Materials Science and Engineering **403** (2018) 012066 doi:10.1088/1757-899X/403/1/012066

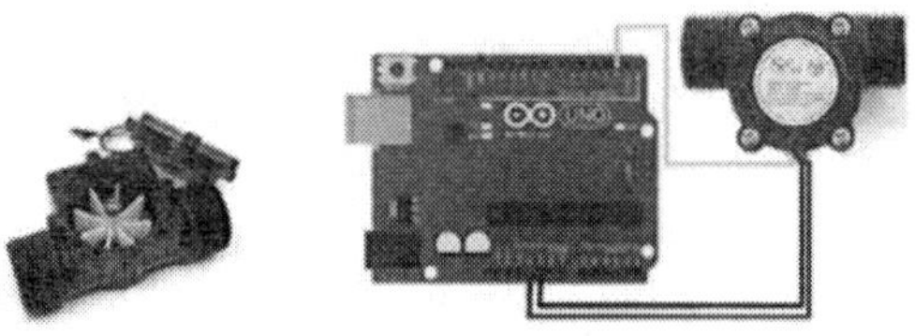

Figure 3. Water flow sensor and arduino

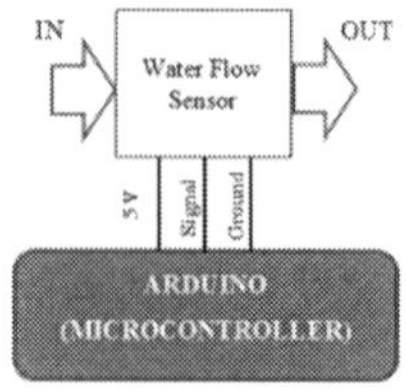

Figure 4. Circuit Diagram

The water speed detection is very significant in the water flooding detection. Accurate flow measurement is an essential step both in the terms of qualitative and economic point of view. Flow meters have proven excellent device for measuring water flow. Water flow sensor and circuit diagram show in Figure 3 and Figure 4. [5]

The position of this sensor is in the line with the water line and inside of sensor contains a pinwheel sensor. The pinwheel sensor has a function to measure how much water has moved through it. In this sensor also has an integrated magnetic Hall-effect sensor with an electrical pulse output every revolution. The rate of water flow will be directly proportional to the number of pulses counted. The pulse signal is converted into liters per minute using the following formula [8].

$$\text{Pulse frequency (Hz)} / 7.5 = \text{flow rate L/min} \quad (2)$$

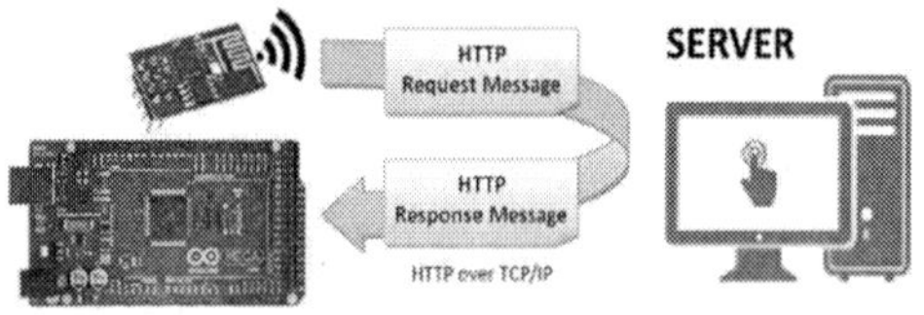

Figure 5. Communication process between The Sensor and Web Server.

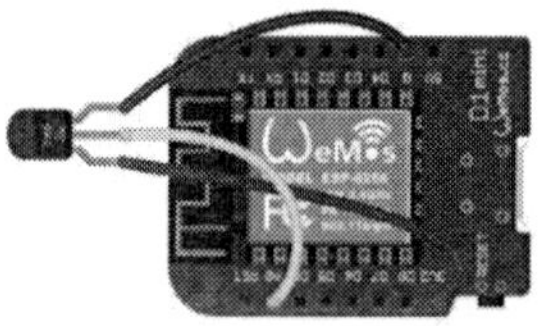

Figure 6. Wemoss d1-mini

As we can see in the Figure 5, The Sensor devices and web server monitoring are interconnected with wireless connection using Wifi shield module. In the process of data transmission or communication between sensors with the web using HTTP protocol and this communication occurs in the form of request and response. To send commands to turn on flood warning alarms that exist in residential areas, the sensor will send a Get-Request command to the web server, while the web when receiving the Get-Request command will respond by sending http-response command. When the water condition is in a normal state between the sensor and the web is still connected.

As we can see in the Figure 6, The Wemos D1 Mini is a mini Wi-Fi device based on ESP8266EX chip. This device is a very compact solution for prototyping small smart objects linked to the World Wide Web, and it have an impressive ESP8266 Wi-Fi functionalities. The sensory device consists of an ultrasonic sensor used to measure the water level, and the waterflow sensor that is used to measure the speed of water flow, while the mini-d-1 wemos are used for water-level data transmission connection via wifi network which also uses wemos d1-mini device.

This research also make an application to inform water flow and water level condition. This information send to BPPD and local community which use this application. If there is critical condition SWOD application will make alert to the people. SWOD application flow shown in Figure 7.

IOP Conf. Series: Materials Science and Engineering **403** (2018) 012066 doi:10.1088/1757-899X/403/1/012066

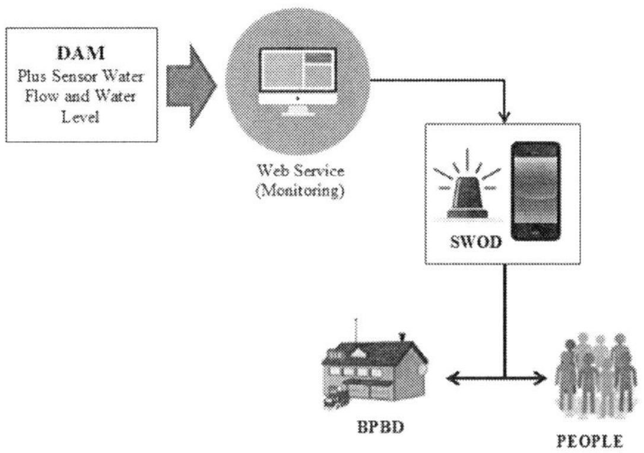

Figure 7. SWOD application flow

3.1. The Experiment

The results of the data monitoring and SWOD aplication will be displayed in the web information systems and specially for android application will be explain in the text below:

- Data Monitoring of DAM Kalimati and Kretek Gantung.

From this system we get the result of data measurement during the time we observed and the level status of DAM kalimati & DAM Kreteg gantung in each our of observation can be seen in Figure 8 and Figure 9.

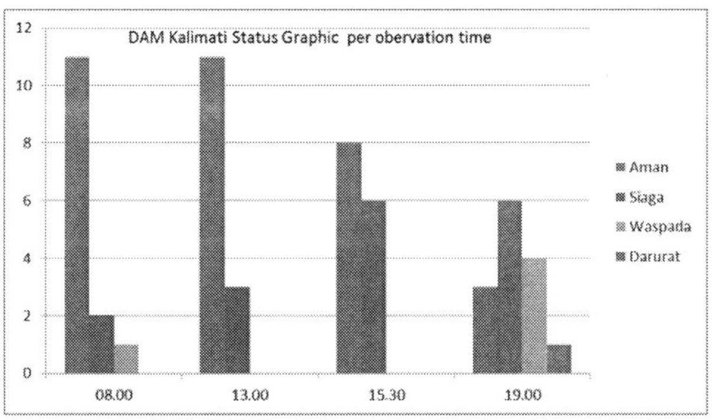

Figure 8. Kalimati DAM

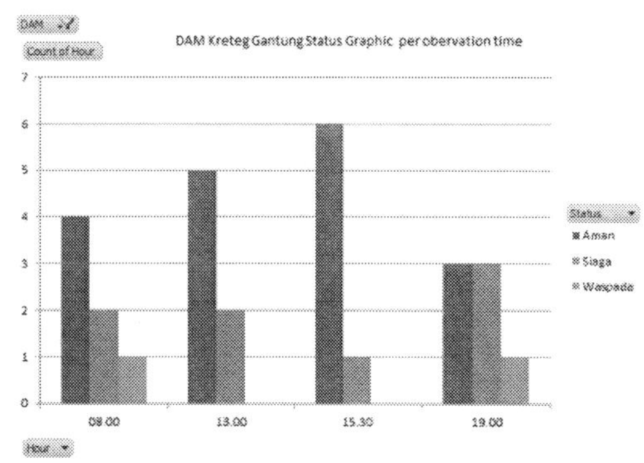

Figure 9. Kretek Gantung DAM

From each observation hour in graphic shown that DAM Kalimati is more dangerous than DAM Kretek Gantung, especially on night, which had "Darurat" status level.

- SWOD Aplication.

The SWOD application is an early warning system (EWS) that is used to monitor the water level in a dam from a distance without the need to come directly to the location of the weir or irrigation building. This app contains altitude information sent from the dam based on the sensor, as well as the change graph that has been completed with the safe boundary status of the dam conditions. In this application is also equipped with tips on handling information ranging from Safe to in case of emergency. This app can provide a notification if things start to standby until emergency even if not continuously in open.

Figure App Login View, in this view the user who can login is only those who have registered as a user in the head office that has been adjusted with the weir or buildings in the monitor. If unregistered then the user cannot see and receive notifications.

(a) (b) (c) (d)

Figure 10. (a)(b) SWOD App Login View; (c) Home View; (d) Emergency Call View

As We can see in Figure 10, the home view in the immediate live view contains real time water status information including the time of change. In this view there is also a safe limit so that when passing through the red line officers can more easily understand the graph, the status can be viewed on a weekly and monthly basis in Figure 10 (c). Emergency calls can be done quickly using the emergency call button connected to the BPBD (Badan Penanggulangan Bencana Daerah) for Kabupaten Banyuwangi.

4. Conclusion

Refer to above data show that application can be used as a solution for real time water level monitoring process along with flood alert status. The system also easy to use and the information of system is very useful for people to prepare any condition from disaster effect. This application is made for users on each dam to facilitate the monitoring process associated with the maximum water heights that potentially flood disaster in each Technical Services Unit (UPT).

5. Acknowledgment

This research supported by LAB Nirkabel - State Polytechnic of Banyuwangi, LAB Elektronika - Brawijaya University and Ristekdikti in PEKERTI 2017 (Penelitian Kerjasama Antar Perguruan Tinggi).

6. References

[1] Gosavi. Amit, Vishnu. S.S. "Disaster Alert and Notification System Via Android Mobile Phone by Using Google Map," IJETAE ISSN 2250-2459,ISO 9001:2008 Certified Journal,Volume 4, Issue 11, November 2014.

[2] HapsarI Ratih Indri, Zenurianto Mohammad, "View of Flood Disaster Management in Indonesia and the Key Solutions", American Journal of Engineering Research (AJER), e-ISSN: 2320-0847 p-ISSN : 2320-0936 Volume-5, Issue-3, pp-140-151, 2016.

[3] Soebroto Arief Andy, Soekotjo D, Suhartanto Ery, "Prototype of the Real Time Decision Support System for Flood Early Warning at Brantas River Basin", International Journal of Engineering Innovation & Research, Volume 2, Issue 3, ISSN: 2277 – 5668, 2013.

[4] Made Saraswati, Endrowednes Kuantama, Pono Mardjoko, "Design and Contructio of water measurement system accesible through SMS", UKSim-AMSS 6th European Modelling Symposium IEEE DOI 10.1109/EMS.2012.60 pp 46-51, 2012.

[5] Ravi Hosamani, Ravi Bagade, "Arduino Based Water Billing System for Domestic Purpose", Internatinal Journal of Modern Trends in Engineering and Research, ISSN(online): 2349-9745, Vol.2, Issue 06, 2015.

[6] Ping ultrasonic distance sensor (online), 2017. Available in https://www.parallax.com/sites/default/files/downloads/28015-PING-Sensor-Product Guide-v2.0.pdf

[7] Kiki Prawiroredjo, Nyssa Asteria, "Detektor Jarak Dengan Sensor Ultrasonic Berbasis Mikrokontroller", JETRI, Vol 7, ISSN 1412-0372,2008

[8] Godhini Prathyusha, B.Ramamurth, "Development of arduino based flow control system implementing fuzzy with PID controller", International Journal of Innovative Research in Computer and Communication Engineering, Vol.3 ISSN : 2320-9801, 2015.

Named entity recognition model for Indonesian tweet using CRF classifier

Y Munarko, M S Sutrisno, W A I Mahardika, I Nuryasin and Y Azhar

Teknik Informatika, Universitas Muhammadiyah Malang

yuda@umm.ac.id

Abstract. Named Entity Recognition (NER) is a part of Natural Language Processing (NLP) that acts to recognize the existing word entity in the document. By using NER, it is possible to perform activities such as information extraction and text summary. One of the data sources for the NLP process is tweets which are real time, occurred frequently, but limited by the number of words per tweet. In Indonesia, twitter is one of the most popular social media with various topics, so, it is necessary to provide models, train data, and test data for Indonesian tweet. In this study, the models were built using Conditional Random Field classification from 8,000 tweets that have been grouped to formal tweets and informal tweets. By testing the models to 2,000 training data, it provided recall and precision results of 62% and 87% respectively for formal tweets, 36% and 90% respectively for informal tweets, and 60% and 86% respectively for mixed tweets. These results indicate that the created Indonesian tweet models can be used for automatic NER.

1. Introduction

Twitter is one of the most popular social media in Indonesia with a large number of users. As a micro-blogger, Twitter provides users with the flexibility to tweet a maximum length of 140 characters. This is quite interesting because users tend to share information frequently with a typical sentence structure which is different with common text documents. Information on Twitter is very useful when extracted appropriately, for example, the information extraction on earthquake detection and sentiment analysis in real time [1-3]. For information extraction needs, Named Entity Recognition (NER) plays a very important role. Early NER research generally used documents in English corpus or traditional text documents, but later research concerned on various corpora, such as non-English language, tweets, user review, and biomedical domain [4-10]. Algorithms used for NER are varied but can be classified into two types, those are machine learning based and rule-based. The example of machine learning based is the use of Conditional Random Fields (CRF) [4], then the example of rule-based is MUSE that utilized GATE [11].

In general, machine learning based such as CRF provides good performance almost for all types of corpora. CRF is based on sequence probability that used to build a model. The model, then, is utilized to detect entities automatically. Different with machine learning based, rule-based is a kind of tailor-made solution which usually only work for a particular corpus. The rule is built by analyzing the nature of the corpus and then employs various features such as part of speech, context, and the lexical word [12].

For the Indonesian language, there were several studies on NER, including InNer [12], the use of association rule [13, 14], and the use of supervised learning [15]. Those studies already shown a promising results with high precision and recall values. However, the data used were only standard documents which were using standard sentence structure and standard words. So, there is no evident that those studies may works well on Twitter which uses non-standard sentence structure and non-standard word. Moreover, the rule-based approach may experience suffer performance when apply for a different corpus domain.

Study about NER for microblog already conduct by several scholars, such as Twiner [8] and supervised learning approach [16]. Those studies were concerning on a domain specific corpus. Twinner target was a stream of tweets in English which firstly tokenized into phrases and then each phrase classified to a suitable entity. The second study was tried to detect words' entity for the tweet about labor strike event in Indonesia.

In this research, we built a NER model that can be used to recognize entities in Indonesia tweets automatically using various corpus, so it can be used as a general model for any kind of corpus. We were also utilizing CRF as a machine learning based solution, which suitable for the non-domain specific corpus.

2. Research method

The research was conducted by several steps, begins by data collection, and then follows by data classification, data preprocessing, entity tagging, model development, and testing.

2.1. Data collection

Data collection was conducted over two months from several popular general users, news agencies, and government agencies. The purpose of choosing a tweet based on the type of user is to get a collection of formal and informal tweets. Popular general users usually make informal tweets, while news agencies and government agencies tend to make tweets with more formal grammar and formal word selection. Tweets collected as many as 10,000 tweets that will be divided into train data and test data, and formal tweet and informal tweet.

2.2. Data classification

For testing purposes, the tweet was divided into two datasets, i.e. train data, and test data. Train data is used to build a model, while test data is used to perform model testing. The two datasets are then grouped by linguists into two groups. The first group contains tweets that use formal Indonesian, while the second group contains tweets that use informal Indonesian. The classification of data is described in table 1.

Table 1. The data used.

Type of data	Indonesian Tweets		#tweets
	Formal	Informal	
Train data	3,814	4,186	8,000
Test data	1,000	1,000	2,000
Total			10.000

2.3. Data processing

Before tagging entities manually, tweets are first preprocessed. Standard preprocessing is to convert characters into lower case and tokenizing. However, these simple preprocessing processes become quite complicated on twitter. For example, the tokenizing process cannot be based on white space only, because there are some words that are not separated by whitespace. To simplify, in this study, the tokenizing process was performed manually. Furthermore, the characters in a tweet are not just letters, such as emoticons, so, non-letter characters were left as is. The entity tagset is shown in table 2.

Table 2. Entity tagset.

#	Tag	Description	Example
1	Person	Name of person	Ahok, Ridwan Kamil
2	Location	Name of location	Jakarta, rumah, teras
3	Organization	Name of organization	BMKG, BNI, BNN
4	O	Other	yang, di, ke

2.4. Entity tagging

Tagging entities for train data and test data were done manually against 10,000 tweets. As can be seen in table 2, there are four entity tag, Person, Location, Organization, and O. All tokens are classified into one of the four entities in table 2. Examples of manual tagging results are presented in table 3.

Table 3. Tweet examples.

#	Tweets	Tweet type	Tagged tweets
1	Masihkah Madrid Merasa Kehilangan Di Maria ?	Formal	Masihkah/O Madrid/ORGANIZATION Merasa/O Kehilangan/O Di/PERSON Maria/PERSON ?/O
2	Densus 88 Tangkap Warga Klaten , Sita Motor dan Buku Bertema Radikalisme https://t.co/lnY8xs6VRF	Formal	Densus/ORGANIZATION 88/ORGANIZATION Tangkap/O Warga/O Klaten/LOCATION ,/O Sita/O Motor/O dan/O Buku/O Bertema/O Radikalisme/O https://t.co/lnY8xs6VRF/O
3	Nyeblak (Evisahara, Avis Yuni)	Informal	Nyeblak/O (/O Evisahara/PERSON ,/O Avis/PERSON Yuni/PERSON)/O
4	3 in 1 bakal dihapus brooo https://t.co/eIsdguQnXO	Informal	3/O in/O 1/O bakal/O dihapus/O brooo/PERSON https://t.co/eIsdguQnXO/O

2.5. Model development

The tagged train data was then processed using CRF classifier, resulting three models. In order to generate these model, we were utilizing Stanford NER which already implemented CRF classifier [4]. The first model is based on format tweets, the second model is based on informal tweets, and the last one is the combination of both formal and informal tweets.

2.6. Testing

The test was performed to measure the recall and the precision values. The recall is the comparison between the token that correctly identified, I_c, and the number of tokens that should be correctly identified, I_r, shown as equation (1). Then, the precision is the comparison between the token that correctly identified, I_c, and the number of tokens that identified as correct, I_{db}, shown as equation (2).

$$Precision = \frac{I_c}{I_r} \tag{1}$$

$$Precision = \frac{I_c}{I_{db}} \tag{2}$$

3. Result and discussion

Each model was tested for three different test data, formal, informal and mixed. The purpose of these tests was to see the significance of classifying tweets to formal or informal classes. In addition, precision and recall measurements were performed with ten-fold cross-validation to determine the performance of the model when it does not consider formal and informal classification. The results of the tests are shown in table 4.

We found that the precision value for the test of all test data using all models is of high value. However, selecting the right type of model against test data will result in better performance. For example, the highest precision value for formal test data was obtained by using a formal model, as well as for informal test data and mixed test data. For those tests, the precision values are 0.8760, 0.9076 and 0.8749 consecutively. This indicates that the use of CRF for NER detection on Twitter in Indonesian is of high accuracy.

Moreover, it seems that the completeness of detection was moderately for formal and mixed test data, but slightly low for informal data. This indicated by the highest recall values which were 0.6229, 0.6221 and 0.3601 for formal, mixed, and informal test data respectively. The problem with twitter data is the vast use of non-standard words, both informal tweets and even more in informal tweets. Thus, it is recommended to normalize the word on Twitter using a dictionary or certain rules, so that detection completeness increases. When compared to the results of the entity detection of Indonesian document in other studies, there is an indication that the use of CRF has a better performance. For example, a study that incorporated context, part of speech, and lexical features to formulate rules for detecting entities in formal Indonesian documents, the recall was 0.6343 and the precision was 0.7184 [12].

Table 4. Test result

Model	Test data	Recall	Precision
Formal	Formal	0.6229	0.8760
	Informal	0.3922	0.8121
	Mixed	0.5400	0.8699
Informal	Formal	0.2866	0.8272
	Informal	0.3601	0.9076
	Mixed	0.3339	0.8749
Mixed	Formal	0.6221	0.8749
	Informal	0.5576	0.8523
	Mixed	0.6080	0.8684
Mixed, 10-fold cross validation	Mixed	0.6906	0.8774

In general, we may agree that in order to detect entity correctly, firstly a tweet should be classified into a formal or informal tweet. Secondly, the entities in the tweet can be detected using a suitable model. In a case that tweet classification is not performed, the mixed model may perform very well since the precision and recall values of the test of the mixed model with the mixed train data are only slightly under the precision and recall values of the test of the formal model with the formal train data. To make sure this condition, we measured precision and recall of mixed data using ten-fold cross-validation, which the results are 0.8774 and 0.6906 respectively, which is higher than other measurements.

4. Conclusion

Based on test results, CRF classifier can be used for NER with good performance, where the recall and precision values are respectively 62% and 87% for formal tweets, 36% and 90% for informal tweets, and 60% and 86% for mixed tweets. Thus, the selection of the model has an effect on automatic recognition accuracy, therefore, it is necessary to classify tweets into a formal or informal then selected the appropriate model for NER. Under conditions where it is not possible to classify tweets, the use of mixed models is adequate, evidenced by 69% of recall value and 87% precision of value in 10-fold cross-validation test.

References

[1] Sakaki T, Okazaki M and Matsuo Y 2010 Earthquake shakes twitter users: real-time event detection by social sensors *Proceedings of the 19th international conference on World wide web. ACM* pp 851–60

[2] Earle P S, Bowden D C and Guy M 2012 Twitter earthquake detection: earthquake monitoring in a social world *Annals of Geophysics* vol 54 no 6

[3] Wang H, Can D, Kazemzadeh A, Bar F and Narayanan S 2012 A system for real-time twitter sentiment analysis of 2012 US presidential election cycle *Proceedings of the ACL 2012 System Demonstrations. Association for Computational Linguistics* pp 115–20

[4] Finkel J R, Grenager T and Manning C 2005 Incorporating non-local information into information extraction systems by gibbs sampling *Proceedings of the 43rd annual meeting on association for computational linguistics. Association for Computational Linguistics* pp 363–70

[5] Leonandya R A, Distiawan B and Praptono N H 2015 A semi-supervised algorithm for indonesian named entity recognition *Computational and Business Intelligence (ISCBI), 2015 3rd International Symposium on. IEEE* pp 45–50

[6] Alfina I, Manurung R and Fanany M I 2016 Dbpedia entities expansion in automatically building dataset for indonesian ner *Advanced Computer Science and Information Systems (ICACSIS), 2016 International Conference on. IEEE* pp 335–40

[7] Suwarningsih W, Supriana I and Purwarianti A 2014 Imner indonesian medical named entity recognition *Technology, Informatics, Management, Engineering, and Environment (TIME-E), 2014 2nd International Conference on. IEEE* pp 184–8

[8] Li C, Weng J, He Q, Yao Y, Datta A, Sun A and Lee B-S 2012 Twiner: named entity recognition in targeted twitter stream *Proceedings of the 35th international ACM SIGIR conference on Research and development in information retrieval. ACM, 2012* pp 721–30

[9] Kazama J, Makino T, Ohta Y and Tsujii J 2002 Tuning support vector machines for biomedical named entity recognition *Proceedings of the ACL-02 workshop on Natural language processing in the biomedical domain-Volume 3. Association for Computational Linguistics* pp 1–8

[10] GuoDong Z and Jian S 2004 Exploring deep knowledge resources in biomedical name recognition *Proceedings of the International Joint Workshop on Natural Language Processing in Biomedicine and its Applications. Association for Computational Linguistics* pp 96–9

[11] Maynard D, Tablan V, Ursu C, Cunningham H and Wilks Y 2001 Named entity recognition from diverse text types *Recent Advances in Natural Language Processing 2001 Conference* pp 257–74

[12] Budi I, Bressan S, Wahyudi G, Hasibuan Z and Nazief B 2005 Named entity recognition for the indonesian language: combining contextual, morphological and part-of-speech features into a knowledge engineering approach *Discovery Science* Springer pp 57–69

[13] Budi I and Bressan S 2003 Association rules mining for name entity recognition *Web Information Systems Engineering, 2003. WISE 2003. Proceedings of the Fourth International Conference on. IEEE, 2003* pp 325–8

[14] Budi I and Bressan S 2007 Application of association rules mining to named entity recognition and co-reference resolution for the indonesian language *International Journal of BusinessIntelligence and Data Mining* vol 2 no 4 pp 426–46

[15] Wibawa A S and Purwarianti A 2016 Indonesian named-entity recognition for 15 classes using ensemble supervised learning *Procedia Computer Science* vol 81 pp 221–8

[16] Purwarianti A, Madlberger L and Ibrahim M 2016 Supervised entity tagger for indonesian labor strike tweets using oversampling technique and low resource features *TELKOMNIKA (Telecommunication Computing Electronics and Control)* vol 14 no 4 pp 1462–71

IOP Conf. Series: Materials Science and Engineering **403** (2018) 012068 doi:10.1088/1757-899X/403/1/012068

Expert system diagnose disease dermatitis using web based certainty factor

Y Findawati[1], A I Afrina[2]

[1,2]Universitas Muhammadiyah Sidoarjo
Jl. Raya Gelam 250 Candi Sidoarjo
+62-031-8945444

[1]yulianfindawati@umsida.ac.id

Abstract. Dermatitis is one of the skin disorders characterized by itching and clinically appearing a rash on the skin. Generally can occur repeatedly to a person in the form of skin inflammation (epidermis and dermis) in response to the influence of exogenous factors or endogenous factors.Dermatitis disease diagnostic expert system is a web application that allows users to diagnose the type of dermatitis disease they are experiencing. The purpose of making this application is to facilitate the user in diagnosing the type of dermatitis disease based on the symptoms that are being felt. In the use of expert system application of dermatitis diagnosis of this disease the user must register first to get a username and password that will be used to log on expert system and do diagnosis dermatitis disease. After diagnosis the user can know the type of dermatitis disease they are experiencing along with the description of the disease type and the solution. Based on testing, the accuracy of this application is 90%. With trial data that is 10 patients.

Keywords: Expert System, Dermatitis, Certainty Factor, Web Application

1. Introduction

Expert System (Expert System) is designed to solve quite complex problems that can only be solved by experts. The making of an expert system is not to replace the expert himself but can be used as a highly experienced assistant (Kusumadewi, 2003). One of the methods available in the expert system is the Certainty Factor method. Certainty Factor expressed confidence in an incident (or facts or hypotheses) based on evidence or expert judgment (Turban, 2005) .One of the disease are still many people are infected in dermatitis, and many people who do not know and understand about Dermatitis disease. So they do not know the right solutions to deal with skin diseases they suffer. Dermatitis is an inflammation of the epidermis and the dermis which is characterized by symptoms such as lesions that are objective and subjective symptoms itching polymorphs, can they are due to endogenous or exogenous factors (Maryunani, 2010).

There are still many people who do not know if they have dermatitis. To be sure they should consult a dermatologist first. It is encouraging in the development of an expert system that can diagnose dermatitis disease to be realized. So in this final project, the author raised the title "Expert System Diagnosis Dermatitis Disease Using Web Based Certainty Factor Method". With this expert system the layman or user can perform diagnosis of dermatitis itself through perceived symptoms. Based on the background that has been mentioned above it can be concluded into the formulation of the problem How to design and build an expert system to diagnose dermatitis by using web based certainty factor method

2. Methodology

2.1. Certainty Factor

The general equation used to calculate the value of Certainty Factors is defined as follow [J.Martin, and S. Oxman,1988]:

$$CF(H,E) = MB(H,E) - MD(H,E) \quad (1)$$

Description:
CF(H,E) : Hypothesis H of the Certainty Factor for the emergence of symptoms E,
MB(H,E) : The measure of increased belief towards hypothesis H that is influenced by symptoms E,
MD(H,E) : The measure of increased disbelief towards hypothesis H that is influenced by symptoms E.

Equation (2) is the basic form of the Certainty Factor formula in a rule in form of "IF E THEN H" [8]

$$CF(H,e) = CF(E,e) * CF(H,E). \quad (2)$$

Description:
CF(H,e) : The measure of increased disbelief towards hypothesis H that is influenced by symptoms e,
CF(E,e) : CF of symptoms E that is influenced by symptoms e,
CF(H,E) : CF of the hypothesis, with the assumption that the symptoms are known for sure when CF(E,e) = 1.

3. Research Method

3.1. Inference Machines

Inference engine is a function of system reasoning pattern used by expert system. In the dermatitis diagnosis system, there are three types of dermatitis disease that will be diagnosed, namely AD (Atopic Dermatitis), ACD (Allergic Contact Dermatitis), and ICD (Irritant Contact Dermatitis). The following are the rules used to diagnose dermatitis:

- **Rule 1:**
  The rules of Atopic Dermatitis (AD)
  IF Itching when sweating
  AND Dry skin, scaly / flaking AND Occurs Eczema (spots / reddish spots) AND Have a history of asthma & Allergy to food AND Allergy to drugs AND Feeling anxious, selfish, frustrated, aggressive or feeling depressed AND Sensitive to wool AND Sensitif Against AND ANDITIFICAL fur on dust AND Skin pale face / dim THEN Atopic Dermatitis
- **Rule 2:**
  The rule of Allergic Contact Dermatitis (ACD)
  IF Skin feels dry, scaly / flaking AND Occurs eczema (spots / reddish spots) AND Itching and redness occur after 24-48 hours after exposure to a substance AND There is a reaction to the skin after exposure to latex / rubber
  AND There is a reaction to the skin after using jewelry, watches, and metal zippers AND There is a reaction to the skin after exposure to dye AND There is a reaction to the skin after using certain cosmetics, perfumes, nail polishers, or hair dye AND There is a reaction to the skin after using Topical medications
  THEN Contact Allergic Dermatitis
- **Rule 3:**
  Irritant Contact Dermatitis (ICD)
  IF Skin feels dry, scaly / flaking
  AND Occurs Eczema (spots / rashes) AND Increases pain, heat and burns AND There is a reaction to the skin after exposure to detergent / soap AND A reaction to the skin after exposure to engine lubricating oil AND A reaction to the skin after exposure to acid and alkali AND There is a reaction On the skin after exposure to the soil AND There is a reaction to the skin after exposure to the wood powder AND There is a reaction to the skin after exposure to antiseptic or antibacterial agent AND There is a reaction to the skin when exposed to water containing chlorine or lime THEN Contact Dermatitis Irritant
  Certainty Factor or certainty factor is the value of clinical parameters that will be given to indicate the magnitude of confidence and the level of confidence in a diagnosis of some of the symptoms given. In the dermatitis disease

consultation session, patients or expert system users will be given several answer choices that each have the following weights (Very Sure = 1; Sure = 0.8; Slightly Sure = 0.5; Not Sure = 0.2; No = 0)
Then it will be determined the weight value of the patient's confidence in each of the symptoms, by way of expert system gives some questions about the symptoms of dermatitis disease. There are about 25 questions to be answered by the patient or expert system user, including:

1. Expert: Does your skin feel itchy when sweating? Slightly Confident (CFuser = 0.5)
2. Expert: Does your skin feel dry, scaly / flaking? Slightly Confident (CFuser = 0.5)
3. Expert: Does eczema (reddish spots) occur? Sure (CFuser = 0.8)
4. Expert: Do you have a history of asthma? Patients: No (CFuser = 0)
5. Expert: Are you allergic to food?Slightly Confident (CFuser = 0.5)
6. Expert: Are you allergic to drugs?No (CFuser = 0)
7. Experts: Do you feel anxious, selfish, frustrated, aggressive or depressed? Not Sure (CFuser = 0.2)
8. Expert: Are you sensitive to wool?No (CFuser = 0)
9. Expert: Are you sensitive to animal fur?No (CFuser = 0)
10. Experts: Are you sensitive to dust?Sure (CFuser = 0.8)
11. Expert: Is your facial skin pale / dim?Slightly Confident (CFuser = 0.5)
12. Expert: Does itchy and redness occur after 24-48 hours after exposure to a substance? Unsure (CFuser = 0.2)
13. Expert: Is there a reaction to the skin after being exposed to latex / rubber? No (CFuser = 0)
14. Experts: Is there a reaction to the skin after using jewelry, watches, and metal zippers? No (CFuser = 0)
15. Experts: Is there a skin reaction after exposure to a dye? No (CFuser = 0)
16. Expert: There is a reaction to the skin after using certain cosmetics, perfumes, nail polish, or hair dye? Slightly Confident (CFuser = 0.5)
17. Expert: Any reaction to the skin after using certain topical medications? No (CFuser = 0)
18. Expert: Is it painful, hot and burning?Slightly Confident (CFuser = 0.5)
19. Expert: Is there a skin reaction after exposure to detergent / soap? No (CFuser = 0)
20. Expert: Is there a skin reaction after exposure to engine oil? No (CFuser = 0)
21. Expert: Is there a reaction to the skin after exposure to acid and alkali? No (CFuser = 0)
22. Expert: Is there a reaction to the skin after exposure to the soil? No (CFuser = 0)
23. Expert: Is there a reaction to the skin after exposure to wood powder? No (CFuser = 0)
24. Expert: Any reaction to the skin after exposure to antiseptic or antibacterial agents? No (CFuser = 0)
25. Expert: Any reaction to the skin when exposed to water containing chlorine or lime? No (CFuser = 0)

The next step is to calculate the CF value of each rule by multiplying the value of CFpakar and CFuser.

Rule 1: Atopic Dermatitis (DA)

CF $= CF_{expert} X\ CF_{user}$
CF 1 = 0,4 x 0,5 = 0,2
CF 2 = 0,8 x 0,5 = 0,4
CF 3 = 0,9 x 0,8 = 0,72
CF 4 = 0,3 x 0 = 0
CF 5 = 0,4 x 0,5 = 0,2
CF 6 = 0,3 x 0 = 0
CF 7 = 0,1 x 0,2 = 0,02
CF 8 = 0,1 x 0,2 = 0,02
CF 9 = 0,1 x 0 = 0
CF 10 = 0,2 x 0,8 = 0,16
CF 11 = 0,2 x 0,5 = 0,1

Rule 2: Allergic Contact Dermatitis Disease (ACD)

CF $= CF_{pakar}\ x\ CF_{user}$
CF 1 = 0,8 x 0,5 = 0,4
CF 2 = 0,9 x 0,8 = 0,72
CF 3 = 0,7 x 0,2 = 0,14
CF 4 = 0,7 x 0 = 0
CF 5 = 0,7 x 0 = 0
CF 6 = 0,6 x 0 = 0
CF 7 = 0,8 x 0,5 = 0,4
CF 8 = 0,4 x 0 = 0

Rule 3: Irritant Contact Dermatitis Disease (ICD)

CF $= CF_{expert}\ x\ CF_{user}$
CF 1 = 0,8 x 0,5 = 0,4
CF 2 = 0,9 x 0,8 = 0,72
CF 3 = 0,9 x 0,5 = 0,45
CF 4 = 0,8 x 0 = 0
CF 5 = 0,7 x 0 = 0
CF 6 = 0,8 x 0 = 0
CF 7 = 0,3 x 0 = 0
CF 8 = 0,7 x 0 = 0
CF 9 = 0,8 x 0 = 0
CF 10 = 0,8 x 0,2 = 0

The next step is to combine the CF values of each rule to determine the diagnosis of dermatitis, as follows:

Atopic Dermatitis (AD)

$CF_{combine}(CF_1, CF_2) = 0{,}2 + 0{,}4 * (1 - 0{,}2) = 0{,}52 \rightarrow (CF_{old1})$
$CF_{combine}(CF_{old1}, CF_3) = 0{,}52 + 0{,}72 * (1 - 0{,}52)$

Allergic Contact Dermatitis (ACD)

$CF_{combine}(CF_1, CF_2) = 0{,}4 + 0{,}72 * (1 - 0{,}4) = 0{,}832 \rightarrow (CF_{old1})$

Contact Irritant Dermatitis (CID)

$CF_{combine}(CF_1, CF_2) = 0{,}4 + 0{,}72 * (1 - 0{,}4) = 0.832 \rightarrow (CF_{old1})$

$= 0,8656 \rightarrow (CF_{old2})$
$CF_{combine}(CF_{old2}, CF_4) = 0,8656 + 0 * (1 - 0,8656) = 0,8656 \rightarrow (CF_{old3})$
$CF_{combine}(CF_{old3}, CF_5) = 0,8656 + 0,2 * (1 - 0,8656) = 0,8924 \rightarrow (CF_{old4})$
$CF_{combine}(CF_{old4}, CF_6) = 0,8924 + 0 * (1 - 0,8924) = 0,8924 \rightarrow (CF_{old5})$
$CF_{combine}(CF_{old5}, CF_7) = 0,8924 + 0,02 * (1 - 0,8924) = 0,8945 \rightarrow (CF_{old6})$
$CF_{combine}(CF_{old6}, CF_8) = 0,8945 + 0,02 * (1 - 0,8945) = 0,8966 \rightarrow (CF_{old7})$
$CF_{combine}(CF_{old7}, CF_9) = 0,8966 + 0 * (1 - 0,8966) = 0,8966 \rightarrow (CF_{old8})$
$CF_{combine}(CF_{old8}, CF_{10}) = 0,8966 + 0,16 * (1 - 0,8966) = 0,9131 \rightarrow (CF_{old9})$
$CF_{combine}(CF_{old9}, CF_{11}) = 0,9131 + 0,1 * (1 - 0,9131) = 0,9217 \rightarrow (CF_{old10})$
Percentage of confidence = $CF_{combine}$ * 100 %
= 0,9217 * 100 = 92,17 % Atopic Dermatitis

$CF_{combine}(CF_{old1}, CF_3) = 0,832 + 0,14 * (1 - 0,832) = 0,8555 \rightarrow (CF_{old2})$
$CF_{combine}(CF_{old2}, CF_4) = 0,8555 + 0 * (1 - 0,8555) = 0,8555 \rightarrow (CF_{old3})$
$CF_{combine}(CF_{old3}, CF_5) = 0,8555 + 0 * (1 - 0,8555) = 0,8555 \rightarrow (CF_{old4})$
$CF_{combine}(CF_{old4}, CF_6) = 0,8555 + 0 * (1 - 0,8555) = 0,8555 \rightarrow (CF_{old5})$
$CF_{combine}(CF_{old5}, CF_7) = 0,8555 + 0,4 * (1 - 0,8555) = 0,9133 \rightarrow (CF_{old6})$
$CF_{combine}(CF_{old6}, CF_8) = 0,9133 + 0 * (1 - 0,9133) = 0,9133 \rightarrow (CF_{old7})$
Percentage of confidence = $CF_{combine}$ * 100 %
= 0,9133 * 100 %
= 91,33 % Allergic Contact Dermatitis

$CF_{combine}(CF_{old1}, CF_3) = 0.832 + 0,45 * (1 - 0.832) = 0,9076 \rightarrow (CF_{old2})$
$CF_{combine}(CF_{old2}, CF_4) = 0,9076 + 0 * (1 - 0,9076) = 0,9076 \rightarrow (CF_{old3})$
$CF_{combine}(CF_{old3}, CF_5) = 0,9076 + 0 * (1 - 0) = 0,9076 \rightarrow (CF_{old4})$
$CF_{combine}(CF_{old4}, CF_6) = 0,9076 + 0 * (1 - 0,9076) = 0,9076 \rightarrow (CF_{old5})$
$CF_{combine}(CF_{old5}, CF_7) = 0,9076 + 0 * (1 - 0,9076) = 0,9076 \rightarrow (CF_{old6})$
$CF_{combine}(CF_{old6}, CF_8) = 0,9076 + 0 * (1 - 0,9076) = 0,9076 \rightarrow (CF_{old7})$
$CF_{combine}(CF_{old7}, CF_9) = 0,9076 + 0 * (1 - 0,9076) = 0,9076 \rightarrow (CF_{old8})$
$CF_{combine}(CF_{old8}, CF_{10}) = 0,9076 + 0 * (1 - 0,9076) = 0,9076 \rightarrow (CF_{old9})$
Percentage of confidence = $CF_{combine}$ * 100 %
= 0,9076 * 100 %
= 90,76 Contact Irritant Dermatitis

So it can be concluded that the patient was diagnosed with Atopic Dermatitis (DA) disease with a percentage of confidence or CF value of 92.17%.

3.2. Data Flow Diagram (DFD)
Application design using data flow diagram. The result of Data flow diagram Level 1 can be found in Figure 1.

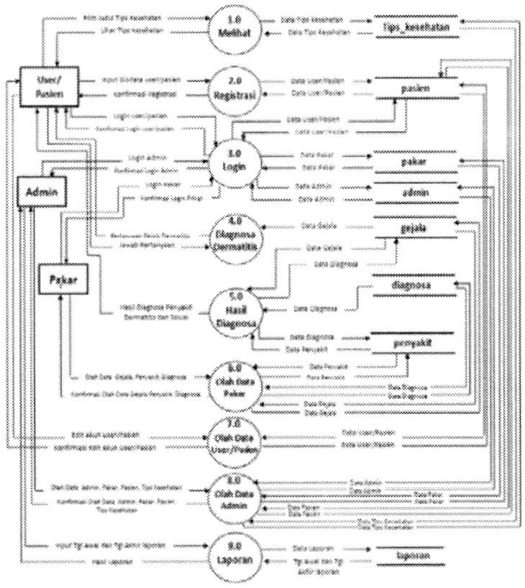

Figure 1. DFD Level 1

The level 1 DFD consists of Process view, Registration process, login process, which is the process to enter the expert system diagnosis dermatitis disease. There are three login process: user login / login, admin login, and expert login, diagnostic dermatitis process, diagnostic process, data processing process, user data processing process,

4. RESULTS AND DISCUSSION
The result of Diagnosis Patient Pageview can be found in Figure 2

Figure 2. Patient Diagnostic Page View

On the patient's diagnostic page there are symptoms of dermatitis symptoms that the patient must fill in to determine the type of dermatitis patient's disease in accordance with the symptoms experienced by the patient.

The test is done with 10 data. Where testing is performed to compare the results of expert system applications and expert physicians. The testing result of case can be found in Table 1.

Table 1. Testing Result of Case

No	Symptom	Application Results	Diagnostic Results
1	G-001, G-003, G-012, G-005, G-004	Atopic Dermatitis (AD) CF = 71.576 %	Atopic Dermatitis (AD)
2	G-003, G-012, G-017	Allergic Contact Dermatitis (ACD) CF = 87.13 %	Allergic Contact Dermatitis (ACD)
3	G-012, G-003, G-005	Atopic Dermatitis (AD) CF = 78.748 %	Atopic Dermatitis (AD)
4	G-002, G-003, G-019, G-018	Contact Irritant Dermatitis (CID) CF = 98.983936 %	Contact Irritant Dermatitis (CID)
5	G-012, G-016, G-003	Allergic Contact Dermatitis (ACD) CF = 80.134 %	Allergic Contact Dermatitis (ACD)
6	.G-012, G-002, G-003, G-010	Atopic Dermatitis (AD) CF = 72.28 %	Atopic Dermatitis (AD)
7	G-016, G-003, G-018	Allergic Contact Dermatitis (ACD) CF = 74.975 %	Allergic Contact Dermatitis (ACD)
8	G-002, G-003	Contact Irritant Dermatitis (CID) CF = 89.11 %	Atopic Dermatitis (AD)
9	G-013, G-003	Allergic Contact Dermatitis (ACD) CF = 79.188 %	Allergic Contact Dermatitis (ACD)
10	G-003, G-018, G-019	Contact Irritant Dermatitis (CID) CF = 99.3903616 %	Contact Irritant Dermatitis (CID)

Based on the test results, 9 data in accordance with doctors data and 1 data is not appropriate. So it can be concluded the accuracy of 90%.

IOP Conf. Series: Materials Science and Engineering **403** (2018) 012068 doi:10.1088/1757-899X/403/1/012068

5. Conclusion

From the results of the design, manufacture, implementation and testing of expert system program diagnosis dermatitis disease using certainty method based on the web, it can be obtained some conclusions and suggestions for the development of expert system program diagnosis of dermatitis disease towards the better. Based on the results of the program testing, expert system diagnosis of dermatitis disease using certain methods of web-based certainty factor is quite helpful to diagnose dermatitis disease based on the symptoms that complained by the patient. Dermatitis disease diagnostic expert system allows users to diagnose their dermatitis disease. In addition users can also obtain various kinds of information about skin health especially related to dermatitis disease through the health tips provided in this expert system. Based on program testing, the dermatitis disease diagnosis expert system has an accuracy value of 90%. And the erroran value of this expert system is 10%. necessary to try to improve the calculations and knowledge base of the expert system, in order to increase the accuracy value and reduce the error value of the dermatitis expert system.

REFERENCES

[1] Ardhie, A.M. (2004), "Dermatitis Dan Peran Steroid Dalam Penanganannya", *DEXA MEDIA*, Vol. 17, No. 4, hal. 157,158.

[2] Desiani, Anita., Arhami, Muhammad. (2006), *Konsep Kecerdasan Buatan*, Penerbit ANDI, Yogyakarta.

[3] Evina, Belda. (2015), "Clinical Manifestations And Diagnostics Criteria of Atopic Dermatitis", *J MAJORITY*, Vol. 4, No. 4, hal. 23.

[4] Ikatan Dokter Indonesia. (2014), *Panduan Praktik Klinis Bagi Dokter Di Fasilitas Pelayanan Kesehatan Primer*, Edisi 2/Revisi, Jakarta.

[5] J.Martin, and S. Oxman, "*Building Expert System a tutorial*", Prentice Hall, New Jersey, 1988.

[6] Kariosentono, Harijono. (2006), *Dermatitis Atopik (Eksema)*, Cetakan 1, LPP UNS dan UNS Press, Surakarta.

[7] Kusrini. (2006), *Sistem Pakar Teori Dan Aplikasi*, Penerbit ANDI, Yogyakarta.

[8] Raharjo, Budi. (2011), *Modul Pemrograman WEB, HTML, PHP, dan MYSQL*, Modula, Bandung.

[9] Saputra, Agus. (2011), *Web Tips PHP, HTML 55, dan CSS*, Jasakom, Jakarta.

[10 Suciani, Wulan. (2016), *Sistem Pakar Mendiagnosa Penyakit Jantung Menggunakan Metode Certainty Factor Berbasis Web*, Tugas Akhir S.Kom., Universitas Muhammadiyah Sidoarjo, Sidoarjo.

[11] Suyanto. (2011), *Artificial Intelligence*, Informatika, Bandung.

[12] Yakub, Suardi. (2008), *Sistem Basis Data Tutorial Konseptual*, Graha Ilmu, Yogyakarta.

Classification of heart signal using wavelet haar and backpropagation neural network

H Hindarto, I Anshory and A Efiyanti

Universitas Muhammadiyah Sidoarjo
Jl. Mojopahit 666B Sidoarjo
Jl. Raya Gelam 250 Candi Sidoarjo
+62-031-8945444

hindarto@umsida.ac.id

Abstract. Researchers used many methods to extract and classify heart signals.In this study wavelet haar is use to extract characteristics of heart signals. Artificial neural networks Backpropagation for the classification of heart signals. The data is taken from Physiobank namely MIT-BIH Arrhythmia Database and MIT-BIH Normal Sinus Rhythm Database. The data is processed using Haar wavelet method for its extraction.The results of feature extraction methods will be use for the classification process. The research found that by using Wavelet Haar feature extraction and classification using Backpropagation obtained classification accuracy rate of 92%.

Keywords: Heart; electrocardiography; Wavelet Haar; backpropagation

1. Introduction

Electrocardiogram (ECG) is the electrical device to detect the activity of the human heart. The ECG is a composite of 5 waves - P, Q, R, S and T. These signals can be measured with electrode, placed on the human body . The signal from this electrode is connect to a simple electrical circuit with amplifier and analog - digital converter. Application of heart rate and heart rate detection by using an electronic circuit with a stethoscope and pulse sensor as input to detect. The pulse sensor is use to determine the heart of a person and a stethoscope is use to detect heart sound. While the series of electronics used ardunio series as an interface to the monitor screen. From Test results on 10 subjects, to measure of heart rate with the pulse sensor obtained the accuracy of the tool sensory pulses by way of 99% [1].

The heart frequency can be detected by many methods and algorithms. Many heart signal detection algorithm is based on the distance between the QRS complexes. Complex QRS algorithms are from the field of artificial neural networks, genetic algorithms, wavelet transforms or bank filters [2]. In addition the next way to detect complex QRS is to use adaptive threshold [3] such as Direct method for detection of heart rate spectral signal spectral ECG [4] and the method of Short-Term Autocorrelation [5]. ECG signals can be used to diagnose heart disease, however, t ECG signals do not fully describe the heart characteristics. It is because the heart is also affected by the opening and closing of the heart valve is a factor in the conscience. In addition, there is a heart damage that is difficult to detect using an ECG, such as natural structural abnormalities or opening and closing imperfect heart valves, as well as heart murmurs or abnormal sounds [6].

There are several studies by taking data from MIT-BIH data base. This research uses Heart Rate Variability (HRV) analysis method to look for feature extraction and classification using artificial neural network classifier. Results obtained with an accuracy of 99.38% [7]. F. Yaghouby et al., In his penisilitan using Generalized Discriminant Analysis (GDA) method for extracting and using Multilayer Perceptron (MLP) method of artificial neural network grouping, the result is 100% [8]. R. Acharya et al., This study took eight classes, feature extractions used by taking from the spectral entropy, the Poincar geometry plot and the largest Lyapunov exponent (LLE). Then classification using artificial neural network method and fuzzy relationship. This research yields value. 80-85 5 [9]. L. Hussain et al., This study to look for feature extraction using HRV analysis includes linear (time and frequency domain) and non-linear techniques and with the result of 92.5% method using LMT (logical tree model) method [10].

This study divides the heart signals into two classes of normal heart and abnormal heart.The first presents a method of Wavelet Haar to as a heart signal feature extraction. The retrieved characteristic is then classified using a backpropagation neural network.

2. Method

Heart signal data is taken from Physiobank database are MIT-BIH Normal Sinus Rhythm Database and MIT-BIH Arrhythmia Database. This database includes for 18 ECG recordings from subjects at the Arithmia Laboratory at Bethlehem Israel Beth (now Beth Israel Deaconess Medical Center). Subjects, included in this database were found to have insignificant arrhythmia; They are 5 men, aged 26-45, and 13 women, aged 20 to 50. The database contains 48 half-hour citations of two-channel ECG recordings of a person treated on the road, obtained from 47 subjects studied by BIH Arrhythmia Laboratory between 1975 and 1979. Twenty-three randomly selected records of a set of 4,000 24-hour ambulatory ECG records are collected from a mixed population of inpatients (about 60%) and outpatients (about 40%) at Home Pain of Israel Beth Boston; 25 remaining records are selected from the same set to include less common but clinically significant arrhythmias that would not be represented in the sample randomly. Digital recording at 360 samples per second per channel with 11-bit resolution at 10 mV range. Two or more cardiologists independently described each record; Disputes were resolved to obtain a readable computer reference explanation for each beat (about 110,000 annotations at all) included with the database [11].

The used data e are 15 files of 1 minute length and 360 Hz sample frequency, normal pulse signal and various types of pulse arrhythmias. All selected files take 6 seconds segmentation to get 150 samples from the experimental data.

Wavelet transformation is an improvement of the Fourier transform. If the Fourier transform only provides information about the frequency of a signal, the wavelet transform provides information about the combination of scale and frequency. Wavelet comes from a scaling function, can be made as a mother wavelet. Other wavelets will come from scaling, dilation and mother wavelet shifts. Wavelet transformation is improvement of the Fourier transform. If the Fourier transform only provides information about the frequency of a signal, the wavelet transform provides information about the combination of scale and frequency. Wavelet comes from a scaling function. From this scaling function can made a mother wavelet. Other wavelets will come from scaling, dilation and mother wavelet shifts.

There are two functions that play a major role in wavelet analysis, which is function of the wavelet scale (wavelet) and wavelet ψ (mother wavelet). The simplest wavelet analysis is base on the Haar function scale [12]. The Haar scale function is define as another using formula (1).

$$\varphi(x) = \begin{cases} 1, & if\ 0 \le x < 1 \\ 0, & OTHER \end{cases} \tag{1}$$

The Haar mother wavelet function is define as formula (2).

$\Psi(x) = \varphi(2x) - \varphi(2x - 1)$

$$\Psi(x) = \begin{cases} 1, & 0 \le x < 1/2 \\ -1, & \frac{1}{2} \le x < 1 \\ 0, & OTHER \end{cases} \tag{2}$$

Like all wavelet transforms, the Haar Wavelet transform is a discrete signal that decomposes into two sub-signals of half its length. One sub signal is averages in a row. Haar Level Wavelet Transformation 1, Haar Wavelet transformation is do in several stages, or levels.

Assume 1 dimension of signal f with signal length equal to N. Haar level transformation 1 for $f = (x_1, x_2, \ldots\ldots, x_N)$.

$$f \overset{H_1}{\rightarrow} (a^1 | d^1) \tag{3}$$

Where :

$$a^1 = \left(\frac{x_1 + x_2}{\sqrt{2}}, \frac{x_3 + x_4}{\sqrt{2}}, \ldots\ldots\ldots, \frac{x_{N-1} + x_N}{\sqrt{2}}\right)$$
$$d^1 = \left(\frac{x_1 - x_2}{\sqrt{2}}, \frac{x_3 - x_4}{\sqrt{2}}, \ldots\ldots\ldots, \frac{x_{N-1} - x_N}{\sqrt{2}}\right)$$

Heart signals that have been selected based on normal and abnormal heart, then each signal is process using Wavelet Haar level 4. Totals of 300 heart signals are extrac to get two special features. 300 cardiac data signals consisting of 150 normal and 150 abnormal heart signals. Extraction of heart signal features is take from the Haar Wavelet process with level 4.

3. Results and Discussion

In this study, Haar Wavelet transformation was use for characteristic extraction of normal and abnormal heart signals. Normal and abnormal heart signals can be seen in Figure 1.

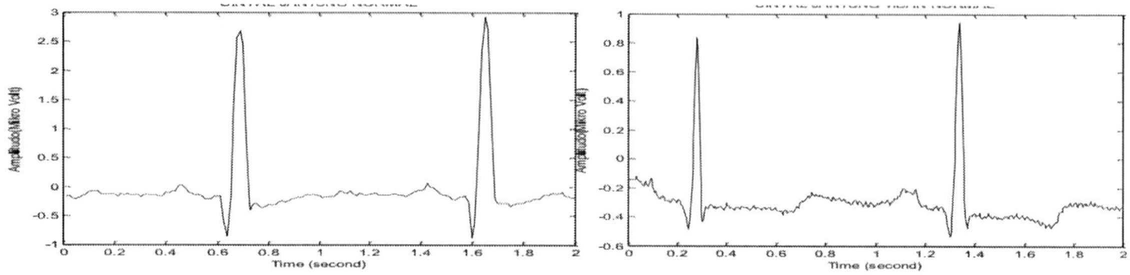

(a) (b)

Figure 1. (a) a normal heart signal (b) an abnormal heart signal

Results Haar Wavelet process in the form of a vector consisting of several data point elements. In this study using Wavelet Haar level 4, the results of the wavelet haar process can seen in Figure 2.

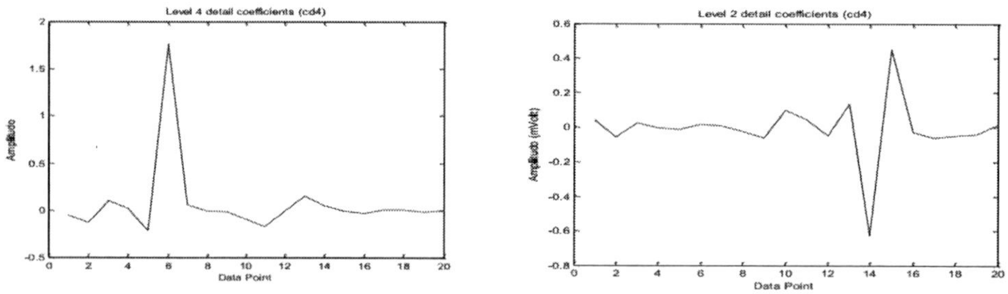

(a) (b)

Figure 2. Wavelet Haar level 4 results (a) normal heart signal (b) abnormal heart signal

The data taken in this study is 300 data files of heart signals. A normal heart signal file has 350 data points and an abnormal heart signal file has 80 data points. The result of the wavelet haar value is show in Table 1 by taking five samples of haar wavelet results on each signal.

Table 1. Result Of Wavelet Haar Of Heart Signal

Signal	**Signal 1**	**Signal 2**	**Signal 3**	**Signal 4**	**Signal 5**
	20 Point data	20 Point data	20 Point data	20 Point data	20 Point data
Normal heart signal	0.03	-0.013	-0.06	-0.02	0.001
	-0.05	-0.05	0.03	0.045	-0.015
	0.02	0.03	0.02	-0.075	0.013
	-0.02	0.01	0.01	-0.02	0.014
	-0.02	0.02	0.03	0.03	0.002
	0.01	-0.01	0.02	-0.02	-0.032
	0.02	0.02	0.01	0.02	-0.08
	-0.02	-0.04	0.03	0.11	0.08
	-0.05	0.01	0.02	0.03	0.04
	0.15	0.04	0.02	0.01	-0.01
	0.06	0.02	0.01	0.62	0.05
	-0.04	0.45	-1.61	-1.58	-0.05

IOP Conf. Series: Materials Science and Engineering **403** (2018) 012069 doi:10.1088/1757-899X/403/1/012069

	0.13	-2.61	1.43	1.08	0.30
	-0.62	2.23	0.02	0.02	0.02
	0.44	0.02	0.01	-0.07	-0.04
	-0.02	-0.02	0.05	-0.03	-0.04
	-0.05	-0.03	0.01	-0.03	-0.13
	-0.04	-0.05	0.01	0.02	0.03
	-0.03	0.01	0.03	0.04	-0.03
	0.01	0.02	0.02	-0.02	-0.02
abnormal heart signal	-0.04	-0.015	-0.07	0.00	-0.02
	-0.12	-0.22	-0.15	-0.06	-0.01
	0.12	0.11	0.12	0.09	-0.07
	0.02	0.03	0.021	0.09	-0.03
	-0.21	-0.61	-0.751	0.05	0.16
	1.75	1.46	1.632	0.26	0.01
	0.05	0.04	0.042	0.22	-2.20
	-0.02	-0.01	0.013	0.00	0.42
	-0.02	-0.02	-0.013	-0.01	0.00
	-0.08	-0.05	-0.074	-0.01	-0.02
	-0.16	-0.13	-0.142	-0.10	-0.02
	-0.02	-0.02	-0.022	-0.15	-0.07
	0.15	0.12	0.174	0.04	-0.16
	0.04	0.02	0.075	0.13	-0.07
	-0.02	-0.01	0.012	0.00	0.17
	-0.02	0.01	0.002	0.02	0.05
	0.012	0.01	0.022	-0.02	-0.04
	0.014	0.02	0.012	0.00	-0.01
	-0.015	0.012	0.013	0.03	-0.03
	0.001	-0.013	0.034	-0.01	0.00

Classification of heart signals processed using neural network Back Propagation as show in Figure 3. Final processing is do after the initial process is feature search.

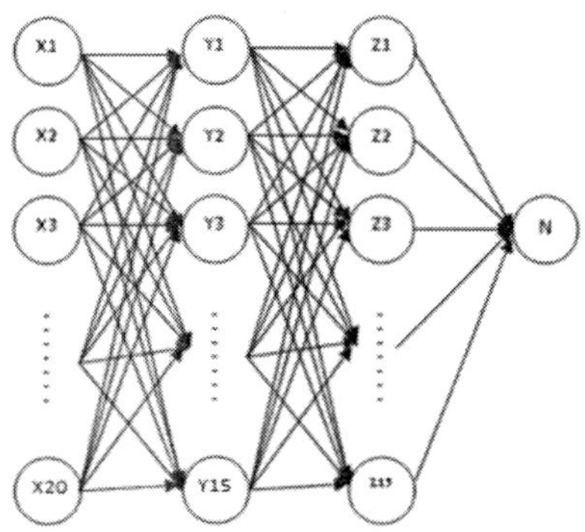

Figure 3. Neural Network Architecture Backpropagation with 2 Hidden layers

Extraction feature of wavelet haar is use for input to Backpropagation, this study using Backpropagation (20-15-15-1) number 20 is the input of the wavelet haar value, the number 15 layer is hidden one to the number of nodes 15, the number 15 represents the number of nodes in the layer Hidden second, and the number 1 is the target (normal heart and abnormal heart).

There are two stages for the classification process that is the learning process and Mapping process. The learning process using learning rate parameter 0.1 and error be achieved 0,00001. The value for weights is random in the range of -1 to 1. In search of optimal parameter performance to produce the best value of the neural network is to assess the size of Mean squared error (MSE) and the number of hidden layer units at trainning. Examples of performance results can found in Figure 4.

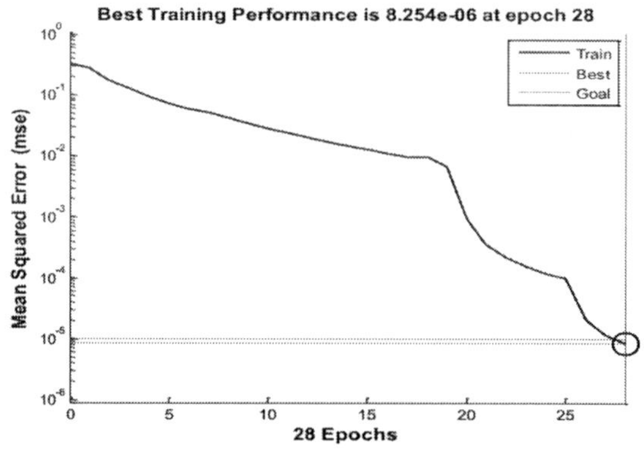

Figure 4. Performance performance results of the identification of Backpropagion (20-15-15-1)

The backpropagation performance of 20-15-15-1 show that the desired target error of 0.00001 has been achieved. In this network come about error 8.25. 10-6 so that the training process for the classification on this heart signal can already reach 100% accuracy. The result of the trainning process for classification of heart signals by using the 20-10-15-1 network has obtained the value of bias and weight. After the trainning process, the next process is the testing process. The data used in this trial process is to retrieve data as much as 100 heart signal data, consisting of 50 data of normal heart signals and 50 data of heart signals upnormal. In this research, data trainning and data testing with total data amount 300 data yield accuracy level 276/300 * 100% = 92%. From Figure 4 shows that by using 2 hidden layers have reached the desired target.

Table 2. The Performance Of The Neural Network To The Different Number Of Hidden Layer

	1 Hidden Layer	2 Hidden Layer	3 Hidden Layer
Time	30	53	62
Iteration	1000	503	407
MSE	$9.9x10^{-4}$	$8.25x10^{-6}$	$3.55x10^{-6}$
Accuracy	89 %	92 %	92 %

5. Conclusion

In this study, researchers introduced Wavelet Transformation by taking a 4th level haar wavelet to extract features. Backpropagation neural network is used to train 200 heart signal data files.. The testing process used 300 data file data signal heart. The precise classification of two hidden layers of backpropagation is 92%. The future research will be focused on exploration of better feature extraction methods.

References

[1] Hindarto, dkk. Aplikasi Pengukur Deteksi Detak dan Suara Jantung. Jurnal Saintek, Volume 13, nomer 2, Desember.
[2] Kohler, B.-U.; Hennig, C.; Orglmeister, R. The principles of software QRS detection. Engineering in Medicine and Biology Magazine IEEE, vol. 21, pp. 42 – 57, January -February 2002.
[3] I. I. Christov. Real time electrocardiogram QRS detection using combined adaptive threshold. BioMedical Engineering OnLine, 2004. [cit: 2011-10-16]. [Online]. Available on internet: http://www.biomedical-engineering-online.com/content/3/1/28.
[4] Surda, J.; Lovas, S.; Pucik, J.; Jus, M. Spectral Properties of ECG Signal. Radioelektronika, 2007. 17th International Conference, Brno, Czech Republic, 24th – 25th April 2007, pp. 1 – 5.
[5] Piotrowskia Z.; Różanowski K. Robust Algorithm for Heart Rate (HR) Detection and Heart Rate Variability (HRV) Estimation. ACTA PHYSICA POLONICA, vol. 118, pp. 131 – 135, No. 1/2010.
[6] Glass L, Mackey MC. *From Clocks to Chaos: The Rhythms of Life.* Princeton, NJ: Princeton University Press; 1988.

[7] Babak Mohammadzadeh-Asl, Seyed Kamaledin Setarehdan, Neural Network Based Arrhythmia Classification Using Heart Rate Variability Signal. 14th European Signal Processing Conference (EUSIPCO 2006), Florence, Italy, September 4-8, 2006, copyright by EURASIP
[8] F. Yaghouby, et al. Classification of Cardiac Abnormalities Using Reduced Features of Heart Rate Variability Signal, orld Applied Sciences Journal 6 (11): 1547-1554, 2009.
[9] R. Acharya U, et al. Classification of cardiac abnormalities using heart rate signals. Medical & Biological Engineering & Computing 2004, Vol. 42.
[10] L. Hussain, et al. Classification Of Normal And Pathological Heart Signal Variability Using Machine Learning Techniques. International Journal Of Darshan Institute On Engineering Research & Emerging Technologies Vol. 3, No. 2, 2014.
[11] PhysioBank, at http://www.physionet.org
[12] Chun-Lin, Liu. A Tutorial of the Wavelet Transform. February 23, 2010.

Very short term load forecasting peak load time using fuzzy logic

J Jamaaluddin, D Hadidjaja, I Sulistiyowati, EA Suprayitno, I Anshory and S Syahrorini

Universitas Muhammadiyah Sidoarjo
Jl. Mojopahit 666B Sidoarjo
Jl. Raya Gelam 250 Candi Sidoarjo
+62-031-8945444

jamaaluddin@umsida.ac.id

Abstract. One of the important things to do in electric power system operation is load forecasting. Load forecasting consists short term load forecasting and very short term forecasting. The very short-term load forecasting is predicting electrical loads in every 30 minutes. This forecasting is done to decide which plant to operate. The capacity of the plant to be operated adjusts to the load plan to be supplied the next day One method utilized in this research is Fuzzy Logic. This method has been applied for short-term load forecasting and will be employed for very short-term forecasting peak load time. Fuzzy logic expected has a small MAPE (0,6244%).

Keywords: Very Short Term Load Forecasting; Fuzzy Logic; Main Average Percentage Error (MAPE)

1. Introduction

Electrical energy is a major requirement at this time. Almost all things are very dependent on electricity [1].One second without electricity, modern society will not be able to conduct its activities [2][3]. Likewise, in Indonesia,. electricity system is divided into several areas; one of which is the Java-Bali electricity system.

Very short term load forecasting can estimate electricity consumption over a certain time span. Accurate forecasting can improve safety and reliability in electric power system operations such as load flow, maintenance unit maintenance and unit commitment [4].

Load characteritics is very important to determine the load forecasting parameters. The power change characteristics received by the load of the electrical power system at any given time interval is known as the daily load curve, as shown in Figure 1.

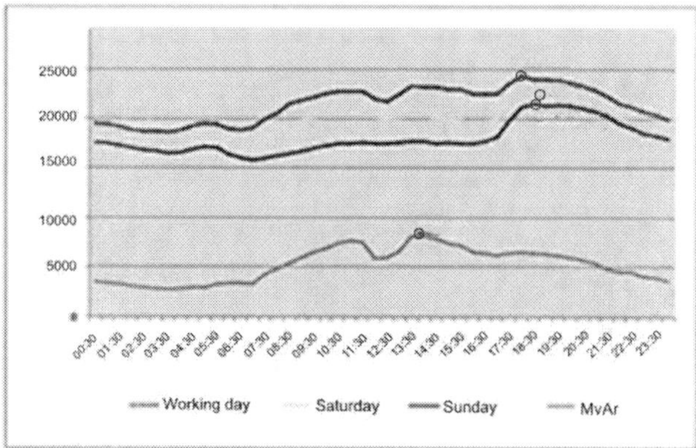

Figure 1. Daily Rhythm of Java Bali System 2015[6]

Forecasting very short term load forecasting required data tape (data logger) because the accuracy of load forecasting depends on the data provided. Therefore we need tools (tools) that can monitor the load with accurate and flexible so that the load data can be anytime required. In this study case study was conducted on Java-Bali electricity system because in that place is the largest system available in Indonesia. More than two decades, widely used fuzzy logic to control, prediction and optimization in power systems [5]. Previous Reshearches about Very Short Term Load using Artificial Neural Network (ANN) have result (MAPE) between 0,89% – 1,25% [6]. Whereas if using Based on Autoregressive Integrated Moving Average Model (ARIMA) have result between 2.62% - 5,27%, If using Adaptive Neuro-Fuzzy Inference System (ANFIS) have result between 10,21% - 18,45%[7]. This research try using Fuzzy Logic to Very Short Term Load Forecasting so get the better MAPE.

Fuzzy logic is one of the methods in forecasting short-term expenses. Therefore, this study developed forecasting for short-term expenses using Fuzzy Logic The period outside the peak load will be forecasted, since at the time burden characteristic is not too big.

2. Methodology

The very short term load forecasting, referred in this study, is the hourly load planning for a certain time on the same day of each year, based on very short run time data at the same time in the previous 4 days in 3 years period., The Flow chart of Methodology Showed Fig.2.

2.1. Calculation of Value of Input Variable X

Preparation stage preparedaily load data in every 30 minutes 4 , with four days earlier in 3 past years. This temporary calculation is used by the working day represented by Friday. This process is done to find the actual Variation Load Deffrence (VLD_{MAX}). It is used to calculate Variable X.

1. Search examples for the load value at peak Load time 22:00 on the first Friday of October 2015, using data on the first Friday of October 2013, 2014 and 2015.
2. Identify the load to search (P_{max}) on the previous four days at the same hour before the load time is analyzed.

$$MaxWD_{(i)} = \frac{WD_{(i)d-4} + WD_{(i)d-3} + WD_{(i)d-2} + WD_{(i)d-1}}{4} \tag{1}$$

3. Calculate the difference in load sought (Load Difference) on the clock to be predicted.

$$LD_{MAX}(i) = \frac{MaxSD(i) - MaxWD(i)}{MaxWD(i)} x100 \tag{2}$$

4. Looking for load difference characteristics on typical load hours or TLD_{MAX} (i) by averaging the same LD_{MAX} peak load (i) in previous years.
5. Search for Variation Loads on the clock to be predicted (Variation Load Difference) at each hour

$$VLD_{\max}(i) = LD_{\max}(i) - TLD_{\max}(i) \tag{3}$$

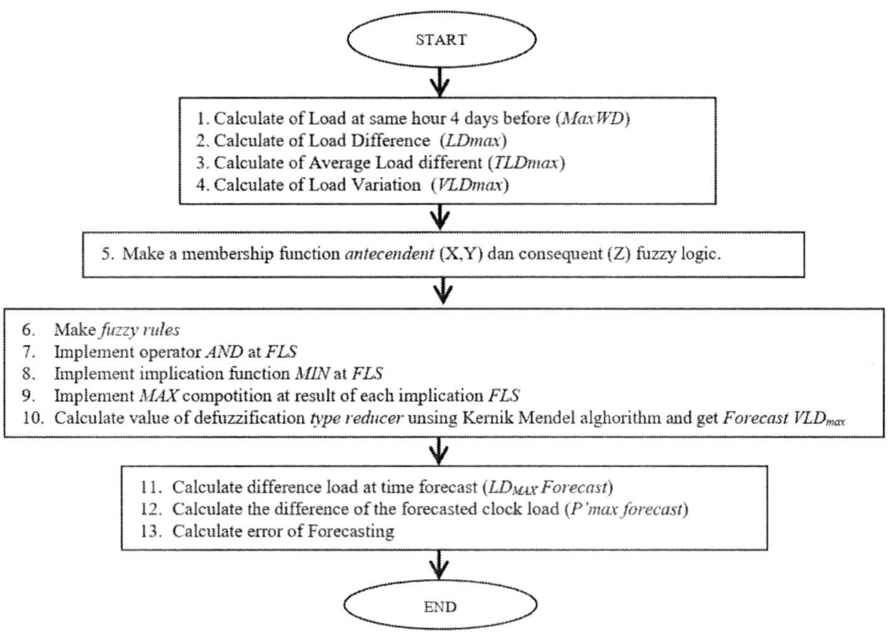

Figure 2. Flow Chart of Research

a) Electrical power at 22.00 First Friday in October 2013 $(R_1\text{-}2013)$

$MaxWD(R_1\text{-}2013)_{d\text{-}4} = 22.345$ MW
$MaxWD(R_1\text{-}2013)_{d\text{-}3} = 20.432$ MW
$MaxWD(R_1\text{-}2013)_{d\text{-}2} = 19.700$ MW
$MaxWD(R_1\text{-}2013)_{d\text{-}1} = 21.735$ MW
$MaxSD = 20.564$ MW

$$MaxWD(R_1\text{-}2013) = \frac{21.735 + 19.700 + 20.432 + 22.345}{4} = 21.053 \text{ MW}$$

$$LD_{MAX}(R_1\text{-}2013) = \frac{MaxSD(R_1\text{-}2013) - MaxWD(R_1\text{-}2013)}{MaxWD(R1\text{-}2013)} \times 100\%$$

$$Max_{WD}(\text{R1-2013}) = \frac{20.564 - 21.053}{21.053} \times 100\% = -0{,}02322$$

b) Electrical power at 22.00 First Friday in October 2014 $(R_1\text{-}2014)$

$MaxWD(R_1\text{-}2014)_{d\text{-}4} = 23.211$ MW
$MaxWD(R_1\text{-}2014)_{d\text{-}3} = 19.887$ MW
$MaxWD(R_1\text{-}2014)_{d\text{-}2} = 23.211$ MW
$MaxWD(R_1\text{-}2014)_{d\text{-}1} = 20.137$ MW
$MaxSD = 21.437$ MW

$$MaxWD(R_1\text{-}2014) = \frac{20.137 + 23.211 + 19.887 + 23.211}{4} = 21.612 \text{ MW}$$

$$LD_{MAX}(R_1\text{-}2014) = \frac{MaxSD(R_1\text{-}2014) - MaxWD(R_1\text{-}2014)}{MaxWD(R_1\text{-}2014)} \times 100\% = \frac{21.437 - 21.612}{21.612} \times 100\% = -0{,}00808$$

$$TLD_{MAX}(R_1\text{-}2014) = \frac{LD_{MAX}(R_1\text{-}2013) + LD_{MAX}(R_1\text{-}2014)}{2} = \frac{-0{,}02322 + (-0{,}00808)}{2} = -0{,}016$$

$$VLD_{MAX}(R_1\text{-}2014) = LD_{MAX}(R_1\text{-}2014) - TLD_{MAX}(R_1\text{-}2014) = -0{,}00808 - (-0{,}016) = -0{,}024$$

c) Electrical power at 22.00 First Friday in October 2015 $(R_1\text{-}2015)$

$MaxWD(R_1\text{-}2015)_{d\text{-}4} = 19.789$ MW
$MaxWD(R_1\text{-}2015)_{d\text{-}3} = 23.789$ MW
$MaxWD(R_1\text{-}2015)_{d\text{-}2} = 20.764$ MW
$MaxWD(R_1\text{-}2015)_{d\text{-}1} = 21.367$ MW
$MaxSD = 22.214$ MW

$$MaxWD(R_1\text{-}2015) = \frac{21.367 + 20.764 + 23.789 + 19.789}{4} = 21.427 \text{ MW}$$

$$LD_{MAX}(R_1\text{-}2015) = \frac{MaxSD(R_1\text{-}2015) - MaxWD(R_1\text{-}2015)}{MaxWD(R_1\text{-}2015)} \times 100\% = \frac{22.214 - 21.427}{21.427} \times 100\% = 0{,}03672$$

$$TLD_{MAX}(R_1\text{-}2015) = \frac{LD_{MAX}(R_1\text{-}2013) + LD_{MAX}(R_1\text{-}2014) + LD_{MAX}(R_1\text{-}2015)}{3} = \frac{-0{,}02322 + (-0{,}00808) + (0{,}03672)}{3} = 0{,}002$$

$$VLD_{MAX}(R_1\text{-}2015) = LD_{MAX}(R_1\text{-}2015) - TLD_{MAX}(R_1\text{-}2015) = 0{,}03672 - (0{,}002) = 0{,}035$$

In the same way we will get the values of LD_{MAX}, TLD_{MAX} and VLD_{MAX} as shown in Table 1, Table 2 and Table 3.

IOP Conf. Series: Materials Science and Engineering **403** (2018) 012070 doi:10.1088/1757-899X/403/1/012070

Table 1. Calculation Forecasting First Friday on October 2013

2	21.00	19.619	19.900	22.173	21.500	20.684	20.798	-0,548
3	20.00	22.987	20.564	19.988	21.800	21.464	21.335	0,606
4	19.00	21.800	20.684	23.341	22.983	22.173	22.202	-0,131

Table 2. Calculation Forecasting First Friday on October 2014

	day / Hour	d-4	d-3	d-2	d-1	d	WDMAX 2014	LDMAX 2014	TLDMAX 2014	VLDMAX 2014
1	22.00	23.211	19.887	23.211	20.137	21.437	21.612	-0,807	-1,565	0,757
2	21.00	21.100	21.900	20.451	21.378	21.404	21.207	0,928	0,190	0,738
3	20.00	20.342	23.898	23.455	22.843	22.452	22.635	-0,806	-0,100	-0,706
4	19.00	22.452	21.404	20.275	22.965	22.483	21.774	3,256	1,563	1,693

Table 3. Calculation Forecasting First Friday on October 2015

	day / Hour	d-4	d-3	d-2	d-1	d	WDMAX 2015	LDMAX 2015	TLDMAX 2015	VLDMAX 2015
1	22.00	19.789	23.789	20.764	21.367	22.214	21.427	3,672	0,181	3,491
2	21.00	20.400	21.324	23.008	21.700	21.286	21.608	-1,490	-0,370	-1,120
3	20.00	21.286	20.477	23.236	23.008	22.357	22.002	1,615	0,471	1,143
4	19.00	23.900	24.112	22.134	23.236	23.008	23.346	-1,446	0,560	-2,006

2.2. Calculation of Value of Input Variable Y

In the same way as finding the value of variable X, then we can the value of variable Y as set out in Table 4, Table 5 and Table 6.

Table 4. Calculation Forecasting second Friday on October 2013

	Day / Hour	d-4	d-3	d-2	d-1	d	WDMAX 2013	LDMAX 2013
1	22.00	19.213	19.111	18.900	20.400	19.332	19.406	-0,381
2	21.00	19.567	19.954	19.265	20.178	19.890	19.741	0,755
3	20.00	20.517	19.998	19.786	21.762	20.700	20.516	0,898
4	19.00	21.138	21.776	21.563	22.321	21.900	21.700	0,924

Table 5. Calculation Forecasting second Friday on October 2014

	Day / Hour	d-4	d-3	d-2	d-1	d	WDMAX 2014	LDMAX 2014	TLDMAX 2014	VLDMAX 2014
1	22.00	20.116	19.800	20.891	21.400	20.200	20.552	-1,712	-1,046	-0,665
2	21.00	20.200	22.911	20.883	21.789	21.400	21.446	-0,213	0,271	-0,484
3	20.00	22.989	19.899	22.911	23.348	22.400	22.287	0,508	0,703	-0,195
4	19.00	22.400	21.400	24.321	23.783	22.833	22.976	-0,622	0,151	-0,773

Table 6. Calculation Forecasting second Friday on October 2015

	Day / Hour	d-4	d-3	d-2	d-1	d	WDMAX 2015	LDMAX 2015	TLDMAX 2015	VLDMAX 2015
1	22.00	21.977	20.864	19.121	19.562	20.450	20.381	0,339	-0,585	0,923
2	21.00	20.450	22.787	20.231	22.113	21.270	21.395	-0,585	-0,015	-0,571
3	20.00	21.966	20.450	23.222	22.114	22.353	21.938	1,892	1,099	0,792
4	19.00	23.322	23.114	21.347	23.222	22.999	22.751	1,089	0,464	0,625

2.3. Calculation of Value of Input Variable Z

The calculation of the variable very short term forecasting at 22.00 The first Friday of October 2015 is to find the value of Variable Load Deference (VLD_{MAX}) forecasting hours. With the same calculation for the second Friday of October between 2013 - 2015 in get the value (VLD_{MAX}) which results can be seen as Table 7.

Table 7. Value of WD_{MAX}, LD_{MAX} and VLD_{MAX} 2013-2015

2	21.00	20.798	(0,548)	21.207	0,928	0,190	0,738	21.608	(1,490)	(0,370)	(1,120)
3	20.00	21.335	0,606	22.635	(0,806)	(0,100)	(0,706)	22.002	1,615	0,471	1,143
4	19.00	22.202	(0,131)	21.774	3,256	1,563	1,693	23.346	(1,446)	0,560	(2,006)
	Second Jumat on October										
1	22.00	19.406	(0,381)	20.552	(1,712)	(1,046)	(0,665)	20.381	0,339	(0,585)	0,923
2	21.00	19.741	0,755	21.446	(0,213)	0,271	(0,484)	21.395	(0,585)	(0,015)	(0,571)
3	20.00	20.516	0,898	22.287	0,508	0,703	(0,195)	21.938	1,892	1,099	0,792
4	19.00	21.700	0,924	22.976	(0,622)	0,151	(0,773)	22.751	1,089	0,464	0,625

2.4. Membership Function for Input and Output Variable

Input variables (X, Y) and output variables (Z) consists of 11 fuzzy sets are described as follows:

Negative Very Big (NVB)	range of values -12 s/d -8
Negative Big (NB)	range of values -10 s/d -6
Negative Medium (NM)	range of values -8 s/d -4
Negative Small (NS)	range of values -6 s/d -2
Negative Very Small (NVS)	range of values -4 s/d 0
Zero (ZE)	range of values -2 s/d 2
Positive Very Small (PVS)	range of values 0 s/d 4
Positive Small (PS)	range of values 2 s/d 6
Positive Medium (PM)	range of values 4 s/d 8
Positive Big (PB)	range of values 6 s/d 10
Positive Very Big (PVB)	range of values 8 s/d 12

The mathematical description of the antecendent membership function (X, Y) and concequent (Z) is used for the manufacture of Rules Base for the Fuzzy Inference System process. The establishment of Fuzzy Rule Base for very short term forecasting for 2015 is shown in Table 8 up to Table 13.

Table 8. Input (X, Y) and output (Z) By VLD_{MAX} in 2014 and 2015

	Hour	*VLDMAX*	*VLDMAX*	Input		output
		2014	*2015*	X	Y	Z
1	22.00	0,757	3,491	0,757	0,923	3,491
2	21.00	0,738	-1,120	0,738	-0,571	-1,120
3	20.00	-0,706	1,143	-0,706	0,792	1,143
4	19.00	1,693	-2,006	1,693	0,625	-2,006

Tabel 9. Process Rules for Input X in 2015

			Derajat Keanggotaan											Himp
	Hour	Nilai X	NVB	NB	NM	NS	NVS	ZE	PVS	PS	PM	PB	PVB	X
1	22.00	0,757						0,737	0,263					ZE
2	21.00	0,738						0,77	0,23					ZE
3	20.00	-0,706					0,21	0,79						ZE
4	19.00	1,693						0,48	0,52					ZE

Tabel 10. Process Rules for Input Y in 2015

			Derajat Keanggotaan											Himp
	Hour	Nilai Y	NVB	NB	NM	NS	NVS	ZE	PVS	PS	PM	PB	PVB	Y
1	22.00	0,923						0,69	0,31					ZE
2	21.00	-0,571					0,18	0,82						ZE
3	20.00	0,792						0,71	0,29					ZE
4	19.00	0,625						0,83	0,17					ZE

Tabel 11. Process Rules for Output Z in 2015

			Derajat Keanggotaan											Himp
	Hour	Nilai Z	NVB	NB	NM	NS	NVS	ZE	PVS	PS	PM	PB	PVB	Z
1	22.00	3,491							0,68	0,32				PVS
2	21.00	-1,120					0,35	0,65						ZE
3	20.00	1,143						0,35	0,65					PVS
4	19.00	-2,006					0,12	0,88						ZE

Tabel 12. Basic Rules table (fuzzy rules) for forecasting the year 2015

NVS											
ZE						PVS/ZE					
PVS											
PS											
PM											
PB											
PVB											

Tabel 13. Conversion Table Basic Rules Forecasting the Year 2015 for Matlab Software Code

no	Antecendent		Cons
rules	X	Y	Z
1	ZE	ZE	ZE
2	ZE	ZE	PVS

no	Antecendent		Cons
rules	X	Y	Z
1	6	6	6
2	6	6	7

Very Short-term load forecasting using Fuzzy Logic Executed by using Matlab software to obtain the va forecasting. by using Equation VLD_{MAX} which has been obtained, it will get the results of the comparison ta shown in Table 14.

Table 14. Comparison Forecasting and Actual load on First Friday October 2015

	pukul	h-4	h-3	h-2	h-1	h	WDMAX	LDMAX	TLDMAX	VLDMAX	Output Forecast	Forecast LDMAX	Forecast P'MAX	Actual (MW)	Error (%)
							2015								
	First Jumat on October 2015														
1	22.00	19.789	23.789	20.764	21.367	22.214	21.427	3,6720	0,1809	3,4911	2,7625	2,943388	22.058	22.214	0,7028
2	21.00	20.400	21.324	23.008	21.700	21.286	21.608	(1,4902)	(0,3702)	(1,1200)	-2,0028	-2,3729899	21.095	21.286	0,8962
3	20.00	21.286	20.477	23.236	23.008	22.357	22.002	1,6146	0,4714	1,1433	0,7683	1,2396907	22.275	22.357	0,3690
4	19.00	23.900	24.112	22.134	23.236	23.008	23.346	(1,4457)	0,5600	(2,0056)	-2,5276	-1,9676388	22.886	23.008	0,5296
													MAPE average		0,6244

In Table 14, We can find the average error value used Fuzzy Logic Load at First Friday October 2015 (; 21.00; 19.00:18) have MAPE arround 0,6244 %. Actual and forecast Peak Load Time First Friday on O 2015 (22.00 ; 21.00; 19.00:18) is shown in Figure 3.

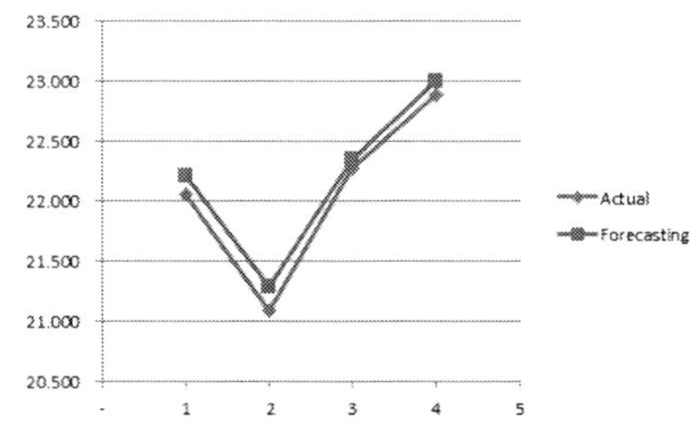

Figure 3. Actual and forecast Peak Load Time First Friday on October 2015 (22.00 ; 21.00; 19.00:18)

3. Conclusion

From the analysis we can conclude that the Very short-term load forecasting at First Friday on October 2015 (: 22.00; 21.00; 20.00) has an error value 0,6244 % of MAPE.

Thus the Fuzzy Logic can be proposed as one of the methods used to conduct very short-term load forec The membership function can be expanded to increase the accuracy of the model.t model. Expanded memb function may shrink the data range for resulting more accurate forecasting results.

References

[1] Jamaaluddin;Imam Robandi, "Short Term Load Forecasting of Eid Al Fitr Holiday By Using In Type – 2 Fuzzy Inference System (Case Study : Electrical System of Java Bali in Indonesia)," in *IEEE Region 10, TENSYMP*, 2016, vol. 0, no. x, pp. 237–242.

[2] I. Robandi, *Desain Sistem Tenaga Modern*. Yogjakarta: Andi Ofset, 2006.

[3] I. Robandi, *Modern Power System Control*, 1st ed. Yogjakarta: Andi, 2009.

[4] M. H. P. Agus Dharma, Imam Robandi, "Application of Short Term Load Forcasting On Special Using Interval Type-2 Fuzzy Inference System; Study Case in Bali Indonesia," *J. Theor. App. Technol.*, vol. Vol.49 no.

[5] S. Ahmadi, H. Bevrani, and H. Jannaty, "A fuzzy inference model for short-term load forecasting," *Second Iran. Conf. Renew. Energy Distrib. Gener.*, pp. 39–44, 2012.

[6] W. Charytoniuk and M.-S. Chen, "Very Short-Term Load Forecasting Using Artificial Neural Network *IEEE Trans. Power Syst.*, vol. 15, no. 1, p. 263, 2000.
[7] L. C. Moreira de Andrade and I. Nunes da Silva, "Very Short-Term Load Forecasting Based on ARIMA Model and Intelligent Systems," *2009 15th Int. Conf. Intell. Syst. Appl. to Power Syst.*, pp. 1–6, 2009.

IOP Conf. Series: Materials Science and Engineering **403** (2018) 012071 doi:10.1088/1757-899X/403/1/012071

Android based Tarjih Muhammadiyah information systems

M Iqbal and Suyono

STIKOM Muhammadiyah Batam
Prof. Dr. Hamka St No. 3 Batu Aji, Batam, Indonesia

rnpanay@yahoo.com

Abstract. Muhammadiyah is a large Islamic organization in Indonesia. Muhammadiyah does da'wah movement in order to realize the main Islamic society. This study aims to know clearly the considerations of the Assembly Tarjih and Tajdid of Muhammadiyah in determining the fatwa and decisions of everything in daily life. The research generates an android application of material viewers of Muhammadiyah in the form of tarjih fatwah and question and answer.. The app has a knowledge and search comparison feature. The application runs well on a tablet or smartphone with the minimum requirement of CPU specification is 600/900 MHz, need a memory space of 111 Mb which consists of 2 files of the installer file (Muhammadiyah Guide.apk of 6.59 Mb) and data files (org.muhammadiyah 105 Mb) and Android Frozen Yoghurt Operating System (Froyo) 2.2.2.

Keywords: Android, Tarjih Muhammadiyah, Mobile.

1. Introduction

Mobile communication technology has created a major change in the world of communication. The rapid development of these technological devices has given rise to a variety of new functions Mobile device which is one of communication media become part of society need even mobile device become a common thing, because the circulation is very wide and easy to use. Formerly this mobile application only created and developed by the telecommunications companies themselves. But now with the standard of Android programming language, everyone can create and develope the desired applications for a mobile device.

As described by Syamsul Anwar, the decisions of the Majelis Tarjih Muhammadiyah are divided into three forms. The forms of the decision are, (1) - the decision of the Tarjih; (2) The Fatwa of the Tarjih Assembly; and (3) Tarjih Discourse. The Tarjih Decision is made in a formal mechanism called the National Deliberation of Tarjih and has a binding force for members. This ruling has a stronger authority based on al-qur'an and hadith. The popular Tarjih verdict known among Muhammadiyah members is the HPT [1].

The Tarjih discourse contains ideas, thoughts, and opinions on contemporary issues. It is open for discussion and can be rejected. Borrow fiqh terminology, 'Tarjih discourse is mubah.

Muhammadiyah have the application of the Tarjih Decision Assemblage (HPT) of Muhammadiyah. However, This application is able to display various kinds of information about the Decision of the Muhammadiyah Tarjih Assembly which is text with pdf format in the form of online book, so user must read or pass through sheets before if we want to read a particular topic, because how to read existing HPT applications such as manuals and not digitally arranged yet.

Therefore, we will make an application of the Tarjih Decision Assemblage (HPT) of Muhammadiyah based on andorid. HPT may facilitate Fatwah tarjih Muhammadiyah always in the application made into various topics of discussion.

2. Theoretical Basis

2.1. *Tarjih Muhammadiyah Forum*

Tarjih in terms of association is a brief description of "Matan Faith and Aims of Muhamadiyah" is to compare opinion in discussion and take which have a stronger reason. In the first stages, task of the Majlis Tarjih was appropiate with the name which merely to choose among some of the opinions that exist in the Khazanah of Islamic Thought, which is considered stronger. The Fatwa of the Tarjih Assembly is a decision made outside the formal mechanism [2].

The fatwa is issued by the Tarjih Assembly to respond to questions raised by members or the wider society, sent to the editors of Muhammadiyah official magazines, Muhammadiyah votes, or direct to the Tarjih Assembly. According to Syamsul Anwar, the fatwa issued by the Tarjih Assembly will be binding on the members and questioners. But unlike Syamsul Anwar's opinion, Muhammad Azhar 'is of the opinion that the Fatwa of the Tarjih Assembly is not too binding on Muhammadiyah members but good if followed.

2.2. *Software Engineering*

According to O'Brien in 1999, Software is the entire command used to process information. The software can be a program or procedure. Program is a collection of commands understood by the computer while the procedure is a command required by the user in processing information [3].

2.3. *Information System*

In fact, what is meant by information system does not have to involve computer. Information System using computer commonly called computer based information system (Computer Based Information System or SBIS). In Practice, the term information most used without computer-based frills although in reality the computer is an important part [4].

2.4. *Android*

Android is an open source operating system, and Google is releasing its code under the Apache License. Open source code and licensing licenses on Android allow software to be freely modified and distributed by device makers, wireless carriers and app developers. In addition, Android has a large number of app developer communities (apps) that extend device functionality, commonly written in customized versions of the Java programming language.

Android architecture consists of these following parts:

a. Applications and Widgets: layer (layer) where the user is only associated with the application only.
b. Applications Framework: the layer where the developers Doing the creation of applications that will run on the operating system Android with its components include views, contents Provider, resource manager, notification manager, activity manager.
c. Libraries: layers where android features reside above the kernel include core C / C ++ libraries such as Libc and SSL.
d. Android Run Time: layers that execute Android is divided into two parts namely Core Libraries and Dalvik virtual machine.
e. Linux Kernel: Layer manages the Processing, memory, resources, drivers, and android operating system Others.

2.5. *Unified Modeling Language (UML)*

UML is a visual language for modeling and communication of a system using diagrams and supporting texts [5] says: In the development of object oriented programming techniques, there emerged a standarized modeling language for software development that was built using object-oriented programming techniques, namely Unified Modeling Languange (UML).

3. Research Methods

This research is using Research and Development method to develop products to have a higher quality. Research and development results can be: Plan, Design, Building / Construction (Construct), or Development Results of a Model/Method/Software/Hardware / System [6]. Plans, designs or constructions of engineering models, systems or products must be tested on the basis of formal methods, computational methods (hard / soft computing).

In writing this research authors collect data using data collection methods as follows :

a. Primary Data, that is data collected by researchers directly from the source.
b. Secondary Data, that is data that first collected by other party. This data is obtained by means of, among others: documentation or research library, as well as books that have to do with research.

Software (software) is a computer program associated with the documentation of needs, design models, and how to use (user manual). A computer program without associated with the documentation then can not be called software (software) [7.

The software characters are as follows:

1. Software is built by engineering (software engineering) instead Manufactured by manufacture or manufacturer.
2. The software is never outdated ("wear out") due to a deep disability Software can be fixed.
3. Production goods manufacturers usually new components will continue to be produced, While the software usually continues to improve as it grows needs.

Creation of android-based applications concerning Information System Tarjih Muhammadiyah, need to make the flow and process in the reciprocal system in the form of diagrams.

Use Case Diagram (Figure 1) is a graphical representation of some or all of the actors, use cases, and interactions that introduce a system. Use case diagrams do not explain in detail the use of use cases, but only give a brief description of the relationship between usecase, actor, and system. In this use case will be known what functions are on the system created.

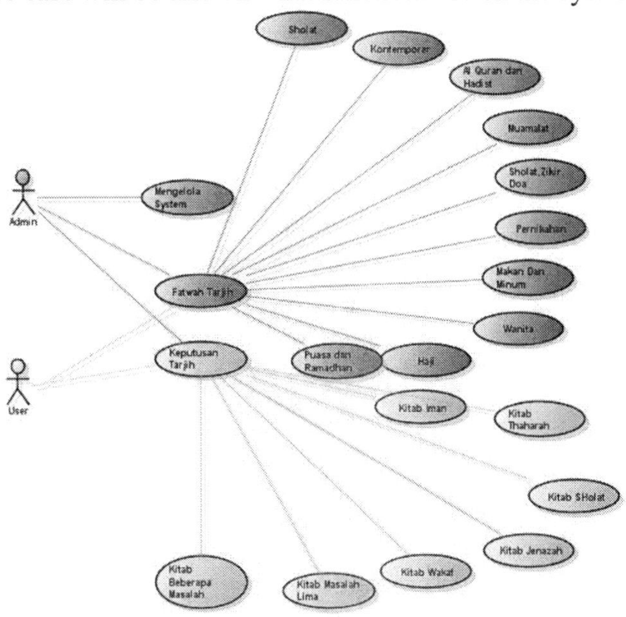

Figure 1. Use Case Diagram Information System Tarjih Muhammadiyah

State Machine Diagram (Figure 2) is a technique commonly used to describe the behavior of a system. It is used to help analysts, designers and developers to understand the behavior of objects on the system.

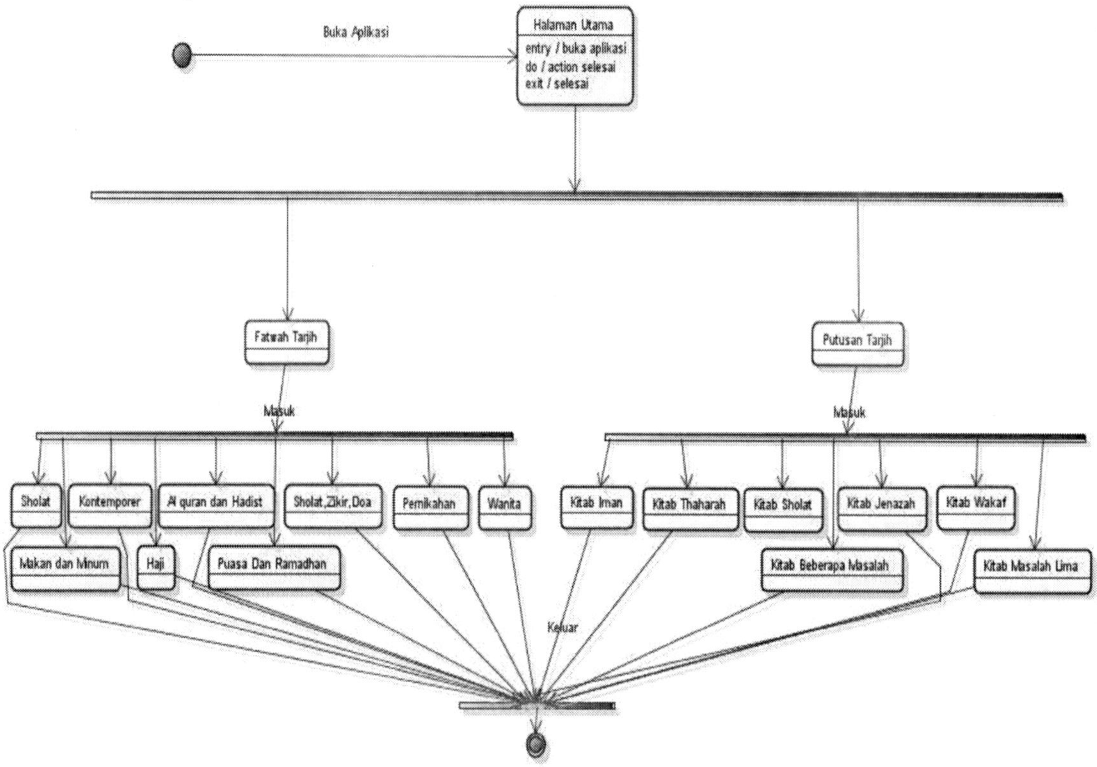

Figure 2. Diagram State Machine Information System Tarjih Muhammadiyah

Activity diagram, as the name implies this diagram describes the activities that occur on the system. From first to last, this diagram shows the steps in the system work process that we make.

4. System Display Results

The main view is the initial view of this tarjih muhammadiyah information system application. Consists of two menus are: Fatwah Tarjih, Decision Tarjih and exit (Figure 3).

Figure 3. Display Main Menu Tarjih Muhammadiyah

After the main view appears, then enter the menu view :

a. Fatwah Tarjih Muhammadiyah Menu View
   Views on fatwah tarjih menu there are sub-sub menus available such as: aqidah, prayer, al-quran and hadith, zadat, prayer, zikr, women, fasting and ramadan (Figure 4).

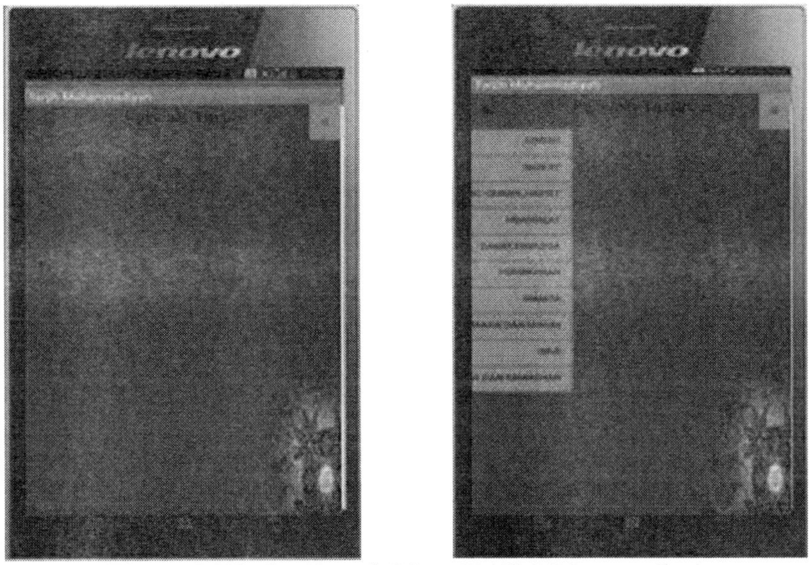

Figure 4. Display Fatwah Menu Tarjih Muhammadiyah

b. Display Menu Sentence Tarjih Muhammadiyah
   In the tarjih decision menu view there are sub-menus that have been available such as: book of faith, book of thaharah, book of prayer, book of corpse, book of waqf, book of problem five, book of some problem (Figure 5).

Figure 5. Display Menu Decision Tarjih Muhammadiyah

Sub menu is useful to inform the information contained in muhammadiyah tarjih according to sub menu choice.

a. Fill Sub Menu Fatwah Tarjih Muhammadiyah
   In the Decision Tarjih menu there is a sub menu selection that contains about Muhammadiyah Tarjih Decision (Figure 6).

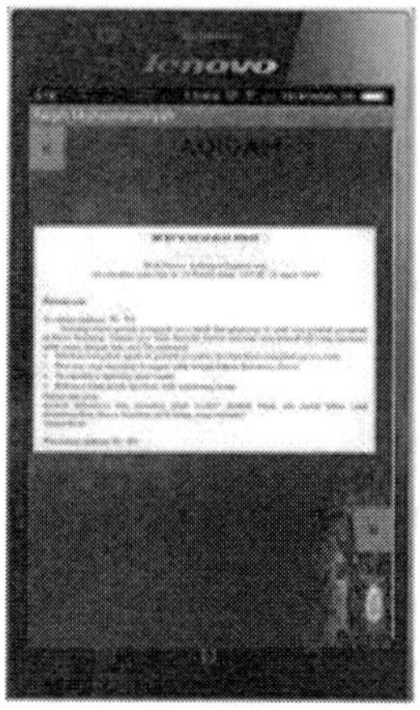

Figure 6. Content of Fatwah Sub Menu Tarjih Muhammadiyah

b. Sub division Sub Menu Tarjih Muhammadiyah
   In the Fatwah Tarjih Menu there is a sub menu collection in the sub menu contains the Fatwa Tarjih Muhammadiyah (Figure 7).

Figure 7. Content of Fatwah Sub Menu Tarjih Muhammadiyah

5. Conclusion

The result has successfully created information system Tarjih Muhammadiyah. This application consists of fatwah and judgment tarjih muhammadiyah.

This Information System has implemented android technology, using an attractive design makes this application look good and this application can be used anytime and anywhere.

References

[1] Qaem Aulassyahied. 2016. Discourse Interconnection Studies Hadith, Tarjih Journal Vol.13(2)
[2] Tarjih PP Muhammadiyah Forum (t.t), Muhammadiyah Tarjih Decision Forum,pp.284-28

IOP Conf. Series: Materials Science and Engineering **403** (2018) 012071 doi:10.1088/1757-899X/403/1/012071

[3] Mulyanto, Aunur R. 2009. Software Engineering, Vol 1. Jakarta: Directorate of Vocational High School Development.
[4] Abdul Kadir, (2014). "Introduction Of Information Systems" Revised Edition.Andi.Yogyakarta
[5] S, Rosa dan M. Shalahuddin. 2015. Structured Software Engineering and Object Oriented. Bandung: Informatika
[6] Rianto Rahadi, Dodi. (2014). Measurement Usability System Using Use Quesionnaire On Android Applications. Postgraduate Program Bina Darma University Palembang.Palembang
[7] Shalahuddin, M., Rosa A.S. 2011. Software Engineering (structured and object-oriented). Modula : Bandung.

IOP Conf. Series: Materials Science and Engineering **403** (2018) 012072 doi:10.1088/1757-899X/403/1/012072

User participation in building language repository: the case of Google Translate

H Thamrin[1*], G Ariyanto[1], E W Pamungkas[1] and Y Sulistyono[2]

[1]Department of Informatics, Universitas Muhammadiyah Surakarta, Jl. A Yani Pabelan, Kartasura, Sukoharjo 57162, Indonesia
[2]Department of Education of Bahasa Indonesia, Universitas Muhammadiyah Surakarta, Jl. A Yani Pabelan, Kartasura, Sukoharjo 57162, Indonesia

[*]husni.thamrin@ums.ac.id

Abstract. A language repository is a place that stores language resources, such as dictionary or corpus. Accurate and large size repository is important to improve the performance of natural language processing applications, especially those that employs repository-based algorithms. This research tries to confirm the possibility of building language repository in Bahasa Indonesia involving user participation. We survey 191 people about their experience and opinion in using Google Translate, at the time they first used the application several years ago and recently. Opinions were then analyzed for sentiment score using lexicon-based sentiment analysis. The shift of sentiment was calculated and confirmed using t-test. There is a change in opinion score, from 0.3174 at first time use of the application to 0.4980 at recent use. The improvement can be attributed to the introduction of better translation algorithm and the improvement of data repository (or dictionary) by crowdsourcing. We conclude that it is possible to involve users to build language repository for Bahasa Indonesia.

1. Introduction

Repository is a place that collects or stores resources, and the term usually has implicit mean of immensity. The field of Natural Language Processing (NLP) has introduced more specific meaning for repository, which is usually a form of digital storage containing dictionary, corpus, or knowledge base. Corpus and knowledge base are collections of words, phrases or sentences that have been annotated with particular information [1]. Repository is important in NLP because many algorithms work upon its existence, for example algorithms that utilize WordNet [2, 3], N-gram[4], or lexicon [5, 6].

Repositories for Bahasa is unfortunately rare. Big dictionary may be the only repository in Bahasa. Thesaurus has been attempted to build, but the size is low. Other type of repositories such as WordNet, do not yet exist although attempts to build one have long been initiated [7]. Hence, implementation of repository based algorithms in Bahasa has not been successful.

Building a repository can be expensive because the work does not only involve digital data collection but also data annotation. Annotation is conducted by language experts and/or native speakers of the specified language. Repositories are traditionally built by a team of experts in a long-term project, which may involve people under supervision. Examples are the development of language dictionaries [8] and WordNet [9]. However, the advances of internet technology and virtual interaction among people have facilitated a way to build repositories involving people in the virtual world. This is commonly called crowdsourcing. Wikipedia has a successful story to build digital repository containing encyclopedic entries. Wikipedia has more than 3 million users that produce 45 million articles in various languages [10]. The number of Wikipedia articles is growing with the rate of 20 thousands articles per month involving more than 100 thousand active users.

To overcome the problem of building language repository for Bahasa, we proposed a solution to involve people in an online system. This article describes the potential to build language repository for Bahasa Indonesia, taking the case of Google Translate (GT). At the current stage, this research tried to confirm whether crowdsourcing can be successful and improve the performance of an application that utilizes the repository. This study reveals that many Indonesian speakers believe that GT has been performing better in recent years. The achievement can be attributed to the introduction of crowdsourcing in building the repository.

2. Related Works

Efforts to build language repositories for Bahasa Indonesia have been conducted by many parties. The earliest attempt was conducted by *Balai Bahasa* which built *Kamus Besar Bahasa Indonesia* (Grand Dictionary of Bahasa). The first edition was printed in 1988 and was available online at 2008 [11]. The same institution has published thesaurus containing synonyms, hyponyms and meronyms [12].

Building word net for Bahasa has also been attempted by [7] replicating the model of English WordNet. It has 1203 synset and 1659 words. However, this process does not seem to be continuing and the repository is hardly accessible. The idea to build Indonesian corpora has been raised several years ago by [13] but the result has not been evident.

In the aforementioned works, repository building is conducted by team members of the corresponding projects. Attempt has been rare to involve public in building an Indonesian language repository. We can mention Oxford Dictionaries and Google Translate among websites that try to attract Indonesian people to contribute in building English-Indonesian dictionary. To the extent of our knowledge, no Indonesian institution has been conducting such an effort.

3. Method

To confirm whether building repository of Bahasa through user participation can be successful, we have selected Google Translate (GT) as a study. This is because GT has started to use crowdsourcing for a while, since 2014 and the resulted repository is used in another part of the application [14].

3.1. Data collection

We surveyed 191 people about their experiences dealing with the GT. Our respondents come from various background including students, civil servants, and private sector employees. Their ages starts from 21 to more than 40.

Table 1. List of questions in our questionnaire

No	Questions
1	Have you ever used Google Translate?
2	When did you first access Google Translate?
3	What was the last time you use Google Translate?
4	What are your reasons in using Google Translate?
5	In the scale of 1 – 10, rate the accuracy of Google Translate in translating a word?
6	In the scale of 1 – 10, rate the accuracy of Google Translate in translating a phrase?
7	In the scale of 1 – 10, rate the accuracy of Google Translate in translating a short sentence? (a complete sentence having less than 5 words)
8	In the scale of 1 – 10, rate the accuracy of Google Translate in translating a complex sentence?
9	How many times in total do you use Google Tranlate?
10	Have you ever recommended your colleagues to try Google Translate?
11	Write your opinion about the translation result by Google Translate at the first time you tried its service.
12	Write your opinion about the translation result by Google Translate at the last time you accessed the service.

We prepared a questionnaire containing 12 questions. The questions are grouped into three parts. The first part is to confirm that the respondent use the software. The second part asks the respondent's experience in using the software. The last part asks the respondent's opinion about the software. (see Table 1). User's opinion about the accuracy of GT translation is probed using 1 – 10 level scoring. On the other hand, user's opinion about the improvement of GT translation after years of usage is probed using free-text questions.

3.2. Data analysis

The main purpose of the research is to indicate whether user involvement and participation in compilation and validation of language resource or repository can improve the usefulness of the application that uses the resource or repository. Usefulness in this context may be proved by the opinion of users about the accuracy of resource and the shift of their opinion about the output of the application.

The opinion shift may be detected by analyzing the sentiment of answers to question 11 and 12 of the survey questionnaire. Both questions ask respondents for opinion about the translation result of GT. However, the first question focuses on the translation result several years ago, while the latter focuses on the translation result recently. As depicted in **Figure** 1, sentiment analysis and t-test are used to detect the opinion shift.

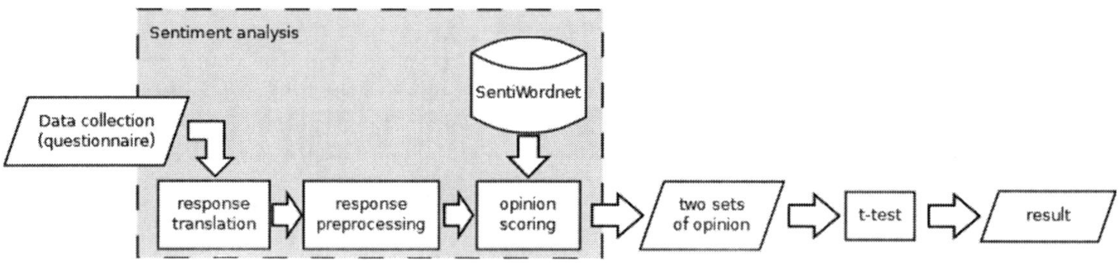

Figure 1. Data analysis

Sentiment analysis of user answers were conducted using a lexicon based method utilizing SentiWordNet [15]. Several steps have to be gone through. Our survey used Bahasa Indonesia and users answer questions in Bahasa, while the lexicon based sentiment analysis utilizes SentiWordNet, which is an English repository. Therefore, prior to conducting sentiment analysis, free-text user responses were translated into English. Thereafter, we analyze the sentiment of each response statement using SentiWordNet. The output of this series of processes is a sentiment score of each response statement.

Translated responses were preprocessed before performing sentiment scoring. There are two steps in preprocessing, i.e. tokenizing and POS Tagging. Tokenization of a response statement divides the statement into a token of words and symbols. Tokenization also helps the step of POS tagging. We use an existing library (Apache Lucene) to perform tokenization.

POS tagging is conducted after tokenization. It aims at getting the exact part of speech based on the context of the sentence. Part of speech for each word is very decisive in the searching process of sentiment score in SentiWordNet. We used Stanford POS Tagger to do POS Tagging process. At the end of this process, we'll get a list of tokens and the part of speech for each token.

SentiWordNet has the sentiment score of each token in the list, either positive or negative sentiment. Sentiment score of a response statement is obtained by summing up the sentiment score of tokens of that statement. Tokens are treated as lexicon in SentiWordNet. Part of speech of words is used to choose relevant synset. For polysemous words, we choose a sense using First Sense method.

Sentiment scores of responses for question 11 and 12 become two sets of scores. The first set resembles the opinion of users several years ago at the first time they used GT and the second set resembles their opinion recently at the last time they used GT. The two sets can be presented into a table (see **Table** 2). The average score can be used to decide whether the opinion of users shifts more positive or more negative. To confirm that there is an opinion shift, we use t-test to examine statistically whether the two sets are similar or not (in other words, the opinion from the two time point is the same or different). A change towards positive opinion suggests that the application is successful in doing what it should be doing.

Table 2. Output of sentiment analysis against respondent opinion

Respondent	**Opinion scores**	
	First time access	Recent access
respondent 1	...	...
...	...	...
Average		

4. Result and Discussion

4.1. Result

Most of our respondents are frequent users of Google Translate (GT), translating text from and to Bahasa Indonesia. Among 191 respondents, 164 people have used the web application frequently, i.e. more than 25 times. About 151 respondents, or 79%, started using the application more than three years ago and 39 respondents started just recently. Among the 151 respondents who started using GT three years ago, 138 (or 91%) are active users who still work with the application during the last three months. The statistics confirms that the respondents are reliable to give opinion about Google Translate in that most of them are active users for several years.

Our survey results in 191 pairs of responses where each pair contains statements about using GT at first time use and at recent use. Sentiment analysis generates sentiment score of each statement hence we've got 191 pairs of opinion scores, which are grouped into two sets of data. The first set is for opinion at first time use and the other set is for opinion at recent use. **Figure** 2 depicts the opinion scores for the two sets. Scores in the first set are depicted into red points and scores in the second set are depicted in into blue points.

Not all respondents state that GT is better recently than it used to be. As depicted in **Figure** 2, several respondents state that GT is not getting better (as indicated by blue point that is lower than red point). However, many respondents believe that the output of GT is better recently than it used to be several years ago (as indicated by blue point that is higher than red point). The average opinion scores of the two sets are 0.317 and 0.498, which means that in average the respondents have more positive sentiment during the use of GT recently than that when they first used the application.

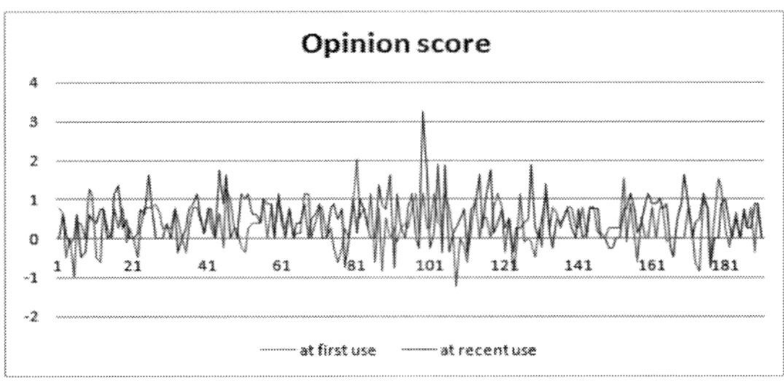

Figure 2. Opinion score for the 191 respondents at first time use (red) and at recent use (blue)

The average opinion scores for the two sets of data suggest that there is a change in user opinion. To prove the significance of the change, however, we have to examine the data using t-test. The t-test checks whether two groups of data are statistically different. Paired t-test was used because each pair of data was obtained from the same respondent and for that purpose we have used the data analysis feature obtained in Microsoft Excel.

Table 3 presents the result of paired t-test for the two sets. Variable 1 is the set of opinion score at first time use, which has a mean score at 0.3174, while variable 2 is the set of opinion score at recent use, which has a mean score at 0.498. The t-test results in t-value as high as 3.5865, which is higher than critical t-value, either for one-tail or two-tail test. The p-value is very low (less than 0.001) which suggests that the test result is not coincidental. The t-test result approves that the two sets of opinion scores are statistically and significantly different. Therefore, we can conclude that users have more positive sentiment on the recent output of GT as compared to their experience several years ago.

Table 3. Result of paired t-test

	Variable 1	*Variable 2*
Mean	0.3174	0.4980
Variance	0.2812	0.3318
Observations	191	191
Degree of freedom	190	
t Stat (t-value)	-3.5865	
P(T<=t) one-tail	0.0002131	
t Critical one-tail	1.6529	
P(T<=t) two-tail	0.0004262	
t Critical two-tail	1.9725	

4.2. Discussion

Google Translate has been improving in recent years as admitted by respondents in our survey. This is particularly true for translation from and to Bahasa Indonesia. Many respondents are apparently happy with the application and this statement is supported by the fact that 133 of respondents, or 70%, have ever recommended colleagues or friends to use GT. Such improvement may come from the improvement of translation algorithm. Translation algorithm of GT reached a new milestone around late 2016 when Google introduced the use of neural machine translation. The method reduces the number of translation error by 60 percent as compared to the previous algorithm called phrase-based machine translation [16]. However, translation process relies heavily on a form of repository, called dictionary. Hence, the size and correctness of repository plays a significant role in the improvement of the translation result. No one can imagine machine translation without the help of dictionary.

We realize that sentiment analysis may provide flawed output because of three reasons, i.e. phenomenon of thwarted expectation, slang and non-standard language, and ambiguous words. The flaw may be exacerbated because our sentiment analysis involves language translation from Bahasa to English which may not result in exact translated words [15], 17, 18]. However, we may expect that the use of the lexicon-based sentiment analysis on a large set of statements increase the variance of sentiment scores but it does not significantly change the mean. Therefore, the t-test result can still be valid.

Natural Language processing for Bahasa Indonesia still lacks in development. Existence of repository is one of the reasons. Bahasa has already had a dictionary and thesaurus but the number of entries is still low. On the other hand, many NLP algorithms require repositories to work, e.g. for [19]. Even the development of NLP algorithm needs repositories as a standard test bed. Therefore, it is imperative for the field of NLP to build repositories, and a way to do it is crowdsourcing.

5. Conclusion

This study confirms that the potential to involve users in building repository, particularly for Bahasa Indonesia, is existent. Our case study suggests that Google Translate has improved and the improvement can be attributed to the introduction of better translation algorithm and better data repository (or dictionary) involving user participation. The improvement is supported by our respondents' opinion. Respondents' statement about the result of GT translation at first time use has an average sentiment score of 0.3174, while their statement about the translation at recent use has an average sentiment score of 0.4980. Hence, it is possible to involve users to build repository of Bahasa Indonesia.

References

[1] J. Pustejovsky and A. Stubbs, *Natural Language Annotation for Machine Learning: A guide to corpus-building for applications*. " O'Reilly Media, Inc.," 2012.
[2] J. J. Castillo and M. E. Cardenas, "Using sentence semantic similarity based on WordNet in recognizing textual entailment," in *Advances in Artificial Intelligence--IBERAMIA 2010*, Springer, 2010, pp. 366–375.
[3] H. Liu and P. Wang, "Assessing sentence similarity using wordnet based word similarity," *J. Softw.*, vol. 8, no. 6, pp. 1451–1458, Jun. 2013.
[4] I. Islam, E. Milios, and V. Keselj, "Text Similarity Using Google Tri-Grams," in *25th Canadian Conference on Advances in Artificial Intelligence*, 2012, pp. 312–317.
[5] T. Wei, Y. Lu, H. Chang, Q. Zhou, and X. Bao, "A semantic approach for text clustering using WordNet and lexical chains," *Expert Syst. Appl.*, vol. 42, no. 4, pp. 2264–2275, Mar. 2015.
[6] H. Thamrin and A. Sabardila, "Utilizing Lexical Relationship in Term-Based Similarity Measure Improves Indonesian Short Text Classification," *ARPN J. Eng. Appl. Sci.*, vol. 11, no. 22, pp. 13141–13145, 2016.
[7] D. D. Putra, A. Arfan, and R. Manurung, "Building an Indonesian wordnet," in *Proceedings of the 2nd International MALINDO Workshop*, 2008.
[8] J. D. Anderson, "The Development of the English-French, French-English Bilingual Dictionary: a Study in Comparative Lexicography.," 1971.
[9] P. Smrž, "Quality control for wordnet development," in *Proceedings of the second international WordNet conference*, 2004, pp. 206–212.
[10] "Wikipedia: Size of Wikipedia." [Online]. Available: https://en.wikipedia.org/wiki/Wikipedia:Size_of_Wikipedia. [Accessed: 24-Aug-2017].
[11] D. Sugono, "Selamat datang di KBBI Daring," 2008. [Online]. Available: http://badanbahasa.kemdikbud.go.id/kbbi/. [Accessed: 01-May-2016].
[12] Pusat Bahasa, *Tesaurus Alfabetis*. Bandung: Mizan, 2009.
[13] R. Manurung, B. Distiawan, and D. D. Putra, "Developing an Online Indonesian Corpora Repository.," in *PACLIC*, 2010, pp. 243–249.
[14] F. Lardinois, "Google Wants To Improve Its Translations Through Crowdsourcing," 2014. [Online]. Available: https://techcrunch.com/2014/07/25/google-wants-to-improve-its-translations-through-crowdsourcing. [Accessed: 24-Aug-2017].
[15] E. W. Pamungkas and D. G. P. Putri, "An experimental study of lexicon-based sentiment analysis on Bahasa Indonesia," in *2016 6th International Annual Engineering Seminar (InAES)*, 2016, pp. 28–31.
[16] Y. Wu *et al.*, "Google's neural machine translation system: Bridging the gap between human and machine translation," *arXiv Prepr. arXiv1609.08144*, 2016.
[17] J. Holmes, *An introduction to sociolinguistics*. Routledge, 2013.
[18] B. Ohana and B. Tierney, "Sentiment classification of reviews using SentiWordNet," in *9th. it & t conference*, 2009, p. 13.
[19] H. Thamrin and A. Sabardila, "Using Dictionary in a Knowledge Based Algorithm for Clustering Short Texts in Bahasa Indonesia," in *2014 International Conference on Data and Software Engineering (ICODSE)*, 2014, pp. 1–4.

IOP Conf. Series: Materials Science and Engineering **403** (2018) 012073 doi:10.1088/1757-899X/403/1/012073

Prediction of population growth using Sugeno and Adaptive Neuro-Fuzzy Inference System (ANFIS)

A H S Jones[1], A Pranolo[2], A Dianto[3] and S Winiarti[4]
Department of Informatics, Universitas Ahmad Dahlan University, Indonesia

[1]annahendri@tif.uad.ac.id, [2]andri.pranolo@tif.uad.ac.id, [3]agusdianto92@gmail.com, [4]sri.winiarti@tif.uad.ac.id

Abstract. Government use population growth data to design sustainable policies frameworks. This research aims to predict the population growth using adaptive neuro-fuzzy inference system (ANFIS) and Sugeno as comparison method. The ANFIS consists of determining layers (1 to 5), system design, implementation, and system testing stage. The results of using ANFIS is 0.44% while prediction test of using Sugeno is 16,09% in year 2010. The Sugeno result is categorized as a negative growth since it is far up from the ideal rate set by the government of 0.5% per year.

1. Introduction

Indonesian government held the population census in every 10 years. Census data is the primary data, used by social, economic, and demographic planners. ANFIS (adaptive neural fuzzy inference system) combined the of fuzzy inference system and neural network adaptive learning ability, which used for optimization, data mining, selection multi criteria, and prediction [1-6].

The selection of the methods was taking based on previous works reported by Oprea *et al.* [4] who used ANFIS as an internal fuzzy rule base in a prediction. When rule base is empty, ANFIS generates the rules and adjust the FIS parameters to match the input output datasets. The results are compared with data mining technique and evaluated by RMSE, MAPE, and simulation time. ANFIS has a small value of RMSE which is indicates that ANFIS is better in prediction than another technique in this research. This is corroborated by Zhang and Lei [7] who reported that ANFIS convergence speed is faster and predictive values are in conformity with the measured values in the case of prediction of laser cutting roughness. In many cases, ANFIS are combined with several methods such as ANFIS-Genetic Algorithm, ANFIS-Differential evolution, ANFIS-particle swarm optimization [8]. ANFIS result is depend on selection of optimization method in training [5].

This research will predict the population growth using a computer based system. The method used in this research are Adaptive Neuro Fuzzy Inference System (ANFIS) to determine the prediction of growth rate of the inhabitants of the per year, where Sugeno [9, 10] is used as a baseline method. This research is designed to ease the government to predict the annual of population growth. The data retrieves from the Central Bureau of the Republic of Indonesia with population census period of 1961, 1971, 1980, 1990, 2000, 2010.

2. Method

2.1. Adaptive Neuro Fuzzy Inference System

Adaptive Neuro Fuzzy Inference System (ANFIS) is a combination of artificial neural networks and fuzzy logic. Fuzzy inference systems trained using artificial neural network systems. Thus, the hybrid possesses all advantages of artificial neural network systems and fuzzy inference systems. From its ability to learn, the neuro-fuzzy system is often referred as adaptive neuro fuzzy inference systems (ANFIS). One of the most well-known structures is the fuzzy inference of the Takagi-Sugeno-Kang model [9, 10], is shown in figure 1.

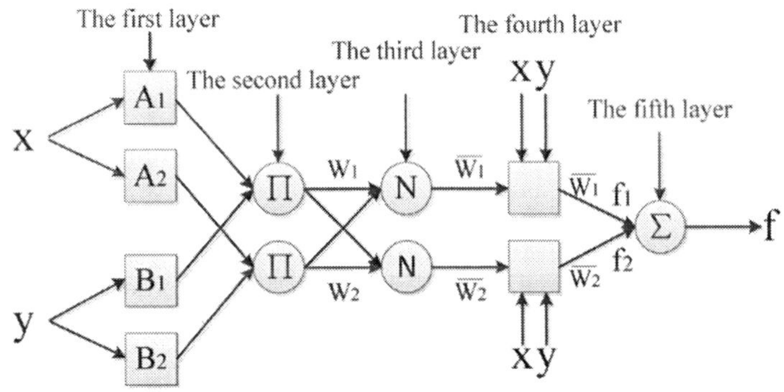

Figure 1. ANFIS structure.

In the Neuro-Fuzzy system, there are five layers of process fuzzification (1st Layer), production (2nd layer), normalization (3rd layer), defuzzification (4th layer), and summarization (5th layer) [10-12].

2.2. Sugeno methods

Fuzzy sugeno is a fuzzy inference technique which can be seen as the form of IF – THEN rules, where the output is a linear equation or constant values [9]. The Sugeno model uses the Singleton membership function that is a membership function that has a membership degree of 1 on a single crisp value and 0 on another crisp value [10].

a. Fuzzy Sugeno Orde-Zero

In general, the form of fuzzy model Sugeno Orde zero is shown in equation (1).

$$\text{IF } (X_1 \text{ is } A_1) \text{o } (X_2 \text{ is } A_2) \text{o} \dots \text{o } (X_n \text{ is } A_n), \text{then } z = k \tag{1}$$

where A_i is the set of fuzzy to-*i* as antecedents and k is a constant as a consequence.

b. Model Fuzzy Sugeno Orde-One

In general, the form of fuzzy model Sugeno Orde one is shown in equation (2).

$$\text{IF } (X_1 \text{ is } A_1) \text{ o } (X_2 \text{ is } A_2) \text{ o} \dots \text{o } (X_n \text{ is } A_n) \text{ THEN } z = p_1 * X_1 + p_2 * X_2 + \dots + p_N * X_n = q \tag{2}$$

where A_i is the set of fuzzy to-i as antecedents and p_i is a constant to i and q is also a constant in consequence.

3. Results and discussion

3.1. ANFIS

The ANFIS of population growth in Sleman regency – Indonesia as data sample is shown in equation (3). Where $\mu A(x)$ is degree of membership, x is population of census data, c is standard deviation, a is mean, and y, z are census time.

$$\mu_a(x) = \frac{1}{1+\left|\frac{x-c}{x}\right|^{2b}} \tag{3}$$

Based on figure 1 can be describes the process for each layer. The first layer, fuzzy layer, converts inputs into a fuzzy set by means of membership functions (MFs). It contains adaptive nodes, that can be seen in table 1.

Table 1. 1st layer.

Data	Y		Z	
	X1	X2	X1	X2
1	516,653	588,313		
2	588,313	677,323		
3			677,323	780,334
4	780,334	901,377		
5			901,377	1,093,110
Mean	628,433.3	722,337.6	789,350	936,722
Standard deviation	64,272.2	76,043.2	79,215.0	110,583.0

From table 1, The fuzzy set by MFs can be find by using describe as $\mu A_1(x) = \frac{1}{1+\left|\frac{x-628433,3}{64272,2}\right|^2}$, $\mu A_2(x) = \frac{1}{1+\left|\frac{x-722337,6}{76043,2}\right|^2}$, $\mu B_1(x) = \frac{1}{1+\left|\frac{x-789350}{79215,0}\right|^2}$, and $B_2(x) = \frac{1}{1+\left|\frac{x-936722}{110583,0}\right|^2}$. The result shows various forms of membership function for fuzzy set A. The parameters in this layer are called premise parameters which degree of its membership can be seen in table 2.

Table 2. Degree of membership.

Data	Degree of Membership			
	μA_1	μA_2	μB_1	μB_2
1	0.248465	0.243529	0.077817	0.091520
2	0.719602	0.740514	0.134395	0.153787
3	0.633468	0.632245	0.333333	0.333333
4	0.151846	0.152828	0.987211	0.907310
5	0.052537	0.040366	0.333333	0.333333

In the 2nd layer, every node is a fixed node, with the function node to be multiplied by input signals to serve as output signal as shown in equation (4).

$$W_1 = \mu A_1(x) * \mu B_1(x);\ W_2 = \mu A_2(x) * \mu B_2(x) \tag{4}$$

where W_1 and W_2 are *firing-strength,* $\mu A_1(x)$ is degree of membership of set Y, $\mu B_1(x)$ is degree of membership of set Z. The 2nd layer results can be seen in table 3.

Table 3. 2nd Layer result.

Data	Output 2nd layer	
	w_1	w_2
1	0.019335	0.022288
2	0.096711	0.113882
3	0.211156	0.210748
4	0.149904	0.138662
5	0.017512	0.013455

Every node in 3rd layer is considered a fixed node, marked by a circle and labelled by *N*, with function node to normalize the firing strength by computing the ratio of the node firing strength to sum of all firing strength rules by using equation (5).

$$\overline{w} = \frac{w1}{w1+w2}; \quad \overline{w} = \frac{w2}{w1+w2} \tag{5}$$

where $\overline{w}$ is normalized firing strength, w_1 and w_2 are the output of 2nd layer. The quantity $\overline{w}$ is known as the normalised firing strength. The output of 3rd layer can be seen in table 4.

Table 4. Normalized firing strength.

Data	**Output layer 3**	
	$\overline{w}_1$	$\overline{w}_2$
1	0.464527	0.535473
2	0.459233	0.540767
3	0.500483	0.499517
4	0.519478	0.480522
5	0.565501	0.434499

The output of the parameter coefficients can be seen in table 5.

Table 5. The parameter coefficients.

Data	**Parameter coefficients**					
	p_1	q_1	r_1	p_2	q_2	r_2
1	0.929053	1.858106	0.464527	2.677367	13.386836	0.535473
2	0.918465	1.836931	0.459233	2.703837	13.519184	0.540767
3	1.000966	2.001932	0.500483	2.497585	12.487924	0.499517
4	1.038956	2.077913	0.519478	2.402609	12.013046	0.480522
5	1.131002	2.262003	0.565501	2.172496	10.862481	0.434499

In the 4th layer, every node is an adjustable node, marked by a square, with node function as in equation (6).

$$o_{4,i} = \overline{w}_i \, . \mathrm{f}_i = \overline{w}_i(p_i x + q_i x + \mathrm{r}_i) \tag{6}$$

where p_i, q_i and r_i are the parameters set, referred to as the consequent parameters.
The last process is finding w_i and y_i for summarization process, $y' = \Sigma\overline{w}_i y_i = \overline{w}_1 y_1 + \overline{w}_2 y_2$, in 5th layer which can be shown in table 6.

Table 6. Output of 4th and 5th layer.

Data	**4th Layer**		**5th Layer**
	$\overline{w}_1 y_1$	$\overline{w}_2 y_2$	$\Sigma\overline{w}_i y_i$
1	10.822204	184.717673	195.539877
2	10.582174	188.275869	198.858043
3	12.523681	161.447750	173.971430
4	13.472641	149.786203	163.258845
5	15.915474	123.342183	139.257657

After finding the 5th layer, the data will be processed under normal circumstances, thus we need to normalize it by using equation (7).

$$result = \left(\frac{a_1}{L_5}\right) + a_1 \tag{7}$$

where a_1 is population growth data, and L5 is output of 5th layer. The results of long-term population forecasts normalization which is conducted using ANFIS, then it compared with the actual data from 1961 to 2000 (table 7).

Table 7. Comparison results between real data of population and ANFIS.

Data	Year	Population	ANFIS	Error
1	1961	516,653	519,295	0.51%
2	1971	588,313	590,911	0.44%
3	1980	677,323	680,292	0.44%
4	1990	780,334	783,498	0.41%
5	2000	901,377	905,087	0.41%
			Average	0.44%

3.2. Sugeno
In this research, Fuzzy Sugeno method will be used as a comparison. The steps of Sugeno method can be described as:
1) Establishment of membership for Increase function and decrease Function (figure 2).

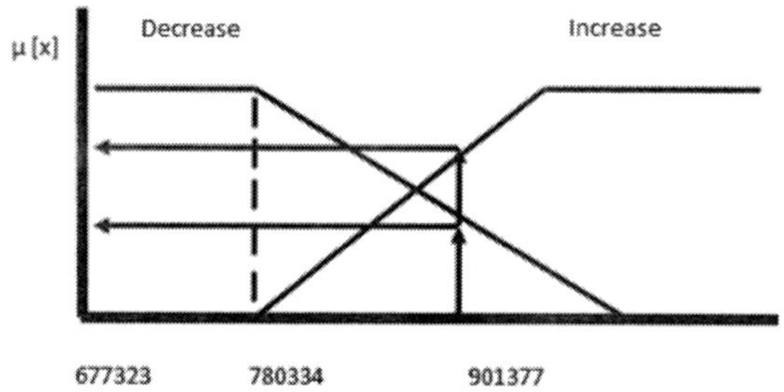

Figure 2. Decrease and Increase Degree of membership function.

Degree of membership value from decrease function and increase function, can be calculated as µptmbh Decrease [780334] = $\frac{901377-780334}{224054}$ = 0.54, and µptmbh Increase [780334] = $\frac{780334-677323}{224054}$ = 0.46.
2) Establishment of membership for Few function and many Function (figure 3).

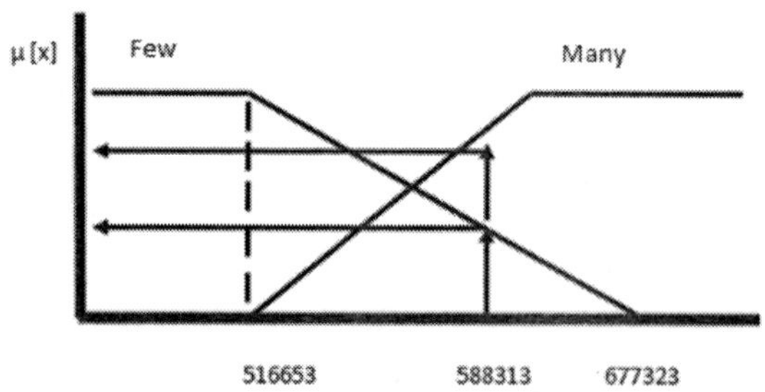

Figure 3. Decrease and Increase Degree of membership function.

Degree of membership value from few function and many function can be calculated as µptmbh Few [588313] = $\frac{677323-588313}{160670}$ = 0.56, and µptmbh Many [588313] = $\frac{588313-516653}{160670}$ = 0.44. Some possible rules which possible can be created by Sugeno for the experimental data, and the population created by Sugeno can be seen in table 8, and the error of prediction of Sugeno is 16.09% (higher than 3%).

Table 8. Comparison of Sugeno and real data population (year 2010).

Real data population	Sugeno
1,093,110	1,269,064

Based on the calculation of Sleman growth rate prediction, it can be said that there is a negative growth due to the occurrence of spike and increase of population equal to 16.09%. This result far from ideal rate which has been set by government which equal to 0.5% per year.

4. Conclusion

The result of population growth prediction is 0.44% from census data, while the prediction data of population growth using Sugeno method is 16.10%. Based on this result, the Sugeno method resulted in negative growth predictions, meaning that this result is far from the tolerance of government prediction that is only 0.5% per year.

References

[1] Yousefi M, Hooshyar D, Remezani A, Sahari K S M, Khaksar W and Alnaimi F B I 2015 Short-term wind speed forecasting by an adaptive network-based fuzzy inference system (ANFIS): an attempt towards an ensemble forecasting method *Int. J. Adv. Intell. Informatics* vol 1 no 3 pp 140–9
[2] Munandar D 2017 Wavelet discrete transform, ANFIS and linear regression for short-term time series prediction of air temperature *Int. J. Adv. Intell. Informatics* vol 3 no 2 pp 68–80
[3] Mohan A, Kuchhal P and Sharma M G 2017 Prediction Models for Global Solar Radiations, Diffused Radiations and Direct Solar Radiations Using ANFIS *Proceeding of International Conference on Intelligent Communication, Control and Devices* pp 27–37
[4] Oprea M, Popescu M, Mihalache S F and Dragomir E G 2017 Data Mining and ANFIS Application to Particulate Matter Air Pollutant Prediction. A Comparative Study *ICAART (2)* pp 551–8
[5] Tabrizi S S and Sancar N 2017 Prediction of Body Mass Index: A comparative study of multiple linear regression, ANN and ANFIS models *Procedia Comput. Sci.* vol 120 pp 394–401
[6] Wang J-S and Ning C-X 2015 ANFIS Based time series prediction method of bank cash flow optimized by adaptive population activity PSO algorithm *Information* vol 6 no 3 pp 300–13
[7] Zhang Y and Lei J 2017 Prediction of Laser Cutting Roughness in Intelligent Manufacturing Mode Based on ANFIS *Procedia Eng.* vol 174 pp 82–9
[8] Chen W, Panahi M and Pourghasemi H R 2017 Performance evaluation of GIS-based new ensemble data mining techniques of adaptive neuro-fuzzy inference system (ANFIS) with genetic algorithm (GA), differential evolution (DE), and particle swarm optimization (PSO) for landslide spatial modelling *Catena* vol 157 pp 310–24
[9] Zeng W, Zhao Y and Yin Q 2016 Sugeno fuzzy inference algorithm and its application in epicentral intensity prediction *Appl. Math. Model.* vol 40 no 13–14 pp 6501–08
[10] Sugeno M and Kang G T 1988 Structure identification of fuzzy model *Fuzzy sets Syst.* vol 28 no 1 pp 15–33
[11] Noori R, Safavi S and Shahrokni S A N 2013 A reduced-order adaptive neuro-fuzzy inference system model as a software sensor for rapid estimation of five-day biochemical oxygen demand *J. Hydrol.* vol 495 pp 175–85
[12] Ahmed A A M and Shah S M A 2017 Application of adaptive neuro-fuzzy inference system

(ANFIS) to estimate the biochemical oxygen demand (BOD) of Surma River *J. King Saud Univ. Sci.* vol 29 no 3 pp 237–43

IOP Conf. Series: Materials Science and Engineering **403** (2018) 012074 doi:10.1088/1757-899X/403/1/012074

Similarity measurement algorithms of writing and image for plagiarism on Facebook's social media

M A Kurniawan[1] and K Surendro[2]

Institut Teknologi Bandung, Jalan Ganesha No 10, Bandung, Indonesia

[1] aghust.kurniawan@gmail.com, [2] surendro@gmail.com

Abstract. Currently, social media is growing very rapidly as a medium of information and interaction. However, these developments commonly not followed by a system that can detect the validity of the information also where the source came from, that resulting a negative potential such as fake news (hoaxes) or plagiarism. As a solution, in this research, will be testing plagiarism of writing and image on social media up by using URL as input test. Methods of measuring similarities that used in this research are Smith-Waterman Algorithm and Latent Semantic Analysis. The research measures the accuracy of both methods with local alignment and term-document approaches on documents and Facebook's social media. The result of the research for Smith-Waterman Algorithm method with local alignment approach is better on document testing, with performance level up to 99.77%. In addition, in the proposed algorithm, image testing performed using Google Image Search can also indicate its relevance to a post. Thus, this algorithm has the opportunity to be develop again in other similarity searches such as authorship identification on a post as well as a hoax analyser.

1. Introduction

Social media is an internet-based media technology that connects every user to participate, share and create content such as blogs, social networks, forums, encyclopedias and virtual worlds. Various kinds of technological devices are currently able to access information on social media with ease due to the availability of an increasingly affordable Internet network. So then, the data spread on the internet more easily accessible and used as a reference. However, sometimes it is become misused by taking the work of others as a property of their own intentionally or not. It is included in the act of plagiarizing without permission or without the inclusion of reference sources commonly known as plagiarism.

Prevention of plagiarism has been widely practiced in previous research. Some tools for identifying plagiarism have been widely developed. Basically comparing only two documents in syntactic resemblance, more precisely looking for similarities based on words, phrases or sentences [1].

Currently, many tools also applied some document similarity detection such as WriteCheck, PlagScan, Turnitin, iThenticate, Viper, etc. In this research, will be submitted an algorithm to test writing and an image on social media as illustration supporting writer. The results of the study expected to provide opportunities for the development of this algorithm with the method of measuring similarity more widely.

2. Plagiarism Detection Methods

2.1. Textual Similarity Approach

The detection of document plagiarism has several approaches in measuring similarity, can be categorized into three types: String-based, Corpus-based, and Knowledge-based [2]. The measurement of textual similarity has an important role in textual relations research and its application in information retrieval, text classification, document clustering, topic detection, topics search, question building, answer questions, essay grading, short answer assessments, translator engines, text summarization and so on.

2.2. Smith-Waterman Algorithm

Smith-Waterman Algorithm uses local alignment techniques to determine similar parts between two sequences of nucleoid acid sequences or protein sequences. Compared to searching through the sequences, this algorithm compares segments with varying lengths that allow and finds optimal alignment. The scheme for obtaining an optimal alignment value can be defined in equation (1).

$$H(i,0) = 0, \quad 0 \le i \le m$$
$$H(0,j) = 0, \quad 0 \le j \le n$$
$$H(i,j) = \max \begin{Bmatrix} 0 & - \\ H(i-1,j-1) + s(a,b) & Match/Mismatch \\ \max\limits_{k\ge1}\{H(i-k,j) + W_k\} & Deletion \\ \max\limits_{l\ge1}\{H(i,j-l) + W_l\} & Insertion \end{Bmatrix}, \quad 1 \le i \le m, 1 \le j \le n \qquad (1)$$

Equation (1) will yield an appraisal matrix from the comparison of two series, so that the optimum value obtained which shows the value of similarity between the two. The flow process of this method illustrated by Figure 1.

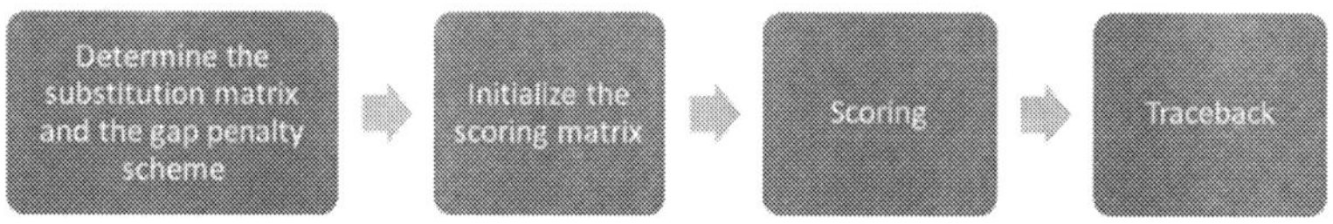

Figure 1. The Process flow of Smith-Waterman Method

2.3. Latent Semantic Analysis

Latent Semantic Analysis (LSA) is a statistical model of language learning and representation that uses vector averages in the semantic space to test the similarity between words or text [3]. The use of LSA was first use by Foltz for grading essays in 1995. In addition, several other examples of use as representative models and measuring human verbal concepts [4].

LSA uses Singular Value Decomposition (SVD) to decompose the input matrix which is the matrix of document-term weights into three matrices [5]. The process flow can be illustrate as shown in Figure 2.

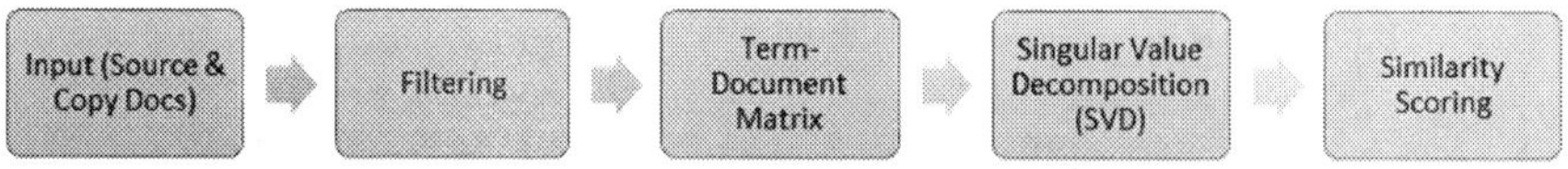

Figure 2. The Process flow of Latent Semantic Analysis

2.4. Performance Measurement

In a documentary measurement system or plagiarism detection, as a classification system determines a pair of documents considered to contain plagiarism. However, the assessment should be support by performance measurement of the method used, to see how accurately it measures the similarity.

Therefore, some performance measurement techniques are need to look at precision, recall and F-Measure values as a measure of an accuracy.

3. Methodology

The design of model architecture in this research use research methodology of Design Science Research Methodology (DSRM) to test case of plagiarism in document and writing in social media. Steps in DSRM include (1) identifying problems and motivations, (2) defining goals for solutions, (3) design, (4) testing, (5) evaluation and (6) communication. These stages are the workflow process conducted on the research and will be combine with the process of filtering and document similarity measurement method.

The design results produce a model that classifies the process into the DSRM step to simplify the testing process. The model shown in Figure 3.

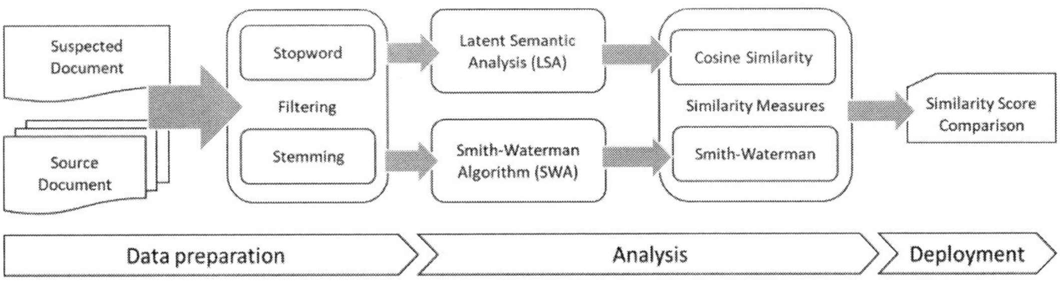

Figure 3. Architecture Model of Plagiarism Testing

In the filtering process, the process of eliminating words that do not have a link of meaning as a whole (Stopword filtering) and eliminate affixes to get the root word (stemming process). Then the measurement of document resemblance done in the analysis phase, to generate the value of similarity.

4. Testing and Implementation

4.1. Testing Similarities on Documents

Tests of similarity in the document process with three scenarios, including testing the same document, testing different documents and examining documents similar to changes in word layout or phrase (paraphrase). In the same document test, both methods show the same results; while for other tests shown in the Table 1.

Table 1. Testing on different documents

No	*Copy Doc*	*Source Doc*	*Similarity Score*	
			LSA	*SWA*
1	A01	A04	75%	4%
2	A01	A12	81%	2%
3	A01	A14	92%	5%
4	A02	A04	70%	3%
5	A02	A05	91%	6%
6	A02	A11	84%	3%
7	A02	A13	65%	6%
8	A02	A17	87%	4%
9	A03	A11	72%	3%
10	A03	A13	63%	3%

No	*Copy Doc*	*Source Doc*	*Similarity Score*	
			LSA	*SWA*
11	A03	A16	83%	3%
12	A03	A20	87%	2%
13	A03	A24	82%	2%
14	A04	A09	64%	3%
15	A04	A12	87%	2%
16	A04	A14	74%	2%
17	A05	A08	71%	3%
18	A05	A11	87%	3%
19	A05	A13	71%	6%
20	A05	A17	85%	4%
21	A06	A07	60%	3%
22	A06	A13	67%	3%
23	A06	A21	80%	3%
24	A06	A23	75%	2%
25	A07	A09	62%	3%
26	A07	A10	91%	6%
27	A07	A23	86%	4%
28	A08	A11	78%	3%
29	A08	A13	76%	3%
30	A09	A18	80%	2%
31	A10	A23	75%	4%
32	A10	A28	71%	4%
33	A11	A13	92%	3%
34	A11	A17	74%	2%
35	A11	A21	69%	2%
36	A12	A14	88%	3%
37	A13	A21	65%	2%
38	A15	A22	70%	5%
39	A15	A27	84%	5%
40	A15	A29	84%	6%
41	A16	A20	90%	5%
42	A16	A24	99%	37%
43	A17	A19	73%	2%
44	A17	A21	61%	2%
45	A20	A24	87%	4%
46	A22	A25	77%	3%
47	A22	A27	97%	6%
48	A25	A27	76%	3%

The percentage of similarity that use as a reference is the result of testing on the Latent Semantic Analysis (LSA) method, with an average percentage of similarity 78% because the term frequency in the compared documents has similarities although the topic of the document is different. While the test results with Smith-Waterman Algorithm (SWA) shows a small alignment value or can be said average percentage of similarity 4%.

Assuming the results of the tests conducted objectively by the authors as a reference level of accuracy, then the results show the average percentage of similarity between the documents close to 0% (no resemblance), can be seen graphically on the Figure 4.

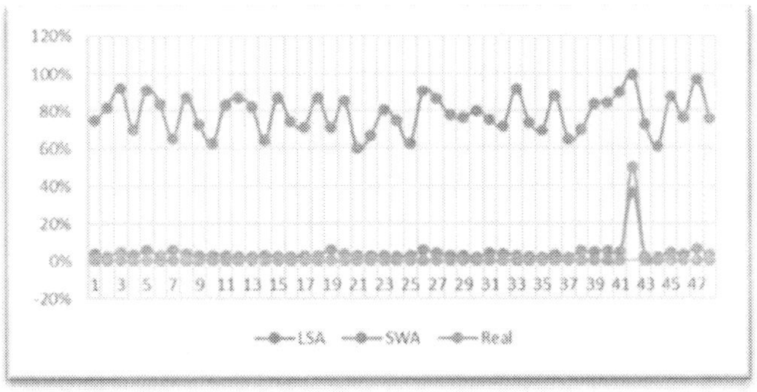

Figure 4. Testing on different documents

In testing similar documents, the test results of both methods are not significantly different, as shown in Figure 5.

	Similarity Score									
	B01		B02		B03		B04		B05	
	LSA	SWA	LSA	SWA	LSA	SWA	LSA	SWA	LSA	SWA
B01	100%	100%	100%	77%	87%	100%	87%	94%	96%	100%
B02			100%	100%	87%	100%	87%	95%	96%	100%
B03					100%	100%	99%	97%	76%	100%
B04							100%	100%	75%	93%
B05									100%	100%

Figure 5. Testing on Similar Documents

From Figure 6 shows that for testing with Smith-Waterman Algorithm, the percentage of similarity is 95.6%. While for Latent Semantic Analysis also shows an overall percentage of 89% average.

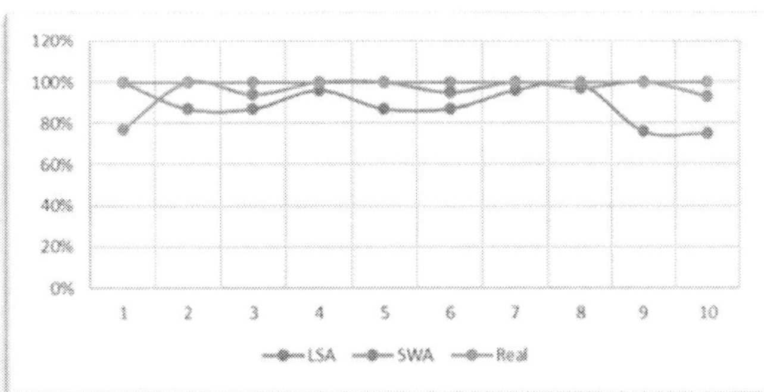

Figure 6. Testing on Similar Documents

4.2. Implementation on Social Media Testing Similarities

Implementation testing of similarity writing and images on Facebook's social media done with some initialization process. The process structured into an algorithm that can extract the content contained in

Facebook's social media. Figure 7 shows an algorithm used to detect plagiarism in Facebook's social media.

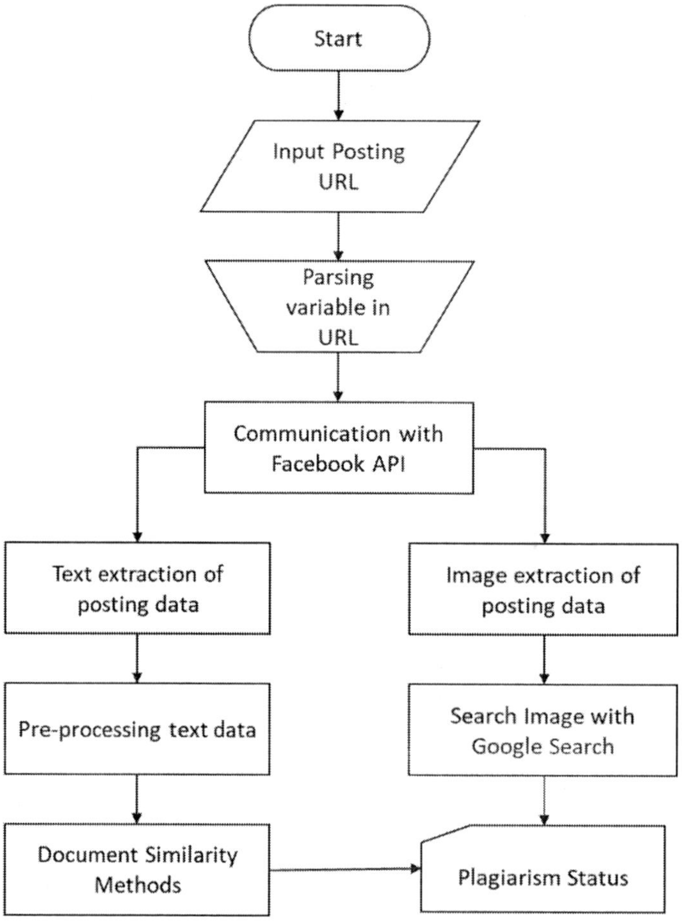

Figure 7. Plagiarism Detection Algorithm on Facebook's social media

Figure 9 is an example of a plagiarism case on Facebook's social media and blogspot. Testing done by initiating the data access of both social media. In this research, the tools used is R, the initiation process, the measurement of similarity and the performance measurement done using the tools. The result of measurement on three cases of plagiarism in social media as shown in Figure 8.

No	*Copy Post*	*Source Post*	*Similarity Score*		
			LSA	*SWA*	*Real*
1	F1	F2	100%	67%	60%
2	F3	F4	100%	27%	40%
3	F5	F6	100%	44%	70%

Figure 8. Testing on social media

F3	Agama Facebook Anggaplah engkau tersesat di sebuah perkampungan primitif yang sangat jauh dari peradaban. Saat itu, di sana sedang terjadi kehebohan diantara penduduknya. Mereka semua saling berdebat tentang "facebook", sebuah kata di secarik surat botol yang terdampar di pesisir laut. Ada yang menganggapnya hantu, ada yang berpendapat sosok dewa, ada yang ngotot itu surga, ada yang mengira sebuah aliran sesat, ada yang bilang sebuah agama baru, bahkan ada yang tidak percaya bahwa facebook ada. Lalu dirimu yang sebenarnya sudah terbiasa menggunakan facebook mencoba menjelaskannya pada mereka. Tapi tetap saja, mereka masih ngotot dengan pendirian mereka masing-masing. Mereka tak percaya padamu. Penjelasanmu tidak masuk akal. Perdebatan terus berlanjut. Bahkan ada yang sampai bermusuhan setelahnya. Kau akhirnya memilih diam. Kau merasa lebih baik menyaksikan tingkah mereka sambil geleng-geleng kepala. Bagaimanapun, penjelasanmu tidak akan berguna selama mereka tidak melihat facebook dengan mata kepala. Kau hanya bisa berharap semoga listrik dan jaringan internet bisa segera masuk ke kampung tersebut, agar mereka bisa melihat sendiri apa yang sedang mereka ributkan ini. - Afi	F4	Agama Facebook [illegible] Hahaha.. [illegible] The Mind Navigator
Tulisan sudah dihapus penulis (menggunakan data file teks) Tanggal Posting: 28 November 2016		[illegible]/2015/02/agama-facebook.html Tanggal Posting: 24 Februari 2015	

Figure 9. Facebook's status and blogspot's writing

From Figure 8, the Smith-Waterman Algorithm method has a high degree of accuracy and precision when paired with actual results. So the category of plagiarism becomes different, whereas the LSA assumes the whole writing as plagiarized, while SWA considers not all plagiarized. The results of these measurements are not yet be accurate. Therefore, it is necessary to measure performance in the case of plagiarism in social media.

In the image testing on Facebook social media, the image extracted the URL data then performed the same image search using Google Image Search, so that the conclusion about the relevance of the article with the image shown. Figure 10 is one example of testing.

Link to Posting	*Picture from Posting*	*Status*
https://www.facebook.com/mochamadridwankamil/photos/a.165770243574805.1073741828.163237587161404/929268827224939/?type=3		Gambar yang dimuat relevan dengan tulisan serta tidak memuat unsur plagiarisme
Result from Image search	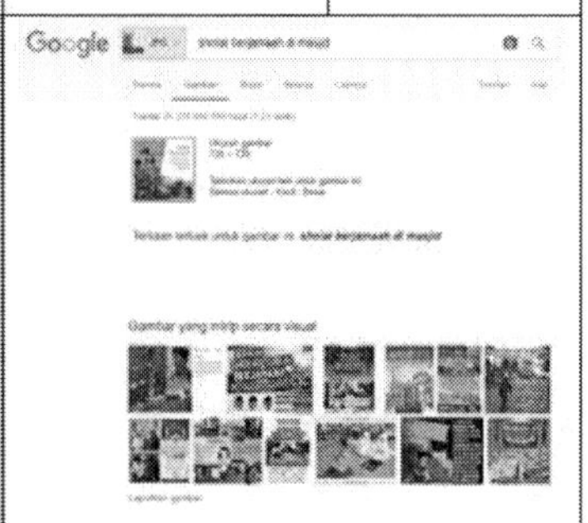	

Figure 10. Testing on Image Relevance

4.3. Performance Measurement

In the examination of document similarity, it can be seen the level of precision, recall and F-Measure for both methods in measuring its performance in Figure 11 and Figure 12.

Latent Semantic Analysis			Smith-Waterman Algorithm		
Pre	Rec	F-Me	Pre	Rec	F-Me
23,67%	104,76 %	38,61%	95,23%	104,76%	99,77%

Figure 11. Performance of Methods on Documents

From the results of performance measurement, shows the level of precision and recall method Smith-Waterman Algorithm higher than Latent Semantic Analysis. The average harmonic value between precision and recall of high F-Measure or F-Score indicates a good degree of accuracy. So then, it can be concluded that Smith-Waterman Algorithm method has better accuracy.

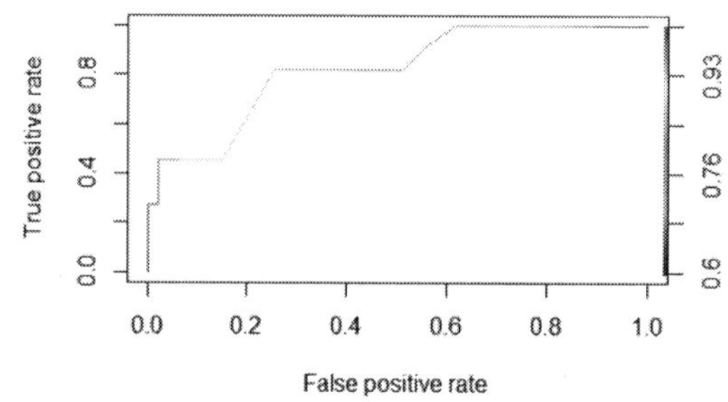

Figure 12. Performance of LSA Method

Performance measurement data are also shown on the graph of true positive rate and false positive rate, where it can be seen that the relevance of document suspected to have similarity/plagiarism is higher in the Smith-Waterman Algorithm method (Figure 13), which is shown with a high true positive rate.

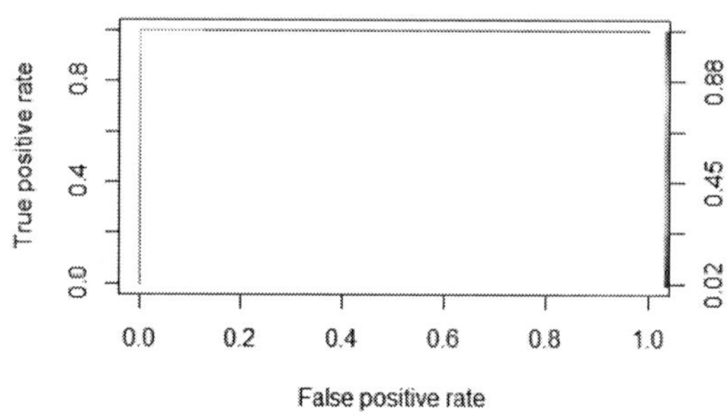

Figure 13. Performance of Smith-Waterman Method

In the performance measurement of similarity test of writing in social media, in Figure 14 show the level of precision and F-Measure Smith-Waterman Algorithm higher, it shows that the accuracy of this method is better.

Latent Semantic Analysis			Smith-Waterman Algorithm		
Pre	Rec	F-Me	Pre	Rec	F-Me
1%	1,76 %	1,27%	2,17%	1,76%	1,93%

Figure 14. Performance of Social Media Plagiarism

5. Results and Suggestions

5.1. Results

Based on the test results, the percentage of high document resemblance does not indicate a good degree of accuracy, a performance measurement process must be performed to determine a good degree of accuracy. After the performance measurement is obtained that the level of precision and high recall indicate a good level of accuracy to the method of measuring similarity of documents.

1. Test results on scientific paper documents and performance measurements, Smith-Waterman Algorithm has a higher accuracy with F-Measure 99.77%.
2. Similar results were obtained on testing similarities of writing data on Facebook social media and performance measurements, Smith-Waterman Algorithm has a higher accuracy with F-Measure value of 1.93% versus 1.27% for LSA.
3. The process of testing the similarity of images based on the extraction of the image is done by the process of execution of the function on R for image search in Google Image Search, because the Google API feature for images is no longer provided by Google. Search function of this image to see the relevance of writing and images on Facebook social media.

5.2. Suggestions

In this study, the authors wanted to show the methods in the measurement of popular document similarities such as Latent Semantic Analysis and Smith-Waterman Algorithm, in addition to having different characteristics, each method has advantages such as the LSA is more efficient when used in testing on large corpus. However, the results obtained show good accuracy on Smith-Waterman Algorithm method by not using corpus.

The suggestion for further research is the development of the Smith-Waterman Algorithm method to perform similarity tests on images in social media as a hoax analyzer. And the implementation of Smith-Waterman Algorithm method in testing the writing pattern with authorship analysis.

References

[1] A. Sarkar, U. Marjit, and U. Biswas, "A conceptual model to develop an advanced plagiarism checking tool based on semantic matching," *Bus. Inf. ...*, pp. 104–108, 2014.

[2] W. Gomaa and A. Fahmy, "A survey of text similarity approaches," *Int. J. Comput. Appl.*, vol. 68, no. 13, pp. 13–18, 2013.

[3] T. K. Landauer and S. T. Dumais, "A Solution to Plato's Problem : The Latent Semantic Analysis Theory of Acquisition, Induction, and Representation of Knowledge," *Psychol. Rev.*, vol. 104, no. 2, pp. 211–240, 1997.

[4] T. K. Landauer, P. W. Foltz, and D. Laham, "An Introduction to Latent Semantic Analysis," *Discourse Process.*, vol. 25, no. 2, pp. 259–284, 1998.

[5] S. Soleman and A. Purwarianti, "Experiments on the Indonesian plagiarism detection using latent semantic analysis," *2014 2nd Int. Conf. Inf. Commun. Technol.*, no. 2010, pp. 413–418, 2014.

Customer trust of online shopping using social media

V E Mariana[1], A P Wibawa[2], J Hammad[3], R Herdianto[4] and Haviluddin[5]

[1,2,4]Universitas Negeri Malang, Malang, Indonesia.
[3]Departement of Computer Information System, Alquds Open University, Bethlehem, Palestine, Department of Electrical Engineering, Institut Teknologi Sepuluh Nopember(ITS), Surabaya, Indonesia.
[5]Department of Computer Science, Universitas Mulawarman, Indonesia.

[1]rianavika@gmail.com, [2]aji.prasetya.ft@um.ac.id, [3]jhammad35@hotmail.com.

Abstract. Recently, people have tendency to purchase goods from online stores. This study aims to determine the level of students trust in social media online stores. In this descriptive qualitative study, data and information gathered by interviewing 50 respondents from various departments in Universitas Negeri Malang. This research revealed that the level of students' trust in online stores on social media is very good, where 20 of them recommend online stores to others. The most purchased goods are clothing, 58% of respondents. This study also showed that 30 respondents used Instagram while 14 others used Facebook.

Keywords: trust, online shopping, social media

1. Introduction

The development of technology is growing rapidly. Smart phones and internet access make people easy to do communication especially on using social media. Social media is now flourishing and plays a more and more important role in most people's lives as a means of information acquirement and sharing [1].

Social media is being widely used by various sized companies in s to advertise their products and services [2]. Small size traders also look social media as a promotion opportunity. Traders can promote their goods by posting the details of product they are selling. Their reasons of using social media as a promotional platform are: (1) the easiness of creating an account, (2) zero cost for posting advertisement, (3) large number of users on social media, especially, and (4) direct interaction between buyer and seller in the transaction process.

The strongest points of using social media for marketing is their ability to generate advertisements that tailored to the likings of each consumer. Social media users continuously declare their likings by joining product groups, following company news, submitting personal information on their online profiles [3].

Apart from the excess advantages, there are many fraud cases in social media online shopping. Most types of fraud are: (1) Fake accounts. (2) Undelivered items (even though the customer did the payment), and (3) Goods sent are different than offered in the post.

Most of Facebook traders have done many ways to minimize the number of online fraud. Some of them make transactions through joint accounts and create lists of deceptive accounts. Another way is watching the product reviews and number of traders' followers. However, still there are frauds cases occurred.

Based on the advantages and disadvantages of online shopping, we want to know the public trust online shopping through Facebook and Instagram. This may also reveal the type of most often purchased goods, and the most popular social media, used by respondents. Facebook, LinkedIn and blogs are the major social media platforms that marketers have employed over the years to advertise their products and services [4].

In recent years, scholars have taken a keen interest in social media advertising, which defined as application of social media tools to create awareness, persuade and educate consumers about a firm's products and services [4]. Recently, there has been an increasing amount of attention been placed on marketing over social media websites as an alternative option for product promotion and customer feedback [3][5]. Facebook is working on a system, targeted its advertising to specific networks of people with similar interest [2]. On the other hand, Pinterest and Instagram are increasingly popular social media platforms [5]. As consequence, shopping with social media is more comfortable for both seller and buyer.

2. Trust in online shopping

Trust linked to uncertainty. The uncertainty prevalent in most online transactions, means that the importance of trust is elevated in e-commerce [6] . Online transactions typically involve two types of trust: trust in the integrity of the transaction (behavioral risk) and trust in the infrastructure supporting the transaction (environmental risk) [6]. A central explanation for the trend described the uncertainties and particular risks, involved in online trading and the resulting lack of trust this engenders among consumers [7]. There are many issues in social media shopping Problem that often occurs are: (1) Packet delays. (2) Suitability of goods. (3) Payment system.

3. Related Work

The studies on commerce-oriented reviewing and sharing platforms (CRSP) primarily addressed the problems of summarizing and analysing online reviews [1] . Some study explore whether awareness and understanding of German trust marks have changed from 2007 to 2012 through increased Internet experience and online purchasing activity as well as through the wider proliferation of Internet trust marks [6], examine how emotions influence trust with the use of new technology [7]. And understanding of consumer online behaviour for future research in e-commerce [8].

Most of the research study above discuss about online shopping in web application like eBay, Amazon, etc, but there still scantily studies about online shopping through social media. The studies about Consequences of customer engagement behaviour: when negative Facebook posts have positive effects Most of the research study above discusses about online shopping in web application like eBay, Amazon, etc. but there still scantily studies about online shopping through social media. The studies about Consequences of customer engagement behaviour: when negative Facebook posts have positive effects which are discuss about the consumer habits on social media Facebook that focus on the impact of posting on the restaurant brand [9].

The studies which identify the extent to how people trust on online shopping through social media Facebook and Instagram is still slightly, therefore we want to explain how the levels of public trust in online shopping through social media Facebook and Instagram. We choose social media Facebook and Instagram because both are very popular social media and have many users from all over the world.

4. Methodology

This study uses a qualitative descriptive approach. We reviewed 50 random respondents from various faculties in Universitas Negeri Malang. Some other is expert player while the rest is newcomers or never do an online shopping.

We asked the reason of respondents who never do an online transaction. For the online shopper, we ask about the goods they used to buy, do they still do online shopping if being frauded, and what is social media platforms they use to shop. We used the results to draw conclusions about their level of trust in social media online shopping, in this case Facebook and Instagram.

5. Discussion and Conclusion

From 50 college respondents, there are 43 students who have been shopping online and 7 others never do it. One who never did shopping online have different reasons. Two students prefer to buy traditionally, while five others do not believe in online transaction. The 43 respondents who have done online shopping grouped into each faculty, shown in the Figure 1.

Figure 1. Number of Respondents.

Based on Fig1, the numbers of respondents who ever shop online were mostly from the Faculty of Literature with 13 respondents, then the Faculty of Economics with 10 respondents, and the Faculty of Engineering with 9 respondents. Respondents who have made an online shopping transaction, have a tendency to shop with certain categories of goods presented in the Figure 2.

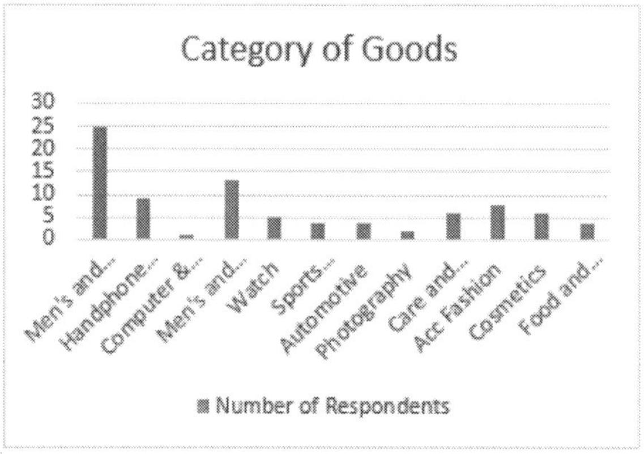

Figure 2. Category of Goods.

Based on the graph Figure 2, the category of goods that are mostly purchased by respondents is men's or women's clothing with 25 buyers. Social media that is often used to shop online by respondents (Figure 3) is Instagram with 30 respondents and Facebook with 14 respondents. Here, each respondent can choose more than one choice of social media.

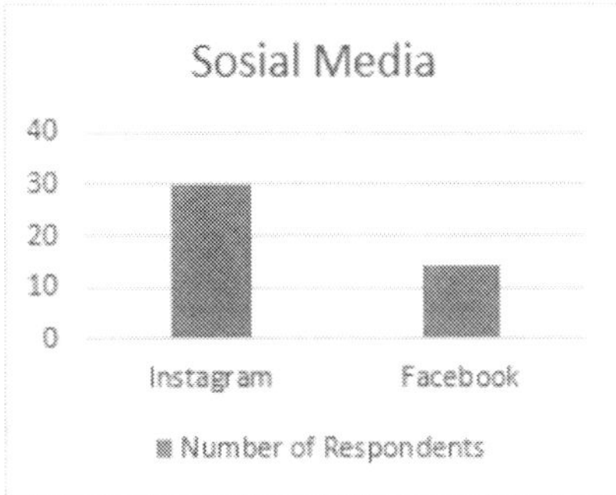

Figure 3. Social media which is used.

The result of this paper found the service quality of online shopping in social media fair enough, twenty-eight (28) respondent argue that seller acted friendly to buyers during the transaction, 21 respondents stated that seller would retake a picture for consumer satisfaction. Twenty-six respondents said that prices at the online shops compete with other shops, while 18 respondents stated that goods at store are always on stock. Seventeen 17 respondents said that the warranty service in online shopping is fair enough. The warranty usually is in form refund or replace good. Twenty-four respondents said that online transaction has fast respond to buyers from transaction process until the delivery of goods. The online shop through social media has fast respond because buyer and seller talk through private message.

In term of reliability, 21 respondents stated that goods purchased in accordance with the description of goods on social media, 27 respondents stated that the order system are easy and simple and 20 respondents stated that the shop/store they usually bought can be recommended to others.

6. Conclusion

The conclusions of this study are (1) the level of student's trust in online stores on social media is very good, 20 of 43 respondents trusted and recommend online stores to others, (2) the categories of goods that were mostly purchased by respondents is men and women clothing, (3) Instagram is the most used social media for online shopping. They happy with Instagram since it is a photo-sharing mobile application that allows users to take pictures, apply filters to them, and share them on the platform itself [10]. Instagram is free and has user-friendly interface.

References

[1] M. Li, J. Zhang, Z. Liu, and G. I. Johnson, "An experimental study of chinese shopping related sharing behaviors," in *IFIP Conference on Human-Computer Interaction*, 2013, pp. 608–615.
[2] M. Saravanakumar and T. SuganthaLakshmi, "Social media marketing," *Life Sci. J.*, vol. 9, pp. 4444–4451, 2012.
[3] P. Mikalef, A. Pateli, and M. Giannakos, "Why are users of Social Media inclined to Word-of-Mouth?," in *Conference on e-Business, e-Services and e-Society*, 2013, pp. 112–123.
[4] H. Boateng and A. F. Okoe, "Determinants of Consumers' Attitude towards Social Media Advertising," *J. Creat. Commun.*, vol. 10, pp. 248–258, 2015.
[5] A. E. Doub, A. Levin, C. E. Heath, and K. LeVangie, "Mobile app-etite: Consumer attitudes towards and use of mobile technology in the context of eating behaviourNo Title," *J. Direct, Data Digit. Mark. Pract.*, vol. 17, pp. 114–129, 2015.
[6] C. Tislar, J. Sterkenburg, W. Zhang, and M. Jeon, "How Emotions Influence Trust in Online Transactions Using New Technology," in *Proceedings of the Human Factors and Ergonomics Society Annual Meeting*, 2014, pp. 1531–1535.
[7] K. Rüdiger and M. J. G. Rodríguez, "Do we need innovative trust intermediaries in the digital

economy?No Title," *Glob. Bus. Perspect.*, vol. 1, pp. 329–340, 2013.
[8] M. Qin, "Consumer behavior towards continued use of online shopping: An extend expectation disconfirmation model," *ntegration Innov. Orient to E-Society*, pp. 400–407, 2007.
[9] S. Bitter and S. Grabner-Kräuter, "Consequences of customer engagement behavior: when negative Facebook posts have positive effects," *Electron. Mark.*, vol. 26, pp. 219–231, 2016.
[10] S. Alhabash and M. Ma, "A Tale of Four Platforms: Motivations and Uses of Facebook, Twitter, Instagram, and Snapchat Among College Students?," *Soc. Media + Soc.*, no. January-March, pp. 1–13, 2017.

Mobile CRM student-parent information system

A Setiawan[1] and R Widaryanto[2]

[1,2]Department of Informatics Engineering, Universitas Muhammadiyah Magelang, Magelang, Indonesia.

[1]setiawan@ummgl.ac.id

Abstract. This paper presents UML (unified modeling language) modeling on mobile CRM for a parent guardian information system. The current problem is that the relationship between parents of students and university is still there. The purpose of this study is to eliminate the distance between parents of students with universities in the aspect of information related to the academic development of their children. Using ones of system development life cycle, UML is used to model the system requirements that will be developed based on an analysis of existing needs. The result of this study is a UML model of mobile CRM for parental guardian information system.

1. Introduction

Utilization of information technology on this day has become an inevitable requirement. According to APJII data of Indonesian internet users in January of 2016 has reached 88 million [1],[2]. Of course, internet usage will increase every time, because the government and the private sector continue to provide facilities increasingly reach new areas in remote locations [3]. Internet users will always learn to utilize the internet to help all things in life activities. The use of emails has replaced conventional letters, e-books have replaced books in libraries, could be corrupted by termites, and e-news has replaced newspapers every morning.

Information technology has been utilized in the world of education to process. One of the benefits in the world of teaching is the existence of e-learning. e-Learning can remove the problem of time and distance to deliver information from educators to learners. The smartphone is an information and communication technology equipment that has been owned by everyone. Smartphone has been able to replace the function as a computer in the form of a PC or laptop. People use longer smartphones than PCs and laptops [2].

However, the use of information systems to foster relationships with parents or guardians of students to improve customer satisfaction by the concept of CRM is still less attention. So it still often happens ignorance of parents or guardians of students with the development of studies and student activities in the college where his college where science demands. UAS value information for each course, IP semester, cumulative IP, college payment status, and other academic activities should not be obtained from his / her child but can be through a direct information system. It can be delivered directly to the parents of the student guardian. One of Industrial application for customer convenience is CRM, in the world of education must also provide satisfaction to all customers called learners and parents guardians of children who have entrusted their children to educational institutions concerned.

Information technology is a fast-growing technology [4]. The world of education is required to use information technology to accelerate the delivery of information because it is essentially education is the transfer of knowledge. Universitas Muhammadiyah Magelang (UMMgl) has started to utilize information technology one of them with KRS online with URL address krsol.ummgl.ac.id. Universities will be able to cut the course time and cost with the utilization of information technology.

One method to improve customer satisfaction is using a concept of customer relationship management. Fostering good relationships with their customers will give us confidence and ultimately customer loyalty to increase greater market share. Loyal customers will be loyal and will indirectly promote to new prospective customers.

The implementation of CRM in the education world is growing stronger with the emergence of several models of distance learning, e-learning, and e-business [5]. Implementation of CRM becomes imperative because CRM is also part of SaaS (Software as a Service) in Cloud Computing technology. Looking at stakeholders as customers will make their feelings more comfortable and loyal to provide a competitive advantage. College by implementing CRM will be able to attract, retain, and serve its customers.

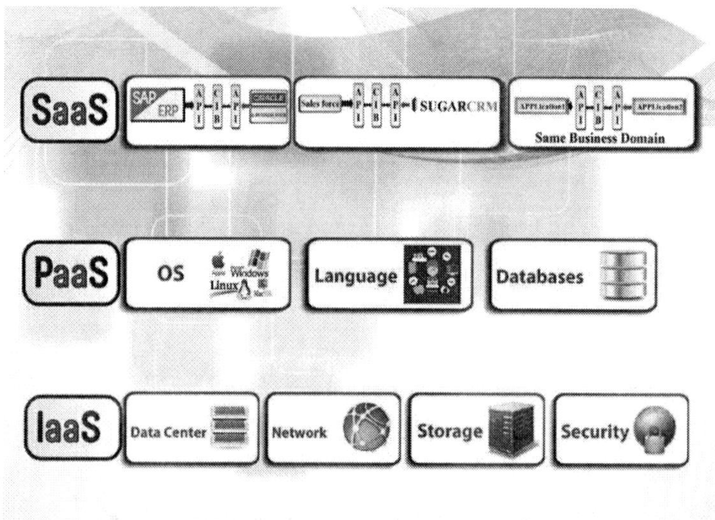

Figure 1. CIB Location inside SaaS layer of Cloud Computing Layers [6]

In the previous study, CRM application is also applied to CRM reporting of enterprise development to provide confidence to customers and improve customer loyalty [7]. Many companies develop and maintain CRM-based data warehouse systems appropriately to get what they expect, that is, to maintain a sustainable customer relationship. The use of CRM information technology makes the service to customers more flexible, especially for multinational companies certainly have different currency, with this system is very easy to convert because based on the system.

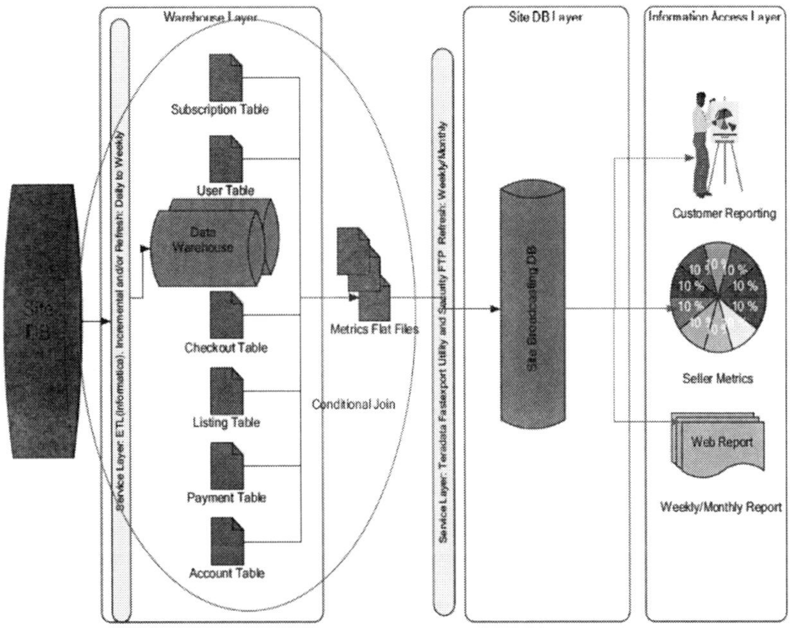

Figure 2. Technical architecture of this CRM/Customer Data Reporting system [7]

Furthermore, some research on CRM still focuses on students as customers but has not considered parents as customers. Therefore, research has focused on parents of student guardians as customers. In this case, mobile CRM developed and designed using UML.

2. Method

In this study, an activity to identify the object of research and related environment further to deepen the situation and condition of the system to be developed. Expectations and constraints on the system that run will be used as a reference in developing the system. The preliminary study was conducted by collecting information about existing educational business processes at universities, consulting with related parties and supporting information about the relationship of parents of guardian students and universities.

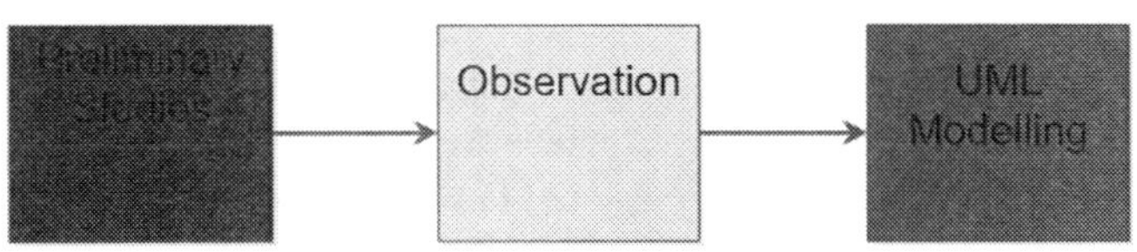

Figure 2. Set up experiment and apparatus

After the data is obtained, the next step is to browse the system that has been running along with its advantages and disadvantages. To accomplish this, the results of this activity are also reinforced by literature and library studies with accountable sources, such as from the literature on mobile customer relationship management, information systems, and academic systems.

Researchers conduct a meeting with student guardians (prospective users of information systems) to identify the purpose of the application or system and determine the information needs to achieve the goal. Researchers analyze aspects of business or management and analysis of technological issues. The business perspective analysis is carried out to obtain information related to the flow and relation of information needed from the interaction of parents of guardians with the education providers (universities). Technological aspect analysis is done to get information related to technology already owned by the college, and the possibility of developing what needs to be done in the future in connection with the implementation of mobile customer relationship management information system of parents of student guardian. This analysis will result in minimal requirements of a system. Requirement Planning will create a minimum specification of a system.

This design stage is done to get the blueprint of system design and prototype from the parent student information system. System design is closely related to the previous process, namely the planning of the needs. These design stages include the design of business process models to be developed and the design of information-based information systems based on mobile customer relationship management. This interface design will be done on the Android and iOS platforms because these two platform models have different characteristics. This design researcher requires the help of the professional UI / UX designer to create an interface that matches the character of each platform mobile phone (smartphone). This stage of the design process also includes the design transformation into the coding of mobile-based programming languages. From the results of this implementation has been obtained mobile-based information system installed on the smartphone and set up a server that is used to serve parent guardian information systems based on mobile CRM.

3. Result and Discussion

3.1. Preliminary Studies

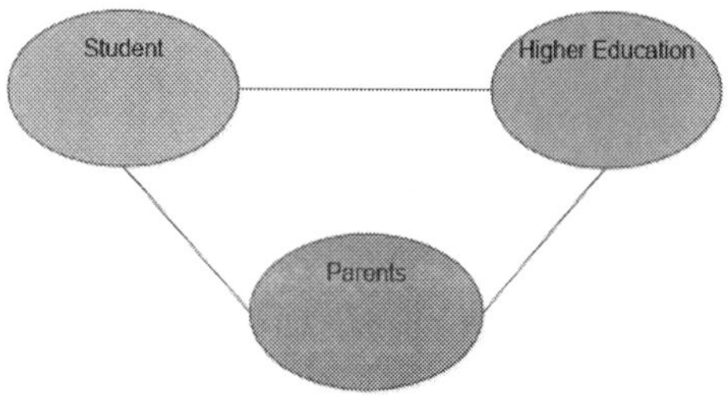

Figure 3. Relationship between Students, University, Parents

Figure 3 shows the relationship between students, university, and parent/guardian, the dashed line of guardian parents towards the higher education shows an indirect relationship, but the student towards the higher education is a direct connection. The inconsequential relationship that occurs is because the parents of the student guardian are less likely to interact with the university than the students with the higher education. Students and parents, will encounter obstacles if their children (students) never provide information about academic development obtained by his son at the university. In the course of the study plan in each semester, students will only fill out online through the krsol.ummgl.ac.id page, unnoticed by the parents of the student guardian. The parents of the student guardian also have an important role in the success of their children in university education, without the supervision and control of higher education parent will also find it difficult to provide academic-related actions. Thus, the role of parents in the success of education in college is also one of the important points.

The researcher conducts a preliminary study of the existing SOP from the educational business process of the student to an academic system of Study Plan Card (KRS). It it is concluded that there is a standard procedure for the registration of the study plan for the next semester with the existing online system also through the krsol.ummgl.ac.id page. This online KRS process hopes to facilitate the student in enrollment of study plan for one semester ahead with more efficient because it can be done according to the schedule specified. Students can also determine according to their ability because students can choose the courses that will be taken during the next semester which will also impact with the bill one semester. Students will surely deliver this to guardian parents as a supporter for the still. Obstacles that arise later then the parents cannot see the detailed resume bill, so it appears nominal so paid for immediately. Current constraints from interviewing with the Guardian's parent, her child sometimes cannot explain about the fee, so parents need to come to the campus for clarification.

3.2. Observation

Mobile application development can be maximized and by its characteristics [8]. Mobile devices have many types of sizes, designs, and layouts, but they have very different features of the desktop system, which are:

a) Small size
Mobile devices have a small size. Consumers want the smallest devices for their convenience and mobility.
b) Limited memory
Mobile devices also have a small memory, namely primary (RAM) and secondary (disk). This restriction is one of the factors that affect the writing of programs for various types of these devices. By limiting the amount of memory, special considerations must be taken to maintain the use of these expensive resources.
c) Limited processing power
Mobile systems are not as tight as their desktop counterparts. Size, technology, and cost are some factors that affect the status of these resources. As with discs and RAM, you can find them in sizes that fit a small package.
d) Low power consumption
Mobile devices spend less power than desktop machines. These devices must conserve power because they are running in a state where the power supplied is limited by batteries.
e) Powerful and reliable
Since mobile devices are always taken wherever they are, they must be robust enough to deal with collisions, movement, and occasional drops of water.
f) Limited connectivity
Mobile devices have low bandwidth, some of them are not even connected. Most of them use the wireless connection.

3.3. UML Modelling

Based on preliminary study and observation then mobile CRM is expected to meet the information needs that no longer pass the student but can be directly enjoyed by guardian parents directly with mobile devices. Based on the summary of requirements analysis, the minimum specification of a system is found:

a) Able to provide KHS information
b) Able to provide KRS information
c) Able to provide billing information and Payment Status

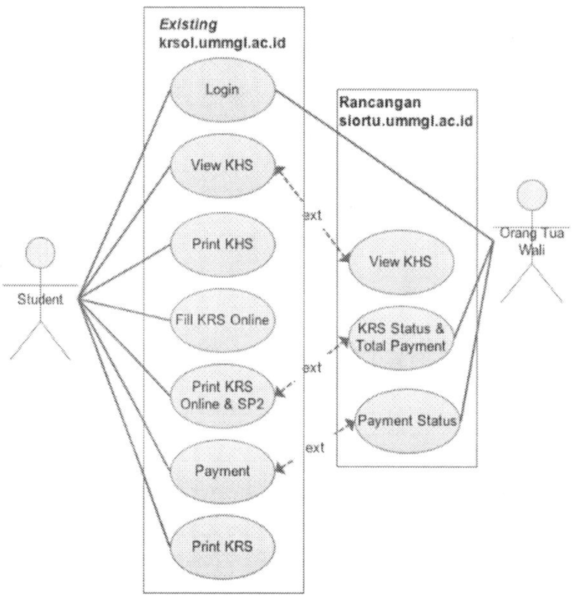

Figure 4. Parents information system [9]

Based on Figure 4, the development of parental guardian information system is developed without changing the existing system. The parent guardian's information system only retrieves data related to the necessary information to be presented to the parents of the student guardian. In figure 4 above also seen the primary function of the parental guardian student information system is to provide three essential information, as well as the expectations and specifications that will be the system to be developed. As shown in Figure 5 below, the system will be designed using the use case diagram from the parent guardian information system. Thus, that in the development will be done on the limits of information that has been set in the needs planning. The language of general facilitated the use of language generally to assist users in understanding the term education. So "KHS" we make comfortable with "Value" and "KRS" with "Study Plan."

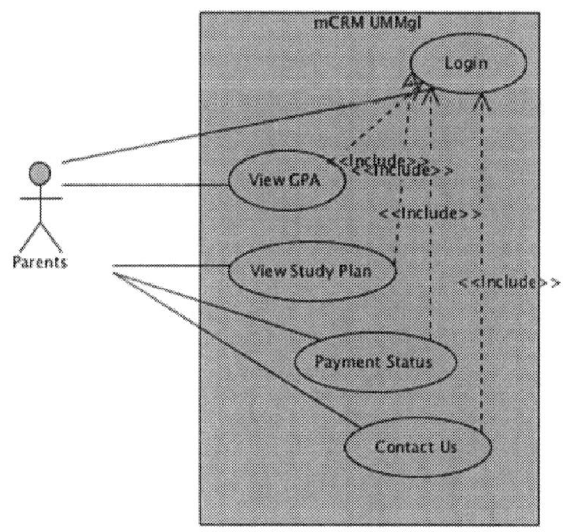

Figure 5. Use Case Diagram for mCRM

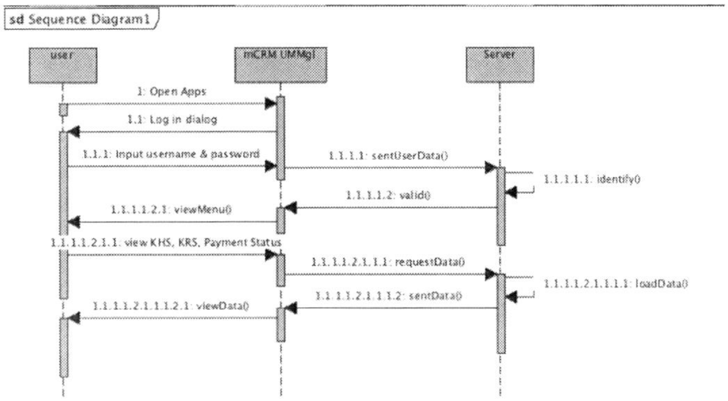

Figure 6. Sequence Diagram for mCRM

4. Conclusion

In this study, UML model usable diagram use case diagram and sequence diagram which can be the basis of further mobile CRM software development. The existing use case diagram already meets the minimal needs of a CRM mobile system for the parent student information system at a college. Mobile CRM has a function to connect the distance between parents of students with universities to provide academic information development of students to their parents.

Acknowledgment

This article is part of research on improving the capacity of Research Beginner Lecturer Directorate of Research and Community Service (PDP DRPM) entitled mobile CRM for Parent Information System Student Guardian at Higher Education. Therefore, thank DRPM KEMENRISTEKDIKTI and LP3M Universitas Muhammadiyah Magelang for their support.

References

[1] APJII, "PROFIL PENGGUNA INTERNET INDONESIA 2016," 2016. [Online]. Available: https://apjii.or.id/content/read/39/27/PROFIL-PENGGUNA-INTERNET-INDONESIA-2016. [Accessed: 23-Jan-2016].

[2] I. Banyumurti, "Statistik Internet Indonesia 2016," 2016. [Online]. Available: http://www.banyumurti.my.id/2016/01/statistik-internet-indonesia-2016.html. [Accessed: 20-Apr-2016].

[3] Y. Paragian, "Dalam 5 tahun terakhir, jumlah pengguna internet Indonesia naik 430 persen (GRAFIK)," 2014. [Online]. Available: https://id.techinasia.com/dalam-5-tahun-terakhir-jumlah-pengguna-internet-indonesia-naik-430-persen-grafik.

[4] N. Karna, I. Supriana, and N. Maulidevi, "Social CRM using Web Mining for Indonesian Academic Institution," *Int. Conf. Inf. Technol. Syst. Innov.*, 2015.

[5] E. D. Seeman and M. O'Hara, "Customer relationship management in higher education," *Campus-Wide Inf. Syst.*, vol. 23, no. 1, pp. 24–34, 2006.

[6] H. Ali, R. Moawad, and A. A. F. Hosni, "A cloud interoperability broker (CIB) for data migration in SaaS," *Proc. 2016 IEEE Int. Conf. Cloud Comput. Big Data Anal. ICCCBDA 2016*, pp. 250–256, 2016.

[7] X. Wei and L. Maoqing, "The research on building a global customer centered data warehouse system," *ICCSE 2011 - 6th Int. Conf. Comput. Sci. Educ. Final Progr. Proc.*, no. Iccse, pp. 971–973, 2011.

[8] F. Liming and Z. Sanyin, "The Application of Mobile Internet in Teaching and Learning Development of University Young Teacher," *2016 Eighth Int. Conf. Meas. Technol. Mechatronics Autom.*, pp. 700–703, 2016.

[9] A. Setiawan, "Analisis Sistem Informasi Orang Tua Wali Mahasiswa berbasis Website Studi Kasus Universitas Muhammadiyah Magelang," *J. Inform. UPGRIS*, vol. 2, no. 1, pp. 8–17, 2016.

IOP Conf. Series: Materials Science and Engineering **403** (2018) 012077 doi:10.1088/1757-899X/403/1/012077

Achievement of quality management system ISO 9001:2015 strategy in Vocational High School

R Irsyada[1], S Isbiyantoro[1], A P Wibawa[2], M F Teng[3], A F O Gaffar[4], R Herdianto[5] and A R Witarsa[6]

[1] Student, Universitas Negeri Malang, Malang, Indonesia

[2] Senior Researcher, Universitas Negeri Malang, Malang, Indonesia

[3]Assistant Professor, American University of the Middle East, Kuwait

[4] Lecture, State Polytechnic of Samarinda, Indonesia

[5]Researcher, Pascasarjana Universitas Negeri Malang, Indonesia

[6] Researcher, PPPPTK BOE Malang, Indonesia

irsyada.rahmat@gmail.com

Abstract. The purpose of this research is to find out the strategy to achieve the ISO 9001:2015 quality management system in Vocational High School in Malang. Strategies found in this research when applied correctly enable the certification of the ISO 9001:2015 standard. In order to further enhance the success of obtaining the certification, commitment and continues support from the school principal, president, as well as forming a plenary, education, and training teams and bringing in external auditors will be required. Implementation of ISO 9001 based quality management has a positive impact on school performance and activities. The strategy of achievement of ISO 9001:2015 version emphasized risk register compared to the previous version, which is the ISO 9001:2008 where risk register is not taking into account. The risk register is a preventive action that reduces the risk that occurs. Therefore, ISO version 2015 becomes a novelty quality management system because the additions of risk register.

1. Introduction

ISO 9001 is a set of standards developed and published by the International Organization for Standardization (ISO), which defines, builds and maintains an effective quality assurance system for manufacturing and service industries [1]. In an educational context, the services are the outputs or quality of the human resources.

The problem that is in Indonesia in terms of education is still the low quality of education at every level as explained by the Ministry of National Education that one of the factors that cause the low quality of education is inadequacy in school participation, especially by teachers and the community, in the implementation of education. Teacher participation in decision-making is often overlooked or ignored yet whether or not changes in schools are largely dependent on teachers [2]

By looking at this, it is necessary to improve and handle the quality of education that is by doing improvement process, security process of human resources especially teachers and other education stakeholders; and school environment that is by applying the quality management system ISO 9001: 2015 which is revision of ISO 9001:2008.

Public Vocational high school 2 Malang is one of the public vocational high schools in Malang City which has implemented ISO 9001:2015 quality management system. However, the main constraint that happened is inconsistency for example teachers have to make lesson planning but many teachers do not make the lesson planning, so teachers become inconsistent in a way there are not in accordance with responsibilities that ultimately lead to achievement of the quality objectives.

The first of these studies [3] is about the ISO 9001:2015 capability to solve the problem of an organization risk. Implementing ISO 9001:2015 has great benefits for an organization because the process of activities is documented. The second study [4] investigates that the ISO 9001:2008 to be the new ISO 9001:2015 is because of good time which relates to better changes. This becomes a new technological advancement for an organization to be competitive as the quality management system evolves.

The third study [5] indicates that the ultimate goal of ISO 9001 still gives confidence in the ability of organizations to consistently provide products and services in increasing customer satisfaction. Therefore, from this vision of ISO 9001:2015, the concept and meaning of "risk" and "risk free" is understand. The Fourth Research [6] concludes about how to properly implement an ISO 9001 quality management system in avoiding errors in the procedure, so the university prepares the proposal in preparation of documents such as requirements and stages of ISO 9001 quality management implementation.

The ISO standard is set of a tools focus on the quality issues, which the primary objective is to ensure applying total quality control is expected in organization and to achieve the effectiveness and efficiency of an organization. The ISO 9001 standard has been implemented by more than one million organizations in 187 countries, and a large number of these organizations aim to achieve product innovation and ISO 9000 series certification system the mean objective is the process standardization on quality of practice tracking and conformity assessment of product quality worldwide [7–9]. Quality Management system 9001 is a quality management system which refers to a set of interrelated or interacting elements to define policies and objectives to achieve those objectives that are used to guide and control an organization quality [10]. QMS is essential improves the quality of Vocational High School education in which educational institutions have implemented ISO Quality Management System to attract customers' attention and satisfaction with quality procedures that are often used as a competitive tool to produce ready-to-work output. In addition, government bodies require the education institutions to apply continuous improvement processes to prevent any risks involved in vocational schooling and improve the quality of education [8].

Improving the quality of education is one of the problems that until now is still under the supervision by the Indonesian government. One of the documented problems affecting the quality of education, for instance is where the, education management has not been improved professionally, this including vocational education as well [11]. The most important thing in the implementation of education is the improvement and quality assurance of education institutions. Guarantee and quality improvement of education conducted to answer competitiveness, imaging to society, and educational accountability. The preparation of qualified human resources can be achieved only with quality education, supported by all components of the education system (inputs, processes, and outcomes), and must meet quality and management standards used.

Strategy definition is a way to achieve long-term goals [12]. Hence, the defined strategies are the steps that must be followed to achieve the desired targets. Vocational education is a secondary education system that prepares students primarily to work in a particular field. Various efforts have been made to improve the quality of vocational education, among others through various training and quality improvement of school management. However, various indicators of educational quality have not shown any significant improvement, in fact, most schools are still in poor conditions [13]. Total quality management is a systematic quality improvement approach for overall enterprise management with the aim of improving performance in terms of quality, productivity, customer satisfaction, and profitability [14]. ISO 9001 created in 1987, has been present in successive revisions in the year 1994, 2000, 2008 and 2015 [15]. ISO 9001 is one of a series of quality management standard for business process system that will help to realize the best of an organization by understanding processes of providing products and or services to customers [16]. Based on the study presented in [17], measuring the effectiveness of implementing the ISO 9001 standard shows it has a positive effect on innovation in the context of education.

Both ISO 9001:2008 and ISO 9001:2015 consist of three similar clauses, there are; 1. Introduction, 2. Scope, and 3. Normative references. Figure 1 shows, in ISO 9001:2008 consist of clauses 4.a. Quality management systems, 5.a. management responsibility, 6.a. resource management, 7.a. Product realization, and 8.a. Measurement, analysis and improvement. Compare to ISO 9001:2015 are point; 4.b. Context of the organization, 5.b. Leadership, 6.b. Planning for the quality management system, 7.b. Support, 9.b. Performance evaluation, and 10.b. Improvement [18].

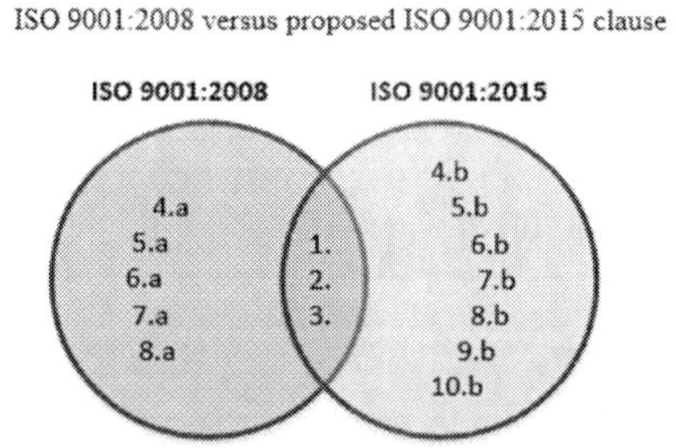

Figure 1. ISO 9001:2008 versus proposes ISO 9001:2015 clause

In essence, the implementation of ISO 9001:2008 all activities be documented while ISO 9001:2015 on the other hand, requires all activities to be documented as well with additional clauses regarding risk register. This risk register documentation is very helpful in risk management where the ISO 9001:2015 is documented in a way to be able to identify the potential risks and problems it may occur in the future. In fact, significant differences between ISO 9001:2008 and 9001:2015 that the standard is the same process from planning, implementation, and evaluation. What distinguishes the ISO 9001:2015 version is that diverts from the risk registers are more emphasized because it will inventory the risk of problems that will occur and is contained under clause 6. Whereas in ISO 9001:2008 version, there are 8 clauses and ISO 9001:2015 version there are 10 clauses. Version 2015 focuses on risk control and identification. In addition, in this case version 2008 requires management to take a more active role in aligning quality policy with business needs [7]. Hence, in the risk registers we will document the risks that occur in the process of activity in both external and internal factors. Therefore, the risks can be identified and solved to reduce the same risks re occur again in the future.

One example, when students are enjoying out Industry practices, they get tools from the place of existence (store) which is risky. With the risk register there, it will try to find a solution or problem solving. With the existence of preventive action, a risky event will be overcome by inventory before the incident. The benefits of an ISO Quality Management System will bring tangible and measurable benefits to all sectors, supportive standards for technology that can be relied on ensuring expected quality of an educational institution [19].

So the existence of an organization in education, especially Vocational High School by applying quality management system ISO 9001: 2015 will change the performance mindset of the education stakeholders so that a quality management in an organizations especially vocational high schools will become systematic and documented.

2. Methodology

Data collected in this research are through interviews and observations on the achievement strategy of ISO 9001: 2015 in Vocational High Schools. Schools interviewed are schools that have been certified with both ISO 9001:2008 and ISO 9001:2015. The purpose of this paper is to know the achievement of ISO 9001:2015 strategy. Thus, this paper will provide information and explanation on the differences and the fundamental achievement of the quality management process in Vocational High School in the Malang Region.

3. School that can implement ISO version 2015

ISO 9001: 2015 standard was launched by the end of the year 2015, and not many schools are implementing this version of the quality management system, as a mere formality. There are some schools that are not international-school oriented but implemented the ISO 9001:2015 and have developed quite well, especially for schools that are still in the pioneering stage. The benefits of ISO are felt in terms of documentation, workflow-coordination-communication, job description and standard operating procedures (SOPs) so that each member of the school organization can understand its work.

Figure 2 shows a general design of quality management system process for Vocational High School. The management section consists of several components, these include: extra-curricular, finance, procurement, and support which supporting the running of business process quality management system in Vocational High school. This process starts from the design of development planning in accordance to ISO 9001:2008 and 2015 quality model system. Therefore, all the components are involved, where on the basis of ISO principles such as planning the implementation evaluation and then back again to component before. In the certification process, the teaching and learning process is an evaluation phase. Figure 3 shows the information about the implementation of ISO 9001 in Malang Region, based of observation of 20 vocational high schools.

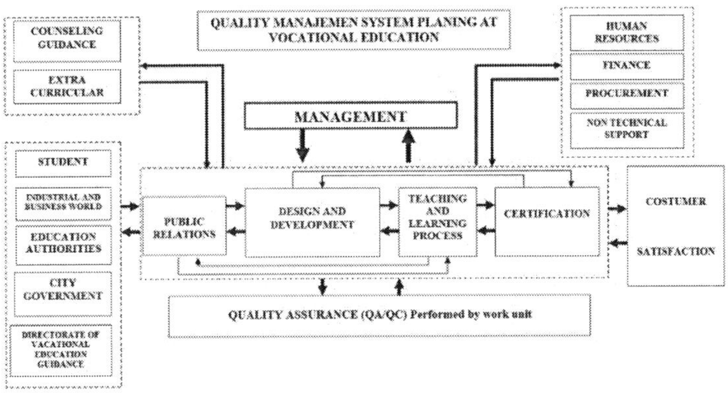

Figure 2. Quality Management System Planning at Vocational Education.

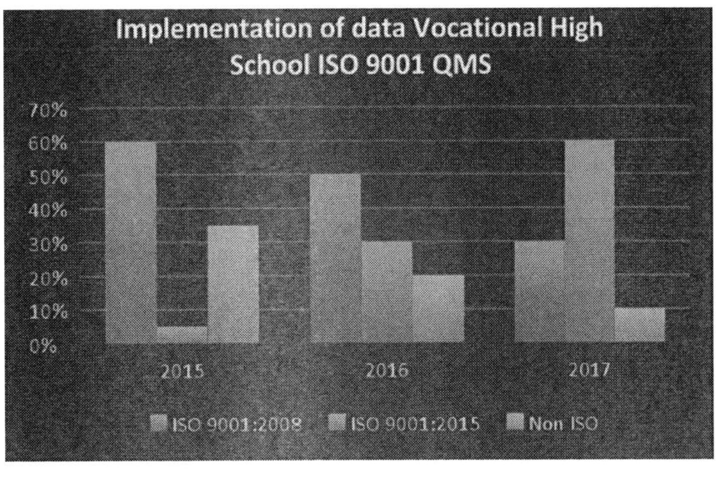

Figure 3. Implementation of data Vocational High School ISO 9001 QMS.

School that implemented the ISO 9001:2008 standard was 60%, 50%, and 30% for the year 2015, 2016, and 2017 respectively. Only 5% of the school in the year of 2015 implemented the ISO 9001:2015 version. The number increases to 30% and 60% for the year 2016 and 2017 respectively for the ISO 9001:2015 standard.

Based on these results, it showed that many schools are willing to upgrade or implementing the ISO 9001:2015 standard because it greatly benefits and affects the performance of the colleagues in the Vocational High Schools. The increase in performance can only be achieved if the school implement the ISO standard in accordance to the standard of ISO 9001:2015 quality management system. This is to ensure to have a systematic procedure according to the activities undertaken by the school community in the institution.

4. School implemented ISO 9001:2015

The strategy for schools to achieve the ISO 9001:2015 standard is presented in figure 4. Strategy achievement of ISO 9001:2015 version basically start with arranging risk register. From the risk register, it will be able to map document to be arranged. The second preparation is the four compulsory documents, which is the quality manual, quality procedures, work instructions and several forms. Where the documents must meet the required format, and will be checked by the auditor from the institution of ISO organizer. Therefore, the school will only receive the ISO 9001:2015 certificate if the compulsory documents are completely presented.

The next step is to prepare the supporting documents, ranging from quality plans to school work programs. This follow by the socialization of all the citizens of the school, because school stakeholders will implement the ISO management and if it is in the socialization, it means that the school residents have implemented procedures in accordance with their respective units. From the implementation, it will result in the documenting, which must be maintained and archived or stored. Hence, from the process there will be an evaluation to determine the progress of certification.

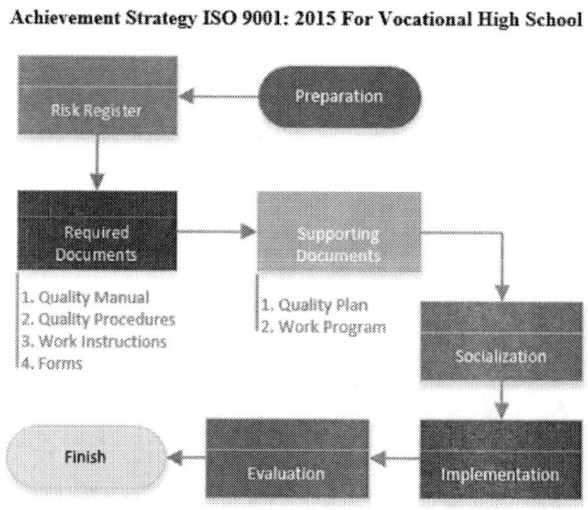

Figure 4. Achievement Strategy of ISO 9001:2015 in Vocational High School.

By carrying out the audit or checks, and if it works, it concludes it has been successfully implemented the ISO 9001:2015 certification. If a school wanted to implement the ISO 9001:2015, it means the school must follow all the four compulsory documents that have been prepared. Where all the citizens of the school have to perform standard SOP (standard operational procedure) that has been set. Therefore, after implementing the standard process that has been established it means implementation strategy of ISO has been completed. In implementing the management of ISO 9001, there is no force that all schools in Indonesia are required to implement the ISO 9001:2015 standard. However, implementing and obtaining the ISO certification will benefit the schools to have a systematic management, will have a positive impact on the schools.

5. Conclusion

Based on the results of observations and interviews conducted in Vocational High School in Malang that implemented the quality management system that based on the ISO 9001:2008 quality management system must make changes to the new ISO 9001:2015 version by conducting a an internal audit and then follow by an external ISO institutions to ensure knowledge management oriented ISO 9001:2015 to the stakeholders in Vocational High School is presented.

As for the interview result, it indicates that there is no significant difference in the clause of ISO version 2008 to 2015. The only changes of ISO 9001:2008 is more emphasize on improvement while ISO 9001:2015 is more emphasize on risk register or risk prevention. A long-term strategy for achieving ISO 9001:2015 standard is a must to upgrade the quality management system with addition of clause of risk register. The implementation of ISO 9001: 2015 quality management system in education institute especially in Vocational High School is a way to improve the quality of education so that the output from Vocational High School can be absorbed in both industry workplaces and in the universities. With the implementation of ISO 9001:2015 will change the performance mindset of the education stakeholders so that a quality management in an organizations especially vocational high schools will become systematic and documented.

Reference

[1] Liao R C 2015 Customers' Perspectives on ISO 9001 QMS Auditors' Personality Traits: A Preliminary Investigation from Taiwan's Certificated Companies *Proc. Int. Conf. Serv. Sci. ICSS* vol. 2015–Octob pp 215–219

[2] Nugroho A W 2017 Strategi Sekolah dalam Menerapkan Sistem Manajemen Mutu (SMM) Berbasis ISO 9001:2015 *Manaj. dan Supervisi Pendidik.* vol 2 no 2 pp 155–166

[3] Fonseca L M 2015 From Quality Gurus and TQM to ISO 9001:2015 : A Review Of Several Quality Paths *Int. J. Qual. Res.* vol 9 no 1 pp 167–180

[4] Othman N, Sanury S, Mokhtar M, Norhasni M and Asaad M 2017 Quality management system ISO 9001 : 2008 and ISO 9001 : 2015 standards within higher education *J. Glob. Bus. Soc. Entrep.* vol 3 no 6 pp 40–46

[5] SITNIKOV C S 2015 The Role of Risk Management In ISO 9001: 2015 *Proceedings of the 9th international management conference* pp 1009–16

[6] Alalfy H R 2015 A Suggested Proposal to Implementation Quality Management System ISO-9001 in Egyptian Universities *Am. J. Educ. Res.* vol 3 no 4 pp 483–9

[7] Manders B, De Vries H J and Blind K 2016 ISO 9001 and product innovation: A literature review and research framework *Technovation* vol 48–49 pp 41–55

[8] Sampaio P, Saraiva P, and Rodrigues A G 2009 ISO 9001 certification research: questions, answers and approaches *Int. J. Qual. Reliab. Manag.* vol 26 no 1 pp 38–58

[9] Dreyfus L P, Ahire S L and Ebrahimpour M 2004 The Impact of Just-In-Time Implementation and ISO 9000 Certification on Total Quality Management vol 51 no 2 pp 125–141

[10] Ed S W 2015 Information Systems : Development, Applications, Education

[11] Susilo P H, Larasati A, Kejuruan P and Malang P N 2016 Sistem Manajemen Mutu Iso 9001 : 2008 Digital vol 1 no 5 pp 971–7

[12] Sari D P 2014 Strategi pencapaian standar mutu dan pengaruhnya terhadap hasil penjualan di pasar eropa vol 16 no 1 pp 47–54

[13] Riban I and Mukhadis A 2011 Implementasi iso 9001:2000 pada pembelajaran produktif bidang keahlian bangunan di sekolah menengah kejuruan *Teknol. dan Kejuru.* vol 34 no 2 pp 141–154

[14] Sadikoglu E and Zehir C 2010 Investigating the effects of innovation and employee

performance on the relationship between total quality management practices and firm performance: An empirical study of Turkish firms *Int. J. Prod. Econ.* vol 127 no 1 pp 13–26

[15] Barry Z, Bensaid Z, Boukhris I, Hamrit S, Mnif F, Soto L, Derathé A and Farges G 2016 Allier qualit?? et performance via l'ISO 9001 version 2015 *IRBM News* vol 37 no 2 pp 68–71

[16] Debnath N, Uzal R, Montejano G and Riesco D 2010 Web application to improve police management performance *2010 Seventh International Conference on Information Technology Web, 2010* pp 32–35

[17] Kafetzopoulos D, Gotzamani K and Psomas E 2013 Quality systems and competitive performance of food companies *Benchmarking An Int. J.* vol 20 no 4 pp 463–83

[18] Fonseca L M 2015 ISO 9001 : 2015 Revision *International Conference on Quality Engineering and Management, 2015* pp 391–400

[19] Vernier 2017 *International Organization for Standardization* [Online]. Available: https://www.iso.org/standards.html.

Development of knowledge management system for determining organizational culture in micro, small and medium enterprises using organizational culture assessment instrument

R Z A Aziz[1], M F Azima[2] and S Y Irianto[3]

Informatics Engineering Department Graduate Program Informatics and Business Institute Darmajaya Bandar Lampung, Indonesia

[1]rz_aziz@darmajaya.ac.id, [2]mfauzanazima@darmajaya.ac.id, [3]suhendro@darmajaya.ac.id

Abstract. Organizations must improve their competitiveness due to global competition and free market. Organizational culture and continuous improvement are one of the main factors used not only to face the competitiveness in the free market but also to improve organizational performance and vision and mission achievement. Moreover, the organizational culture also has an important element for determining the direction of Micro, Small and Medium Enterprises (MSMEs). However, many MSMEs currently do not pay attention to their organizational culture so that difficulties emerge in determining the right direction and policy. To solve the problems, this research aimed to determine the companies' organizational culture by using organizational culture assessment instrument (OCAI) based on knowledge management system (KMS) in order to provide recommendations. In the developing the system, the researchers conducted surveys to several MSMEs and invite experts and practitioners to conduct the process of knowledge capture and literature study and to achieve the best organizational culture recommendations on MSMEs. This research produced a software used for determining the organizational culture based on KMS and OCAI principles. The developed software was tested to several companies and MSMEs to obtain results with a very high level of accuracy in providing recommendations.

Keywords. Organizational Culture, Knowledge Management System, MSMEs, OCAI.

1. Introduction

The use of information technology in companies is indicated by the use of software. Software used in companies is in the form of electronic business, electronic commerce, electronic budgeting, and electronic marketing [1]. The use of software is not only used for business activities but also for organizational culture. The organizational culture is one of the variables closely related to a corporate or an organizational performance improvement [2].

The organizational culture is very important because each organization or company has its own different organizational culture from the other organizations [3]. Moreover, the organizational culture is closely related to the knowledge management system because it is a part of the technologies and mechanisms used to support the knowledge management in the organizational culture [4]. Besides, the knowledge management is the combination of planning, organizing, motivating, and controlling people, processes, and systems in the organization used to ensure that the knowledge-related assets improve [5]. In addition, the knowledge management involves the data and information systems used to codify, categorize, and transfer asset [6]. The organizational culture can also influence the success of the TQM implementation and organizational performance. Therefore, knowing the organizational culture is important for the company [7] because the organizational culture is recognized as an important determinant of the quality management success and organizational performance [8,9].

Each company has different leaderships and organizational cultures which influence particular organizational cultures [10]. According to Cameroon and Quinn 2011, there are four types of the organizational culture embedded in the company e.g., clan culture, adhocracy culture, market culture, and hierarchy culture [3]. Problem statement of this research is a lack of the information about the current organizational condition e.g., the condition for MSMEs, a gap among leaderships, middle manager and employee condition, and limited communication between the work culture and organizational culture from managers to each division and section. To solve these problems, a tool as a policy maker must be made and is used to make decisions in implementing the organizations. From the background of this study, the researchers are interested in developing the knowledge management system (KMS) used for determining organizational culture on micro, small medium enterprise (MSMEs) by using the organizational culture assessment instrument (OCAI) as a measurement model.

2. Research Method

2.1. Knowledge Mapping

Organizational culture in MSMEs was very important for managers so that this research was conducted through organizational culture assessment instrument (OCAI) model which was used to determine the culture. The result of this instrument was in the form of a competing value which influenced organizational culture profile. Knowledge mapping played an important role as a recommendation for the types of organizational culture so that the system was able to categorize and recommend the best organizational culture for the company. Stages on the knowledge mapping were adopted from the knowledge management system life cycle [11] illustrated in Figure 1.

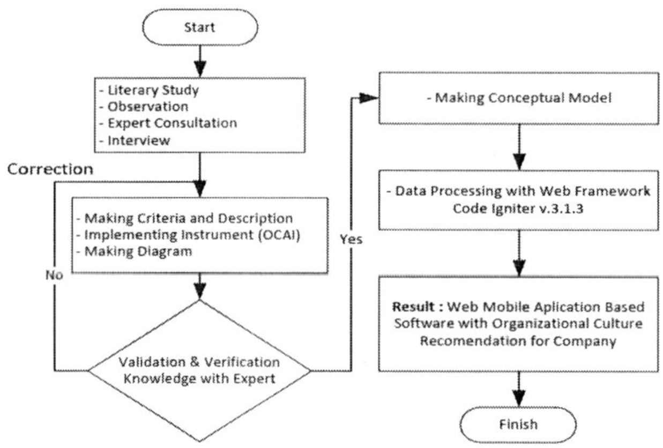

Figure 1. Knowledge Mapping Flowchart

2.2. Research Instrument

The instrument of this research was organizational culture assessment instrument [3] consisting of 6 dimensions: dominant characteristics, organizational leaderships, management of employees, organizational glues, strategic emphasis, and criteria of success. Each organizational culture dimension had 4 questions leading to the alternative cultural profiles (clan culture, adhocracy culture, market culture, and hierarchy culture). Respondents were asked to provide an assessment of each alternative answer. OCAI instrument had 2 assessment columns (now and preferred columns). Now column explained an assessment on the current cultural conditions but preferred column represented the expected culture conditions. Figure 2 showed the competing values of organizational culture [3].

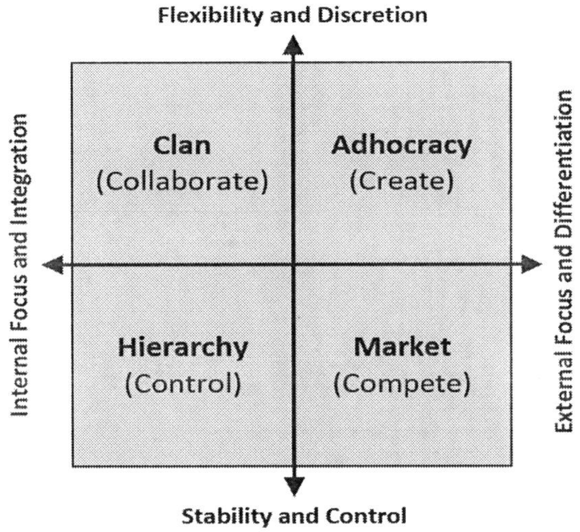

Figure 2. Competing Values Framework of An Organizational Culture

2.3. Instrument Assessment Scale

This research used quantitative data so that data analysis was done by using quantitative approach (inferential statistics). The questionnaire used a likert scale from 1-7. Score 7 was given at Very Strongly Agree (VSA), score 6 was given on Strongly Agree (SA), score 5 was given on Agree (A), score 4 was given on Simply Agree (SA), score 3 was given on Less Agree (LA), score 2 was given on Disagree (DS), and score 1 was given on Strongly Disagree (SDS).

2.4. Determining Algorithm of Organizational Culture

According to [3], OCAI calculation was that respondents were asked to fill out the questionnaire in 6 dimensions from which each dimension had 100 as a maximal score so that respondents were not confused in filling the questionnaire. Filling the questionnaire with direct numbers (0-100) made respondents difficult to determine how many numbers should be filled to determine 100 in each of OCAI dimensions. Therefore, this research developed likert scale conversion algorithm from scale 1-7 and converted this number to scale 100 on each OCAI dimension. The algorithms developed for determining the organizational culture were as follows:

- Respondents filled scores on OCAI questionnaire with 1 – 7 as scales
- Converting each score into 100 on each OCAI dimension with formula (1).

$$X_{ij\,new} = \frac{X_{ij\,old}}{\sum X_{ij\,old}} x100\% \tag{1}$$

  where i indicates OCAI dimension (1 – 6) and j indicates questions on each OCAI dimension (1 – 4)
- Summing up each dimension and finding the mean score in columns A, B, C, and D in *now* and *preferred* columns for each record of respondents.
  Summing up each dimension in columns A, B, C and D using formula (2)

$$t_{kr} = \sum_{i=1}^{6} X_{ji\,\text{new}} \tag{2}$$

  Finding the mean score of each dimension in columns A, B, C and D using formula (3).

$$\overline{X}_{kr} = \frac{t_{kr}}{i} \tag{3}$$

where k indicates organizational culture (1 – 4) and r shows number of respondents

- Summing up all mean score of respondents, clan, adhocracy, market, and hierarchy and dividing them by a number of respondents in each of the *now* and *preferred* columns using formula (4).

$$T_{kr} = \sum_{r=1}^{R} \frac{\overline{X}_{kr}}{R} \tag{4}$$

Where R shows total of respondents.

After summing up all mean score of respondents, the organizational culture profile was created in the percentage using formula (5).

$$BO_k = \frac{T_{kr}}{\sum_{k=1}^{j} T_{kr}} \tag{5}$$

where BO is Organizational Culture.

- Classifying 4 largest organizational culture profiles to the smallest organizational culture profiles and obtaining the greatest organizational culture using formula (6).

$$T_k = \text{arsort}(BO_k) \tag{6}$$

Culture profile changes in order (BO1, BO2, BO3, BO4) from largest to smallest.

After culture profile was being in order, the first largest organizational culture was compared with the second largest. If they were included in the range of ≤ 5% of the first culture, 2 pieces of dominant organizational culture (BO1 & BO2) was obtained; if they were more than ≥ 5% of the largest culture, the dominant organizational culture was only 1 (BO1).

- Creating an organizational culture profile diagram.
- Providing the best recommendations based on knowledge capture and organizational culture profile data.

2.5. *Algoritm Calculation*

Respondents filled out questionnaires on a scale 1-7 with Strongly Disagree (STS) to Very Strongly Agree (SSS) shown in Table 1.

Table 1. OCAI Instrument Score in Scale 7

No	**Dimension** $X_{ij\,old}$									
	1				**2**				**...**	**6**
	A	B	C	D	A	B	C	D		
r1	$X_{ij\,old}$	...	...	...	...	...	...	...	...	...
r2	...	...	...	...	...	...	...	...	...	...
r n..	...	...	...	...	...	...	...	...	...	...

Converting scale 7 questionnaire data into scale 100 using equation (1).

Table 2. Scale 7 Conversion to Scale 100

No	Dimension $X_{ij\ new}$									
	1				**2**				**...**	**6**
	A	B	C	D	A	B	C	D		
r1	$X_{ij\ new}$	...	...	...	...	...	...	...	...	...
r2	...	...	...	...	...	...	...	...	...	...
r n..	...	...	...	...	...	...	...	...	...	...

Summing up the columns a = clan, b = adhocracy, c = market, d = hierarchy, using equation (2) and finding the mean score using equation (3).

Table 3. Score of Organizational Culture per Respondents

No	**Clan**	t_{kr}	$\overline{X}_{kr}$	**Adhocracy**	t_{kr}	$\overline{X}_{kr}$	... 4
r1	$X_{ji1}+X_{ji2}+\ldots X_{ji6}$	t_{kr}	$\overline{X}_{kr}$	$X_{ji1}+X_{ji2}+\ldots X_{ji6}$	t_{kr}	$\overline{X}_{kr}$	
r2	$X_{ji1}+X_{ji2}+\ldots X_{ji6}$	t_{kr}	$\overline{X}_{kr}$	$X_{ji1}+X_{ji2}+\ldots X_{ji6}$	t_{kr}	$\overline{X}_{kr}$	
r n..	$X_{ji1}+X_{ji2}+\ldots X_{ji6}$	t_{kr}	$\overline{X}_{kr}$	$X_{ji1}+X_{ji2}+\ldots X_{ji6}$	t_{kr}	$\overline{X}_{kr}$	

Summing up the total score of respondents, clan, adhocracy, market, and hierarchy; and dividing them by the number of respondents using equation (4). And use equation (5) to make precentage of organizational cultue profile.

Table 4. Organizational Culture Profile

Clan	Adhocracy	Market	Hierarchy
$T_{clan}=\sum_{r=1}^{R}\frac{\overline{X}_{kr}}{R}$	$T_{adhocracy}=\sum_{r=1}^{R}\frac{\overline{X}_{kr}}{R}$	$T_{market}=\sum_{r=1}^{R}\frac{\overline{X}_{kr}}{R}$	$T_{hierarchy}=\sum_{r=1}^{R}\frac{\overline{X}_{kr}}{R}$

After organizational culture profile score was obtained, finding the dominant organization culture by classifying the organizational culture profile use equation (6).

Table 5. Finding the largest organizational culture

No	Organizational Culture (*Now*)	Organizational Culture (Preferred)
1st	$BO_{dominant1}$	$BO_{dominant1}$
2nd	$BO_{dominant2}$	$BO_{dominant2}$
3rd	$BO_{dominant3}$	$BO_{dominant3}$
4rd	$BO_{dominant4}$	$BO_{dominant4}$

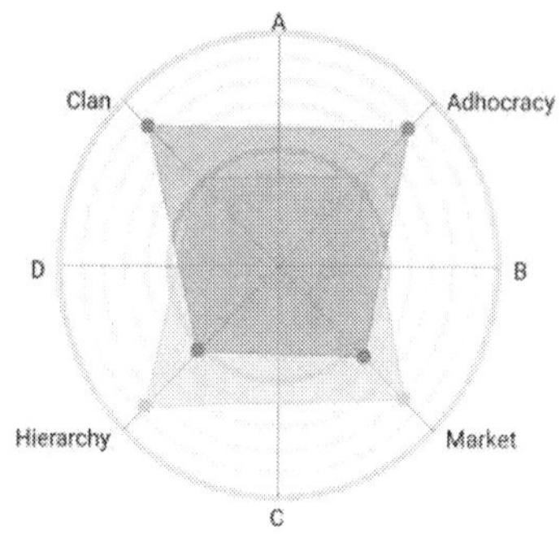

Figure 3. Competing Value Framework Of AnOrganizational Culture Profile Diagram

According to the organizational culture profile, the best recommendation was showed through the organizational culture profile data which were appropriate with the knowledge capture data.

2.6. *Knowledge Capture*
Knowledge capture was useful to capture the knowledge from experts, practitioners, and organizational culture profile form. The development of this system was carried out by surveys and interviews with experts and practitioners used to conduct the knowledge capture on good recommendations on each type of organizational culture profile. This knowledge was a recommendation on the MSMEs organizational profile.

2.7. *Use-case Diagram*
The design of a use-case diagram system for determining organizational culture was done in the process of software development. The design of this system consisted of several actors who played a role in the system, illustrated in Figure 4.

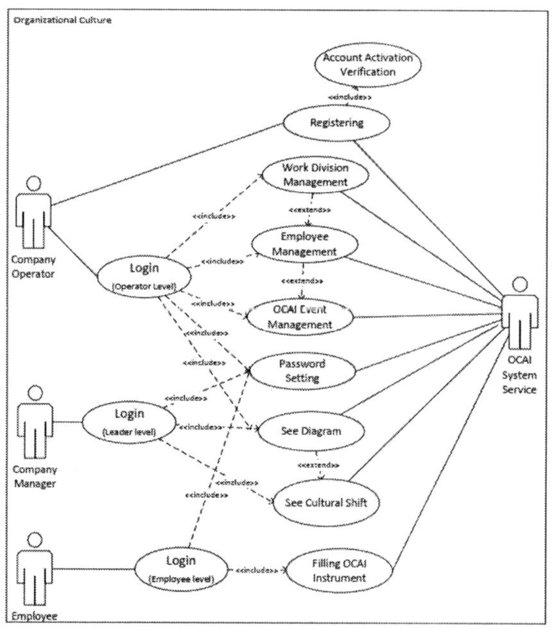

Figure 4. Use-Case Diagram of Determining Organizational Culture System

Operator has several tasks covering Sign up, verify account, and manage data. Manager has several tasks include gets report description of OCAI diagram, organizational culture profile, and recommendation, and Employee filling OCAI Instrument.

3. Research Findings And Discussion

This study produced useful software to assist managers in determining good organizational culture in MSMEs. The following paragraphs described the appearance of the organizational culture software.

3.1. Menu Page and Quistionnare

The menu page was a link provided for the user to access the system based on the user level. Questionnaire was a page for employees to fill out the OCAI questionnaire. This questionnaire was an instrument that played a role in getting a profile of organizational culture (Figure 5).

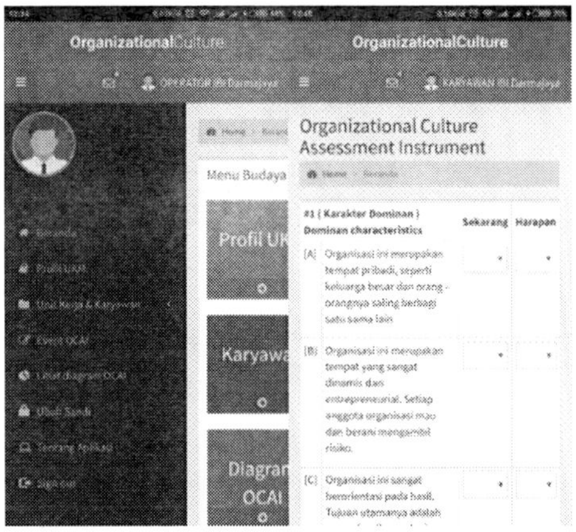

Figure 5. Menu Page andQuestionnaire

3.2. OCAI Diagram Page and Details

OCAI diagram page was a page describing the organizational culture profile in the form of a diagram. There were types of organizational culture e.g., clan, adhocracy, market, and hierarchy. The detail page was the result of the organizational culture calculation so that the dominant organizational culture number was showed in Figure 6.

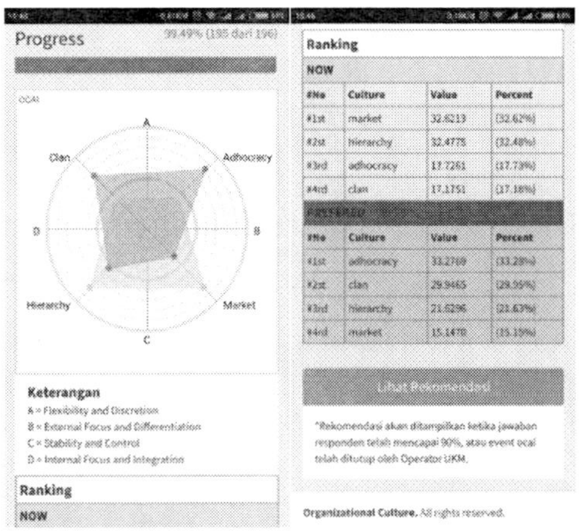

Figure 6. OCAI Diagram Page and Details

3.3. Organizational Culture Recommendation Page

Recommendation page was a useful page to show the best organizational culture recommendation to MSMEs. Recommendations were obtained through the process of knowledge capture and its result was adjusted to the nearest score on the profile of organizational culture (Figure 7).

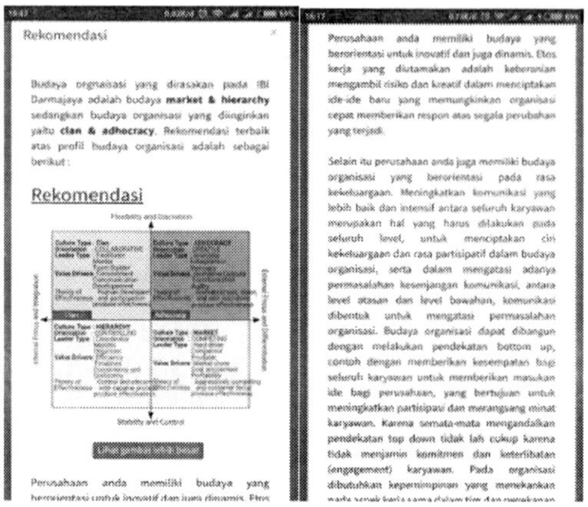

Figure 7. The Best Organizational Culture Recommendation Page in MSMEs

4. Conclusion

The development of algorithms for determining organizational culture was by Likert scale 1-7. It allows respondents to fill out OCAI questionnaires. Moreover, respondents do not have to count the number of figures up to 100 in each OCAI dimensions. This software is useful for the managers to know the organizational culture condition, to get the best recommendations on the organization, and to follow up the company based on organizational culture diagrams, recommendations, and profiles. Knowledge capture provides the organizational culture decision recommendation which is appropriate with OCAI's organizational culture profile picture. Organizational culture can be determined by using knowledge management system (KMS) combined with organizational culture assessment instrument (OCAI). Organizational culture software can be used to find out the profile of organizational culture and provide appropriate recommendations to MSMEs.

5. Significance Of The Research

In this research, the basis of knowledge as the database is still incomplete so that it is expected that further researches can increase or develop more varied capture knowledge for some cultural conditions in order to maximize the recommendation organizational culture, and to find correlation between organizational culture and organizational performance.

This study has important practical and academic implications. Prior to the implementation of TQM, managers must determine the dominant organizational culture in their company. Management needs to assess the culture using Organizational Culture software in order to develop step to implement TQM.

References

[1] Marius, P., & Anggoro, S. Profil Pengguna Internet Indonesia 2014. APJII, Jakarta. 2015.
[2] Megawati, & Nashri, M F. Evaluasi Budaya Organisasi Dalam Penerapan Teknologi Informasi Menggunakan Organizational Culture Assessment Instrument (Ocai) Pada PT. Perkebunan Nusantara V Pekanbaru. Jurnal Ilmiah Rekayasa Dan Manajemen Sistem Informasi, 2015. 1(1), pp 17-30.
[3] Cameron, K S., & Quinn R E. Diagnosing and Changing Organizational Culture: Based on The CompetingValues Framework-3/E. 2011.

[4] Fernandez IB, Sabherwal R. Knowledge Management: System and Processes. England: M.E. Sharpe. 2010.
[5] García-Holgado, Alicia, et al. "Analysis and improvement of knowledge management processes in organizations using the Business Process Model Notation." Annual Conference of the Global Innovation and Knowledge Academy. Springer, Cham, 2015.
[6] Murphy, Glen, and Sonia Salomone. Using social media to facilitate knowledge transfer in complex engineering environments: a primer for educators. European Journal of Engineering Education 38.1 (2013): 70-84.
[7] Aziz, RZ Abdul, and Hiroshi Morita. National culture, organizational culture, total quality management implementation, and performance: an empirical investigation. International Journal of Productivity and Quality Management, 2016. 19(2), pp 139-159.
[8] Prajogo, D.I. and McDermott, C.M. (2005) 'The relationship between total quality management practices and organizational culture', International Journal of Operations and Production Management, Vol. 25, No. 11, pp.1101–1122.
[9] Prajogo, D.I. and McDermott, C.M. (2011) 'The relationship between multidimensional organizational culture and performance', International Journal of Operations and Production Management, Vol. 31, No. 7, pp.712–735.
[10] Alvesson, M., & Sveningsson, S. Changing organizational culture: Cultural change work in progress. Routledge. 2015.
[11] Awad, E.M dan Ghaziri, H.M. Knowledge Management. Person Education Inc., New Jersey. 2010.

IOP Conf. Series: Materials Science and Engineering **403** (2018) 012079 doi:10.1088/1757-899X/403/1/012079

Analytical hierarchy process for mode choice model at Perumnas Palur, Karanganyar

N Hidayati, S Sunarjono and A Putri

Center for Transportation Studies, Engineering Faculty, Universitas Muhammadiyah Surakarta

nurul.hidayati@ums.ac.id

Abstract. The increasing need for transportation is influenced by the development of an area, including what happened at Perumnas Palur, Karanganyar. This residential area has a strategic location with many public facilities that easier people to do various movements with various modes of transportation. What mode is used by the Perumnas Palur society is inseparable from the criteria of mode choice. The linkage between the user reason criteria and the selection of these modes can be solved with an analytical hierarchy process (AHP). This study uses questionnaires that have been filled by the community selected as a sample. The questionnaire component includes travel characteristics and economic characteristics that are tailored to the conditions of the object. The dominance of the travel characters was obtained: motorcycle users (61%), work trips (44%), more than 5 trips per day (43%), 10-15 minutes of travel time (26%) and travel cost less than IDR 10,000 (43%). The percentage of mode types chosen according to the criteria of the reasons are Motorcycle (27%), Walking (26.6%), Car (18.5%), Bicycle (15%) and Bus (12.4%). The percentage of the vector priority each criterion of the reason is Safe (47.5%), Time (24%), Easy (17%), and Cost (11.5%).

1. Introduction

A model is a tool that can be used to accurately reflect and simplify a reality. The mode choice model is one of the four stages of the planning process, besides trip generation, distribution and route choice [1]. According to [2], there are several factors influencing the selection of modes, namely: characteristics of travelers, travel characteristics made, and characteristics of available transportation systems. Travel characteristics included travel destinations, distance, time, and travel expenses, departure times, weather conditions, and interconnection of different modes of transport [3].

The movements of people from origin to destination have different characteristics, generally affected by land use activities, and also car ownership. Sometimes, mode choice is also influenced by qualitative factors that are difficult to quantify. In this case, it could be analyzed by using the Analytical Hierarchy Process (AHP) method to describe multi-factor or criteria problems into a hierarchy [4]. Xi *et al.* [5] stated that AHP was a subjective assessment of decision makers using a combination of data between qualitative and quantitative analysis to help make decisions. Based on the above background, then the study of mode choice model at Perumnas Palur, Karanganyar is necessary.

Perumnas Palur is the first Perumnas in Central Java established around the 1975s. This study aims to determine the characteristics of travel, and socioeconomic conditions of people in the housing. In addition, the study also aims to create a mathematical model of AHP and determine the percentage of mode selection based on the criteria of reason. This study is expected to provide an overview of the use of AHP in modal mode analysis. This is evidence that factors influencing the selection of modes can be expressed either qualitatively or quantitatively.

2. Method

Decision-making with AHP can be divided into several steps, namely: define the problem, determine the desired solution, and create a hierarchical structure. This hierarchical structure was the composition of the network begins with the general goal, followed by the criteria and alternative options to be ranked. Networks were built on a comprehensive structure with regard to the issues to be developed [6]. The advantage of this method is the ability to combine qualitative and quantitative elements. Quantification of qualitative elements is done by providing a pair-wise comparison scaled perception. Comparisons are made on the judgment of the decision maker by assessing the level of importance of an element over other elements. The hierarchical structure used in this study can be seen in figure 1, whereas the rating scale is shown in table 1 and the process is shown in figure 2.

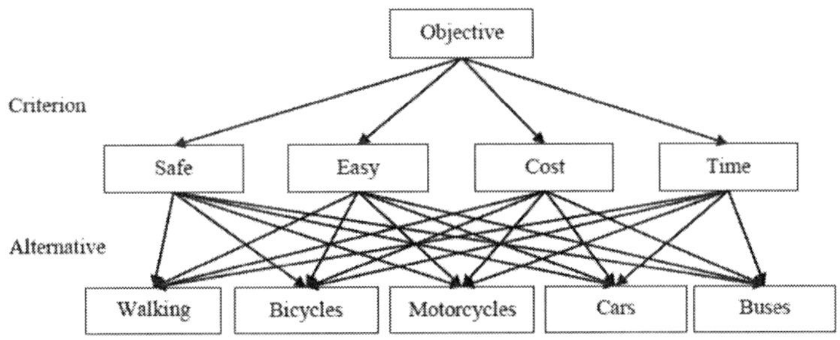

Figure 1. The hierarchical structure.

Table 1. Scale of hierarchy assessment.

The intensity of interest	Definition of variable	Explanation
1	Equally important	Both elements have the same effect
3	Slightly more important	Assessment is slightly in favor of one element.
5	More important	Assessment is one-sided by one element.
7	Very important	Evidence that one of the elements is very influential and dominant
9	Absolutely more important	Evidence that one element is more important than the others and very clear
2, 4, 6, 8	Middle rating between two scale	A scale is given if there is doubt between two options
Reciprocal	If the comparison of elements i and j produces one of the values above then the comparison between the elements j to i will yield the inverse value	

The calculation process of the AHP method uses a comparison matrix (reciprocal) as shown in equation (1), ie if $A_{ij} = a$, then $A_{ji} = 1/a$. If A_i has the same interest with A_j then $A_{ij} = A_{ji} = 1$, for the spesial case, $A_{ij} = 1$ for all i.

$$A = \begin{bmatrix} 1 & a_{(1,2)} & \cdots & a_{(1,n)} \\ {}^{1}/_{a_{(1,2)}} & 1 & \cdots & a_{(2,n)} \\ \cdots & \cdots & \cdots & \cdots \\ \cdots & \cdots & \cdots & \cdots \\ {}^{1}/_{a_{(1,n)}} & {}^{1}/_{a_{(2,n)}} & \cdots & 1 \end{bmatrix} \quad (1)$$

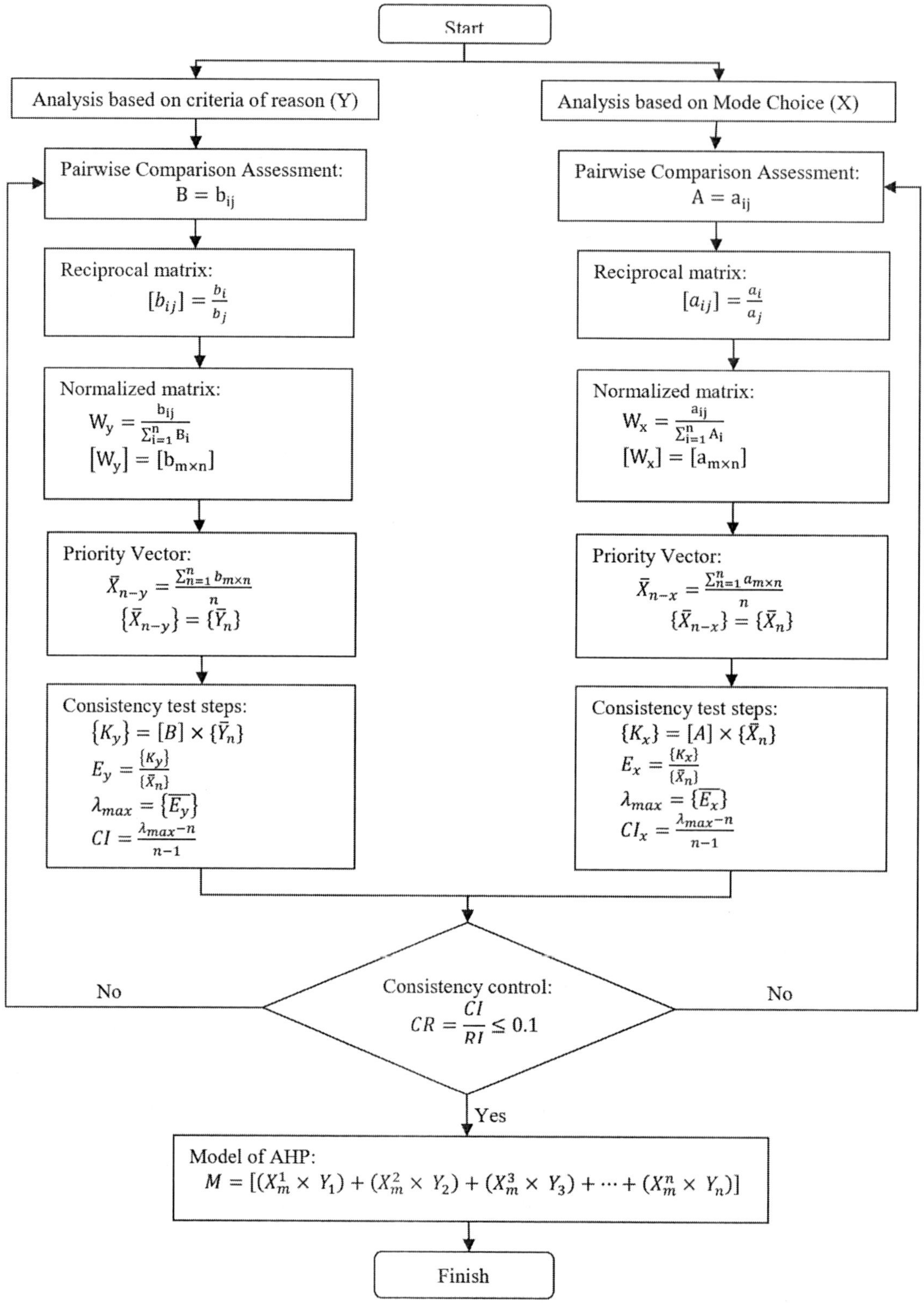

Figure 2. Analytical hierarchy process procedure.

The matrix in equation (1) is then normalized by dividing each element in a matrix into pairs with the total value of each column. The next step in the process of this analysis is: calculating the priority vector and checking for consistency. The procedure of AHP analysis can be seen in the following figure 2.

The data used in this study were collected from respondents in Perumnas Palur. Based on the data of *Kelurahan Ngringo*, the population of Perumnas Palur is about 1927 household. Referring to Slovin Formulas with 7.5% error rate, a minimum sample size was obtained. In this research, the total sample was 195 respondents.

$$n = \frac{N}{1 + N(e)^2} = \frac{1927}{1 + (1927(0.075)^2)} = \frac{1927}{11.839} = 162.762$$

Where n = number of sample, N = number of population, and e = error tolerance.

3. Result and discussion

The characteristics of travel to be discussed are a number of the trip, type of destination trip, travel time, type of mode used, and travel cost. It was found 43% respondents made more than five trips a day, and around 44% dominant on a work trip. Travel time are dominated by the respondent with 10-15 minutes (26%), and majority using a motorcycle for their activities (61%). Besides above characteristic, this study also found that 14% of respondents have the lowest revenue (less than IDR 500,000), and 29% of them have more than IDR 2,500,000. The first step of the AHP process for the criterion is to set-up the matrix in pairs and then proceed it with the reciprocal matrix as shown in table 2. The values in table 2 are then normalized by dividing the elements of each column by the number of matrix columns (see table 3).

Table 2. Reciprocal matrix of criteria of reasons

	Safe	Easy	Cost	Time
Safe	1.000	3.289	2.789	2.728
Easy	0.304	1.000	2.316	0.483
Cost	0.359	0.432	1.000	0.468
Time	0.367	2.068	2.138	1.000
Total	2.029	6.789	8.243	4.679

Table 3. Normalized matrix

	Safe	Easy	Cost	Time
Safe	0.493	0.484	0.338	0.583
Easy	0.150	0.147	0.281	0.103
Cost	0.177	0.064	0.121	0.100
Time	0.181	0.305	0.259	0.214
Total	1.000	1.000	1.000	1.000

The next step after finding normalized matrix is to calculate the vector priority value of each criterion as follows:

$$\text{Safe} = \frac{0.493 + 0.484 + 0.338 + 0.583}{4} = 0.475$$

$$\text{Easy} = \frac{0.150 + 0.147 + 0.281 + 0.103}{4} = 0.170$$
$$\text{Cost} = \frac{0.177 + 0.064 + 0.121 + 0.100}{4} = 0.115$$
$$\text{Time} = \frac{0.181 + 0.305 + 0.259 + 0.214}{4} = 0.240$$

Data of the vector priority is then used to determine the weighting of each element. However, this value should be tested in advance as following steps:
Determining vector [K]

$$[K] = \begin{bmatrix} 1.000 & 3.289 & 2.789 & 2.728 \\ 0.304 & 1.000 & 2.316 & 0.483 \\ 0.359 & 0.432 & 1.000 & 0.468 \\ 0.367 & 2.068 & 2.138 & 1.000 \end{bmatrix} \times \begin{bmatrix} 0.475 \\ 0.170 \\ 0.115 \\ 0.240 \end{bmatrix}$$
$$= \begin{bmatrix} (1.000 \times 0.475) + (3.289 \times 0.170) + (2.789 \times 0.115) + (2.728 \times 0.240) \\ (0.304 \times 0.475) + (1.000 \times 0.170) + (2.316 \times 0.115) + (0.483 \times 0.240) \\ (0.359 \times 0.475) + (0.432 \times 0.170) + (1.000 \times 0.115) + (0.468 \times 0.240) \\ (0.367 \times 0.475) + (2.068 \times 0.170) + (2.318 \times 0.115) + (1.000 \times 0.240) \end{bmatrix} = \begin{bmatrix} 2.010 \\ 0.698 \\ 0.471 \\ 1.013 \end{bmatrix}$$

Finding the Eigen Value (E) by dividing vector [K] with the vector priority of reason criteria.

$$[E] = \begin{bmatrix} 2.010/0.475 \\ 0.698/0.170 \\ 0.471/0.115 \\ 1.013/0.240 \end{bmatrix} = \begin{bmatrix} 4.236 \\ 4.096 \\ 4.083 \\ 4.226 \end{bmatrix}$$

Above value then is used to calculate the average value of all elements, as follows:

$$\lambda_{max} = \frac{4.236 + 4.096 + 4.083 + 4.226}{4} = 4.160$$

Determining the index of consistency (CI).

$$CI = \frac{\lambda_{max} - n}{n - 1} = \frac{4.160 - 4}{4 - 1} = 0.054$$

Calculating the ratio of consistency (CR). To determine the value of CR, it is needed to know about the random index (RI) value that can be found from table 4 as follow.

Table 4. The value of RI [4]

Matrix size (n)	1, 2	3	4	5	6	7	8	9	10	11
Random index (RI)	0.00	0.58	0.90	1.12	1.24	1.32	1.41	1.45	1.49	1.51

$$CR = \frac{CI}{RI} = \frac{0.054}{0.9} = 0.059 \leq 0.1$$

Based on the value of CR, it can be stated that the value is adequate. Therefore, the vector priority can be used in the analysis. The percentage of the vector priority each criterion of the reason is Safe (47.5%), Time (24%), Easy (17%), and Cost (11.5%). Besides the AHP analysis conducted on the criteria of reason, it was also conducted on the Mode Choice used. Using the same process of analysis, the results of mode selection are shown in table 5 – table 8.

Table 5. The AHP analysis of mode choice based on safe criteria

	Walking	Bicycle	Motorcycle	Car	Buses	Vector Priority
Walking	1.000	3.035	2.841	3.128	2.665	0.353
Bicycle	0.329	1.000	2.807	0.327	0.339	0.139
Motorcycle	0.352	0.356	1.000	2.991	3.177	0.202
Car	0.320	3.061	0.334	1.000	3.148	0.182
Buses	0.375	2.949	0.315	0.318	1.000	0.124
$\lambda_{max} = 5.200$	CI = 0.050		RI = 1.12		CR = 0.04	

Table 6. The AHP analysis of mode choice based on easy criteria

	Walking	Bicycle	Motorcycle	Car	Buses	Vector Priority
Walking	1.000	3.007	0.305	3.426	3.421	0.241
Bicycle	0.333	1.000	0.266	3.310	3.073	0.157
Motorcycle	3.281	3.762	1.000	4.219	4.512	0.443
Car	0.292	0.302	0.237	1.000	3.210	0.100
Buses	0.292	0.325	0.222	0.312	1.000	0.059
$\lambda_{max} = 5.470$	CI = 0.118		RI = 1.12		CR = 0.10	

Table 7. The AHP analysis of mode choice based on cost criteria

	Walking	Bicycle	Motorcycle	Car	Buses	Vector Priority
Walking	1.000	3.227	3.528	3.713	3.532	0.416
Bicycle	0.310	1.000	3.419	3.830	3.429	0.262
Motorcycle	0.283	0.292	1.000	3.327	3.017	0.161
Car	0.269	0.261	0.301	1.000	0.395	0.063
Buses	0.283	0.292	0.331	2.532	1.000	0.098
$\lambda_{max} = 5.467$	CI = 0.117		RI = 1.12		CR = 0.10	

Table 8. The AHP Analysis of Mode Choice Based on Time Criteria

	Walking	Bicycle	Motorcycle	Car	Buses	Vector Priority
Walking	1.000	0.170	0.142	0.148	0.250	0.038
Bicycle	5.897	1.000	0.358	0.230	0.273	0.114
Motorcycle	7.026	2.794	1.000	2.070	2.027	0.356
Car	6.756	4.342	0.483	1.000	3.016	0.311
Buses	4.002	3.669	0.493	0.332	1.000	0.182
$\lambda_{max} = 5.457$	CI = 0.114		RI = 1.12		CR = 0.10	

All values of vector priority each type of mode are recapitulated in a new matrix. This matrix is then multiplied by the result of vector priority of reason criteria as shown in table 9.

Table 9. Recapitulation of the AHP Analysis of Mode Choice Based on Reason Criteria

	Safe	Easy	Cost	Time		Vector Priority
Walking	0.353	0.241	0.416	0.038		0.475
Bicycle	0.139	0.157	0.262	0.114		0.170
Motorcycle	0.202	0.443	0.161	0.356	×	0.115
Car	0.182	0.100	0.063	0.311		0.240
Buses	0.124	0.059	0.098	0.182		

The result of calculation on Table 9 can be presented as follows:

Walking = 0.353 Safe + 0.241 Easy + 0.416 Cost + 0.038 Time = 26.6 %
Bicycle = 0.139 Safe + 0.157 Easy + 0.262 Cost + 0.114 Time = 15%
Motorcycle = 0.202 Safe + 0.443 Easy + 0.161 Cost + 0.356 Time = 27%
Car = 0.182 Safe + 0.100 Easy + 0.063 Cost + 0.311 Time = 18.5%
Buses = 0.124 Safe + 0.059 Easy + 0.098 Cost + 0.182 Time = 12.4%

Above result shows that the ranked of mode choice based on reason criteria are Motorcycle (27%), Walking (26.6%), Car (18.5%), Bicycle (15%) and Bus (12.4%).

4. Conclussion

The dominance of the travel characters in Perumnas Palur was obtained: motorcycle users (61%), work trips (44%), more than 5 trips per day (43%), 10-15 minutes of travel time (26%) and travel cost less than Rp10,000.00 (43%). The majority of respondents are housewives (26%) and the highest percentage of income is greater than Rp2,500,000.00 (29%). The percentage of mode types chosen according to the criteria of the reasons are Motorcycle (27%), Walking (26.6%), Car (18.5%), Bicycle (15%) and Bus (12.4%). The percentage of the vector priority each criterion of the reason is Safe (47.5%), Time (24%), Easy (17%), and Cost (11.5%).

References

[1] Tamin O Z 2000 Perencanaan *dan Pemodelan Transportasi*, Edisi Kedua. (Bandung: Institut Teknologi Bandung)
[2] Warpani S 1990 *Merencanakan* Sistem *Perangkutan* (Bandung: ITB)
[3] Schiebel J, Omrani H and Gerber P 2015 Border Effects on The Travel Mode Choice of Resident and Crossborder Workers in Luxembourg *EJTIR Eur. J. Transp. Infrastruct. Res.*, vol 15 no 4 pp 570–96
[4] Saaty T L 2008 Decision Making with the Analytic Hierarchy Process *Int. J. Serv. Sci.* vol 1 pp 83–98
[5] Xi J, Zhao Z, Li W and Wang Q 2016 A Traffic Accident Causation Analysis Method Based on AHP-Apriori *Procedia Eng.* vol 137 pp 680–7
[6] Baidya R and Ghosh S K 2015 Model for a Predictive Maintenance System Effectiveness Using the Analytical *Hierarchy* Process as Analytical Tool *IFAC Int. Fed. Autom. Control* vol 48 no 3 pp 1463–8

IOP Conf. Series: Materials Science and Engineering **403** (2018) 012080 doi:10.1088/1757-899X/403/1/012080

Lack of knowledge matching algorithms using distance measurements on brainwave features

A Azhari[1] and DP Ismi[2]

[1,2]Informatics Department, Universitas Ahmad Dahlan, Yogyakarta, Indonesia.

[1]ahmad.azhari@tif.uad.ac.id, [2]dewi.ismi@tif.uad.ac.id

Abstract. All human activities are controlled centrally through the brain. The brain produces electrical forces that transmit information between neurons that are represented in wave form. Brainwaves can be used as a medium for identifying and authenticating users because individual brain waves are different and have distinctive characteristics. Characteristics of brainwaves can be triggered using a pattern that is protruding and constant. The application of cognitive tasks developed based on the incorporation of human perceptions and psychological characteristics can serve as a stimulus. Feature extraction is done by applying statistics in the form of mean value, standard deviation, skewness, kurtosis, and entropy. These features are analyzed and compared using a distance measurement matching algorithm. The results obtained from this test indicate that the Minkowski algorithm is capable of providing optimal results in data matching.

1. Introduction

Biometrics is a method used to recognize humans based on the size of physiological characteristics and human behavior [1]–[7]. The use of Biometrics to recognize a person is widely used in previous studies [5],[6],[8]–[19]. A person's face has been used for recognition, both for recognizers and unrecognized. A person's voice can also be recognized without seeing the person directly. Step motion or gesture of a person can also be recognized from the way it goes. Physiological characteristics using the physical parts of a person's body include fingerprint, face image, hand vessels, retina, iris, palms, teeth and body odor / sweat. While the introduction based on behavioral characteristics using one's behavior include gesture, signature, and sound [3],[4],[20].

There are six biometrics commonly used for self-recognition systems, including: fingerprint, iris, face, voice, hand geometry, and signature [3]. The latest development of the field of biometrics is cognitive biometrics based on specific responses of the human brain, including: brainwaves, odors, and face monitors [4],[5].

The purpose of application of Biometrics is to identify and verify human characteristics [1]. The verification system aims to accept or deny an identity claimed by a person, while the identification system aims to resolve a person's identity [2],[5],[6],[20],[21]. Various approaches have been developed for self-introduction. The approaches can basically be grouped into three, namely: first, based on possessions-based, such as keys (physical keys) and cards. Second, based on knowledge-based, such as user identity, PIN, and password, and third, biometrics (biometrics-based), such as fingerprint, face, sound, and so on [1],[3],[4],[17]. Advantages of biometrics system that uses attributes attached to the body (something what you are) and attributes attached to the behavior of someone (something what you do). This advantage can reduce duplication, shared, and forgotten usage as it has in traditional recognition systems [3],[5],[7], [20]–[22].

To get a specific response from the human brain needed a trigger. Reference [1],[5],[17] mentions that the characteristics of brainwaves become very strong when a person is exposed to specific stimuli. In cognitive biometrics, stimuli can be derived from the human cognitive aspect. To record and measure the brainwave activity required electroencephalogram.

Electroencephalogram (EEG) is one of the biomedical instruments that function to obtain, process, and display the wave activity generated by the brain in the form of electrical potential, with a small electrical voltage. By using EEG, brainwave activity can be analyzed to find different work levels in each part of the brain [1],[6],[7],[16],[17].

2. Related Work

Reference [7] using EEG features to authenticate user on biometrics systems. The issues raised are about Vulnerabilities threatening user in Authentication. This study used Common Spatial Patterns (CSP) Values as Main Features and Linear Discriminant Analysis (LDA) as Classification algorithm for set of user data. EEG data used wirelessly improvement accuracy can be expected by avoiding electrodes. The result showed that Beta Band is the most suitable band for imagination and active thinking with 90,91% accuracy better than Alpha Band with 81,81% accuracy. Combination of Both Band (Alpha and Beta Band) showed better accuracy than individual frequency band with 96.96% Accuracy.

Reference [17] proposed feature extraction using cross-correlation method to find specific characteristics of short-wave signals applied specifically to assist patients with disorder neuromuscular channel, in moving the wheelchair to turn right and turn left. The cross-correlation is done by finding the average of the maximum peak point and the minimum peak point coming from the correlation signal. The characteristic of the signal used as a reference to make a special short-wave is the signal feature that has the lowest correlation coefficient. The lowest correlation coefficient of the signal characteristic pair indicates the greatest difference between the signal pair turning right and left turns. The matching used in each signal characteristic is the curve fitting method. The average accuracy at the testing phase of 22 participants for 4 types of movement showed a similar average of 76% for right turn movement, 76% for left turn movement, 74% for right turning imagination and 72% for left-handed imagination.

Reference [5] conducted a study which aims to classify Brainwaves feature using K-Means clustering. EEG single-sensor is utilizing in this study. Several feature statistics such as mean, standard deviation, skewness, kurtosis, and entropy use as a EEG feature. Cognitive activity divided into activity of right brain and the left brain (so-called cognitive task). Cognitive task shows different cluster based on two times test on mathematical activities. This study showed that signal characteristic based on math activity can be used as basic for BCI (brain-computer interface) applications development.

3. Methodology

Data collection is conducted through biometric data acquisition method retrieved from some people. Data collection was collected by applying an EEG single-sensor headset, named NeuroSky Mindset. EEG Electrodes are placed by applying 10-20 placement electrode system [23], right above the FP1 (frontal temporal) or frontal lobe position. EEG retrieval process is done gradually and separately obtained from 6 subjects. Each subject performed two times data retrieval by doing some cognitive task assigned. EEG retrieval process generate 2,560 data along for 20 seconds using sampling frequency of 128 Hz per second. EEG signals are generated by applying stimulus of cognitive tasks. Nine type of EEG stimulus of cognitive task include breath color, face, fingers, counting, objects, pass-thought, singing, and sports.

Based on previous research [13],[19],[24],[25], cognitive task stimulus retrieve by applying several psychological perceptions to get a natural responses of the brain. The nine types of cognitive task are described on Table 1.

Table 1. Nine types of cognitive task

Category	Task	Description
Cognitive	Breathing Task (Breath)	Subject focuses on breathing for 20 seconds.
	Object Counting Color Task (Color)	Subjects are asked to remember the color which appears alternately for 20 seconds.
	Mathematical Task (Math)	Subjects are asked to focus on calculating a simple two-numbered multiplication for 20 seconds. False and True answers are ignored in this task.
	Song Recitation Task (Song)	Subjects are asked to imagine a song or sound with lyrics.
Imaginative	Simulated Movement Finger (Finger)	Subjects are asked to focus on simulating by moving a finger without for 20 seconds.
	Simulated Facial Reconstruction (Face)	Subjects are asked to focus reconstructs a person's face for 20 seconds.
	Simulated Object Reconstruction (Object)	Subjects are asked to focus reconstructs the object for 20 seconds.
	Simulated Sport Task (Sport)	Subjects are asked to imagine the exercise movement for 20 seconds.
	Simulated Password Recall Task (Pass-thought)	Subjects are asked to imagine the sentence of password provided for 10 seconds.

In terms of signal processing, biometric data is obtained from data acquisition. Data separately will be grouped into 4 category including subject, cognitive task, and time data retrieval. Furthermore, the feature data will be extract to get difference characteristics that represent unique characteristics of the signal. Feature extraction retrieve from by performing mean, standard deviation skewness, kurtosis, and entropy.

Mean measure of data distribution, standard deviation measures variations of data distribution, skewness measures the asymmetric level of data distribution, kurtosis measures how flat or high the distribution of data is to a normal distribution, and entropy is used to measure the randomness of the data distribution. If mean ($\bar{x}$), standard deviation (σ), skewness (s), kurtosis (k), entropy (H) has the sum of data N and x_i is data i then all data can be calculated by equation (1-5).

$$mean = \bar{x} = \frac{1}{N}\sum_{i=0}^{N} x_i \tag{1}$$

$$standard\ deviation = \sigma = \sqrt{\frac{1}{(N-1)}\sum_{i=0}^{N}(x_i - \bar{x})^2} \tag{2}$$

$$skewness = s = \sum_{i=0}^{N}(x_i - \bar{x})^3/(N\sigma^3) \tag{3}$$

$$kurtosis = k = \sum_{i=0}^{N}(x_i - \bar{x})^4/(N\sigma^4) \quad (4)$$

$$entropy = H = E[-log_2P(x)] = -\sum_{i=0}^{N} P(x)log_2P(x)\ bits \quad (5)$$

Figure 1 shows the general procedure of proposed method. Data from feature extraction will be normalized before matching algorithm. Matching algorithm using distance measurement. The smallest value of each distance measurement method will be selected and then using the calculation score will be selected the largest value.

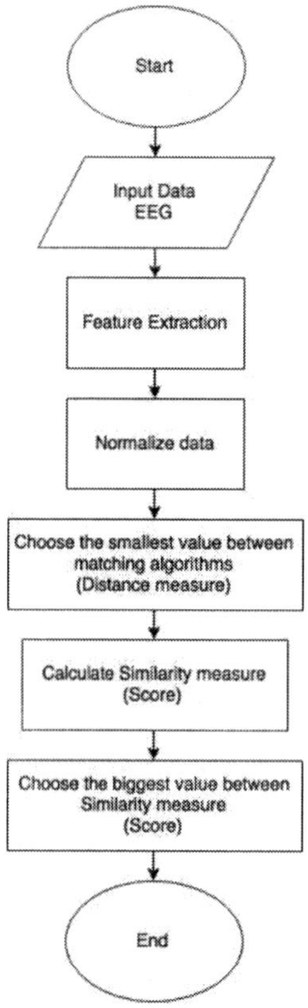

Figure 1. General Procedure of Proposed Method

3.1. Euclidean Distance

Euclidean Distance is the most commonly used metric for calculating the similarity of two vectors. Euclidean distance calculates the root of the square of the difference of two vectors. Euclidean Distance using formula (6).

$$d_{ij} = \sqrt{\sum_{k=1}^{n}(x_{ik} - x_{jk})^2} \quad (6)$$

The smaller d_{ij} score the more similar the two vector features are matched. Conversely, the greater d_{ij} score the more different the two character traits. The nature of the Euclidean distance is that the result is in the range of $0 \leq d \leq 1$. Notice that the Euclidean distance is a special case of the Minkowski metric, where $\lambda = 2$.

3.2. City Block / Manhattan Distance

Manhattan distance calculates the value of the absolute difference of the two vectors. Manhattan distance using formula (7).

$$d_{ij} = \sum_{k=1}^{n} |x_{ik} - x_{jk}| \tag{7}$$

3.3. Chebyshev Distance

Chebyshev distance is also called the maximum value distance that serves to check an absolute magnitude of two vector difference. From each value of difference will be selected the greatest value to be made Chebyshev distance. Distance Chebyshev distance using formula (8).

$$d_{ij} = max|x_{ik} - x_{jk}| \tag{8}$$

3.4. Minkowski Distance

This Minkowski distance with the λ generalize some of the previous metrics, where $\lambda = 1$ is expressed as the distance of City Block and $\lambda = 2$ is expressed by Euclidean distance. Chebyshev distance is a special case of Minkowski distance with $\lambda = \infty$ (infinity) using formula (9).

$$d_{ij} = \sqrt[\lambda]{\sum_{k=1}^{n} |x_{ik} - x_{jk}|^{\lambda}} \tag{9}$$

3.5. Canberra Distance

For each value of the two vectors to be matched, the Canberra distance divides the absolute difference of two values by the sum of the absolute two values. The results of each of the two matching values are then summed to gain distance to Canberra. If the two coordinates are zero-zero, then defined by $^0/_0 = 0$. This distance is very sensitive to slight changes with the two coordinates close to zero. Distance Canberra using formula (10):

$$d_{ij} = \sum_{k=1}^{n} \frac{|x_{ik} - x_{jk}|}{|x_{ik}| + |x_{jk}|} \tag{10}$$

3.6. K-Nearest Neighbors

The next stage after the EEG signal performs feature extraction and dimensional reduction that is classified by K-Nearest Neighbour (KNN) classification method. The KNN data sample is based on distance measurement with the number of patterns stored in memory. Testing data was calculated to each training data using distance method the smallest distance of training data to its testing was assumed the correct class.

In the training phase, the algorithm is just doing the feature vector storage and classification of training sample. In the classification phase, the same features are calculated for testing data. The distance from the new vector of the entire sample training vectors are calculated and the amount of the closest k taken.

The way to calculate the KNN algorithm:

- Determining *k* parameter (the number of nearest neighbors)

IOP Conf. Series: Materials Science and Engineering **403** (2018) 012080 doi:10.1088/1757-899X/403/1/012080

- Calculate the distance Euclidean distance of each object on the data samples provided
- Sort the objects into groups that have the smallest distance.

3.7. Similiarity Measure

The decision-making process will use the formula (11). The higher the score obtained then it can be assumed that the task will be more dominant.

$$Score = (1 - d_{ij})/2 \tag{11}$$

4. Results

EEG data acquisition results will be extracted to generate the feature. Feature extraction results are obtained using statistical features. Statistical features used are mean values, standard deviation values, skewness values, kurtosis values, and entropy values. These features are grouped according to three categories: subject, cognitive task, and data retrieval time.

Furthermore, Matching features using distance measurement methods. Among the matching methods used were Euclidean distance, Manhattan distance, Chebyshev distance, Minkowski distance, and Canberra distance shown in Table 2 - Table 6.

Table 2. Euclidean Distance

	Subject_1	**Subject_2**	**Subject_3**	**Subject_4**	**Subject_5**	**Subject_6**
Breath	0,29	0,17	0,08	0,01	0,02	0,01
Color	0,06	0,11	0,20	0,01	0,02	0,02
Face	0,02	0,24	0,16	0,02	0,02	0,11
Finger	0,03	0,04	0,03	0,04	0,02	0,01
Math	0,34	0,10	0,70	0,02	0,19	0,03
Object	0,03	0,02	0,10	0,05	0,05	0,00
Pass-thought	0,03	0,02	0,06	0,03	0,01	0,01
Sing	0,01	0,02	0,04	0,08	0,03	0,05
Sport	0,02	0,09	0,13	0,01	0,06	0,01

Table 3. Manhattan Distance

	Subject_1	**Subject_2**	**Subject_3**	**Subject_4**	**Subject_5**	**Subject_6**
Breath	0,42	0,24	0,13	0,02	0,03	0,02
Color	0,10	0,19	0,29	0,02	0,03	0,02
Face	0,03	0,34	0,21	0,04	0,03	0,14
Finger	0,06	0,06	0,04	0,07	0,03	0,01
Math	0,44	0,15	1,05	0,03	0,27	0,05
Object	0,04	0,04	0,20	0,08	0,08	0,01
Pass-thought	0,04	0,03	0,08	0,04	0,03	0,02
Sing	0,01	0,03	0,06	0,14	0,04	0,09
Sport	0,03	0,15	0,19	0,02	0,10	0,02

IOP Conf. Series: Materials Science and Engineering **403** (2018) 012080 doi:10.1088/1757-899X/403/1/012080

Table 4. Chebyshev Distance

	Subject_1	Subject_2	Subject_3	Subject_4	Subject_5	Subject_6
Breath	0,23	0,14	0,07	0,07	0,01	0,01
Color	0,05	0,08	0,18	0,18	0,01	0,01
Face	0,01	0,19	0,16	0,16	0,02	0,10
Finger	0,03	0,04	0,02	0,02	0,02	0,01
Math	0,33	0,10	0,52	0,52	0,14	0,02
Object	0,02	0,02	0,08	0,08	0,05	0,00
Pass-thought	0,02	0,01	0,05	0,05	0,01	0,01
Sing	0,01	0,02	0,04	0,04	0,03	0,04
Sport	0,02	0,08	0,11	0,11	0,04	0,01

Table 5. Minkowski Distance

	Subject_1	Subject_2	Subject_3	Subject_4	Subject_5	Subject_6
Breath	0,26	0,15	0,07	0,01	0,01	0,01
Color	0,05	0,10	0,19	0,01	0,01	0,01
Face	0,01	0,22	0,16	0,02	0,02	0,10
Finger	0,03	0,04	0,02	0,04	0,02	0,01
Math	0,33	0,10	0,62	0,02	0,17	0,03
Object	0,02	0,02	0,09	0,04	0,05	0,00
Pass-thought	0,03	0,01	0,06	0,03	0,01	0,01
Sing	0,01	0,02	0,04	0,07	0,03	0,05
Sport	0,02	0,08	0,11	0,01	0,05	0,01

Table 6. Canberra Distance

	Subject_1	Subject_2	Subject_3	Subject_4	Subject_5	Subject_6
Breath	1,34	1,87	0,89	0,27	0,54	0,10
Color	2,10	2,26	1,52	0,12	1,18	0,22
Face	2,10	2,04	1,43	0,27	1,13	0,90
Finger	1,62	2,04	0,29	1,19	1,08	0,10
Math	2,31	1,23	1,45	0,19	1,67	1,27
Object	2,20	0,87	0,98	0,69	1,16	0,08
Pass-thought	1,94	0,84	1,18	0,55	0,69	0,15
Sing	2,02	1,46	0,74	0,90	1,11	0,83
Sport	0,42	2,15	0,88	1,12	0,88	0,36

The next step, based on distance measurement using several methods, is to take the mean value of each distance measurement result. The data with the mean value are then grouped into one table and made a comparison. Comparison is done by using the threshold value. The threshold value determines the matching success rate. The threshold value is set based on the smallest value of the overall value in the table. Using KNN classification method, take the smallest value using range $0 \leq d \leq 1$. Table 7 shows the comparison results between the matching algorithm based on the mean value.

Table 7. Comparison of matching algorithms

Task	Euclidean	Manhattan	Chebyshev	Minkowski	Canberra	Threshold
Breath	0,0960	0,1429	0,0908	**0,0869**	0,9850	0,0869
Color	0,0700	0,1089	0,0867	**0,0635**	1,1900	0,0635
Face	0,0946	0,1309	0,1081	**0,0885**	1,1850	0,0885
Finger	0,0286	0,0439	**0,0240**	0,0268	1,0700	0,0240
Math	0,2296	0,3343	0,2717	**0,2100**	1,2500	0,2100
Object	0,0418	0,0735	0,0407	**0,0369**	1,1400	0,0369
Pass-thought	0,0266	0,0417	0,0281	**0,0245**	1,0450	0,0245
Sing	0,0384	0,0619	**0,0277**	0,0349	1,3800	0,0277
Sport	0,0541	0,0857	0,0604	**0,0488**	1,2550	0,0488

The decision-making stage is the last stage of the verification process. At this stage a decision is produced which method is best suited to use. To obtain such a decision, a comparison with the threshold value is required. To facilitate the decision-making process used the similarity measure method. The smaller the score then it can be assumed that the method is suitable. Table 8 shows the results of decision making using similarity measurement methods.

Table 8. Decision-making process

Task	Minkowski	Threshold	Score
Breath	0,0869	0,0869	0,4565
Color	0,0635	0,0635	0,4683
Face	0,0885	0,0885	0,4557
Finger	0,0268	0,0240	0,4880
Math	0,2100	0,2100	0,3950
Object	0,0369	0,0369	0,4815
Pass-thought	0,0245	0,0245	0,4877
Sing	0,0349	0,0277	0,4862
Sport	0,0488	0,0488	0,4756

5. Conclusions

Based on the measurement of similarity between features using distance measurement methods, Minkowski's distance method shows the smallest and near-zero scores. To facilitate the decision-making process used the method of measurement of similarity. The range of reference range used between 0 and 1. The smaller the score indicated by the distance between features shows the higher similarity level, and vice versa, the bigger the lower the similarity level.

From the cognitive task used as a stimulus, it is found that "Finger", "Pass-thought", "Sing", and "Object" tasks show the highest similarity scores with respective values of 0.4880, 0.4877, 0.4862, and 0.4815. The highest score is shown by two cognitive activities namely "Finger" and "Pass-thought" with values of 0.4880 and 0.4877 with the lowest distance range 0.0268 and 0.0245. These results indicate that cognitive activity "Finger" and "Pass-thought" can provide more dominant signal characteristics.

Acknowledgments
This research is supported by The Indonesian Ministry of Research, Technology and Higher Education (RISTEK-DIKTI) research grant no. 118/SP2H/LT/DRPM/IV/2017.

6. References

[1] A. Azhari, "Ekstraksi Ciri Gelombang Otak Menggunakan Alat Neurosky Mindset Berbasis Korelasi Silang.pdf," Universitas Gadjah Mada, Yogyakarta, 2015.
[2] U. S. M. A. West Point Information and Educational Technology Division, "Biometrics Metrics Report." United States Military Academy West Point Information and Educational Technology Division, 2012.
[3] D. Putra and K. Gede, "Sistem Verifikasi Biometrika Telapak Tangan dengan Metode Dimensi Fraktal dan Lacunarity," *Maj. Ilm. Teknol. Elektro*, vol. 8, no. 2, 2012.
[4] D. Putra, *Sistem Biometrika*. Andi Offset, 2009.
[5] A. Azhari and L. Hernandez, "Brainwaves feature classification by applying K-Means clustering using single-sensor EEG," *Int. J. Adv. Intell. Inform.*, vol. 2, no. 3, pp. 167–173, Nov. 2016.
[6] J. Klonovs, C. K. Petersen, H. Olesen, and J. S. Poulsen, "Development of a Mobile EEG-Based Feature Extraction and Classification System for Biometric Authentication," Aalborg University Copenhagen, 2012.
[7] I. Jayarathne, M. Cohen, and S. Amarakeerthi, "BrainID: Development of an EEG-based biometric authentication system," in *Information Technology, Electronics and Mobile Communication Conference (IEMCON), 2016 IEEE 7th Annual*, 2016, pp. 1–6.
[8] P. Ackermann, C. Kohlschein, J. Á. Bitsch, K. Wehrle, and S. Jeschke, "EEG-based automatic emotion recognition: Feature extraction, selection and classification methods," in *e-Health Networking, Applications and Services (Healthcom), 2016 IEEE 18th International Conference on*, 2016, pp. 1–6.
[9] A. Patil, C. Deshmukh, and A. R. Panat, "Feature extraction of EEG for emotion recognition using Hjorth features and higher order crossings," in *Advances in Signal Processing (CASP), Conference on*, 2016, pp. 429–434.
[10] Y. Zhong and Z. Jianhua, "Recognition of Cognitive Task Load levels using single channel EEG and Stacked Denoising Autoencoder.pdf," in *Proceedings of the 35th Chinese Control Conference*, 2016, pp. 3907–3912.
[11] C. F. H. Putra, "Analisis Citra Otak Pada Color-Task Dan Word-Task Dalam Stroop Task Dengan Menggunakan Electroencephalography (EEG)," Universitas Gadjah Mada, Yogyakarta, 2014.
[12] A. A. Ghodake and S. D. Shelke, "Brain controlled home automation system," in *Intelligent Systems and Control (ISCO), 2016 10th International Conference on*, 2016, pp. 1–4.
[13] B. Johnson, T. Maillart, and J. Chuang, "My thoughts are not your thoughts," 2014, pp. 1329–1338.
[14] R. F. Ahmad, A. S. Malik, N. Kamel, F. Reza, and A. H. A. Karim, "Simultaneous EEG-fMRI data acquisition during cognitive task," in *Intelligent and Advanced Systems (ICIAS), 2014 5th International Conference on*, 2014, pp. 1–4.
[15] M. Diykh, Y. Li, and P. Wen, "EEG Sleep Stages Classification Based on Time Domain Features and Structural Graph Similarity," *IEEE Trans. Neural Syst. Rehabil. Eng.*, vol. 24, no. 11, pp. 1159–1168, Nov. 2016.
[16] B. H. Tan, "Using a low-cost eeg sensor to detect mental states," Carnegie Mellon University, 2012.

IOP Conf. Series: Materials Science and Engineering **403** (2018) 012080 doi:10.1088/1757-899X/403/1/012080

[17]E. Yulianto and A. Susanto, "Spesifikasi Ciri Sinyal EEG Untuk Perintah Pergerakan Motorik Berbasis Gelombang Singkat Khusus Berdasarkan Hasil Ekstraksi Ciri," Universitas Gadjah Mada, 2013.

[18]M. Varela, "Raw EEG signal processing for BCI control based on voluntary eye blinks," in *2015 IEEE Thirty Fifth Central American and Panama Convention (CONCAPAN XXXV)*, 2015, pp. 1–6.

[19]A. Saha, A. Konar, P. Das, B. S. Bhattacharya, and A. K. Nagar, "Data-point and feature selection of motor imagery EEG signals for neural classification of cognitive tasks in car-driving," in *2015 International Joint Conference on Neural Networks (IJCNN)*, 2015, pp. 1–8.

[20]N. K. Ratha, A. Senior, and R. M. Bolle, "Automated biometrics," in *International Conference on Advances in Pattern Recognition*, 2001, pp. 447–455.

[21]A. Azhari, "Analisis Pengaruh Cognitive Task Berdasarkan Hasil Ekstraksi Ciri Gelombang Otak Menggunakan Jarak Euclidean," *SEMNASTEKNOMEDIA ONLINE*, vol. 5, no. 1, pp. 3–1, 2017.

[22]A. Azhari, A. Susanto, and I. Soesanti, "Studi Perbandingan: Cognitive Task Berdasarkan Hasil Ekstraksi Ciri Gelombang Otak.pdf," presented at the Seminar Nasional Teknologi Informasi dan Multimedia, Yogyakarta, 2015, vol. 3.1, p. 7.

[23]Trans Cranial Technologies ldt, "1020 System Positioning." Trans Cranial Technologies ldt, 2012.

[24]A. Azhari, "Ekstraksi Ciri Gelombang Otak Menggunakan Alat Neurosky Mindset Berbasis Korelasi Silang.pdf," Universitas Gadjah Mada, Yogyakarta, 2015.

[25]R. Zafar, A. S. Malik, H. U. Amin, N. Kamel, S. Dass, and R. F. Ahmad, "EEG spectral analysis during complex cognitive task at occipital," in *Biomedical Engineering and Sciences (IECBES), 2014 IEEE Conference on*, 2014, pp. 907–910.

A profile analysis of blacksmith in Desa Limbang Jaya I Kecamatan Tanjung Batu Kabupaten Ogan Ilir - Indonesia

Irmeilyana[1], Ngudiantoro[2], A. Desiani[3]
[1,2,3]Department of Mathematics, Faculty of Mathematics and Natural Sciences
University of Sriwijaya, Indralaya, Indonesia

[1]e-mail address: imel_unsri@yahoo.co.id

Abstract. This paper aims to analyze the relationship of factors showing the profiles of blacksmiths in Desa Limbang Jaya I in terms of social demography, average monthly income, and productivity of craftsmen. The productivity studied includes the average number of products for 1 month, the average length of working time per day, the average duration of the work of 1 kodi of product (in hours), and the price of 1 kodi of product. Based on the description analysis, artisan of blacksmith in Desa Limbang Jaya I majority aged 25-55 years, married status with majority of his wife works as a songket craftsman. The results of a biplot analysis with a representative level of 50.6% indicate that the income of the craftsmen depends only on the price of the product, but not on the number of products. Working hours of craftsmen in the day is quite high, but the effect is not significant enough to increase the number of products produced. The result of cluster analysis is found that majority of craftsmen have high age, high experience, and have the highest work day, but lower income. While most of the craftsmen have low productivity features. Only a small percentage of craftsmen whose age and work experience are low but they have higher income.

Keywords: Desa Limbang Jaya I, profiles of the blacksmith, biplot analysis, cluster analysis, productivity of blacksmith.

1. Intoduction

The potential of craftsmen from Desa Limbang Jaya can be said the largest in South Sumatra Province. The village also has the most number of blacksmiths. Other villages with fewer blacksimths are Desa Tanjung Pinang (as a neighboring village directly adjacent to Desa Limbang Jaya), Desa Tanjung Laut, and Desa Merangin. Production of, both artisans of songket and blacksmith in Limbang Jaya using raw materials imported from outside the region. Raw materials for the craft of iron in the form of iron that the steel level is good enough. The iron material is mostly sent from Surabaya. Charcoal fuel for this handicraft sent from Jambi.

The level of population migration in Desa Limbang Jaya is quite high, especially blacksmith. This is because the insufficient natural potential and also the artisans do not use the potential of their own region. They mostly migrated to Jambi, whereas a small number of people migrated to Sulawesi and Papua. They make and develop business as a blacksmith. They wander not to bring their family and some are permanently moved to other areas. Although permanent, they often return home at least once a year. Their wife usually work as songket artisans.

The Blacksmith's products of Limbang Jaya artisans are marketed to smallholder plantations in Kecamatan Tanjung Batu and also out of areas with many plantation areas, such as Kabupaten OKI, MUBA, and even out of provinces, namely Bengkulu and Jambi. The products of the craftsmen are good enough for the traditional craftsmen class at the national level.

Reference [1] researched the factors that affect the productivity of lontar women craftsmen in Desa Bona, Kabupaten Gianyar. This study used multiple linear regression analysis, with the variables studied are social demographic and economic factors of craftsmen, including: age, education, work experience, and marital status. The results showed that factors affecting productivity were age, work experience and marital status. The level of education does not affect productivity. Age is an influential dominant factor.

The purpose of this paper is to present the character, profile, and potential of blacksmith in Desa Limbang Jaya in terms of social demography factors including age, education level, work experience as craftsman, marital status and wife profession, average income per month, and productivity of craftsmen. The productivity studied includes the average number of products produced in 1 month (in kodi), the average length of working time per day, the average duration of workmanship of 1 kodi product (in hours), and the price of the product. The monthly income is income after deducting the cost of raw materials.

By profile analysis of blacksmith, we can analyze the factors that affect productivity of blacksmith, so we can formulate the need of craftmen to increase their productivity. The high productivity has strong relationship with income.

2. Literature Review

Reference [2] suggested the importance of government participation to contribute to how to increase the income of songket craftsmen in the industrial center of songket in Kelurahan 30 Ilir Palembang. Maternal income, maternal education, the number of family members, and investment influence the pattern of food consumption is greater than non-food.

Reference [3] discussesed the profile of the bamboo handicraft creative industry and how innovations can improve the performance of artisans in Desa Sumber, Kecamatan Trucuk, Kabupaten Klaten. Research showed that there was a need for handycraff bamboo craftsmen to get training on entrepreneurship.

Reference [4] researched the work culture of female songket weavers in Desa Bukit Batu, Kabupaten Bengkalis. The business of the artisans had made no significant progress, even though the government had provided donation. This phenomenon occurs because of the low work culture (average 44.33%) and the absence of entrepreneurial mentality in the songket weavers in Desa Bukit Batu. The factors that cause the low work culture of songket weavers can be divided into two parts, namely (1) internal factors consisting of low levels of education and insight, the attitude of the songket weavers, and low motivation in work; (2) external factors consisting of customs and culture prevailing in society and government policies (especially during the New Order period).

The motivation of weavers is doing work more based on the material, not the non-material. They do this work more to economic demands or fulfilment of household needs [5].

Reference [6] examined the inhibiting factors and efforts made by craftsmen in maintaining the sustainability of the blacksmith industry in Desa Tumbukan Banyu and Desa Sungai Pinang, Kecamatan South Daha, Kabupaten Hulu Sungai Selatan. The results of this study showed that inhibiting factors affecting the blacksmith industry include expensive raw materials, limited capital, unsuitable labor with his education and employment backgrounds, the use of traditional technology, marketing that is only sold to pensions, less preferred institutional functions, lack of interest participate in government activities and the disappearance of regional cultural features. Efforts are made to maintain the sustainability of the industry by purchasing raw materials wholesale, adding capital by individual loans, training and internships, machine use, marketing at industrial sites or outlets and improving the function of business groups.

Several methods in multivariate analysis relating to the simplification of data cluster problems with many variables and objects are the principal component analysis (as the initial method), biplot analysis (as graphical analysis), correspondence analysis, and cluster analysis. These analysis can reduce the variables by identifying smaller groups of variables [7].

3. Research Methods

This research is a case study. The population in this study is the community of blacksmith craftsmen in Desa Limbang Jaya I, Kecamatan Tanjung Batu, Kabupaten Ogan Ilir. The sampling technique used is purposive sampling technique. The data used in the form of primary data based on observation and direct interviews or through questionnaires.

The steps in this research are:

1. Identification of the variables studied include factors that can characterize the profile and potential of artisans of the blacksmith; which includes social demography factors that are limited to age, education level, length of time to profession as craftsmen, marital status (also wife profession), average monthly income, and productivity of craftsmen. The productivity studied includes the average number of product produced in 1 month (in kodi units), the average length of working time per day, the average duration of the processing of 1 kodi product, and the price of the product. The identification of variables is accommodated in the form of quessionary questions. The type of product made determines the wage (income after deducting the raw material) per kodi and also the length of time of workmanship.
2. Quantitative descriptive analysis; to give an overview of the characteristics of the variables.
3. Perform multivariate analysis, i.e. biplot analysis and cluster analysis.
4. Interpretation of results.

4. Results and Discussion

Based on information from the Village Head and his apparatus, there are about 100 - 150 highly productive blacksmiths in the village. Blacksmiths who only cut iron raw materials into standard sizes, or suppliers of raw materials, collectors or containers of products, and also wandering blacksmith (but not permanent, it's means not to bring the family to go wander) are not included as a population of skilled artisans iron. In this case, with Slovin formula obtained a sample of 34 respondents.

The manufacture of blacksmith products through several stages; namely sketching, forging ("memandai"), sharpen, and finishing (smoothing and installation of accessories). All the blacksmiths can perform at every stage. Especially for the manufacture of wooden handles in knives, they usually work together with wooden craftsmen and knife accessories. The selling price and the profit earned depend on the type of product made. The products made are rubber tapping knives, dodos for palm oil, machetes, and various types of knives including kitchen knives. The economics of artisans of blacksmith craftsmen along with the high price of rubber, because the products they produce the majority are used by rubber farmers.

Work as a craftsman is the main job, because the natural conditions are less support for the creation of employment. Iron marking skills are usually obtained from generation to generation from parents. The tools used are still traditional in the form of charcoal stoves and "asahan" for sharpeners, as well as semi-modern tools by using electricity in the form of blowers, grinding machines, and welding machines.

According to the Village Head of Limbang Jaya who is also a blacksmith businessman (he employ several craftsmen with its own brand), Limbang Jaya' blacksmith products are better quality than artificial craftsmen in Medan. But they are lost to compete with products from outside, especially Malaysia. It is viewed from materials and sharpening technology. Table 1 shows statistics descriptive of the respondents.

Tabel 1. Statistics descriptive of the respondents

Variable	Mean	StDev	Minimum	Q1	Median	Q3	Maximum
Age (in years)	41.74	12.78	18.00	31.50	43.00	48.50	67.00
Work experience	22.15	13.50	2.00	9.50	23.00	30.50	50.00
Income (in Rp)	2960294	1389341	1200000	2000000	2750000	3050000	7200000
Number of product (in kodi)	35.76	26.41	10.00	20.00	30.00	40.00	150.00
Work hour per day	7.691	1.206	4.000	7.000	8.000	8.000	10.000
Duration of work (in hours)	6.956	3.207	1.600	5.000	7.000	8.000	20.000
Product price	388529	182610	60000	275000	400000	500000	750000

Based on statistical descriptions, we obtain the average of the variables and quartile limits of the respondents, so that in addition to using a histogram (with curve normal), the statistics descriptive can also be a benchmark in compiling the categories of variables.

Figure 1 is a percentage graph of the composition of respondent categories in terms of social demography factors, productivity, and income.

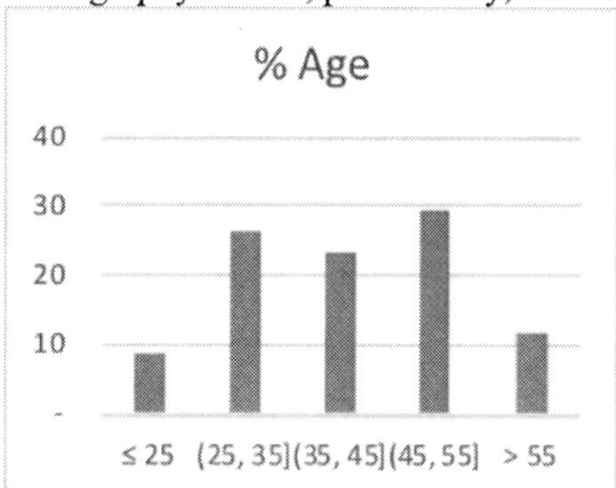

(a) Age

(e) Work experience as a blacksmith

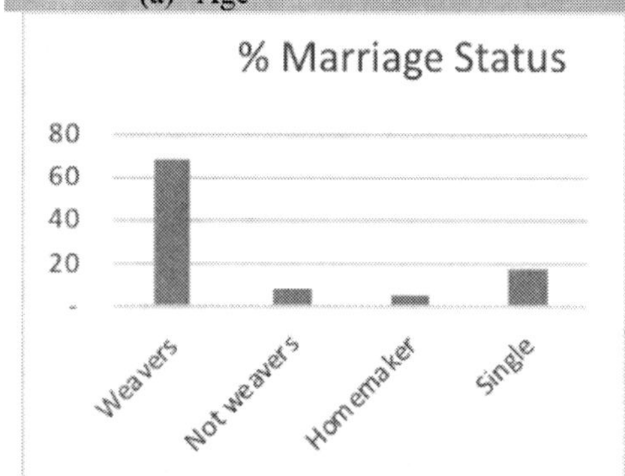

(b) Marital status and wife's profession

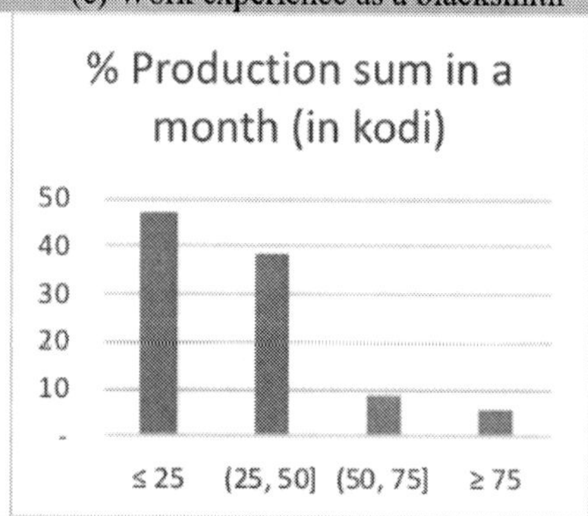

(f) Number of products per month

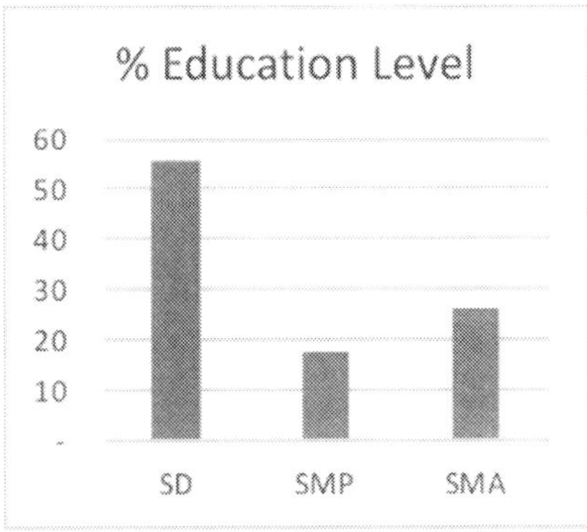

(c) Education level

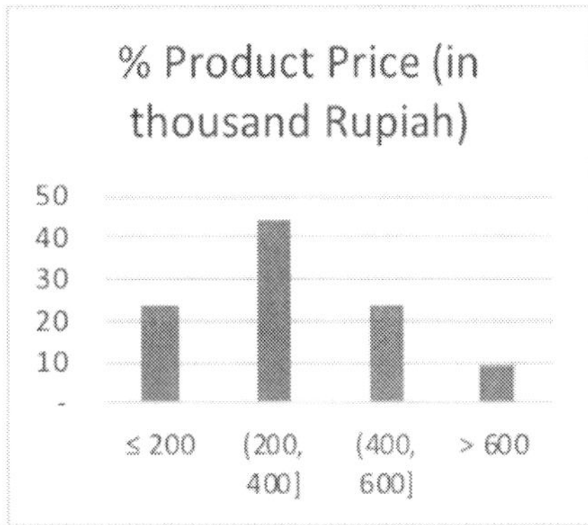

(g) Product price per kodi

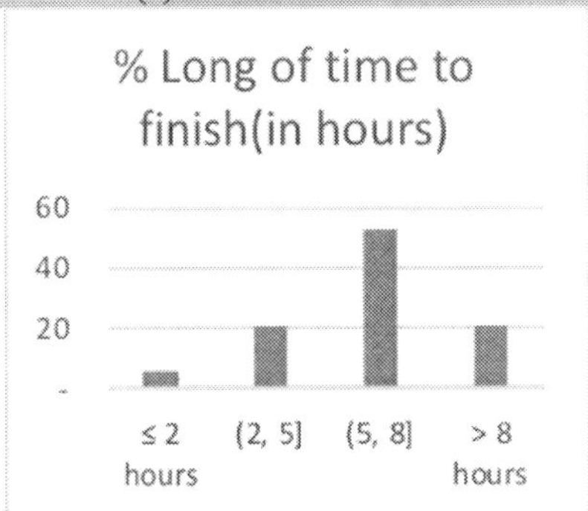

(d) Duration of work per 1 kodi product

(h) Work hour per day

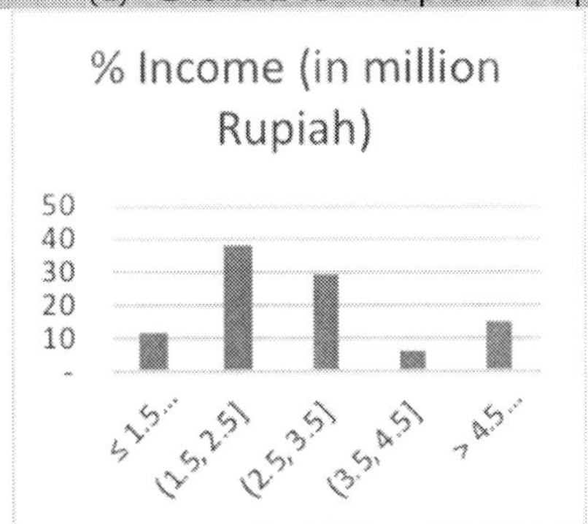

(i) Netto Income per Month

Figure 1. Percentage of respondent composition based on (a) - (i)

According to Figure 1, the majority of respondents are 25-55 years old (79%), with age ranges (25, 35], (35, 45], and (45,55] respectively 26%, 24% and 29%. The 82% of respondents are married, with 82% of their wives working as songket weavers. Single status who has a high net income per month is 18%.

Respondents had low education, namely elementary level (56%). The majority of them have undergone the profession as craftsmen for 20-30 years (29%), even> 30 years (24%). Their product yields are average of ≤25 kodi per month (47%), with their product prices are Rp 200,000-Rp 400,000, - per kodi. Working hours per day from craftsmen are mostly full day, which are 6-8 hours (65%). The duration of workmanship of 1 kodi product is ranged from 5-8 hours (53%). The monthly net income of the respondents is average Rp 1,500,000 untill Rp 2,500,000 (about 38%) and even up to Rp 3,500,000 (29% of the respondents).

Based on value of variables from 34 respondents, it can be formed a data matrix. Since the units of the variables are different, the data matrix is standardized, so the principal component analysis is done by using the correlation matrix. The correlation between variables can be seen in Table 2.

Table 2. Correlations among variables

	Edu.	**Ages**	**Work Exp.**	**Income**	**Number. of product**	**Work day**	**Duration of work**
Ages	-0.302						
Work experience	-0.311	**0.845**					
Income	0.043	-0.082	-0.148				
Number of product	0.272	-0.009	-0.028	0.180			
Work hour/day	0.267	0.083	0.195	0.269	0.206		
Duration of work	-0.111	-0.019	0.072	- 0.062	**-0.623**	0.083	
Product price	0.082	-0.290	-0.166	0.410	-0.282	0.044	0.161

Description: The coefficient in bold states there is a fairly close correlation between 2 variables.

Based on Table 2, there is low correlation among the level of education with work experience and the income level. Age is positively correlated closely with work experience. Income is most related to product prices, with a correlation of +0.41. The number of products is negatively correlated quite closely with the length of workmanship per kodi.

Correlations among variables and relative relationships between objects (respondents) and variables can be seen from biplot which is a graphical representation of the group of variables studied, as in Figure 2. This biplot can only represents variance of the data with goodness of fit as 50.6%, so biplot is not enough to represent information. It can also be seen from the length of the segment of the variable lines that are not the same length. Since the first four principle components (PC) have a goodness of fit 82% (> 75%), so PC analysis results can be used.

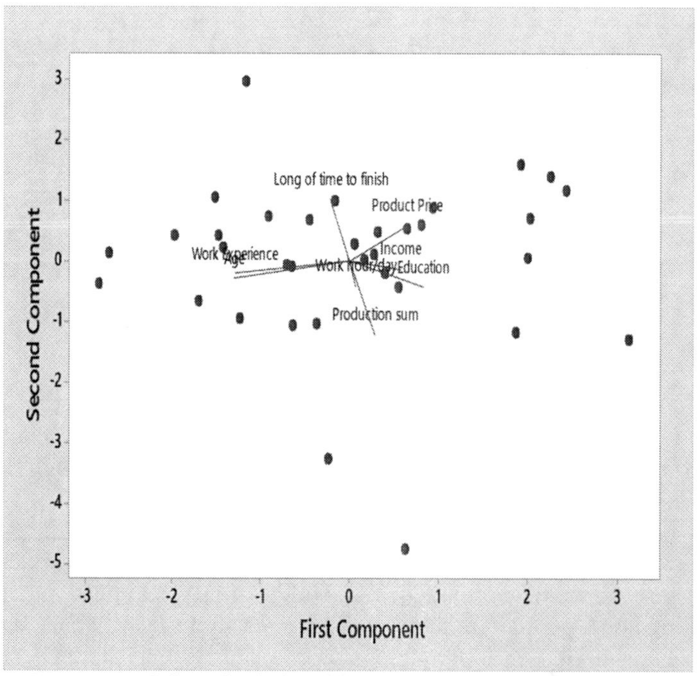

Figure 2. Biplot of relationship among objects with variables

The eigenvectors obtained from the correlation matrix can be seen in Table 3.

Table 3. The coefficient value of the principle components

Variable	PC1	PC2
Education	0.371	-0.233
Age (in years)	**-0.587**	-0.156
Work experience	**-0.583**	-0.104
Income (Rp)	0.255	-0.070
Number of product (in kodi)	0.127	**-0.660**
Work Hour per day	0.030	-0.224
Duration of work (in hours)	-0.097	**0.557**
Product price	0.292	0.332

Based on Table 3, the dominant variables representing the principle components of the biplot are age, work experience, number of products, and duration of product work.

Based on Figure 2, respondents whose age is old has a high working period (work experience) and low education. In this case, the profession as craftsmen started from the teenage years. Working hours of craftsmen in the day is quite high, but the effect is not significant enough to increase the number of products produced (low positive effect or correlation), and does not affect the decrease in the duration of workmanship for 1 kodi of product.

Income depends only on the price of the product, but not on the number of products. This can be due to the product being produced at various prices. The higher the price of the product, the greater the profit.

If the number of products produced is high, it will require higher processing time and is not affected by working hours/days. Because the average working hours / days of the craftsmen is 6-8 hours (all day).

If the average working hours in a day is high, then it should be able to decrease the duration of processing 1 kodi of product becomes shorter and the number of products produced in 1 month more, so as to increase revenue. If it is analyzed based on observations, answers to questionnaires, and direct interviews, it is influenced by work culture, dependence on raw material and capital availability, lack of appropriate targeted craftsmanship, and marketing dependence on collectors. They market their products so they can earn money to meet their daily needs. Furthermore, it affects the entrepreneurial spirit and low performance.

Cluster analysis is a technique used to classify objects or cases (respondents) into groups that are relatively homogeneous and have similar characteristics. Figure 3 and Figure 4 below are dendograms of cluster analysis results with a complete linkage method of the respondents.

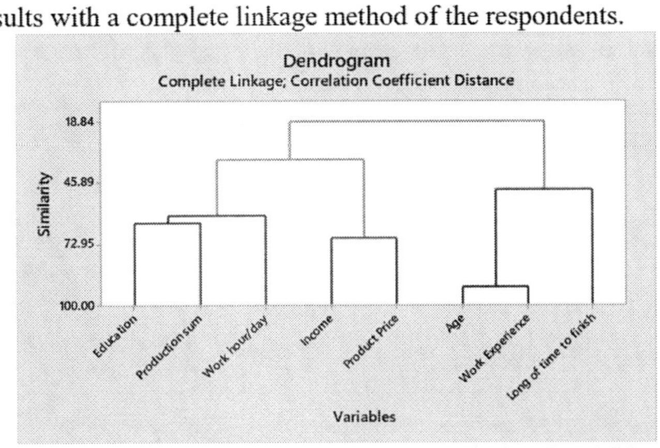

Figure 3. Dendogram on cluster analysis of variables

Based on Figure 3, there are 3 clusters of variables formed. If they are viewed from the meaning of the problem, then clusters can be interpreted as: cluster 1 is characterized by income and product price variables; cluster 2 is characterized by age and work experience variables; and cluster 3 has relatively no informative features although work hour/day is related to production sum. The results of cluster analysis of observations are as in Table 4 to Table 6.

Table 4. Numbers of Observations and average distance from centroid

	Number of observations	Within cluster sum of squares	Average distance from centroid	Maximum distance from centroid
Cluster1	14	48.2012	1.75526	3.56189
Cluster2	11	68.7610	2.38247	3.98700
Cluster3	7	33.0952	2.10862	3.23410
Cluster4	2	2.9325	1.21090	1.21090

Table 5. Cluster centroid

Variable	Cluster1	Cluster2	Cluster3	Cluster4
Education	**-0.709156**	0.005918	0.99049	**1.46481**
Age	0.015122	**0.710660**	**-1.23115**	0.29456
Work experience	-0.100840	**0.904931**	**-1.30190**	0.28540
Income	-0.256757	-0.102288	**0.76891**	-0.33130
Production sum	-0.312945	-0.166648	-0.02355	**3.18959**
Work hour/day	**-0.632270**	**0.745962**	0.01916	0.25604
Long of time to finish	-0.104279	0.334047	0.14292	**-1.60755**
Product price	-0.160143	0.102641	0.64955	**-1.71694**

Based on Figure 4, there are 4 clusters of respondents that can be formed. The numbers of observations in each cluster can be seen in Table 4.

Based on Table 5, cluster 1 consisted of the most (respondent) objects is characterized by the lowest education and work hour variables values, and also had low income, production sum and work hour. So, cluster 1 is dominantly characterized by craftsmen whose productivity is low. Cluster 2 consisting of 11 respondents is characterized by high age, work experience, and work hour variables. While cluster 3 is characterized by the lowest age and work experience variables and highest income. Cluster 4 consisting of 2 respondents is characterized by highest education and production sum variables, but lowest long of time to finish product and product price variables. So, cluster 4 can be interpreted as a cluster of minority craftsmen who has special skills to work on products in large quantities, but have a low price and the fastest processing time.

Based on Table 6, cluster 4 is very different from other clusters.

Table 6. Distances between cluster centroids

	Cluster1	Cluster2	Cluster3	Cluster4
Cluster1	0.00000	2.05247	2.85681	4.76437
Cluster2	2.05247	0.00000	3.35462	4.61766
Cluster3	2.85681	3.35462	0.00000	5.03265
Cluster4	4.76437	4.61766	5.03265	0.00000

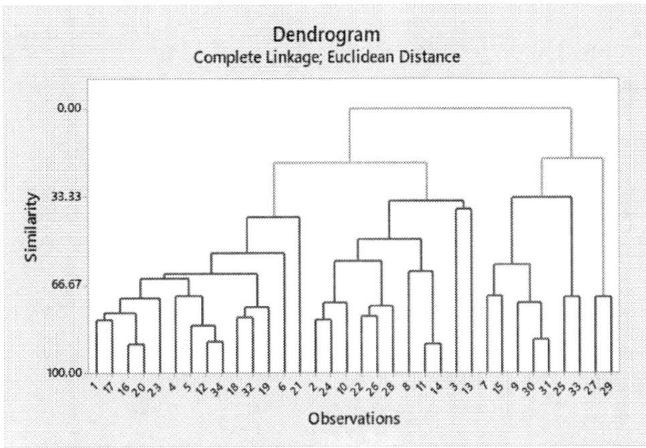

Figure 4. Dendogram on cluster analysis of observations

5. Conclusion

From the result we can conclude that Artisan blacksmith in Desa Limbang Jaya I majority of 25-55 years old, married status with majority of his wife work as songket weavers. Craftsmen are also mostly low-educated, that is elementary level and have undergone the profession as craftsmen for 20-30 years. The product yields an average of 25 kodi per month with working hours per day 6-8 hours and the duration of the work of 1 kodi of product is average 5-8 hours. The monthly net income of the average craftsmen is less than Rp 3,500,000.

The result of a biplot analysis with a representative level of 50.6% indicates that the income of the craftsmen depends only on the price of the product, but not on the number of products. If the number of products produced is high, it will require longer processing time and is not related to working hours/days. Old-age artisans have a high work experience and low education. Working hours of craftsmen in the day is quite high (all day), but the effect is not significant enough to increase the number of products produced (low positive effect).

The result of the cluster analysis obtained by the majority of craftsmen profiles of high age have high work experience as well, and have the highest work day, but lower income. While most of the craftsmen have low productivity features. Only a small percentage of craftsmen whose age and work experience are low but they have higher income.

Acknowledgment

The research leading to this study financially supported by PNPB University of Sriwijaya for support through "Hibah Penelitian Unggulan Kompetitif", 2017.

References

[1] Andari, N.P.U. dan L. P. Aswitari, 2010. Pengaruh sosial demografi terhadap produktivitas tenaga kerja perempuan pengrajin lontar di Desa Bona, Gianyar. Denpasar.

[2] Lisnini dan Purwati, 2012. Analisis pola konsumsi rumah tangga pengrajin songket di Kota Palembang. J. Orasi Bisnis 7, 55–58.

[3] Nurlaela, S., 2010. Profil industri kreatif pengrajin handycraff di Desa Sumber Kecamatan Trucuk Kabupaten Klaten. GEMA 683–694.

[4] Hasbullah, 2011. Budaya kerja kaum perempuan Melayu (studi terhadap perempuan pengrajin songket di Bukit Batu Kabupaten Bengkalis).

[5] Hasbullah dan Jamaluddin, 2013. Pengrajin Songket di Bukit Batu Kabupaten Bengkalis). Sos. Budaya Vol. 10 No, 1–13.

[6] Azmi, A. R., K. P. Hastuti, P.A., 2015. Upaya pengrajin pandai besi dalam menjaga keberlangsungan industri kerajinan rumah tangga di Desa Tumbukan Banyu dan Desa Sungai

IOP Conf. Series: Materials Science and Engineering **403** (2018) 012081 doi:10.1088/1757-899X/403/1/012081

Pinang Kecamatan Daha Selatan Kabupaten Hulu Sungai Selatan. JPG (Jurnal Pendidik. Geografi) 2, 66–80.
[7] Johnson, R.A. and D.W.Wichern., 2007. Applied Multivariate Statistical Analysis, 6th ed. Prentice-Hall, Inc., New Jersey.

IOP Conf. Series: Materials Science and Engineering **403** (2018) 012082 doi:10.1088/1757-899X/403/1/012082

A rule-based method for living organisms classification

A Desiani[1], M Arhami[2], Firdaus[3] and S I Maiyanti[4]

[1,4]Mathematics Departement, Faculty of Mathematics and Natural Science Universitas Sriwijaya,
[2]Informatics Engineering Department, Politeknik Negeri Lhokseumawe
[3]Faculty of Computer Science Universitas Sriwijaya

[1]anita.arhami@gmail.com, [2]Muhammad.arhami@gmail.com

Abstract. The implementation of an expert system for classification of a living organism could help researchers to classify an organism without having to face the complexities of a dichotomous key. This paper proposes a new system that can inference and give a solution for classification living organisms based on rule-based in taxonomy biology. The knowledge in taxonomy is represented by the rules-based method as a set know ledge for the system. the system is named as the expert system for living organisms. the system can be a solution to overcome obstacles in biological research caused by the problem of identification of living organisms. The system opens the possibilities to integrate other automated taxonomy systems because It can be used not only for a specific species but for all of other living organisms. The expert system of the classification is applied Peffer's DSRCM as a framework and guidelines. Forward chaining modified with backward chaining used as reasoning technique in the system. The expert system has tested on some groups respondent where all of the respondents give scores around 80-90 for the performance system. It shows that the system works well for classification living organims.

Keywords: Taxonomy, Classification, Living Organism, expert system, design search research

1. Introduction

An expert system is a concluded solution, derived from many solutions for solving a problem.The expert system is able to learn from some facts and decide right conclusion based on the given facts. An expert system is a part of the artificial intelligence that is able to make a conclusion, giving advice, explaining like a consultant, as well as having the ability to think logically[1]. An expert system (ES) is a application computer which has a special design to simulate an expert's abilities into the computer to solve a problem[2]. An expert system consists of 5 components[3], there are; 1) Graphical user interface, 2) Knowledgebase or learning base, 3) Database of facts, 4) Inference/reasoning engine, 5) Working memory(Figure 1). The expert system has the ability to capture, collect and compile knowledge from experts to an integrated system, sometimes some people name ES as Knowledge-based systems [4]. Expert systems have been extensively applied in various fields of research such as agriculture, medical, education, nanotechnology, control, scheduling, Geoscience, mineral identification, etc[2][3][5]–[12].

In the expert system, the technique of reasoning is a most crucial part of the system. Reasoning technique is a technique which is run by the engine so the machine can be an inference engine that has the ability to draw conclusions based on the facts. One of the familiar and famous reasoning technique is production rules (rule-base). production rules have been widely used in developing the expert system inference engine. Production rule has a simple form but powerful for knowledge representation. The production has flexible ability to combine declarative and procedural rules representation that is integrated into one form of a cognitive process's model of a set of rules [13].

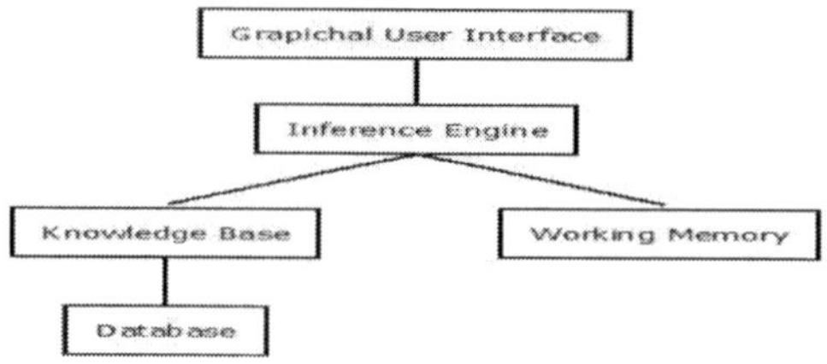

Figure 1. Expert System Components

Production rule or rule-based system can display human performance that the knowledge is seemingly got through experience[12]. Forward chaining and backward chaining are two methods of production rules for reasoning [12]. Forward chaining method (or bottom-up reasoning) is an inference that works from antecedents toward consequents. Otherwise, backward chaining begins work with a set of hypothesis or goals and performs to look for antecedents based on given consequents that support the consequents[14].

Classification is a technique in Taxonomy to classify living organisms as specified with their anatomical similarities, if the organisms have similar anatomical, they are put in one group.The organisms are grouped ranging from bigest organisms groups to smallest organisms groups. The orders of the groups are a kingdom(the greatest group), phylum (plural phyla), class, order, family (familia), genus (plural genera) and species (the smallest group)[15]. Classification is used for many applications. The first, the classification helps researchers to order organisms sequence. The Second, it is able to recognize newly discovered organisms based on the group that has similar with it. Third, it may ease the understanding process while learning the sorted organisms[16].

The method of organism classification is named as Taxonomy [16]. Taxonomy uses a dichotomous key for organism identification. A dichotomous key is a way of reasoning that provides some questions series to identify the organism that has only two choice answer yes/no. Every answer of the question will lead to next question until we get a conclusion about the organism. Reasoning in living organism needs precision and much time because living organisms in the world are very much and have a lot of variety[15]. .

The principle of classification process in Taxonomy has been used in several other filed such as software engineering (Design and Development)[17], Character recognition[18], fuzzy system and soft-computing[19][20], searching, and employee recruitment[21]. Learning application for the English Language[22]. In the field of Biology, taxonomy application has been developing on several specifics tasks, such as Dinoflagellates Classification using Neural network[23], species identification of fungi[24], automated insect identification system[25][26], Plant identification using image processing and computer vision[27][28], Snake identification[29].

The knock about identification of living organism that owns so many characters, becomes a major hindrance in biological research [30]. Reference [31], applied artificial neural networks to seek evolutionary relationships between living organisms but only for certain species., has developed machine to identify In the taxonomy, there are some disagreements about the kingdom system. There is a claim that in the world there is only 2 kingdoms system, but there are also other claims that there are 3 kingdoms system, or 6 kingdoms system, 7 kingdoms system, and 8 systems kingdom[32]–[36]. this paper will propose a new system for taxonomy that can inference the problem and give a solution based on rule-based in taxonomy biology. the knowledge in taxonomy biology is represented by the rules-based method as the knowledge for the system. the system is named as the expert system for living organisms. the system can be a solution to overcome obstacles in biological research caused by the problem of identification of living organisms. The system in this study will open the possibilities for other automated taxonomy systems because It can be used not only for a specific species but for all of other living organisms, although they are on the different kingdom systems.

Implementation of the expert system for the classification of living organisms can help people to classify an organism without having to face the complexities of a dichotomous key. The expert system classification of the living organisms also can be as a basic component to construct a larger system such

IOP Conf. Series: Materials Science and Engineering **403** (2018) 012082 doi:10.1088/1757-899X/403/1/012082

as eco-informatics. The expert system classification of the living organism will use a production rule as inference engine system and forward chaining as an inference method.

2. Method

This expert system of living organism classification is built by using Peffer's as based framework and guidelines. DSRCM has six phases that should go through from the identification of the problem phase until the communication phase[37]. Design-science research helps research paradigm pragmatic to find artifacts as real-life problem solving [38][39]. There are; concepts, models, methods, and instantiations[39][40]. DSRCM includes 6 phase:

1. Identity problem, this phase is to review the literature on the classification of living organisms and interview with various groups that have a direct relation to the classification of living organisms problem.
2. Establish the objectives for solutions. The Second phase is to identify the objects involved in the system to design solutions that are needed in the system. If the objects can be derived from the definition of the problem or the knowledge which obtained in phase 1. The selected object is an object that has a high rational in the problem. The results of this phase will be a guideline in designing an expert system for classification of living organisms.
3. Design and development. The third phase is built artifacts. In this paper, the obtained artifacts will become objects that have a contribution to the system. Some artifact will be a use case for the user in the system, a rational database, and knowledge inference models. in this phase, the system has just started to develop.
4. Demonstration. In the fourth phase, the artifacts that obtained in phase 3 is started to be implemented to develop an expert system for classification of living organisms. the system will be simulated and tested to see whether it is appropriate to user requirements or no and whether it provides solutions for users or no. This phase focuses on implementation of all the features in the design phase to build an expert system classification of living things using the programming language and database
5. Evaluation. In this phase, the artifacts in an expert system classification of living organisms will be observed and measured its performance if it can be completed and provide solutions to the problems that exist or no, whether all existing artifacts have features that correspond to the problem at an early stage. In addition, the user is prompted to interact directly with the expert system classification of living organisms to measure the performance of the system. This phase compares the solution of interest with the results obtained in the demonstration phase.
6. Communication. In this phase, the results obtained should be communicated between users, namely the public and researchers, to assist the further development system. The expert system for classification of living organisms can be integrated with a larger another system like information systems such as eco-informatics and bio-diversity system.

3. Result and Discussion

In taxonomy, classification of living organisms are divided into some levels depends on certain groups. the classification is arranged from the highest level of the kingdom to the lowest level of species. To facilitate the grouping of living organisms that are enormous and varied, so the classification will be grouped by some rules as in Figure 2.

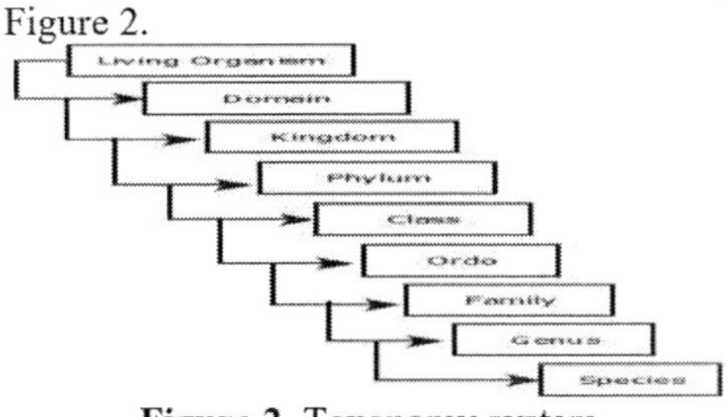

Figure 2. Taxonomy system

3.1. Database

From the process of data collection and interviews can be concluded that each of the characteristics of living organisms put them in different groups. The higher the level will have less in similar characteristics that must be considered, conversely the lower level of grouping, it will be more detail and more similar characteristics that must be considered. This classification system is useful for the identification of living organisms based on common characteristics in the taxonomy system. In the analysis phase, this phase obtains some Entities data and their relation to the system. It is explained by using the ERD (Entity Relationship Diagram) that can be seen in Figure 3.

The ERD will be implemented to database programming. The database is a basic component of knowledge in an expert system.

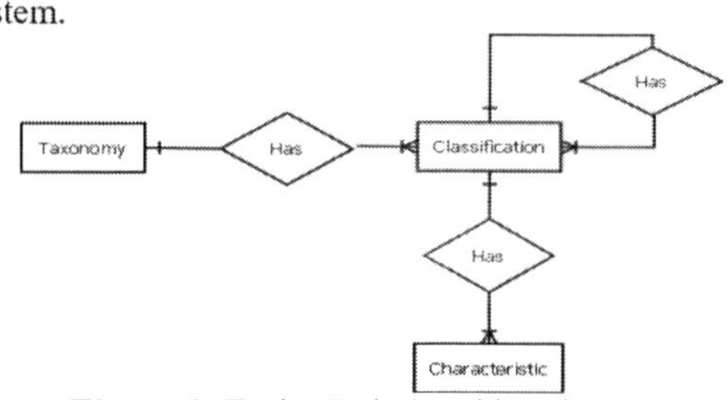

Figure 3. Entity Relationship Diagram

3.2. Knowledge Base

Designing knowledge has some several ways such as the semantic net, procedural representation, production rules, and frames[4]. For this research, the knowledgebase will be designed by production rule sometimes called IF-THEN Rule or Rule-based. Some paper used Production rule or Rule-base as a knowledgebase on their system. Reference [41] used production rule in an expert system for tourism in Jordan. Reference [42] applied production rule for nutrition care process of older adults expert system.

The premise of IF-Then Rules condition can combine attribute and logical expression like 'or' and 'and'. When Clause IF is true, so the clause after THAN will be run. When the clause IF is false, so it should check the other rules. All of IF-Then rules in the classification of the living organism have two general form :

- IF FACT 1: class of taxon found in database AND FACT 2: character of organism is found in database THEN CONCLUSION 1: show name of organisms in the taxon
- IF FACT 3: Conclusion 1 AND FACT 4: new character of organism is found in database THEN CONCLUSION 2: show name of organism and its taxon

3.3. Inference Engine

The inference engine is The essential component in the expert system Inference Engine (IE) allows the ES to communicate the knowledge stored in the KB and to run the system, as it draws an inference by relating user-supplied facts to KB rules. This paper uses forward chaining that modified with backward chaining. Forward chaining is a bottom-up model where the solution to some problems naturally starts with the collection of information. Otherwise, backward chaining is a top-up model where is the model start from goal[43]. Inference uses this information to obtain logical conclusions for the solution[14]. Pseudo code for inferencing the classification of living organisms in the system is:

```
REPEAT
   FOR every rule do
         IF antecedents match assertions in the working memory and consequents would change
         the working memory THEN
         Create triggered rule instance
   END IF
   END FOR
Pick one triggered rule instance, using conflict resolution strategy if needed, and fire it (throw away
other instances)
UNTIL no change in working memory, or no STOP signal
```

For the example :
IF F1: taxon is class and F2: is Fishes **THEN**
(C1: Ordo is Primata OR C2: Ordo is Proboscidea) OR (C3: Ordo is Lagomorpha OR C4: Ordo is Marsupialia) OR (C5: Ordo is Rodentia OR C6: Ordo is Artiodactyla) OR (C7: Ordo is Dermoptera OR C8: Ordo is Cetacea) OR (C9: Ordo is Perissodactyla OR C10: Ordo is Insectivora) OR C11: Ordo is Carnivora

IF F3: ORDO is Carnivora AND (CF1: puny and cobby animals; AND CF2: cobby legs, globular ears, woolly) AND (CF3: deliver a strong-putrid secretion for sexual cueing and mark out area AND CF4: owns rectal whiff glands)
THEN C12: Familia is Mustelidae

IF F4: Familia is Mustelidae AND (FC4: its long body is between 173-217 mm or 6.8 to 8.5 inc AND FC5: the females are shorter than the males) AND (FC6: owns brown or red top layers and white maw AND FC6: some species of the population molt to a fully their white mantles in wintertime) AND (FC7: the population own lengthy and slim body, that permit them to adhere the prey into the hole AND FC8: They have tail with long are around 34-52 mm or 1.3-2.0 inc long)
THEN C13: Genus Mustela

IF F5: Genus Mustela AND (FC9: it has large and long tail with a protruding sooty tip AND FC10: its original from North America and Eurasia)
THEN C14: The Species of the organism is Stoat.

where, F_i : living organism class
FC_i are Characteristics of living organism
C_i : Conclusion

3.4. Demonstration

The expert system was built to identify living organisms based on their taxonomy. Database implementation on the system used Mysql which still keep normalization rule and relational database. Expert system for classification is developed based on internet, so the system can be accessed anytime and anywhere. Expert system for classification of living organisms has some interface. Figure 4 is a user interface to input the level of taxonomy. This interface manages the level of taxonomy data for every kingdom systems. The interface of the system can provide two kingdoms system, five kingdoms system, etc. The other interface has a facility to manage characteristics of living organisms for every taxonomy level. This data will be used to identify organism based on their characteristic. This interface can see in Figure 5.

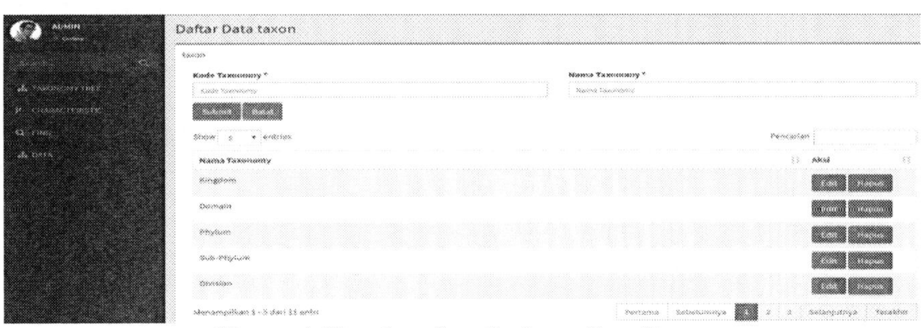

Figure 4. User Interface for Input Data Taxon

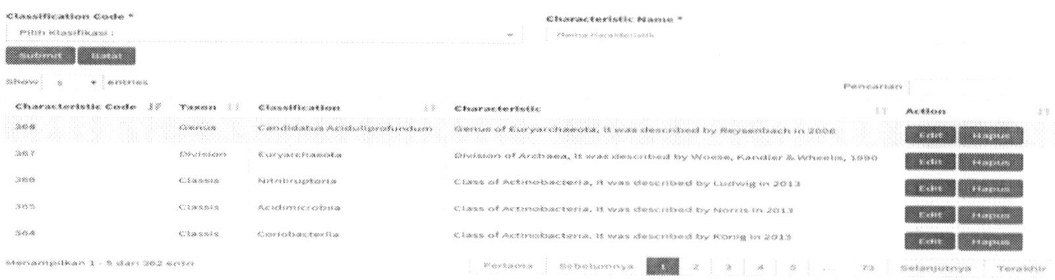

Figure 5. User Interface for Input Characteristic

For output, the system can draw the hierarchy classification tree of living organisms based on their group or based on their characteristic. So the user can look the three hierarchy of one various of living organisms or more (Figure 6) and also the user can see detail characteristic of organisms (Figure 7).

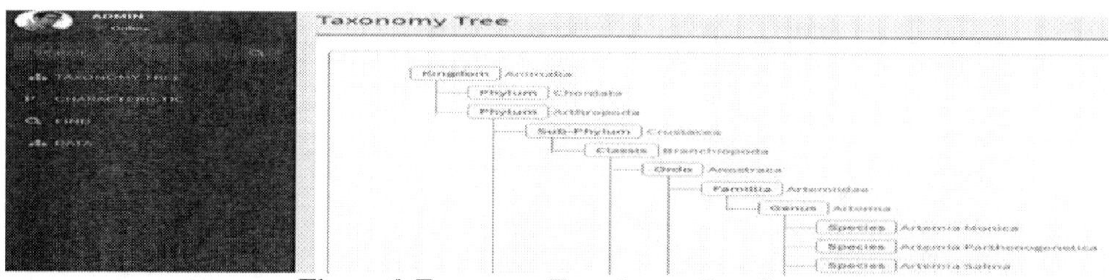

Figure 6. Taxonomy Tree Output Interface

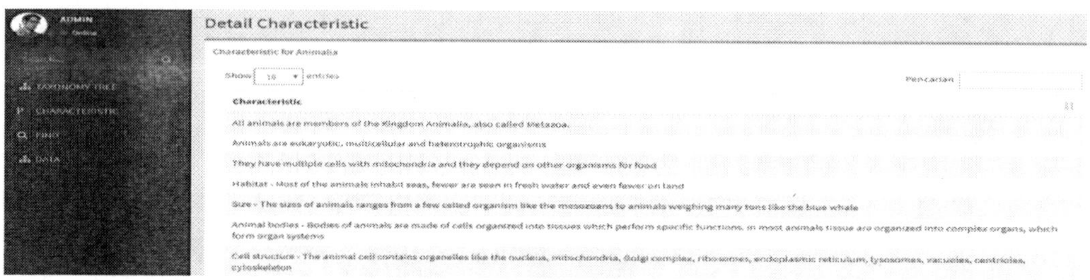

Figure 7. Characteristic Interface

4. Testing and Result

The Expert system application for classification of living organism has been testing on some groups respondent. The experiment is performed on 28 test subjects consist of 10 students from the different background, 10 undergraduate students from the biology department, 5 lecturers, and 3 researchers. The experiment subjects were tested with summated rating algorithms where the total rating is count. Summated rating is a quantitative method which the test subjects will give their response on favorable and unfavorable.

The result of the testing application gives the conclusion that expert system for classification of the living organism works well because all of the respondents give scores around 80-90 for the performance system. The respondents give suggestion the expert system for classification of living organisms can provide the better image with bigger size than existing image in the application.

5. Conclusion

The expert system application for living organisms used Peffer's DSRCM as a framework and guidelines to develop an application. The application can adopt every kingdom systems such as 2 kingdom 3 kingdom , and other system. Applied reasoning technique uses modified forward chaining with backward chaining because of the detail information in the classification of living organisms. The expert system for classification of the living organism is a quick, easy and precise solution to identify an organism without having to face the complexities of the dichotomous key.It has been shown by the respondents scores while they tested the application. An expert system classification of living organisms is a basic component to construct a larger system such as an ecoinformatics.

References

[1] E. Turban, *Decision Support System and Expert Systems*. USA: Prentice Hall International Inc, 1993.
[2] S. P. R. P. Isakki, "The Expert System Designed To Improve Customer Satisfication," *Adv. Comput. Int. J. ACIJ*, vol. 2, no. 6, pp. 69–84, 2011.
[3] K. . olorunso, I. O, Abikoye, O. C, 2 Jimoh, R. G. and 2 Raji, "A Rule-Based Expert System for Mineral Identification," *J. Emerg. Trends Comput. Inf. Sci.*, vol. 3, no. 2, pp. 205–210, 2012.
[4] S. Liao, "Expert system methodologies and applications — a decade review from 1995 to 2004," *Expert Syst. Appl. Elsevier*, vol. 28, no. June 2004, pp. 93–103, 2005.
[5] H. Ahmadi, M. Nilashi, L. Shahmoradi, and O. Ibrahim, "Computers in Human Behavior Hospital Information System adoption : Expert perspectives on an adoption framework for Malaysian public hospitals," *Comput. Hum. Behav.*, 2016.
[6] J. P. Astruc and R. Tufeu, "Knowledge Base Specification to Automate The Fluid," *Applied Artificial Intelligence*, no. scholarship 18860. pp. 453–471, 2001.
[7] R. E. Plant and M. P. Vayssie, "Combining expert system and GIS technology to implement a state-transition model of oak woodlands," *Elsevier Comput. Electron. Agric.*, vol. 27, pp. 71–93, 2000.
[8] B. Prasadl, P. E. S. N. K. Prasad, and Y. Sagar, "An Approach to Devepol Expert System In Medical Diagnosis Using Machine Learning Algorithms and A Perpormance Study," *Int. J. Soft Comput. IJSC*, vol. 2, no. 1, pp. 26–33, 2011.
[9] L. Salekhova, A. Nurgaliev, R. Zaripova, and N. Khakimullina, "The Principles of Designing an Expert System in Teaching Mathematics," *Univers. J. Educ. Res.*, vol. 1, no. 2, pp. 42–47, 2013.
[10] A. M. Salem, "Reasoning Techniques for Diabetics Expert Systems," *Procedia - Procedia Comput. Sci.*, vol. 65, pp. 813–820, 2015.
[11] G. W. Sasmito, "Application Expert System of Forward Chaining and The Rule Based Reasoning For Simulation Diagnose Pest and Disease Red Onion and Chili Plant," in *Proceedings of The 1st International Conference on Information Systems For Business Competitiveness (ICISBC)*, 2011, pp. 392–398.
[12] P. T. Sharma, P. N. Tiwari, and P. K. Shah, "Student Counseling System : A Rule-Based Expert System based on Certainty Factor and," *Int. J. Appl. Innov. Eng. Manag.*, vol. 2, no. 1, pp. 371–375, 2013.
[13] G. Giarattano, Joseph, Riley, *Expert Systems Principles and Programming*. Boston: PWS-KENT Publishing Company, 2005.
[14] D. K. P. Sharma, Tilotma, N. Tiwari, "Study of Difference Between Forward and Backward Chaining," *Int. J. Emerg. Technol. Adv. Eng.*, vol. 2, no. 10, pp. 271–273, 2012.
[15] M. A. Ruggiero *et al.*, "A Higher Level Classification of All Living Organisms," *Plus One Open Access J.*, pp. 1–60, 2015.
[16] S. Rudyshyn and V. Samilyk, "Development Of Knowledge of The Taxonomy and Phylogeny Of Living Organisms for Future Biology Teachers," *Adv. Sci. J.*, vol. 2015, no. 1, pp. 75–82, 2015.
[17] M. Usman, R. Britto, J. Börstler, and E. Mendes, "Taxonomies in software engineering : A Systematic mapping study and a revised taxonomy development method," *Inf. Softw. Technol.*, vol. 85, pp. 43–59, 2017.
[18] T. Bouwmans, L. Maddalena, and A. Petrosino, "PT US CR," *Pattern Recognit. Lett.*, 2016.
[19] A. Saeed, S. Hafizah, A. Hamid, and M. B. Mustafa, "The experimental applications of search-based techniques for model-based testing: Taxonomy and systematic literature review," *Appl. Soft Comput. J.*, vol. 49, 2016.
[20] A. Fernández, V. López, M. José, and F. Herrera, "Knowledge-Based Systems Revisiting Evolutionary Fuzzy Systems : Taxonomy , applications , new trends and challenges," *J. Knowl.-Based Syst. KNOSYS*, vol. 80, pp. 109–121, 2015.
[21] G. Michel, "Levelized Taxonomy Approach for The Job Seeking / Recruitment Problem," in *IEEE International Conference on Embedded and Ubiquitous Computing, and International*

Symposium on Distributed Computing and Applications to Business, Engineering and Science, 2016, pp. 448–451.

[22] E. Yadegaridehkordi and M. Alaa, "Mobile Learning for English Language Acquisition: Taxonomy , Challenges , and Recommendations," *IEEE Access*, vol. 5, 2017.

[23] T. A. Meehan-Andrews, "Teaching mode efficiency and learning preferences of first year nursing students," *Nurse Educ. Today*, vol. 29, no. 1, pp. 24–32, 2009.

[24] M. Hsu *et al.*, "Species identification of medically important fungi by use of real-time LightCycler PCR," *J. Med. Microbiol.*, no. 2003, pp. 1071–1076, 2017.

[25] P. J. D. W. M. A. O. N. K. J. Gaston and I. D. Gauld, "Automating insect identification : exploring the limitations of a," *J. Appl. Entomol.*, vol. 123, no. 1, 1937.

[26] R. Gurgel-gon, E. Komp, L. P. Campbell, A. Khalighifar, H. L. Owens, and K. De, "Automated identification of insect vectors of Chagas disease in Brazil and Mexico : the Virtual Vector Lab," *PeerJ*, vol. 5, pp. 1–23, 2017.

[27] J. Joaci, D. M. Sá, P. César, R. Marta, and A. Ricardo, "images of palisade parenchyma," *Ecol. Inform.*, vol. 39, no. April, pp. 119–122, 2017.

[28] P. Barré, B. C. Stöver, K. F. Müller, and V. Steinhage, "A computer vision system for automatic plant species identi fi cation," *Ecol. Inform.*, vol. 40, no. May, pp. 50–56, 2017.

[29] V. H. K. R, MAHESHWARI. V, KUMAR V. K, "A Neural Network Based Specie Identification Technique in Venom Interacted Cases in India," *J Venom Anim Toxins Incl Trop Dis*, vol. 13, pp. 766–781, 2007.

[30] K. J. Gaston and M. A. O. Neill, "Automated species identification : why not?," *Philos. Trans. R. Soc. B Biol. Sci.*, vol. 359, no. March, pp. 655–667, 2004.

[31] A. Kasperski and R. Kasperska, "A new approach to the automatic identification of organism evolution using neural networks," *BioSystems*, vol. 142–143, pp. 32–42, 2016.

[32] T. Cavalier-smith, "Only six kingdoms of life," in *Proceedings of the Royal Society B: Biological Sciences*, 2004, no. May, p. 1251.

[33] O. R. Anderson *et al.*, "The New Higher Level Classification of Eukaryotes with Emphasis on the Taxonomy of Protists," *J. Eukaryot. Microbiol.*, vol. 52, no. 5, pp. 399–451, 2005.

[34] G. E. Woese, C. R., & Fox, "Phylogenetic structure of the prokaryotic domain: the primary kingdoms.," in *Proceedings of the National Academy of Sciences*, 1977, pp. 5088–5090.

[35] M. G. Simpson, "Plant systematics," in *Science and Technology Book*, Second., Academic Press is an imprint of Elsevier, 2010.

[36] T. S. R. Gayathri, *Biology (Zoology)*. Chennai, India: Tamil Nadu TextBook Corporation, 2003.

[37] K. Peffers, "A Design Science Research Methodology for Information Systems Research," *J. Manag. Inf. Syst.*, vol. 24, no. 3, pp. 45–78, 2007.

[38] S. Y. R. Esearch, B. A. R. Hevner, S. T. March, J. Park, and S. Ram, "D ESIGN S CIENCE IN I NFORMATION," *MIS Q.*, vol. 28, no. 1, pp. 75–105, 2004.

[39] P. Offermann, S. Blom, M. Schönherr, U. Bub, and D. T. Laboratories, "Artifact Types in Information Systems Design Science – A Literature Review," pp. 77–92, 2010.

[40] O. Wongso, Y. Rosmansyah, and Y. Bandung, "Gamification Framework Model , Based on Social," in *International Conference on Technology, Informatics, Management, Engineering & Environment*, 2014, pp. 10–14.

[41] N. S. Hussein and M. J. Aqel, "ESTJ : An Expert System for Tourism in Jordan," *Procedia - Procedia Comput. Sci.*, vol. 65, no. Iccmit, pp. 821–826, 2015.

[42] T. Cioara *et al.*, "Expert System for Nutrition Care Process of Older Adults," *Future Gener. Comput. Syst.*, 2017.

[43] A. Al-ajlan, "The Comparison between Forward and Backward Chaining," *Int. J. Mach. Learn. Comput.*, vol. 5, no. 2, 2015.

Planting time determination for food crops using decision tree

[1]T Setiadi, [2]A Tarmuji, [3]B Suhendra, [4]F Noviyanto and [4]S L Khasbullah
[1,2,4] Department of Informatics Engineering, Universitas Ahmad Dahlan, Indonesia
[3] Department of Accounting, Universitas Ahmad Dahlan, Indonesia
E-mail: [1]tedy.setiadi@tif.uad.ac.id.com, [2]alitarmuji@tif.uad.ac.id, [3]benisuhendra@uad.ac.id
[4]fiftin.noviyanto@tif.uad.ac.id, [4]sigitlutfy@gmail.com

Abstract. Before the world experienced climate change,people can predict rainy and dry seasons easily. It is because the period is fixed throughout the year. The industrial revolution has caused global warming and affect climate change resulting in climate anomalies may decrease or increase in air temperature extreme, rainfall and seasons are shifting from the usual pattern and not erratic as well as rising sea levels and the occurrence of rob in some areas including in Indonesia. As consequence, the planting period is become difficukt to predict. Climatic factors, especially temperature, humidity, and rainfall strongly influences the process of determining the planting period for food crops. The domino effect of this mistake stimulates pest explosions, increased production costs as well the difficulty of providing food in the market. One of the technologies to assist farmers in determining the planting period is based on climate predictions from BMKG and data on food crops from the Dinas Pertanian. The process of extraction of climate data from BMKG can be applied with webservice technology. The process of determining the planting period and selection of the appropriate plant species is the application of data mining using the method of decision tree. The data in the process in this study is weather and harvest data in 2014 to 2016. The results of this study is an application that can facilitate farmers in determining the right planting or type of seeds.

1. Introduction

Farmers use the Java planting calendar or prey order for farming guidance. The determination of the type of food crop follows the prey calendar. It is not precise anymore since the current climate that tends to be unpredictable. Therefore, there is a need for a solution to increase farmers' productivity by minimizing errors in the selection of plant species.

Data Service provided by BMKG contains climate data on certain areas including temperature, rainfall, humidity. The data can be publicly accessed and updated regularly by BMKG. The Agricultural Service collects annual crop data for food crops. The yield data will then be combined with climate data obtained from BMKG to be stored in the application data warehouse.

The data collected in the data warehouse then performed the process of data mining using the decision method. After the data mining process will beknown what type of plant is suitable to be planted in a certain period. In order not to widen the problem there needs to be a systematic discussion of the problem, so the limitation on this research is to help for the farmers to determine the appropriate cropping pattern with the weather. The location of research in this research is in Sleman Regency Yogyakarta Indonesia.

2. Literature Review

Preliminary studies conducted by climate-related and weather reporters have been reported by Mass Clifford (Mass., 2012). In this study it was submitted that Nowcasting combines descriptions of the current state and short-term estimates of how the atmosphere will evolve over the next several hours. Problems resolved in this study include: addressing weather mesoscale information, quick fixes in numerical modelling and assimilation data. The results are presented in the form of graphic and data information.

Furthermore in 2015 conducted weather prediction research using Deep Hybrid method by Groffer and Horvitz (Grover & Horvitz, 2015). Predictions are made with the Deep Hybrid approach that incorporates discriminatory models and trained neural networks. Test results from this study indicate that the new methodology yields better data based on the NOAA benchmark, as well as showing improvement over the benchmark.

A study that combines weather prediction in the determination of the period planting done in 2012 by [8]. This study examined the effect change of date of Rice planting in dry season simulated with using Decision Support Software of Agrotechnology Transfer System (DSSAT 4.5).

Setting cropping pattern that is guided by hereditary habits on dry land, currently at risk of failure due to climate change. This research performed by [9].

3. Method

a. Data collection

At this stage will be data collection service API (Application Programmable Interface) about climate and weather from BMKG in order to receive data in real time. In addition, the necessary data from the Agriculture Service related to the suitability of the type of plants with weather conditions as well as historical data of the harvest. After the data collected then performed data selection and continued with data transformation process.

b. Web service Design

Based on the data service that has been collected and extraction process, then designed the process of data synchronization of API service into the server system to be developed.

c. Implementation of Data Service Integration BMKG and Datamining: Learning

At this stage will be implemented data integration with datamining method, namely Decision Tree. Web-developed system to communicate with data service and accessed further in the mobile native application development stage.

4. Results and Discussion

4.1. System Analysis

Climatological search procedures in accordance with climatology to be planted in a certain period can be seen in Figure 1. Based on the results of data synchronization BMKG and observation of data from the Department of Agriculture has been done, the data is stored in the database.

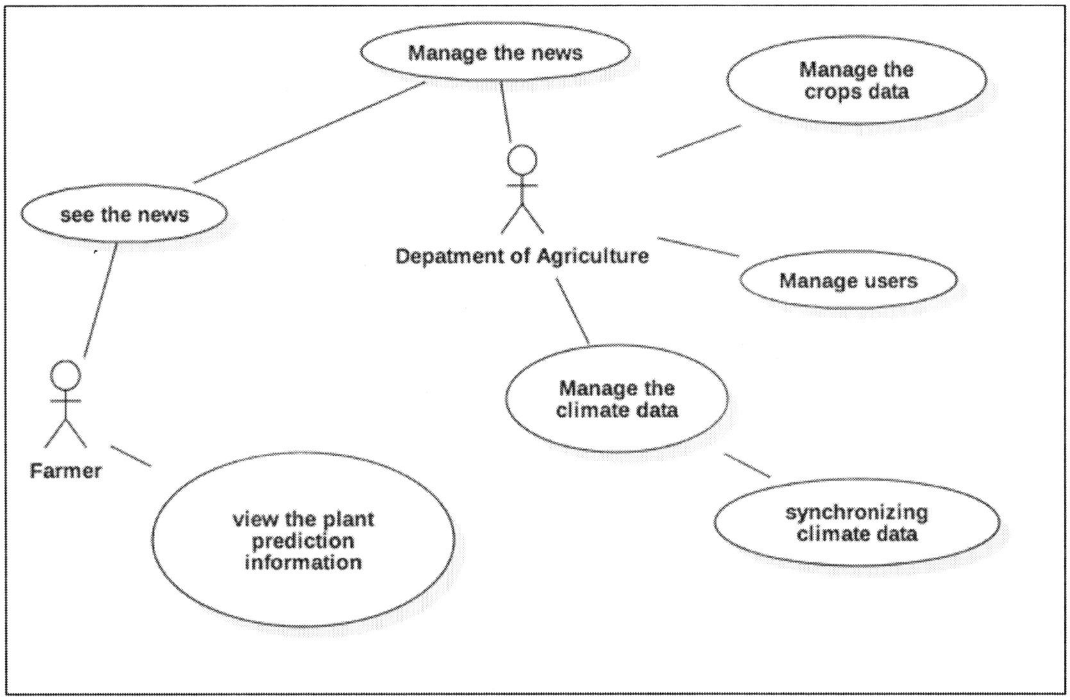

Figure 1. Use Case Diagram System

The application flow of the use case consists of:

- Climate Data Synchronization
  Climate data used for mining process obtained from public data BMKG through address http://data.bmkg.go.id. From the public data is then performed the process of synchronization into the application using the method -simple_load_xml. The picture synchronization process is as Figure 2.

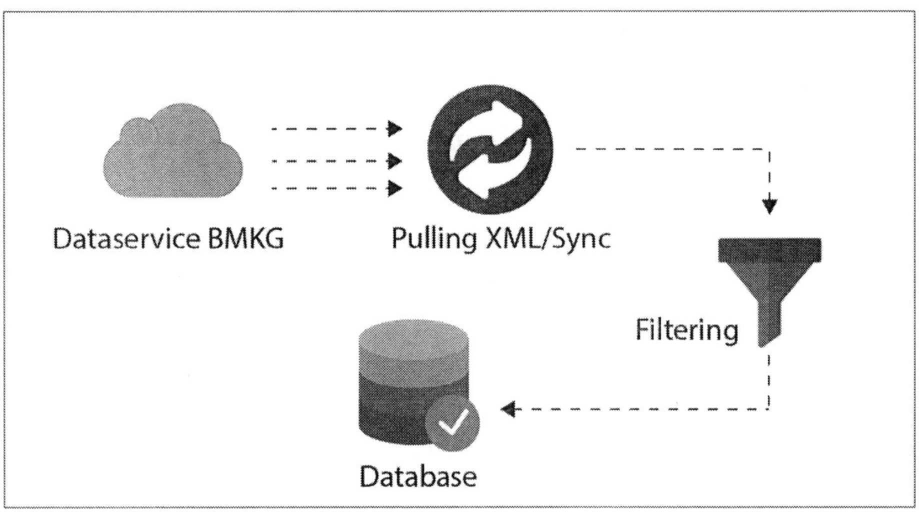

Figure 2. Synchronization Process

After xml data obtained from BMKG then done process of data filtering. Filtering data aims to filter what data will be used in the next process. After the filtering process is completed proceed by storing data in the database.

- Data collection
  The data collection process is done after obtaining climate data from the synchronization process and the average data every four months of food crops in Sleman, Yogyakarta.
- Making Decision Tree
  At this stage crop and climate data are combined to create a data integration table. Data integration that has been made into materials to do the mining process using decision tree method. The structure of the integration data table are: ID as primary key, Temperature, Humidity, Rainfall, rice, corn, beans, soybean, cassava. From the data integration then made a decision tree that will be applied into the application. Examples of decision trees that have been made can be seen in Figure 3.
  After the decision tree is created the next step is to implement the rule into the application.

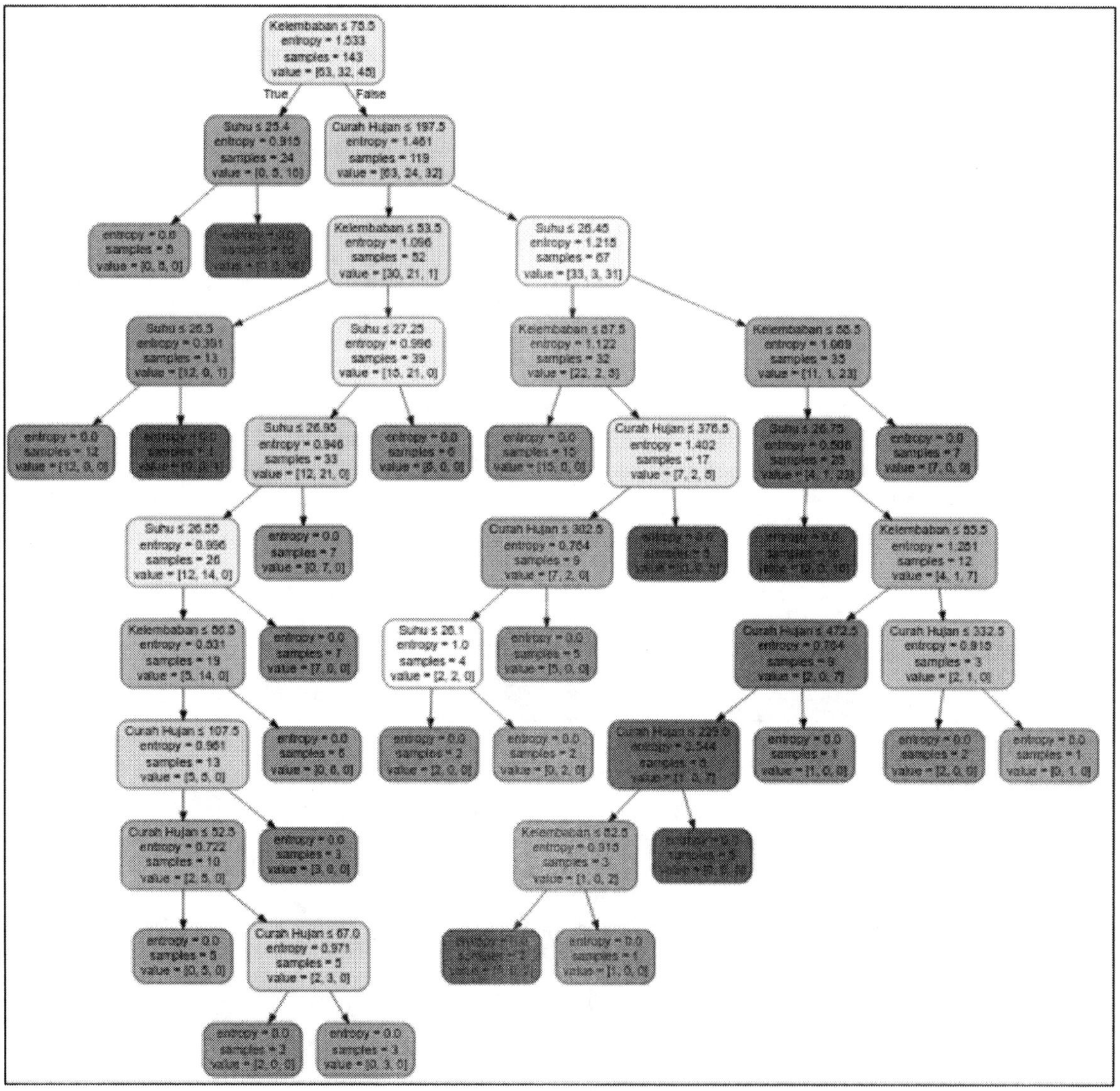

Figure 3. Decision Tree

4.2. Implementation

Users of this application are Farmers and Agriculture Service. Department of Agriculture in charge of input basic data which then processed by the system. Processed results are then displayed to be absorbed Farmers who access this application. Interface design created among others:

1) Weather Sync

The synchronization process begins by pulling the last XML data from the BMKG data service (Figure 4). After withdrawal then filtered as needed. The last step of the sync process is to save the results into the database.

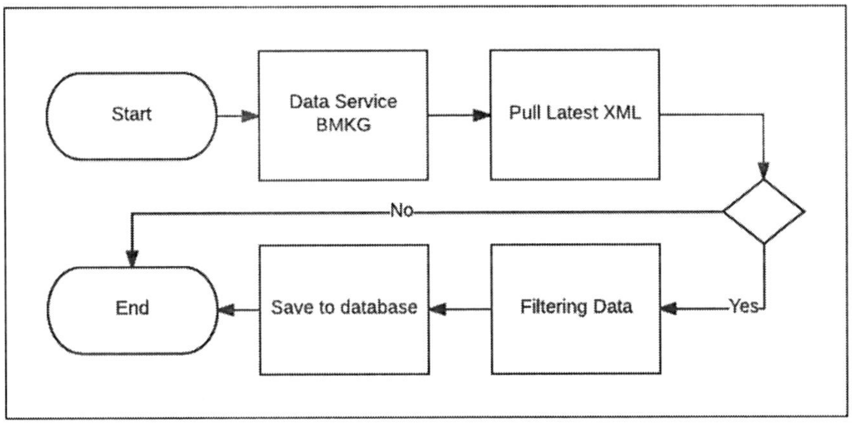

Figure 4. Synchronization

2) Option Choice for prediction

   To be able to predict farming patterns, farmers are faced with two choices. First prediction of planting based on the type of seeds that will predict the future is the appropriate planting period. The second is the prediction of planting based on the planting period that the result is the type of seed.

3) Predicted Results

   Figure 5 contains sample predictions for January. January is included in the category of quarter 1. The exact crops for January are predicted to be rice.

Figure 5. predicted planting and cropping result.

4.3. Testing

After the implementation of the application proceed with the testing process. The testing process is done in 2 stages of system test and acceptance.

1. System Test
   In testing system test done by developer using BlackBox test method. From blackbox testing to the created application, the whole function can be operated well. The conclusion of the blackbox test on the application is said to be successful.
2. Acceptance Test
   Acceptance Testing on application is performed by the end user, that is Agriculture Department and farmer. This test aims to measure the usefulness of the application made. The test results are shown in Figure 6 and Table 1.

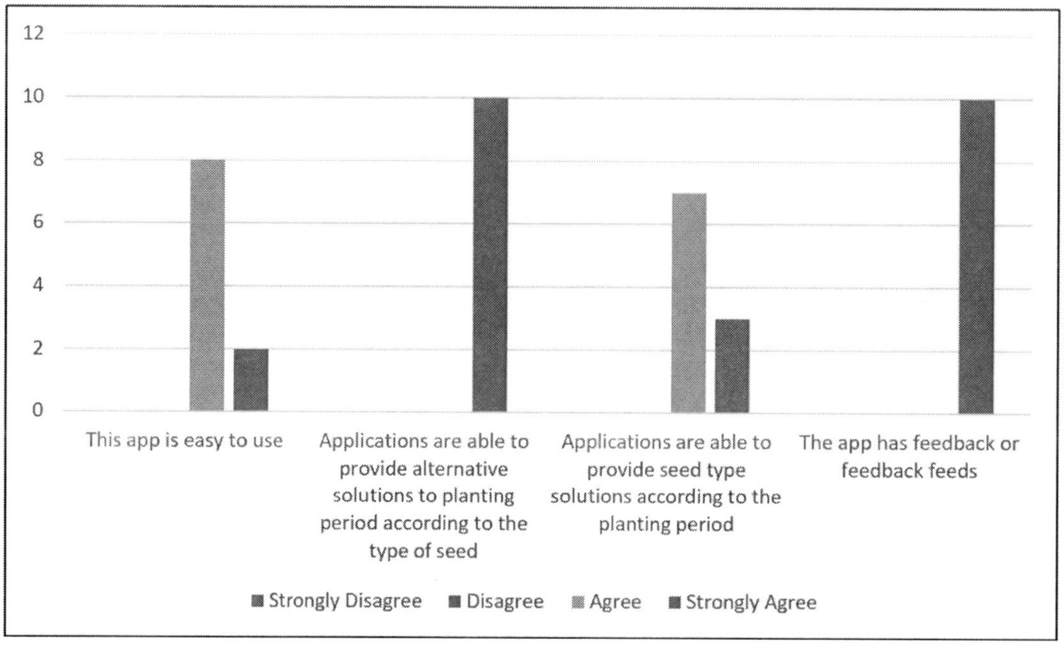

Figure 6. Acceptance Testing Results

Table 1. Acceptance Testing by Farmers

No	Question	Answers SA	D	A	SD
1	This app is easy to use			8	2
2	Applications are able to provide alternative solutions to planting period according to the type of seed				10
3	Applications are able to provide seed type solutions according to the planting period			7	3
4	The app has feedback or feedback feeds				10
	Total			15	25

Description: SA: Strongly Agree D: Disagree A: Agree SD: Strongly Disagree

5. Conclusions

After analysing, designing, implementing and testing stages, it can be concluded that the built application can assist the farmers in determining the planting period according to the type of seed, or determine the type of seeds in certain planting period according to the climate. For future development, the error in determining the type of plant can be minimized to avoid the risk of crop failure.

Acknowledgments
The research was funded by the directorate general of higher education, the Ministry of technology research and higher education Indonesia through Competitive Research Grant scheme 2017r.

References
[1] Sze S M 1969 *Physics of Semiconductor Devices* (New York: Wiley–Interscience)
[2] Dorman L I 1975 *Variations of Galactic Cosmic Rays* (Moscow: Moscow State University Press) p 103
[3] Caplar R and Kulisic P 1973 *Proc. Int. Conf. on Nuclear Physics (Munich)* vol 1 (Amsterdam: North-Holland/American Elsevier) p 517
[4] Szytula A and Leciejewicz J 1989 *Handbook on the Physics and Chemistry of Rare Earths* vol 12, ed K A Gschneidner Jr and L Erwin (Amsterdam: Elsevier) p 133
[5] Kuhn T 1998 Density matrix theory of coherent ultrafast dynamics *Theory of Transport Properties of Semiconductor Nanostructures (Electronic Materials* vol 4*)* ed E Schöll (London: Chapman and Hall) chapter 6 pp 173–214
[6] Hosang, P. R., Tatuh, J., & Rogi, J. E. X. (2012). Analisis Dampak Perubahan Iklim Terhadap Produksi Beras Provinsi Sulawesi Utara Tahun 2013 – 2030 Analysis of The Climate Change Impact on Rice Production of The Province of North Sulawesi In 2013 - 2030. Eugenia, 18(3).
[7] Nurdin. (2010). Antipasi perubahan iklim untuk keberlanjutan ketahanan pangan. Gorontalo
[8] Dharmarathna, W. R. S. S., Herath, S., & Weerakoon, S. B. (2014). Changing the planting date as a climate change adaptation strategy for rice production in Kurunegala district, Sri Lanka. Sustainability Science, 9(1), 103–111. http://doi.org/10.1007/s11625-012-0192-2.
[9] Dwiratna, Nawawi, & Asdak. (2013). Analisis Curah Hujan dan Aplikasinya dalam Penetapan Pola Tanaman Pertanian Lahan Kering di Kabupaten Bandung. Bionatura Jurnal IlmuIlmu Hayati Dan Fisik, 15(1), 29–34.

Grouping the community health center patients based on the disease characteristics using C4.5 decision tree

N Anwar[1], A Pranolo[2] and R Kurnaiwan[3]
Informatics Department, Universitas Ahmad Dahlan, Yogyakarta, Indonesia

nuril.anwar@tif.uad.ac.id[1], andri.pranolo@tif.uad.ac.id[2], riza.kur93@gmail.com[3]

Abstract. Community health centers (Puskesmas) is one of the important public health service facilities in Indonesia. Puskesmas serves many patients on performing the examination or treatment in every day. Accumulated medical record data is not utilized to generate new information or knowledge. One existed datamining techniques is the process of grouping an object with unknown label into a class. The C.45 algorithm is used to mine the patients diagnosis data available on 2015 2016. As a result, C4.5 algorithms can be applied for grouping disease. The first test using 85 training data has 78% accuracy level, while the second test of 115 training data reaches 88% accuracy rate.

1. Introduction

Community health centers (Puskesmas) in Indonesia is a health center owned by government. Puskesmas always improve the quality of service to the patient through the way of involving technological progress in health world. So the current government of Indonesia immediately take action such as in the form of Social Health Insurance Provider Body, called BPJS. As a service in the form of BPJS Health it can be ascertained the number of patients increases. Activities at this community health centers can generate and collect a lot of medical record data every day. Heaps of medical record data are used for operational needs. Daily medical record data is always increasing. It can be explored to be used as a source of historical data to find a new pattern and knowledge for the community health centers, community and related agencies. This way can be called as a data mining technique [1-3]. Data mining is a term used to describe the discovery of knowledge in a database. Data mining is a process that uses statistical, mathematical, artificial intelligence, and machine learning techniques to interact and identify useful information and related knowledge from large databases [1, 4, 5].

Previous studies have utilized the data mining for many purposes and techniques, such as early prediction of heart diseases [6], to predict liver diseases progress [7], gut microbiota profiles characterization in coronary artery disease patients [8], breast cancer detection [9, 10], and mind performance in Alzheimer's disease [11]. One of techniques, which used for this conducted research, is C4.5 algorithm [12]. C4.5 algorithm is based on decision tree form [12-14].

This research aims to mine valuable information from the historical patient data by using C.45 algorithms. The data will be grouped based on disease characteristics. Therefore, a system that can help the community health centers in determining the number of patients is badly needed.

2. Method

2.1. Dataset

We have collected 150 patient data from Community health centers (Puskesmas) Jetis 1 Bantul district, Indonesia. The data is an attribute owned by the patient, the data in question is the data that has at least

IOP Conf. Series: Materials Science and Engineering **403** (2018) 012084 doi:10.1088/1757-899X/403/1/012084

two columns of attributes. One column as the insert attribute column and another as the target attribute column. From each column there are values to be used for calculation, and the value of each attribute must be discrete. The application will read the input with the target attribute located in the last column of the table. Therefore, from the last column the system will recognize it as the input attribute of the system. Some components of variables were:

1) Age. This variable contains the age of each data held by the patient to be filled in the program input process. The pre-defined values for this program are 15-24 years, 25-44 years, 45-64 years, 65+ years.
2) Sex. This variable contains sex data from the patient. The existing groupings are based on the provisions made by the program: L (male) and P (female).
3) Check year. This variable contains the checking years of each patient. The predefined variables in the app based on groupings are 2015 and 2016.
4) Symptoms. This variable contains patient data based on the symptoms that occur in the patient. Symptom data obtained include atrophy, cough, shortness of breath, nausea and vomiting, fever, headache, chills. Grouping based on the provisions made by the program has 2 values that is yes and no.
5) Type of Illness. This variable is data that serves to determine the outcome of the decision. In the grouping of data has been fixed permanently to avoid errors in the calculation process of the program. Decision data has two values: "Infection" and "Degenerative"

2.2. Research design

The main purpose of design is to provide a design description to be built, as well as to understand the flow of information and processes within the system. Figure 1 determined the stages to be performed in system design. The calculation process is done by C4.5 algorithm method, to get the entropy value and gain value, which will be made a decision tree with node and node.

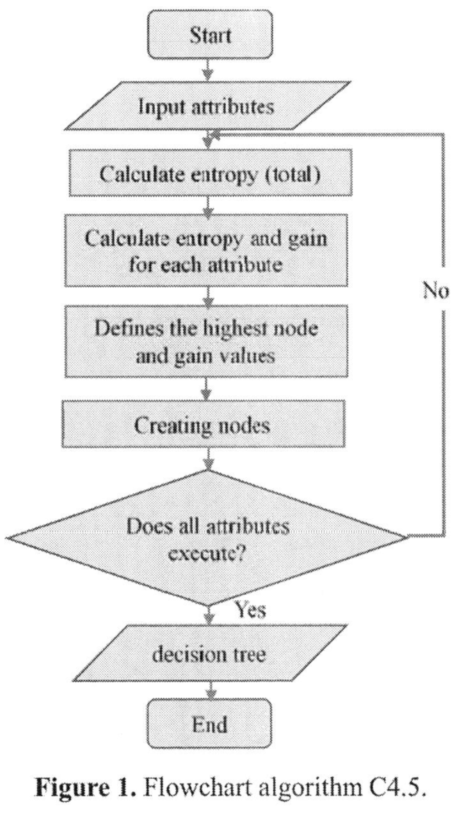

Figure 1. Flowchart algorithm C4.5.

2.3. C4.5 algorithm

In the C4.5 algorithms, decision trees are formed based on the decision-making criteria. The decision tree is a very powerful and well known method of classification and prediction. The decision tree method transforms a very large fact into a decision tree that represents the rule. Rules can be easily understood with Natural language. They can also be expressed in the form of database languages such as Structured Query Language to search for records in certain categories. In general, the C4.5 algorithm constructs a decision trees by selecting attribute as root, creating a branch for each value for the case in the branch, and the process will be repeated for each branch until the case on the branch has the same class [12, 13]. The attribute selection is based on the highest gain value, using equation (1).

$$Gain\,(S,A) = Entropy(S) - \sum_{i=0}^{n} \frac{|S_i|}{|S|} * Entropy(Si) \quad (1)$$

where S are case set, A are attributes, N is a number of attribute partition A, |Si| is the number of cases on the i-th partition, and |S| are the number of cases in S. The value of entropy is calculated before getting a gain value. Entropy is used to determine how informative an attribute is to generate an attribute. The basic formula of entropy is as in equation (2).

$$Entropy(S) = \sum_{i=1}^{n} - pi * log_2\, pi \quad (2)$$

where S are case set, A are features, N is a number of partition S, p_i is a proportion from S_i to S.

3. Results and discussion

To perform data mining process, we start on data preprocessing which includes the steps of data cleaning, data integration, data selection, and data transportation. Table 1 shows the transformation result of age. In the data mining stage we calculate an $Entropy(Total)$ by using equation (2), for example $\left(-\frac{60}{150} * log_2\left(\frac{60}{150}\right)\right) + \left(-\frac{90}{150} * log_2\left(\frac{90}{150}\right)\right)$, then we get an entropy (total) is 0.970950. The gain (total, age) can be calculated by using equation (1), for example $0.970950 - \left(\left(\frac{20}{150} * 0\right) + \left(\frac{80}{150} * 1\right) + \left(\frac{20}{150} * 0\right) + \left(\frac{30}{150} * 0\right)\right)$, then we got $Gain(Total, Age)$ is 0.437617.

The final decision result can be seen from the result of entropy that the result is zero. If we have obtained the result of entropy zero, then next we see the result of the decision variable with the most value.

Table 1. Node 1 of C4.5.

		Number of Case (S)	Infection (S1)	Degenerative (S2)	Entropy	Gain
Total		150	60	90	0.970950	
Age						0.437617
	15-24	20	20	0	0.000000	
	25-44	80	40	40	1.000000	
	45-64	20	0	20	0.000000	
	65+	30	0	30	0.000000	
Atrophy						0.419973
	Yes	60	0	60	0.000000	
	No	90	60	30	0.918295	
Cough						0.278259
	Yes	70	50	20	0.863120	
	No	80	10	70	0.543564	
Nausea and Vomiting						0.219493
	Yes	76	50	26	0.926819	
	No	74	10	64	0.571354	
Fever						0.067272

		Number of Case (S)	Infection (S1)	Degenerative (S2)	Entropy	Gain
	Yes	115	55	60	0.998635	
	No	35	5	30	0.591672	
Headache						0.019973
	Yes	60	30	30	1.000000	
	No	90	30	60	0.918295	

Table 1 shows that the highest gain value is age variable (0.437617). Thus, age can be used as a root node. There are four attributes of age which have decisions, except the category 25-44 age that has not produced yet a decision. Furthermore, we need to process for it age category (table 2).

Table 2. Node 1.1. of C4.5.

		Number of Case (S)	Infection (S1)	Degenerative (S2)	Entropy	Gain
Age						
25-44		80	40	40	1.000000	
Atrophy						0.548795
	Yes	30	0	30	0.000000	
	No	50	40	10	0.721928	
Cough						0.323073
	Yes	45	35	10	0.824587	
	No	35	5	30	0.430827	
Nausea and Vomiting						0.454819
	Yes	56	30	26	0.558268	
	No	24	10	14	0.979869	
Fever						0.311278
	Yes	60	40	20	0.918296	
	No	20	0	20	0.000000	
Headache						0.048795
	Yes	30	20	10	0.918296	
	No	50	20	30	0.970951	

From the table 1 and table 2, we can create a knowledge representation which represented by a decision tree (figure 2). In table 2, four attributes age has each decision, 15-24 is infection and 45-64 is degenerative. Hence, no further calculation is required, but for 15-24 attribute is still needed further process. In Table 2, both Atrophy attributes have degenerative decision, so no further calculation is required. Figure 2 shows the final result of the decision tree which is containing the rules, as follows:

1) When the age is "15-14 years", then the decision is infection.
2) When the age is "45-64 years", then the decision is degenerative.
3) When the age is "65+ years", then the decision is degenerative.
4) When the age is "25-44 years", then the decision is atrophy.
5) When the atrophy is "Yes", then the decision is degenerative.
6) When the atrophy is "No", then the factor is cough.
7) When cough is "Yes", then the decision is infection.
8) When the cough is "No", then the factor is a headache.
9) When the headache is "No", then the decision is degenerative.
10) When the headache is "Yes", then factor is fever.
11) When the fever is "Yes", then the decision is infection.
12) When the fever is "No", then the factor nausea-vomiting.
13) When nausea-vomiting is "Yes", then the decision is degenerative.
14) When nausea-vomiting is "No", then the decision is infection.

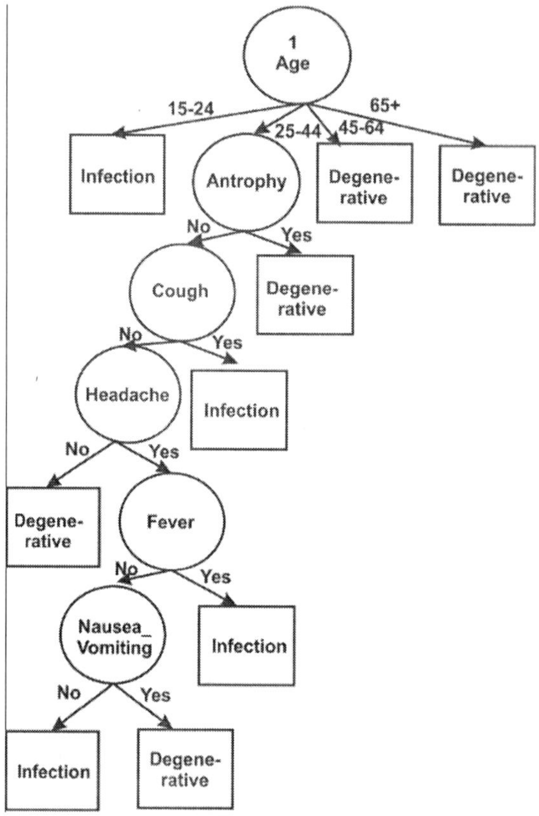

Figure 2. Final result of the decision tree.

4. Conclusion

It can be concluded that Conclusions that data mining can be used to help provide useful information in predicting the number of groupings diagnosis of patient where in this thesis use one of algorithm from data mining that is C4.5 algorithm. The result of accuracy testing with confusion matrix method, test one with amount of training data 85 and data testing 33 produce 78% accuracy with error 22%. The two tests with the amount of training data 115 and the 18 test data yielded 88% accuracy with 12% error.

References

[1] Han J and Kamber M 2006 Data Mining: Concepts and Techniques 2nd ed, The Morgan Kaufmann Series in Data Management Systems (Morgan Kaufmann)

[2] Berkhin P 2006 A survey of clustering data mining techniques *Group. Multidimens. data* vol 25 pp 25-71

[3] Cohen W W and Richman J 2002 Learning to match and cluster large high-dimensional data sets for data integration *Proc. eighth ACM SIGKDD Int. Conf. Knowl. Discov. data Min.* pp 475–80

[4] Hall M, Frank E, Holmes G, Pfahringer B, Reutemann P, Witten I H 2009 The WEKA data mining software: an update *ACM SIGKDD Exploration Newsletter* vol 11 issue 1 pp 10-18

[5] Hand D J 2007 Principles of data mining *Drug Saf.* vol 30 no 7 pp 621–622

[6] Chaurasia V and Pal S 2017 Early prediction of heart diseases using data mining techniques

Carribean Journal of Science and Technology vol 1 pp 208-217

[7] Saad Y, Awad A, Alakel W, Doss W, Awad T and Mabrouk M 2018 Data mining of routine laboratory tests can predict liver disease progression in Egyptian diabetic patients with hepatitis C virus (G4) infection: a cohort study of 71 806 patients *Eur. J. Gastroenterol. Hepatol.* vol 30 no 2 pp 201–206

[8] Emoto T, Yamashita T, Kobayashi T, Sasaki N, Hirota Y, Hayashi T, So A, Kasahara K, Yodoi K, Matsumoto T, Mizoguchi T, Ogawa W and Hirata Kl 2017 Characterization of gut microbiota profiles in coronary artery disease patients using data mining analysis of terminal restriction fragment length polymorphism: gut microbiota could be a diagnostic marker of coronary artery disease *Heart Vessels* vol 32 no 1 pp 39–46

[9] Chaurasia V and Pal S 2017 A novel approach for breast cancer detection using data mining techniques *International Journal of Innovative Research in Computer and Communication Engineering* vol 2 issue 1

[10] Chaurasia V and Pal S 2017 Data mining techniques: To predict and resolve breast cancer survivability *International Journal of Computer Science and Mobile Computing IJCSMC* vol 3 issue 1 pp 10–22

[11] Ramanan S, de Souza L C, Moreau N, Sarazin M, Teixeira A L, Allen Z, Guimaraes H C, Caramelli P, Dubois B, Hornberger M and Bertoux M 2017 Determinants of theory of mind performance in Alzheimer's disease: A data-mining study *Cortex* vol 88 pp 8–18

[12] Li Y, Jiang Z L, Yao L, Wang X, Yiu S M and Huang Z 2017 Outsourced privacy-preserving C4. 5 decision tree algorithm over horizontally and vertically partitioned dataset among multiple parties *Cluster Comput.* pp 1–13

[13] Tsai C-H, Weng S-J, Chou C-A, Wu H-H and Hung W-Z 2017 Applying C4. 5 decision tree to analyze insomnia symptoms *Taiwan Gong Gong Wei Sheng Za Zhi* vol 36 no 5 pp 449–60

[14] Sharma N and Dubey S K 2014 Semantic based Web Prefetching using Decision Tree Induction *2014 5th* Int. *Conf. - Conflu. Next Gener. Inf. Technol. Summit* pp 132–7 Sep. 2014

IOP Conf. Series: Materials Science and Engineering **403** (2018) 012085 doi:10.1088/1757-899X/403/1/012085

Implementation of use case point as software effort estimation in Scrum Framework

H Yuliansyah[1], S N Qudsiah[2], L Zahrotun[3] and I Arfiani[4]
[1,2,3,4] Department of Informatics, Universitas Ahmad Dahlan
Jl. Prof Soepomo, Janturan, Umbulharjo, Yogyakarta, DI. Yogyakarta, Indonesia

[1]herman.yuliansyah@tif.uad.ac.id

Abstract. Estimation is a mechanism to estimate and provide a value for software development projects. Use Case Point (UCP) is a method of software estimation effort in software development based on Use Case model. Scrum is a framework for software development that can solve complex problems and always change. The aims of this study are finding the implementation of UCP as a method of estimating efforts software development with Scrum framework. The methodology of this research is identifying the estimation process with UCP and examine collaboration with Scrum Framework. The results of UCP and Scrum Framework collaborations will be simulated using a web-based project management system. The result of this research is an implementation of UCP and Scrum can be done by associate the User Stories on the Product Backlog with the Use Case on UCP and by modifying User Story attribute to Functionality and adding transaction attribute when defining the User Storie on Product Backlog.

1. Introduction

Estimation is an anticipation of the resource that requires to finishing the software development by observing the environment and the method [1]. With software estimation, we can find software size estimates, software development effort, software development cost and software development schedule for the software project in an environment by using defined methods, tools, and techniques [1]. A good software estimation method will produce optimal resource by utilization and improve the quality of the final products to reach higher level of customer satisfaction [2].

Use Case Point (UCP) is a method to estimate the required effort (Estimation Effort) in completion of software development [3][4]. Modeling software with Use Case is the basis of Estimation Effort used in UCP method [5]. Estimation Effort calculation is multiplication between UCP value and value of Effort Rate (ER) [6]. The value of ER is a prediction value that calculated based on previous project statistics and this is an important value to identify the number of human resources for a project [6]. Several previous studies have defined ER values in UCP. Karner proposed an ER value of 20 Man Hours per UCP [6]. Schneider & Winters has proposed a value of 20, 24, 36 Man Hours Per UCP [7]. Clemmons has proposed a value of 18 Man Hours per UCP [4]. Ochodeck has proposed a value of 4 to 25 Man Hours per UCP [3]. Subriadi has proposed a value of 8.2 Man Hours per UCP [8]. Sholiq has proposed a value of 4.41 Man Hours per UCP [9].

Scrum is a framework with a very simple process to manage complex projects [10]. Besides Extreme Programming (XP) and Kanban, Scrum is also included in Agile Software Development (ASD) [11]. Some methods of effort estimation on Scrum framework include Expert Judgment, Planning Poker [12][13][14], Story Point [15][16], Bucket System [17], T-shirt Size [17], Relative Mass Valuation [17], and Dot Voting [17].

Planning poker introduced by James Grenning, described further by Cohn, is a game that team members can play to estimate the tasks that exist in software development [13][18][19]. Planning poker uses cards to estimate stories with values of 1, 2, 3, 5, 7, 10. This value represents the number of days required to be ideal for programming. The poker planning estimation results from experts will be more accurate with the actual effort required and will be more accurate than the combination of statistics and individual estimates [14].

Story Point is an effort estimation based on the size of a task to represent consensus between team and task effort [16][20]. A point is assigned to each Story Point. This point is relative. This is because a two-pointed story will assume to take twice the effort of a story. Story Points can be assigned based on the effort involved, complexity and inherent risks of developing features [15].

Bucket system is a way to estimate the number of tasks in a group of people in small to medium quantities and done quickly. T-shirt size is an estimation method that uses non-numeric estimation technique/not with numbers, this is done to prevent confusion in interpreting the number of hours is required for an attempt. T-shirt size will estimate a task with the title extra-small, small, medium, large, extra-large, or double extra-large. Relative mass valuation is a quick way to get through the task and estimate everything because it is related to each other. To use this approach, first, write a card for each task. Then, you have to create a large table to move tasks each other. Dot Voting is a technique that allows participants to choose preferences among a set of items. This is done by placing a dot on an item higher than the other items. Items, with higher priority dots, are higher than items with fewer dots. [17].

A variety of estimation methods have been applied in the context of ASD, although most of these methods do not yield good predictive accuracy values. However, Story Point and Use Case Point is a technique approaching with software requirements specification method such as user stories and use case. This specification is the most commonly used [12]. Based on the description, the story point is a method that has been widely implemented in Scrum Framework minus Use Case Point [20]. Therefore, he novelty in this article is how to implement Use Case Point with Scrum Framework. This is an improvement to utilize the software effort estimation in software development.

2. Method

This research will be done in 3 stages. First stage is to identify the estimation process with UCP. This identification is done by reviewing the related literature in the estimation process with UCP. This identification will know the components required in the estimation process with UCP.

The second stage is to examine the mechanism of implementation of Scrum framework to find UCP and Scrum *integration*. The result of this stage is to find the part that can be an intersection and can be associated with UCP and Scrum. These results will be the system needs to be developed into a simulation application.

The last stage is to develop a simulation application. This simulation application is a web-based project management system. This application will give an overview of the implementation process of UCP as Software Effort Estimation method in Scrum Framework.

3. Results and Discussion

The UCP estimation process begins by measuring system functionality based on the use case model in a count called Unadjusted Use Case Point (UUCP). To calculate UUCP is done by giving the value of each actor on Use Case and give weighting to each Use Case based on the complexity of the function to be implemented from Use Case. The results of the actor rating are simple, average or complex values with reference to Table 1 and each Use Case is the reference to Table 2.

Table 1. Assessment of Weighted Actor [9].

Complexity	Definition	Weight
SIMPLE	If it represents with application programming interface.	1
AVERAGE	If it has interaction through a protocol OR human with a line terminal.	2
COMPLEX	If it interaction through GUI.	3

Table 2. Assessment of Weighted Use Cases [9].

Complexity	Definition	Weight
SIMPLE	If it has 3 or less transactions.	5
AVERAGE	If it has 3 to 7 transactions	10
COMPLEX	If it has more than 7 transactions	15

The next estimation process is to calculate the Technical Complexity Factor (TCF) and the Environmental Factor (EF) that will affect the completion of the software process. EF will help to estimate how efficient the project. The TCF component will contribute to the complexity shown in Table 3 and the EF components that will contribute to the efficiency shown in Table 4.

Table 3. Assessment of TCF - Technical Complexity Factor [9].

F_i	Factors Contributing to Complexity	W_i
F_1	Distributed systems.	2
F_2	Application performance objectives, in either response or throughput.	1
F_3	End user efficiency (on-line).	1
F_4	Complex internal processing.	1
F_5	Reusability, the code must be able to reuse in other applications.	1
F_6	Installation ease.	0.5
F_7	Operational ease, usability.	0.5
F_8	Portability.	2
F_9	Changeability.	1
F_{10}	Concurrency.	1
F_{11}	Special security features.	1
F_{12}	Provide direct access for third parties	1
F_{13}	Special user training facilities	1

Table 4. Assessment of EF - Environmental Factor [9].

F_i	Factors Contributing to Complexity	W_i
F_1	Familiar with Objectory	1.5
F_2	Part time workers	-1
F_3	Analyst capability	0.5
F_4	Application experience	0.5
F_5	Object oriented experience	1
F_6	Motivation	1
F_7	Difficult programming language	-1
F_8	Stable requirements	2

The value of UCP can be calculated by multiplying between UUCP, TCF, and EF values. After that multiply, the number of UCP with Mean Resources needed per UCP (MR) or Effort Rate (ER).

Based on the calculation of the software development estimation process with UCP, it can be seen that from the process UUCP calculation is related to the analysis and system design with UML Use Case. This relation is the value of Weighted actors derived from the level of complexity of the actor type and Weighted use cases obtained from the level of complexity of the number of transactions that may arise from a Use Case. The calculation of UUCP will be focused later on in integration with Scrum.

To find the integration of UCP with Scrum, the first step is to examine the process model or process sequence of the Scrum framework. Figure 1 describes the Scrum framework process model. In Figure 1, there are several activities that occur in Scrum: determining product backlog, determining sprint backlog, implementation (coding) and daily Scrum, sprint review, and sprint retrospective [21].

Product Owner defines the product backlog [22]. The Product Owner arranges and lists all the goals that will be produced in the software. The List of Product Backlog is sorted by priority. Product Backlog is not limited and will be the main reference or source of all the features that must exist in the software to be developed. Product Backlog will be used during Sprint Planning Meeting to determine the tasks that will be done as Sprint Backlog.

The determination of the Sprint Backlog is determined by reference to a predefined Product Backlog. The Developer Team implements the Sprint Planning Meeting to determine the Sprint Backlog to be performed by the Developer Team. In Sprint Planning Meeting also determine the start time and estimated time of each Sprint work by the Developer Team in hours and determine Sprint Goal in this application. Sprint Planning Meeting will generate Sprint Backlog, which will be ready to be completed.

After Sprint Planning Meeting is completed and generates Sprint Backlog and Sprint Goal, the Developer Team will complete the Sprint Backlog that has been obtained from the Sprint Planning Meeting. During the software development process, the Developer Team meets every day for a maximum of 15 minutes to deliver and discuss the performance of Sprint Backlog that has been done and what will be done and the constraints that occur during the completion of the performance. Daily Scrum is done every day and at the same hour. Scrum Master and Developer Team attended daily Scrum.

Sprint Review is done at the end of Sprint Backlog completion. Sprint Review activity is carried out to review work that has been done by the Developer Team within a certain time. The Sprint Review was attended by the Scrum and Stakeholder Teams and was held for a maximum of four hours. The Scrum teams and Stakeholder collaborate to discuss Sprints that have been worked on and completed by the Developer Team. Sprint Retrospective is done to allow the Scrum Team to review itself and make plans for future work.

Based on the process can be concluded that the determination of Product Backlog and Sprint Backlog in Sprint Planning Meeting is a planning phase in Scrum, while the work of Sprint Backlog and Daily Scrum is the implementation phase of software development and Sprint Review is the evaluation phase of software development. So in this analysis found that estimation with Use Case Point (UCP) will be able to be integrated in Scrum framework at Product Backlog and Sprint Backlog determination phase.

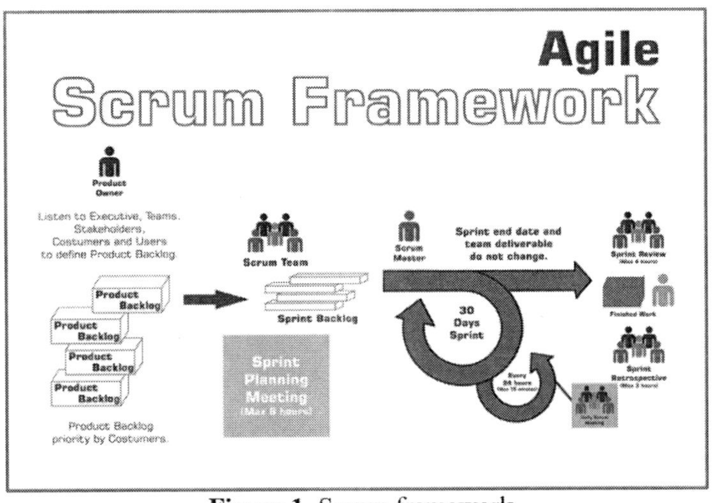

Figure 1. Scrum framework

To clarify the integration process, then I will make a simulation of the integration process. Figure 2 is a simple Use Case example illustrating an application. According to Alan Dennis, Barbara Haley Wixom and David Tegarden, a Use Case is a major part of system functionality and is labelled with a descriptive verb phrase or noun [23].

In Product Backlog determination in Scrum, Product Backlog list can be written by creating User Story. According to Mike Cohn writing User Story can explain a function that will be something of value to the user of the software[18]. Alistair Cockburn in Mike Cohn's book proposes the Use Case writing template among the most commonly used as in Figure 3 [18]. Figure 4 is the equivalent of Writing User Story.

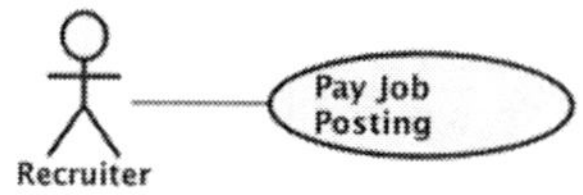

Figure 2. Use case example

Based on Ivar Jacobsen in Mike Cohn's book mentioned that Use Case is a general description of a set of interactions between the system and one or more actors, in which an actor is a user or another system. The Use cases can be written in unstructured text or to adapt to structured templates [18]. Mike Cohn also stated that User Stories is not a Use Case. One of the most obvious differences between User Stories and Use Case is the scope. Both are sized to deliver business value, but Use Stories remain smaller within scope limits. Use Case almost always includes a much larger scope than a User Stories [18].

Use Case Title: Pay for a job posting
Primary Actor: Recruiter
Level: Actor goal
Precondition: The job information has been entered but is not viewable.
Minimal Guarantees: None
Success Guarantees: Job is posted; recruiter's credit card is charged.
Main Success Scenario:

1. Recruiter submits credit card number, date, and authentication information.
2. System validates credit card.
3. System charges credit card full amount.
4. Job posting is made viewable to Job Seekers.
5. Recruiter is given a unique confirmation number.

Extensions:
2a: The card is not of a type accepted by the system:
  2a1: The system notifies the user to use a different card.
2b: The card is expired:
  2b1: The system notifies the user to use a different card.
2c: The card is expired:
  2c1: The system notifies the user to use a different card.
3a: The card has insufficient available credit to post the ad.
  3a1: The system charges as much as it can to the current credit card.
  3a2: The user is told about the problem and asked to enter a second credit card for the remaining charge. The use case continues at Step 2.

Figure 3. Example of writing Use Case with Alistair Cockburn template [18]

"A Recruiter can pay for a job posting with a credit card."

Figure 4. Example case for writing User Story

Based on the statement, in this article proposes a new method that can be used. While it still complies that Use Case is a larger scope of User Stories and still ensures that Use Case can explain system functionality. Use Case can be used as a list of statements in Product Backlog and Sprint Backlog in Scrum framework and able to adapt to Use Case Point especially on Unadjusted Use Case Point (UUCP).

The proposal to be offered is to modify the way of writing Product Backlog and Sprint Backlog by making 3 factors that must be written in this writing. These three factors are Product Backlog and Sprint Backlog must include:

- Actor Name that has been set by actor type based on Weighted actors,
- The Functionality statement that will represent Use Case and will be the goal to be achieved and written on the Product Backlog in the Scrum framework,
- Transaction Statement which will represent a more detailed scope of Functionality. This transaction will also be used as a statement for Sprint Backlog and will be a parameter to calculate the rating of Weighted use cases.

For further details of this proposal will be made a simulation in the form of web applications. This application aims to provide a comprehensive overview of the effect of implementation of the proposed offer.

3.1 Web-based application project management system

The implementation of web-based project management system simulation proposed are:

- Figure 5 is the dashboard page used to view existing projects and to create new projects by selecting the "Project" menu,
- Figure 6 is the page to create a new project. There are several fields in the form of project name, filler position, working hours of the day, project start date and description,
- Figure 7 is the Product Backlog stuffing page. This field is a proposal in this research to integrate UCP and Scrum framework,
- Figure 8 is an entry page for the Technical Complexity Factor. This page related to Table 3,
- Figure 9 is the contents page for Environmental Factor. This page related to Table 4,
- Figure 10 is the calculated page of UUCP, TCF and EF. On this page, user must be filled is the value of the Effort Rate (ER) which is used as a reference in determining the estimated value of UCP,
- Figure 11 is the result of UCP calculation. This value is the final value of the estimated estimate,
- Figure 12 is an estimation graph depicted in the form of a Burndown Chart. This graph is obtained from the calculation of estimation by UCP which has been converted to day unit based on UCP value and project hours.

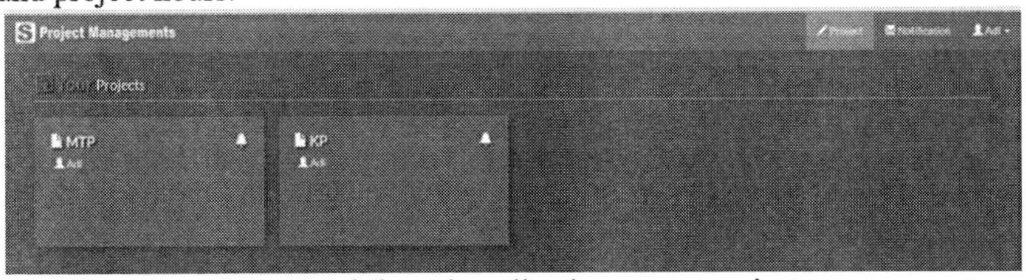

Figure 5. Dashboard page web-based application scrum project management system

Figure 6. The new project form

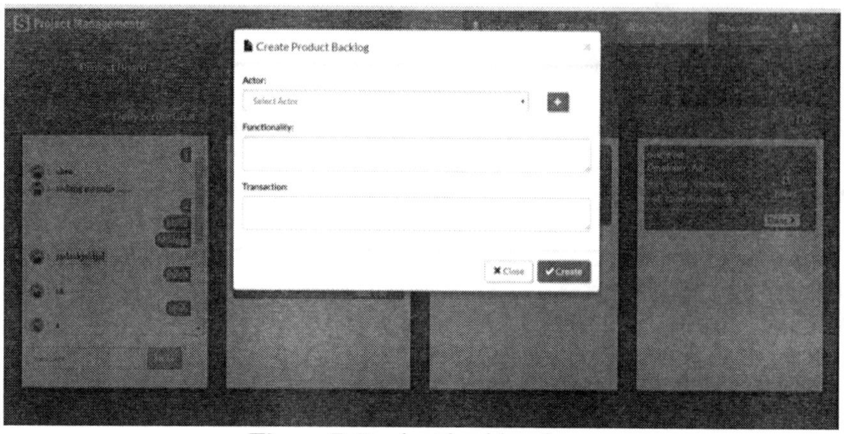

Figure 7. Product backlog form

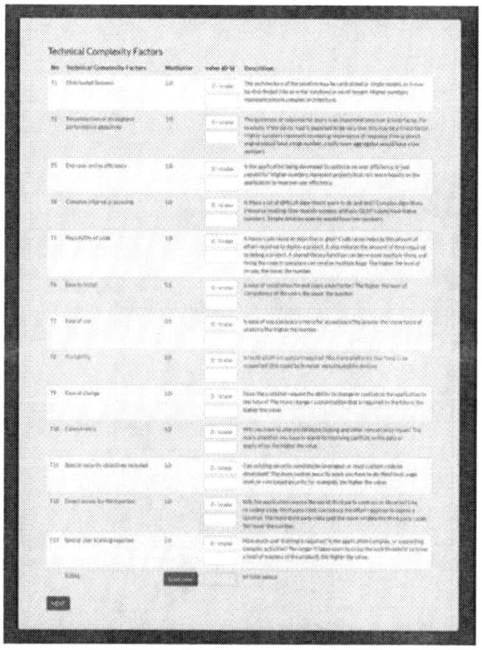

Figure 8. Technical complexity factor form

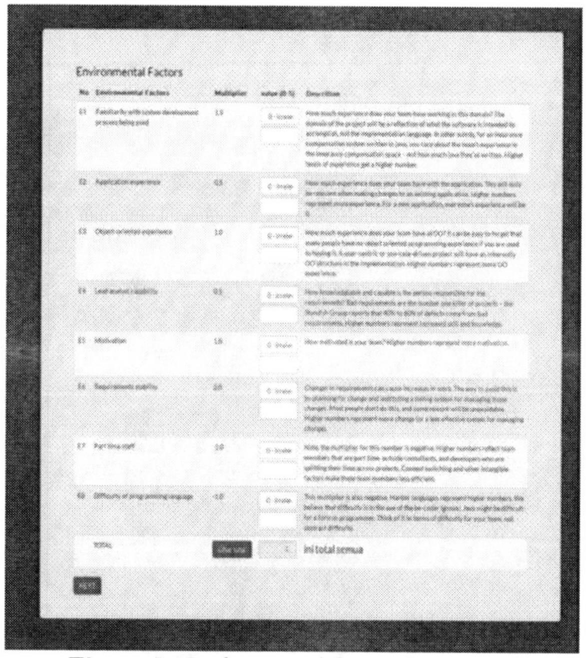

Figure 9. Enviromental factor form

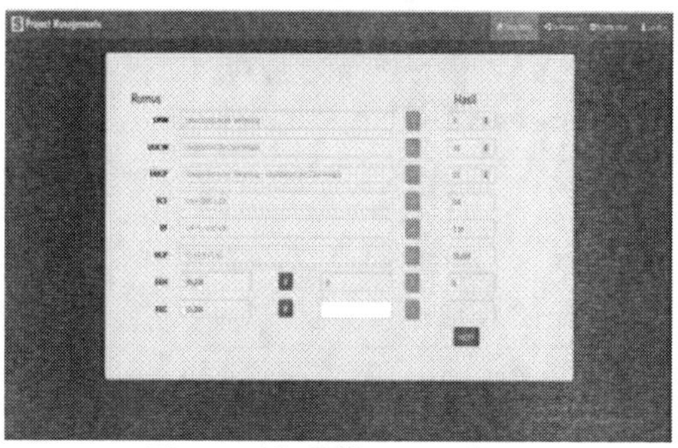

Figure 10. Effort rate form

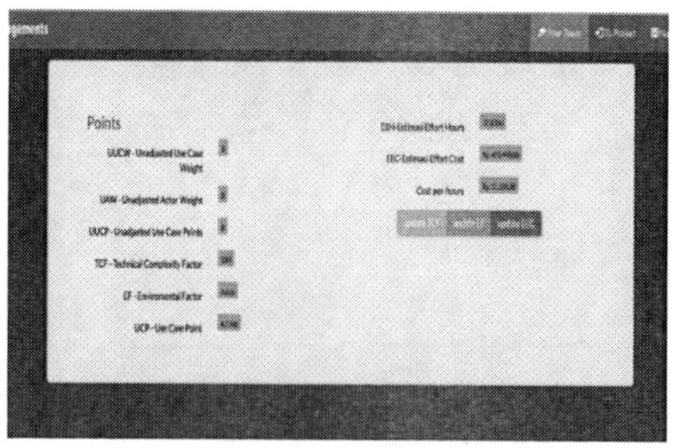

Figure 11. UCP calculation result

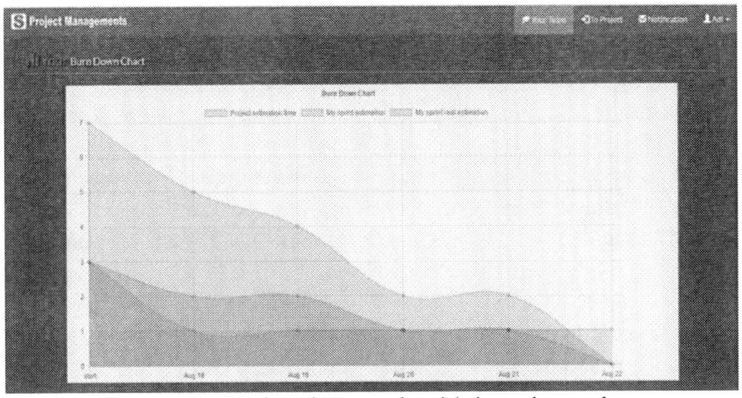

Figure 12. Estimation graph with burndown chart

4. Conclusion

This research proposes the integration of UCP and Scrum framework by associating User Stories on Product Backlog with Use Case on UCP by modifying the User Story attribute into Functionality and adding transaction attribute when defining Product Backlog. Simulations of web-based applications that appear capable of manipulating Product Backlog statements that have been filled to be the components required in calculating UUCP is the assessment of Weighted Use Cases and assessment of Weighted actors.

For future works, this proposal needs to be tested by an expert who has a field of competence in the Scrum framework. This needs to be done because to test that these proposed things do not conflict with the implementation of Scrum framework. The estimates undertaken in this article are global estimates, which means estimates of overall software development work. A further study of the utilization of these estimates can be used to determine the estimates needed to accomplish the tasks in the Sprint Backlog statement. From these results, it is necessary to review the values of the Effort Rate (ER) that can be established and related to the topics of the field of software work to be built.

5. Acknowledgment

Universitas Ahmad Dahlan has supported this research in the scheme Fundamental Research with grants number No: PF-081/SP3/LPP-UAD/IV/2017.

6. References

[1] M. K. Chemuturi, *Software Estimation Best Practices, Tools & Techniques*. USA: J. Ross Publishing, 2009.

[2] M. M. Kirmani and A. Wahid, "Use Case Point Method of Software Effort Estimation : A Review," *Int. J. Comput. Appl.*, vol. 116, no. 15, pp. 43–48, 2015.

[3] M. Ochodek, J. Nawrocki, and K. Kwarciak, "Simplifying effort estimation based on Use Case Points," *Inf. Softw. Technol.*, vol. 53, no. 3, pp. 200–213, 2011.

[4] R. K. Clemmons, "Project Estimation With Use Case Points," *J. Def. Softw. Eng.*, pp. 18–22, 2006.

[5] P. P. Jena, P. Jha, P. P. Jena, and R. K. Malu, "Estimating Software Development Effort using UML Use Case Point (UCP) Method with a Modified set of Environmental Factors Estimating Software Development Effort using UML Use Case Point (UCP) Method with a Modified set of Environmental Factors," *Int. J. Comput. Sci. Inf. Technol.*, vol. 5, no. 3, pp. 2742–2744, 2014.

[6] G. Karner, "Resource estimation for objectory projects," *Object. Syst. SF AB*, pp. 1–9, 1993.

[7] G. Schneider and J. P. Winters, *Applying use cases: a practical guide*. Pearson Education, 1998.

[8] A. P. Subriadi, Sholiq, and P. A. Ningrum, "Critical review of the effort rate value in use case point method for estimating software development effort," *J. Theroretical Appl. Inf. Technol.*,

vol. 59, no. 3, pp. 735–744, 2014.

[9] Sholiq, T. Sutanto, A. P. Widodo, and W. Kurniawan, "Effort Rate on Use Case Point Method for Effort Estimation of Website Development," *J. Theroretical Appl. Inf. Technol.*, vol. 63, no. 1, pp. 209–218, 2014.

[10] K. Schwaber and M. Beedle, *Agile software development with Scrum*, vol. 1. Prentice Hall Upper Saddle River, 2002.

[11] R. Hoda, N. Salleh, J. Grundy, and H. M. Tee, "Systematic literature reviews in agile software development: A tertiary study," *Inf. Softw. Technol.*, vol. 85, pp. 1339–1351, 2017.

[12] M. Usman, E. Mendes, F. Weidt, and R. Britto, "Effort Estimation in Agile Software Development: A Systematic Literature Review," in *Proceedings of the 10th International Conference on Predictive Models in Software Engineering*, 2014, pp. 82–91.

[13] J. Grenning, "Planning poker or how to avoid analysis paralysis while release planning," *Hawthorn Woods Renaiss. Softw. Consult.*, pp. 1–3, 2002.

[14] V. Mahnič and T. Hovelja, "On using planning poker for estimating user stories," *J. Syst. Softw.*, vol. 85, no. 9, pp. 2086–2095, 2012.

[15] E. Coelho and A. Basu, "Effort Estimation in Agile Software Development using Story Points," *Int. J. Appl. Inf. Syst.*, vol. 3, no. 7, pp. 7–10, 2012.

[16] S. Porru, A. Murgia, S. Demeyer, M. Marchesi, and R. Tonelli, "Estimating Story Points from Issue Reports," *Proc. 12th Int. Conf. Predict. Model. Data Anal. Softw. Eng. - PROMISE 2016*, pp. 1–10, 2016.

[17] V. Singh, "Agile Estimation Techniques," 2016. [Online]. Available: https://www.scrumalliance.org/community/articles/2016/january/agile-estimation-techniques. [Accessed: 27-Aug-2017].

[18] M. Cohn, *User stories applied: For agile software development*. Addison-Wesley Professional, 2004.

[19] K. Moløkken-Østvold, N. C. Haugen, and H. C. Benestad, "Using planning poker for combining expert estimates in software projects," *J. Syst. Softw.*, vol. 81, no. 12, pp. 2106–2117, 2008.

[20] E. Ungan, N. Cizmeli, and O. Demirors, "Comparison of functional size based estimation and story points, based on effort estimation effectiveness in SCRUM projects," *Proc. - 40th Euromicro Conf. Ser. Softw. Eng. Adv. Appl. SEAA 2014*, pp. 77–80, 2014.

[21] P. Rola, D. Kuchta, and D. Kopczyk, "Conceptual model of working space for Agile (Scrum) project team," *J. Syst. Softw.*, vol. 118, pp. 49–63, 2016.

[22] H. S. Sverrisdottir, H. T. Ingason, and H. I. Jonasson, "The Role of the Product Owner in Scrum-comparison between Theory and Practices," *Procedia - Soc. Behav. Sci.*, vol. 119, pp. 257–267, 2014.

[23] A. Dennis, B. H. Wixom, and D. Tegarden, *Systems analysis and design: An object-oriented approach with UML*. John Wiley & Sons, 2015.

Fracture toughness characterization of polymers-based composites using essential work of fracture method

Purnomo[1], P H Setyarini[2] and D Cahyandari[1]

[1]Mechanical Engineering Department, Universitas Muhammadiyah Semarang, Indonesia
[2]Mechanical Engineering Department, Brawijaya University, Malang, Indonesia

purnomo@unimus.ac.id

Abstract. Fracture toughness of polymer-based composites in plane stress condition has been successfully evaluated using latest method namely essential work of fracture (EWF). This method has been extensively performed in order to assess the performance of two parameters in fracture, i.e. essential work of fracture (w_e) and non-essential work (βw_p) ones. The use of EWF method is widespread due to the preparation of uncomplicated specimens, easy examination and easy data processing procedures but accurate in data analysis. This review is focused on the EWF application for the characterization of fracture toughness of polymer-based composites in which the fracture occurs in plane stress state. The EWF testing procedure is in accordance with the recommendations of the ESIS protocol TC-4. Literature reviewed includes all the major areas of testing and simplification of associated data for both quasi-static and dynamic condition. The latter is an EWF application classified by the type of load applied.

1. Introduction

A polymer is made of large molecules composed of many repeated subunits that are connected by covalent bonds. Polymer have many advantages such as light weight, good corrosion resistance and low cost, and have become the important commercial material such as construction, automotive, biomedical, household appliances, etc. An important parameter in evaluating the materials is fracture toughness. This parameter is very important to know especially for materials that are subjected to impact load in its application. In the case of biomaterials for load bearing implant application such as skull bone implant, it is important to examine the various fracture parameters with the consideration that they protect a critical part of human brain.

Since the wide use of polymers, their mechanical properties such as fractures are very importance. Fracture behavior, i.e. brittle or ductile, is categorized by the amount of deformation that occurs during the fracture process. Brittle fractures are characterized by very limited or no plastic deformation involved. As for the ductile fracture, the polymer undergoes plastic deformation prior to and during the fracture occurred. The fracture toughness behavior of materials can be evaluated using fracture mechanics methods [1-4]. In order to analyze the material fracture, a linear elastic fracture mechanics (LEFM) approach widely used to assess the fracture mechanics of material under plane-strain condition that mostly occurs in extremely thick specimens and satisfy small-scale yield conditions [5, 6]. The LEFM is particularly suitable to be applied to brittle materials. In ductile materials, fracture toughness is evaluated through the crack tip opening displacement (CTOD) parameter, also can be evaluated using the J-Integral method pioneered by [7] in which the plane-strain fracture toughness is obtained at relatively low specimen thickness compared to the LEFM approach [6]. In a plastic deformed material prior to failure, the CTOD testing can be accurately applied to characterize fracture toughness.

Recently, the fracture toughness behavior in ductile materials including thermoplastics and polymer matrix composites widely be assessed using the EWF approach that divided the fracture energy in two different parts, the w_e and βw_p [8-10]. The use of this method has evolved to characterize the fracture toughness in ductile materials. The EWF test is performed using a double edge-notched tension (DENT) subjected to tensile load in quasi-static conditions as well as single-edge notched bend in impact load (dynamic conditions) [8, 11, 12]. In quasi static state, the test is performed in accordance with the standards set forth in the ESIS protocol which will be explained later in this paper. In the case of fracture due to impact load causing failure, the test is performed according to Charpy impact tester.

2. The EWF method

2.1. Theory

The concept developed in the EWF method is an alternative solution to respond the weaknesses in the previous method in term of the assessment of the fracture parameters of the ductile materials. This method is able to provide clarity of energy partition during the fracture process of materials. Many researchers [13, 14] have developed the EWF method as a testing standard for evaluating fracture behavior in ductile materials including polymers and composites. In the EWF testing, the DENT specimens as shown in figure 1(a) are used as a standard specimen shape. The EWF concept was first introduced by [15] and is based on a study conducted by [16], i.e. the stability of crack growth occurs as a result of autonomous input energy during the fracture process. Consequently, when a DENT specimen is subjected to a tensile load up to total fracture, the total fracture energy (W_f) is consumed by material in two distinct zones (see figure 1(a)). Firstly, the fracture process zone where the fracture takes place and energy is used to create a new crack surface during crack propagation occurs. Second, the plastic deformation zone that is around the ligament length. The energy consumed in both first and second zones is the essential (W_e) and non-essential (W_p) work of fracture, respectively. Consequently, the W_f can be divided into two distinct parts namely W_e and W_p. In mathematical form, the W_f is the sum of both energy and can be written as equation (1) [17, 18].

$$W_{\mathrm{f}} = W_{\mathrm{e}} + W_{\mathrm{p}} \tag{1}$$

According to the EWF method, the crack resistance parameter of the material represents the fracture energy required to form a new cracked surface. Therefore, W_e value are strongly associated with the ligaments length (ℓ) for fixed material thickness, so essential energy also called the surface energy. The plastic work W_f is corresponding to volume of outer zone in which proportional to the quadratic of the ligament length, therefore the non-essential work is the volume energy.

From figure 1, (a) is the schematic illustration diagram for DENT shape showing the location of fracture process zone (FPZ) and outer process zone (OPZ), (b) typical of load-displacement curve and (c) data reduction of EWF. Based on the fracture zone separation as explained, the two zones are assumed to be inside the ligament of the specimen as illustrated in figure 1(a), in specific term, equation (1) can be written into equation (2).

$$W_{\mathrm{f}} = w_{\mathrm{e}}.t.\ell + \beta.w_{\mathrm{p}}.t.\ell^2 \tag{2}$$

By dividing the parameters in equation (2) by $\ell.t$, the following specific fracture parameters are obtained as shown in equation (3).

$$w_f = w_{\mathrm{e}} + \beta.w_{\mathrm{p}}.\ell \tag{3}$$

The shape factor denoted by β is a factor that characterizes the dimension of the plastic zone and depends on the OPDZ shape on the length of ligaments on the DENT specimen with a thickness t. By using figure 1b, the total work of fracture is equal to the area under load-displacement curve obtained by experimentally from the DENT shape under tensile loading in quasi static condition. In according to the equation (3), the w_f is a linear function of ligament length in plane stress state. By plotting the w_f value against the ligament length (see figure 1(c)), the specific essential work of fracture (w_e) value is the intercept at zero ligament length ($\ell = 0$) and the specific non-essential work of fracture (βw_p) is the slope of the line formed by the linear equation of w_f (see figure 1(c)). From figure 1(c), it can be explained that when the ℓ value is lower than (3-5)t, the materials undergoes a plane strain fracture or in a mixed-mode fracture state. The same thing will also be occurred when the ℓ value is less than or equal to (W/3 or $2r_p$).

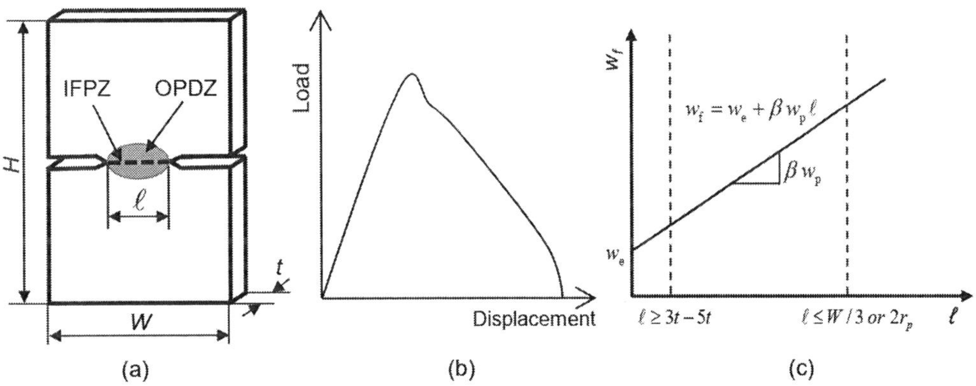

Figure 1. Schematic illustration diagram for (a) DENT shape showing the location of fracture process zone (FPZ) and outer process zone (OPZ), (b) typical of load-displacement curve and (c) data reduction of EWF.

2.2. *Experimental Consideration*

The requirements that must be met prior to applying the EWF approach to evaluate the plane stress fracture toughness as recommended by European Structural Integrity Society (ESIS) – TC 4 protocol [18] are as follows:

a. Under tensile load, the fully ligament yielded should be reached before the crack is initiated
b. Equation (3) can be applied if the notched specimen (DENT) is in a plane stress fracture condition. Furthermore, the fracture condition be verified according to the Hill's criterion [19].
c. The load-displacement of all specimens tested having different ligament length should clearly exhibit a self-similarity for each ligament length. In this term, the data is obtained from experimental series in DENT shape.

2.2.1. Specimen geometry and testing condition. In materials with ductile fracture, the fracture process zone undergoes a necking process which then breaks into a fracture surface. According to ESIS – TC 4 protocol and many researchers [16, 20], it has suggested that the specimen geometry must meet the specific conditions required so that the fracture occurs in plane stress manner. The required conditions involve the specimen thickness (t), the width of DENT specimen (W), and the plastic zone size (r_p). The relationship between parameters of those requirements can be expressed in equation (4).

$$(3-5)t \leq \ell \leq \min\left(\frac{W}{3} \text{ or } 2r_p\right) \tag{4}$$

The parameter r_p (see equation (4)) is related to the essential work of fracture (w_e) as well as the tensile properties of the materials test including the Young's modulus (E), and the tensile yield stress (σ_y). The r_p can be estimated using equation (5).

$$2r_p = \frac{1}{\pi}\frac{E w_e}{\sigma_y^2} \tag{5}$$

Note that if the left side of equation (5), i.e. $2r_p$ is equal to the ligament length means the ligament of DENT specimen tested is completely full-yielding before the fracture is initiated and subsequently undergoes propagation. To ensure the plane stress fracture was occurred in the test, the ligament length should be met to the value for left hand-side of equation (4). If ligament length is smaller than the lower bound, the material will fall in plane-stress to plane-strain fracture transition. Schematic illustration of the requirement is presented in figure 1(c). Nevertheless, the threshold value is only discussed in a little literature so that the boundary value is not supported by strong theoretical studies. The ESIS protocol was revised [20, 21] with emphasis that Hill's criterion could be improve by the similar stress condition, verified by the highest tensile stresses of all specimens tested. According to this revised protocol, the validity of ℓ under plane stress fracture conditions are checked using formula $0.9 < \sigma_{max}/\sigma_m < 1.1$, in which the σ_{max} is the net section stress while σ_m is the mean value of σ_{max}. This means that if $\sigma_{max} < 0.9\ \sigma_m$ or $\sigma_{max} > 1.1\ \sigma_m$ then the data is declined because it does not match from the terms required in the EWF.

2.2.2. Energy partitioning. Total energy consumed throughout the fracture occurs has been studied and formulated in various expression by many researchers. An approach proposed Mai and co-workers [13] partitioned the total work (W_f) into two distinct parts: (i) work or energy for initiate fracture (W_i), and (ii) the fracture work for crack growth including crack propagation and necking (W_g), as schematically depicted in figure 2(a). The w_e parameter is independent to ℓ while the βw_p one is corresponding to ℓ. A similar approach was also introduced by [12, 22] that divided the W_f into the energy dissipated for yielding (W_y) and the energy for necking and tearing (W_n), as schematically depicted in figure 2(b). Consequently, the energy for yielding is also divided into two parts, the essential and the non-essential energy. This logic applies also to the energy used for the necking and tearing process. The load-displacement curve is partitioned only to explain how stored elastic work is considered. The elastic work is estimated based on the concept of energy initiation that the energy released in the subsequent fracture process as a result of elastic work contributed significantly to the overall energy consumed. Based on energy-partitioned work of fracture methods, it can be written as equation (6) and (7).

$$W_f = W_y + W_n \tag{6}$$

$$w_f = (w_{e,y} + \beta_y . w_{p.y} .\ell) + (w_{e.n} + \beta_n w_{p,n} . \ell) \tag{7}$$

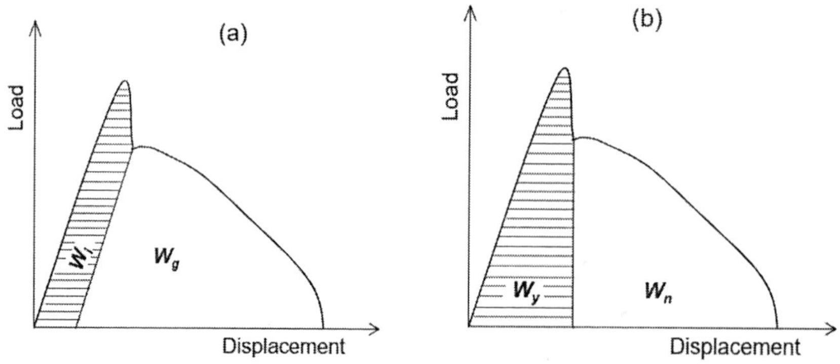

Figure 2. Schematic diagram showing the energy partitioning for initiate fracture (W_i) and growth (W_g) (a), and energy for yielding (W_y) and necking subsequent tearing (W_n) (b).

From equation (6), the specific essential is sum of the essential energy of the composites at the initiation/yielding stage and the necking one. While the non-essential energy is also the sum of both energy components at the yielding stage and necking as in the essential energy. So, it can be expressed in equation (8) and (9).

$$w_e = w_{e,y} + w_{e.n} \tag{8}$$

$$\beta w_{p\cdot} = \beta_y w_{p,y} + \beta_n w_{p,n} \tag{9}$$

in which the $w_{e,y}$ is the specific essential work for yielding, and the $w_{e,n}$ is the specific essential fracture work for necking and tearing. The parameter $w_{p,y}$ and $w_{p,n}$ are specific work for plastic deformation during yielding and necking subsequent tearing, respectively. Many ductile materials exhibit delayed yielding conditions, therefore the yielding part can be divided into two distinct parameters, i.e., crack tip blunting and ligament yielding [23].

2.2.3. EWF Under Impact Loading. The behavior of deformation and impact fracture toughness are very important to understand in order to gain a deep knowledge on fracture mechanics of ductile materials, especially polymer-based composites. Deformation and composites fracture behavior in this condition can behave somewhat brittle at high strain rates. It should be considered that the experimental fracture toughness in ductile material is not suitable as a basis for the assessment of fracture strength parameters because the specimens tested are often not fully fractured. The use of EWF method has been developed by [24]. They have been developed the application of EWF concept for characterized the fracture of ductile polymer under impact loading by using single edge notched bend (SENB) specimen. Furthermore, Martinatti and Ricco [25] confirmed the validity of the use of this method to evaluate the fracture toughness of polypropylene-based materials. The development for impact loading application, equation (3) is then modified with another terminology [26] into equation (10) as follows:

$$\frac{U}{A} = \mathrm{u}_0 + \mathrm{u}_{\mathrm{d}}.\ell \tag{10}$$

in which U/A is total impact fracture energy per unit area where the fracture occurred, u_o is the limiting specific fracture work, and u_d is the dissipative energy density. In relation to equation (10), Kudva [27] considers that equation (10) is precisely inappropriate, that equation (3) should be more appropriate for dynamic condition (impact loading) or high speed rate. For the validity of the used the EWF method, Fasce *et al.* [28] clarified its use for impact load or other load type mentioned. Finally, they concluded that this method is well suited to characterize the fracture toughness and behavior in polymeric composites and other ductile materials.

2.2.4. EWF Application. At present, the fracture mechanics of various thermoplastic matrix composites and other ductile materials have been effectively assessed using the EWF concept rather than the J-Integral and CTOD. It can be understandable because it provides many advantages over the others. The selected references exploring the application of EWF method for both static and dynamic loading in different polymer matrix composites can be elucidate below. The EWF method has been widely applied to characterize the fracture toughness of various polymer-based materials including polyethylene, polypropylene and polyamide based materials. Among of them, the fracture toughness of polyethylene-based materials most widely investigated using that method. Costa *et al.* [29] have successfully evaluated the fracture behavior of nanocomposites composed of polyethylene incorporated with Mg/Al LDH. In a quasi-static state, the micro composite of HDPE combined with natural zeolite has been effectively assessed by [30, 31] with using the partitioning energy based on the EWF concept. Similar to that done by them, many other researchers [32-35] also conduct the assessment of both the HDPE and MDPE. For the materials in high impact rate test, the EWF concept has been extensively used for assessing the performance of mechanical fractures of the ductile materials, especially polymer-based materials. The researchers [36-39] used a single edge notched bend (SENB) shape to evaluate the impact fracture toughness and behavior of ductile materials. In the case of impact fracture, they have been reported that the EWF method is suitable and should be further developed in assessing the robustness and mechanics of materials fracture widely.

3. Conclusion

The EWF method has grown and is well suited for evaluating the polymer-based composites fracture toughness behavior. The EWF testing parameters can be used as standard testing for fracture toughness of the polymers matrix composites. In fact, EWF techniques show a deeper interpretation of the fracture behavior. The superiority of the EWF method compared with the others, especially the J-Integral and CTOD, is that the fracture energy can clearly be separated into the essential and non-essential parts of the energy during the fracture process. The EWF method has been shown to be effective for use primarily in thermoplastic polymer, allowing the study of the effect of various variables on fracture mechanics that are not available in the other methods. In the last decade, this method continuous to be developed by many researchers and they strongly recommend for the EWF method for assessing the fracture mechanics especially the fracture toughness behavior of the ductile materials widely. This recommendation is reasonable because there are so many advantages offered by that method in solving various problem in mechanical analysis and fracture of ductile materials.

References

[1] Reis J M L and Jurumenha M A G 2011 *Mat Res* **14** 326-30
[2] Crouch B A and Huang D D 1994 *J Mater Sci.* **29** 861-4
[3] Seidler S and Grellmann 1993 *J Mater Sci.* **28** 4078-84
[4] Zhou Z, Landes J D and Huang D D 1994 *Polym Eng Sci.* **34** 128-34
[5] Ritchie R O, Koester K J, Lonova S, Yao W, Lane N E and Ager J W *Bone* **43** 798 - 812
[6] Ural A, Krishnan V R, Papoulia K D *Int J Solid Struct* 2009 **46** 2453 - 62
[7] Rice J A 1968 *J Appl Mech* **35** 279-86
[8] Mazidi M M, Aghjeh M K and Abbasi 2012 J Mater Sci **47** 6375 - 86
[9] Yilmaz S, Kodal M, Yilmaz T and Ozkoc G 2013 *Compos Part B Eng* **56** 527 - 35
[10] Karger-Kocsis J and Moskala E J 2002, SPE-ANTEC paper, **60** 1751-55
[11] Mohsenzadeh M A, Mazinani M and Zebarjad M M 2015 *nanocomposite* **1** 27 - 36
[12] Karger-Kocsis J, Czigany T and Moskala E J 1998 *Polymer* **39** 3939-44
[13] Mai Y-W and Cotterell B 1986 *Int. J. Fract.* **32** 105-25
[14] Hashemi S 1993 *Plast Rub Compos Process Appl* **20** 229-37
[15] Cotterel B and Reddel J K 1977 *Int. J. Fract.* **13** 267-77
[16] Broberg K B 1975 *J. Mech. Phys. Solids* **23** 215-37
[17] Yang J L, Zhang Z and Zhang H 2005 *Compos Sci Technol* **65** 2374-79
[18] Gray A 1993 Testing protocol for essential work of fracture, ESIS-TC4
[19] Hill R 1952 *J. Mech. Phys. Solids* **1** 19-30
[20] Clutton E 1997 Testing protocol for essential work of fracture, editor, ESIS - TC4
[21] Williams J G and Rink M 2007 *Eng Fract Mech* **74** 1009-17
[22] Karger-Kocsis J, Czigany T and Moskala E J 1997 *Polymer* **38** 4587
[23] Karger-Kocsis J and Barany T 2002 *Polym Eng Sci* **42** 1410-19
[24] Wu J and Mai Y-W 1993 *J Mater Sci* **28** 3373-84
[25] Martinatti F and Ricco T 1995 ESIS **19** 83
[26] Yoo Y J, Shah R K and Paul D R 2007 *Polymer* **48** 4867 – 73
[27] Kudva R A, Keskula H and Paul D R 2000 *Polymer* 335
[28] Fasce L, Bernal C, Frontini P and Mai Y W 2001 *Polym Eng Sci* **41** 1-14
[29] Costa F R, Satapathy and Wagenknecht 2006 *Eur Polym J* **42** 2140-52
[30] Purnomo, Soenoko R, Irawan Y S and Suprapto A 2017 *Journal of Engineering Science and*

Technology **12** 1191-2203
[31] Purnomo, Soenoko R, Irawan Y S and Suprapto A 2015 *International Journal of Applied Engineering Research* **10** 28001-12
[32] Kwon H J and Jar PYB 2007 *Eng Fract Mech* **74** 2471-80
[33] Elmegueni M, Nait-Abdelaziz M, Zairi F and Gloaguen JM 2013 *Int J Fract* **83**(2) 119-33
[34] Peres F M, Tarpani J R and Scon C G 2013 *Eng Fract Mech* **105** 136-51
[35] Wetherhold R C and Mouzakis D E 1992 *J Eng Mater Trans ASME* **121** 483-7
[36] Purnomo, Soenoko R, Irawan Y S and Suprapto A 2014 *International Journal of Applied Engineering Research* **9** 28737-47
[37] Purnomo and Subri M 2017 *International Review of Mechanical Engineering* **11** 87-93
[38] Charief , Elmeguenni M and Benguediab 2017 *J. Appl. Mech. Phys* **58**(2) 335-41
[39] Purnomo, Soenoko R, Suprapto and Irawan Y S 2016 *FME Transactions* **44** 180-7

Refrigeration effect and energy efficiency ratio (EER) calculation of 1/2 cycle refrigeration system on LPG-fueled vehicles

M Setiyo[1], B C Purnomo[1], B Waluyo[1], D R B Syaka[2] and N Hamidi[3]

[1]Department of Automotive Engineering, Universitas Muhammadiyah Magelang, Magelang, Indonesia.
[2]Department of Mechanical Engineering, State University of Jakarta, Jakarta, Indonesia.
[3]Department of Mechanical Engineering, Brawijaya University, Malang, Indonesia.

setiyo.muji@ummgl.ac.id

Abstract. This paper presents a lab-scale experiment on the prototype of the ½ cycle refrigeration system of an LPG-fueled vehicle. LPG before being sent to the engine is used as a heat absorbing medium. The experiment was carried out at 0.15 MPa evaporation pressure and 1-6 g/s of LPG flow rate which representing of Light Duty Vehicle (LDV) fuel consumption. The results of this study indicate that ½ cycle refrigeration system capable of producing refrigeration effect of more than 1.2 kW with EER of 2.72.

1. Introduction

In recent decades, Liquefied Petroleum Gas (LPG) has become a substitute alternative to gasoline that is widely for SI engines, in addition to alcohol and Compressed Natural Gas (CNG). In fact, the utilization of LPG as a vehicle fuel has been more than 100 years [1]. Although 60% of the world's LPG is produced from oil refineries [2, 3], the trend of LPG usage for transportation sector shows a significant increase. In 2015, based on World LPG Association's annual report, global LPG consumption reached 26.4 million tons with an increase of 24% when compared to 2009. Until now, the number of LPG vehicles operating in the world is not known with certainty, but WLPGA reported more than 27 million [4].

LPG has an energy content (High Heating Value, HHV) which is slightly higher than gasoline, which is 46.23 MJ/kg (depending on percentage propane to butane). LPG has a relatively high energy content per unit mass, although per unit volume is low [5]. As a result, LPG tanks have larger volumes than gas tanks to accommodate the same energy. LPG car trials show about 15% reduction in greenhouse gas emissions (per unit distance), fewer NOx emissions, and almost zero particulate emissions when compared to gasoline [6]. However, due to its physical properties, LPG engine performance is generally below the performance of gasoline engines [7-10].

With the latest generation of converter kits, where LPG is inserted into the intake manifold or even directly into the combustion chamber in the liquid phase, and with advanced injection settings, LPG engine performance is equivalent to gasoline [4]. In LPG engines with first-generation converter kits (converters and mixers), the best performance only can be obtained under partial conditions even with the unique design of mixers and ignition timing [11, 12]. However, along with the development of research activities, recently, found a hybrid system. LPG not only functions as fuel but serves as a refrigerant before it supplied to the engine (figure 1). Cooling effects generated through lab-scale tests show promising results, which more than 1.2 kW [13, 14].

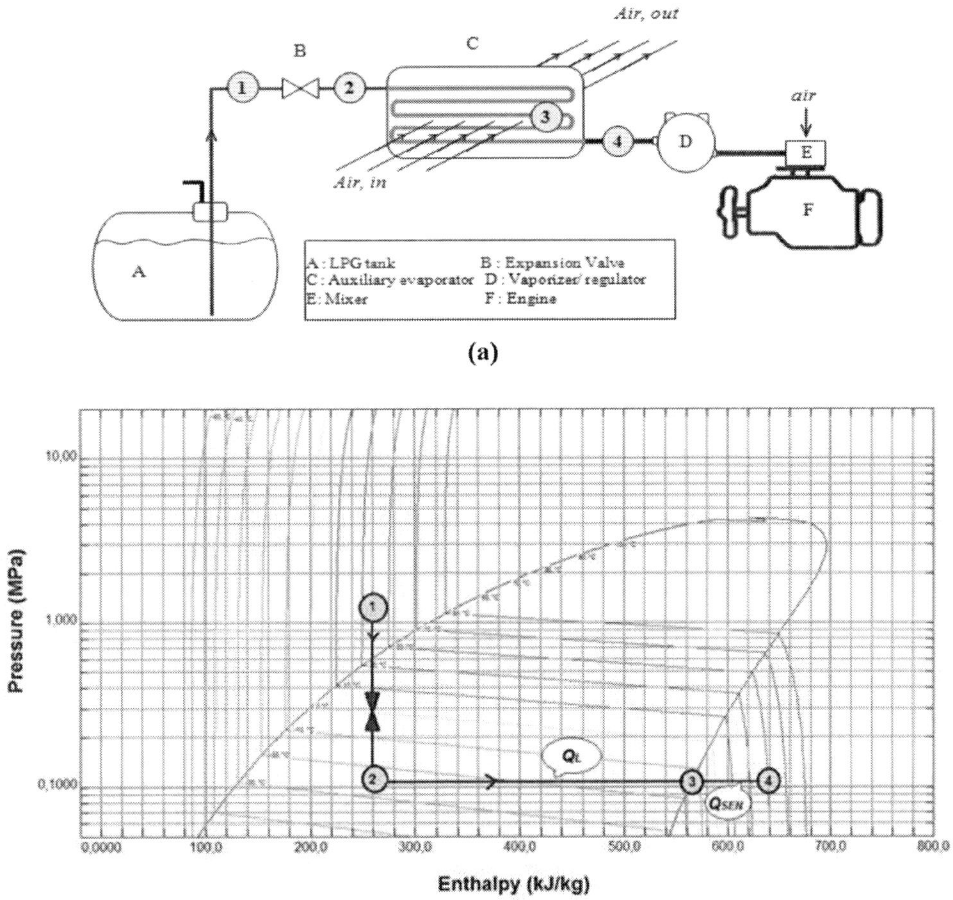

(a)

(b)

Figure 1. Illustration of hybrid system on LPG fueled vehicle (a) and P-h diagram on ½ cycle refrigeration concept (b).

A half cycle refrigeration system on LPG-fueled vehicles is a harvesting cooling effect without compressor and condenser. As an advantage, LPG in a cylinder is a pressurized liquid. Then LPG is expanded in an evaporator. LPG evaporates with the heat derived from the airflow that crosses the evaporator. Finally, LPG flows the engine under low-pressure vapor. In this case, the price per liter of LPG in the tube already includes the cost of compression on the refinery.

In the previous study [13], a 4 g/s LPG flow rate expanded from 0.6 MPa to 0.15 MPa was capable of producing a cooling effect of 1.2 kW with a humidity reduction of about 10%. This test can generate COP up to more than 6.0. Compression work to calculate COP is considered as energy to increase the pressure of LPG from atmospheric to 0.7 MPa (gauge). Meanwhile, studies conducted by Shah and Gupta [15] in a domestic refrigerator resulted in a maximum COP of 5.08. Denominators for calculating COP by Shah and Gupta use total energy to compress LPG per unit mass obtained from PCRA energy audit.

Furthermore, since there is no compressor work in ½ cycle refrigeration, the energy to compress the LPG into the cylinder is neglected. Instead, the performance of the system is calculated by the Energy Efficiency Ratio (EER) method. In this case, the input work is considered as electrical energy to flow the air across the evaporator by an electric blower.

2. Method

In this study, an LPG tank was removed from the vehicle then assembled with an expansion valve, an evaporator, and supporting equipment as shown in figure 2 [13].

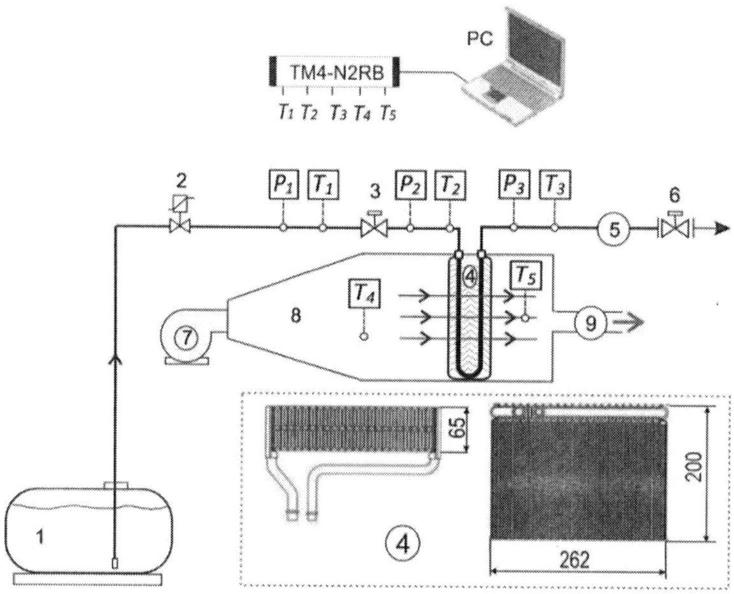

1. LPG tank,
2. Shut-off valve
3. Adjustable expansion valve
4. Evaporator
5. LPG flow meter
6. Hand operated valve
7. Electric blower
8. Box,
9. Air flow meter.

Dimension evaporator in mm

Figure 2. Set up experiment and apparatus.

This study uses LPG tanks with a maximum capacity of 40 liters equipped with multiple valves and shut-off valve. The expansion valve used is an adjustable expansion valve type to adjust the flow rate. In according to minimize heat loss in the body expansion valve, the expansion valve is made of nylon that has a low thermal conductivity (0.24 W/m•K). The compact evaporator is chosen because it has the smallest pressure drop compared to the other types. In this study, used evaporator air conditioner of Peugeot 206.

Selection of electric blower is based on the estimated of maximum potential heat transfer [14]. It is assumed that the LPG mass flow rate for a passenger vehicle is a maximum of 6 g/s and the difference in enthalpy value at the outlet and inlet evaporator is 298.6 kJ/kg, then the total absorption of latent heat is 1.79 kW. Assuming the air enters the evaporator at 38 °C and 90% RH and out of the evaporator at 15 °C and 50% RH, the enthalpy at the inlet and outlet of the airbox are 143 and 29 kJ kg^{-1}, respectively (Δh = 114 kJ kg^{-1}). Thus, the need for maximum air flow rate is 1.79 kW/114 kJ kg^{-1} = 15.7 g/s. Initial test results obtained that electric blower of 2A/3600 rpm produces air flow rate on the outside box of 16 g/s. Then, this value is taken because it is above the maximum requirement of 15.7 g/s.

3. Result and discussion

This section explains the refrigeration effect and discusses the energy efficiency ratio of the experiment results.

3.1. Refrigeration effect

During the air passes through the evaporator, the heat from the air will be absorbed by LPG for evaporation. As a result, the air becomes colder and drier. Dehumidifying occurs because some moisture in the air condenses. Therefore, the actual refrigeration effect is calculated based on equation (1).

$$\dot{q}_{ev} = \dot{m}_a \cdot \Delta h \tag{1}$$

Where $\dot{q}_{ev}$ is the refrigeration effect, $\dot{m}_a$ is the air mass flow rate, and Δh is the difference of specific enthalpy of air on the inlet and outlet airbox.

In this study, the specific enthalpy (h) obtained from the online psychometric calculator. Location of data collection at altitude 380 m above sea level with the barometric pressure of 96.842 kPa. The changes in air temperature during the test are presented in figure 3. Meanwhile, the relative humidity of air during the test is presented in figure 4. The air at the inlet of the airbox is 32°C and 92 ± 1% RH. Then, the refrigeration effect at various LPG flow rates is presented in figure 5.

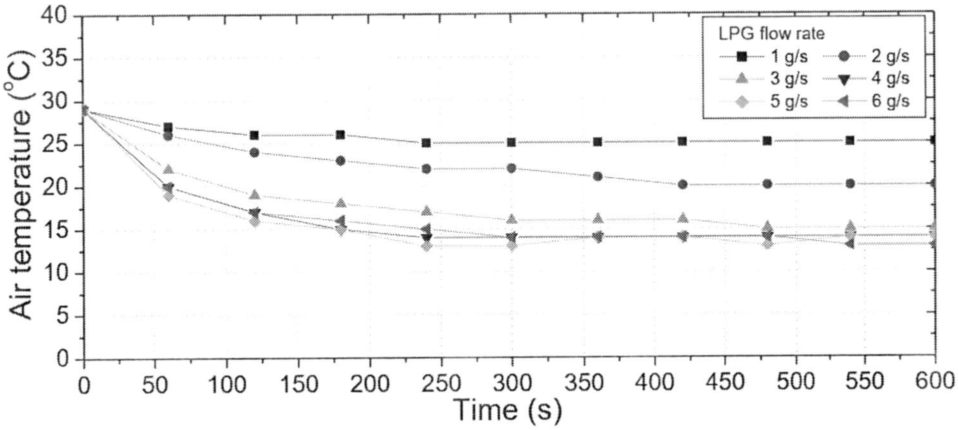

Figure 3. Temperature drop during data collection at various LPG flow rate.

From figure 3, information can be given that the greater the flow rate of LPG the more energy is absorbed for evaporation. As a result, the cooling effect will increase as can be seen in figure 5. Then, due to cooling and decreasing the pressure of the airflow, it will cause a drop in the humidity as shown in figure 4.

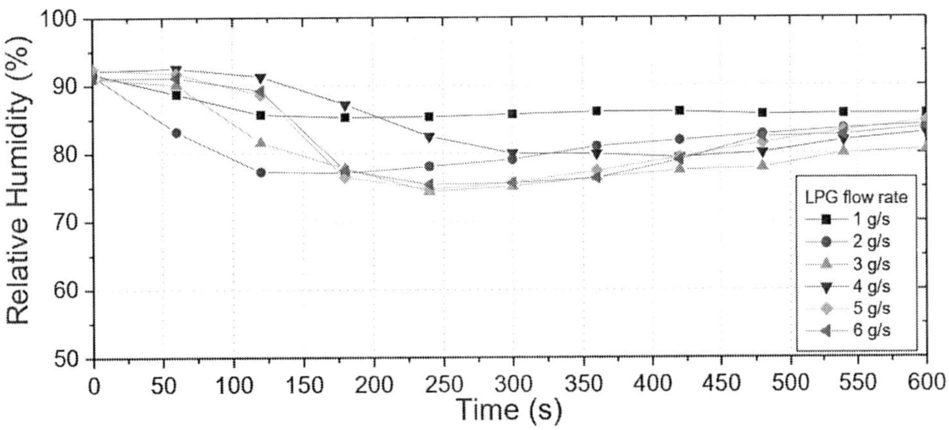

Figure 4. Relative humidity during data collection at various LPG flow rate.

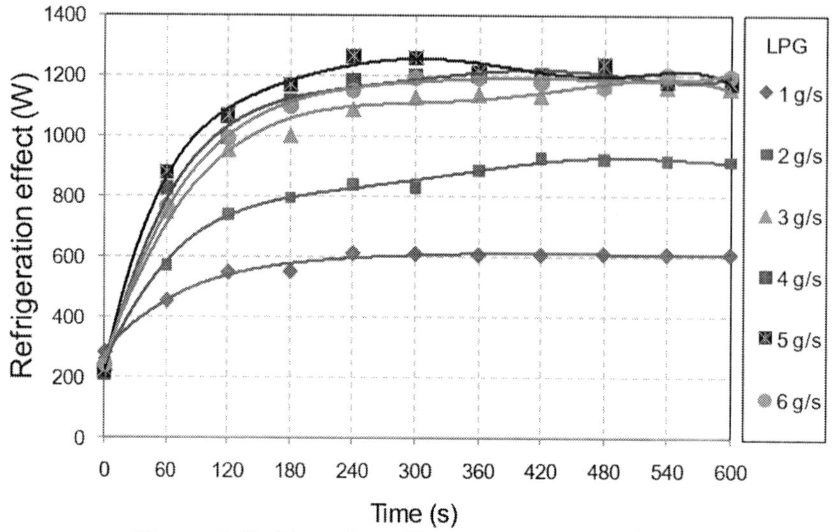

Figure 5. Refrigeration effect at various LPG flow rate.

3.2. Energy Efficiency Ratio (EER)

In addition to COP, the performance of the refrigeration system is also calculated by the Energy Efficiency Ratio (EER) [16]. EER is the ratio of actual refrigeration effect ($\dot{q}_{ev}$) to the total power input to produce refrigeration effect (equation 2). In the case of 1/2 cycle refrigeration, the power input is the electricity load to drive the electric blower (w_b). The concept of EER calculation on 1/2 cycle refrigeration system is presented in figure 6.

$$EER = \frac{q_{ev}}{w_b} \tag{2}$$

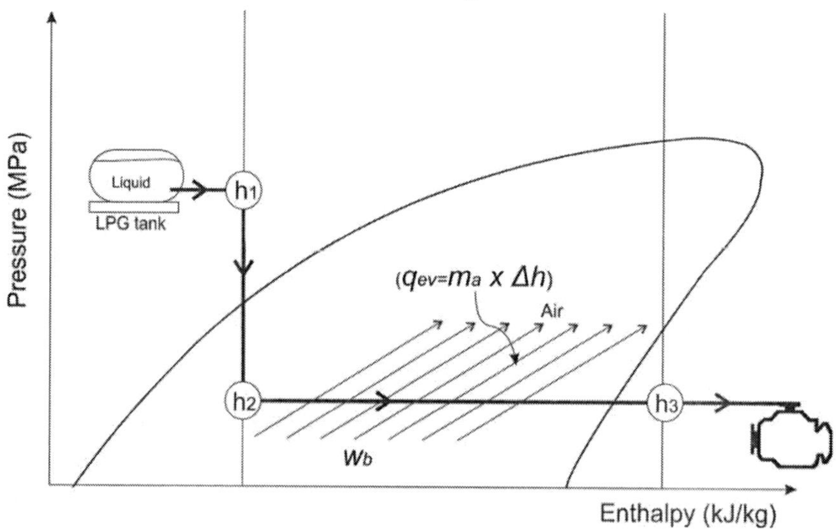

Figure 6. The concept of EER calculation on 1/2 cycle refrigeration system.

The actual refrigeration effect (cooling and dehumidifying) of airflow on various of LPG flow rate have been presented in figure 5. In this study, the electric blower used is 220V, 2.0A. Thus, the electric blower (w_b) is 440 Watt. The EER data is presented in figure 7.

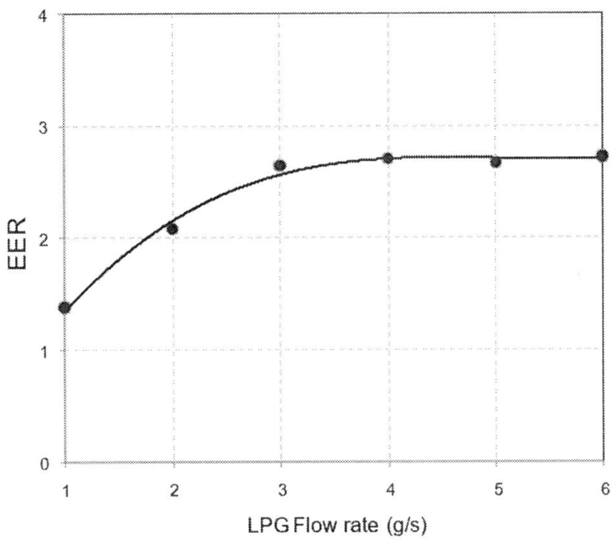

Figure 7. The EER of 1/2 cycle refrigeration system at evaporation pressure of 0.15 MPa (gauge).

4. Conclusion

Lab-scale experiments on the prototype of the ½ cycle refrigeration system of an LPG-fueled vehicle have proven to produce refrigeration effect of more than 1.2 kW. Then, the EER calculation results show a non-linear curve to the LPG flow rate because the input work is constant. The actual EER value obtained from this study is 2.72. In conclusion, the ½ cycle refrigeration system promises to be applied to LPG-fueled vehicles as an effort to reduce fuel consumption and exhaust emissions.

Acknowledgment

This paper is part of the project of hybrid A/C system development on LPG-fueled vehicles which is conducted at the Automotive Laboratory of Universitas Muhammadiyah Magelang. The researchers are grateful to the technicians who have helped data collecting and also the reviewers.

References

[1] Werpy M R, Burnham A and Bertram K 2010 *Propane Vehicles : Status, Challenges, and Opportunities* Argonne National Laboratory
[2] IEA 2010 *Natural Gas Liquids Supply Outlook 2008-2015* Paris
[3] Adolf J, Balzer C, Joedicke A and Schabla U 2015 *Shell LPG Study* Shell Deutschland Hamburg
[4] World LPG Association 2016 *Autogas Incentive Policies* Neuilly-sur-Seine
[5] ETSAP 2010 *Automotive LPG and Natural Gas Engines* IEA ETSAP - Technology Brief T03 April pp 1–5
[6] Norris J 2010 *Light Goods Vehicle – CO2 Emissions Study* AEA Energy and Environment Didcot Oxfordshire Final Report
[7] Masi M and Gobbato P 2012 Measure of the volumetric efficiency and evaporator device performance for a liquefied petroleum gas spark ignition engine *Energy Conversion and Management* vol 60 pp 18–27
[8] Ceviz M A and Yüksel F 2006 Cyclic variations on LPG and gasoline-fuelled lean burn SI engine *Renewable Energy* vol 31 pp 1950–60
[9] Campbell M, Wyszyński Ł P and Stone R 2004 *Combustion of LPG in a Spark-Ignition Engine* SAE Technical Paper 2004-01-09

[10] Manzie C, Watson H and Palaniswami M 2002 Air fuel ratio control in liquefied petroleum gas Injected SI engines *15th Triennial World Congress* Barcelona, Spain
[11] Setiyo M, Waluyo B, Anggono W and Husni M 2016 Performance of Gasoline/LPG Bi-Fuel Engine of Manifold absolute Pressure Sensor (MAPS) Variations Feedback *ARPN Journal of Engineering and Applied Sciences* vol 11 no 7 pp 4707–12
[12] Setiyo M, Waluyo B, Husni M and Karmiadji D W 2016 Characteristics of 1500 cc LPG fueled engine at various of mixer venturi area applied on tesla A-100 LPG vaporizer *Jurnal Teknologi* vol 78 no 10 pp 43–9
[13] Setiyo M, Soeparman S, Hamidi N and Wahyudi S 2017 Cooling effect characteristics of a ½ cycle refrigeration system on an LPG fuel system *International Journal of Refrigeration* vol 82 pp 227–237
[14] Setiyo M, Soeparman S and Hamidi N 2017 Numerical Study on Cooling Effect Potential from Vaporizer Device of LPG Vehicle *Journal of Engineering Science and Technology* vol 12 no 7 pp 1766–79
[15] Shah I H and Gupta K 2014 Design of LPG Refrigeration System and Comparative Energy Analysis with Domestic Refrigerator *International Journal Of Engineering Sciences & Research Technology* vol 3 no 7 pp 206–13
[16] Wang S K 2000 *Handbook of Air Conditioning and Refrigeration, 2nd Edition* (New York: The McGraw-Hill Companies, Inc.)

IOP Conf. Series: Materials Science and Engineering **403** (2018) 012088 doi:10.1088/1757-899X/403/1/012088

Donut mixer using bevel gears transmissions with 18 kg/hour in capacity

F Maghfurah and Windarta
Mechanical Engineering Department, Universitas Muhammadiyah Jakarta

windarta@ftumj.ac.id

Abstract. Small scaled donut productions use hand mixer to mix flours, eggs and another dough contents. The usage of this mixer was followed by some obstacles, such as the plastic material astigmatism worm gearbox that resulting the edge of gears has friction loss. Metal material bevel gearbox may overcome the dough stirring problem. This gearbox is able to overcome big load and high spin. As a result, the capacity of stirring container is able to be maximized that increasing working productivity result. There are a few things must be designed in scheming of the design process, such as the scheming of shaft calculation, bearing, bevel gear transmissions and frame. The results show that the production capacity increase 225 % from 7 kg/hour in capacity using hand mixer to 18 kg/hour in capacity.

1. Introduction

In general, small scale industry uses hand mixer device. The device capacity is small and difficult when the mixing of density donut. It is because of the gearbox made of plastic material. The hand mixer will prone damage because of wear out in spur. Furthermore, the usage of hand mixer is less effective for time production efficiency.

The problem could be solved by the usage of metal gearbox for bevel gear type. As a result, the metal gearbox can handle the high burden and high turning, which may increase the result of job productivity. The design of mixer has been proposed by many researchers [1-7]. Ardiansyah *et al.* [1] state that the same kind of device ever be made in the usage for the dough mixing where kind of density of the dough was the same to the donut resistance. Furthermore, the speed of mixing of dough can be attempted constantly in order that create the qualify product both level of ripeness evenly. As the solution, job principle of the set performance of controlled direction of a full wave which combined with rotary encoder as velocity sensor to produce constant velocity of motor and adaptive torsion.

2. Method

The design concept of the mixer in this research is described in the following procedure:

1) Design the mixer with capacity 7 kg/hour.
2) Calculate the shaft, pin, bearing, bevel gear transmission and frame.
3) Assembling all the components.
4) If the result of the design is 20kg/hour the do the analysis. Otherwise repeat step 2.

IOP Conf. Series: Materials Science and Engineering **403** (2018) 012088 doi:10.1088/1757-899X/403/1/012088

3. Result and discussion

The performance design uses bevel gear, by the mover of gear about 10 gear and 16 moved gears. The calculation for finding out the comparison of gear wheel or gear ratio, can be calculated by the equation (1).

$$i = \frac{z_2}{z_1} = \frac{16}{10} = 1.6 \tag{1}$$

Surely, the comparison of gear wheel or gear ratio in this device is about 1.6. From the specification of electric motor that be used, the speed of motor is about 1400 rpm. So the speed of output axis from this mixer device (figure 1) is formulated as equation (2).

$$\frac{n_1}{n_2} = \frac{z_1}{z_2} = n_2 = n_1 \cdot \frac{z_1}{z_2} = 1400 \cdot \frac{10}{16} = 875 \text{ rpm} \tag{2}$$

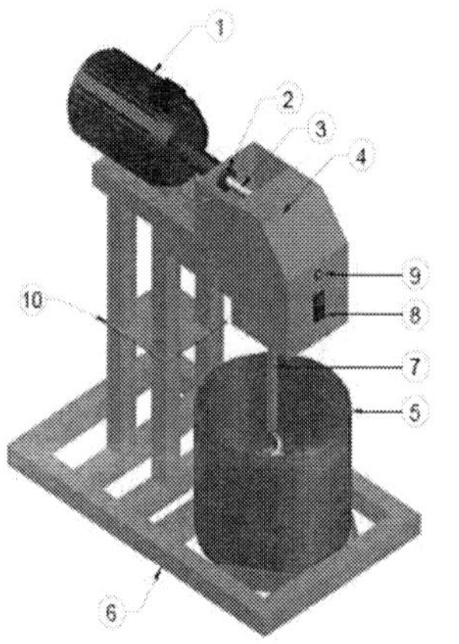

1. Motor
2. Bearing
3. Shaft
4. Gear box
5. Bucket
6. Frame
7. Stir
8. On/Off
9. Indicator
10. Hinge

Figure 1. Design of mixer device.

Based on calculation the speed of mixer is 875 rpm. This result will be compared to measurement using digital tachometer (figure 2(a) and figure 2(b)). The result can be presented in table 1.

(a) (b)

Figure 2. The speed of (a) shaft measurement (b) mixer measurement using digital tachometer.

Table 1. The result of speed comparison of mixer.

	Motor speed	Output speed
Theory	1400 RPM	875 RPM
Measured	1454 RPM	911 RPM
Error	3.86 %	4.21 %

It can be observed that the error based on theory and measured the motor speed is 3.86 %. The error for output (mixer) speed is 4.21 %. It is indicating that this result is secure since the error under 10 %.

3.1. Output result of donut cake dough

From the speakers who the writer makes an interview, the usage of hand mixer only can be used small measure device that only have capacity about 2 liters or 1.6 kg. Hand mixing process about 7 minutes, and 5 minutes in entering cake dough composition to the media of mixing. As a result, the output is about 8 kg/hour

Bevel gear mixer has the capacity of mixing media about 7 kg. From the data testing result, the time for the mixer media and mixing process until evenly be gotten is about 8 minute. The mixer output is then about 26.25 kg/hours or 26 kg/hour. The result of time of mixing between hand mixer and mixer bevel gear device is presented in figure 3 (a) and (b).

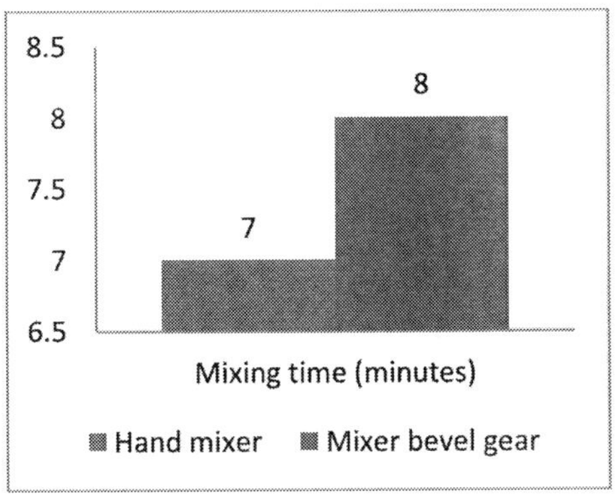

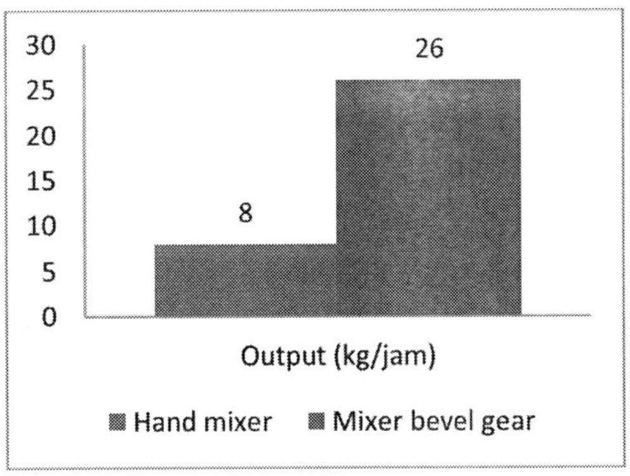

Figure 3. Hand mixer and mixer bevel gear (a) mixing time (b) output.

Mixing process in mixer bevel gear needed longer time because the bigger capacity compared by hand mixer. However, output result from mixer bevel gear create more output, namely 26 kg/hour.

3.2. Break even point

The price of component for 1 mixer vertical device of donut cake dough by gearbox for bevel gear type by capacity about 7 kg will be presented in table 2. In making of 1 mixer bevel gear device, the workshop imposes the project service about IDR 800,000. So that, the price of mixer device that the writer makes in designing for by the expense of IDR 3,800,000 If this mixer device will be sold, according to [2], the price is about 10% + IDR 3,800,000 = IDR 4,180,000.

Table 2. The price of component.

No	Component name	Price (IDR)	Quantity	Total (IDR)
1	Electric motor 0.5 HP	700,000	1	700,000
2	Bevel Gear	325,000	2	650,000
3	Pivot ǿ 20mm	300,000	1	300,000
4	Baring	20,000	5	100,000
5	Iron plate 10 mm	550,000	1	500,000
6	Hollow (40 x 40) 1.2 mm	300,000	1	300,000
7	Stainless steel container	170,000	1	170,000
8	ON-OFF switch	50,000	1	50,000
9	Clutch	50,000	1	50,000
10	Stand by light	20,000	1	20,000
11	Cable	20,000	1	20,000
12	Bolt	5,000	4	20,000
13	Nut	5,000	4	20,000
14	Spray paint	25,000	2	50,000
			Total	3,000,000

If it will be mass production, the use of BEP formula is necessary. There are permanent expense and its variable have been gotten by the leader in workshop which is the place of the making pf such mixer. Thus, by using the calculation of BEP formula, its detail is viewed as in table 3 and table 4.

Table 3. Fixed cost.

No	Fixed Cost	Price (IDR)
1	Electric	800,000
2	Rent warehouse	1,000,000
3	Rent Car/month	1,500,000
	TOTAL	3,300,000

Table 4. Variable cost.

No	Variable Cost	Price (IDR)
1	Salary	2,000,000
2	Carrying cost	100,000
	TOTAL	2,100,000

The price of mixer that planned to be sold. The calculation of Break Even Point rupiah is done as follows:

$$\text{BEP in IDR (Rupiah)} = \frac{fixed\ cost}{1-\left(\frac{variable\ costs}{Selling\ price}\right)} = \frac{3{,}300{,}000}{1-\left(\frac{2{,}100{,}000}{3{,}000{,}000}\right)} = \text{IDR } 11{,}000{,}000$$

BEP Unit. From data of calculation for the value of *Break Even Point* can be concluded whereas BEP (IDR) will be reached for 4 units production.

4. Evaluation

The design mixer device is vertical for donut cake dough by gearbox t bevel gear type for capacity about 7 kg use electricity motor by 0.5 HP by round of motor 1400 rpm. The used gearbox is about 1,6 of mixer bevel gear ratio to reduce the output round to 875 rpm. The mixer uses the media with 7 kg capacity, so that can create output about 26 kg/hour.

The result of output of hand mixer that only 8 kg/hour, so that mixer bevel gear device can increase the result of productivity with more than 18 kg/hour output. In other words, 225% from capacity of hand mixer.

5. Conclusion

The use of hand mixer only can be used in the media that have capacity about 2 litre or 1.6 kg. Process of mixing by hand mixer is about 7 minutes, and 5 minutes in entering the composition of cake dough in the media of mixing. To do so, the result of output in the usage of hand mixer kind is 8 kg/hour. Meanwhile, mixer bevel gear capacity is about 7 kg. By the time about 8 minutes in entering the composition of cake dough in entering the media of mixer and the process of mixing by the time about 8 minutes. The result of output that be gotten by this mixer device is about 26.25 kg/hour or 26 kg/hour. By the result of capacity from hand mixer that only 8 kg/hour, so mixer bevel gear device can increase the result of productivity by capacity more than 18 kg/hour to increase 225% than the capacity of hand mixer device. The Break Even Point will be reached in 4 units production.

References

[1] Ardiansyah R, Hasanah R N and Wijono 2013 Perancangan dan Pembuatan Alat Pengaduk Adonan Dodol dengan Kecepatan Konstan dan Torsi Adaptif *Jurnal Mahasiswa TEUB* vol 1 no 4 pp 1–6

[2] Suyadi S and Rachman F N 2014 Rancang Bangun Mesin Pembuat Es Puter Dengan Pengaduk dan Penggerak Motor Listrik *Jurnal Rekayasa Mesin* vol 9 pp 41–6

[3] Ichwanudin N, Arifin M and Wardoyo 2011 Rancang Bangun Alat Mixer Roti Bandung Dengan Variasi Percepatan *PKMM*

[4] Prabowo K S 2013 Perancangan Mesin Mixer Pengadon Kue Bolu Guna Meningkatkan Produktivitas *Jurnal Mahasiswa Teknik Industri*

[5] Sri W and Rusiyanto 2010 Penerapan Adsorpsi dan Modifikasi Mixer Untuk Meningkatkan Kualitas Produk Industri Kecil Kecap *Rekayasa* vol 8 no 1

[6] Subiyono and Karyono S 2011 Pengaduk Adonan Untuk Pengrajin Seni Fiberglass vol 15 no 2 pp 172–81

[7] Tiwan 2010 Rancang Bangun Mesin Pengaduk Adonan Bahan Dodol Batang dan Buah Pepaya Untuk Meningkatkan Kapasitas Produksi Industri Kecil di Pedesaan vol 14 pp 11–24

Comparison of plastic and stainless-steel as solar still material

D Mugisidi[1], O Heriyani[2] and H Fathurahman[1]

[1]Mechanical Engineering Department, [2]Electrical Engineering Department
University of Muhammadiyah Prof. Dr. HAMKA, Jl. Tanah Merdeka no 6 Kampung Rambutan Ciracas Jakarta Timur, Indonesia

dan.mugisidi@uhamka.ac.id

Abstract. The purpose of this study was to select the most suitable material to be used as solar still material. Materials compared in this study were plastic and stainless-steel because of their resistance to corrosion. The tested material was utilized as a container to accommodate sea water to be evaporated. The experiment carried out in Jakarta, Indonesia. Material temperature profile similar with the ambient temperature profile. When the sun is shining brightly, the material temperature is higher than the ambient temperature but in the afternoon and night the material temperature falls to ambient temperature. Therefore, it can be said that plastic and stainless-steel materials do not store heat. As a material for solar still, stainless-steel has higher evaporation rate than plastic, which is 7.66 ml/h for stainless-steel and 7.09 ml/h for plastics. So, based on this research, it can be concluded that stainless-steel is better to be used as a solar still material.

1. Introduction

Water is the decisive material in life. Humans, plants, and most animals will not survive without water. Along with the growing population in the world, water consumption is increasing. The addition of human population as much as 15% will reduce the water source and increase the water shortage by 40% [1], while the amount of fresh water on the earth surface, only available 2.8% and the rest is sea water [2]. Therefore, sea water is a potential source of water.

Because of its enormous potential, various methods and research have been done to convert sea water into fresh water. Distillation is one of the most widely used methods. Evaporation processes that occur slowly make the contaminants left behind, so that the resulting water becomes pure. The distillation process becomes cheaper by utilizing solar energy, although the production is not high. The direct heating method with solar heat is the most suitable way to produce fresh water up to 200 m^3/day [3].

The process of distillation by solar heating is generally done by using solar still. Various methods and materials are used to improve the performance of solar still. It seems that evaporation rate affected by its solar still. The use of Portland cement black-blackened, as heat storage, increase freshwater production by 39% [4]. Concave-shaped solar stills are used to increase evaporation by [5]. The perforated aluminum is mounted on the water's surface on the solar still to increase evaporation [6]. The use of plastics for easy cleaning of salt deposits is done by [7]. The tilted-wick was placed on the surface of the solar still spreading water to the entire surface of the wick through capillary fibers. So, the temperature becomes higher in the thin layer [8]. Five trays made of aluminum and stainless steel AISI 304 (polished, chrome plated, and non-polished) were tested to study its performance as a function of different tray materials. The result showed that the aluminum tray had the highest desalinated water production, while the polished steel tray had the lowest production [9].

The performance of a solar still is proportional to distillate output and indicated by the internal heat transfer coefficient. There are three parameters based on internal heat transfer coefficients: convective heat transfer coefficient, radiative heat transfer coefficient and evaporative heat transfer coefficient [10]. The material that holds sea water also receives heat from solar energy. Some of the heat in material flows into the water and some loss to ambient. Beside that some material able to store heat and release it when the heat source is not available. Then the material becomes a contributing factor in the evaporation of sea water.

Sea water is a very corrosive substance, the material used in desalination process must be corrosive resistance material. So, it is necessary to conduct research to do experiment with material which has beneficial effect on evaporation rate, but corrosion resistance. Nevertheless, still no related research has examined the effect of material on evaporation of seawater in solar-still. Then, this research examines the plastic and stainless steel container material to evaporate the sea water and their heat storage ability is compared.

2. Material and methods

The experiment uses a pan made of plastic and stainless steel (SUS 316). The pan of plastics and stainless-steel has same shape and size. Five pieces pan for each type of material, filled with 250 ml sea water which taken from Ujung Kulon, Indonesia.

3. Results and discussion

In figure 1, it shows that the material and water temperature have the same profile as the ambient temperature. When the temperature is about 31° – 33°C, temperature of material is almost equal to the ambient temperature but in higher temperatures, temperature ambient about 37°C, the temperature of materials rises above the ambient temperature. Plastic has the highest temperatures because the slow propagation and release of heat. Plastic is non-metallic materials that do not have free electrons. Therefore, the heat distribution occurs with lattice vibration [11] so the heat distribution becomes slow and the thermal conductivity becomes low. The slow distribution of heat from the plastic surface causes heat accumulation and the heat become higher than stainless steel. Even though temperature of plastic is higher than stainless steel, the temperature of water inside the stainless-steel pan is higher than the water inside plastic pan.

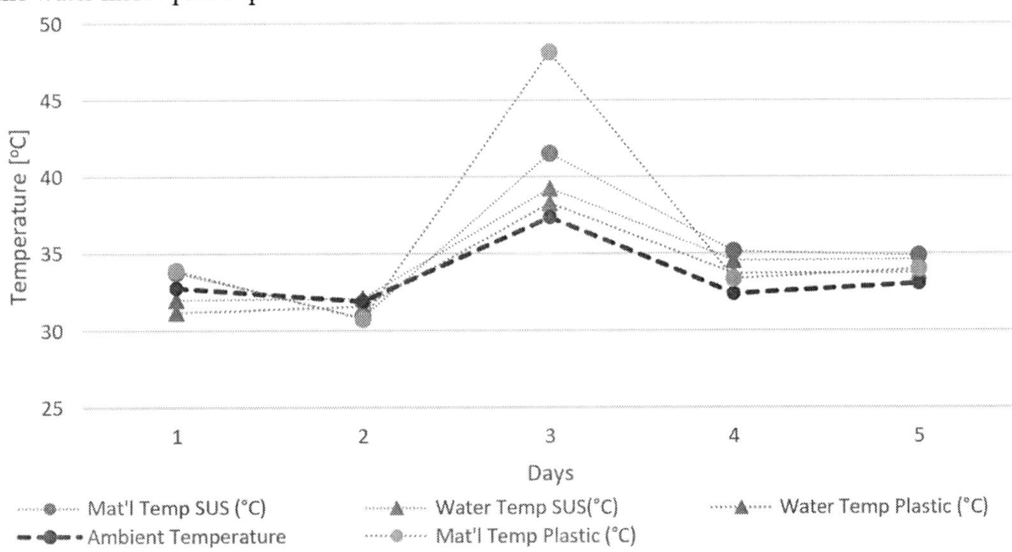

Figure 1. Temperature profile of material, water and ambient temperature during the experiment.

As the result, evaporation of seawater inside the plastic pan is lower than stainless steel pan, as shown in figure 2. In figure 2, the evaporation of sea water in plastic pan is 28.36 ml and stainless-steel pan is 35.22 ml. The conductivity of the stainless steel (16.95 W/m°C) is higher than plastics (0.15 W/m°C). This result along with the result of [12] that the heat transfer increases with decrease in thickness and increase in thermal conductivity and thermal conductivity of the material has a significant effect on the convective heat transfer coefficient [13]. The heat from solar radiation is absorbed and accumulated by the material. Some of the heat of material flows into water and some of it flows to the ambient. Heat in the water, which come from material and sun radiation, balance with evaporation energy [14]. Since the heat propagation in stainless-steel is better than plastic, heat in stainless-steel flows into water and elevate the evaporation. Furthermore, thermal conductivity of material plays important role in evaporation.

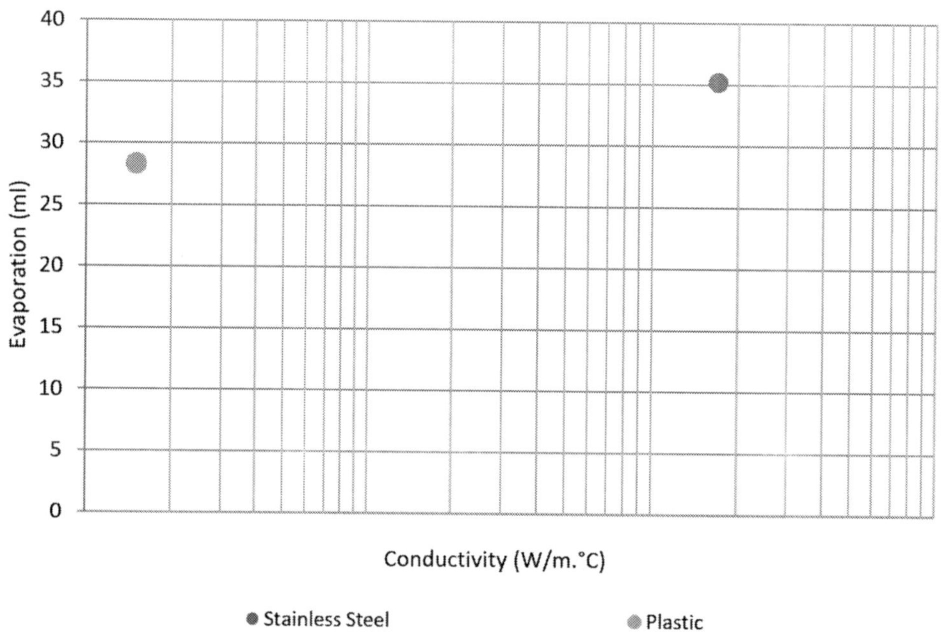

Figure 2. Effect of container material conductivity to evaporation.

In figure 3, as previously stated, the material temperature is higher than the ambient temperature in bright sunlight. In the afternoon, after 15:00, the material temperature falls below the ambient temperature. As the sun sets and the ambient temperature drops to 27°C, the temperature of stainless-steel and plastic, starts from 31°C, follow drops to about 27°C. When the ambient temperature is lower than material, heat flows from material to ambient. So that the material temperature drops to about ambient temperature. This proves that these materials do not have the ability to store heat. Nevertheless, through in the very small gap, temperature of stainless-steel is lower than temperature of plastic. Since conductivity of stainless steel is higher than plastics so the heat in stainless-steel release faster.

Figure 4 shows evaporation and evaporation rate in extended time. Horizontal axes show duration of evaporation. For example, 11-14 means that the time starts is at 11.00 AM and stop at 14.00 (2 PM). Evaporation is checked at 14.00 (2 PM). Evaporation of sea water in material stainless-steel and plastic in extended time shows that the evaporation in stainless steel is higher than in plastic. Even though evaporation increase as the time increase, evaporation rate decrease. Interestingly, rate of evaporation in plastic material decrease slower than evaporation rate of stainless-steel, though still lower.

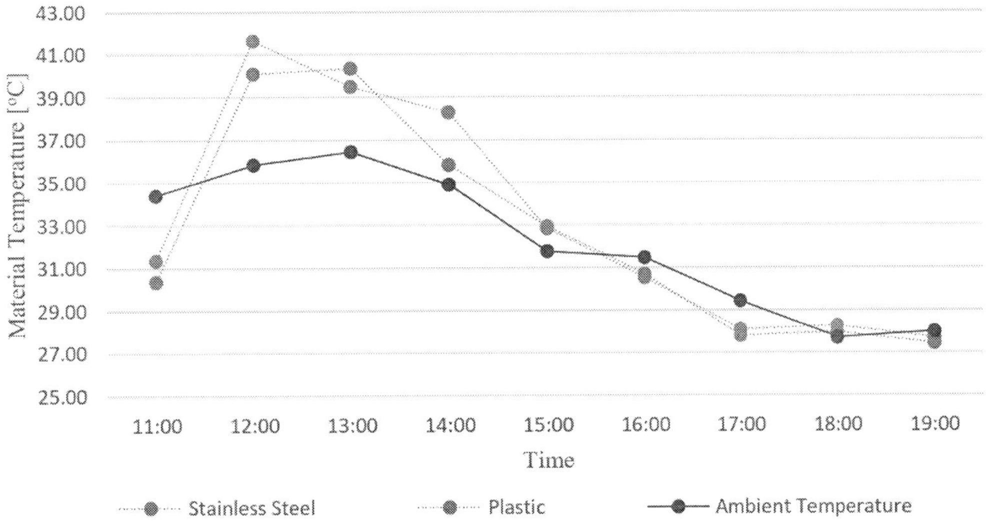

Figure 3. Ability of material to store the heat.

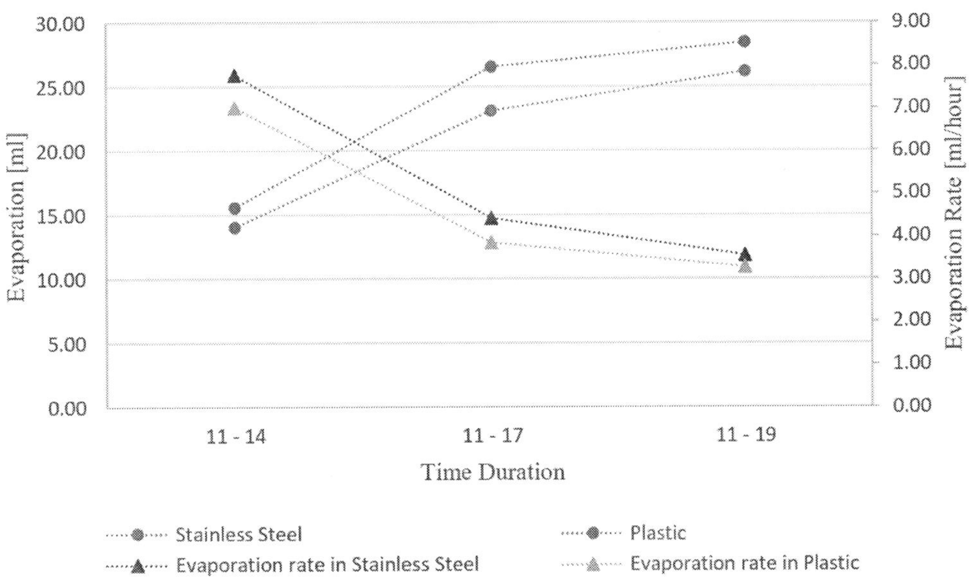

Figure 4. Evaporation and evaporation rate for extended time.

4. Conclusion

The experiment results show that the material affected the temperature of the water inside. Temperature of plastic is higher than the stainless steel but water temperature inside is lower. The temperature gap between water and its container affected by the conductivity of material. In the afternoon and night, the material temperature falls close to ambient temperature since the heat in materials flows into ambient. Therefore, it can be said that plastic and stainless-steel materials do not store heat. As a material for solar still, stainless-steel has higher evaporation rate than plastic, which is

7.66 ml/h for stainless-steel and 7.09 ml/h for plastics. It can be concluded that stainless-steel is better to be used as a solar still material than plastic, since ability of plastics to store heat can be neglected.

References

[1] Schewe J, Heinke J, Gerten D, Haddeland I and Arnell N W 2014 Multimodel assessment of water scarcity under climate change *Proceedings of the National Academy of Sciences of the United States of America* vol 111 no 9 pp 3245-50
[2] Belessiotis V, Kalogirou S and Delyannis E 2016 *Thermal solar desalinisation - methods and systems, 1st ed.* Academic Press Elsevier
[3] García-Rodríguez L 2002 Seawater desalination driven by renewable energies: a review *Desalination* vol 143 issue 2 pp 103-113
[4] Sellami M H, Guemari S, Touahir R and Loudiyi K 2016 Solar distillation using a blackened mixture of Portland cement and alluvial sand as a heat storage medium *Desalination* vol 394 pp 155–61
[5] Kabeel A E 2008 Performance of solar still with a wick concave evaporation surface *Twelfth Int. Water Technol. Conference, IWTC12 2008* Alexandria, Egypt pp 1137–46
[6] Valsaraj P 2002 An experimental study on solar distillation in a single slope basin still by surface heating the water mass *Renew. Energy* vol 25 issue 4 pp 607–12
[7] Tarawneh M S K 2007 Effect of water depth on the performance evaluation of solar still *Jordan Journal of Mechanical and Industrial Engineering (JJMIE)* vol 1 no 1 pp 23–29
[8] Al-Karaghouli A, Renne D and Kazmerski L L 2009 Solar and wind opportunities for water desalination in the Arab regions *Renew. Sustain. Energy Rev.* vol 13 issue 9 pp 2397–407
[9] Da Silva M E V, Schwarzer K, Pinheiro F N, Rocha P A C and De Andrade C F 2015 Experimental study of tray materials in a thermal desalination tower with controlled heat source *Desalination* vol 374 pp 38–46
[10] Panchal H N and Shah P K 2011 Modelling and verification of single slope solar still using ANSYS-CFX *Int. J. Energy & Environ.* vol 2 issue 6 pp 985–98
[11] Hahn D W and Özişik M N 2012 *Heat conduction*, 3rd ed. (John Wiley & Sons, Inc.)
[12] Aboul-Enein S, El-Sebaii A A and El-Bialy E 1998 Investigation of a single-basin solar still with deep basins *Renew. Energy* vol 14 issue 1–4 pp 299–305
[13] Tiwari A K and Tiwari G N 2005 Effect of the condensing cover's slope on internal heat and mass transfer in distillation: an indoor simulation *Desalination* vol 180 issue 1-3 pp 73–88
[14] Qiblawey H M and Banat F 2008 Solar thermal desalination technologies *Desalination* vol 220 issue 1–3 pp 633–44

The effect of the ball size on the product characteristics of shaker HEM to produce nano particle from bamboo charcoal

Supriyono[a], Ngafwan[b] and J W Joharwan[c]

[a, b]Dept. of Mechanical Eng. Universitas Muhammadiyah Surakarta, Jl. A Yani Tromol Pos 1 Pabelan Kartosuro, Surakarta, Indonesia
[c]Dept. of Mechanical Eng. Akademi Teknik Warga, Jl. Raya Solo Baki km 2, Sukoharjo, Indonesia

supriyono@ums.ac.id

Abstract. The aim of this research is to study the effect of the ball size on characteristics of the product of high energy milling (HEM) of shaker type to produce nano particle from bamboo charcoal. A new shaker mechanism is developed. The characteristics of the particle are represented by particle size, surface morphology, and the substances presence in the product. Particle size analyzer (PSA) is conducted to have the particle sizes, whereas SEM and EDX are used to have surface morphology and substances presence in the products respectively. The mixtures of 11 gr of bamboo charcoal powder of 200 mesh and about 299 gr of steel ball are placed in vials. The vial is from stainless steel cylinder with 11/4 inch diameter and 120 mm length. The stroke of the shaker is 54 mm at 233 rpm. The number of cycle is 4 million cycles. The steel ball sizes are 1/8, 5/32, 3/16 and 1/4 inch. PSA results show that there is no certain correlation between the steel ball size and particle size. The final shape of the particles is determined by fracture mechanism. The highest substance presence in the result is C.

1. Introduction

High energy milling (HEM) is a production method where a blend of powder put in the ball mill and is exposed to high impact energy of collision from the balls. HEM often called as mechanical alloying is one of the most broadly method used to produce ultrafine materials. Mechanical alloying was established in the mid 1960 by John Benjamin to have nickel-based super alloys of oxide dispersion strengthened (ODS) for the applications of gas turbine. HEM process is commonly employed in the industries of ceramic and metal processing [1]. The process in HEM comprises of repetitive fracture, mixing, and cold welding of a fine blend powder (alloy particles) causing in size reduction of the particles and sometimes in chemical reactions. Currently, the HEM is used to make nanostructured materials which are intensively researched. HEM equipment are available in different types both for mechanical alloying and nanoparticle formation. They are different in their size and or milling efficiency.

Examples of research in preparing nanostructured material using HEM can be seen in [2-6]. Li and Ramanujan [2], studied the microstructure evolution and formation of nanocrystal line FeCo based alloys. Mishra *et al.* [3] conducted a research to study the magnetic characteristics of nano-composite materials of iron nitride–alumina made by HEM. Hamzaoui *et al.* [4] studied the magnetic properties and structure of nanocrystalline of mixture of Fe-10wt % Ni and Fe-20 wt% Ni alloys. Pilarczyk *et al.* [5] did a research to examine the structure and characteristics of Fe-Co-Ni-B-Si-Nb alloy. El-

Eskandarany *et al.* [6] conducted a research to study the effect of milling speed in ball milling process. They have a conclusion that by increasing the speed of rotation, it will decrease the time necessary for the development of an amorphous phase in the Co-Ti system.

Nanotubes, nanowires and nanorods are one-dimensional (1D) nano materials. It has been showed by many researchers that nanomaterials have many advantages compared to the bulk material in terms of properties, functionalities and a wide range of promising applications. However, for the future industrial usages there is a major obstacle for the large-quantity production of the nanomaterials. Chen *et al.* [7] reported that there is potential process to achieve mass production of nanomaterials by having ball milling and annealing process. To show such capability, several examples are presented in their work. The examples included C, BN nanotubes and SiC, Zn nanowires. Vertical rotating ball mill and a planetary ball mill are used in this Study. From this research, it can be stated that the formation of the nanotubes and nanowires is determined by the prior HEM process. The nanotubes and nanowires cannot be produced by annealing process without the prior ball-milling process.

In this research, shaker type of HEM is used to produce nano particle from bamboo charcoal. A new shaker mechanism is developed. This research is part of the examination of the new mechanism to examine some variables which influence the nano particle results. This paper reports the effect of the ball size to the characteristics of the results.

On the other hand, the demand for nano particle of pure carbon in large quantities and low costs is a lot. Charcoal is one of the alternative of carbon sources. In Indonesia the used of bamboo is very much which results in a lots of unused material. Bamboo has carbon content more than 90% of its weight [8]. It is a potential carbon source. The unused material of bamboo can be made to become charcoal. As far as, author's knowledge, there is very limited research on the bamboo as carbon source for nano-carbon particle. This research tries to have carbon nano-particle at large quantity and low cost production from bamboo.

2. Experiments

Different types of high-energy mills (HEM) have been created. In this research, a new shaker mechanism is developed (see figure 1(a)). The shaker mechanism is obtained from rotational motion changed to translational motion by connecting rod. The length of the stroke is determined by the length of the connecting rod. The speed of the motor can be controlled and the ration of the motor pulley to the connecting rod pulley is 1:3.

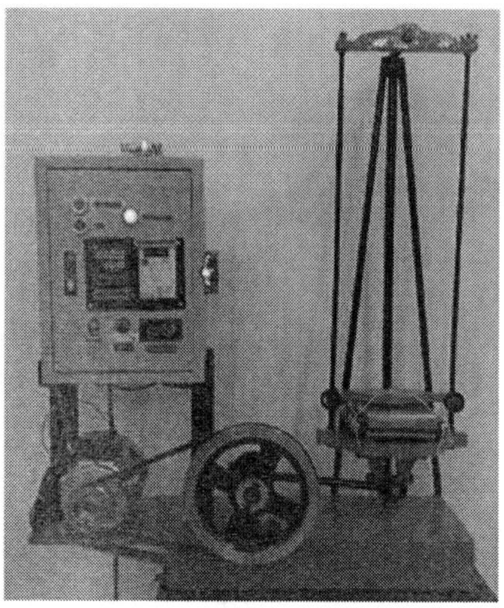

(a)

(b)

Figure 1. Materials used in this research (a) shaker type of HEBM (b) vials.

IOP Conf. Series: Materials Science and Engineering **403** (2018) 012090 doi:10.1088/1757-899X/403/1/012090

The vial is from stainless steel cylinder with 1¼ inch diameter and 120 mm length (see figure 1(b)). The contents of the vial are 11 gram of bamboo charcoal powder of 200 meshes and about 299 gram of steel ball with diameter of 1/8, 5/32, 3/16 and 1/4 inch. A total of four vials can be run at the same time to produce large quantities of materials. The stroke of the shaker is 54 mm at 233 rpm. Characterization of the particle is conducted by Particle Size Analyzer (PSA), Scanning Electron Microscope (SEM) and Energy Dispersive X-Ray (EDX). The characteristics of the particle are represented by particle size, surface morphology, and the substances presence in the particle.

3. Results and discussion

The results of particle size analyzer are shown in table 1. It can be seen that the increase in diameter of the steel ball does not have any certain correlation to the decrease of the particle size. The smallest particle having average size of 273.8 nm is produced by the ¼ inch ball diameter. The sizes of 490.1 nm, 573.3 nm and 515.2 nm are produced by the 1/8, 5/32, and 3/16 inch ball diameter respectively. The size reduction process is explained as follows: at the time of the milling process, materials are repetitively crushed, broken and welded [7]. Collisions happen in between the balls or the balls and the vial wall. The collisions setup some particles between their surfaces. The high impacts energy from the collision severely deform the particles and create atomically fresh, new surfaces, as well as a high density of dislocations and other structural defects. Moreover, the deformation and breaking of particles make the continuous size reduction.

Table 1. Particle size analyzer results.

Steel ball diameter (inch)	Average Particle Size (nm)
1/8	490.1
5/32	575.3
3/16	515.2
1/4	273.8

The increase in diameter means the increase in mass. At the same speed, it also means that the impact energy is increased. If the impact energy is higher, it should reduce the particle size smaller. In this case, the 5/32 inch and 3/16 inch steel ball diameter have higher impact energy than 1/8 inch. They should produce smaller particle than 1/8 inch. However, it does not happen. According to [9] and [10] too long milling time cause agglomeration. Under SEM examination, the morphology of the agglomerated particles seems bigger than un-agglomerated particles. In this research, the agglomeration of the particles occurred for the ball sizes of 1/8, 5/32, and 3/16 inch. It can be seen in figure 2 (a), (b), and (c). The particles produced by the three balls appear in sheet due to the agglomeration. Whereas, the particle resulted from the ¼ inch ball did not agglomerate. It can be seen on figure 2(a).

If the impact energy is enough to prevent agglomeration, according a research conducted by [11], there will be grinding limit. Grinding limit is condition at which the increase number of milling time has no effect on the particle size. Kim reported that after HEM process for 30 hours of Fe-Co powders, the milling process attained a steady state. In this condition, the particles have become homogenized in size and shape. Other research, showing the grinding limit is done by [12]. After milling of Ni–Al powder for 100 hours, the particle size and shape become homogenous. Umemoto *et al.* [13] conducted research on milling of Fe-C. After 500 hours milling time the grinding limit is reached and the particle size is 4.7 nm with homogenous shape.

The agglomeration did not happen for the ¼ ball after 4 million cycles. But grinding limit is not reached yet. SEM morphology of the particles as in figure 2(a) shows this fact. It can be seen that the particle size is not homogenous yet. Even though the average size of the particle is 273.8 nm (as in PSA result), it can be seen that there are still particles with more than 1 μm. In this research, 4 million cycles are equivalent to 288 milling hours. The particles have fractional shape resulted from fracture mechanism.

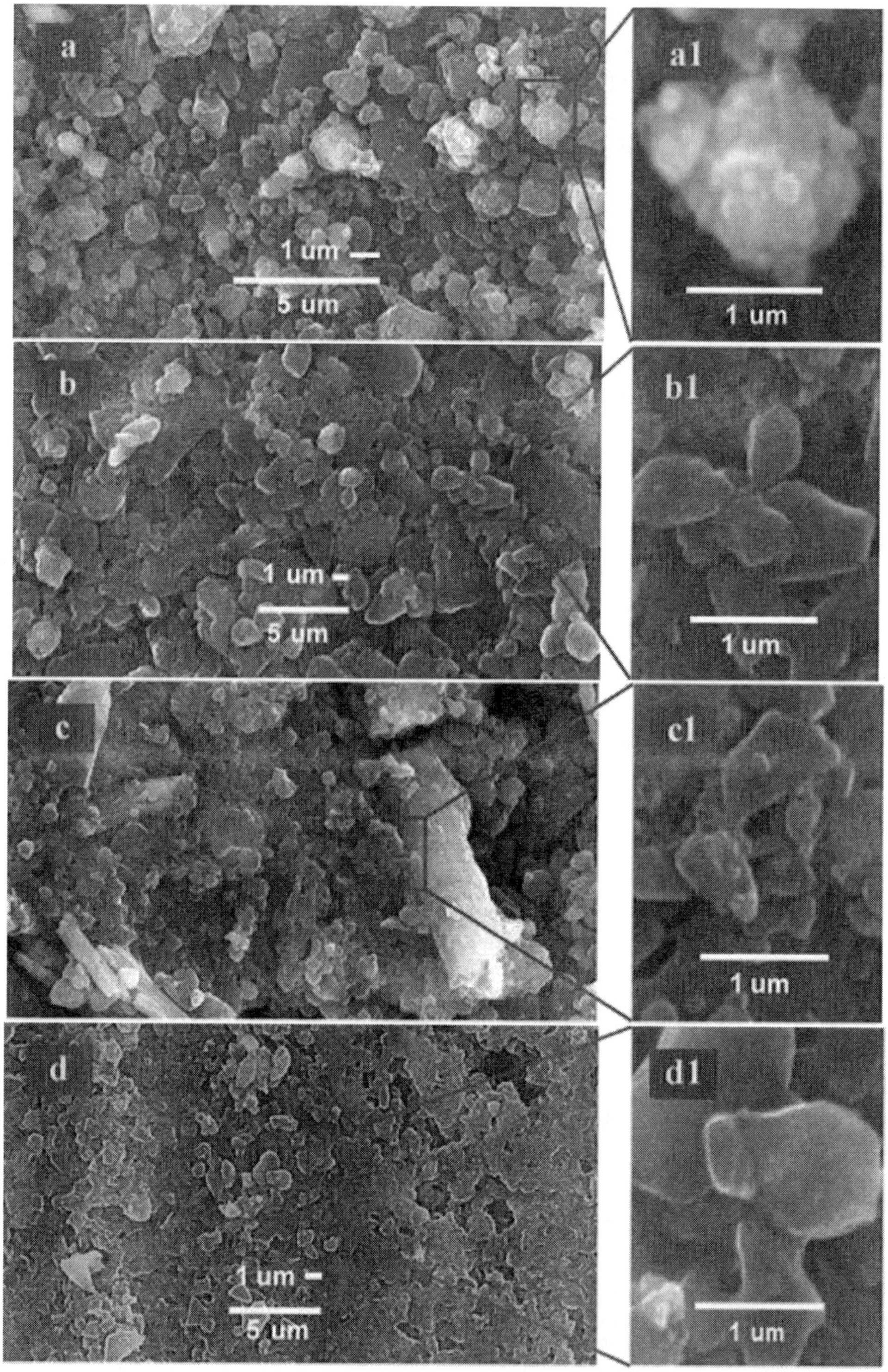

Figure 2. SEM morphology of 4 million cycles, (a) is for ¼ inch, (b) is for 3/16 inch, (c) is for 5/32 inch and (d) is for 1/8 inch).

Table 2 is the substances presence in the particles as a result of EDX. It can be seen that the highest component is carbon. This result is not in line with the result of [8]. They reported that the carbon content of bamboo is more than 90% of its weight. In this research the only one having carbon more than 90% is the result of ¼ inch ball diameter which is 93.03%. The results of 1/8 inch, 5/32 inch, and 3/16 inch are 80.27%, 59.94% and 66.39% respectively.

K_2O, SiO_2 and P_2O_5 are the dominant components after carbon. K_2O has percentage of 0.55%, 8.07%, 14.47%, and 20.41% respectively for ¼ inch, 1/8 inch, 3/16 inch, and 5/32 inch. P_2O_5 has percentage of 0%, 1.37%, 2.64%, and 6.19% respectively for ¼ inch, 1/8 inch, 3/16 inch, and 5/32 inch. It can be seen that the increase percentage of carbon will have the decrease percentage of K_2O and P_2O_5. This finding needs to be studied more to have the answer. However, it can be said that bamboo is the right alternative carbon producer because it is a renewable natural resource. Bamboo is a kind of biomass that has been widely planted. Currently, bamboo resources are plentiful and the area of bamboo is about five million hectares. Similar to wood, bamboo is composed of hemicellulose, cellulose and lignin. It has great potential as a future bio-energy resource.

Table 2. Particle composition.

Element	Composition (% weight)			
	Steel ball 1/8 inch	Steel ball 5/32 inch	Steel ball 3/16 inch	Steel ball 1/4 inch
C	80.27	59.94	66.39	93.03
K_2O	8.07	20.41	14.47	0.55
SiO_2	5.66	7.90	10.99	4.19
P_2O_5	1.37	6.19	2.64	0.00
Na_2O	0.26	1.02	0.44	0.00
SO_3	0.40	0.00	2.64	0.00
Cl	1.07	3.28	0.93	0.00
FeO	0.76	0.87	0.89	0.57
CuO	0.68	0.00	0.59	0.95
ZnO	0.62	0.00	0.00	0.55
ZrO_2	0.84	0.00	0.00	0.00
MgO	0.00	0.20	0.00	0.00
Al_2O_3	0.00	0.20	0.00	0.16

4. Conclusion

HEM can be used to prepare nano particle from bamboo charcoal. The size reduction depends on the impact energy. The final shape of the particle is determined by fracture mechanism. The most dominant element presence is carbon. It seems that bamboo charcoal is a potential source to produce nano carbon.

Acknowledgments
This work is supported by Research Institute of Universitas Muhammadiyah Surakarta.

References

[1] Nagesha K V, Rajanish M and Shivappa D 2013 A review on mechanical alloying *Int. J. Eng. Res. Appl.(IJERA)* vol 3 issue 3 pp 921–24
[2] Li H F and Ramanujan R V 2005 Microstructural evolution and nanocrystalline formation kinetics in FeCo based alloys during mechanical alloying *J. Metastable Nanocrystalline Mater.* vol 23 pp 187–90
[3] Mishra S R, Long G J, Grandjean F, Hermann R P, Roy S, Ali N and Viano A 2003 Magnetic

properties of iron nitride-alumina nanocomposite materials prepared by high-energy ball milling *Eur. Phys. J. D - At. Mol. Opt. Phys.* vol 24 issue 1 pp 93–6

[4] Hamzaoui R, Guessasma S and ElKedim O 2008 Analysis of structure and magnetic properties of nanocrystalline milled alloys *J. Alloys Compd.* vol 462 issue 1–2 pp 29–37

[5] Pilarczyk W, Nowosielski R and Pilarczyk A 2008 Structure and properties of Fe-Co-Ni-B-Si-Nb alloy prepared by mechanical alloying method *J. Achiev. Mater. Manuf. Eng.* vol 30 issue 2 pp 141-46

[6] El-Eskandarany M S, Aoki K, Sumiyama K and Suzuki K 2002 Cyclic phase transformations of mechanically alloyed $Co_{75}Ti_{25}$ powders *Acta Materialia* vol 50 issue 5 pp 1113–23

[7] Chen Y, Li C P, Chen H and Chen Y 2007 One-dimensional nanomaterials synthesized using high-energy ball milling and annealing process *Sci. Technol. Adv. Mater.* vol 7 issue 8 pp 839–46

[8] Sa'diyah H, Nurhimawan S, Fatoni S A, Irmansyah and Irzaman 2016 Ektraksi silikon dioksida dari daun bambu *Pros. Semin. Nas. Fis.* vol V no. Oktober pp 13–16

[9] Ikono R, Nofrizal, Nugroho D W, Rahman T P, Widyaningrum R, Widayanto W B, Sukarto A, Siswanto and Rochman N T 2012 Sintesis nanopartikel ZnO dengan metode mechanochemical milling *Prosiding Pertemuan Ilmiah Ilmu Pengetahuan dan Teknologi Bahan* Serpong Oktober 2012 pp 60–62

[10] Yan J W, Liu Y, Peng A F and Lu Q G 2009 Fabrication of nano-crystalline W-Ni-Fe pre-alloyed powders by mechanical alloying technique *Trans. Nonferrous Met. Soc. China (English Ed.)* vol 19 suppl 3 pp s711–s717

[11] Kim Y D, Chung J Y, Kim J and Jeon H 2000 Formation of nanocrystalline Fe – Co powders produced by mechanical alloying *Mater. Sci. Eng.: A* vol 291 issue 1-2 pp 17–21

[12] Eckert J and Börner I 1997 Nanostructure formation and properties of ball-milled NiAl intermetallic compound *Mater. Sci. Eng.: A* vol 239-240 pp 619–24

[13] Umemoto M, Liu Z G, Masuyama K, Hao X J and Tsuchiya K 2001 Nanostructured Fe-C alloys produced by ball milling *Scr. Mater.* vol 44 issue 8-9 pp 1741–45

IOP Conf. Series: Materials Science and Engineering **403** (2018) 012091 doi:10.1088/1757-899X/403/1/012091

Variation of sitting and standing position at beverage packing factories

R Septiari, Pratikto, B S Purnomo and P T Ishardita

Mechanical Engineering Department, Brawijaya University, Malang, Indonesia

rennyseptiari@gmail.com

Abstract. This was preliminary study in granting the variation sitting and standing position at two different factories in the habit of packing process. The duration of time and pattern of sitting and standing was proposed based on this study to the two beverage packing factories in which the packing process of the drinks with the sitting and stand pattern. Determination of fatigue was by the heart rate, the length of time that showed productivity, job burnout was performed at 6 operators for 4 hours at each firm. The purpose of this study was to assess the pattern of time sitting and standing or walking postures of which may cause fatigue and also observe productivity based on the duration of packing from the variation of sitting, standing, and sitting-standing.

1. Introduction

In the marketing or distribution process of products, the packing process becomes a major instrument for the processed products. Packing itself is an attempt to organize and measure the product in a certain packaging according to certain the size and shape. A sitting position in work is done all the time and for a long time results can increase employee packing work discomfort [1]. Most of these employees work all the time with the same style [2], where they spend an average of more than 8 hours per day on the job.

Packing work carried out at a beverage company located in the Kudus area of Central Java belongs to one of the repetitive and monotonous works. Where during the work time the activity is to enter the cup drinks into the dos-dos that have been provided with a sitting position. This activity takes place continuously for one shift work that is 7 hours. Repeated activities performed on average the feeling of boredom and fatigue will appear after they work ± 2-3 hours, this is known by the decreasing level of packing speeds made by operators. Therefore, people who pack the products have a variety of shapes and types according to product type, size, shape, economical and tastes of the consumers. In the event of repeated packing process, it is common to cause back pain easily. This pain is often experienced by workers in the range of 75% up to 85% [3]. The symptom of back pain might be caused by several factors in the industrial world. In the beverage packing industry by using paper carton is still done manually. This is not done with the automation machine for it is still done by human power or manual; the costs incurred by the company can be more economical.

Packing the beverage means doing repetitive work and monotonous onset may lead back pain that indicates fatigue [4]. The tasks of packing the beverage is done by sitting can lead back pain. There is also the use of the foot as a support stand is often the case when there is no possibility for visual control in progress, such as when a motion made between the accelerator and the brake pedal of a car

[5]. This activity may cause fatigue as well. The process itself, the activity is done by either standing or sitting beside the conveyer where glasses of drinks may run over conveyer toward the workers packing. Then the workers will take the beverage one by one for the next set in the boxes having been provided.

In Pati, Central Java, Indonesia, there are two beverage packaging factories. Both factories have a different tradition of packing. For Factory A whose products are favored by children in Indonesia, the beverage packing process is done while standing. Instead, Factory B which also produces competitor product from Factory A, the beverage packing process is done while sitting. Repetitive and monotonous task is done by standing or sitting position and performed for nearly an 8 hours work shift reduced by one hour for a break.

In this research, various treatments were employed. The treatments were given by combining the packing process while sitting and standing. This study was expected to give recommendation for the factories of an efficient packing process and may increase productivity of the factories

2. Research method

The study was conducted in two beverage packing factory in Pati, Central Java, Indonesia. Research was conducted on the operator of both factories. Data were obtained for seven hours. The data obtained were in the form of the number packed boxes, heart rate, and level of fatigue due to the job. For the data related to the number of packed boxes was the amount collected boxes during 4 working hour of the operator. The 4 working hour was used since it was adjusted to the position change of work that was being tested at the time the research was being conducted. It was expected that outcome of packed boxes for 4 hours may represent the overall performance of each operator packing during working hour.

The data related to the heart rate were obtained from the calculation of the heart rate at the six operators, and for the work fatigue data were obtained by questionnaire given to the six operators. Rating the level of work fatigue was based on Likert's scale with the following assumptions:

1 = No	3 = Often
2 = Sometimes	4 = Always

The results of the work fatigue level were based on the answer choice of each worker which was the most preferred among proposed 17 questions. The chosen number was the presentation of the most common fatigue experienced by the workers ranged between 60% to 100%.

The change of the heart rate indicates the change of workload. The heart rate of each operator was taken after they had performed the work in accordance with the change of position that had been set. For the duration of action following the change in the time schedule was for 4 hours. Heart rate data were taken due to the changes that occurred in each operator.

3. Findings

The findings of this research is explained in the following subsections.

3.1. *Number of boxes*

Based on figure 1(a) and figure 1(b), it is known that an increasing number of boxes packed after a given treatment sitting-standing. At the Factory A, the operator had the habit of standing, while for Factory B the operator was accustomed to sitting. As it is presented in figures 1(a) and figure 1(b), there was increased number of boxes packed while variation sitting-standing position for both operators at the Factories A and B. Work while prolonged sitting may cause inconvenience working position against given time [6]. Sitting position is the position where the feet are not weighed down with body weight and a stable position during work. Adjustment posture with an increasing frequency throughout the working day is used to reduce discomfort during work [7]. The variation sitting-standing may provide comfort condition to the operators, so that it may improve their work performance [8]. In addition, the productivity can be assessed in various ways, one of them is with the quality of work produced [9]. In the packing process, it can be assessed from the number of boxes having been packed by the workers. For variation sitting-standing was out of the habit of the operators is believed to make increase in productivity compared with if their position was monotonously either sitting alone or standing alone. This position will provide relaxation for the body of the operators, especially on the coccyx and spine [10].

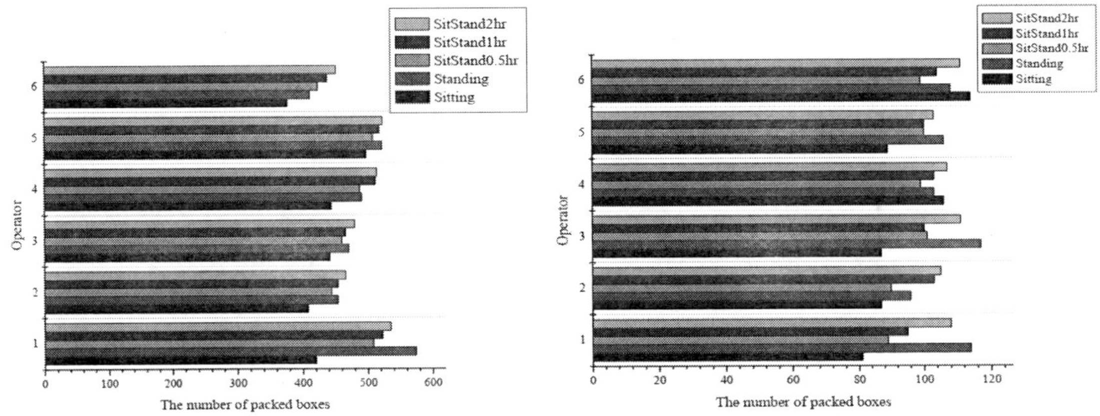

(a) (b)

Figure 1. The number of packed boxes produced by the operators at the time of various treatments at (a) Factory A and (b) Factory B.

3.2. Heart rate

Based on figure 2(a) and figure 2(b), it was shown that the variation of the sitting-standing position gave lower heart rate of the workers. If someone feels the workload is high, and then the change in heart rate will also happen which is beating faster. Sitting and standing positions have become the habit of the operators in each factory. The change of position habit will cause discomfort in work [11]. There was a decrease in heart rate in the sitting-standing position variation, where it was known that work fatigue is synonymous with an increase in heart rate [9]. At work habits where the operator only either sits or stand during the working hour may be risky to the health of your back. This is due to manual handling, repetitive movements, twisting and bending, often lifting and whole body vibration [3]. Variations of the sitting-standing will reduce repetitive movements, giving relaxation and vibration for the whole body.

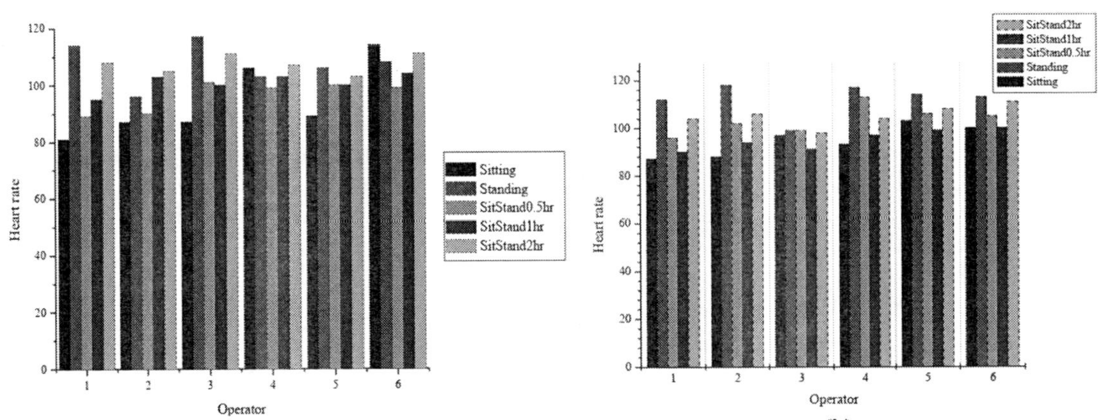

(a) (b)

Figure 2. Heart rate during the various treatments at the (a) Factory A and (b) Factory B.

3.3. Fatigue level

Figure 3(a) and figure 3(b) showed the fatigue level based on interviews with the operators. The treatment of variation sitting-standing position may reduce work fatigue felt by the operators. This study showed the tendency of inconvenience of working in one position only either sitting or standing alone. Monotonous and repetitive position may lead into the inconvenience [12]. In other words, one position either just sitting or standing is likely more to inconvenience. In the case of beverage packing process, the inconvenience is likely to occur due to fatigue on the wrist and hand muscles. In addition, either only sitting or standing would easily cause spinal shrinkage [11]. Analysis of the posture for each body movement and variation can give detailed calculation of the level of fatigue due to the one static position.

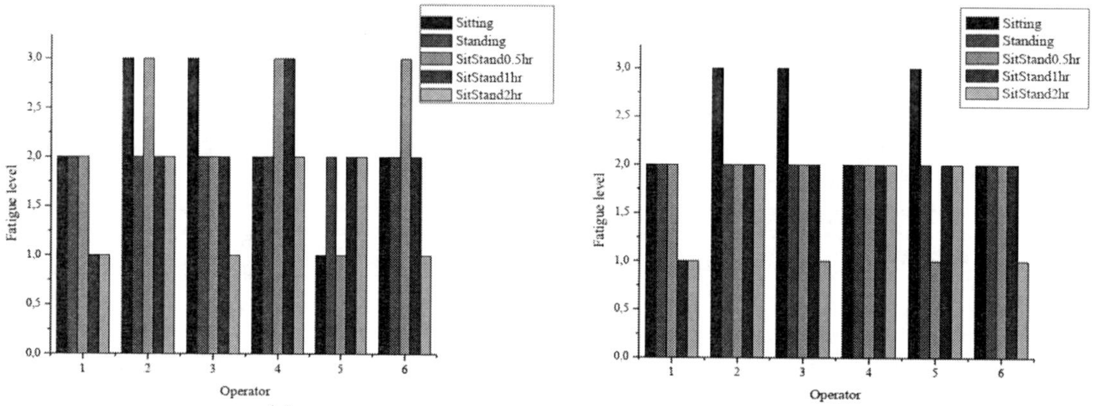

(a) (b)

Figure 3. Fatigue level of various treatments at the (a) Factory A and (b) Factory B.

4. Conclusion

Variation of sitting-standing position gives increased productivity as it was shown from the number of boxes packed at the two factories. The boxes were produced in a sitting-standing position for two hours. There was a decrease in heart rate with the variation of sitting-standing position. Variation of sitting-standing position may lower the levels of fatigue and heart rate experienced by the operators. The initial assessment of this variation may proceed to the calculation of the change in position by the angle formed from the packing tasks.

References

[1] Fenety A and Walker J M 2002 Short-term effects of workstation exercises on musculoskeletal discomfort and postural changes in seated video display unit workers *Phys. Ther.* vol 82 issue 6 pp 578–89

[2] Brown W J, Miller Y D and Miller R 2003 Sitting time and work patterns as indicators of overweight and obesity in Australian adults *Int. J. Obes.* vol 27 issue 11 pp 1340–6

[3] Boos N and Aebi M 2008 *Spinal disorders : fundamentals of diagnosis and treatment* (Berlin, Heidelberg: Springer)

[4] Chan A H S and Chan K W L 2010 Three-dimensional spatial stimulus–response (S–R) compatibility for visual signals with hand and foot controls *Appl. Ergon.* vol 41 issue 6 pp 840–8

[5] Hoffmann E R 1991 Accelerator-to-brake movement times *Ergonomics* vol 34 pp 277–87

[6] Callaghan J P, Gregory D E and Durkin J L 2010 Do NIRS measures relate to objective low back discomfort during sedentary tasks? *Int. J. Ind. Ergon.* vol 40 issue 2 pp 165–70

[7] Haynes S and Williams K 2008 Impact of seating posture on user comfort and typing performance for people with chronic low back pain *Int. J. Ind. Ergon.* vol 38 issue 1 pp 35–46

[8] Vink P, Konijn I, Jongejan B and Berger M 2009 Varying the office work posture between standing, half-standing and sitting results in less discomfort *International Conference on Ergonomics and Health Aspects of Work with Computers (EHAWC 2009)* pp 115–20

[9] Husemann B, Von Mach C Y, Borsotto D, Zepf K I and Scharnbacher J 2009 Comparisons of musculoskeletal complaints and data entry between a sitting and a sit-stand workstation paradigm *Hum. Factors* vol 51 issue 3 pp 310–20

[10] Roelofs A and Straker L 2002 The experience of musculoskeletal discomfort amongst bank tellers who just sit, just stand or sit and stand at work *Ergonomics* vol 14 issue 2 pp 11-29

[11] Zhang L, Helander M G and Drury C G 1996 Identifying factors of comfort and discomfort in sitting *Hum. Factors J. Hum. Factors Ergon. Soc.* vol 38 issue 3 pp 377–89

[12] Ebara T, Kubo T, Inoue T, Murasaki G, Takeyama H, Sato T, Suzumura H, Niwa S, Takanishi T, Tachi N and Itani T 2008 Effects of adjustable sit-stand VDT workstations on workers' musculoskeletal discomfort, alertness and performance *Ind. Health* vol 46 issue 5 pp 497–505

Design and experimental test for solar chimney power plant: case study in Riau Province, Indonesia

A Ridwan[1], H Hafizh[1] and M R Fauzi[2]

[1]Mechanical Engineering Department, [2]Otomotive Department
Universitas Muhammadiyah Riau, Indonesia

abrar.ridwan@umri.ac.id

Abstract. Solar chimney is an energy generator that utilizes solar thermal energy. The working principle air is heated in the solar collector, the increased temperature causes downward pressure, and the air density becomes reduced to create a buoyance force or air moving upward through the chimney. This research aimed to investigate the theoretical design in determining dimension of solar chimney and experimental data test. The solar chimney made with collector roof area of 1.3225 m^2 with clear glass as thick as 5 mm. Chimney is made from PVC pipe with 83 mm inner diameter and high 193 cm. The material of solar chimney made of aluminum pipe. Experimental data was taken on March 2017 from 11:00 AM to 15:00 (3 PM). The design power output was 0.03 watt and the actual values ranged between 0.0133 watt and 0.0317 watt The design of chimney inlet air velocity was 2.15 m /s while the actual value varied from 1.4 m/s to around 2.2 m/s. These results show that the design is able to assess the potential of solar chimney system in Riau Province.

1. Introduction

Nowadays, refer to the increasing rate of environmental pollution and limitation on fossil fuel resources, the use of sustainable energy seems to be inevitable and an absolute necessity for the world. Recently the impact of environmental pollution caused by fossil fuels is one of huge concern. Solar chimney is the energy generation that utilizes solar thermal radiation energy. This energy generator consists of a solar collector and a tower or chimney placed at the center of the collector. Solar energy is used to heat the air inside the solar collector. The air flow to the chimney is formed due to the temperature difference outside and inside the collector. On a scale power plant kinetic energy in the form of speed from the high air to the chimney is used to drive the wind turbine coupled to the generator which will generate electricity.

The greatest advantage of solar thermal energy compared to other energies because it is clean without pollution environment, and its abundant availability in the world. Indonesia as the part of the world is a region which has a considerable intensity of solar thermal radiation. Compared with the monthly rate energy production of 126.83 kWh per day which has been built in Manzanares, Spain, the capacity of energy production in Riau Province, Indonesia almost doubled with 202.72 kWh per day [1]. Another fact, Riau Province, Indonesia also get sun radiation intensity with 6.06 kWh per m^2 in March [2]. This is a potential for Riau Province to utilize the heat of solar energy into a clean energy source.

Solar chimney is one of the energy generation that utilizes solar thermal radiation energy. This energy generator consists of a solar collector, a tower or chimney placed at the center of the collector and turbine. The speed from the high air to the chimney is used to drive the wind turbine coupled to the generator which will then generate electricity. To design a large-scale solar chimney power plant, it is necessary to ensure that the power plant works as expected. Some of the part needs to prepare like: height chimney, collector area, the intensity of solar radiation, the height of collector roof, heat storage material and others. Therefore, this study aims to design a prototype of solar chimney with experimental test in Riau Province, Indonesia.

2. Mathematical model

Solar chimneys are energy plants that use solar radiation to raise air temperature, buoyancy of warm air to accelerate the flow of air throughout the system. The main parts of the solar chimney are shown in figure 1.

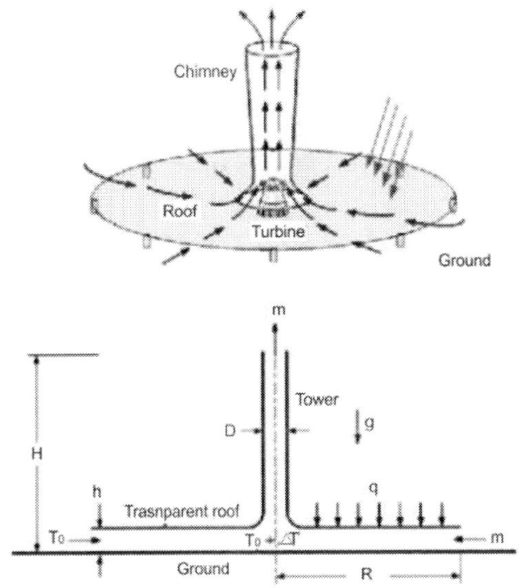

Figure 1. Dimension of solar chimney [3].

The air gets hot as a result of the greenhouse effect under a transparent roof (collector). Because the roof is open at the edges, the buoyancy of hot air forms a continuous flow from the edge of the roof to the chimney. A turbine is placed on the passage of the air stream to convert kinetic energy from the air flowing into electrical energy. Based on the scheme of figure 1, the increasing temperature can be estimated from the energy equation in the roof section as shown in equation (1) – (6).

$$\dot{m}C_p(T_2 - T_1) + \frac{1}{2}\dot{m}(V_2^2 - V_1^2) = q^{"}A_r \quad (1)$$

$$\dot{\mathbf{m}} = \boldsymbol{\rho}\mathbf{A}_{\mathbf{ch}}\mathbf{V} \quad (2)$$

$$\dot{m}c_p\Delta T = q"A_r \quad (3)$$

$$T_2 = T_1 + \frac{q"A_r}{\dot{m}c_p} \quad (4)$$

$$q" = \alpha I - U\Delta T \quad (5)$$

$$\Delta T = \frac{\alpha I}{\frac{\dot{m}c_p}{A_r}+U} \quad (6)$$

The efficiency collector η_{coll} can be expressed using equation (7) and (8).

$$\eta_{coll} = \frac{q"}{I} \quad (7)$$

$$A_r = \frac{\dot{m}c_p\,\Delta T}{\eta_{coll} I} \quad (8)$$

Without the turbine, maximum airflow velocity V is reached and all pressure differences are used to accelerate the air so that it becomes kinetic energy as shown in equation (9) and (10).

$$P_{tot} = \frac{1}{2}\dot{m}V^2 \quad (9)$$

$$V = \sqrt{2gH\frac{\Delta T}{T_a}} \quad (10)$$

The coefficient losses at top collector can be expressed in equation (11) - (16).

$$U_t = (h_w + h_{rs}) \quad (11)$$

$$h_w = \frac{k}{L}Nu \quad (12)$$

$$h_{rs} = \frac{\sigma\varepsilon(T_1+T_s)(T_1^2+T_s^2)(T_1-T_s)}{(T_1-T_\sim)} \quad (13)$$

$$Nu = 1 + 1.44\left[1 - \frac{1708(sin1.8\beta)^{1.6}}{Ra\,cos\beta}\right]\left[1 - \frac{1708}{Ra\,cos\beta}\right]^+ + \left[\left(\frac{Ra\,cos\beta}{5830}\right)^{1/3} - 1\right]^+ \quad (14)$$

$$Ra = \frac{g\beta'\Delta T L^3}{\nu\alpha} \quad (15)$$

$$T_s = T_\sim\left[0.711 + 0.0056(T_{dp} - 273.15) + 0.000073(T_{dp} - 273.15)^2 + 0.013\cos(15t)\right]^{1/4} \quad (16)$$

Meanwhile for the equation of losses coefficient under collector is shown in equation (17).

$$\frac{1}{U_b} = \frac{L_1}{k_1} + \frac{L_2}{k_2} + \frac{L_3}{k_3} \quad (17)$$

All are above those equations taken from [4].

3. Method of design

To define the design dimension of solar chimney, first we have to decide several assumptions, where the solar radiation determined after calculation is 918 w/m^2 while ambient temperature T_a is 30 and dew point temperature T_{dp} = 25.7°C. For the coefficient losses of collector U, 14 W/m^2 K. The collector installed in this solar chimney is glass which its emissivity ε_c, is 0.9 and ground absorptivity is α, 0.98, while emissivity of ground absorber ε_p, is 0.88. In order to achieve 50 m watt power without turbine dimension of glass 115 cm x 115 cm (available dimension in Indonesia market), meanwhile different temperature between inlet and outside ΔT, 37 C. With those above equations found the required air velocity based on equation (9) is 2.15 m/s. The height of the required chimney based on equation (10) is 1.93 m. The chimney diameter is 0.086 m which calculated based on equation (2). Collector height taken 6 cm based on experimental test conducted by [5] and it was the best height.

Clear glass applied as collector with thickness 1.5 mm is good material for its transmissivity value 0.9. The dimension of clear glass was selected considering availability of collector material in the local market. The points of measurement were adjusted for temperature inlet collector, plat absorber, inlet chimney and ambient utilized by Advantech 4718. Krisbow KW06035 anemometer was chosen to collect data for air velocity of chimney inlet. The data collected for every 15 minute from 10:00 AM until 15:00 (03:00 PM).

4. Result and discussion

4.1. Radiance intensity in Riau Province

Daily solar radiations of direct and diffuse for clear sky day were calculated based on a mathematical models equation from [6], due to solar irradiance instrumental very expensive so that these approach equations are matched. In other hand this value has to be determined to calculate air mass flow rate and power output. The solar irradiance in Riau Province is shown in figure 2.

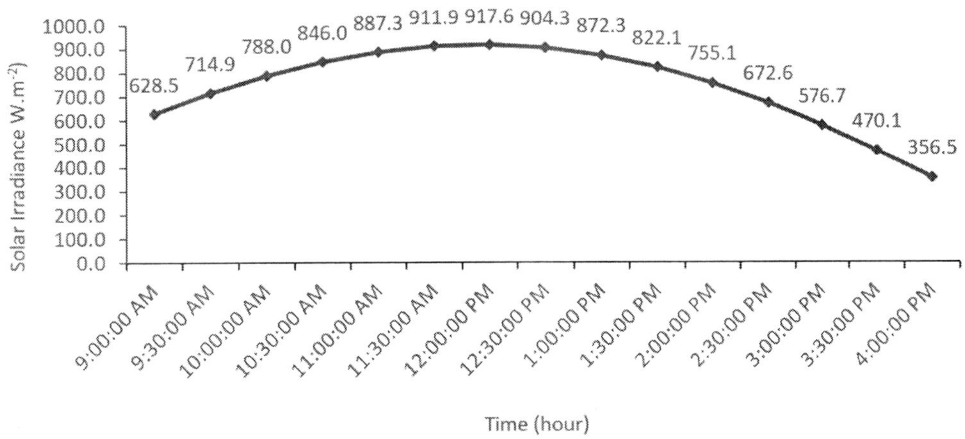

Figure 2. Solar irradiance in Riau Province.

The peak solar irradiance is 918 W.m^{-2} occurred in 12:00 and this intensity larger than other country such as Manzanares, Spain which only 800 W.m^{-2}, because this region located in the equator line.

4.2. Solar chimney dimension

The whole equations above used to design the prototype of solar chimney. There several parameters to be as assumption such as the power 30 mWatt, diameter of clear glass 1.15 m and differential temperature of ambient and plat absorber as 37 ^{0}C. To calculate coefficient of losses at top collector side by its value 12.4 W.m^{-2} K and also there are many parameters found by all above equations such ash_w = 2.28 W.m^{-2} h_{rs} = 10.14 W.m^{-2}.K with Nu= 5.1 and Ra = 284686. Hopefully by these values can helps in solar chimney's design.

4.3. Absorber and collector

The selected heat absorber floor material is an aluminum plate with a thickness of 0.02 mm and is dyed black to maximize the absorption of solar radiation the aluminum is better of reflectivity value than stainless steel plate. The position of solar chimney made from aluminum hollow 1x1 or 23 mm x 23 mm and 5 mm, in other hand the clear glass is selected as collector's roof with its transmissivity value 0.9, whereas high roof collector is 0.06 m this is the best height of inlet collector. The view of frame collector and absorber is shown in figure 3.

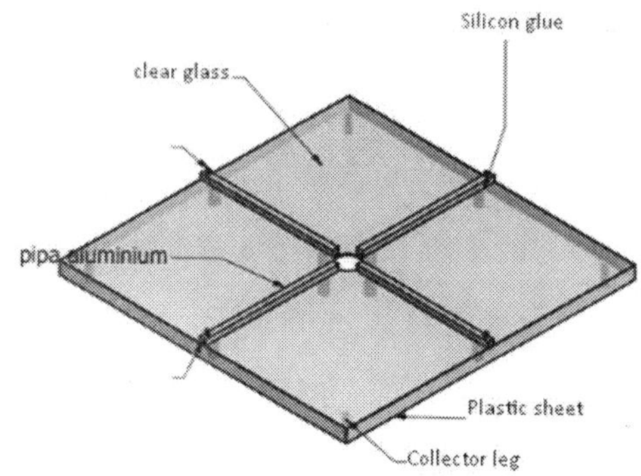

Figure 3. The view of frame collector and absorber.

4.4. Chimney

PVC pipe is selected as chimney material is selected due to its low conductivity thermal as well as it considered to be adiabatic system inside chimney. From equation (10) it is found that the height of chimney 2 m while the air mass flow rate 0.013 kg/s. The adiabatic system of chimney will create the pressure drop from the inlet into outlet of chimney, by these the force buoyancy occurred. The cross wind ambient measured around 5 – 8 m/s therefore need a rope to tie the chimney. Inner diameter of chimney supposed to be 0.083 which has constant diameter. Figure 4 shows the result of solar chimney design (a) side-view of solar chimney design (b) top-view over solar chimney (c) solar chimney.

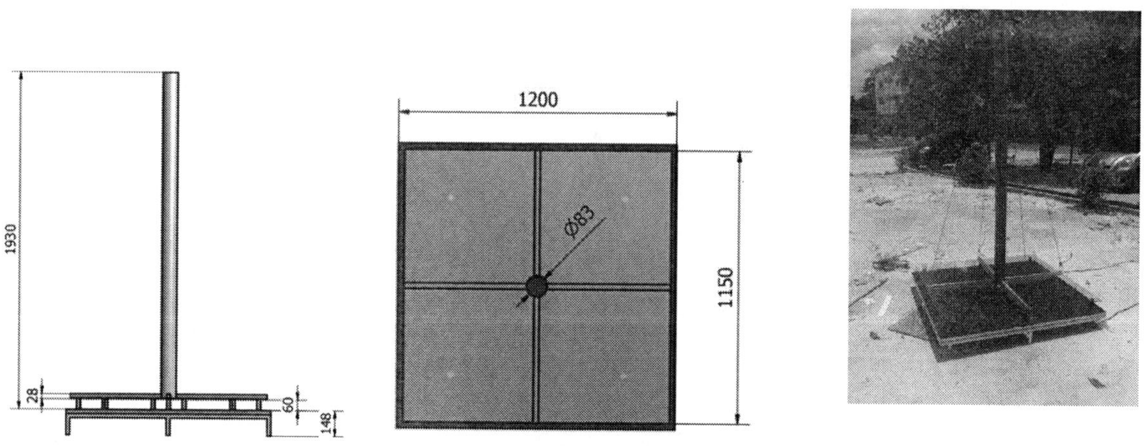

(a) (b) (c)

Figure 4. Results of solar chimney design.

The experimental was conducted in the yard of Universitas Muhammadiyah Riau on March 21, 2017. Which graphical represented latitude 0.49989036^0 and longitude 101.41544804^0, where 27 m above sea level based on Google maps. Data collected for every 15 minutes from 11:00 AM to 14:00 (02:00 PM). The measuring instrument used thermocouple K type, and an anemometer placed at the base of the chimney. The measured data are ambient air temperature, air temperature at the bottom of chimney, heat absorber plate temperature, and air velocity at the base of chimney. The data is then analyzed to obtain power without turbines by using equation (9).

4.5. Absorber insulation

In order to isolate heat losses comes from the bottom of basement into atmosphere, it is necessary utilizing aluminum foils, Styrofoam and carpet by their thickness 2 mm, 7 mm and 20 mm respectively, while the aluminum also become support the equipment of solar chimney. Figure 5 depicts how the system isolated.

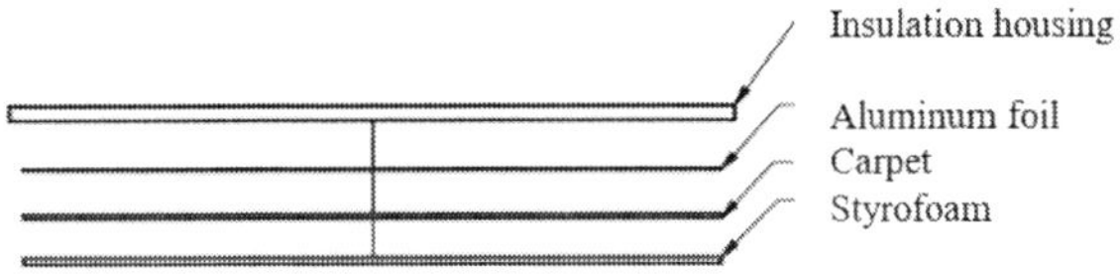

Figure 5. The composition of material insulation.

By these above isolators the heat only flow over the plate and going into chimney inlet, even though there are still not still good isolator to be applied.

4.6. Result test

After manufacturing and assembling the solar chimney by its instrumental measurement figure 6 illustrates temperature of ambient, plat and chimney inlet. It indicates that the highest temperature of plate 57.8 ^{0}C occurred in 11:00 AM while the maximum temperature inlet chimney also at that time. It concludes that temperature chimney inlet was effected by absorber temperature.

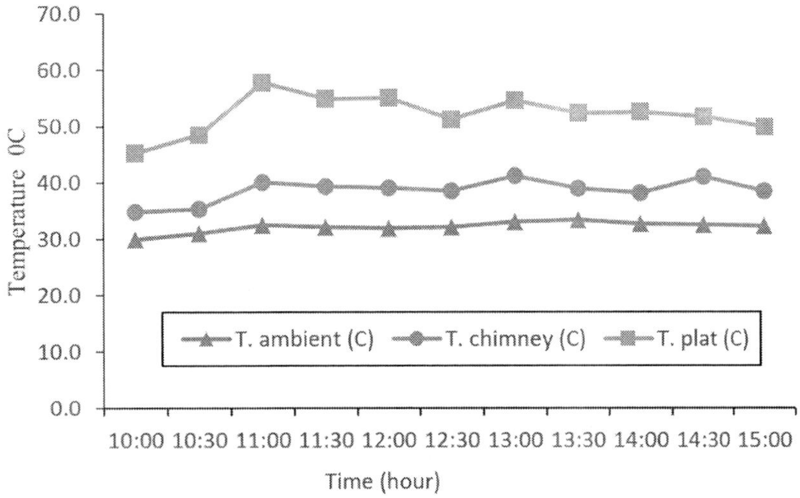

Figure 6. The graph relationship between ambient, chimney inlet, plat and time.

From the graph in figure 6, the differential temperature between ambient and plat really significant and will lead to higher power output. The ambient wind influenced the experimental test involved to power fluctuation. Hence the effect of wind velocity also caused by ambient wind velocity. The air velocity inlet chimney graph is shown in figure 7.

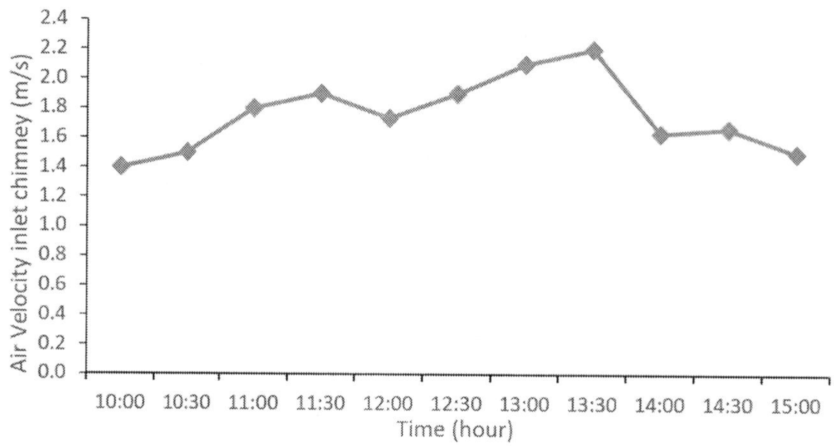

Figure 7. Graphic trend line time versus air chimney inlet velocity.

From the graphs in figure 7, the air velocity inlet chimney starts in 10:00 until 15:00, the air velocity tends to increase and stops at 11:30 because there is cross wind ambient moving over plat absorber that is why absorber become colder and after 12:00 rises again until 13:30 and then decreased because of ambient temperature dropped. The highest air velocity inlet chimney is 2.2 m/s in 11:30 meanwhile the lowest air velocity 1.4 m/s happened in 10:00.

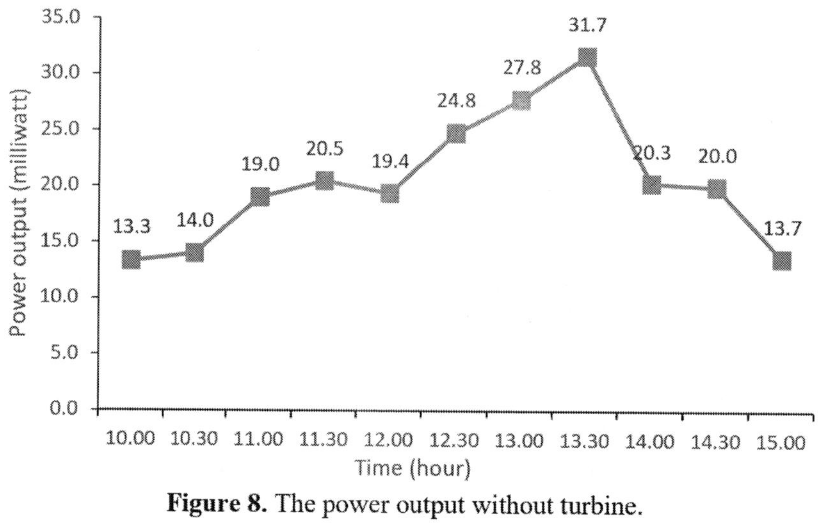

Figure 8. The power output without turbine.

The measured power output without turbine is shown in figure 8. Based on the measurement, temperature and air velocity, the resulting power without turbine reaches 0.0317 Watt at 13:30. And the lowest is 0.0133 Watt at 10:00. Data collected 23, 26 and 28 August 2017. The change of power caused by air velocity and temperature at the base of the chimney varies, in this study observed due to various factors such as ambient cross wind going into the collector that would increase the solar chimney's power. By this condition also enlarge the coefficient loss of collector.

It is clearly seen that the power output along with the parameters which are homogeneous to the power like velocity inlet chimney, mass flow rate approaches their peak almost always at mid-day approximately as supposed because at that time solar radiation becomes maximum due to itis position.

5. Conclusion

Solar chimney that has been made has height of the chimney is 200 cm, the size of the collector roof is 115 cm x 115 cm, the inner diameter of the chimney is 8.3 cm, the collector roof height is 6 cm. Material insulation selected are very satisfactory even the efficiency of collector 41%. In the data analysis, solar chimney produces the lowest power 0.0133 Watt and the highest 0.0317 Watt. Average power generated is 0.020 W. The design power output was 0.03 Watt and the actual values found ranged between 0.0133 Watt and 0.0317 Watt The design of chimney inlet air velocity was 2.15 m /s while the actual value varied from 1.4 m/s to around 2.2 m/s. These results show that the design is able to assess, with a fairly good with experimental and Riau province has great prospect to generate electricity by solar chimney system.

References

[1] Hafizh *et al.* 2015 Solar Chimney *Disertation* Kyoto University

[2] Rumbayan M, Abudureyimu A and Nagasaka K 2012 Mapping of solar energy potential in Indonesia using artificial neural network and geographical information system *Renewable and Sustainable Energy Reviews* vol 16 issue 3 pp 1437-49

[3] Koonsrisuk A, Lorente S and Bejan A 2010 Constructal solar chimney configuration *International Journal of Heat and Mass Transfer* vol 53 issue 1-3 pp 327-33

[4] Koonsrisuk A and Chitsomboon T 2013 Mathematical modeling of solar chimney power plants *Energy* vol 51 pp 314-22

[5] Ghalamchi M, Kasaeian A, Ghalamchi M and Mirzahosseini A H 2016 An experimental study on the thermal performance of a solar chimney with different dimensional parameters *Renewable Energy* vol 91 pp 477-83

[6] Duffie J A and Beckman W A 2013 *Solar engineering of thermal processes*, 4th ed. (John Wiley & Sons, Inc.)

IOP Conf. Series: Materials Science and Engineering **403** (2018) 012093 doi:10.1088/1757-899X/403/1/012093

Potential of ½ cycle refrigeration system for food transport application

M Setiyo[1], B Waluyo[1], B C Purnomo[1], D W Karmiadji[2,3] and I C Setiawan[3]

[1] Department of Automotive Engineering, Universitas Muhammadiyah Magelang, Magelang, Indonesia.
[2] Agency for the Assessment and Application of Technology, PUSPITEK Serpong, Indonesia.
[3] Master of Mechanical Engineering, Pancasila University, Jakarta, Indonesia.

setiyo.muji@ummgl.ac.id

Abstract. Maintaining the quality of food products from an agricultural land and farms to the final consumer is usually conducted through the car-cooling box. However, as long as the refrigeration equipment takes energy from the car engine, it will significantly increase fuel consumption. Therefore, this paper presents the concept of 1/2 cycle refrigeration system on LPG-fueled vehicles to reduce fuel consumption due to the load of the refrigeration system. From the numerical study, the ½ cycle refrigeration system on LPG-fueled vehicles is potential to be applied on the car-cooling box as the secondary refrigeration system. This system provides a considerable potential cooling effect of 0.8 kW at LPG flow rate of 2.4 g/s.

1. Introduction

Food processing and distribution industries take full responsibility for maintaining the quality of agricultural products and farms from producers to consumers, or even to the dining table. The need for fresh vegetables and fruits is a great market demand [1]. In fact, it is impossible to provide agricultural land and settlements in one area. The need for large amounts of fresh food takes place in urban centers, industrial centers, and residential centers. Meanwhile, agricultural land to produce vegetables, fruits, or even farms is commonly far from the city. In some archipelago countries such as Indonesia, the availability of land supporting agricultural and farms activities is outside the needy island [2]. In fact, for a long time, fresh food has been transported from one country to several other countries, from one continent to another [3]. This is not only sufficient availability of land but more on geographical factors, where specific vegetable and fruits commodities can only be produced optimally on a certain high plateau or rainfall [4].

Currently, food preservation technology through refrigeration to withstand a few days or through freezing for several months becomes the need to support the distribution system. In order to maintain the quality of the product, a modern thermo-box not only regulates temperature, but has come to the arrangement of the relative humidity, the air velocity, and perhaps [1]. This is because the supply chain of foodstuffs from farmland, plantations, or farms can form short chains or very long chains according to the type and pattern of distribution. Figure 1 presents the complexities of cold chain development to maintain the quality of long-chain foodstuffs [5].

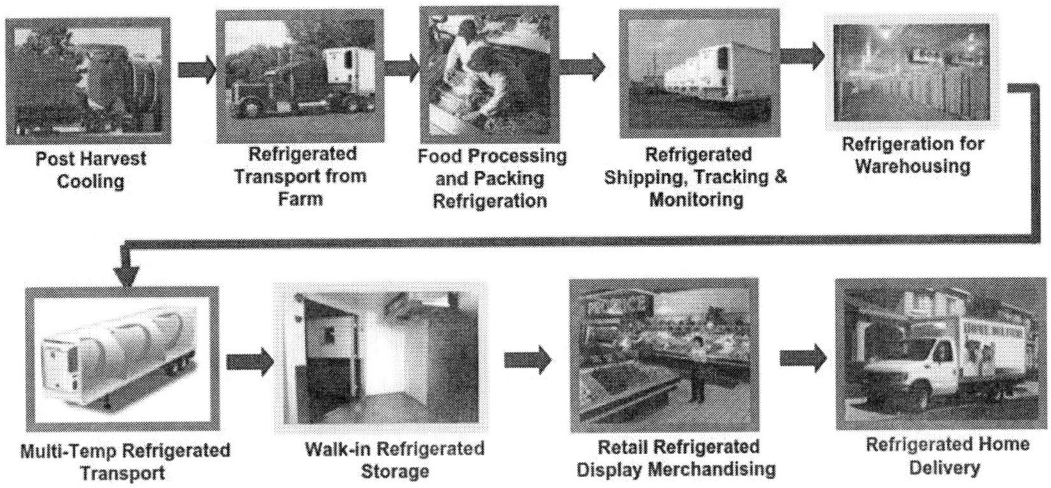

Figure 1. Cold chain development to maintain the quality of the farming product.

The standard transport of foodstuffs and agricultural products was governed by an agreement which was subsequently withdrawn by The Inland Transport Committee of the United Nations Economic Committee for Europe in 1970-1971 [6] which was amended several times and the last of 2016 [7]. This agreement is to facilitate international traffic related to food security by establishing recognized standards. The agreement was known as an ATP deal and was adopted in Britain in 1980 which was then used by many countries. This standard provides general information for controlled vehicles with controlled temperatures, such as road vehicles, trains, and marine containers [8]. Several types of transport equipment of agricultural and farms products are presented in figure 2 [5].

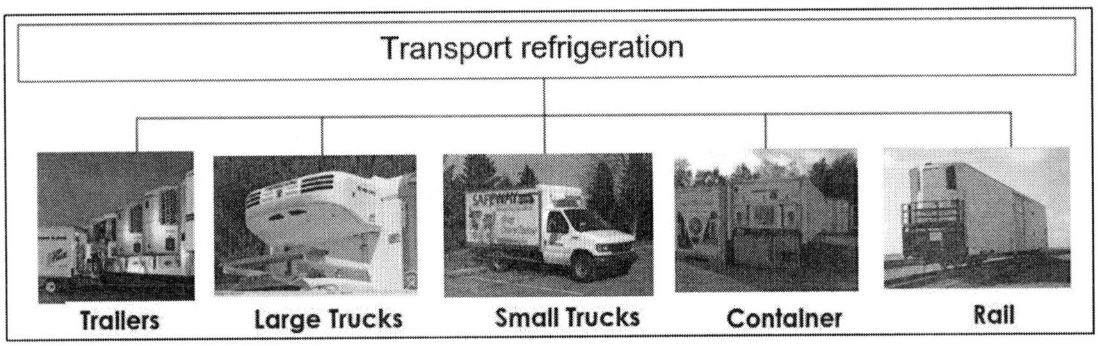

Figure 2. Transport refrigeration provides temperature control solutions during distribution of agricultural product.

On the one hand, a cooling capacity is required to transport foodstuffs to ensure their quality is maintained. However, as long as the refrigeration system is driven by a car engine with vapor compression system, it will take significant fuel and produce more exhaust emissions [9–11]. Waste of fuel is not only an effect on the increase in operational costs that cause the price of products to be more expensive but also gives an adverse effect on increasing global warming. This is also a concern for environmental practitioners, engineers, and stakeholders. In addition to the vapor compression system, the air cycle system has been piloted in food transport systems but has low performance [8].

IOP Conf. Series: Materials Science and Engineering **403** (2018) 012093 doi:10.1088/1757-899X/403/1/012093

2. Opportunity from LPG fueled vehicle

In previous studies, we have successfully harvested the cooling effect of the LPG evaporation process during LPG flowing in the fuel line from the tank to the engine through an additional evaporator. A numerical simulation has been done to calculate the potential cooling effect that can be obtained and the result is quite large reaching 1.5 kW at 3000 rpm for 2000 cc engine [12]. Then, the results of this simulation are re-verified more closely and the result is about 1.2 kW [13]. Finally, the results of the simulation were validated by lab-test and proven to produce of 1.2 kW cooling effect at LPG flow rate of 3 g/s representing the average consumption of 2000 cc eco-driving car [14]. The ½ cycle refrigeration concept of an LPG-fueled vehicle allows for a hybrid system with an air-conditioning system installed on the car as shown in figure 3.

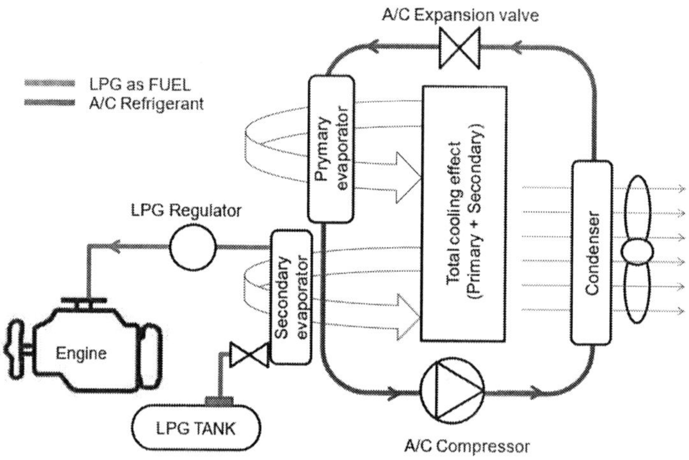

Figure 3. Concept of Hybrid A/C system on LPG fueled vehicle.

3. Simulation on 1500 cc SI Engine

In this study, a 1500 cc LPG fueled vehicle was simulated for the transportation of agricultural products and food products with cooling boxes as shown in figure 4. In this case, there are two refrigeration systems, the fully cycle system driven by engine and 1/2 cycle system from LPG evaporation.

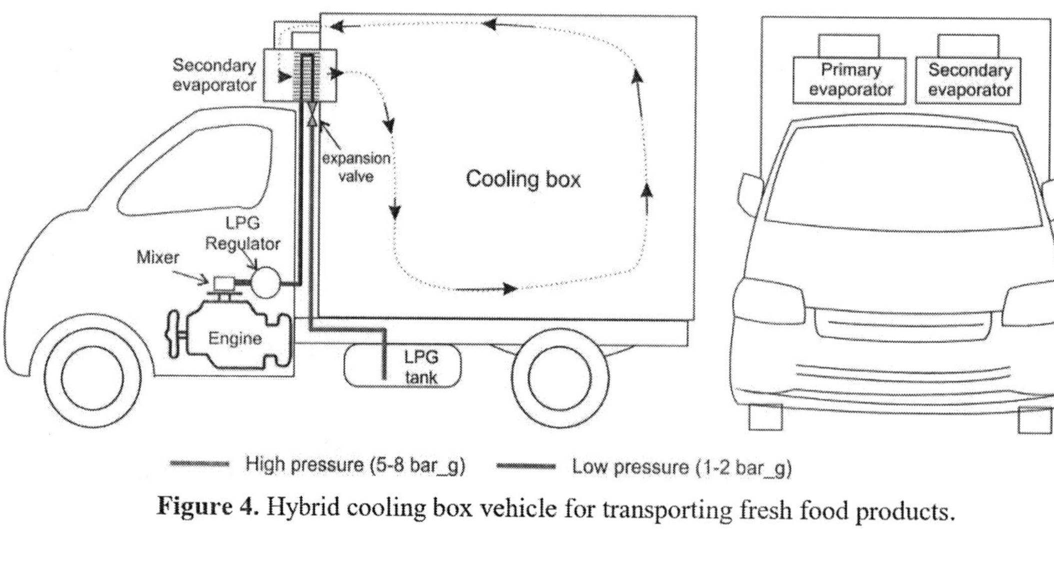

Figure 4. Hybrid cooling box vehicle for transporting fresh food products.

4. Result and discussion

This section discusses the simulation result on potential cooling effect and contribution to the refrigeration load in the following subsections.

4.1. Potential cooling effect

In this study, the potential cooling effect calculated by the following assumptions: (a) The LPG used is a mixture of 50% propane and 50% butane; (b) The expansion process takes on iso-enthalpy; (c) LPG before expansion valves is 30 °C, 8 bar_g and exits the evaporator at 24 °C, 2 bar_g; (d) LPG mass flow rate is estimated at 2.4 g/s, calculated on stoichiometric mixtures for 1500 cc engine and 3000 rpm with volumetric efficiency is estimated from previous studies [12]; and (e) LPG properties are obtained from REFPROP-NIST. The P-h diagram of 1/2 of the refrigeration cycle based on the assumption of assumption is shown in figure 5.

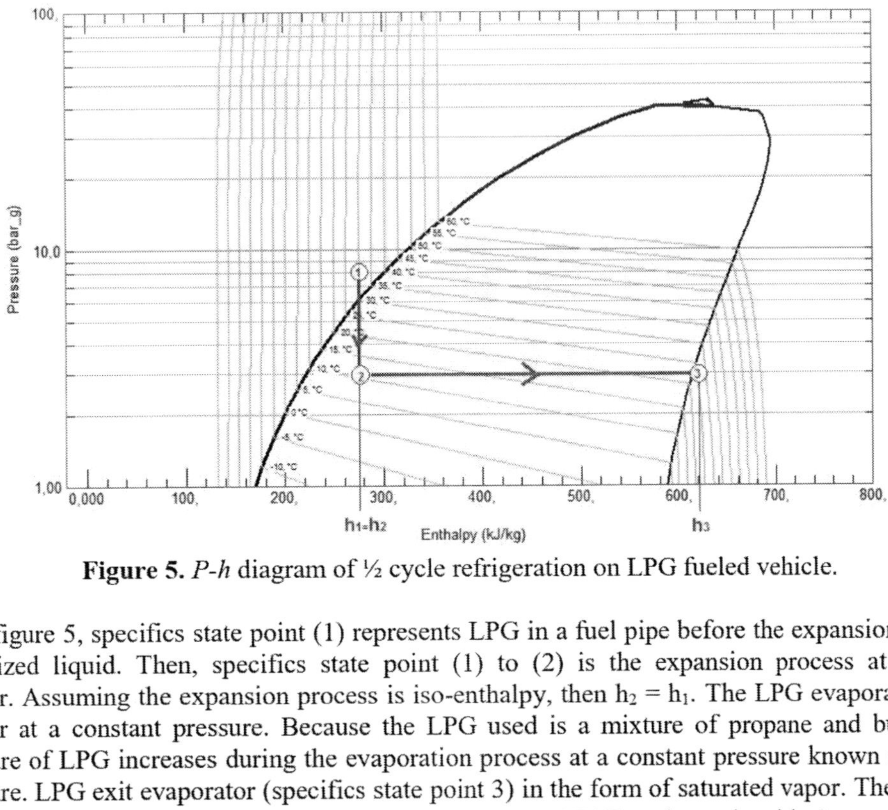

Figure 5. *P-h* diagram of ½ cycle refrigeration on LPG fueled vehicle.

From figure 5, specifics state point (1) represents LPG in a fuel pipe before the expansion valve as a pressurized liquid. Then, specifics state point (1) to (2) is the expansion process at the inlet evaporator. Assuming the expansion process is iso-enthalpy, then $h_2 = h_1$. The LPG evaporates in the evaporator at a constant pressure. Because the LPG used is a mixture of propane and butane, the temperature of LPG increases during the evaporation process at a constant pressure known as gliding temperature. LPG exit evaporator (specifics state point 3) in the form of saturated vapor. The enthalpy value for each state point (h_1, h_3) is obtained from REFPROP-NIST as shown in table 1.

Table 1. Enthalpy data at each specifics state point (h_1, h_3).

REFPROP - NIST Reference Fluid Properties
File Edit Options Substance Calculate Plot Window Help Cautions
2: butane/propane: Specified state points [Barometric ...

	Temperature (°C)	Pressure (bar_g)	Density (kg/m³)	Enthalpy (kJ/kg)	Entropy (kJ/kg-K)
1	30,000	8,0000	526,33	274,85	1,2560
2	24,000	2,0000	6,5587	620,65	2,5020
3					

Using the first law of thermodynamics, the heat released by air (cooling effect) is the same as the heat used to evaporate the LPG. Therefore, the potential cooling effect that can be harvested is defined in equation (1) and equation (2) [13, 15].

$$q_{into\ LPG} = q_{loss\ by\ air} \quad (1)$$

$$q_{ev} = \dot{m}_{LPG}\ (h_3 - h_1) \quad (2)$$

Where, q_{ev} is the potential cooling effect (kW), $\dot{m}_{LPG}$ is flow rate of LPG (kg/s), and h_3, h_1 are the specifics enthalpy of LPG in the outlet and inlet evaporator, respectively. Using the enthalpy data from table 1 and the LPG flow rate of 3 g/s, the potential cooling effect that can be harvested can be calculated as $q_{ev} = \frac{0.024kg}{s}(620.65 - 274.85)\left(\frac{kJ}{kg}\right)\ \ [kW] = 0.83\ [kW]$.

4.2. Contribution to the refrigeration load

We found that refrigeration load for a small-size cooler box car is using Hwa Sung Thermo HT-100 II with refrigeration cooling capacity of 3,224 kW [16]. This means that the potential cooling effect harvested from the LPG fuel system provides approximately 25.7 % of the refrigeration capacity requirement for medium-size cooler boxes. However, this potential varies according to the mass flow rate of LPG and the quality of the vapor formed in the evaporator. With one condition, LPG entering the expansion valve must be liquid.

5. Conclusion

From the numerical study, the ½ cycle refrigeration system on LPG-fueled vehicles is opportunities to be applied on the food transportation box as the secondary refrigeration system. This system provides a considerable potential cooling effect of more than 0.8 kW at LPG flow rate of 2.4 g/s. However, this number is an estimated number at a stationary condition of 3000 rpm. In fact, it is impossible to drive a vehicle at a constant high speed so that the potential cooling effect that can be harvested depends on driving mode, in this case, the LPG flow rate from the tank to the engine. This work will be continued by creating a prototype that resembles the original to test the reliability of the system.

Acknowledgment

This paper is part of the project of hybrid A/C system development on LPG-fueled vehicles which is conducted at the Automotive Laboratory of Universitas Muhammadiyah Magelang. The researchers are grateful the reviewers.

References

[1] Stoecker W F 1998 *The Industrial Refrigeration Handbook*, 1 st (New York: McGraw-Hill)
[2] Bhermana A, Sunarminto B H, Nuryani S, Utami H and Gunawan T 2013 The combination of land resource evaluation approach and gis application to determine prime commodities for agricultural land use planning at developed area (a case study of central Kalimantan province, Indonesia) *ARPN Journal of Agricultural and Biological Science* vol 8 no 12 pp 771–84
[3] European Commission 2004 *From farm to fork: Safe food for Europe ' s consumers* (Brussels: Office for Official Publications of the European Communities)
[4] Geospatial World 2011 Referral agricultural commodities mapping in Indonesia - Geospatial World [Online] Available: https://www.geospatialworld.net/article/referral-agricultural-commodities-mapping-in-indonesia/ [Accessed: 13-Aug-2017]
[5] Fick T 2007 *Transport Refrigeration* (Dakota: Upper Great Plains Transportation Institute, North Dakota State University) pp 1–21
[6] United Nations Economic Commission for Europe (UNECE) 1970 *Agreement on the*

International Carriage of Perishable Foodstuffs and on the Special Equipment to be used for such Carriage (ATP) [Online] Available at: http://www.unece.org/fileadmin/DAM/trans/main/wp11/ATP_publication/Ch_XI_B_22p.pdf

[7] United Nations Economic Commission for Europe (UNECE) 2016 *Agreement on the international carriage of perishable foodstuffs and on the special equipment to be used for such carriage (ATP)* [Online] Available at: http://www.unece.org/fileadmin/DAM/trans/main/wp11/ATP_publication/ATP-2016e_-def-web.pdf

[8] Tassou J L S A and De-Lille G 2012 *Food transport refrigeration* (UK: Centre for Energy and Built Environment Research, School of Engineering and Design, Brunel University) pp 1–25

[9] Farrington R and Rugh J 2000 Impact of Vehicle Air-Conditioning on Fuel Economy, Tailpipe Emissions, and Electric Vehicle Range *Earth Technologies Forum* [Online] Available at: http://www.nrel.gov/docs/fy00osti/28960.pdf

[10] Huff S, West B and Thomas J 2013 Effects of Air Conditioner Use on Real-World Fuel Economy *SAE Technical Paper* 2013-01–05

[11] Lee J, Kim J, Park J and Bae C 2013 Effect of the air-conditioning system on the fuel economy in a gasoline engine vehicle *Proceedings of the Institution of Mechanical Engineers, Part D: Journal of Automobile Engineering* vol 227 pp 66–77

[12] Setiyo M, Soeparman S, Wahyudi S and Hamidi N 2016 A simulation for predicting potential cooling effect on LPG-fuelled vehicles *AIP Conference Proceedings* vol 1717 no 1 p 30002

[13] Setiyo M, Soeparman S and Hamidi N 2017 Numerical Study on Cooling Effect Potential from Vaporizer Device of LPG Vehicle *Journal of Engineering Science and Technology* vol 12 no 7 pp 1766–79

[14] Setiyo M, Soeparman S, Hamidi N and Wahyudi S 2017 Cooling effect characteristics of a ½ cycle refrigeration system on an LPG fuel system *Journal of Refrigeration* vol 82 pp 227-37

[15] Price P, Guo S and Hirschmann M 2004 Performance of an evaporator for a LPG powered vehicle *Applied Thermal Engineering* vol 24 no 8–9 pp 1179–94

[16] Hwa Sung Thermo Co. ltd. 2012 Refrigeration Unit for Truck and Trailer *Hwa Sung Thermo Product Information* [Online] Available at: http://www.hwasungthermo-africa.com/docs/Product Information.pdf [Accessed: 16-Aug-2017]

Application of cold dry cutting and coated carbide tool on high speed turning process against tool wear and surface roughness of Brass MS58

T Sugiarto[1a], B Sugiantoro[1b], N Supriyana[1], K Setyawan[1] and Y Praharto[2]

[1]Mechanical Engineering Department, [2]Electical Engineering Department
STT Wiworotomo Purwokerto,
Semingkir 1[ist] Street, Purwokerto, Central Java, Indonesia, 53352

[a)]trismesinsttw@gmail.com,[b)]biotech.machining@gmail.com

Abstract. The development of cooling process its necessary to aim develop green machining requirement because it still contains subtances that being harmful for the operator and environmental effect. In this study, high speed turning process is done by using insert chisel and cold air (dry cutting) and Cubic Boron Nitride (CBN) with insert tool on Brass Type MS58. The research method refers to Taguchi methods that using four factors (spindle speed, depth of cuth, feed rate and cold dry cutting with defferen pressure). From the test results to get the optimization of machinery Turning on Brass with cooling dry with cold air using Taguchi methods can be assumme that the cooling process that has the most effect on brass surface roughness level is on cutting condition at 50 Psi (pressure), 10 °C (air jet temperature). The most optimum roughness in brass that the cutting conditions used are at spindle speed at 790 Rpm, feed rate 0.0034 mm /min, depth of cut 1.5 mm and good cutting condition at 50 Psi pressure. The highest level of sculptural wear on the brass turning process at the edges and the carving crater. The research results prove that the cooling with cold air can be an option for sustain machining process.

1. Introduction

The development of machining process according to green machining target is to use environmentally friendly material, easy to decompose, safe for operator, and economical, as green machining goal [1, 2]. The target of machining cooling according to an environmentally friendly process as well as cost considerations that must be economical [3]. The development of pressurized air use as coolant has been carried out by various positions of jet cooling improve surface roughness [4]. The usage vortex tube to decrease air temperature [5]. The use of air as an environmentally friendly machining targets encourages process innovation and a cooling air cooler model.

In the machining process the main indicator is seen from surface roughness of workpiece [6]. Pressurized air or Air Jet cooling has the advantage of being one of the cooling mediums in the machining process but still has not produced good surface roughness and the wear is still high, proved the lowest surface roughness palm kernel oil still better than dry cutting [7]. The resulting surface roughness in the dry cutting process is still lower than the use of oil cooling, but high economic value and potential to develop [8]. Temperatures affect the cooling performance, air jet cooling with an uncooled compressor pressure having an average temperature of 28-32°C, to improve performance of air cooling in dry cutting need innovation to decrease temperature cooling under 15°C. With these conditions still need to be developed air brealing technology that is able to produce the output temperature adjusted constant at (10°C). The technology of dry cutting has developed from vortex tube to cold dry cutting, limitations of vortex tube is low reduction of air/gas temperature, set up can be seen at figure 1.

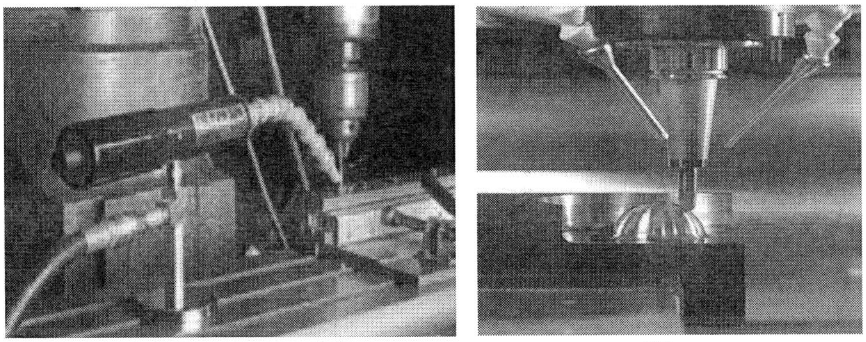

(a) **(b)**

Figure.1. Dry cooling process with (a) vortex tube (b) dry cutting (air jet cooling).

According to the previous research, to decrease air temperature obtained from heat exchanger by utilizing refrigeration process to reduce significant air temperature from compressor. The temperature drop depends on the cross-sectional area of the spiral pipe on the evaporative pipe. To set the air temperature used micro control for air has constant pressure and temperature required during the machining process.

The requirement of industrial machinery demands a fast process with high quality, so that machining process is directed to be executed at high speed machining process. According to industry quality constraint (ASME standard), HSM process range in figure 2.

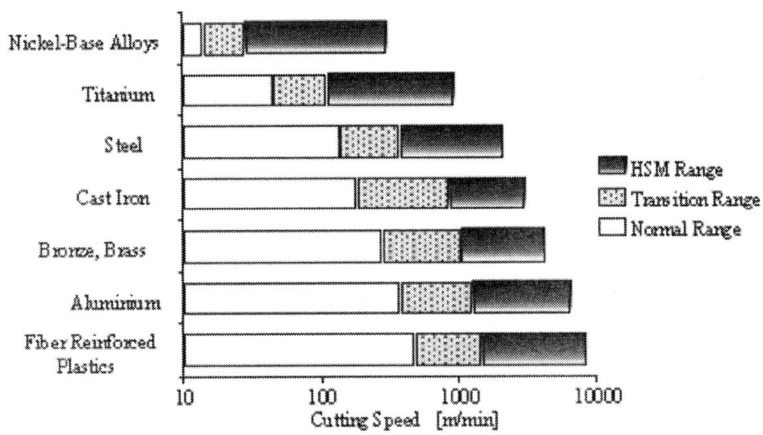

Figure 2. Comparison of cutting speed of material on high speed machining [9].

The choice of feed rate is very important, as it affects material deformation [10]. Pressure influenced residul stress of material. The choice of pressure on dry cutting is essential to prevent increasing temperature and stress in the material and chisels [11]. Parameters affect to the forces and surface roughness [12]. The effect of the feed rate, the spindle speed on the machining process influences the surface roughness of workpiece quality [13]. Cooling with cold air at temperatures below 15°C becomes the dominant factor against temperature and surface roughness [14].

Utilization of high performance refrigeration system becomes an option and not yet used commercially in ASME industry. Performance of cold dry cuttings is measured by looking at the turning machinery process conditions, including temperature process, engine power, tool wear, engine vibration. The lowest roughness condition will be the reference according to industry grouping.

2. Metodology

2.1. Materials

Material used in this research is Brass type (MS58, Code DIN CW614N), good machinability. The result of hardness testing can be seen at table 1. Tools used in this research is CBN Insert (Coated Carbide tool material). The use of chisel with carbide coating improves process speed and tool life and improves the quality of machining surfaces [15, 16]. Cooler air adjustment uses output air temperature, constant at 10°C.

Table 1. The result of hardness testing of Brass type (MS58).

No	Hardness Number (HRB)					Means (HRB)	Hardness HRC	Standart of deviation
1	84.1	87.1	87.5	92.2	91.9	88.65	7.42	3.320

Type of hardness testing: Rockwell Hardness (HRB),
Load: 100 N,
Penetrator: Ball

2.2. Taguchi methods

In this work, parameters are based on the most optimum results, the Taguchi methods used to design efficient experiments and is used to analyze experimental data. Ortogonal arrays $L_9(3)^4$ Desain of experiment can be seen at table 2.

Table 2. Planning of Taguchi experimental on HSM turning on Brass (MS58).

Parameters	Factor	Level Desain Variations		
		1	2	3
Spindle Speed (rpm)	A	790	950	1300
Feed Rate (mm/rev)	B	0.0034	0.0069	0.0137
Depth Of Cut (mm)	C	1	1,5	2
Cutting Condition (cold air injections (psi, °C)	D	P Compresor = 30 Psi, T ; 10 °C	P Compresor = 40 Psi, T ; 10 °C	P Compresor = 50 Psi, T ; 10 °C

Whereas: P: pressure adjustment air cooling, using constant temperature at 10°C, obtained at 15 minute operations of refrigeration process.

2.3. The cooling working principle

The cooling working principle of the cooling process uses compressed air, which requires only the compressor as a pressurized air-generator, effect of evaporation usage to lowering the air temperature. The Pressure Gauge as a valve to adjust the nozzle output air pressure, and the nozzle as the air spray guide. The cooling process is carried out through refrigeration use cooper pipe (diameter 8 mm), using placed on the round of evaporator. The heat exchange fluid aim to reduce air at a temperature of 10°C. Principle of temperature reducer with counter flow as shown figure 3.

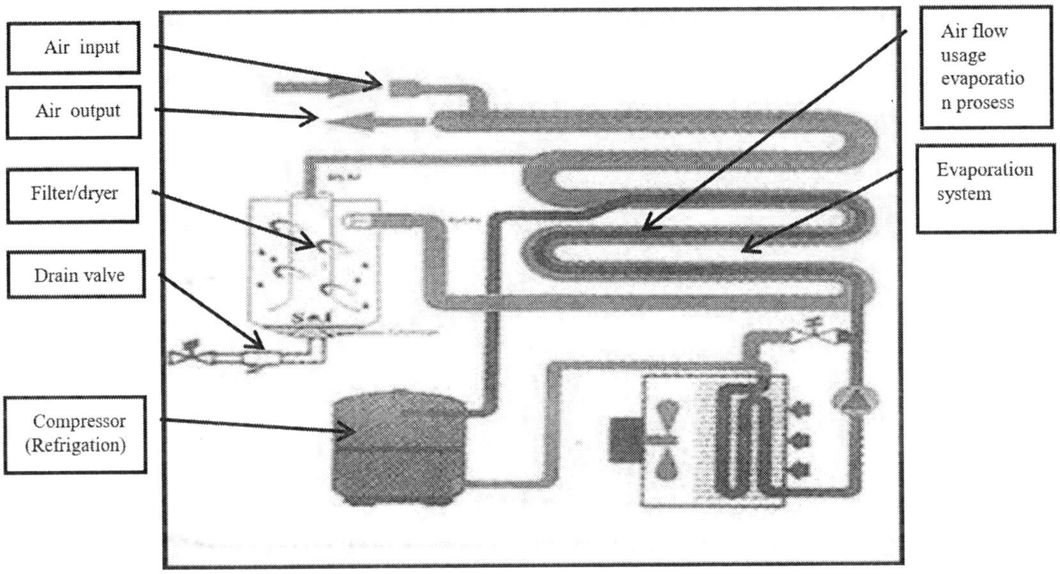

Figure 3. Principle of heat excange to obtained cold air cooling by usage refrigeration process.

2.4. Set-up experimental (turning process using cold dry cutting)

The data collection of turning machine with dry cooling at high speed turning is done with the set up of machining according to figure 4.

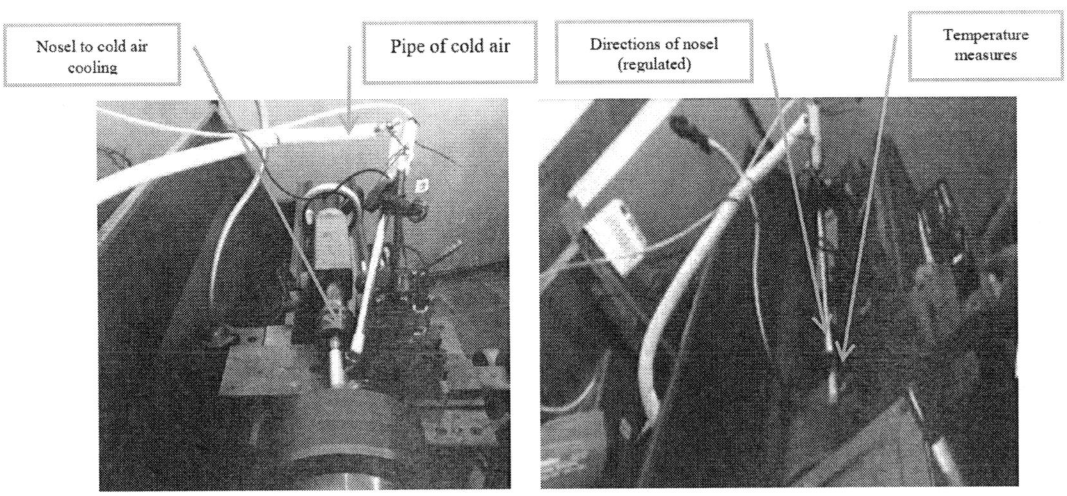

Figure 4. Set-up of experimental cold dry cutting on brass MS58.

2.5. Interpretation of experimental results

Measures for interpreting experimental results using the Taguchi Methods can be done by determining percent contributions, which is a significant portion of each factor interaction to the total observed variation. Quality characteristics used are Smaller-the-Better (STB). Characteristic quality where the lower the value, the better quality, as the equation (1) [17].

$$S/N_{STB} = -10 \log \left[1/n \sum_{i=n}^{n} y^2\right] \quad (1)$$

Vibration in the machining process is very influential on the results, the smaller the vibration indicates engine rigidity and chisel, a small frictional force. Chatter conditions occur because of small stiffness, for that vibration quantity is measured to be able to detect the forces that occur in the machining process. Achievement of process optimization will be known from motor power, vibration, temperature of air cooling pressure, chisel temperature, chip type. To know the phenomenon and condition of data process measured with (clam ampere), vibrationmeter, infrared / thermocouple sensor, pressure gauge. Lowest power, lowest temperature, smallest vibration and best roughness target as target of optimum condition determination.

3. Results and discussion

Roughness test results of the workpiece surface, measured using surface roughness type Surfcorder SE300. The result of measurement of surface roughness of MS58 brass can be seen in table 3.

Table 3. The result of surface roughness of brass (MS58).

Experiments	Specimen	Roughness	means	S/N
	Specimen 1	5.2910		
1	Specimen 2	5.4710	5.3810	-14.6181
	Specimen 3	5.3810		
	Specimen 1	7.5530		
2	Specimen 2	7.8340	7.6440	-17.6678
	Specimen 3	7.5450		
	Specimen 1	5.7930		
3	Specimen 2	5.8730	5.7840	-15.2453
	Specimen 3	5.6860		
	Specimen 1	1.9780		
4	Specimen 2	1.9540	1.9553	-5.8248
	Specimen 3	1.9340		
	Specimen 1	11.1730		
5	Specimen 2	11.1820	11.1730	-20.9634
	Specimen 3	11.1640		
	Specimen 1	28.5920		
6	Specimen 2	28.8910	28.7910	-29.1852
	Specimen 3	28.8900		
	Specimen 1	2.3910		
7	Specimen 2	2.4980	2.4950	-7.9463
	Specimen 3	2.5960		
	Specimen 1	2.8210		
8	Specimen 2	2.9310	2.8310	-9.0421
	Specimen 3	2.7410		
	Specimen 1	15.5010		
9	Specimen 2	15.5010	15.6010	-23.8634
	Specimen 3	15.8010		
		y	81.6553	-144.3564

	ȳ	9.0728	-16.0396

The effect of the level/factors on surface roughness workpiece, temperature of tools, and power consumption on spindle drive motor can be seen in tables 4.

Table 4. Effect of level and factor on surface roughness factor.

Faktor	Level			defference (maks - min)	Ranking
	1	2	3		
Spindel speed	-15.8437	-18.6578	-13.6173	5.0405	3
Feed rate	-9.4631	-15.8911	-22.7647	13.3016	1
Depth of cut	-17.6151	-15.7853	-14.7183	2.8968	4
Cutting Condition	-19.8150	-18.2664	-10.0374	9.7776	2

From table 4 shows that the feed rate is ranked first for the surface roughness response of the turning results region, this means that the feed rate has the greatest effect on the surface roughness of the result of the process. Cutting condition using cold dry cooling with cool air with 10°C occupying the 2nd rank, it shows that the use of cold air can reduce roughness compared to depth of cut and spindle speed, it means the cooling factor promised to develop the sustain machining. Result of high speed turning experiment with Taguchi methods to see the effect and condition of machining process (turning) on brass MS58. Effect level and factor S/N of surface roughness of product at figure 5.

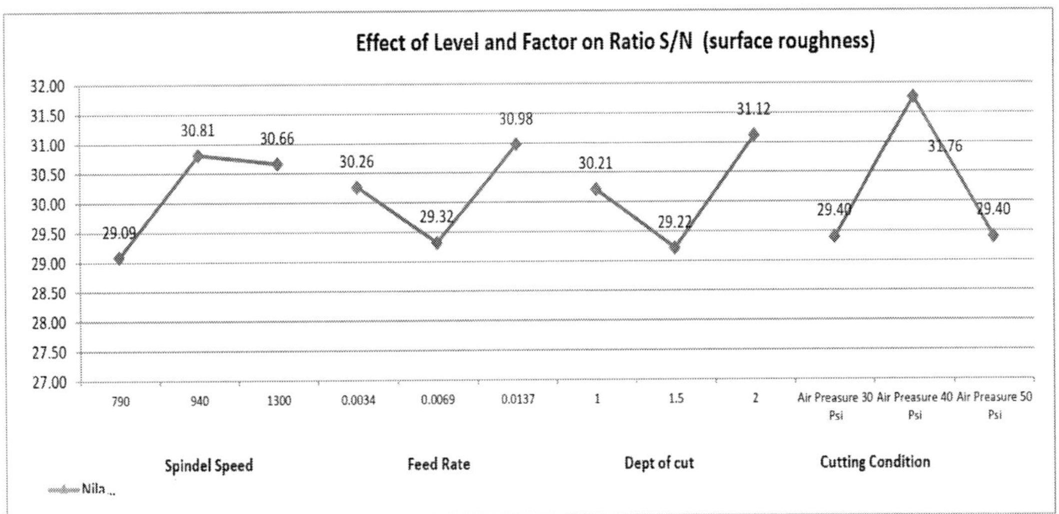

Figure 5. Graph of Effect Level and factor S/N of surface roughness of product.

The influencing factor on the mean value of CBN tools temperature can be seen on table 5.

Table 5. The ranking of effect of factor and level on tool temperature.

Factors	Level			difference (maks - min)	Ranking
	1	2	3		
Spindel speed	-28.72	-29.78	-29.72	1.06	2
Feed rate	-29.05	-29.34	-29.83	0.78	4
Dept of cut	-29.05	-29.31	-29.85	0.80	3
Cutting Condition	-28.82	-30.04	-29.37	1.22	1

From the table 5 it can be seen that, the most influencing factor on the mean value of cutting zone temperature is cutting condition with contribution 42% and the smallest contribution is belong tofeed rate with contribution of 15%. Graph of effect of level and factor ratio S / N temperature of cutting zone is shown in figure 6 and the influence of level and power factor S / N ratio is shown in figure 7.

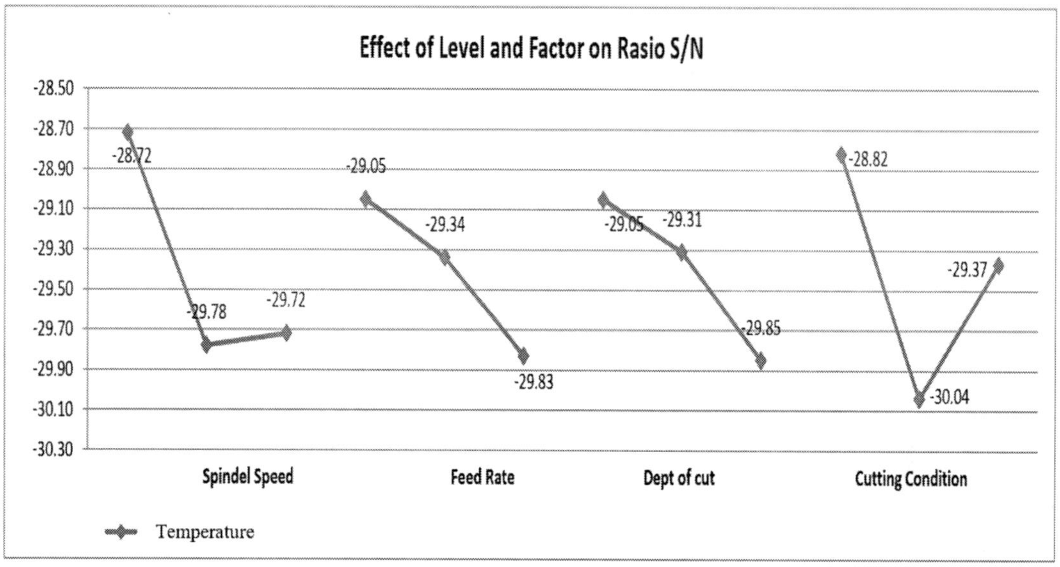

Figure 6. Graph of effect of level and factor ratio S / N temperature of insert tool (CBN).

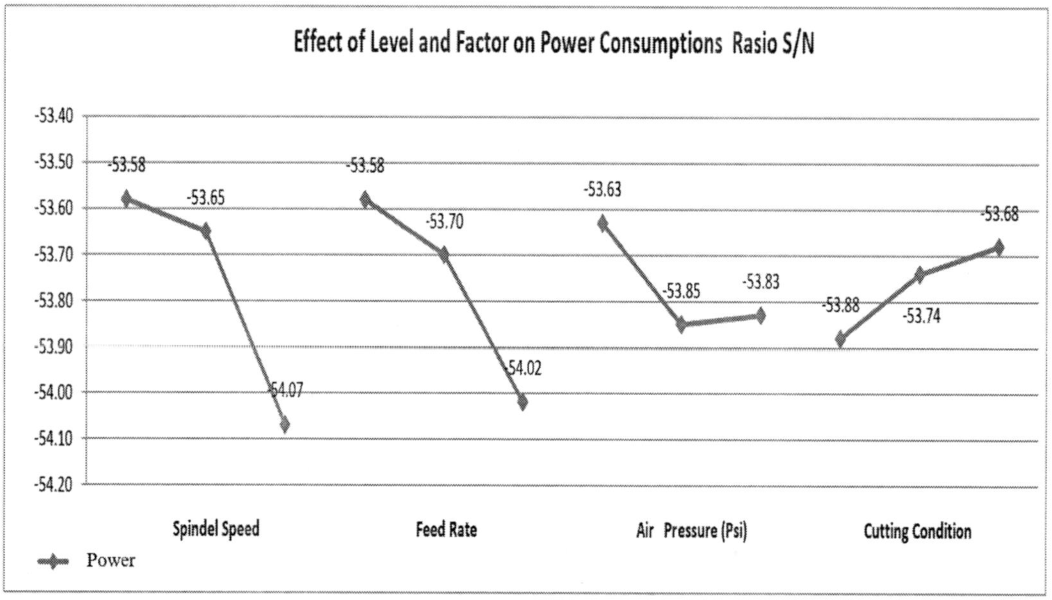

Figure 7. Graph of influence of level and power factor S / N ratio.

Experimental results show for all dry cooling process (cold air) produces roughness identification with the smallest value 1.955 μm, with the smallest temperature 28.40^{0}C and highest 33.23^{0}C. The power consumption depends on highest spindle speed, and highest depth of cut. The biggest wear on the side end of the cutting process turning after 30 minute operation at figure 8. Cutting condition is ranked first for carving temperature response, this means that cutting condition has the greatest effect on chisel temperature yield.

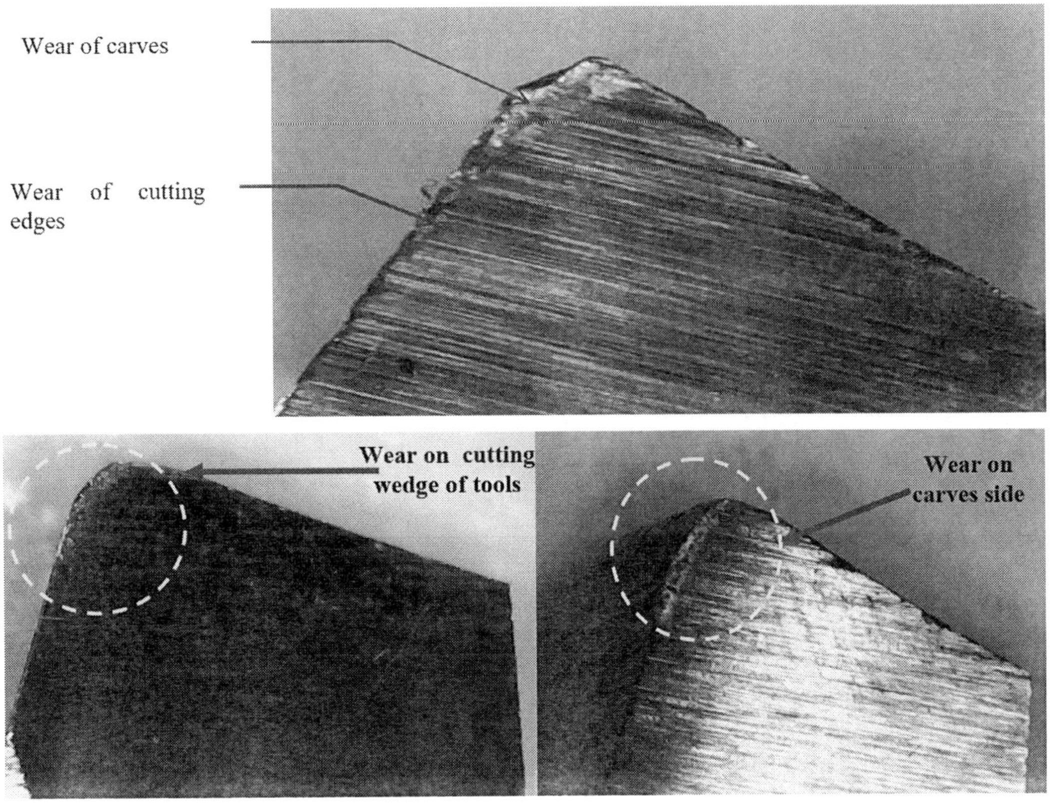

Figure 8. The wear positions on carving insert/cutting edge of CBN tool.

The results of brass research for all cooling processes with as in the above table resulted in a good roughness with the smallest value 1.955 ra, and the largest 28.791 ra. The most optimum roughness in brass then the cutting conditions used are at spindle speed at 790 rpm, Feed rate 0.0034 mm /min, dept of cut 1.5 mm and cold cutting condition at 50 Psi pressure. The lowest temperatures in the smallest number 28.40^{0}C and the highest 33.23^{0}C. Cutting condition is ranked first for the chisel temperature response, this means that the cutting condition has the greatest effect on the chisel temperature results. The cooling with cold air by utilizing refrigeration with constant temperature at 10°C occupies the 2nd rank for surface roughness.

From the above conditions, the most dominant factor on brass MS58 machining is the feed rate because it is ranked first for the surface roughness response of the resultant turning region, this means that the feed rate has the greatest effect on the surface roughness result [18, 19], This result is in accordance with the results of study [20, 21] that carbide-based sculpting improves machining performance, better than vortex tube [22], in other studies spindle speed becomes the most dominant factor [23, 24]. The wear rate dominant occurs on the CBN insert sculpture side of the crater and the chisel side.

4. Conclusion and future work

From the test results to get the optimization of machinery Turning on Brass with cooling dry with cold air using Taguchi methods can be summed up that the cooling process that has the most effect on brass surface roughness level is on cutting condition at 50 Psi (pressure), at constant temperature at 10°C, the most optimum roughness in brass then the cutting conditions used are at spindle speed at 790 rpm, feed rate 0.0034 mm /min, dept of cut 1.5 mm and cold cutting condition at 50 Psi pressure and the highest level of sculptural wear on the brass turning process at the edges and the carving crater.

The use of cold air can be optimized by increasing the lubrication properties with the minimal addition of lubricating oil (very little oil mist). In accordance with the development of green maschining, cooling with air is the target of future cooling. The potential of cold air becomes the potential to be developed for environmentally friendly machining in the future.

Acknowledgments
Head of Research and Community Service, P3M STT Wiworotomo Purwokerto, For Founding Basic Research conducted in 2016-2017.

References

[1] Jain A and Kansal H 2017 Green machining–machining of the future *4th National Conference on Advancements in Simulation & Experimental Techniques in Mechanical Engineering (NCASEme-2017)*
[2] Dahmus J B and Gatowski T G 2004 An environmental analysis of machining *Proceeding of IMECE, ASME International Mechanical Engineering Congress and Expo* Anaheim, California, USA, November 12-19 pp 643-52
[3] Atisková D M 2012 The methodology of economic costs influential on automation of component production *American Journal of Economics* vol 2 no 7 pp 164-70
[4] Rusnaldy P and Utomo T S 2012 Effect of air jet cooling on surface roughness and tool wear *Jurnal Tekno Sains* vol 1 no 2
[5] Rusnaldy P, Utomo T S and Umardani Y 2010 Effect of air jet cooling on surface roughness and tool wear *Proceeding of 2nd International Graduate Student Conference* Yogyakarta, Indonesia.
[6] Waydande R and Ghatge D 2016 Performance evaluation of different types of cutting fluids in the machining of hardened steel–A review *International Journal of Mechanical and Production Engineering* vol 4 issue 3 pp 34-9
[7] Adegbuyi P A O, Lawal G, Oluseye O and Odunaiya G 2010 Analyising the effect of cutting fluids on the mechanical properties of mild steel in a turning operation *American Journal of Scientific and Industrial Research* vol 2 issue 1 pp 1-10
[8] Grover M and Khan Z A 2014 The comparison on tool wear, surface finish and geometric accuracy when turning EN8 steel in wet and dry conditions *Proceedings of the World Congress on Engineering* Vol 2 pp 2-4
[9] ASM Handbook 2009 *13A-Corrosion-Fundamentals* Testing, and Protection
[10] Vilches F J T, Hurtado L S, Fernández M F and Gamboa C B 2017 Analysis of the chip geometry in dry machining of aeronautical aluminum alloys *Appl. Sci.* vol 7 issue 2 p 132
[11] Palakudtewar R K and Gaikwad S V 2014 Dry machining of superalloys: Difficulties and remedies *International Journal of Science and Research (IJSR)* vol 3 issue 7
[12] Rao C J, Rao N and Srihari P 2013 Influence of cutting parameters on cutting force and surface finish in turning operation *Procedia Engineering* vol 64 pp 1405-15
[13] Lawrence I D, Pandiarajan M and Kaviprakash G 2015 Prediction of machining parameters in turning on EN36 *International Journal of Applied Engineering Research* vol 10 no 55 pp 4070-75
[14] Sugiantoro B, Rusnaldhy and Widyanto S A 2014 Optimalisasi parameter proses milling terhadap kualitas hasil permesinan Aluminium dengan Metode Taguchi *TRAK SI* vol 14 no 1
[15] Srithar A, Palanikumar K and Durgaprasad B 2014 Experimental investigation and surface roughness analysis on hard turning of AISI D2 steel using coated carbide insert *Procedia Engineering* vol 97 pp 72-7
[16] Ramanujan R, Venketesan K, Saxena V and Joseph P 2014 Modeling and optimization of

cutting parameters in dry turning of Inconel 718 using coated carbide inserts *Procedia Materials Science* vol 5 pp 2550-9
[17] JMP 2014 *Design of experiments guide* (Cary, NC, USA: SAS Institute Inc.)
[18] Sulaiman S, Ariffin M K A and Norsat R M 2015 Influence of dry machining parameters in minimization of tool wear *3rd International Conference on Advances in Engineering Sciences & Applied Mathematics (ICAESAM'2015)* March 23-24 London (UK) vol 3
[19] Sargade V, Nipanikar S and Meshram S 2016 Analysis of surface roughness and cutting force during turning of Ti6Al4V ELI in dry environment *International Journal of Industrial Engineering Computations* vol 7 issue 2 pp 257–66
[20] Davis R, Madhukar J S, Rana V S and Singh P 2012 Optimization of cutting parameters in dry turning operation of EN24 steel *International Journal of Emerging Technology and Advanced Engineering* vol 2 issue 10 pp 559-63
[21] Andriya N, Rao P V and Ghosh S 2012 Dry machining of Ti-6Al-4V using PVD coated TiAlN tools *Proceedings of the World Congress on Engineering* July 4 – 6, London, UK pp 4-6
[22] Taha Z, Salaam H A, Ya T T, Phoon S Y, Tan C F and Akiah M A 2013 Vortex tube air cooling: The effect on surface roughness and power consumption in dry turning *International Journal of Automotive and Mechanical Engineering (IJAME)* vol 8 p 1477
[23] Ojolo S J and Ogunkomaiya O 2014 A study of effects of machining parameters on tool life *International Journal of Materials Science and Applications* vol 3 no 5 pp 183-99
[24] Okokpujie I, Okonkwo U and Okwudibe C 2015 Cutting parameters effects on surface roughness during end milling of Aluminium 6061 alloy under dry machining operation *International Journal of Science and Research (IJSR)* vol 4 issue 7 pp 2030-36

IOP Conf. Series: Materials Science and Engineering **403** (2018) 012095 doi:10.1088/1757-899X/403/1/012095

Continuous and uncontinuous gasification systems of rice husk using variation modification of burner

Subroto[1], Wijianto[2], Sarjito[2] and D A Himawanto[3]

[1]Department of Mechanical Engineering, [2]Department of Mechanical Engineering, Faculty of Engineering, Universitas Muhammadiyah Surakarta, Jl. A.Yani Pabelan Tromol Pos 1 Kartosuro Surakarta, Jawa Tengah, Indonesia 57102
[3]Department of Mechanical Engineering, Faculty of Engineering, Universitas Sebelas Maret Surakarta, Jl. Ir. Sutami, Surakarta, Jawa Tengah, Indonesia 57612

subroto@ums.ac.id

Abstract. The aim of the research is to know differences between continuous and uncontinuous system of gasification of rice husk with attach the modification of burner. The differences can be measured upon time flame effectiveness and average temperatures that are produced by two systems. Methodology of this research is started by applied all of burners variation such as burner without cap and with cap, burner without reflector and with reflector, after that comparison all variations to continuous and uncontinuous systems. The result of this research shows that continuous gasification system with variation of burner reach better average temperature compare than uncontinuous system. On the other hand, time flame effectiveness tends to the same.

1. Introduction

Recently, World Bank releases their investigated that the world currently generates around 4 billion tons of all types of waste every year. Furthermore, the world's cities can generate approximately 1.5 billion tons of solid waste per year. This volume is showed that occur enhancement of waste about 2.4 billion tons by 2025. In developing countries and lower income countries, waste generation will be increase over the next 25 years [1]. In addition, waste of biomass such as rice husk, in development countries some time being a problem, but that is can be used as source of energy.

Susilo [2] in his research stated that there is a high number of consumption of crude oil Indonesia meanwhile Indonesia's fuel reserves are less than 0.4% of world's fuel reserves. Crude oil is not renewable energy, so this condition is forcing to government and researchers to explore another energy resources which is easy to find and apply in Indonesia.

Biomass, such as rice husk, corn cob, sawdust is available abundantly in Indonesia and having big potential to become source of renewable energies. This is could be the most reasonable solution for our energy crisis. Furthermore, biomass has another benefit to obtain the heat as source of energy that can be used for to cooking, and other activities which using flame as source of energy.

Almost all agricultural country such as Indonesia, Thailand, India, Philippines and more than 75 countries in the world produced rice husk as waste. Nordin *et al.* [3] identifies that around the world, annual rice husk output is around 80 million tons. That is potential to produce energy of 1.2×10 GJ or can be converted to become a heating value of 15 MJ/kg. Another research which done by [4] shows that every year there is 13 million tons of husk is not used in Indonesia. He claims the potential of rice husk in Indonesia can be used as diesel power plant (PLTD), it could be equal to 2.1 million kiloliter or 3.5 thousand billion rupiahs per year. PLTD 100 kilowatt operated with diesel oil, it would need 0.3 litter of diesel oil per kilowatt hour (kWh). If PLTD operated husk, it's only need 0.06 litter of diesel oil per kWh. Nevertheless, diesel oil is needed but it can be more effective till 80 percent.

Gasification is defined as conversion process of solid fuel become gas fuel with high temperature. Gasification process produces environmentally friendly gas such as hydrogen and carbon monoxide. furthermore, gasification process is easier to control than another biomass direct combustion because it is gasiform. Nevertheless, Indonesia only has a few researchers that research of gasification of biomass or utilizing of gasification to become source of energy. Gasification product is usable for cooking, rotate turbine, power up internal combustion engine, and so on. with this fact, gasification technology is suitable to be developed in Indonesia and can produced with mass production.

Another research is done by [5]. They investigate energy supply and the potential of meeting the annual energy requirement with rice husk. The potential of rice husk is widely examined under different utilization scenarios using energy replacement value. Their research is built upon a heating value, greenhouse gas (GHG) reduction potential and annual energy expenditure. Result of their research showed that rice husk could efficiently and effectively to replace wood which achieving up to 67% wood savings, 62.94% GHG emission reduction and an annual parboiling energy expenditure reduction of 73%.

Rice Husk Gasification Furnace with updraft system has investigate by [6]. He designed the simple process gasification by linking up the reactor, burner, ash chamber, and blower into one body. Pathak *et al.* [7] already done another gasification type that is downdraft gasification. From their research, it can be concluded that performance and the relation between air flow velocity from blower and temperature that are produced. When the air flow is faster, temperature will be higher as well.

2. Research methodology

This section explains the research methodology. This research utilizes modification of burner that can be seen in figure 1, figure 2 and figure 3.

2.1. Material

Rice husk is used as main material with a water content of +10%. Rice husk is destroyed by a grinding machine, and then be sifted by size of mesh 10. Then 1.5 kg of the shifted rice husk was put into the gasifier tube. Continuous system done by add rice husk a little bit through the hole on the funnel on the burner modification (figure 3).

2.2. Equipment

The main equipment use type gasifier updraft (TLUD) with an outer diameter of 200 mm and an inner diameter of 160 mm. Refractory clay gasifier is used as an insulator, and placed between an inner and an outer diameter of the gasifier tube. Height of the reactor tube is 600 mm. In the burner attached the reflector and cap on top of burner modification (figure 3).

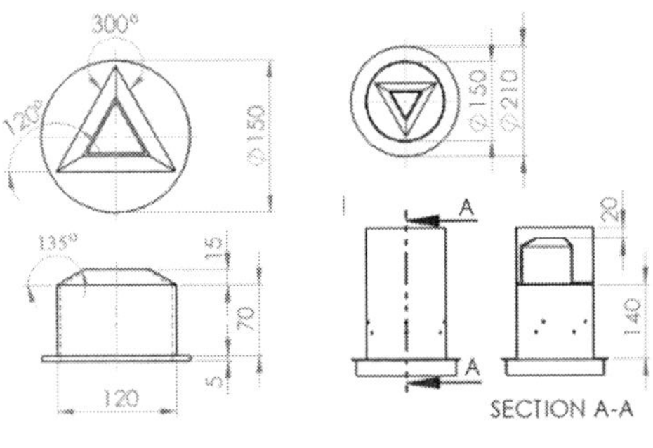

Figure 1. Burner with reflector.

IOP Conf. Series: Materials Science and Engineering **403** (2018) 012095 doi:10.1088/1757-899X/403/1/012095

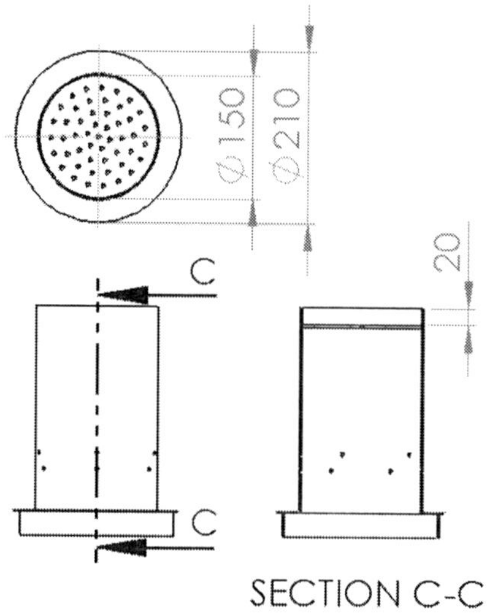

Figure 2. Burner with cap.

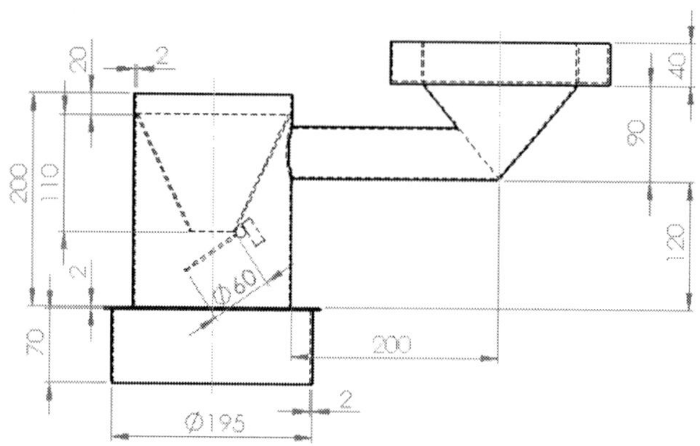

Figure 3. Modification of burner with additional funnel.

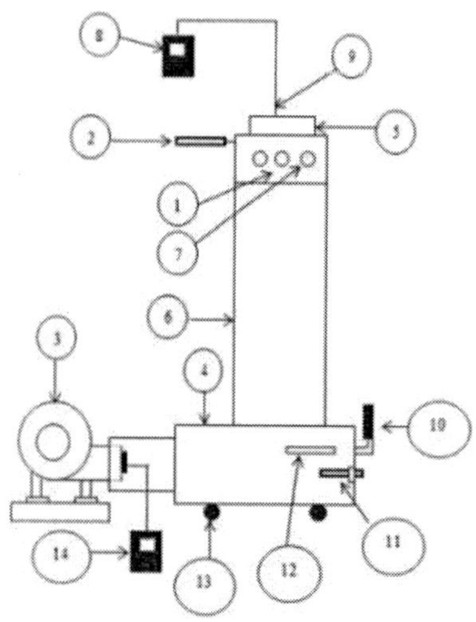

1. Burner
2. Handle of Burner
3. Blower
4. Ash Chamber
5. Reflector
6. Gasifier tube
7. Secondary air flow burner
8. Thermocouple reader
9. Thermocouple
10. Ash discharge lever
11. key
12. handle of door lever
13. wheels

Figure 4. Research installation.

Figure 4 shows the installation of the research. Initially, the gasifier tube (6) was filled with 1.5 kg rice husk. Secondly, the process of gasification was started with burning the surface of rice husk on the gasifier using the lighter. Following this process, the blower (3) was turned on at a speed of 10 m/s. After gasification process began, the temperature at the burner was measured using thermocouple (9) while the time process was noted using a stopwatch. The steps were done repeatedly with time span of 30 seconds until the fire off. Additional rice husk to continuous process system passing to the hole on the funnel a little by little with mass around 250 grams as long as two times.

3. Results and discussion

Figure 5 represents result of gasification process without reflector and without cap, burner with reflector only and burner with cap only at uncontinuous gasification type. Obviously burner with reflector having the highest temperature is around 700°C with average temperature approximately 655°C, and decline gradually. On the other hand, burners with reflector on uncontinuous gasification type having length of flame as long as 34 minutes. Comparing with gasification using burner without reflector and without cap time flame effective almost the same, that is approximately 34 minutes as well. Moreover, burner with reflector having the best of average temperature compare with burner without reflector and burner with cap only.

Figure 6 shows results of the research using continuous gasification type. It was apparent that the continuous gasification type which was reflector on the burner reaches length of flame as long as 45 minutes, burner without reflector and without cap reaches time flame around 39 minutes and burner with cap only reaches 41 minutes to flame. As can be seen on figure 6, burner with reflector gain the average temperature more stable which is around 600°C.

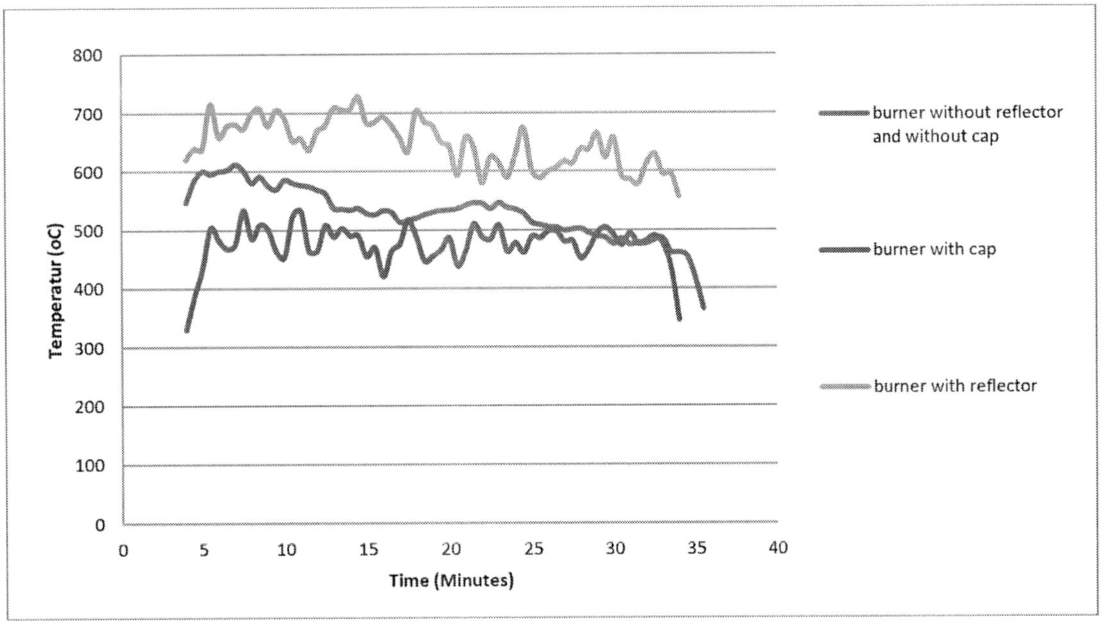

Figure 5. Comparison result between gasification without reflector and without cap and with reflector and with cap at uncontinuous gasification type.

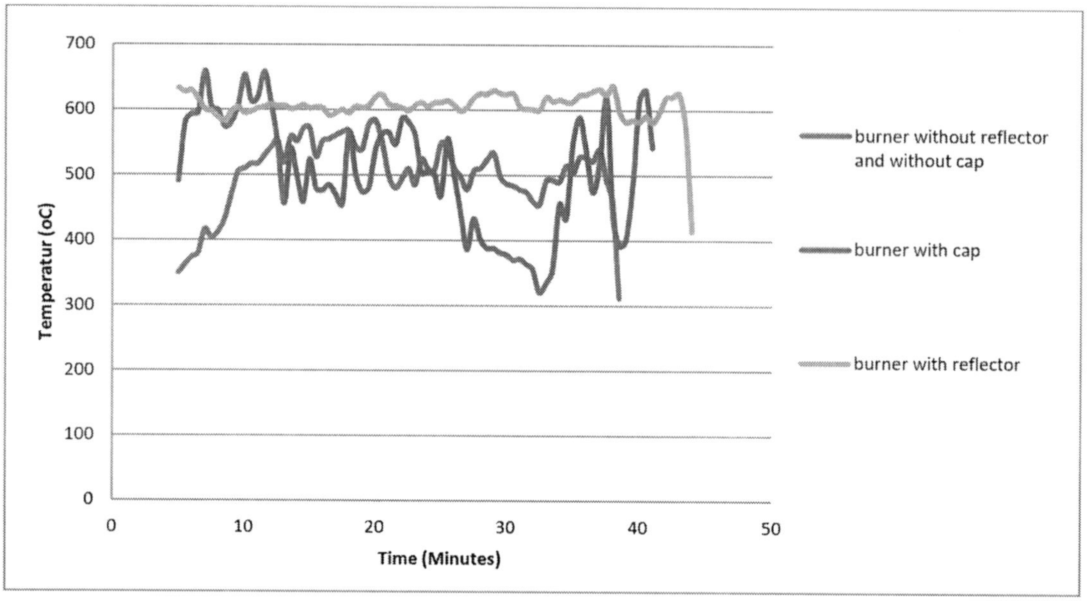

Figure 6. Comparison result between gasification without reflector and cap and with reflector and cap at continuous gasification type.

4. Conclusions

According on analysis of data result, there are some particulars about correlation between variation of burner and modification that burner with combustion temperature and effective time flame on rice husk gasification, such as the average highest temperature is gained on burner with reflector. Burner with reflector both continues and uncontinuous type are the best modification and the longest gasification process without modification burner recorded reach around 34 minutes and the longest gasification process in continue system with reflector on the burner obtain the effective time flame until 44 minutes and the average temperature is around 600oC. It means that continuous system gasification with reflector can be used to prefer in develop model and manufacturing of gasification stove.

Acknowledgement

Authors would like to appreciate and thank to Directorate General of Research, Technology and Higher Education, Ministry of Research, Technology and Higher Education of the Republic of Indonesia. This Research is financially supported by *Hibah Bersaing* Project at the fiscal year 2017-2018 with contract number 211.47/A3.III/LPPM/V/2017.

References

[1] Hoornweg D and Bhada-Tata P 2012 What a waste: A global review of solid waste management *Urban Development Series Knowledge Papers* no 15, The World Bank, Washington, DC

[2] Susilo S 2013 *Cadangan Minyak Kita Cuma 1/100 Venezuela* Article [Online] Available at https://www.esdm.go.id/en/media-center/arsip-berita/cadangan-minyak-kita-cuma-1-100-venezuela.

[3] Natarajan E, Nordina A and Rao A N 1998 Overview of combustion and gasification of rice husk in fluidized bed reactors *Biomass and Bioenergy* vol 14 issues 5-6 pp 533-46

[4] Djuwarno 2003 *Menggantikan Solar, Sekam Bisa Diubah Jadi Sumber PLTD* Article [Online] Available at http://www.energi.lipi.go.id/utama.cgi?cetakartikel&1104525263

[5] Kwofie E M and Ngadi M 2016 Sustainable energy supply for local rice parboiling in West Africa: The potential of rice husk *Renewable and Sustainable Energy Reviews* vol 56 pp

IOP Conf. Series: Materials Science and Engineering **403** (2018) 012095 doi:10.1088/1757-899X/403/1/012095

1409-18
[6] Belonio A T 2005 *Rice Husk Gas Stove Handbook* (Philippines: Central Philippine University)
[7] Pathak P S, Khan T A and Sharma P 2004 Biomass production, its utilization and surplus for energy generation in India *Proceedings of the National Seminar on Biomass Management for Energy Purposes-Issues and Strategy: Eds PS Pathak and NSL Srivastava* pp 11-12

IOP Conf. Series: Materials Science and Engineering **403** (2018) 012096 doi:10.1088/1757-899X/403/1/012096

Fabrication and characterization of the porous titanium alloy by argon filled pore expansion technique

A W Nugroho[1], G Leadbeater[2] and I J Davies[2]
[1]Mechanical Engineering Department, Universitas Muhammadiyah Yogyakarta, Yogyakarta, Indonesia
[2]Mecahnical Engineering Department Curtin University, Perth, Western Australia, Australia

ariswidyo.nugroho@umy.ac.id .

Abstract. The characterization of fabricated porous β titanium alloy using powder metallurgy technique from elemental powders, i.e. Ti, Nb, Ta and Zr has been carried out. Those elemental powders were blended and filled in steel container. The container was vacuumed and pressurized with argon gas subsequently. The pressurized container was subjected for hot isostatic pressing (HIP-ing) at temperature of 1100°C under 100 MPa of argon pressure subsequent furnace cooling. The HIP-ed container was cut into cubic shape followed by placing in vacuum furnace at high temperatures of 1100°C and 1125°C for foaming process. While microstructure and hardness examination by electron microscope and Vickers indentation were respectively conducted, chemical composition of the porous alloy was determined using X-ray fluorescence (XRF). Their corrosion behavior was evaluated by using potentiodynamic polarization technique. The result shows that whilst the porosity level and the hardness increased with the increase of foaming temperature, the specimen foamed at the lower temperature exhibited higher pitting and corrosion resistance.

1. Introduction

Titanium and its alloys are widely used for many applications because of their advantages such as high specific strength and excellent corrosion resistance. Metal in porous structure demonstrates significant advantages for examples low density and novel acoustic, physical, electrical, thermal, and mechanical properties. Due to high melting point (1670°C) and high reactivity of pure titanium, some fabrication techniques in solid route have referred to produce porous titanium alloys [1-3] compared to the melting route. From those, the argon filled pore expansion technique exhibits no contamination and porosity c.a. 40% in closed pore structure.

Whilst pure titanium powder or prealloyed titanium alloy powder are commonly used as starting material, the use of elemental powder which lowers in cost is rarely reported. Titanium alloying elements: Ti, Zr, Nb and Ta have been reported showing excellent biocompatibility, lowering the Young's modulus and also possessing a considerably superior corrosion resistance. This paper presents the characterization of fabricated porous titanium alloy based on the pressurized argon pore expansion process by powder metallurgy technique using the elemental powders.

2. Methodology

Ti, Nb, Ta and Zr elemental powders in angular shape with the purity and size of the elemental powders being > 99.5% and < 44 μm, respectively were supplied from Cerac Inc.The powders were weighed in accordance to a desired composition of Ti-29Nb-13Ta-4.5Zr, and were then mixed for 30 minutes in a roller bottle mixer at 50 rpm. While figure 1(a) depicts the morphologies of the elemental powders after the mixing process, figure 1(b) presents their energy spectra from Energy Dispersive Spectrometer (EDS) measurement.

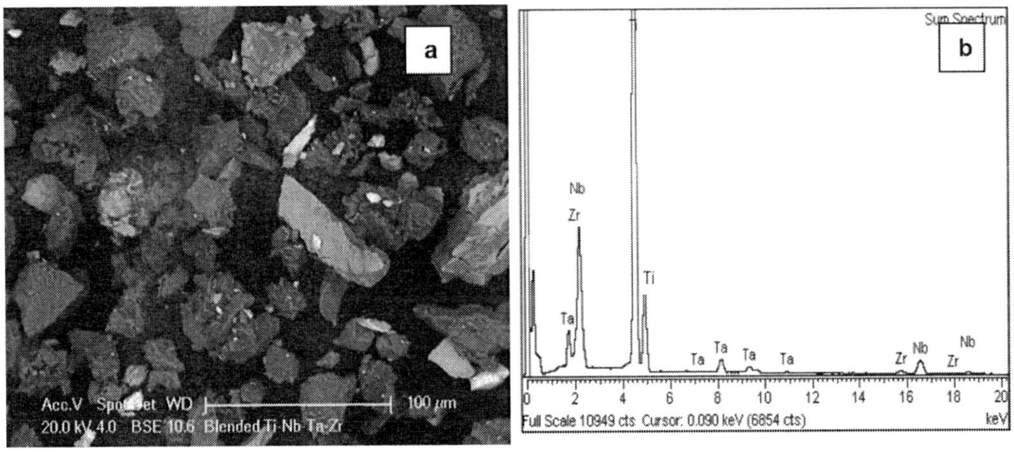

Figure 1. Morphologies of the mixed powder from backscattered scanning electron micrograph (a) the energy spectres from EDS measurement (b).

Afterward, the mixed powder was filled in 30 mm diameter steel containers. The containers were vacuumed and pressurized with argon gas at 0.68 MPa and air tighten by welding subsequently. The pressurized containers were put in hot isostatic press (HIP) machine and heated up at temperature of 1100°C for 1 hours under 100 MPa of argon pressure and slow cooled resulting high pressure argon entrapped in titanium matrix. The HIP-ed billet were cut to cubical shape with dimensions of 10 x 10 x 10 mm by wire electrical discharge machining (WEDM). The cube specimens were then placed in a vacuum furnace for expansion of the pressurized argon or foaming process at 1100° and 1225° for 10 hours. An optical microscopy and an electron microscopy-coupled by EDS were used to examine morphology and microstructure of the prepared foamed specimens. Quantitative analysis of the alloy constituent and fraction porosity were determined using a XRF and open source software digital image analyzer (ImageJ) respectively. Microvickers hardness testing was carried out to reveal hardness values of the porous alloy. Their corrosion behavior was evaluated by open-circuit potential (OCP) and potentiodynamic polarization measurements.

3. Results and discussion

Figure 2 depicted the comparison between a polished HIP-ed specimen and as-foamed specimens before evaluation. The foamed specimens showed darker surface due to pores formed. Generally, the pores expanded in a uniform manner and the extent of pore expansion was determined by the foaming condition. It can be clearly observed that the dimensional changed after foaming process. Foaming at higher temperature appeared in larger side length. This may reflect that porosity level of the foamed specimen at higher temperature (1225°C) is higher than that of the lower one (1100°C).

Barreling shape of the foamed specimen was also noted in which the length of the edge of the foamed specimen was found slightly shorter than that of the middle because of the higher expansion of pressurized argon in metal matrix in the middle. After sectioning by WEDM, surface damage layer of the foamed specimen was removed and polished subsequently. The prepared specimens were then

IOP Conf. Series: Materials Science and Engineering **403** (2018) 012096 doi:10.1088/1757-899X/403/1/012096

examined for pore morphology and pore size. The examination area mostly near to the specimen center, otherwise it would be specifically stated.

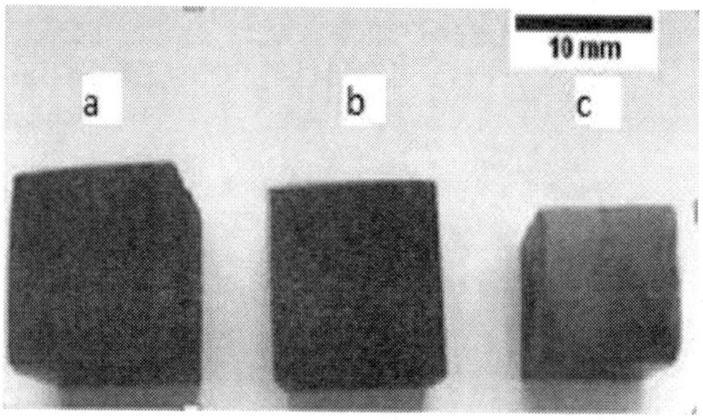

Figure 2. The comparison of foamed titanium alloy specimen at temperature of 1225°C (a), 1100°C (b) and polished HIP-ed billet (c).

The foaming process at 1100°C for 10 hours with pressurized argon gas of 0.68 MPa resulted in porosity level in the range of 10.7-14.5%, while that of 1225°C reached higher porosity level in the range of 21.8-27.0% as shown in figure 3(a).

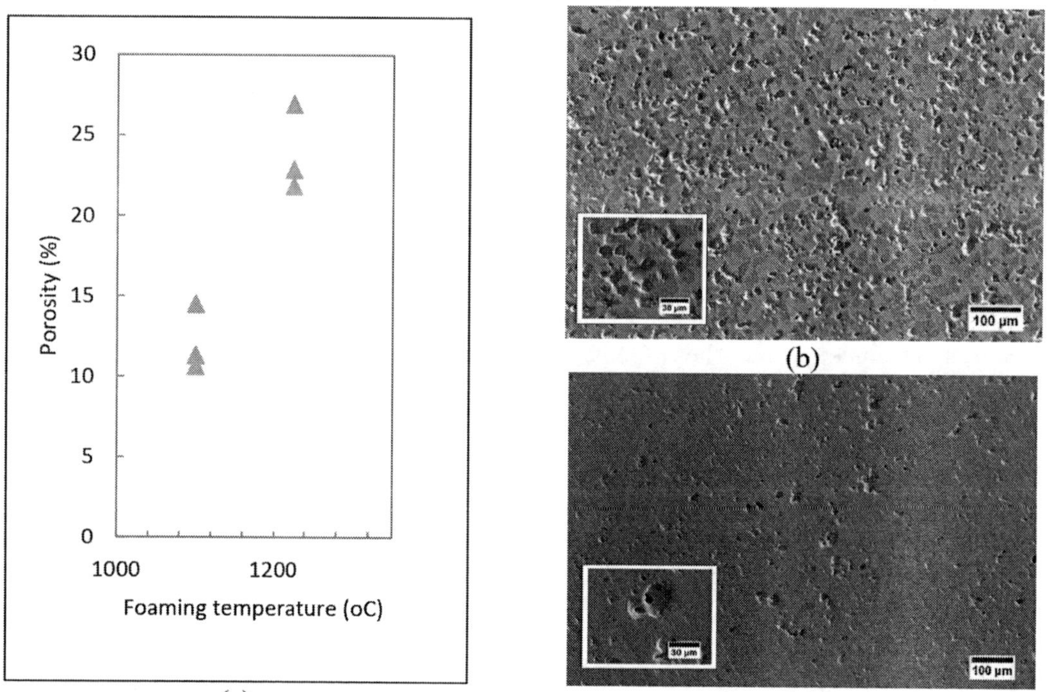

Figure 3. The extent of porosity attained for the specimens being foamed at 1100°C and 1225°C (a), secondary electron microscope image presenting morphology of the pore of a specimen being foamed at 1100°C/0.68 MPa/10 hours with porosity of 10.7% (b) and a specimen being foamed at 1225°C/0.68 MPa/10 hours with porosity of 21.6% (c).

Figure 3(b) shows a typical microstructure depicting pore morphology of a specimen being foamed at 1100°C/0.68 MPa/10 hours. The image also illustrates that a low porosity (10.7%) of porous titanium alloy was obtained by the foaming process. At the porosity level, the pores were distributed unevenly in the specimen with very small number of pore coalescence and the pores being very discrete.

Equiaxed pore shape was mainly observed with probably in 3 dimensions being spherical in shape as depicted by the inset picture. On the other hand, un spherical pore shape (angular) presumably being formed from the expanded interpowder-spacings were also observed.

The extent of porosity is considered equal to past report researches on other porous titanium alloys which stated the porosity of around 12% [4]. In those researches, porous titanium alloys were developed by means of pre-alloyed master powders heated up at 1200oC and held for 10 hours. The pressurized pores of the specimen foamed at 1225°C expanded more obvious because of the metal matrix creeping resulting in specimen with porosity of 21.6% as illustrated in figure 3(b). Morphology of the pores mostly stayed discrete, with more number of pore coalescence being observable (see the inset) than that of the specimen foamed at 1100°C. The enhancing pore connectivity becomes noticeable as well.

In addition, more uniform pore expansion in specimens being fabricated by foaming temperature at 1225°C was found as appeared in figure 4. Since fewer constrain being present, the pores in the area near to surface of the specimen demonstrated higher expansion level than that of the deeper one resulting a raise in porosity level. Yet, pores with thick in wall and discrete in structure were still observed. The number of pore coalescence just increase very slight. In those regards, the pores were still categorized as closed pores. However, some pores had coalesced and connected to the surface of specimen subsequently escaping pressurized argon gas to the atmosphere (see the arrow in figure 4).

The creep deformation that increase with higher foaming temperature was found to attribute the increase of porosity level. Pore morphologies were confirmed in equiaxed shape and elongated shape because of pore coalescence with in general the pore walls remaining thick and the pore connectivity frequently being visible. Therefore, generally, the foaming temperature clearly affects the extent of porosity of the foamed titanium alloys.

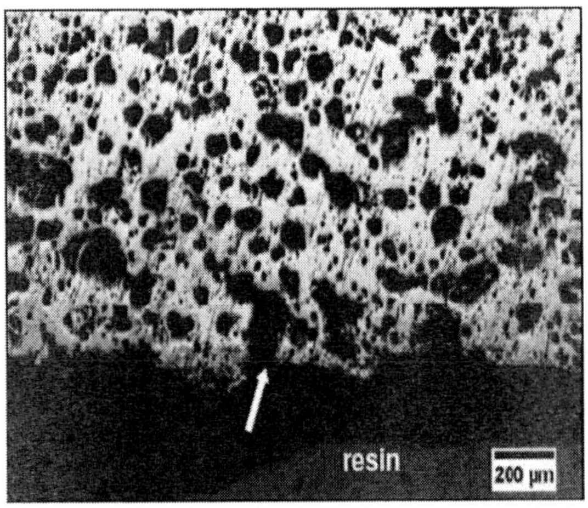

Figure 4. Optical microscope image of the area near to the surface of specimen being foamed at 1225°C showing a relatively uniform pore expansion.

Three specimens were examined by using a XRF to obtain the alloy composition. An XRF spectra of the selected sample is shown in figure 5. The typical chemical composition of Ti, Nb, Ta, Zr and others in the alloy are of 49.6±3.13%, 30.6±2.14%, 13.47±1.47%, 5.97±0.62% and 0.36±0.22 respectively. Whilst this result indicated that the chemical composition of the fabricated alloy is comparable to that of the elemental powders, other elements were detected which may due contamination during alloy processing and samples preparation.

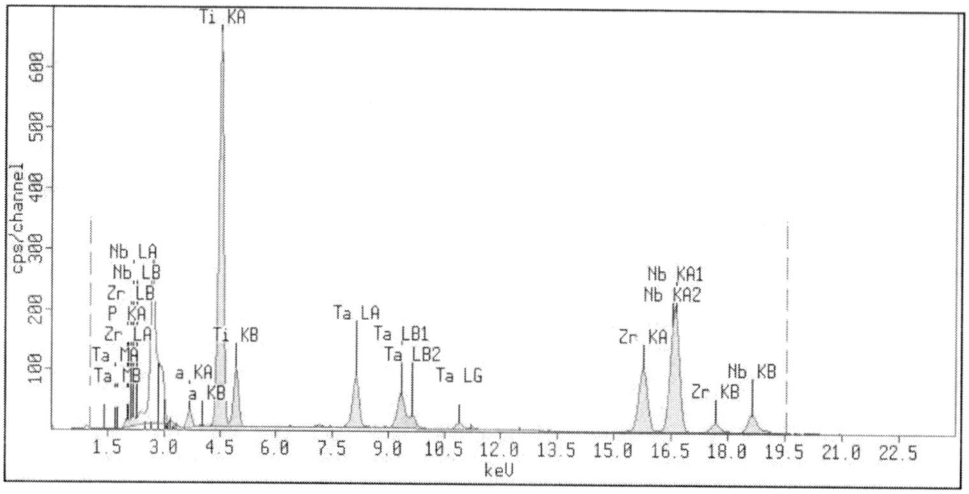

Figure 5. XRF spectra of the titanium alloy.

The hardness testing was conducted on polished specimen surface by means of microvickers hardness testing machine. A load of 100 g was used to carry out the experiment. Figure 6 illustrated the relation between foaming temperature and hardness. Figure 6 presents an increasing foaming temperature increased the hardness.

As expected, the hardness increase with the increase of foaming temperature due to the increase of fraction of α–phase which has higher Young's modulus than that of β–phase as previously revealed [5]. At the higher foaming temperature, fraction of the α-phase rapidly grew rather than that of β-phase resulting in the closer the fraction of the both phases. This allows higher opportunity to obtain individual hardness values of the phases presenting a higher scattered hardness values. In contrast, the specimen being foamed at 1100°C showed a smaller scattered and lower values of hardness value that of at 1225°C. The narrower difference of the hardness for the lower temperature may be suggested by the evident that higher fraction of β-phase was produced by foaming at the lower temperature. Thus, foaming temperature at 1100°C resulted in more homogenous in β–phase than the other. The porous titanium alloy being fabricated consisting of high fraction of β-phase shows hardness which is comparable to the hardness of various biomedical titanium alloys reported in the range of 300-400 HV [6-8].

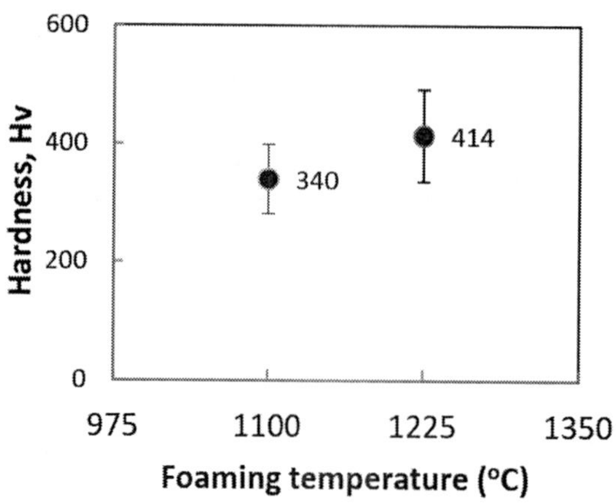

Figure 6. Hardness values of specimen foamed at 1100°C and 1225°C.

In the research, corrosion behavior was assessed using open circuit potential (OCP) method. The OCP measures passivation and layer development of the alloy that occurs in a solution. In addition, potentiodynamic polarization testing was used to identify electrochemical behavior of the material in certain environment. The OCP measurement method's results can be seen in figure 7.

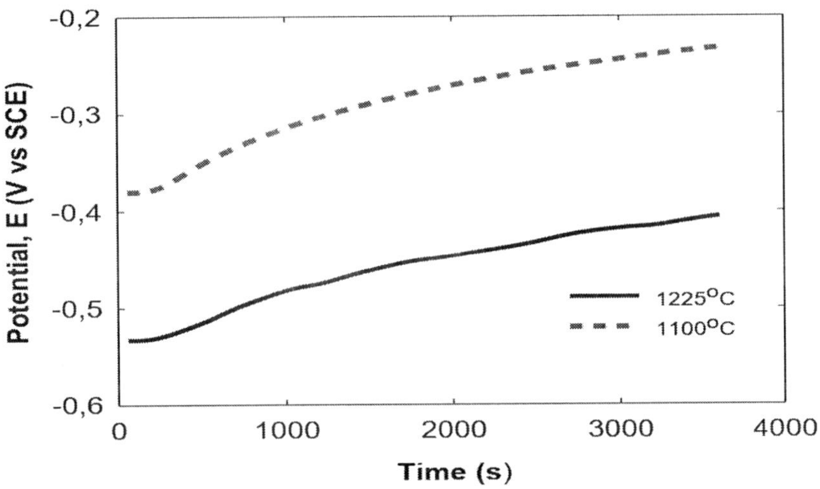

Figure 7. The relation of time to the potential during the OCP measurement in Hank's solution at 37°C of the specimens foamed at 1100°C and 1225°C.

As seen in figure 7, the OCP measurement as a function of time for foaming temperature at 1100°C and 1225°C of the porous alloys. These results can be relatively easy to be reproduced, and all of the curves runs in similar way to each other, in addition to being comparable. Initially, there is a minor upsurge of potential in the first few minutes, followed by a constant rise towards noble direction before becoming steady. The specimen foamed at 1225°C showed the initial E_{corr} of 533 mV. It experienced a moderate increase in increment rate which reached -406 mV after an hour. In addition, the initial corrosion potential (E_{corr}) for the specimen that foamed at 1100°C was found of -380 mV and moderately increase to -232 mV after an hour. The latter specimen had more noble potentials than the former specimen. This increase of potential in the positive way is an indication of a passive film forming, while the steady state implies the film being intact and protective. This signifies a competition between film dissolution and formation, which results in a potential indicating the thickness of surface film and its self-healing properties. The potential variation rate which signifies the formation of oxide was high after a few minutes of immersion, decreasing over time before reaching a steady state.

The potentiodynamic polarization curves for the both foaming temperatures measured in Hank's Solution at 37°C are presented in figure 8. The E_{corr} varied from -445 to -365 mV. This difference among those may be caused by the differing surface condition [9], even though the specimens' preparations used the same procedure.

The corrosion current densities (i_{corr}) were resulted from the polarization's curves by the Tafel extrapolation technique. The breakdown potentials of specimens foamed at 1225°C was of 899 mV. Comparatively, the other specimen which foamed at 1100°C showed neither potential for breakdown nor hysteresis loop (figure 8).

This reveals that the latter specimen has a higher resistance to pitting and crevice corrosion because of its composition, which predominantly comprised the β phase previously reported higher corrosion resistance. In addition, material with higher porosity was also stated susceptibility to crevice corrosion

[10]. Yet, their i_{corr} and E_{corr} in Hank's solution at 37°C could be compared to those of solid CP Ti, Ti 6Al 4V.

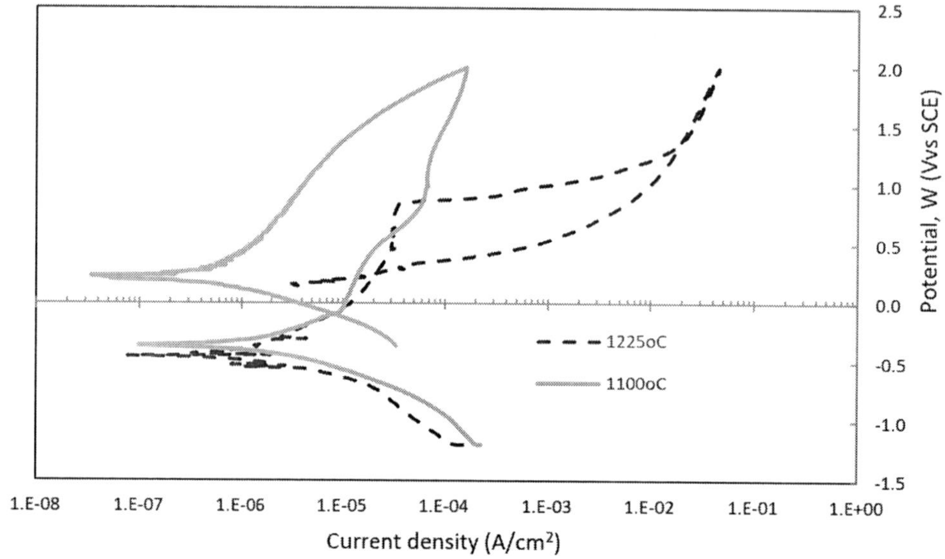

Figure 8. Potentiodynamic polarization curves for specimens foamed at 1100°C and 1225°C in Hank's solution at 37°C.

4. Conclusion

It can be concluded that the porosity level and the hardness increased with the increase of foaming temperature (1100°C and 1225°C) achieving 26.7% and 417mV respectively. The specimen foamed at the lower temperature exhibited higher pitting and corrosion resistance.

References

[1] Oh I-H, Nomura N, Masahashi N and Hanada S 2003 Mechanical properties of porous titanium compacts prepared by powder sintering *Scripta Materialia* vol 49 issue 12 pp 1197-1202

[2] Wen C E, Yamada Y and Hodgson P D 2006 Fabrication of novel TiZr alloy foams for biomedical applications *Materials Science and Engineering: C* vol 26 issue 8 pp 1439-44

[3] Murray N and Dunand D 2006 Effect of initial preform porosity on solid-state foaming of titanium *Journal of Materials Research* vol 21 issue 5 pp 1175-1188

[4] Oppenheimer S 2007 Proccessing and charaterization of porous Ti-6Al-4V and NiTi *Doctor of Philosophy*, Materials Science and Engineering, Northwestern University, Illinois

[5] Nugroho A W, Leadbeater G and Davies I J 2011 Processing and properties of porous Ti-Nb-Ta-Zr alloy for biomedical applications using the powder metallurgy route *Australian Journal of Mechanical Engineering* vol 8 issue 2 p 169

[6] Taddei E B, Henriques V A R, da Silva C R M and Cairo C A A 2007 Age-hardening of Ti-35Nb-7Zr-5Ta alloy for orthopaedic implants *Materials Research* vol 10 no 3 pp 289-92

[7] Majumdar P, Singh S B and Chakraborty M 2008 Wear response of heat-treated Ti-13Zr-13Nb alloy in dry condition and simulated body fluid *Wear* vol 264 issues 11-12 pp 1015-25

[8] Niinomi M, Akahori T, Takeuchi T, Katsura S, Fukui H and Toda H 2005 Mechanical properties and cyto-toxicity of new beta type titanium alloy with low melting points for dental applications *Materials Science and Engineering: C* vol 25 issue 3 pp 417-25

[9] Kim H-K and Jang J-W 2004 Electrochemical corrosion behavior and MG-63 osteoblast-like cell response of surface-treated titanium *Metals and Materials International* vol. 10 issue 5 pp 439-46

[10] Fojt J, Joska L and Málek J 2013 Corrosion behaviour of porous Ti-39Nb alloy for biomedical application Corrosionce *Science* vol 17 pp 78-83

The effect of variation number of holes on burner cap of TLUD gasification stove

Wijianto[1], Sarjito[1], Subroto[1] and D A Himawanto[2]

[1]Department of Mechanical Engineering, Faculty of Engineering UniversitasMuhammadiyah Surakarta, Jl. A.Yani Pabelan Tromol Pos 1 Kartosuro Surakarta, Jawa Tengah, Indonesia 57102
[2]Department of Mechanical Engineering, Faculty of Engineering Negeri Sebelas Maret Surakarta, Jl. Ir. Sutami Surakarta, Jawa Tengah Indonesia 57612

wijianto@ums.ac.id

Abstract–The objective of this research was to investigate the influences of number of holes on the burner cap to performance of TLUD (top lit up draft) gasification. Number of holes on the burner cap usually used to distribute methane gas that comes out from gasifier and then to be flamed become fire. Variation of number of holes on the burner cap was intended to get flame effective and the best temperature at TLUD gasification stove. This TLUD gasification stove use rice husks biomass as source of fuel with a mass of 1.5 kg and mesh 10. The method was by varying number of hole of the burner cap of 20, 30, 40 and 50 holes with a diameter of every hole was 5 mm. Water boiling test was used to prove which hole produced the fastest time to boil 2 liter of water. The results show that the burner cap with number of holes as much as 50 produce a longer time flame and stabilize the combustion temperature produced by TLUD gasification stove.

1. Introduction

Waste of biomass can be used as source of energy to substitute the fossil based fuel. Fuels such as kerosene, LPG, gasoline and others that come from the fossil fuel are very limited and tend to be exhausted. The population also rapidly growing, and increasing the use of fossil fuel consumption. Recently, fossil fuel crisis has drawn attention to develop viable and sustainable alternatives to substitution of kerosene and LPG. In the developed country, some attempt was done to reduce fuel consumption by using alternative energy. They exploit natural resources that are unlimited in numbers such as wind, solar, and ocean waves become source of energy. In some developing countries, waste biomass such as rice husk, sawdust, rice straw and others can take a position as substitution of fossil fuel.

Gasification is conversion process of solid fuel become gas fuel with high temperature. Gasification produces environmentally friendly gas product such as hydrogen and carbon monoxide. Moreover, gasification is easier to control than another biomass direct combustion because it is gasiform. Gasifier is the technology for Gasification (for example TLUD gasifier stove/wood gas stove). Gasification is a process that converts organic or fossil-based carbonaceous materials into carbon monoxide, hydrogen and carbon dioxide. This is achieved by reacting the material at high temperatures (>700°C), without combustion, with a controlled amount of oxygen and/or steam [1].

In most developing countries, the type of fuel that is used depends on household income. To cook, usually using kerosene or LPG. Utilization of biomass waste as a fuel source can reduce costs to be incurred by each household. One of the alternative energy sources that can be developed as a replacement for fossil fuels is rice husk. In an agricultural country such as Indonesia, Philippine, India and in some developing countries is increasing the production of rise husk. The abundance of rice husks continuously can be used as a source of raw materials for TLUD gasification stove.

Examples of research about alternative energies to substitute fossil fuel can be found in the works by [2, 3], and briquette using biomass by [4, 5]. Biomass such as chopped wood, rice husks, straws of rice is not only directly used to fuel material, but also to improve efficiency and calor. It can be made briquette or burned by gasification method. Therefore, this study was made to improve combustion efficiency by using gasification. Rice husk can produce the largest methane gas that can be used as a substitution for LPG gas. To improve the performance of the gasification furnace of TLUD, be attached of reflectors on the burner with a variety of shapes and angles.

Research for the best insulators on the gasification furnace are use clay insulation materials fire-resistant (refractory clay) [6]. Rice husk biomass waste with a capacity of 1.5 kg can produce effective flame for 47 minutes. Air flow rate at the gasification furnace influences the combustion temperature [7].

Study about the influence burner on combustion system worked by [8, 9] and showed that burner give the influences to the maximum temperature. Optimal ratio between height of gasifier and burner and air flow distribution on the TLUD gasification stove of biomass has done by [10], and showed that air flow distribution on burner is very important to get ratio of burner. Optimal ratio is around 1:3 to produces stabilize of flame and blue fire.

2. Methodology

This section explains the methodology used in this research.

2.1. Material

Rice husk used as main material with a water content of $\pm$ 10%. Rice husk is grinding by a machine, and then be sifted by size of mesh 10. Then 1.5 kg of the shifted rice husk was put into the gasifier tube

2.2. Equipment

The main equipment use type gasifier updraft (TLUD) with an outer diameter of 200 mm and an inner diameter of 160 mm (see figure 1). Refractory clay gasifier is used as an insulator, and placed between an inner and an outer diameter of the gasifier tube. Height of the reactor tube is 600 mm.

The variation of burner and burner cap with variation number of hole is shown in figure 2. Number of holes on the burner cap using variation 20, 30, 40 and 50 holes as shown in figure 3(a)-(d) respectively. Every holes having 5 mm of diameter.

Figure 4 shows the installation of the research. Firstly, the gasifier tube (6) was filled with 1.5 kg rice husk. Secondly, the process of gasification was started with burning the surface of rice husk on the gasifier using the lighter. Following this process, the blower (3) was turned on at a speed of 10 m/s. After gasification process began, the temperature at the burner was measured using thermocouple (9) while the time process was noted using a stopwatch. The steps were done repeatedly with time span of 30 seconds until the fire stops.

IOP Conf. Series: Materials Science and Engineering **403** (2018) 012097 doi:10.1088/1757-899X/403/1/012097

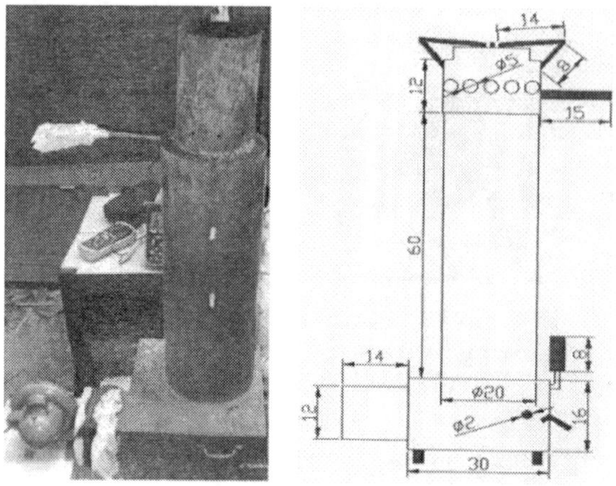

Figure 1. Gasifier tube.

Figure 2. Burner and burner cap with variation number of holes.

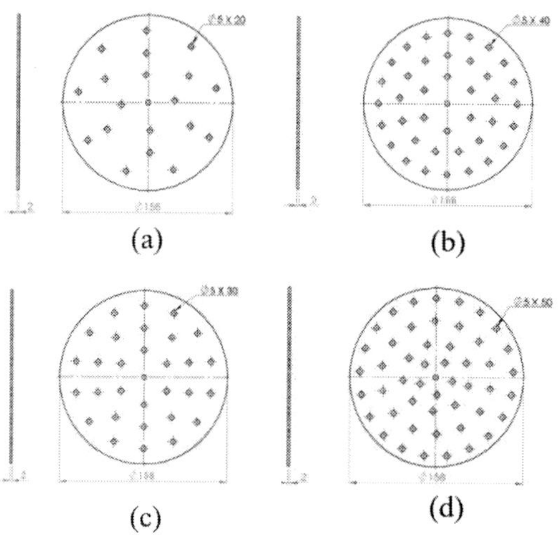

Figure 3. Number of holes (a) 20 holes (b) 30 holes (c) 40 holes and (d) 50 holes.

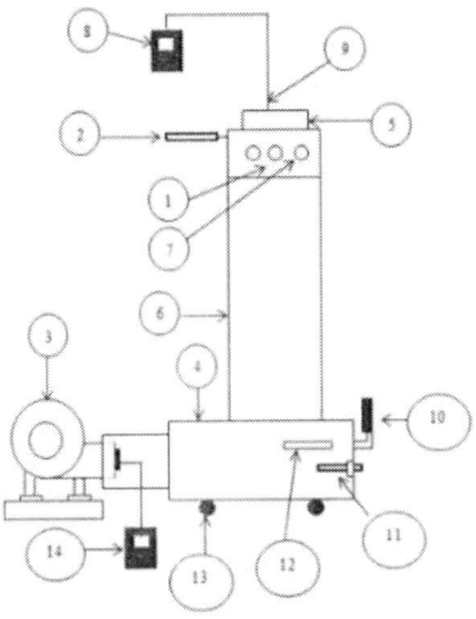

1. Burner
2. Handle of Burner
3. Blower
4. Ash Chamber
5. Reflector
6. Gasifier tube
7. Secondary air flow burner
8. Thermocouple reader
9. Thermocouple
10. Ash discharge lever
11. key
12. handle of door lever
13. wheels

Figure 4. Research installation.

3. Results and discussion

As can be seen on figure 5, which represent result of gasification process with variation of number of holes on burner cap, obviously, that the most stable temperature occurs when using burner cap with number of hole are 50. It reaches average temperature 746°C and length of flame is around 46 minutes. On the other hand, burner with number of holes 20 give the shortest length of flame, and average temperature is not quite high as well. Increasing number of hole on burner cap indicate that average temperature are better than small number or holes. Furthermore, more number of holes produces length of flame better than small number of holes.

Figure 6 represents time to boil of 2 liter of water using water boiling test. Number of holes on burner cap 50 can be boiled water on 7 minutes. That is the fastest time to boil water compare to the other burner with number of hole 20, 30 and 40. This study show that increasing number of hole on the burner cap give influence to the time flame effective and time to boil of water.

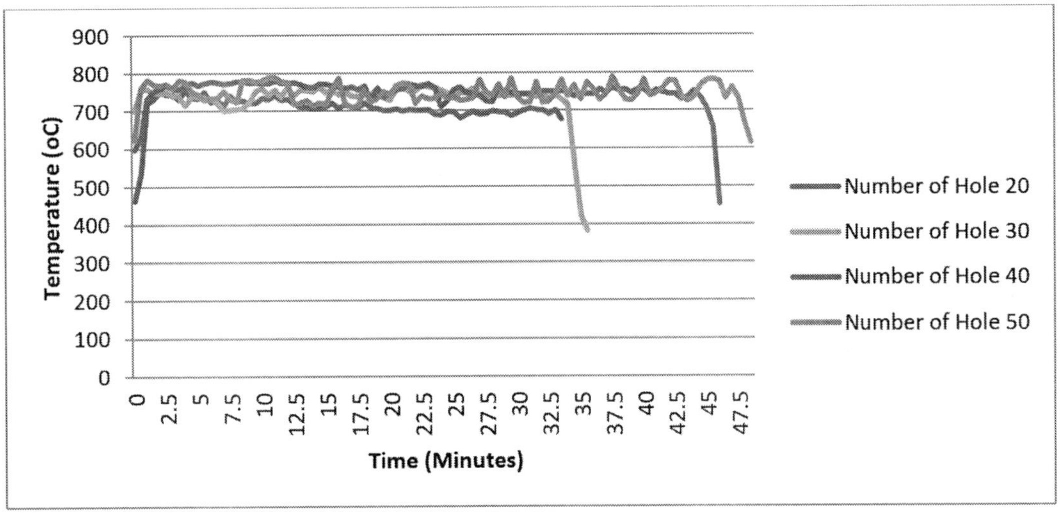

Figure 5. Comparison result of gasification with variation of number of holes on burner cap.

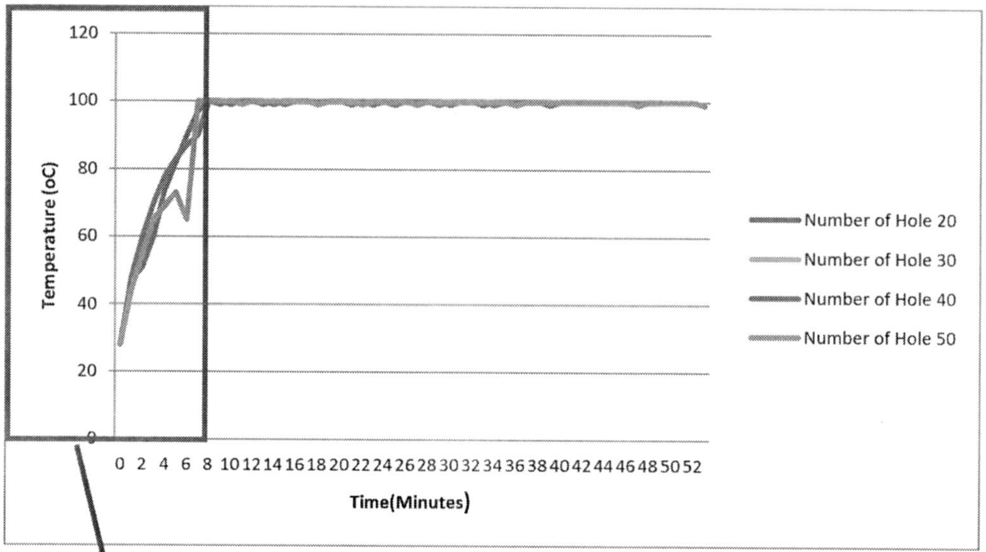

(a)

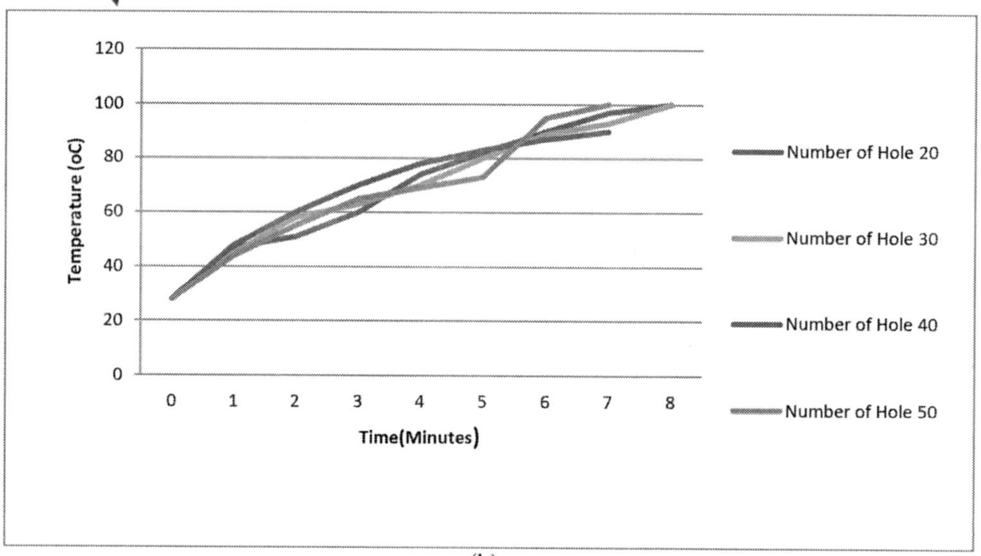

(b)

Figure 6. Comparison result of water boiling test with variation of number of holes on burner cap (a) original comparison graph (b) zoomed version of (a).

4. Conclusions

The result shows gasification processes using variation of number of hole on burner cap give effects to the effective time flame, combustion temperature and time to boil of water. Burner with number of holes 50 produces the best average temperature, the longest time flame and the shortest time to boil of water. The average temperature is 746^{0}C, length of flame is around 46 minutes and approximately 7 minutes to boil of water.

Acknowledgements

Authors would like to thanks to Ministry of Research, Technology and Higher Education of the Republic of Indonesia. This Research is financially supported by *Hibah Bersaing (Penelitian Terapan*

Perguruan Tinggi PTUPT) Project at the fiscal year 2015 and 2016 with contract number 135.40/A3.III/LPPM/IV/2015 and 186.5/A3.III/LPPM/V/2016.

References

[1] Reddy S B N 2012 *Understanding stoves for environment and humanity* (Hertogenbosch, Netherlands: MetaMeta)
[2] McKendry P 2002 Energy production from biomass (part 2): conversion technologies *Bioresource technology* vol 83 issue 1 pp 47–57
[3] McKendry P 2002 Energy production from biomass (part 1): overview of biomass *Bioresource technology* vol 83 issue 1 pp 37-46
[4] Wijianto 2012 Study alternatif penggunaan tar sebagai perekat briket kokas lokal *RAPI* pp M1–M6
[5] Muazu R I and Stegemann J A 2017 Biosolids and microalgae as alternative binders for biomass fuel briquetting *Fuel* vol 194 pp 339–47
[6] Wijianto, Subroto and Sarjito 2015 Analysis of combustion temperature in the choppedwood gasification process with air velocity variation *International Conference on Science, Technology and Humanity (ISETH 2015)* Solo, Indonesia pp 28–35
[7] Arif M J 2013 High temperature air / steam gasification (HTAG) of biomass – influence of air/steam flow rate in a continuous updraft gasifier *Master Thesis* Royal Institute of Technology Stockholm
[8] Dehaj M S 2017 Experimental analysis of natural gas combustion in a porous burner *Experimental thermal and fluid science* vol 84 pp 134-43
[9] Reis L C B S 2014 Numerical modelling of flow through an industrial burner orifice journal of applied thermal egineering vol 67 no 1-2 pp 201-13
[10] Wang X, Niu B, Deng S, Liu Y and Tan H 2014 Optimization study on air distribution of an actual agriculture up-draft biomass gasification stove *Energy Procedia* vol 61 pp 2335-8

Performance evaluation of using water and bio oil-based nanocutting fluids under minimum quantity lubrication with compressed cold air during milling operations of steel

[1]B Sugiantoro, [2]Sakuri and [1]Sutarno

[1]Mechanical Engineering Department, [2]Industrial Engineering Department
STT Wiworotomo Purwokerto, Indonesia
Semingkir Street1[th], Purwokerto, Central Java, 53352

biotech.machining@gmail.com

Abstract. The process machining of metal occurs increasing friction force and high temperature, consequently will affect the wear of tool and product quality. The cooling method contributes significantly to reduce heat, friction force and surface roughness of workpiece. The cooling method that used must develop to aim environmental machining process with minimum quantity lubrications (MQL) and dry cutting, but the dry cutting has limited performance because of low lubrications. The MQL using nanofluid base natural oil/water is major challenge in cooling method. In this study in order to aim high performance of cooling, the bionanofluid formed from high thermal conductivity of nanoparticles (50 nm) emulsifier by water and bio oil as base fluid in MQL process by injection of low temperature of air. To improve the cooling performance, it used cool air as injection pressure, designed by utilizing the evaporative process refrigerant flow with the counter flow direction. The experimental results show the lowest roughness was obtained from Titanium Dioxide TiO_2 with palm oil-based fluid, followed Aluminum Oxide (Al_2O_3) based on coconut oil occurring, the last CuO in a water-based fluid. TiO_2 most dispersing compared to others and correlates to cooling performance, it can be candidates for coolants for machining.

1. Introduction

The process of forming workpiece with machining occurs in the friction and high temperature, consequently will affect the wear of chisels and product quality. The solution to reduce the frictional force, wear and improve surface workpiece, is the cooling method. According to the last development of cooling technology is use of nanocutting based natural fluid with purpose for safe the ecologic and operator health [1]. The optimization of the milling process, the highest quality products and lowest roughness were achieved by the use of vegetable oil based fluids with minimum quantity lubrications (MQL) [2] with purpose for friendly environment [3, 4].

Nanofluids formed from base fluid (dispersants) and particles under dimension of 100 nm (disperse), nanofluid (suspense) can be perfectly dispersed and are permanent due to the effect of motion of Brownian [5]. Mixing Al_2O_3 (2.5% weight) improves convective properties up to 79% [6]. The choice of mixing is based on the fundamental nature of nanoparticles, dispersing nanoparticles with the appropriate base fluid will improve the thermo-physical properties especially the thermal conductivity that impacts on the convective properties of nanofluid [7, 8]. The application of nanofluid based 50 nm coconut oil, can reduce significant friction between tools and the workpiece surface. The coconut oil has heat and stable oxidation properties in the machining process [9], the use of coconut oil with 0.5% weight of AL_2O_3 performs better and is capable of reducing cutting temperature [10], AL_2O_3 and TRIM E709 emulsifier reduce heat generated and improve surface finish [11]. The cooling by the MQL method with nanofluids below 1.5% Al_2O_3 coconut oil-based, reduces machining force [12, 13], with ingredients added Molybdenum Sulfide with 0.5% volume MoS_2 reduces force [14].

The biocutting fluid for machining process influenced by fiber content [15]. The nanofluid characteristics influenced by the time of mixing process, temperature, variation of these two factors will affect the characteristics and performance of nanofluid [16]. The formation of nanofluid to obtain a stable suspension, permanent (durable), free agglomeration (sedimentation) in order to avoid chemical changes, temperature and rotation speed affect the time of synthesis. The ratio of nanoparticle masses per 100 mL (% v/v) is described by equation (1).

$$\varphi = \frac{V_{np}}{V_{bf}+V_{np}} = \frac{V_{np}/V_{np}}{100\ ml\ suspensi} x100 \quad (1)$$

where: np (nano particle), bf (base fluid), φ; Indicates volume fraction.

The cooling performance of bionanofluid in process machining generalized in MQL system. The MQL methods are always developed to improve performance to increase quality and precision high hardness material. The different MQL processes used in the study compared to other MQL methods, is the use of low-temperature air to pressure bionanofluid in mist formations. The cold air used is obtained by lowering the air of the compressor by utilizing the evaporation process of refrigeration system with the Freon R 32 with counter flow heat exchange. Evaluation of performance of bionanofluid under MQL process using Taguchi method with four factors machining parameters (spindle speed, feed rate, depth of cut and bionanofluid), each in three levels. The most optimum process and best nanofluid performance can be seen from its ability to reduce process temperature, power consumption, chatter/vibration and surface roughness of the results as requirement ASTM 2009 in N5 classification.

In this study the effects of changes in thermal conductivity of nanoparticles and nanofluid, and the degree of dispersion will be tested on the milling process. The effect of these two properties on the cooling performance of the milling operation on steel will be analyzed for each type of nanofluid used by looking at temperature, power consumption, vibration, and especially on the machining surface roughness of the milling process.

2. Methodology

This section explains the methodology of the proposed method.

2.1Materials

Materials used in this research are:

a. Nanoparticles: Cu O; AL_2O_3, and TiO_2 (powder dimensions according to MSDS: 50 nm)
b. Basic fluid: Water and bio oil (coconut oil and palm oil).
c. Metals: Steel ST 60 and Stainless Steel 304.
d. Tool: Nachi (HSS) end mill 8 mm, 4 flute.

2.2 Design experiment of bionanofluid formation

The determination parameters process formatting nanofluid using three factors and levels can be seen in table 1. The use of nanofluids made with fiber-containing materials can facilitate decomposition reactions [17]. The addition of nanoparticle elements with high thermal conductivity properties will improve heat transfer capability. The process of making nanofluid aims to produce a nanofluid capable of minimizing the style, the cultivation of the carving and producing high quality machining. CuO can be well ionized if reacted with waterbase [18], Cu dispersing and stable in water [19]. TiO_2 using palm oil base fluid dispersing well, and Al_2O_3 coconut oil base fluid.

Table 1. Factors and levels of formatting of nanofluid ($CuO/Al_2O_3/TiO_2$).

Parameters Formation	Level (1)	Level (2)	Level (3)
Mixing time (minute)	60	120	180
Nano Particle (% weight)(gr)	0,5	1	1,5
Temperature mixing (°C)	25	30	35

2.3 Test of the characteristics of nanofluids

The formation of nanocutting fluids by 3 factors: percentage of weight (0.5, 1.0, and 1.5% weight: base fluid), temperature variation (25°C, 30°C, 35°C), and mixing time (60, 120 and 180 minute) at rotation stirrer (50, 75, 100 rpm), using magnetic stirrer arranged with microcontroller. Test the characteristics of nanofluid to identify thermal conductivity, viscosity, density, and to know crystal structure form of nanofluid, testing by Transmission Electron Microscopy (TEM).

2.4 The principle of reducing temperature of air compressed base refrigeration system

Minimum Quantity Liquid (MQL) with cold air injection at temperature below 15°C. Portable process temperature regulator using the concept of heat exchanger with a system counter flow, when the air compressor pipe (inside) then the cooling evaporation pipe with high performance refrigerant, this process also reduces the humidity of compressed air, the illustration can be seen in figure 1.

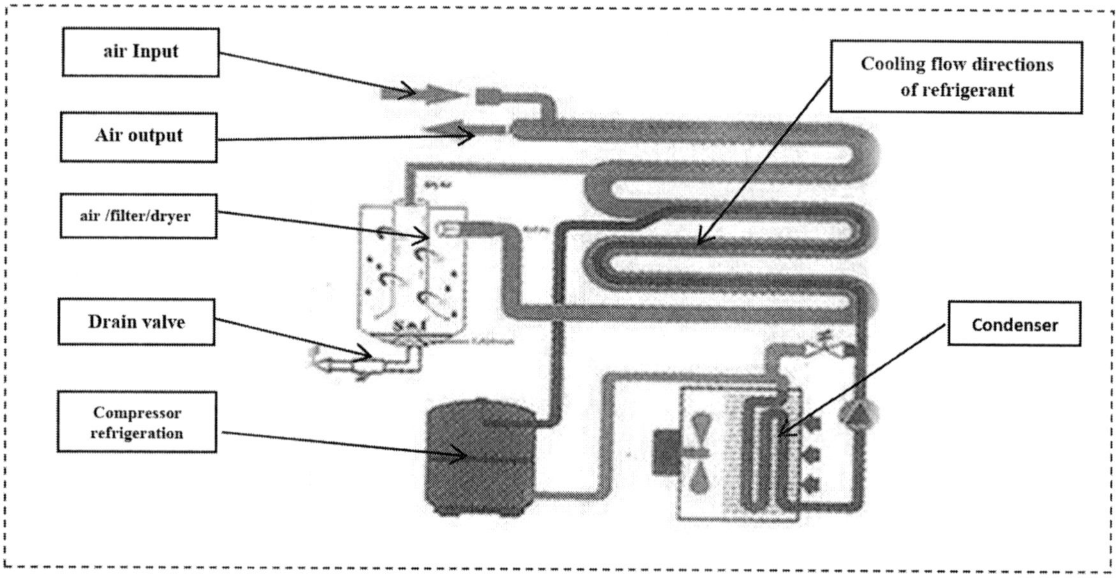

Figure 1. Principle of heat exchange with counter flow direction.

The principle of heat-based heat exchanger by placing a pipe with air flow to be cooled by refrigeration system, to reduce air temperature until continuous at a temperature below 15°C. The exchange of heat aims to reduce the air temperature of the compressor by utilizing the evaporative process of the Freon in the refrigeration system with the flow direction of the counter flow is the most optimum direction in the heat release process. The material used as a compressor air conveyor tube of copper is positioned on the evaporative pipe. To prevent leakage and heat loss on the output side the cold air carrier pipe in isolation is fraught with the objective of minimizing loses.

A refrigeration-based temperature-reducing device can constantly supply air with constant pressure easily, efficiently and optimally. Especially for MQL the fluid used is bio oil based so that the biodegradable index is in accordance with ASTM 2009 requirements. This tool can be used almost for all types of machining process both dry cutting with cold air and MQL for all types of commercial cooling fluid (conventional).

2.5 Experimental design of the milling process under MQL with cooled air pressure
Distribution and data analysis using Taguchi method, to determine the most dominant fluid effect on the results. Performance of nanofluid in machining on milling process materials used as test specimens in this research are ST 60 steel, Stainless steel 304, with length 60 mm, and 40 mm thick, and width 60 mm, and the tools used is a cutter end mill (8 mm, 4 flute/knives). Experimental design for application of cooling performance by MQL method on each nanofluid can be seen in table 2.

Table 2. Factors and levels for the nanolfuid cooling performance on milling.

Parameter of milling	Factors	Level 1	Level 2	Level 3
Spindle speed (rpm)	A	360	565	950
Feed rate (mm/min)	B	98	132	200
Depth of cut (mm)	C	0,75	1,0	1,25
Cutting Condition	D	Nanofluid (mixing time 60 minutes)	Nanofluid (mixing time 120 minutes)	Nanofluid (mixing time 180 minutes)

The cooling performance in milling operations using nanofluid CuO water base, TiO_2 base palm oil and Al_2O_3 base coconut oil performed by the same method on factors (D). Characteristics of quality used is Smaller-the-Better (STB). Characteristic quality where the lower the value, then the better quality. S/N value for the type of STB characteristics shown in the equation (2) [20].

$$S/N_{STB} = -10\,log\ [1/n \textstyle\sum_{i=n}^{n} y^2] \tag{2}$$

Where (S/N): Identify control factor;
n: number of experiment,
y: square of mean

Achievement of process optimization will be known from power consumption, vibration, temperature/injection pressure, tool temperature. The most optimum results are based on the value of the power of spindle, lowest temperature process, lowest vibration and best roughness target as determination of optimum target condition.

3. Results and discussion

This section presents and discuss the experiment results.

3.1. Nanofluids characteristic (thermophysical properties)
The nanoparticles are mixed in order to be suspended, the perfect suspense proof is the mixture become homogeneous (minimal sediment). Thermophysical properties of nanofluids can be seen in table 3. In the cooling process with MQL the process used as the reference is the thermal conductivity value of nanaofluid. The change in the K value on the particle shape and the nanofluid is taken in an average of 3 preparation samples. The result (comparative) on the rate of thermal conductivity of nanoparticles and nanofluid after forming can be seen on figure 2.

The characteristic test results show that in the water-based CuO, there was a 20% decrease in the thermal conductivity value of CuO nanoparticles after dispersed with water from 48.4 W/m.C to 39.62 W/m.C. If compared with Al2O3 the powder conductivity value after dispersed with coconut oil decreased by 15%, from 36.15 W/m.°C to 28.28 W/m.°C. While the increased conductivity value occurs only in TiO2 nanoparticles with palm oil, after being formed into nanofluid the conductivity value increased significantly from 8.74 W/m.°C to 27.56 W/m°C. The specimen with the highest conductivity value tested its crystal arrangement by Transmission Electron Microscopy (TEM) testing. The TEM images can be seen in figure 3. From figure 3(a) is the nanofluids TiO_2 (50 nm), (b) CuO on water (50 nm) and (c) vegetable oil (palm oil) with Al_2O_3.

The TEM test results on three nanofluids describe the dispersion rate of nanoparticles in the base fluid. Nanofluids with base oil of vegetable oil with TiO2 additive material shows very good dispersion rate compared with crystal density and bonding with basic fluid is very good, while at Cu O with water visible process of disperse less visible from crystal structure that loose, at Al2O3 with coconut oil. The level of disperse is quite good and the distribution of crystals quite good. From the comparison, the disperse will improve performance for cutting fluid [21].

Table 3. The result of properties of nanoparticle and nanofluid (CuO with water base fluid).

No	Properties of Nanoparticle and Nanofluid	Measurement (CuO with water base fluid)	Measurement (TiO_2 palm oil base fluid)	Measurement (Al_2O_3) with coconut oil base fluid
1	Thermal conductivity (W/m.°C)	48.4	8.47	36.15
2	Nanoparticles dimension (nm)	50	50	50
3	Density(gr/cm^2)	2.93	4.15	3.64
4	Cristal Structure form	Cubic	Anatase	γ
5	Viscosity (cp) in nanofluid	16.9	29.5	27.8
6	Average K in nanofluids (W/m.°C)	39.62	27.56	28.28

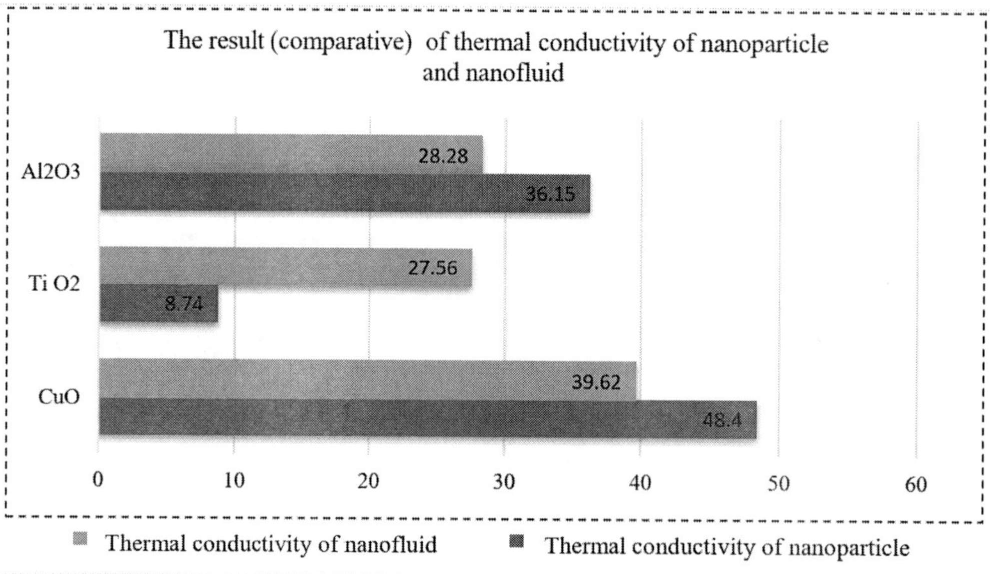

Figure 2. The result (comparative) of thermal conductivity of nanoparticle and nanofluid.

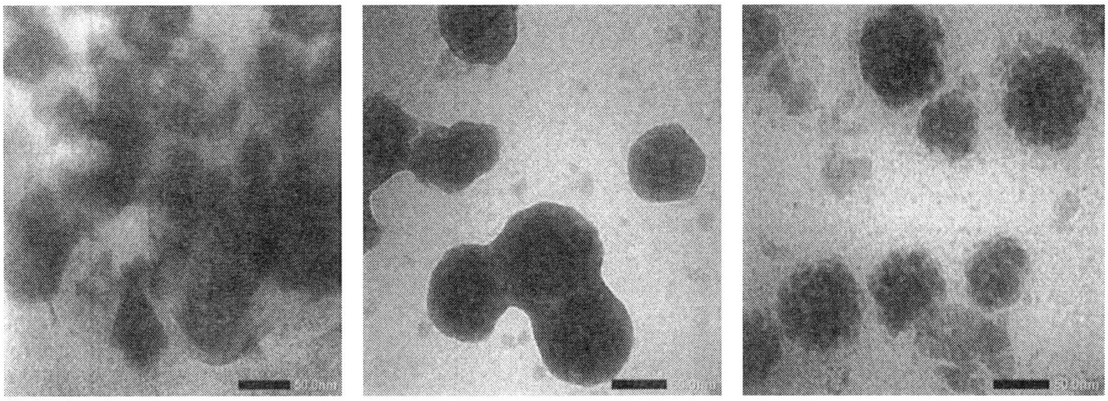

(a) (b) (c)

Figure 3. TEM image (a) TiO_2 palm oil (b) CuO in water and (c) Al_2O_3 coconut base fluid.

3.2. Set-up application of nanofluid under Minimum Quantity Lubrications (MQL) with compressed cool air in milling operations of steel

The nanofluid performance test performed on milling operations in each specimen measured process temperature (tool), vibration, power consumption and surface quality of workpiece can be seen in figure 4. The experimental design uses the Taguchi distribution to detect the most dominant fluid with the specification of the formation process against surface roughness, temperature of cutting zone, tool wear and machining characteristics (vibration, and power consumption).

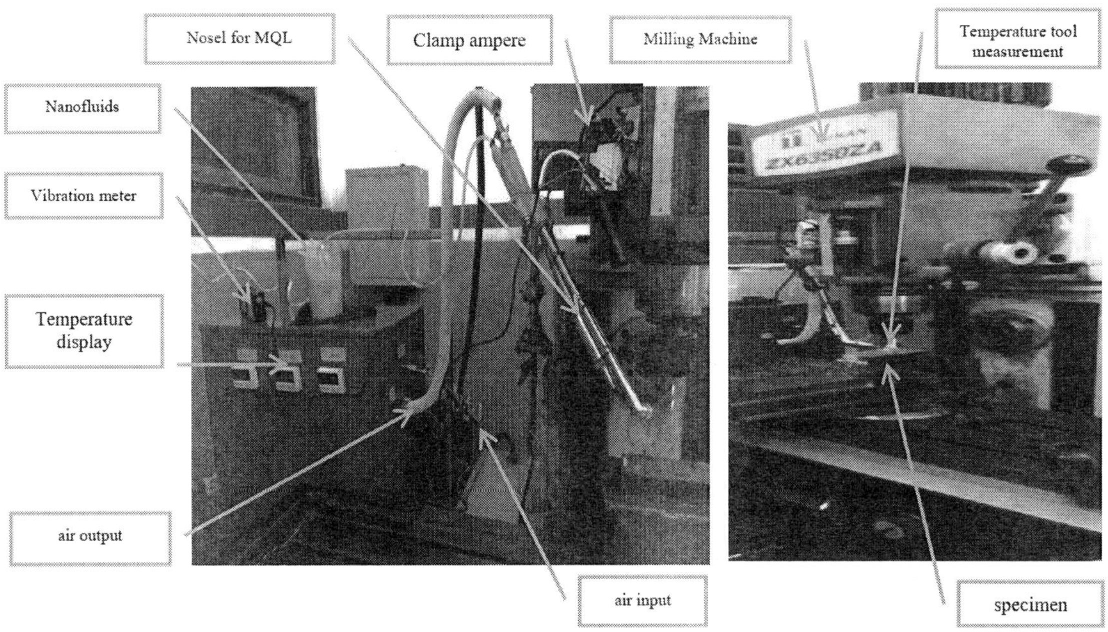

Figure 4. Set-up application nanofluid in the process of milling with Minimum Quantity Lubrications(MQL)with compressed cool air.

3.3. The result of nanofluid performance in the milling process on steel

Table 4 shows parameters obtained bionanofluids under MQL in milling operations on steel.

Table 4. Parameters obtained bionanofluids under MQL in milling operations on steel.

Parameters obtained in milling process Measurement (lowest/highest)	Bio-nanofluids formation (water/bio-oil base)		
	Palm oil based nanofluids Titanium Dioxide (TiO_2)	Coconut oil based nanofluid with (Al_2O_3)	Water based nanofluids water fluid with (CuO)
Lowest temperature cutting zone °C	27.36	30.033	28.00
Highest temperature cutting zone °C	34.10	37.733	35.900
Lowest Surface roughness (ra) µm	0.34	2.058	2.388
Highest Surface roughness (ra) µm	3.2730	6.516	6.628
Power consumptions (lowest) (Watt)	363.73	418.733	388.911
Power Consumptions (highest) (Watt)	427.53	440.733	418.733

The graph of S/N ratio roughness the process and temperature can be seen in figure 5(a) and 5(b).

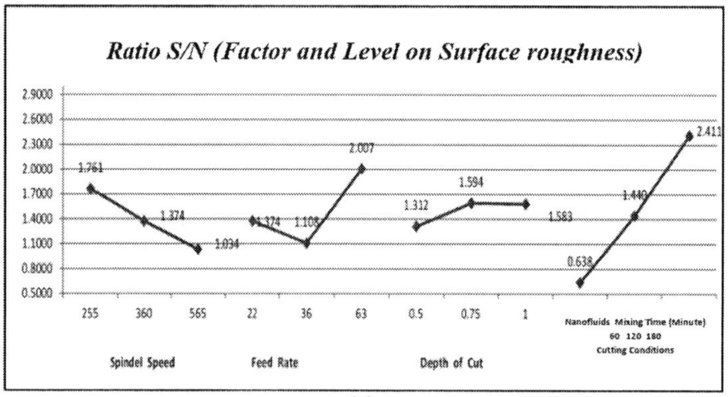

(a)

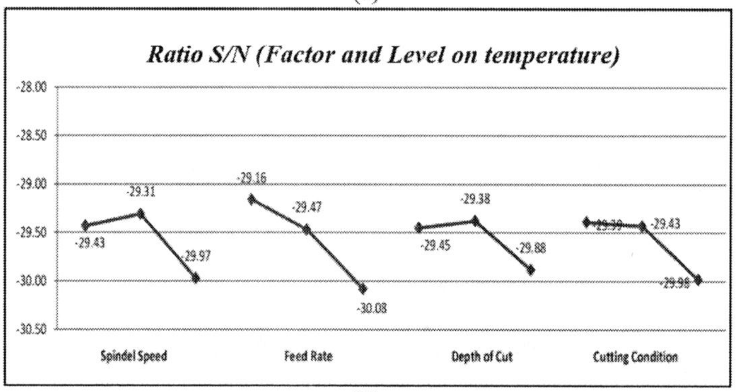

(b)

Figure 5. Graph (a) S/N ratio surface roughness (b) S/N ratio temperature on milling process with nanocutting fluid palm oil with Titanium Dioxide Particles (TiO_2).

The result of milling operations on steel with MQL shows that the use of nanofluid from palm oil with titanium dioxide particles (TiO_2) on stainless steel yielded roughness with the smallest value 0.534 ra, with very low temperature at the smallest temperature at 27.36° C and highest 34.10° C, temperature condition. The low cutting zone is able to reduce the wear of the tool significantly. In accordance with the graph shows that the best roughness is obtained at spindle speed 950 rpm, depth of cut 0.5 mm mixing time 60 minutes, while for the lowest temperature obtained at 180 minutes mixing, this shows that the formation of nanofluid effect on the performance. The characteristics of milling process using nanofluid (coconut oil with Al_2O_3) can be seen on figure 6(a) and figure 6(b).

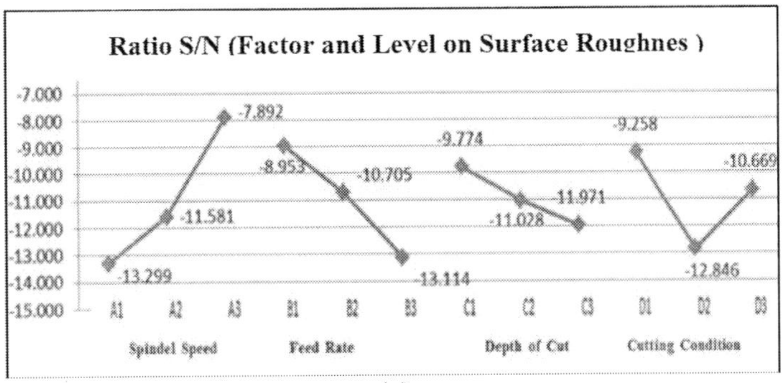

(a)

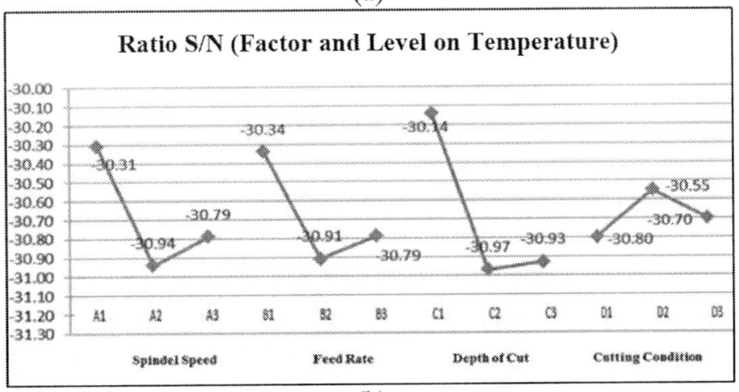

(b)

Figure 6. Graph (a) S/N ratio of surface roughness (b) S/N ratio temperature in the process of milling with nanofluid (coconut oil with Al_2O_3).

The natural (coconut) oil-based nanocutting fluid aluminum oxide (Al_2O_3) applied on ST 60 material on the milling process shows the result of the smoothest roughness at 2.058 ra, and the largest at 6.516 ra, with the smallest cuttingzone temperature at 30.033°C and the highest 37.733°C. With a mixing time of 120 minutes. Characteristics of milling process results with nanofluid base water with (CuO), shown in S/N Roughness ratio of Surface and Temperature in Figure 7(a) and 7(b).

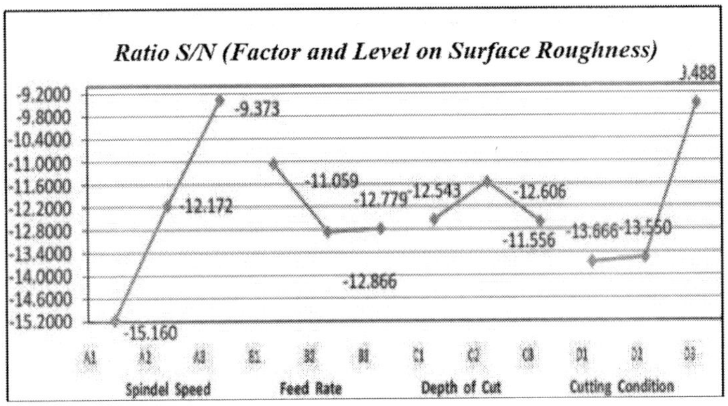

(a)

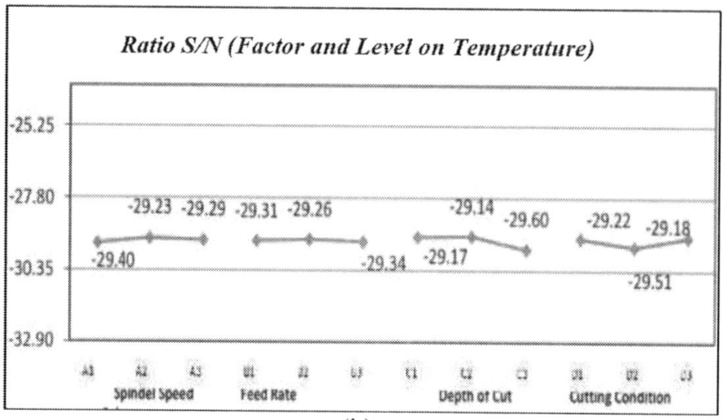

(b)

Figure 7. Graph (a) S/N ratio surface roughness (b) S/N Ratio oftemperature cutting zone in the milling process with nanofluid base water with (CuO).

The nanofluid with added ingredient (CuO) water base fluid shows the lowest roughness is 2,388 Ra and the lowest temperature is 28°C. Length of mixing is not correlated with performance, it can be seen on the graph that the best conditions occur at the time of the mixing, the number of nano particles less effect on changes in nanofluid properties. In terms of wear and tear wear, it can be reduced well. The wornest area is on the edge and side of the end mill chisel. In general, the wear of the tool (end milling) reduce significantly. After 9 times milling process shows that the mean wear of tool reduces significanly, positions and type of tool wears can be seen in figure 8.

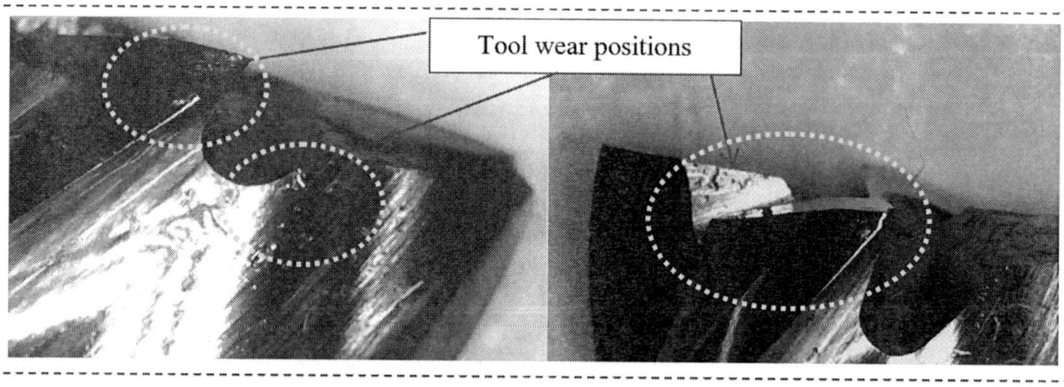

Figure 8. Mean wear sculpting occurs on the side of the cut end mill.

The MQL using bionanofluid with cool air injections increased tool life, it is evident that at 8 mm (diameter) end mill can be used for the formation of SS 304 with a depth of cut up to 1.0 mm, at low temperatures, with minimal wear and low surface finish.

3.4. Discussion

The thermal conductivity value is added to the base of the coconut oil/palm oil fluid with Al_2O_3 the powder conductivity value after dispersed with coconut oil decreased by 15%, from 36.15 W/m.° C to 28.28 W/m.° C. While the increased conductivity value occurs only in TiO_2 nanoparticles with palm oil, after being formed into nanofluid the conductivity value increased significantly from 8.74 W/m°C to 27.56 W/m° C. The result is similar to the research trends [22].

The formation of nanofluid with coconut oil and palm oil shows the process temperature should be below 35°C, because above the temperature began to occur early boiling process. The addition of nanoparticles has an effect on cooling performance, because it can increase enhanced penetration but not directly proportional to the same trend with the research [23]. Bionanocutting fluid with cold air injections can reduce the temperature at the cutting zone and friction force that occurs tool life increased, suitable at the result previous study [24-26]. The nanofluid TiO_2 base palm oil minimize frictions, reductions cutting wear and surface roughness as research trends [27]. The comparison of bio-oil and water base in MQL with micro additives / nano additives shows that palm oil with TiO_2 give the best performance than others, as previous research result [28].

4. Conclusion and future work

Nanofluids used to produce the lowest roughness obtained from nanofluid; titanium dioxide (TiO_2) based palm oils, In sequence Aluminum oxide (Al_2O_3) base coconut oil, and the last (CuO) base water. TiO_2 with palm oil as base oil can be used as a good coolant for machining process It shows that the most disperse level as shows in crystal images has an effect on cooling performance, the lowest temperature even for the hardest material that is stainless steel. The level of dispersion and thermal conductivity of nanofluid into factors affecting cooling performance in milling operations, proven titanium dioxide (TiO_2) based palm oils, became the best coolant in milling of SS 304with reduce surface roughness significantly. Effect of air with cold temperature below 15°C, at pressure 1 bar is most effective as bionanofluid suppressor in MQL process.

Bionanofluid research is strategically developed in the future, because it is environmentally friendly and biodegradable. Subsequent research is indispensable in automated machining and high speed machining to obtain more complete bionanofluid performance characteristics as reference for industrial scale applications.

Acknowledgments

Thank to Ministry of Research, Technology and Higher Education of the Republic of Indonesia, which has provided funding for fundamental Research of fiscal year 2016-2017.

References

[1] Shokoohi Y, Khosrojerdi E and Shiadhi B H R 2015 Machining and ecological effects of a new developed cutting fluid in combination with different cooling techniques on turning operation *J. of Cleaner Prod.* vol 94 pp 330-9
[2] Sugiantoro B, Rusnaldhy and Widyanto S A 2014 Optimization of parameter of milling process on quality of aluminum machinery result with taguchi method *TRAK SI* vol 14 no 1
[3] Shashidhara Y M and Jayaram S R 2010 Vegetable oils as a potential cutting fluid—An evolution *Tribology International* vol 43 pp 1073–81
[4] Ozcelika B, Kuram E, Demirbas E and Şik E 2013 Effects of vegetable based cutting fluids on the wear in drilling *Sadhana* vol 38 issue 4 pp 687–706
[5] Suhanan, Kamal S, Prayitno Y A K, Wiranata A and Pradecta M R 2016 Experimental study of thermophysical properties of nano fluid TiO2/TermoXT-32, *Proceeding of National Symposium on Thermofluids VIII* Yogyakarta, Indonesia
[6] Putra N, Maulana S, Koestoer R A and Danardono A S 2005 Measurement of coefficient of heat transfer convection of nano particle fluid suspended water (Al_2O_3) on fintube heat exchanger *Journal of technology* vol 2 no XIX pp 116-25
[7] Naina H K, Gupta R, Setia H and Wanchoo R K 2012 Viscosity and specific volume of TiO2/water nanofluids *J. Nanofluids* vol 1 no 2 pp 161–5
[8] Vasheghani M, Marzbanrad E, Zamani C, Aminy M, Raissi B, Ebadsadeh T and Barzegar-Bafrooei H 2011 Effect of Al_2O_3 phases on the enhancement of thermal conductivity and viscosity of nanofluids in engine oil *Heat Mass Transf.* vol 47 pp 1401–06
[9] Rao G K M, Padmini R and Krishna P V 2008 Performance evaluation of eco-friendly nano fluids in machining *Recent Advances in Robotics, Aeronautical and Mechanical Engineering* pp 221-9
[10] Sodavadia K P and Makwana A H 2014 Experimental investigation on the performance of coconut oil based nano fluid as lubricants during turning of AISI 304 austenitic stainless

steel *International Journal of Advanced Mechanical Engineering* vol 4

[11] Vasu V and Kumar K M 2011 Analysis of nanofluids as cutting fluid in grinding EN-31 Steel *Nano-Micro Lett.* vol 3 issue 4 p pp 209-14

[12] Shokoohi Y and Shekarian E 2016 Application of Nanofluids in Machining Processes - A Review *Journal of Nanoscience and Technology* vol 2 no 1 pp 59–63

[13] Bose P S C, Rao C S P and Jawale K 2014 Role of MQL and nano fluids on the machining of Nicrofer C263, *5th International & 26th All India Manufacturing Technology, Design and Research Conference (AIMTDR 2014)* Guwahati, Assam, India pp 1 - 4

[14] Padmini R, Krishna V P and Rao G K M 2015 Effectiveness of vegetable oil based nanofluids as potential cutting fluids in turning AISI 1040 steel *Tribiology International* vol 94 pp 490-501

[15] Mirjalili S A, Muirhead J C and Stringer M D 2011 Ultrasound visualizations of the spinal accessory *Journal of surgical research* vol 175 issue 1 pp e11-6

[16] Keblinski P, Eastman J A and Cahill D G 2005 Nanofluids for thermal transport materials *Materials Today* vol 8 issue 6 pp 36-44

[17] El-Nahhal I M, Zourab S M, Kodeh F S, Selmane M, Genois I and Babonneau F 2012 Nano-structured copper oxide-cotton fibers: synthesis, characterization and applications *Int. Nano Lett.* vol 2 issue 1 pp 62-7

[18] Manimaran R, Palaniradja K, Alagumurthi N, Sendhilnathan S and Hussain J 2013 Preparation and characterization of copper oxide nanofluid for heat transfer applications *Applied Nanoscience* vol 4 issue 2 pp 163-7

[19] El-Maghlany W M, Hanafy A A, Hassan A A and El-Magid M A 2016 Experimental study of Cu–water nanofluid heat transfer and pressure drop in a horizontal double-tube heat exchanger *Experimental Thermal and Fluid Science* vol 78 pp 100–11

[20] JMP 2014 *Design of experiments guide* (Cary, NC, USA: SAS Institute Inc.)

[21] Prasad M M S and Srikant R R 2013 Performance evaluation of nano graphite inclusions in cutting fluids with MQL technique in turning of AISI 1040 steel *IJRET: International Journal of Research in Engineering and Technology* vol 2 issue 11

[22] Baby T T and Ramaprabhu S 2011 Enhanced convective heat transfer using graphene dispersed nanofluids *Nanoscale Research Letters* vol 6 p 289

[23] Nam J S, Lee P-H and Lee SW 2011 Experimental characterization of micro-drilling process using nanofluid minimum quantity lubrication *Int J Mach Tools Manuf.* vol 51 pp 649–52

[24] Zhang S, Li J F and Wang Y W 2012 Tool life and cutting forces in end milling Inconel718 under dry and minimum quantity cooling lubrication cutting conditions *Journal of Cleaner Production* vol 32 pp 81–7

[25] Hadad M and Sadeghi B 2013 Minimum quantity lubrication-MQL turning of AISI 4140, steel alloy *Journal of Cleaner Production* vol 54 pp 332–43

[26] Rahim E A and Sasahara H 2011 A study of the effect of palm oil as MQL lubricant on high speed drilling of titanium alloys *Tribology International* vol 44 issue 3 pp 309–17

[27] Belluco W and Chiffre L D 2004 Performance evaluation of vegetable-based oils in drilling, austenitic stainless steel *Journal of Material Processing Technology* vol 148 issue 2 pp 171–6

[28] Padmini R, Krishna P V and Rao G K M 2016 Effectiveness of vegetable oil based nanofluids as potential cutting fluids in turning AISI 1040 steel *Tribology International* vol 94 pp 490–501

IOP Conf. Series: Materials Science and Engineering **403** (2018) 012099 doi:10.1088/1757-899X/403/1/012099

Application pin-on-disc method for wear rate prediction on interaction between rail and wheel

Windarta

Mechanical Engineering Department, Universitas Muhammadiyah Jakarta,
Jl. Cempaka Putih Tengah 27, Jakarta Pusat, Indonesia 10510

windarta@ftumj.ac.id

Abstract. This paper aim to predict wear rate on interaction between rail and wheel using pin-on-disc method. The wear tests were performed on Ducom multispecimen test machine designed according to ASTM G99 standards. The applied loads selected were 40 N, 60 N, 80 N, and 100 N with a common rotating speed of 100 rpm. Pin-on-disc result was used to predict wear rate at real condition. The results show that stress distribution can be used to predict rail and wheel surfaces evolution. It also can be used in a specific condition for prediction at a real applied load. Depth of wear curvature indicate parabolic curve. Depth of wear are 89.33 μm, 127.25 μm, 263.01 μm, and 395.67 μm for 40 N, 60 N, 80 N, and 100 N applied loads respectively in steady state condition. Based on tangential contact stress, the surface shear stress increase in magnitude from leading edge to reach maximum value 668 MPa at longitudinal axis x = -0.73 mm. The adhesion part dominated the contact patch area. The maximum temperature 828.58 °C occurs at the contact surface.

1. Introduction

Interaction between rail and wheel can be divided into rolling-sliding contact and purely sliding contact [1]. All of contact produced damage, such as wear and fatigue. The maintenance cost of wear and fatigue of wheel and rail contacts is about 1.2 billion USD in China annually [2]. Wear prediction is important to be determined. It is well-known fact that sliding contact contributes more to wear than rolling-sliding contact. Sliding contact can be simulated using pin-on-disc.

Actual tests need huge resources. Since actual testing for rail and wheel contacts are expensive and time consuming, therefore simulation studies are selected well than actual experiment. Wear modelling in metals has been conducted by many researchers [3, 4]. Many models have been proposed for different materials and applications. Several models are co relational in nature and specific system. The model is based on the wear data of that system and only applied to that data set [5-8]. The present paper aimed to determine wear rate using pin-on-disc method and wheel and rail contact temperature.

2. Methodology

2.1. Interaction between rail and wheel

Interaction between rail and wheel can be divided onto two kinds of mechanism (figure 1(a)). The first one occurs on railhead and wheels tread contact, it sometimes called rolling-sliding contact. The second occurs on rail edge and wheel flange contact, it is pure sliding contact. In this research, contact between rail and wheel can be assumed non conformal contact.

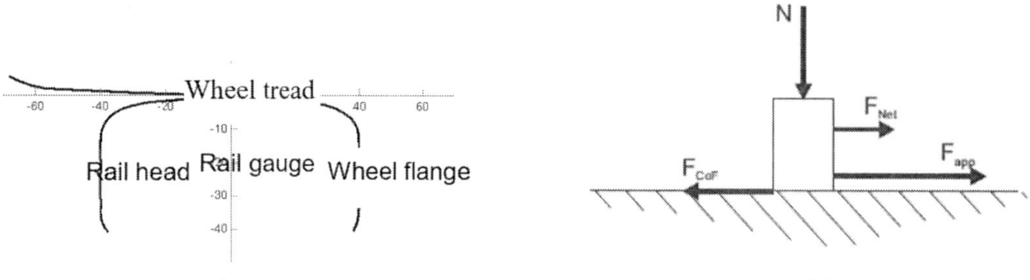

(a) (b)

Figure 1. Overview of wheel-rail contacts (a) Pin-on-disc scheme (b).

A non-conformal contact of two bodies of revolution is considered in the general elastic contact. Depending on their modulus elasticity, the two bodies will deform to certain degrees. The problem may be reduced to that of a single equivalent elastic body of revolution with a plain infinitely rigid body. This assumption is widely adopted in the theory and application of Hertzian contacts so that the conversion formulas of geometry and elasticity are well established [9, 10].

When at the top of an elastic pin of radius r and equivalent modulus of elasticity E_e (figure 1(b)) is pressed with a load N against a rigid plane disc surface, there is a contact area mark [11]. The equivalent modulus elasticity is given by equation (1).

$$E_e = \frac{E}{1-v^2} \tag{1}$$

The pin is rotating with a fixed angular speed ω. Due to the normal force, angular velocity, and coefficient of friction μ, a reaction force will occur. This reaction force is frictional force F_f.

$$F_f = N.\mu \qquad F_{Net} = F_{app} - F_f \tag{2}$$

The net force F_{Net} is equal applied force F_{app} minus frictional force F_f (equation (2)). The contact patch is defined as the region A in the xy-plane: $A = \{(x, y): x^2 + y^2 \leq r^2\}$. The normal stress P_z is given by equation (3).

$$P_z(x,y) = P_0\sqrt{1-\left(\frac{x}{r}\right)^2-\left(\frac{y}{r}\right)^2} \quad \text{with} \quad P_0 = \frac{3N}{2\pi r^2} \tag{3}$$

Since pin-on-disc contact can be categorized as sliding contact, the Archard's wear equation as shown in equation (4) can be applied.

$$V_w = k_{(v_s,P_z)}\frac{N.s}{H} \tag{4}$$

where Vw is the wear volume (m3); s is sliding distance (m); N is normal force (N); H is hardness of materials (Pa); k is wear coefficient (–); vs (x, y) is relative sliding velocity (m/s); Pz (x, y) is contact pressure (Pa).

In the numerical analysis, Archard's wear model is applied locally by partitioning the contact area into a calculation mesh and dividing equation (5) by the area increment, dx dy. The wear depth, Δz, at the center of the mesh element with co-ordinate (x, y) thus becomes:

$$\Delta z(x,y) = k_{(v_s,P_z)}\frac{P_z(x,y).s(x,y)}{H} \tag{5}$$

The wear depth is preferred to determine rail-wheel profile evolution [10]. Wear coefficients (k) are determined by experiment. Wear coefficient values are presented in a maps as a function of contact pressure and sliding velocity. Based on available experimental results for relevant material combinations, a wear map for dry conditions has been defined. Jendel [11] presented the most important regimes for railway operation are the k2 and k3 areas although both higher contact pressures and larger slip may occur in narrow curves or at poor adhesion conditions, respectively.

A non-linear creep force theory has been proposed and developed [10, 12, 13]. The spin contribution in Kalker's linear theory is neglected. The resultant creep force F was assumed to approach the traction bound μWx following a polynomial approximation in equation (6).

$$F = \mu . W_x \left(k - \frac{1}{3} k^2 + \frac{1}{27} k^3 \right) \quad \text{for k} \leq 3 \qquad k = \frac{F}{\mu W} \tag{6}$$

Many researchers investigated contact temperatures between rail and wheel using analytical method in numerous papers [14, 15]. Since pin diameter less than its length, one-dimensional approach can be used. In one-dimensional approach, the heat conduction with included heat transfer into the surrounding can be formed as in equation (7) [15].

$$\frac{\partial T}{\partial \tau} = \kappa \frac{\partial^2 T}{\partial z^2} - b(T - T_a) \qquad T = T(z, \tau);\ z \in (0, \infty);\ \tau \in (0, t) \tag{7}$$

where, T is the pin temperature, z is the pin axis coordinate, Ta is the ambient temperature (20°C). K is the thermal diffusivity coefficient.

Based on the one-dimensional heat conduction problem, the initial and boundary conditions are determined by equation (8) and equation (9).

$$T(\tau = 0) = T_a \qquad k \frac{\partial T}{\partial z}(z = 0) = q'' \tag{8}$$

$$\lim_{z \to \infty} T(z, \tau) = T_a;\ q'' = \mu P_z v \tag{9}$$

where, μ is the coefficient of frictional. v is traveling velocity. The temperature distribution along the pin axis and the contact temperature are given by equation (10).

$$T_{pin} = T(z,t) = T_a + \frac{\mu P_z v}{k} \sqrt{\frac{\kappa}{\pi}} \int_0^t \frac{1}{\tau} e^{-\left(b\tau + \frac{z^2}{4\kappa\tau}\right)} d\tau \ ; \quad T_{contact} = T(z = 0, t) \tag{10}$$

where k is thermal conductivity (ksteel = 41 J/msK) [14].

3. Material and experimental

The rail steels were used to represented real materials in this research. The chemical composition of material is shown in table 1.

Table 1. Nominal chemical composition of the studied steel.

	C	Mn	P	S	Si	Fe
Rail	0.71	0.87	0.02	0.01	0.29	Balance
Wheel	0.53	0.69	0.02	0.01	0.28	Balance

Pin-on-disc is commonly used in wear test. The tests use Ducom multi specimen testing machine designed according to ASTM G99 standards. The wheel material was cut to form disc specimen, which was 42 mm in diameter and 5 mm in width. The rail material has been cut to form pin specimen. The pin samples were prepared as 6 mm in diameter and 12 mm in length. During wear tests, the normal forces were applied. The normal force of 40 N, 60 N, 80 N, and 100 N were selected. Both pin and disc sample were polished using 120, 220 and 500 grit abrasive papers, and cleaned with alcohol and dried.

4. Result and discussion

4.1. Normal load analysis

The samples used in this research to form pin materials and disc materials are rail steels UIC60 and wheel. It is assumed that normal load influence along contact area only. There is not normal load influenced outside contact area. It proved that wheel-rail contact is line contact. The dry sliding tests reported in Figure 2(a) are conducted using Ducom multi specimen tribometer. Rail steel has been cut to form disc specimen, which is 42 mm in diameter and 6 mm in thickness. Pin samples are prepared as 6 mm in diameter and 12 mm in length. The normal force of 40 N, 60 N, 80 N, and 100 N are selected during wear tests. Both pin and disc sample are polished using 120, 220 and 500 grit abrasive papers, and cleaned with alcohol and dried.

Figure 2(a) shows depth of wear are 89.33 μm, 127.25 μm, 263.01 μm, and 395.67 μm for 40 N, 60 N, 80 N, and 100 N applied loads respectively in steady state condition. Depth of wear curvature is exponential curve for transient condition.

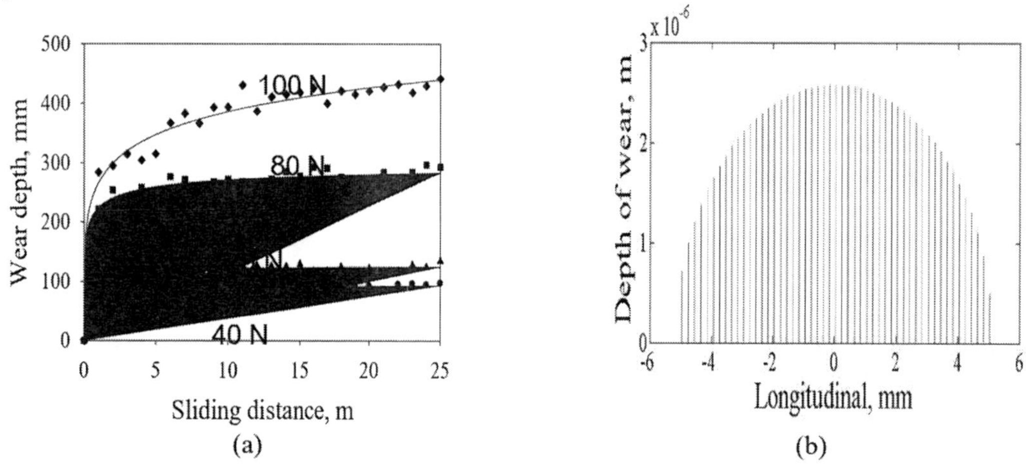

Figure 2. Depth of wear, (a) Wear depth of rail and wheel material for various load after pin-on-disc results; (b) depth of wear prediction for 100 kN applied load.

The total weight of a boogie and passengers in a real operating railway-wheel system conditions are usually used 80 – 120 kN applied load [10]. In this simulation, normal load used 100 kN applied load to predict real condition. The prediction was used to fit the results of experiment, where k was identified to be 5.10-4 mm3/N.mm. Prediction depth of wear for 100 kN applied load is presented in figure 2(b). It is shown that depth of wear has parabolic curvature with maximum value at the center of applied load. This results can be used to predict rail and wheel profile evolution.

4.2. Tangential load analysis

The tangential contact problem is usually connected with friction, and a solution is needed for surface shear stress distributions and related area of adhesion and slip in the contact patch [16]. The slip-stick area distributions in contact patch when the traction coefficient 0.3 is shown in figure 3(a). The stick

IOP Conf. Series: Materials Science and Engineering **403** (2018) 012099 doi:10.1088/1757-899X/403/1/012099

area is located at the leading part of contact area and decreases in size with the increase of tangential load. The surface shear stress (figure 3(b)) increases in magnitude from the leading edge to reach maximum value 668 MPa at longitudinal axis x = -0.73 mm.

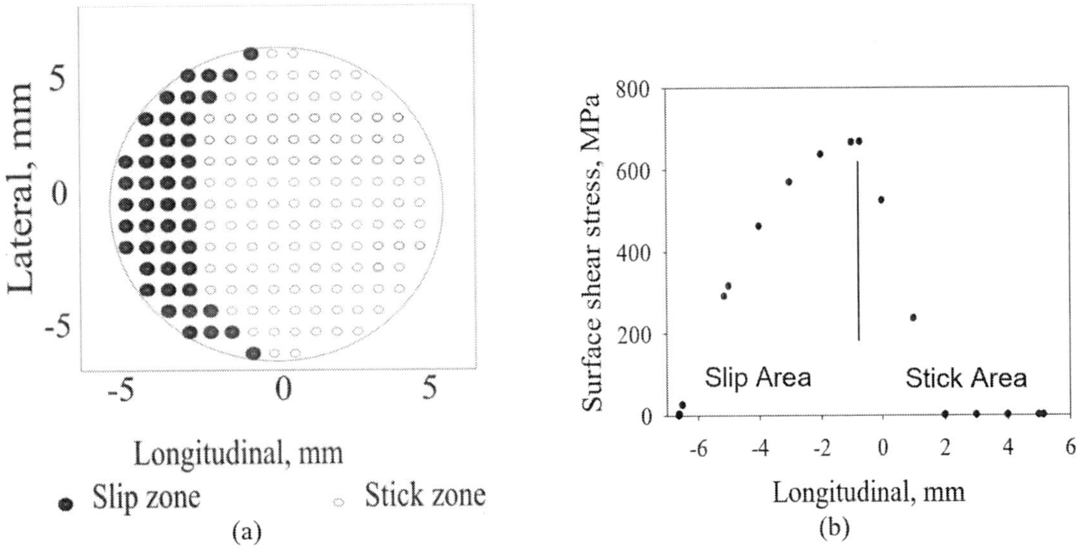

Figure 3. (a) Slip-Stick area distribution for μ=0.3; (b) Surface shear stress.

4.3. Contact temperature

The distribution of contact temperature was calculated using equation (10). The results are shown in figure 4. The results show that the temperatures strongly decrease along the pin axis since the heat is conducted into the surrounding. The highest temperature 828.58°C occurs at the contact surface. This results do not consider cooling effects which caused by environmental contact and heat transfer between pin and disc materials.

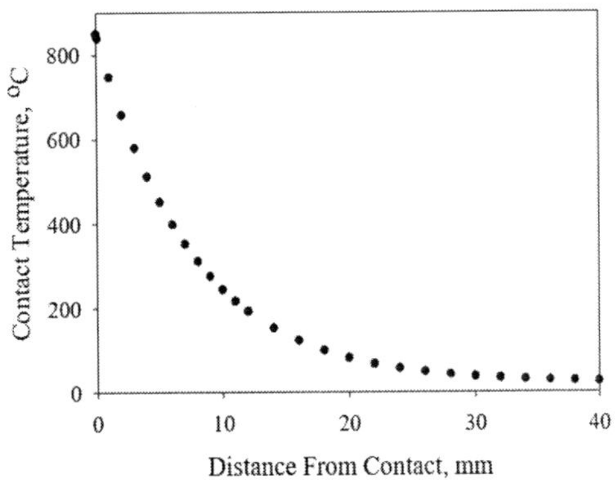

Figure 4. Comparison between experimental results and predicted results.

5. Conclusion

The sliding contact between rail gauge and wheel flange interaction have been investigated using pressure distribution study. The results showed that stress distribution can be used to predict rail and wheel surfaces evolution. It also can be used in a specific condition for prediction at a real applied load. The pin-on-disc method can be used to predict the depth of wear experimentally. The maximum pressures are 2.12 MPa, 3.18 MPa, 4.25 MPa, and 5.31 MPa for 40 N, 60 N, 80 N, and 100 N applied loads respectively. Depth of wear curvature indicate parabolic curve. Depth of wear are 89.33 μm, 127.25 μm, 263.01 μm, and 395.67 μm for 40 N, 60 N, 80 N, and 100 N applied loads respectively in steady state condition. The surface shear stress reach maximum value 668 MPa at longitudinal axis x = -0.73 mm. The adhesion part dominated the contact patch area. The maximum temperature 828.58 °C occurs at the contact surface.

Reference

[1] Sundh J and Olofsson U 2011 Relating contact temperature and wear transitions in a wheel–rail contact *Wear* vol 271 issues 1–2 pp 78–85
[2] Jin X, Wen Z, Xiao X and Zhou Z R 2007 A numerical method for prediction of curved rail wear *Multibody Syst. Dyn.* vol 18 no 4 pp 531–57
[3] Lewis S R, Lewis R, Evans G and Buckley-Johnstone L E 2014 Assessment of railway curve lubricant performance using a twin-disc tester *Wear* vol 314 issues 1–2 pp 205–12
[4] Hsu S M and Shen M 2004 Wear prediction of ceramics *Wear* vol 256 no 9–10 pp 867–78
[5] Vidaud M and Zwanenburg W J 2009 Current situation on rolling contact fatigue – a rail wear phenomenon *9th Swiss Transport Research Conference 2009* pp 1–27
[6] Wang Y and Yuan Q 2013 Contact Force Distribution and Static Load-carrying Capacity of Large Size Double Row Four-point Contact Ball Bearing *Def. Technol.* vol 9 no 4 pp 229–36
[7] Jin Y, Ishida M and Namura A 2011 Experimental simulation and prediction of wear of wheel flange and rail gauge corner *Wear* vol 271 issues 1–2 pp 259–67
[8] Lewis R and Dwyer-Joyce R S 2006 Wear at the wheel/rail interface when sanding is used to increase adhesion *Proc. Inst. Mech. Eng. Part F J. Rail Rapid Transit* vol 220 no 1 pp 29–41
[9] Al-Bender F and De Moerlooze K 2008 A Model of the Transient Behavior of Tractive Rolling Contacts *Adv. Tribol.* vol 2008 no 3 pp 1–17
[10] Enblom R and Berg M 2008 Proposed procedure and trial simulation of rail profile evolution due to uniform wear *Proc. Inst. Mech. Eng. Part F J. Rail Rapid Transit* vol 222 no 1 pp 15–25
[11] Jendel T 2002 Prediction of wheel profile wear—comparisons with field measurements *Wear* vol 253 issues 1–2 pp 89–99
[12] Wang W J, Shen P, Song J H, Guo J, Liu Q Y and Jin X S 2011 Experimental study on adhesion behavior of wheel/rail under dry and water conditions *Wear* vol 271 issues 9–10 pp 2699–705
[13] Polach O 2011 Wheel profile design for target conicity and wide tread wear spreading *Wear*, vol 271 issues 1–2 pp 195–202
[14] Podra P and Andersson S 1999 Simulating sliding wear with finite element method *Tribol. Int.* vol 224 pp 13–21
[15] Zmitrowicz A 2006 Wear Patterns and Laws of Wear – A Review *J. Theor. Appl. Mech.* vol 44 no 2 pp 219–53
[16] Zhao X and Li Z 2011 The solution of frictional wheel–rail rolling contact with a 3D transient finite element model: Validation and error analysis *Wear* vol 271 issues 1–2 pp 444–52

Use of bamboo fiber as a brake pad lining material and the influence of its portion on hardness and durability

P I Purboputro, M A Hendrawan and A Hariyanto

Mechanical Engineering Department, Universitas Muhammadiyah Surakarta, Indonesia

pip272@ums.ac.id

Abstract. This study aims to optimize non asbestos brake pad lining component to find their dynamic characterization. The brake pad component material was made from bamboo fiber, aluminum powder, fiber glass and polyester as matrix. The initial characterization consists of its hardness and wearness in the water and dry condition. The result is highest hardness 14.47 BHN more than existing brake pad 13.7 BHN. Wearness on Ogoshi in dry condition is lowest 0.00041mm^2/kg, while on water wearness variation 1 has lowest wearness 0.0062 mm^2/kg, whether in oil wet condition variation 2 has lowest wearness 0.0003 mm^2/kg. The characterization shows that the brake pad lining that we were made are close to existing characterization.

1. Introduction

This research was dominant to use natural fiber as green technology that recently issues, to substitute asbestos fiber that can cause lung disease. This research that using bamboo fiber, glass fiber and aluminum powder, and polyester as matrix, its wide to use natural fiber to substitute the asbestos component.

These materials were tested on strength with friction Ogoshi wear test method, and Wipro wear test Method. Its hardness tested with Brinell Hardness Test, followed by macro photographs, and testing of braking dynamics characteristics.

The strength of the composite material of these brake particles, greatly influenced by particle size, matrix material and the manufacturing process. Strength of the particle composite is obtained maximally in sizes from 0.01 to 0.1 mm, and also depend on the strength of surface bonding, pressing, and sintering [1]. The process is very rarely exposed to the producers of brake lining, so the need for research on materials and processes are need scientifically standard for brake lining. We hope this field of studying composite base natural fiber can be developed continuously.

The use of brake lining is usually no more than 10,000 contacts, especially for braking in daily use in city travel may be no more than 6 months [2]. In other uneven wear may result from less uniform pressure, due to improper mounting [3], so we hope to find materials that better on breaking characterization and healthy environment in this study.

2. Research purposes and virtue

The purpose of this research is to study on brake pad lining material, using variation of composition from bamboo fiber, glass fiber, aluminum powder, with polyester as matrix, for fair result of its hardness and wear and to characterize prototype of motorcycle brake lining by dynamometer test, to obtain the parameters of braking, absorption of braking, braking time, temperature of braking, and also coefficient of friction brake pad in the variation of dry and wet condition (water and oil). The variation of composition is:

a. 40% bamboo fiber + 10% fiber glass + 10% aluminum (Al) + 40% polyester, as variation 1.
b. 30% bamboo fiber + 15% fiber glass + 15% aluminum (Al) + 40% polyester, as variation 2.
c. 20% bamboo fiber + 20% fiber glass + 20% aluminum (Al) + 40% polyester, as variation 3.

The virtue of this research are conducting applied research for the same field study using natural fiber on brake pad lining materials. Giving alternative of material using natural fiber for better health and environmental consideration and supports local contain materials and home industries among Surakarta.

3. Literature review

The research on friction brake pad by giving sintering time at compaction pressure of 10 minutes is conducted by [4]. Wear of a composite material is getting bigger or more easily worn can be affected by the amount of time given in the compaction process. When the emphasis time is greater than the wear rate is also getting bigger. The hardness value of a material is also affected by the amount of time the compression stresses are given in the process of making the brake lining materials. In making the brake pads, the value of the hardness of the canvas also affects the greater the compaction charged the harder the composite. Composite itself is influenced by several factors in the process of making from material to composite and several causes are: material variation, compaction load given and duration of compaction load, and sintering.
Low wear and high Brinell Hardness Number (BHN) are obtained by providing relatively lower formation pressures than other specimens. This is due to two possibilities [5]:

a. The separation of the resins as the binder of the composite mixture.
b. The resin bond strength is less than the formation pressure applied to the specimen at the time of the specimen preparation process.

The research on friction brake pads by providing improved sintering is conducted by [6]. With the higher sintering temperatures affect the wear rate. High temperature sintering causes the value of wear increases. Then the wear is higher. Increased sintering temperature also affects the hardness of the canvas. The higher the sintering temperature the hardness value will decrease.

Composite material is actually a lot in use of nature fiber, because the composite material can consist of organic and inorganic such as bamboo, wood, leaves, and so on [6]. We are unconsciously familiar with various types of composites.

4. Method

The tools that used in this research are measuring cup, electrical mixer, pressing load machine, thermometer, electrical oven, stopwatch, hardness tester, Ogoshi wear tester, Wipro wear tester and microscopic structural photograph. The procedure of the proposed method is explained as follow:

a. Bamboo fiber cut into less than 2 mm in order make discontinuous fiber.
b. Aluminum powder would be 60 mesh.
c. Fiberglass are also cut for 2 mm length.
d. Preparing mixing specimen conduct based on variation in table 1.
e. Mixed specimen pressed under 2.000 kg load, then sintered among 180°C.
f. Tested on Brinell.
g. Tested on Wearness.
h. Find friction coefficient.

Table 1. Variation of mixing specimen.

Bamboo fiber	Fiber glass	Aluminum Powder	Polyster 157 BQTN/ Phenolic
40%	10%	10%	40%
	15%	15%	30%
	20%	20%	20%
30%	15%	15%	40%
	20%	20%	30%
	25%	25%	20%
20%	20%	20%	40%
	25%	25%	30%
	30%	30%	20%

5. Result and discussion

5.1. Hardness test

Figure 1 shows that variation 1 (40 % bamboo fiber + 10 % fiber glass + 10 % aluminum (Al) + 40 % polyester) is the highest hardness of Brinell Number. This variation is 7,4% harder than exiting material that we find. It tends have good wearness resistance, high hardness materials tend more less in wearness. It shows, that bamboo fiber is good contributed in hardness for this material. The material will be more using time in braking operation, but severe in temperature rising. Variation 2 (30 % bamboo fiber + 15 % fiber glass + 15 % aluminum (Al) + 40 % polyester) is also higher than existing part, although more aluminum, it tends good in heat reducing during friction.

$$BHN = \frac{2P}{\pi D(D-\sqrt{D^2-d^2})} \quad (1)$$

Where BHN is the Brinell Hardness Number, P is the load (15.625 kg) and d is the penetrator diameter (2.5 mm).

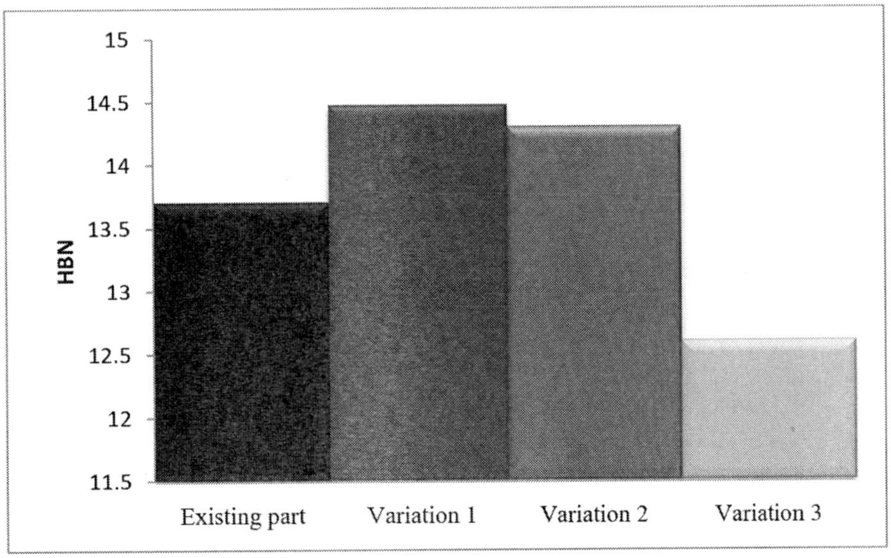

Figure 1. Variation of Brinell Hardness Braking Lining.

5.2. Oghosi wear test

Figure 2 shows that variation 1 is the best in wearness on dry condition, but less than existing material. It not more than 0,0002 mm^3/kg, so the breaking lifetime is not to different. Variation 2 shows better in water and oil wet condition. It shows that bamboo fiber is not affected in wet condition breaking operation, because of Ori bamboo fiber is solid and harder than other natural fiber. Variation 3 is less in hardness and wearness.

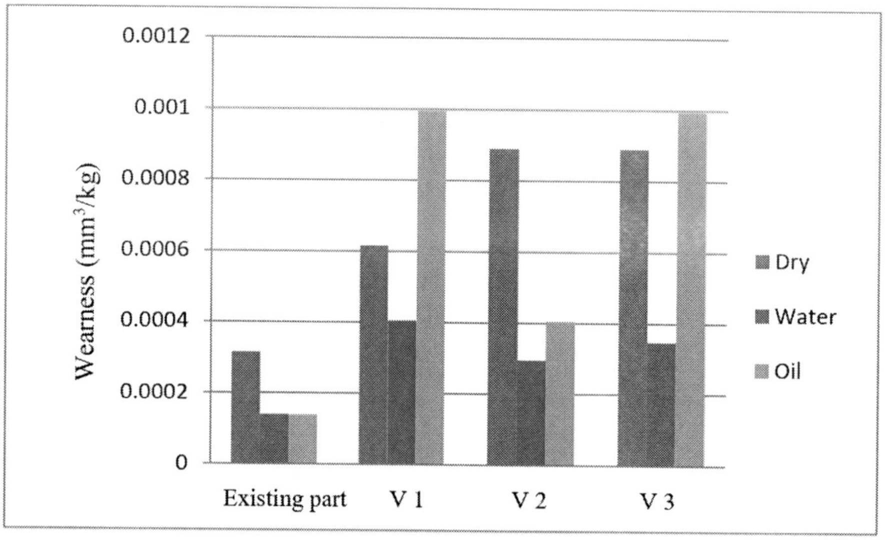

Figure 2. Oghosi wearness for braking wet and dry condition.

5.3. Wipro wear test

Figure 3 shows that the best wearness is variation 3 (20% bamboo fiber + 20% fiber glass + 20% aluminum (Al) + 40% polyester). It is more in fiber glass and aluminum, make more tight in compaction load and shearing force resistance. Variation 3 is not wide difference with existing breaking properties.

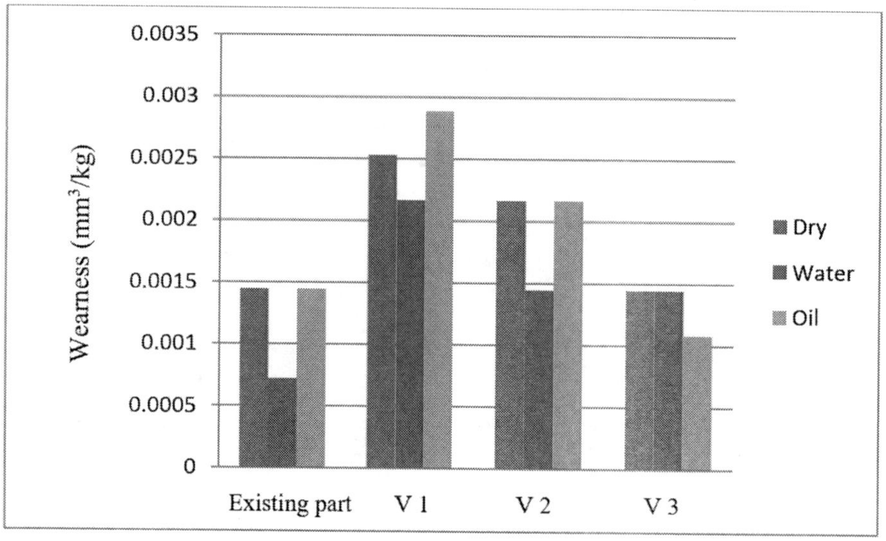

Figure 3. Comparison of Wipro wearness test for braking wet and dry condition.

Figure 4 shows that high power has high torsion resistance. It means high torsion has high coefficient of friction on these braking pad materials. Furthermore, all of these variation is not wide difference on coefficient of friction.

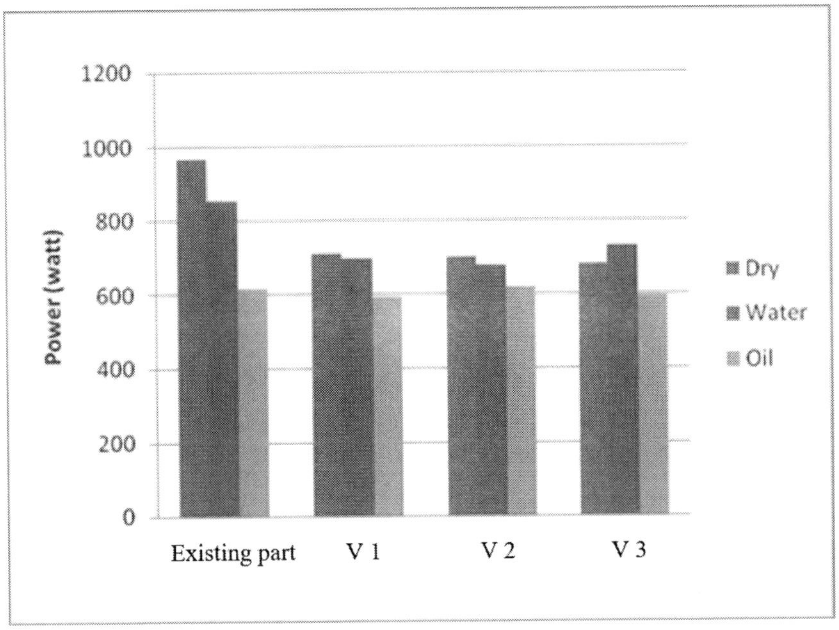

Figure 4. Electrical Power needed for braking.

6. Conclusion

From the discussion in section 5, we can conclude that for Brinell hardness, the material of brake lining with the composition of Variation 1 is the hardest, with hardness price of 14.47 BHN harder than the product on the market with the value of hardness 13.7 BHN, although for Ogoshi Wear Testing on dry conditions, the brake lining material with Variation 2 has the lowest wear value of 0.00041 mm^2/kg, which is slightly larger than the product on the market with 0.00014 mm^2/kg wear.

For Ogoshi wearing tests wet conditions with water, it is found that, brake lining material with the lowest Variation 1 of its wear is equal to 0.0062 mm^2/kg, but still slightly higher than the brake lining of the market that is equal to 0.0032 mm^2/kg Waste Testing Ogoshi Wet condition with oil, it is found that, the material of brake lining with the lowest 2 of its wear is 0.0003 mm^2/kg, but still slightly higher than the brake lining material of the market that is equal to 0.00014 mm^2/kg.

In Wipro Wear Tests on dry conditions, the brake lining material with Variation 3 has the lowest wear value of 0.00014mm2 / kg, which is the same from the product on the market with 0.00014 mm2 / kg wear. For Wipro Wear Tester Wetro be conditions with water, it was found that the brake lining materials with the lowest 2 and 3 variations were 0.144 mm^2/kg, but still slightly higher than the brake pads of the market at 0.0007 mm2 / kg. Wipro Wear Testing Wetro condition with oil, it is found that, brake lining material with the lowest 3 of its wear is 0.0011 mm^2/kg, but still slightly higher than the brake pads material of the market that is equal to 0.00014 mm^2/kg.

Acknowledgement

Ministry of Research and Higher Education of Indonesia, DP2M and *Lembaga Penelitian dan Pengabdian Masyarakat,* University of Muhammadiyah Surakarta.

References

[1] Cândido G M, Rezende M C, Almeida S F M D 2000 Hygrothermal Effects on the Tensile Strength of Carbon / Epoxy Laminates with Molded Edges *Materials Research* vol 3 no 2 pp 11–17

[2] Ilmu P 2009 Influence of Holding Tim on Hardness of Particulate Composite *Media Mesin* vol 2

[3] Ashby M F and Jones D R 1995 Engineering Materials 1 - An introduction to their properties & applications *Fuel and Energy Abstracts* vol 4 no 36

[4] Ilmu P 2009 Influence of Sintering Temperatur on Particulate Coconut Fiber Composite *Media Mesin* vol 1

[5] Harris B 1999 *Engineering composite materials* (London: Institute of metals)

[6] Pradhan S K, Dwarakadasa E S and Reucroft P J 2004 Processing and characterization of coconut shell powder filled UHMWPE *Materials Science and Engineering: A* vol 367 no 1-2 pp 57–62

Institute of Physics
Dirac House, Temple Back
Bristol BS1 6BE UK

ISSN: 1757-8981
ISBN 978-1-5108-7297-4

9 781510 872974

CURRAN ASSOCIATES INC.
proceedings
.com